Progress in Mathematics
Volume 65

Gerd Grubb

Functional Calculus of Pseudo-Differential Boundary Problems

1986

Birkhäuser
Boston · Basel · Stuttgart

Gerd Grubb
Københavns Universitets Matematiske Institut
2100 København Ø
Denmark

Library of Congress Cataloging in Publication Data
Grubb, Gerd.
Functional calculus of pseudo-differential boundary problems.
(Progress in mathematics ; v. 65)
Bibliography: p.
Includes index.
1. Pseudodifferential operators. 2. Boundary value problems. I. Title. II. Series: Progress in mathematics (Boston, Mass.) ; vol. 65.
QA329.7.G78 1986 515.7'242 86-20769

CIP-Kurztitelaufnahme der Deutschen Bibliothek
Grubb, Gerd:
Functional calculus of pseudo-differential boundary problems / Gerd Grubb. —Boston ; Basel ; Stuttgart : Birkhäuser, 1986
(Progress in mathematics ; Vol. 65)
ISBN 3-7643-3349-9 (Basel...)
ISBN 0-8176-3349-9 (Boston)
NE: GT

ISBN 0-8176-3349-9
ISBN 3-7643-3349-9

Printed in the U.S.A.

PREFACE

The theory of pseudo-differential operators has been developed through the last three decades as a powerful tool to handle partial differential equations. Here the pseudo-differential operators, and more generally the Fourier integral operators, include as special cases both the differential operators, their solution operators (integral operators), and compositions of these types. For equations on manifolds with boundary, Eskin, Vishik and Boutet de Monvel invented in particular the calculus of *pseudo-differential boundary operators*, that applies to elliptic boundary value problems.

The aim of the present book is to develop a *functional calculus* for such operators; i.e. to find the structure and properties of functions of these operators defined abstractly by functional analysis.

We consider in particular detail the exponential function of the operators, which leads to a treatment of parabolic evolution problems, and the complex powers of the operators, with applications to spectral theory; and we determine trace formulas and index formulas. The basic tool is a study of the *resolvent* of the operator, and this is worked out in the framework of a calculus of *pseudo-differential boundary problems depending on a parameter* $\mu \in \mathbb{R}_+$. The original parameter-independent theory is included as a special case, and our presentation may be used as an introduction to that theory. A further application of the theory is the treatment of singular perturbation problems; they contain a small parameter ε going to zero, corresponding to $\mu = \varepsilon^{-1}$ going to infinity.

The work was begun during a stay at the Ecole Polytechnique in 1979. At that time, we expected the resolvent analysis to take a few months (- with a sound knowledge of the Boutet de Monvel theory, it should be an easy matter to establish corresponding results in cases with a parameter -), but the task turned out to be not quite so simple. A first version of our calculus was written up in a series of reports from Copenhagen University in 1979-80 [Grubb 11], and much of that is used here (in shortened form). However, it also had some flaws: On one hand, the hypotheses needed to go beyond the most classical boundary conditions were incomplete, and on the other hand, we later found a way to eliminate a certain "loss of regularity $\frac{1}{2}$". Brief accounts of the theory have been given in [Grubb 12-15], where [Grubb 15] corrects earlier defects.

The present work contains much more, both in the form of explicit information on the structure of the boundary problems to which the theory applies, an amelioration of the calculus, and developments of consequences of the theory. It has taken a long time to complete (partly because of the author's other University duties), but we hope that this has led to a maturing of the contents and elimination of disturbing errors.

Various people have been helpful to us during the process. The start of the work benefited from conversations with Charles Goulaouic and Louis Boutet de Monvel in Paris. In 1981, Denise Huet in Nancy told us of the connection with singular perturbation theory, which led to a clarification of the hypotheses; and Bert-Wolfgang Schulze and Stefan Rempel in Berlin showed much interest in the work. When it was in a final stage, Lars Hörmander in Stockholm helped us greatly with criticism and suggestions for improvements. We are very thankful to these and other colleagues that have shown interest, and we likewise thank the editor and referees of the Birkhäuser Progress in Mathematics Series for their encouragement.

The author is deeply grateful to Jannie Larsen and Ulla Jakobsen at the University of Copenhagen Mathematics Department for their efficiency and patience with the task of typing the manuscript and its alterations.

Copenhagen in April 1986,

Gerd Grubb

CONTENTS

I N T R O D U C T I O N

"Vetenskaperna äro nyttiga därigenom att de hindra människan från att tänka"
Citeret i Doktor Glas af Hjalmar Söderberg.

The main purpose of this work is to set up an operational calculus for operators defined from differential and pseudo-differential *boundary value problems*, via a resolvent construction, and to present some applications of this to evolution problems, fractional powers, spectral theory and singular perturbation problems.

1. Functions of an operator.

On the abstract level, there are several well-known methods to define "functions of an operator". For one thing, when A is a selfadjoint operator in a Hilbert space H, $f(A)$ can be defined via the spectral resolution, for any measurable function $f(t)$ on $\mathbb{R}$. Another method, that works when A is not necessarily selfadjoint but has a sufficiently large resolvent set $\rho(A)$, is to define functions $f(A)$ by use of a Cauchy integral formula

(1) $$f(A) = \frac{i}{2\pi} \int_C f(\lambda)(A-\lambda)^{-1}\, d\lambda \ ,$$

when $f(\lambda)$ is holomorphic on the spectrum $\mathbb{C} \setminus \rho(A)$, and the integral converges in a suitable sense; here C is a curve going around the spectrum in the positive direction. We use the latter method, and here our point is not merely to define $f(A)$, but rather to *analyze its structure* in the framework of pseudo-differential boundary operators; to determine its properties in detail. In this method, the fundamental object to analyze is *the resolvent* $R_\lambda = (A-\lambda)^{-1}$.

As particular applications, we consider the "heat operator" $\exp(-tA)$ and the fractional powers A^z. The heat operator is a basic tool in the solution of evolution problems

(2) $$\frac{\partial u}{\partial t} + Au = g \quad \text{for } t > 0 \ , \qquad u|_{t=0} = u_0 \ ,$$

that we shall also discuss.

2. Pseudo-differential operators.

The concept of pseudo-differential operators was invented in the 1960's as a class of operators that includes both *differential operators*, their *solution operators* (or approximate solution operators, called *parametrices*) in the elliptic case, and certain other *integral operators* and *integro-differential operators* (especially the so-called *singular integral operators*). For example, the Laplace operator

(3) $$\Delta = \frac{\partial^2}{\partial x_1^2} + \dots + \frac{\partial^2}{\partial x_n^2} \quad (= -D_{x_1}^2 - \dots - D_{x_n}^2)$$

on $\mathbb{R}^n$ is included as a differential operator, whereas the solution operator $(-\Delta + 1)^{-1}$ (easily defined via the Fourier transform) is a *pseudo-differential operator* (ps.d.o.). Also the solution operator (for $n > 2$)

(4) $$Qf = c_n \int \frac{f(x)dx}{|x-y|^{n-2}}$$

for the equation $-\Delta u = f$, is a ps.d.o. The Laplace operator is of order 2, and $(-\Delta + 1)^{-1}$ and Q are of order -2. Actually, the theory of ps.d.o.s allows operators of any real order; for example, $(-\Delta + 1)^s$ (defined in $L^2(\mathbb{R}^n)$ by spectral theory) is a ps.d.o. of order $2s$ for any $s \in \mathbb{R}$. (Introductions to ps.d.o.s are given e.g. in [Seeley 2]*, [Nirenberg 1], [Hörmander 6,8], [Taylor 1], [Treves 2], and we take them up in the present text in Sections 1.2 and 2.1.)

An important point in the calculus is the relation between a pseudo-differential operator and its *symbol* $p(x,\xi)$ (a function of $(x,\xi) \in \mathbb{R}^n \times \mathbb{R}^n$) where, roughly speaking, *composition* of two ps.d.o.s P and P' corresponds to *multiplication* of their symbols $p(x,\xi)$ and $p'(x,\xi)$ (modulo errors of lower order), and *inversion* of a ps.d.o. P corresponds to *inversion* of the symbol (again modulo lower order errors); the inversion can be carried out when P is *elliptic*, i.e. p has an invertible highest order part $p^0(x,\xi)$. (Here x represents the space variable (position), and ξ is the "dual variable" (momentum) appearing by the Fourier transform. When x runs in a manifold Ω, (x,ξ) should be considered as a point in the cotangent space $T^*(\Omega)$.)

By the symbolic calculus, one has good control over *polynomial* functions of P, $f(P) = \Sigma_{k\leq N} c_k P^k$, and over suitable rational functions $f(P)$ in case P is elliptic. The study of general functions of P was initiated by R. Seeley, who analyzed the resolvent $Q_\lambda = (P - \lambda I)^{-1}$ in [Seeley 1] and used it to prove that P^z (for any $z \in \mathbb{C}$) is a classical ps.d.o. when P is elliptic and classical. Further developments have been given in [Strichartz 1], [Dunau 1],

---

*) [Author X] refers to a reference given in the bibliography.

$p(x,\xi)$ of P for $x \in \overline{\Omega}$, $\xi \in \mathbb{R}^n$; and the other one is the "boundary symbol", which is in reality a family of operators $a(x',\xi',D_n)$ of the form (9) with $\overline{\Omega}$ replaced by $\overline{\mathbb{R}}_+$, and parametrized by $x' \in \partial\Omega$, $\xi' \in \mathbb{R}^{n-1}$ (more precisely by $(x',\xi') \in T^*(\partial\Omega)$). The compositions of Green operators (9) are now reflected in compositions of these one-dimensional *boundary symbol operators*, besides multiplication of interior symbols. Similarly, inversion or construction of parametrices involves not just inversion of a (possibly matrix formed) *function* of (x,ξ), but also the inversion of boundary symbol *operators* on $\overline{\mathbb{R}}_+$, parametrized by (x',ξ').

This kind of complexity is of course well-known also from studies of partial differential operators prior to the systematic pseudo-differential calculus, where one succesful technique was to do the following (expressed in a few words): Freeze the coefficients at a boundary point and reduce to an ordinary differential equation (with boundary conditions) by Fourier transformation in the tangential variables. Then build up the general solution from the solutions of the ordinary differential equations at each boundary point.- In the modern method, the boundary symbol operator $a(x',\xi',D_n)$ corresponds to the Fourier transformed frozen coefficient case, and the construction of a true parametrix from the symbolic cases (or "model cases") is systematized.

In the present work we devote much attention to those pseudo-differential boundary problems that lie fairly close to differential boundary problems. For instance, one can make (5) non-local in the boundary condition only, by taking T of the form

(10) $$Tu = \gamma_1 u + S\gamma_0 u ,$$

where S is a first order ps.d.o. on $\partial\Omega$; or one can take

(11) $$Tu = \gamma_0 u + T_0' u ,$$

or

(12) $$Tu = \gamma_1 u + T_1' u ,$$

where T_0', T_1' are a kind of integral operators going from Ω to $\partial\Omega$, of suitable orders (fitting together with the order of γ_0 resp. γ_1). Here (10) represents a very mild form of non-localness that is often easily handled along with local problems, whereas (11) and (12) are somewhat more different from differential trace operators since they contain a link between the boundary value and the interior behavior of u. Such conditions were called "lateral conditions" in [Phillips 1], and they occur as "boundary renewal conditions" in population theory.

As for the first line in (5), one is not so far from differential operator problems if $-\Delta$ is replaced by, say

$$P = (D_{x_1}^4 + \ldots + D_{x_n}^4)(D_{x_1}^2 + \ldots + D_{x_n}^2)^{-1} \tag{13}$$

(where $(D_{x_1}^2 + \ldots + D_{x_n}^2)^{-1}$ is a parametrix of $D_{x_1}^2 + \ldots + D_{x_n}^2 = -\Delta$). The addition of a singular Green operator G to P will give further links between the interior and the boundary; typically, G can be of the form

$$G = K_0\gamma_0 + K_1\gamma_1 + G' \tag{14}$$

where G' is an integral operator on Ω , and K_0 and K_1 are integral operators going from $\partial\Omega$ to Ω (defining "boundary feedback" terms). We assume that the system $\{P_\Omega + G , T\}$ satisfies suitable ellipticity (or invertibility) conditions.

Along with the stationary problems generalizing (5) we also consider time-dependent evolution problems

$$\begin{aligned} \partial_t u + P_\Omega u + Gu &= f(x,t) \quad &&\text{for } x \in \Omega \;, t > 0 \;, \\ Tu &= \varphi(x,t) \quad &&\text{for } x \in \partial\Omega \;, t > 0 \;, \\ u|_{t=0} &= u_0(x) \quad &&\text{for } x \in \Omega \;, \end{aligned} \tag{15}$$

where P , G and T can be as described above (such problems have been considered by methods of functional analysis e.g. in [Nambu 1], [Triggiani 1]).

An object of importance, both in the study of (5) and its generalizations, and in (15), is the *realization* associated with P , G and T , namely the operator

$$B = (P + G)_T \;, \tag{16}$$

acting like $P_\Omega + G$ and with domain

$$D(B) = \{u \in H^d(\overline{\Omega}) \mid Tu = 0\} \;, \tag{17}$$

here $d > 0$ is the order of P (equal to 2 in the above examples); B is a closed unbounded operator in $L^2(\Omega)$. The discussion of (15) can be carried out on the basis of a discussion of the operator function $\exp(-tB)$, that we call the "heat operator" in view of the resemblance with the case where $P = -\Delta$ and $G = 0$.

Problems like (15) can arize as models of concrete problems, but also as technical tools in studies of differential boundary problems (e.g. by factorization and other reductions). For example, boundary problems for the non-stationary Stokes equation can be reduced to this form [Grubb-Solonnikov 1].

4. Functional calculus.

In the questions of functional calculus of pseudo-differential boundary problems, there are several possible lines to follow. On one hand, one can study operators A as in (9) in the square matrix case where $N = N'$ and $M = M'$, so that one can define $f(A)$ as an operator acting in $C^\infty(\overline{\Omega})^M \times C^\infty(\partial\Omega)^N$ (and in suitable Sobolev spaces). Neither (6) nor (7) belongs to this case. But they do define the unbounded realization $B = P_T$ acting in $L^2(\Omega)$ (and its bounded inverse R); and it is of great interest to study functions of B, e.g. in view of applications to (15). The theory we present in this work permits a study, both of square matrix formed systems A as in (9) and of realizations B as in (16); and we focus particularly on the latter case that we find most interesting.

The basic step in the calculus is a study of the resolvent $R_\lambda = (B - \lambda I)^{-1}$. This study is imbedded in a calculus of *parameter-dependent pseudo-differential boundary problems*. In the differential operator case, the parameter $\mu = |\lambda|^{1/d}$ is easily absorbed as an extra cotangent variable, because of the polynomial nature of the symbols. In the ps.d.o. case, however, the parameter gives severe extra trouble. This is felt already in the case without boundary: The strictly homogeneous principal symbol $p^h(x,\xi)$ of a ps.d.o. P of order $d \geq 1$ is continuous at $\xi = 0$ with locally bounded derivatives in ξ up to order d only; the next derivatives are unbounded at $\xi = 0$ unless P is a differential operator. The ps.d.o. calculus handles easily such an irregularity in a compact neighbourhood of $\xi = 0$ in $\mathbb{R}^n$, but when an extra cotangent variable $\mu \in \mathbb{R}$ is adjoined, the irregularity extends to a noncompact neighbourhood of the full μ-axis in $\mathbb{R}^{n+1}$. (One does not just get anisotropic ps.d.o.s, as in [Fabes-Rivière 1], [Lascar 1], [Rempel-Schulze 1, 4.3.6], cf. Section 1.5.) Now when boundary conditions are included, the strictly homogeneous principal boundary symbol operator $a^h(x',\xi',D_n)$ is generally irregular at $\xi' = 0$, and the addition of an extra cotangent variable μ gives similar trouble as for $p(x,\xi)$, or even more so, because we are here treating operator families depending on the parameter (x',ξ'), not just functions. To clarify these phenomena we introduce the *regularity number* ν for the system $\{P_\Omega + G, T\}$, defined, roughly speaking, as the highest order of derivatives bounded at $\xi' = 0$, possessed by the strictly homogeneous symbols (with respect to suitable symbol norms). It is necessary for the calculus to introduce also noninteger and negative regularity numbers. Keeping account of the regularity numbers is all-important for the strength of the calculus. The parametrix construction works best for systems with strictly positive regularity in the principal part, for this assures a good principal parametrix symbol as well as a decrease of the lower order terms with respect to the parameter. The parameter-elliptic systems occuring in our resolvent study have regularity in the interval $[\frac{1}{2},d]$.

It is perhaps surprising, that the three conditions (10), (11) and (12) give *different regularities* ν for the problem (5), namely $\nu = 1$ for (10), $\nu = \frac{1}{2}$ for (11), and $\nu = \frac{3}{2}$ for (12). What was most surprising to the author was that the Dirichlet-type condition (11) with $T_0' \neq 0$ has a very low regularity $\frac{1}{2}$, barely missing 0. (It would be interesting to know whether non-local Neumann-type conditions (12) are more natural than non-local Dirichlet-type conditions (11); in fact [Nambu 1] and [Triggiani 1] only consider Neumann-type conditions.)

A large part of the book, namely Chapters 2 and 3, is devoted to the systematic parameter-dependent calculus and its application to the resolvent study. The development of the calculus required a very thorough analysis of the Boutet de Monvel calculus; for instance there are several systems of symbol seminorms that are equivalent in the non-parametrized case, but have quite different qualities when the parameter is included (see the discussions of (2.2.82), (2.2.85), (2.4.24), (2.4.26)). This investigation led to improvements also in the non-parametrized case, see [Grubb 17].

The operator-theoretic approach in [Grubb 5] (applied to strongly elliptic ps.d.o.s resulting from reductions of matrix-formed differential operators) gave results on the principal symbol level; whereas the present calculus treats the full asymptotic expansions. It was first written up in the prepublication [Grubb 11], the results were announced in two short notes [Grubb 13,14], and an account of the consequences for the heat equation was given in [Grubb 15], with some corrections to the notes. - Also Rempel and Schulze have taken up the resolvent studies and operational calculus for their operator class, giving principal estimates in [Rempel-Schulze 3,4]. - Elements of parameter-dependent calculi of pseudo-differential boundary problems moreover enter in [Eskin 1] and [Frank-Wendt 1-3], [Wendt 1]. - [Cordes 1] treats the functional calculus from a more abstract point of view; using Banach algebra techniques in a framework of L^2 estimates.

5. Outline.

Some prerequisites for our presentation are collected in the Appendix at the end of the book, which the reader is invited to consult for the notation.

Chapter 1 starts with some examples, and then gives a survey of the known theory of parameter-independent pseudo-differential boundary problems. The new material begins with our study of *Green's formula* and *realizations* of ps.d.o. boundary problems. Here the realizations defined by *normal* boundary conditions (that we introduce as a natural generalization of the concept for differential operators) are especially interesting. On one hand, the class of these realizations is closed under *composition* and passage to *adjoints* (in a precise sense when ellipticity holds), and moreover, the normality is necessary for the type of ellipticity with a parameter that we assume for our resolvent construction. The *parameter-ellipticity* and *parabolicity* concepts are introduced in Section 1.5, where we also briefly discuss the so-called *regularity number* ν . In order to provide concrete examples, we hereafter treat some special cases, in Section 1.6 the *selfadjoint* realizations, and in Section 1.7 the *semibounded* and *coercive* realizations (in particular, we show hos the Friedrichs extension, alias the Dirichlet realization, fits in).

Chapters 2 and 3 give the details of the parameter-dependent calculus and resolvent construction, on which the rest of the book depends. We here go through the full program of building up the calculus: Chapter 2 gives the definition of *parameter-dependent ps.d.o.s* and their *regularity*, the appropriate version of the *transmission property* and its background (sometimes called the Wiener-Hopf calculus), the introduction of *boundary symbol* classes, the associated *operators* and *kernels* (in particular the *negligible* ones) and their behavior under *coordinate changes*, *norm estimates* in Sobolev spaces, rules for *adjoints* and *compositions* (in particular the difficult case of fully x_n-*dependent symbols*). The choice of symbol classes is aimed towards a good *parametrix construction*. (A reader who wants a complete explanation of the parameter-independent theory can actually find it in Chapters 2-3 by disregarding the effects of the parameter.)

Finally, Chapter 4 gives some *applications*. For one thing, we can discuss the solvability of (15) as in classical works on parabolic differential operators, on the basis of the estimates of the resolvent obtained in the preceding chapter. This is done rather briefly in Section 4.1. Secondly, we discuss the *heat operator* exp(-tB) defined by (1), in much detail, in particular its kernel properties. Here the heat operator exp(-tP) for the boundaryless case is also discussed, for comparison and for completeness' sake. However, the largest efforts are devoted to the term

$$W(t) = \exp(-tB) - \exp(-tP)_\Omega \quad , \tag{18}$$

which is of a "singular Green operator" type. An interesting consequence of the study of exp(-tB) is the trace formula

$$\operatorname{tr}\exp(-tB) = c_{-n}(B)t^{-n/d} + c_{1-n}(B)t^{(1-n)/d} + \dots + c_{\nu'}(B)^{\nu'/d} + O(t^{(\nu-\frac{1}{4})/d}) \quad \text{for} \quad t \to 0+ , \tag{19}$$

where ν *is the regularity of* $\{P_\Omega + G, T\}$, and ν' is the largest integer in $[0,\nu[$. For $\exp(-tP)$ one has a more complete asymptotic expansion of the trace, in terms of powers $t^{(j-n)/d}$ with $j \in \mathbb{N}$ and logarithmic expressions $t^{k+1} \log t$ with $k \in \mathbb{N}$, cf. [Duistermaat-Guillemin 1] and [Widom 1,2]; we show how it follows from our resolvent calculus in Section 4.2. Now the regularity number ν in (19) is always $\leq d$ when $\{P_\Omega + G, T\}$ is genuinely pseudo-differential, so (19) does not get near the first logarithmic term $t \log t$, one would expect to have. For $\exp(-tP)$, one can overcome this by composition of the resolvent with high powers of P, which improves the regularity; but for $\exp(-tB)$, the composition with high powers of B does not improve the regularity (this is discussed at the end of Section 3.3 and in Section 4.2).

Formula (19) is accurate enough (*just* accurate enough, for Dirichlet-type trace operators (11)!) to lead to a new *index formula* (Section 4.3) for general normal elliptic realizations $\overline{B}$ of elliptic ps.d.o. boundary problems:

$$\text{index } \overline{B} = c_0(\overline{B}^*\overline{B}) - c_0(\overline{B}\,\overline{B}^*) \tag{20}$$

(previously known for differential boundary problems, cf. [Atiyah-Bott-Patodi 1], [Greiner 1]). It involves slightly fewer terms from the symbols of T and G than the formula of [Rempel 1] for general elliptic ps.d.o. problems. (We have not analyzed possible reductions of the formula, as done for other formulas in [Fedosov 1], [Hörmander 7], [Rempel-Schulze 1].)

Section 4.4 discusses another operator function, namely the *complex powers* B^z, again defined by use of (1). Here

$$B^z = (P^z)_\Omega + G^{(z)} \tag{21}$$

where the ps.d.o. P^z is well studied (cf. [Seeley 1]), so that the main interest lies in the study of $G^{(z)}$. It is usually not a singular Green operator but a *generalized* kind, the symbol satisfying certain but not all of the usual estimates. The symbol is analyzed for $\operatorname{Re} z < 0$, and we furthermore show that the *trace*, which is traditionally defined for large negative values of $\operatorname{Re} z$, extends to a meromorphic function of z in the region $\{z \in \mathbb{C} \mid \operatorname{Re} z < (\nu-\frac{1}{4})/d\}$, with poles in the set

$$\{z = (j-n)/d \mid j = 1,2,\dots,n+\nu',\ j \neq n\}. \tag{22}$$

(The obstacles to extension above $\operatorname{Re} z = (\nu-\frac{1}{4})/d$ are the same as those for getting more terms in (19).) - For some cases where B is selfadjoint positive, we characterize the *domains of fractional powers* B^θ, $\theta \in]0,1[$, as in [Grisvard 1], [Seeley 6] for differential operators.

Section 4.5 is devoted to *spectral theory*. Asymptotic eigenvalue estimates with remainder estimates were obtained for pseudo-differential boundary problems already in [Grubb 8-10] (independently of the present resolvent calculus) and we begin by recalling these and some corollaries, e.g. the estimate

(23) $$N(t;B) = C(p^0,\Omega)t^{n/d} + O(t^{(n-\theta)/d}) \quad \text{for } t \to \infty ,$$

that holds for general $B = (P+G)_T$ with $\theta < \frac{1}{2}$, and with $\theta < 1$ in some special cases. Here $N(t;B)$ denotes the number of characteristic values $s_k(B)$ (eigenvalues of $|B| = (B^*B)^{\frac{1}{2}}$) in the interval $[0,t]$. (It should be recalled here that the finer estimates of [Hörmander 5], [Demay 1], [Seeley 7], [Ivrii 3], [Métivier 1], [Vasiliev 1,2]..., with $\theta = 1$ in the error term, or with a precise asymptotic estimate replacing the error term, are essentially concerned with either boundary value problems for *differential operators*, or with ps.d.o.s on *boundaryless* manifolds.) One consequence of (23) is the estimate for elliptic ps.d.o.s Q of negative order $-d$ on E, having the transmission property at Γ, and considered <u>without</u> boundary conditions:

(24) $$N'(t;Q_\Omega) = C'(q^0,\Omega)t^{n/d} + O(t^{(n-\theta)/d}) \quad \text{for } t \to \infty ;$$

here θ is as above, and $N'(t;Q_\Omega)$ is the number of characteristic values $\geq 1/t$. Next, the symbol estimates for complex powers derived in Section 4.4 are used to show that the generalized s.g.o. term $G^{(z)}$ (cf. (21)) satisfies

(25) $$s_k(G^{(z)}) \leq c_z\, k^{-d|\mathrm{Re}\, z|/(n-1)} \quad \text{for all } k ,$$

when $\mathrm{Re}\, z < -(2d)^{-1}$ and $n > 2$ (a slightly weaker estimate is shown for $n = 2$), by an application of the method in [Grubb 17]. In particular, $G^{(z)}$ is of *trace class* if $\mathrm{Re}\, z < -(n-1)/d$. The estimate (25) is furthermore used to obtain spectral estimates for the *complex powers* B^z, and for the positive and negative part of B in the case where P is not assumed strongly elliptic (as it was in [Grubb 8-10]).

In Section 4.6, we show how the preceding results can, by use of some special reductions, be applied to *implicit eigenvalue problems* (of Pleijel type)

(26) $$\lambda A_1 u = A_0 u \quad \text{on } H^{r+d}(E) , \quad \gamma_j u = 0 \quad \text{for } j < (r+d)/2 ,$$

where A_1 is selfadjoint elliptic and of higher order $(r+d)$ than A_0 (symmetric and of order $r \geq 0$); here A_1 and A_0 are of the form $A_1 = P_{1,\Omega} + G_1$ resp. $A_0 = P_{0,\Omega} + G_0$. It is found that the eigenvalues λ behave like the eigenvalues of an operator of order $-d$, with a principal asymptotic estimate in general cases, and with a remainder estimate as above when A_0 is also elliptic. The results extend to *multi-order systems*. (They have some overlap with results of [Kozlov 1,2], [Levendorskii 1,2], [Ivrii 3].)

Finally, in Section 4.7 we show how our parameter-dependent theory can be applied to the study of *singular perturbation problems*. One considers the perturbed problem

(27) $$\varepsilon^d A_1 u_\varepsilon + A_0 u_\varepsilon = f \text{ in } E, \quad T_0 u_\varepsilon = \varphi_0 \text{ at } \Gamma, \quad T_1 u_\varepsilon = \varphi_1 \text{ at } \Gamma,$$

in relation to the unperturbed problem

(28) $$A_0 u = f \text{ in } E, \quad T_0 u = \varphi_0 \text{ at } \Gamma;$$

where A_1 is of order $r+d$ and A_0 is of order r ($r \geq 0$ and $d > 0$), like in the preceding application, and T_1 resp. T_0 are trace operators, with T_1 normal and formed of strictly higher order operators than those in T_0. We describe a new method to treat this problem (in the differential operator case as well as for pseudo-differential generalizations), consisting of a reduction by use of parameter-independent operators to a situation where the results of Chapter 3 can be applied, with $\mu = \varepsilon^{-1}$. This gives a simple and straightforward representation of the solution u_ε and its relation to the unperturbed solution u, with natural estimates of the convergence as $\varepsilon \to 0$. The problem has been studied earlier by [Vishik-Lyusternik 1], [Huet 1-5], [Greenlee 1], [Demidov 1], [Eskin 1], [Frank 1], [Frank-Wendt 1-3], [Wendt 1] and many others. In comparison with the treatment of (27)-(28) in the works of L. S. Frank and W. D. Wendt using pseudo-differential considerations, our method has the advantage of *avoiding* the problem of *negative regularity*; and it extends readily to general (non-rational) pseudo-differential boundary problems in vector bundles. (The results were published in [Grubb 18].)

6. Further perspectives of the theory.

The detailed presentation of the parameter-dependent calculus in Chapters 2 and 3 opens up for generalizations and refinements of the theory. For one thing, one may try to improve (19), possibly by using more of the special properties of the resolvent in comparison with the general operators in the parameter-dependent calculus. It is not obvious whether methods like those in [Melrose 2] could be of use here. [Widom 4] gives some conjectures concerning full asymptotic expansions in a related, but somewhat different problem.

Another, more applications oriented question, is the extension of the theory to the situation of L^p spaces and Hölder spaces, which is of fundamental interest for *non-linear problems*. The symbol-kernel point of view ((2.3.25)ff.) seems promising here, leading to a reasonable L^p-calculus, that we expect to develop elsewhere. (There are some observations in [Rempel-Schulze 1] for the case without a parameter.)

Thirdly, the questions around the evolution problem (15) have a certain interest for *control theory* (as indicated in [Nambu 1] and [Triggiani 1]); this connection seems well worth developing.

A fourth observation is that L.S. Frank and W.D. Wendt have defined elements of a parameter-dependent version of the Boutet de Monvel calculus, that they use in their treatment of singular perturbation problems in [Frank-Wendt 1-3], [Wendt 1]; here the results of the present book may be useful for the development of a complete presentation.

As a fifth point, let us mention the possibility of including t-*dependent* operators in the evolution problems (which apparently just requires a minor generalization of the symbolic calculus), and problems where ∂_t enters in more general ways, as in [Solonnikov 1], [Eidelman 1] and subsequent works on differential operators (see Sections 3.4 and 4.1 for some generalizations of (15)).

Finally, there is the question of the inclusion of problems where the *regularity number is nonpositive*; here [Rempel-Schulze 3,4] and [Frank-Wendt 1-3], [Wendt 1] treat various interesting cases, particularly for operators with constant coefficients; we discuss these cases in several remarks and examples in the following text. The negative regularity means that the parameter-ellipticity fails in some sense.

In this connection, it is of interest to investigate the possible application of the general methods for *non-elliptic problems* that have been found in recent years, and of which a great deal are collected and developed in the volumes [Hörmander 8], to the study of pseudo-differential boundary problems.

CHAPTER 1

STANDARD PSEUDO-DIFFERENTIAL BOUNDARY PROBLEMS AND THEIR REALIZATIONS.

1.1. Introductory remarks.

The typical class of boundary problems, that we shall be concerned with, lies fairly close to boundary problems for differential operators, but is large enough to include interesting non-local phenomena, as in the following example.

1.1.1 Example. Let Ω be a smooth open subset of $\mathbb{R}^n$ with boundary $\partial\Omega = \Gamma$, and consider the problem

$$(1.1.1)\qquad \begin{aligned} -\Delta u + Gu &= f \quad \text{on } \Omega \\ Tu &= \varphi \quad \text{at } \partial\Omega . \end{aligned}$$

Here Δ is the Laplace operator $\partial_{x_1}^2 + \ldots + \partial_{x_n}^2$, and for G we take an operator of the form

$$(1.1.2)\qquad Gu = K_0\gamma_0 u + K_1\gamma_1 u + G' ,$$

where $\gamma_j u = (-i\partial_n)^j u|_\Gamma$ (∂_n being the interior normal derivative), K_0 and K_1 are (integral) operators going from $\partial\Omega$ to Ω, and G' is an integral operator over Ω. The trace operator T can either be taken as a variant of the Dirichlet trace operator

$$(1.1.3)\qquad T_0 u = \gamma_0 u + T_0' u ,$$

where T_0' is an integral operator going from Ω to Γ; or of the Neumann trace operator

$$(1.1.4)\qquad T_1 u = \gamma_1 u + S_0\gamma_0 u + T_1' u ,$$

where S_0 acts on functions on Γ, and T_1' is an integral operator going from Ω to Γ. Boundary condition like (1.1.3) and (1.1.4), containing interior contributions, were considered already in [Phillips 1], where they were called "lateral conditions".

The terms $K_0\gamma_0$ and $K_1\gamma_1$ in (1.1.2) are sometimes called "boundary feedback" terms (information from the boundary is fed back to the interior). Conditions like (1.1.3) and (1.1.4) can be seen as "boundary renewal" conditions, with a terminology inspired by population theory, where the condition

$$u(0) = \int_0^\infty u(t)f(t)dt \ ,$$

expressing the number $u(0)$ of newborn individuals as a function of the age profile $u(t)$, is a special case of the homogeneous condition (in case $\Omega = \mathbb{R}_+$),

$$\gamma_0 u + T_0' u = 0 \ , \qquad u \in \mathscr{S}(\overline{\mathbb{R}}_+) \ ,$$

with $-T_0'u = \int_0^\infty u(t)f(t)dt$. It is also easy to think of physical situations with non-local phenomena of this kind. Consider for example the temperature distribution in a house, where the walls have temperatures imposed from the exterior (changing with the time of the day or year). Here one could set up further wall heating governed by a temperature measuring system distributed in the interior, in such a way that the boundary temperature is modified proportionally to an integral (weighted average) over the interior. It should be noted that the theory allows integral operators that are rather "singular" near the boundary.

The solvability properties of the system (1.1.1) can be discussed within the framework of the Boutet de Monvel theory [Boutet 1-3], when the operators belong to his class (wider classes have been considered in [Vishik-Eskin 1,2], [Eskin 1], [Rempel-Schulze 3,4] and elsewhere).

One can more generally replace $-\Delta$ in (1.1.1) by a pseudo-differential operator P, or more precisely, a certain restriction to Ω, P_Ω, of a ps.d.o. P defined in $\mathbb{R}^n$. Also G and T can be taken more general, but usually adapted to the order of P.

Let us consider the evolution equation ("heat equation")

$$\text{(1.1.5)} \quad \begin{aligned} &\text{(i)} \quad \partial_t u(x,t) + P_\Omega u(x,t) + Gu(x,t) = f(x,t) \quad \text{for } x \in \Omega \, , \ t > 0 \, , \\ &\text{(ii)} \quad Tu(x,t) = \varphi(x,t) \quad \text{for } x \in \partial\Omega \, , \ t > 0 \, , \\ &\text{(iii)} \quad u(x,0) = u_0(x) \quad \text{for } x \in \Omega \, , \ t = 0 \, . \end{aligned}$$

(Cf. e.g. [Nambu 1] and [Triggiani 1] for various cases of this problem.) As is usual in this kind of problem one may start out by considering cases where part of the data are zero, e.g. where f and φ are 0. The problem can then be written more abstractly

$$(1.1.6)\qquad \begin{aligned} \partial_t u(t) + Bu(t) &= 0 \quad \text{for} \quad t > 0 \\ u(0) &= u_0 , \end{aligned}$$

where B is an operator acting like $P_\Omega + G$ and with a domain defined by the boundary condition $Tu = 0$ (B is a so-called realization of P); and it is natural to search for the solution operator of (1.1.6) on the form

$$(1.1.7)\qquad u(t) = \exp(-tB)u_0 ,$$

with a suitable definition of the operator function $\exp(-tB)$.

One of the purposes of the functional calculus that we shall set up here, will be precisely to give a good sense to $\exp(-tB)$ under reasonable hypotheses on B. Of course, $\exp(-tB)$ has a meaning in the traditional theory of semigroups ([Hille-Phillips 1]), when B has certain elementary properties. What we want to do here is to go much further: investigate the expression for $\exp(-tB)$ in terms of symbolic calculus and pseudo-differential (and Fourier integral) operator techniques, in such a way that we can get detailed information on the solutions in terms of their data, and detailed information on the kernel of the solution operator and its trace.

The basic tool here will be an analysis of the resolvent $R_\lambda = (B-\lambda I)^{-1}$, and this analysis will permit the investigation also of other functions of B, defined by a Cauchy integral formula

$$(1.1.8)\qquad f(B) = \frac{1}{2\pi}\int_C f(\lambda)$$

where the integration is over a curve C in the complex plane going around the spectrum of B.

The road to the precise analysis of R_λ is quite long. Before going deeply into the λ-dependent calculus we shall need, we give an account of the parameter-independent calculus, with a special aim towards those operators that do admit a sensible resolvent. The present chapter is concerned with that. First we recall the essential ingredients in the Boutet de Monvel calculus (Section 1.2), explained with relatively few technicalities (the complete story is told in connection with parameter-dependent symbols later anyway). The standard terminology and definitions of function spaces etc. are collected in the Appendix.

In the rest of this chapter, the focus is on ps.d.o.s P admitting solvable realizations B in L^2, i.e., for which P can be supplied with a singular Green operator G and a system T of trace operators so that the operator B in L^2, acting like $P_\Omega+G$ and defined for functions u with $Tu = 0$, is close to being a bijection of $D(B)$ onto L^2. This holds if the system

$$A = \begin{pmatrix} P_\Omega + G \\ \\ T \end{pmatrix}$$

is elliptic in Boutet de Monvel's sense. Here the systems where T is normal (Section 1.4) are of special interest; on one hand because the corresponding class of realizations is closed under composition and passage to adjoints (in the sense of unbounded operators, not just for the bounded case considered in [Schulze 1], [Rempel-Schulze 1]); and on the other hand because the normality is necessary for the property of "ellipticity with a parameter" that we use in the present work. The concepts of parameter-ellipticity and parabolicity are discussed in Section 1.5. In Section 1.6 we analyze the normal realizations more deeply, determining the adjoint and giving criteria for selfadjointness, and presenting some fundamental examples.

An interesting subset of the systems satisfying the requirement of parameter-ellipticity consists of those that define positive realizations (or m-coercive realizations, satisfying the "Gårding inequality"). Because of their importance for the applications of the theory, we study these systems at some length in Section 1.7, showing how certain ideas from [Grubb 1-4] generalize to the present case. The discussion is carried far enough to explain how the Dirichlet-like and Neumann-like boundary conditions (1.1.3) and (1.1.4) fit into the contect; in fact we often return to Example 1.1.1 to illustrate the general principles.

Occasionally, the later theory returns to parameter-independent cases that do not need the strong assumptions presented in Section 1.5, e.g. the index formula in Section 4.3 and some results on spectral estimates in Section 4.5.

Let us finally mention that certain versions (or variants) of the integro-differential problem (1.1.1) have been studied by [Bony-Courrège-Prioret 1] and [Cancelier 1]; they consider second order cases where they can obtain a maximum principle (Cancelier also treats cases without ellipticity), and they discuss the associated semigroup in spaces of continuous functions. [Widom 4] presents a functional calculus for ps.d.o.s of negative order, restricted to a bounded domain; here a certain extendability condition replaces in some sense the need for boundary conditions.

1.1.2 Remark. One of the immediate questions one could ask concerning the boundary problems considered in Example 1.1.1 is whether the non-localness could be eliminated by a change of independent variable. Indeed, it is possible to remove the non-local terms from the boundary conditions (1.1.3) resp. (1.1.4) by insertion of $u = \Lambda^{-1}v$ for a suitable bijective operator Λ, cf. Lemma 1.6.8 below. This leads to a change of the first line of (1.1.1) to another equation

$$-\Delta v + G_1 v = f$$

with another G_1 (nonzero in general). Although such a change of variables can be useful for some questions (e.g. the discussion of positivity at the end of Section 1.7), it does not in general simplify our problem, since the non-local terms that can occur in the first and second line of (1.1.1) have a closely related nature - one might even say that the G terms are more complicated than the T terms. And of course, in questions of operational calculus, the replacement of u by v would lead to the construction of functions of another operator than the given one.

1.2 The calculus of pseudo-differential boundary problems.

In this section, we briefly recall the essential ingredients in the Boutet de Monvel calculus of Green operators [Boutet 1-3]. Accurate details are given e.g. in [Grubb 17] and [Rempel-Schulze 1], and we shall later in this book give a precise account of the parameter-dependent case, where the original operator classes are included as those operators that are independent of the parameter μ, so for the moment, an outline will suffice. (The reader who wants a self-contained presentation of the theory can actually find this in Chapters 2 - 3, by reading the parameter μ as a constant and neglecting the complications due to the presence of μ.)

The basic notions are defined relative to $\mathbb{R}^n$ and the subset $\overline{\mathbb{R}}^n_+ \subset \mathbb{R}^n$; then they are carried over to manifold situations by use of local coordinate systems.

Pseudo-differential operators (ps.d.o.s) were invented as a class of operators that was rich enough to encompass both differential operators and those (singular) integral operators that appear as inverse operators (parametrices) for elliptic differential operators. One can write a differential operator on $\mathbb{R}^n$ of order m

$$A(x,D_x) = \sum_{|\alpha|\le m} a_\alpha(x)D^\alpha$$

by the help of the Fourier transformation as follows:

$$A(x,D_x)u(x) = (2\pi)^{-n}\int e^{ix\cdot\xi}a(x,\xi)\hat{u}(\xi)d\xi ,$$

where $a(x,\xi)$ is the function

$$a(x,\xi) = \sum_{|\alpha|\le m} a_\alpha(x)\xi^\alpha ,$$

called the symbol. The definition of pseudo-differential operators simply allows a more general function (symbol) to be put in the place of $a(x,\xi)$ - for example, the inverse of the operator $1-\Delta$ on $\mathbb{R}^n$ is expressed by

$$(1-\Delta)^{-1}u = (2\pi)^{-n}\int e^{ix\cdot\xi}\frac{1}{|\xi|^2+1}\hat{u}(\xi)d\xi$$

(where $u\in\mathcal{S}(\mathbb{R}^n)$, or the formula is generalized to larger spaces). Various classes of ps.d.o.s are obtained, according to what conditions are imposed on the function entering as $a(x,\xi)$. We shall work here only with (parameter-dependent generalizations of) the most classical ps.d.o. symbols, because the boundary conditions create sufficient complications; but we note in passing that the same questions are of great interest also for more general symbol classes.

So, the pseudo-differential operator P with symbol $p(x,\xi)$ is the operator defined by

$$(1.2.1)\qquad (Pu)(x) = (2\pi)^{-n}\int_{\mathbb{R}^{2n}} e^{i(x-y)\cdot\xi}p(x,\xi)u(y)dy\, d\xi \quad ;$$

one also denotes $P = OP(p(x,\xi))$. The formula is valid for $u \in \mathcal{S}(\mathbb{R}^n)$ and extends by continuity (or by consideration of the adjoint) to more general u , when $p(x,\xi)$ satisfies suitable hypotheses. We shall mainly be concerned with the case where $p \in S^d_{1,0}$, for some $d \in \mathbb{R}$, i.e., p is a C^∞-function satisfying

$$(1.2.2)\qquad |D^\beta_x D^\alpha_\xi p(x,\xi)| \le c(x)\langle\xi\rangle^{d-|\alpha|} \quad \text{for all indices } \alpha \text{ and } \beta$$

(with a continous function $c(x)$ depending on the indices), and the polyhomogeneous case, i.e. the case where $p \in S^d_{1,0}$ and moreover has an asymptotic expansion

$$(1.2.3)\qquad p(x,\xi) \sim \sum_{\ell\in\mathbb{N}} p_{d-\ell}(x,\xi) ,$$

where the $p_{d-\ell}$ are C^∞-functions that are homogeneous of degree $d-\ell$ in ξ for $|\xi| \ge 1$, and $p - \Sigma_{\ell<M}\, p_{d-\ell} \in S^{d-M}_{1,0}$ for each $M \in \mathbb{N}$. The class of polyhomogeneous symbols is called S^d . P and p are said to be of order d , and p_d (also denoted p^0) is called the principal symbol (of order d). The function $p(x,\xi)$ is called the symbol of P , sometimes denoted $\sigma(P)$. The strictly homogeneous function coinciding with p^0 for $|\xi| \ge 1$ will be called $p^h(x,\xi)$.

The function p in formula (1.2.1) can be allowed to depend on (x,y) instead of just x , in which case the operator is denoted $P = OP(p(x,y,\xi))$. $p(x,y,\xi)$ is again called a symbol of P (with a loose terminology, since $\sigma(P)(x,\xi)$ is essentially determined from P , whereas $p(x,y,\xi)$ is not).

The classes of $p(x,y,\xi)$ in $S^d_{1,0}$, resp. polyhomogeneous, are defined as above, with x replaced by (x,y) .

The operator $P = OP(p(x,y,\xi))$ has the distribution kernel

$$(1.2.4)\qquad K_p(x,y) = (2\pi)^{-n}\int_{\mathbb{R}^n} e^{i(x-y)\cdot\xi}p(x,y,\xi)d\xi = F^{-1}_{\xi\to z}\, p(x,y,\xi)\Big|_{z=x-y} .$$

The middle expression here is understood in the sense of oscillatory integrals [Hörmander 6]. The kernel $K_p(x,y)$ is a C^∞ function of (x,y) outside the diagonal $\{x=y\}$, for we have when $2M > n+d+2N$, that $|z|^{2M}(1-\Delta_z)^N F^{-1}_{\xi\to z}p = F^{-1}_{\xi\to z}(-\Delta_\xi)^M\langle\xi\rangle^{2N}p$, the inverse Fourier transform of an L^1 function in ξ , so that $(1-\Delta_z)^N F^{-1}_{\xi\to z}p$ is continuous in z for $z \neq 0$. This calculation also shows that $F^{-1}_{\xi\to z}p$ is $O(|z|^{-2M})$ for $M > n+d$, so $F^{-1}_{\xi\to z}p$ is rapidly decreasing for $|z| \to \infty$. However, $F^{-1}_{\xi\to z}p$ generally has an important singularity at $z = 0$.

Symbols $p(x,y,\xi)$ in the space $S^{-\infty}_{1,0} = \cap_{d\in\mathbb{R}} S^d_{1,0}$ define ps.d.o.s that are in fact integral operators with C^∞ kernel; such operators are called negligible (the symbols also sometimes called negligible). They are smoothing operators, i.e. send distribution spaces into C^∞ spaces, and so are their adjoints.

About the calculus of ps.d.o.s, which is amply treated elsewhere in articles and textbooks (see also our Chapter 2), let us for the moment just mention the following facts.

1.2.1 Proposition. 1^0 *When* $p \in S^d_{1,0}$, *the associated operator is continuous*

$$OP(p)\colon H^s_{comp}(\mathbb{R}^n) \to H^{s-d}_{loc}(\mathbb{R}^n)$$

for any $s \in \mathbb{R}$.

2^0 *When* P *is a ps.d.o., the adjoint* $P^*\colon H^{d-s}_{comp}(\mathbb{R}^n) \to H^{-s}_{loc}(\mathbb{R}^n)$ *is likewise a ps.d.o., and if* $P = OP(p(x,y,\xi))$ *then* $p^* = OP(\overline{p}(y,x,\xi))$. *When* P *has the principal symbol* $p^0(x,\xi)$, P^* *has the principal symbol* $\overline{p^0(x,\xi)}$.

3^0 *Any ps.d.o.* P *can be written as a sum* $P = P_1 + R$, *where* R *is negligible and* P_1 *is proper, i.e.* P_1 *and its adjoint* P_1^* *map compactly supported functions into compactly supported functions (the supports can be further controlled).*

4^0 *When* P *and* P' *are ps.d.o.s of order* d *resp.* d', *and one of them is proper, then* $P'' = PP'$ *is a ps.d.o. of order* $d+d'$. *If* P *and* P' *are polyhomogeneous, so is* P'', *and the principal symbols satisfy*

$$p''^0(x,\xi) = p^0(x,\xi)p'^0(x,\xi) .$$

In the classical theory of elliptic differential operators A, the theory becomes particularly interesting, when the operators are considered on domains with boundary, so that boundary conditions (representing various physical situations) have to be adjoined to get wellposed problems. The calculus of Boutet de Monvel [Boutet 1-3] is a solution to the problem of establishing a class of operators encompassing the elliptic boundary value problems as well as their solution operators; moreover his class of operators is closed under composition (it is an "algebra").

In the case of a differential operator A, the analysis of the various boundary conditions is usually based on the polynomial structure of the symbol of A; in particular the roots of the polynomial $a^0(x',0,\xi',\xi_n)$ in ξ_n (in the situation where the domain is $\mathbb{R}^n_+$) play a rôle. When pseudo-differential operators P are considered, the principal symbol p^0 is generally not a polynomial. It may be a rational function (this happens naturally when one makes reductions in a system of differential operators), in which case one can consider the roots and poles with respect to ξ_n. But then, even when P is elliptic, there is much less control over how these behave than when a^0 is a polynomial; roots and poles may cancel each other or reappear, as the coordinate ξ' varies. For a workable theory, a much more general point of view is needed.

Vishik and Eskin (see [Vishik-Eskin 1] and [Eskin 1]) based a theory on factorization of symbols. This works well in the scalar case, but can be problematic in the case of matrix formed operators (since the factorization here is generally only piecewise continuous in ξ'). - They mainly consider ps.d.o.s of a general kind, with much fewer conditions on the ξ-dependence than (1.2.2) states.

Boutet de Monvel worked out a calculus (that we shall describe below) for a special class of ps.d.o.s; and one of the advantages of that theory is that it replaces the factorization by a projection procedure, that works equally well for scalar and matrix formed operators (depends smoothly on ξ'); it is linked in a natural way with the projections of $L^2(\mathbb{R})$ onto $L^2(\mathbb{R}_+)$ and $L^2(\mathbb{R}_-)$ obtained by restriction (in the x_n-variable). The description that now follows is given in relation to the latter projections, and the Fourier transformed version (of interest in connection with symbols) will be taken up in full detail in Chapter 2.

When P is a ps.d.o. on $\mathbb{R}^n$, its "restriction" to $\Omega = \mathbb{R}^n_+$ is defined by

$$P_\Omega u = r^+ P e^+ u \tag{1.2.5}$$

where r^+ and e^+ are the restriction and extension-by-zero operators, cf. (A.30-31). When P is proper and of order d, this operator is continuous from $L^2_{\mathrm{comp}}(\mathbb{R}^n_+)$ to $H^{-d}_{\mathrm{comp}}(\overline{\mathbb{R}}^n_+)$, but in general does not map $H^m_{\mathrm{comp}}(\overline{\mathbb{R}}^n_+)$ into $H^{m-d}_{\mathrm{comp}}(\overline{\mathbb{R}}^n_+)$ for $m > 0$; the discontinuity of e^+u at $x_n = 0$ causes a singularity. Boutet de Monvel singled out a class of ps.d.o.s where the mapping properties of P_Ω are nice [Boutet 1, 3], namely the ps.d.o.s having the <u>transmission property</u>. Let us just consider operators of integer order $d \in \mathbb{Z}$, then the transmission property at $x_n = 0$ means that the inverse Fourier transform in ξ_n,

$$(1.2.6)\qquad \tilde{p}_{\alpha,\beta}(x',x_n,\xi',z_n) = F^{-1}_{\xi_n\to z_n} D^\beta_x D^\alpha_\xi p(x',x_n,\xi',\xi_n)$$

satisfies, for all α and $\beta \in \mathbb{N}^n$,

$$(1.2.7)\qquad \tilde{p}_{\alpha,\beta}(x',0,\xi',z_n) \text{ is } C^\infty \text{ for } z_n \to 0+ \text{ and for } z_n \to 0- \ .$$

We denote $\tilde{p}_{0,0} = \tilde{p}$. The arguments given further above show that $\tilde{p}_{\alpha,\beta}$ is always a C^∞ function of z_n for $z_n \neq 0$ (it even goes to zero rapidly for $|z_n| \to \infty$), but (1.2.7) is a particular condition on the behaviour at $z_n = 0$. It does not exclude distributional terms supported by $z_n = 0$ (in fact when P is a differential operator, $\tilde{p}$ is supported by $z_n = 0$). When p depends also on y, the transmission property at $x_n = 0$ follows if for all α,β,

$$\tilde{p}_{\alpha,\beta}(x',0,y',0,\xi',z_n) \text{ is } C^\infty \text{ for } z_n \to 0+ \text{ and for } z_n \to 0- \ ,$$

where $\tilde{p}_{\alpha,\beta} = F^{-1}_{\xi_n\to z_n} D^\beta_{x,y} D^\alpha_\xi p$. (Fourier transformed versions are given below.)

Observe that the transmission property at $x_n = 0$ puts no restrictions on the behavior of p and $\tilde{p}_{\alpha,\beta}$ at the points (x',x_n) with $x_n \neq 0$.

When $p \in S^d_{1,0}$ has the transmission property at $x_n = 0$, the operator $P_\Omega = r^+ \operatorname{OP}(p)e^+$ is seen to be continuous:

$$(1.2.8)\qquad \begin{aligned} &P_\Omega\colon H^s_{\mathrm{comp}}(\overline{\mathbb{R}}^n_+) \to H^{s-d}_{\mathrm{loc}}(\overline{\mathbb{R}}^n_+) \quad \text{for } s > -\tfrac{1}{2}\ ,\\ &P_\Omega\colon H^{(s,t)}_{\mathrm{comp}}(\overline{\mathbb{R}}^n_+) \to H^{(s-d,t)}_{\mathrm{loc}}(\overline{\mathbb{R}}^n_+) \quad \text{for } s > -\tfrac{1}{2}\ ,\ t \in \mathbb{R}\ ,\end{aligned}$$

and it maps $C^\infty_{(0)}(\overline{\mathbb{R}}^n_+)$ into $C^\infty(\overline{\mathbb{R}}^n_+)$ (and $\mathring{\mathcal{E}}'(\overline{\mathbb{R}}^n_+)$ into $\mathring{\mathcal{D}}'(\overline{\mathbb{R}}^n_+)$ if $d \le 0$). In the case $n = 1$ these properties are not hard to see from the condition (1.2.7) on the kernel $\tilde{p}(x_n,x_n-y_n)$, and this gives the essential step in the proof. (Note that the analogous operator for $\Omega = \mathbb{R}^n_-$ will have similar continuity properties in view of the symmetry of (1.2.7); this is particular for the integer-order case - in case $d \in \mathbb{R}\setminus\mathbb{Z}$ one has a one-sided kind of transmission property, cf. [Boutet 1].) Invariance is shown in Theorem 2.2.12 3° later.

The transmission property can also be expressed directly in terms of the symbol $p(x,\xi)$ (or $p(x,y,\xi)$). Here it means that for any α and $\beta \in \mathbb{N}^n$, $D^\beta_x D^\alpha_\xi p(x',0,\xi)$ has an asymptotic expansion when $|\xi_n| \to \infty$

$$(1.2.9)\qquad D^\beta_x D^\alpha_\xi p(x',0,\xi) \sim \sum_{-\infty<\ell\le d-|\alpha|} s_{\ell,\alpha,\beta}(x',\xi')\xi_n^\ell\ ,$$

where the $s_{\ell,\alpha,\beta}(x',\xi')$ are <u>polynomials</u> in ξ' of degree $d-|\alpha|-\ell$. In particular, writing $s_{\ell,0,0} = s_\ell$, we have that

(1.2.10) $$p(x',0,\xi) = \sum_{0\le\ell\le d} s_\ell(x',\xi')\xi_n^\ell + p'(x',\xi) ,$$

where p' is $O(\langle\xi_n\rangle^{-1})$, s_ℓ is of order $d-\ell$, and the first coefficient s_d is a <u>function</u> of x' ,

(1.2.11) $$s_d(x',\xi') = s_d(x') .$$

(More precisely, one has that $p'(x',\xi)$ is $O(\langle\xi'\rangle^{d+1}\langle\xi\rangle^{-1})$.)

There is a very important observation on the transmission property in the polyhomogenous case, namely that here a symbol $p(x,\xi) \in S^d$ <u>has the transmission property at</u> $x_n = 0$ <u>if and only if the homogeneous terms</u> $p_{d-\ell}$ <u>have the symmetry property</u>

(1.2.12) $$D_x^\beta D_\xi^\alpha p_{d-\ell}(x',0,0,\xi_n) = (-1)^{d-\ell-|\alpha|} D_x^\beta D_\xi^\alpha p_{d-\ell}(x',0,0,-\xi_n) \quad \text{for} \quad |\xi_n|\ge 1 ,$$

for all indices α , β , ℓ . The necessity of (1.2.12) is deduced from the fact that when p is homogeneous of degree d and satisfies (1.2.9), then

(1.2.13) $$D_x^\beta D_\xi^\alpha p(x',0,0,\xi_n) = s(x')\xi_n^{d-|\alpha|} , \quad \text{for} \quad |\xi_n| \ge 1 ,$$

where the coefficient $s(x')$ is the function $s_{d-|\alpha|,\alpha,\beta}$ in (1.2.9), which is a polynomial in ξ' of degree 0; the other terms vanish at $\xi' = 0$. For the sufficiency we refer to the proof of Theorem 2.2.5 later, carried out in the (more difficult) parameter-dependent case. The equations (1.2.12) can then be taken as <u>definition</u> of the transmission property.

We see from (1.2.10) that <u>if</u> p <u>is constant in</u> x_n , P is the sum of a differential operator and an operator with a certain L^2-continuity in x_n . Generally, p can be quite nasty for $x_n \neq 0$, but here a Taylor expansion in x_n gives a number of good terms behaving as in (1.2.10) and a term with a factor x_n^M that makes it harmless (there is a precise statement in Lemma 1.3.1 further below).

The transmission property implies that one can build up a theory of boundary value problems for P , with many of the same features as the theory for differential operators (and some differences); much more will be said about this in the rest of the present chapter.

Let us now introduce the other ingredients in the Boutet de Monvel calculus. (The precise symbol definitions are listed in Definition 2.3.13.)

Let P be $N'\times N$-matrix formed. Along with P_Ω, that operates on Ω, we shall consider operators going to and from the boundary, forming together with P a system

$$A = \begin{pmatrix} P_\Omega + G & K \\ & \\ T & S \end{pmatrix} : \begin{matrix} C^\infty_{(0)}(\overline{\Omega})^N \\ \times \\ C^\infty_0(\Gamma)^M \end{matrix} \to \begin{matrix} C^\infty(\overline{\Omega})^{N'} \\ \times \\ C^\infty(\Gamma)^{M'} \end{matrix} . \tag{1.2.14}$$

Here T is a so-called trace operator, going from Ω to Γ ; K is a so-called Poisson operator, going from Γ to Ω ; S is a pseudo-differential operator on Γ ; and G is an operator on Ω called a singular Green operator, it is a non-pseudo-differential term that has to be included in order to have good composition rules. The full system A is called a Green operator. We shall usually take $N = N'$, whereas the dimensions M and M' can have all values, including zero. When P is a differential operator, it is classical to study systems of the form

$$A = \begin{pmatrix} P_\Omega \\ T \end{pmatrix} ; \tag{1.2.15}$$

here $M = 0$ and $M' > 0$. The terms in (1.2.14) will now be explained.

The trace operators we consider are defined in such a way that

$$T = \gamma_0 P_\Omega \tag{1.2.16}$$

is a trace operator, whenever P is a ps.d.o. having the transmission property. Here, when $P = OP(p(x,\xi))$, then $\gamma_0 P_\Omega u = \gamma_0 r^+ OP(p(x',0,\xi))e^+u$, where we can insert (1.2.10). This gives a sum of differential trace operators plus a term where the symbol is $O(\langle\xi_n\rangle^{-1})$. The general definition goes as follows:

A trace operator or order d $(\in \mathbb{R})$ and class r $(\in \mathbb{N})$ is an operator of the form

$$Tu = \sum_{0\le j\le r-1} S_j\gamma_j + T' , \tag{1.2.17}$$

where γ_j denotes the standard trace operator $(\gamma_j u)(x') = D^j_{x_n} u(x',0)$; the S_j are ps.d.o.s in $\mathbb{R}^{n-1}$ of order $d-j$, and T' is an operator of the form

$$(T'u)(x') = (2\pi)^{-1-n} \int_{\mathbb{R}^{n-1}} e^{ix'\cdot\xi'} \int_0^\infty \tilde{t}'(x',x_n,\xi')\acute{u}(\xi',x_n)dx_n \, d\xi' , \tag{1.2.18}$$

with $\tilde{t}' \in \mathcal{S}(\overline{\mathbb{R}}_+)$ as a function of x_n, satisfying estimates for all indices:

$$\| x_n^\ell D^{\ell'}_{x_n} D^\beta_{x'} D^\alpha_{\xi'} \tilde{t}'(x',x_n,\xi') \|_{L^2_{x_n}(\mathbb{R}_+)} \le c(x')\langle\xi'\rangle^{d+\frac{1}{2}-\ell+\ell'-|\alpha|} ; \tag{1.2.19}$$

we denote $F_{x'\to\xi'}u = \acute{u}$ (A.18).

This is the definition for the $S_{1,0}$ type of operator, and one can show that for any such T' there exists a ps.d.o. Q of $S_{1,0}$ type, having the transmission property at $x_n = 0$, such that

$$T' = \gamma_0 Q_\Omega . \tag{1.2.20}$$

The subclass of polyhomogeneous trace operators are those where the S_j are polyhomogeneous, and $\tilde{t}'$ has an asymptotic expansion

$$\tilde{t}'(x',x_n,\xi') \sim \sum_{\ell\in\mathbb{N}} \tilde{t}'_{d-\ell}(x',x_n,\xi') , \tag{1.2.21}$$

where each $\tilde{t}'_{d-\ell}$ is C^∞ and quasi-homogeneous in the sense that

$$\tilde{t}'_{d-\ell}(x',\tfrac{1}{\lambda}x_n,\lambda\xi') = \lambda^{d-\ell+1}\,\tilde{t}'_{d-\ell}(x',x_n,\xi') \quad \text{for } \lambda \geq 1 \text{ and } |\xi'| \geq 1 , \tag{1.2.22}$$

$\tilde{t}' - \Sigma_{\ell<M}\,\tilde{t}'_{d-\ell}$ satisfying the estimates (1.2.18) with d replaced by $d-M$, for any $M \in \mathbb{N}$. The function (distribution when $r > 0$)

$$\tilde{t}(x',x_n,\xi') = \sum_{0\leq j<r} s_j(x',\xi')D^j_{x_n}\,\delta(x_n) + \tilde{t}'(x',x_n,\xi') \tag{1.2.23}$$

is called the symbol-kernel of T; its Fourier transform

$$t(x',\xi) = F_{x_n\to\xi_n}\tilde{t}(x,\xi') \tag{1.2.23'}$$

being the symbol of T. (The symbol spaces and symbol-kernel spaces are taken up again in Chapter 2, see e.g. Definition 2.3.13.) We also set $t'(x',\xi) = F_{x_n\to\xi_n}\tilde{t}'(x,\xi')$. In the polyhomogeneous case, t has an expansion corresponding to (1.2.21)

$$t(x',\xi) \sim \sum_{\ell\in\mathbb{N}} t_{d-\ell}(x',\xi) , \quad \text{with}$$

$$t_{d-\ell}(x',\lambda\xi) = \lambda^{d-\ell}t_{d-\ell}(x',\xi) \quad \text{for } \lambda \geq 1 ,\ |\xi'| \geq 1 . \tag{1.2.24}$$

We often denote t_d and $\tilde{t}_d$ by t^0 resp. $\tilde{t}^0$, the principal symbol and symbol-kernel. Application of the operator definition with respect to the x_n variable only, gives the boundary symbol operator $t(x',\xi',D_n)$ (resp. principal boundary symbol operator $t^0(x',\xi',D_n)$) from $\mathscr{S}(\overline{\mathbb{R}}_+)$ to $\mathbb{C}$,

$$t(x',\xi',D_n)u = \sum_{0\leq j<r} s_j(x',\xi')\gamma_j u + \int_0^\infty \tilde{t}'(x',x_n,\xi')u(x_n)dx_n ; \tag{1.2.25}$$

it is also denoted $\mathrm{OPT}_n(t)$. We can then write

(1.2.26) $$Tu = OP'(t(x',\xi',D_n))u\ , \quad \text{also denoted} \quad OPT(t(x',\xi))u\ ,$$

where OP' stands for application of the ps.d.o. definition with respect to the x' variable. Here $T' = OPT(t')$ is of class 0. The trace operator (1.2.16) has precisely the symbol-kernel $\tilde{p}(x',0,\xi',x_n)|_{x_n\geq 0}$.

A trace operator T of order d and class r has the continuity properties

(1.2.27)
$$\begin{aligned} &T\colon H^s_{\mathrm{comp}}(\overline{\mathbb{R}}^n_+) \to H^{s-d-\frac12}_{\mathrm{loc}}(\mathbb{R}^{n-1}) && \text{for } s > r-\tfrac12\ ,\\ &T\colon H^{(s,t)}_{\mathrm{comp}}(\overline{\mathbb{R}}^n_+) \to H^{s+t-d-\frac12}(\mathbb{R}^{n-1}) && \text{for } s > r-\tfrac12\ , \text{ all } t\in\mathbb{R}\ ,\end{aligned}$$

it maps $C^\infty_{(0)}(\overline{\mathbb{R}}^n_+)$ into $C^\infty(\mathbb{R}^{n-1})$, and, if $r = 0$, $\mathring{\mathcal{E}}'(\overline{\mathbb{R}}^n_+)$ into $\mathcal{D}'(\mathbb{R}^{n-1})$.

Just as the general definition of trace operators was motivated by (1.2.16), one can introduce the class of Poisson operators with the motivation that K defined by

(1.2.28) $$(Kv)(x) = r^+ P(v(x')\otimes\delta(x_n))$$

should be in this class, whenever P is a ps.d.o. having the transmission property. In general, a Poisson operator of order d is defined by the formula

(1.2.29) $$(Kv)(x',x_n) = (2\pi)^{1-n}\int_{\mathbb{R}^{n-1}} e^{ix'\cdot\xi'}\tilde{k}(x',x_n,\xi')\hat{v}(\xi')d\xi'$$

where the symbol-kernel $\tilde{k}$ satisfies just the same estimates as $\tilde{t}'$ in (1.2.19), except that d is replaced by $d-1$. This gives the $S_{1,0}$ type of operators; the polyhomogeneous symbol-kernels $\tilde{k}$ moreover have asymptotic expansions in quasi-homogeneous terms, and the corresponding symbols

(1.2.30) $$k(x',\xi) = F_{x_n\to\xi_n}\tilde{k}(x,\xi')$$

have expansions in homogeneous terms in ξ (for $|\xi'|\geq 1$) of degree $d-1-\ell$. We often denote $\tilde{k}_{d-1} = \tilde{k}^0$ and $k_{d-1} = k^0$; the principal symbol-kernel or symbol. Again, one can view K defined in (1.2.29) (also denoted $OPK(k)$ or $OPK(\tilde{k})$) as an operator

(1.2.31) $$K = OP'(k(x',\xi',D_n))$$

where $k(x',\xi',D_n)$ is the boundary symbol operator from $\mathbb{C}$ to $\mathcal{S}(\overline{\mathbb{R}}_+)$

(1.2.32) $$k(x',\xi',D_n)a = \tilde{k}(x',x_n,\xi')\cdot a \quad \text{for } a\in\mathbb{C}\ ,$$

also denoted $OPK_n(k)$.

One can show that any $S_{1,0}$ type Poisson operator can be written in the form (1.2.28) with a ps.d.o. P of $S_{1,0}$ type.

The Poisson operators of order d are continuous

$$(1.2.33)\qquad \begin{aligned} K:\ & H^{s}_{comp}(\mathbb{R}^{n-1}) \to H^{s-d+\frac{1}{2}}_{loc}(\overline{\mathbb{R}}^n_+) \\ K:\ & H^{s}_{comp}(\mathbb{R}^{n-1}) \to H^{(m,s-d-m+\frac{1}{2})}_{loc}(\overline{\mathbb{R}}^n_+) \end{aligned}$$

for any $s \in \mathbb{R}$, any $m \geq 0$, besides mapping $C_0^\infty(\mathbb{R}^{n-1})$ resp. $\mathcal{E}'(\mathbb{R}^{n-1})$ into $C^\infty(\overline{\mathbb{R}}^n_+)$ resp. $\mathcal{D}'(\overline{\mathbb{R}}^n_+)$.

The order convention may seem a bit strange (polyhomogeneous Poisson operators of order d have principal symbols homogeneous of degree $d-1$), but this fits the purpose that the composition of two operators of order d resp. d' will be of order $d+d'$ (valid e.g. for the ps.d.o. TK on $\mathbb{R}^{n-1}$).

Now there is a very important observation concerning adjoints: The trace operators T' of class 0 (and order d) have as adjoints precisely the Poisson operators (of order $d-1$), and vice versa:

$$(1.2.34)\qquad \begin{aligned} T':\ & L^2_{comp}(\mathbb{R}^n_+) \to H^{-d-\frac{1}{2}}_{loc}(\mathbb{R}^{n-1}) \quad \text{and} \\ K = T'^*:\ & H^{d+\frac{1}{2}}_{comp}(\mathbb{R}^{n-1}) \to L^2_{loc}(\mathbb{R}^n_+) \quad \text{, are each others adjoints.} \end{aligned}$$

This is quite obvious on the boundary symbol operator level, and follows in general after application of Proposition 1.2.1 2° to OP'. In fact, when the OP' notion is extended to symbols depending on (x',y') instead of x', then if T' has symbol-kernel $\tilde{t}'(x,\xi')$, then T'^* has symbol-kernel $\overline{\tilde{t}}'(y',x_n,\xi')$ (the complex conjugate). -Trace operators of class $r > 0$ do not have adjoints within the present calculus.

We now get to the more peculiar element G in A (1.2.14). A singular Green operator (s.g.o.) G arizes for instance when we compose a Poisson operator K with a trace operator T as $G = KT$; this operator acts in $\Omega = \mathbb{R}^n_+$ but is not a P_Ω. Another situation where s.g.o.s enter is when we compose two ps.d.o.s P_Ω and Q_Ω (having the transmission property); then the "leftover" operator

$$(1.2.35)\qquad L(P,Q) \equiv (PQ)_\Omega - P_\Omega Q_\Omega = r^+PQe^+ - r^+Pe^+r^+Qe^+ = r^+P(I-e^+r^+)Qe^+$$

is an operator, acting in Ω, that is not a ps.d.o. It turns out that these cases are covered by operators of the following form (on the boundary symbol level it is a completed tensor product of Poisson and trace symbols).

A singular Green operator G of order d $(\in \mathbb{R})$ and class r $(\in \mathbb{N})$ is an operator

(1.2.36) $$G = \sum_{0\leq j\leq r-1} K_j \gamma_j + G' ,$$

where the K_j are Poisson operators of order $d-j$, the γ_j are standard trace operators, and G' is an operator of the form

(1.2.37) $$(G'u)(x) = (2\pi)^{1-n} \int_{\mathbb{R}^{n-1}} e^{ix'\cdot\xi'} \int_0^\infty \tilde{g}'(x',x_n,y_n,\xi')\acute{u}(\xi',y_n)dy_n \, d\xi' .$$

Here $\tilde{g}'$, the symbol-kernel of G', is in $\mathscr{S}(\overline{\mathbb{R}}^2_{++})$ as a function of (x_n,y_n) satisfying estimates for all indices

(1.2.38) $$\| x_n^k D_{x_n}^{k'} y_n^m D_{y_n}^{m'} D_{x'}^\beta D_{\xi'}^\alpha \tilde{g}'(x',x_n,y_n,\xi)\|_{L^2_{x_n,y_n}(\mathbb{R}^2_{++})} \leq c(x')\langle\xi'\rangle^{d-k+k'-m+m'-|\alpha|} .$$

This is the definition for operators of $S_{1,0}$ type; for the subclass of polyhomogeneous s.g.o.s it is furthermore required that the K_j are polyhomogeneous, and the symbol-kernel $\tilde{g}'$ of G' has an asymptotic expansion in quasi-homogeneous terms

(1.2.39) $$\tilde{g}'(x',x_n,y_n,\xi') \sim \sum_{\ell\in\mathbb{N}} \tilde{g}'_{d-1-\ell}(x',x_n,y_n,\xi')$$

where

(1.2.40) $$\tilde{g}'_{d-1-\ell}(x',\tfrac{1}{\lambda}x_n,\tfrac{1}{\lambda}y_n,\lambda\xi') = \lambda^{d+1-\ell}\tilde{g}'_{d-1-\ell}(x',x_n,y_n,\xi')$$
$$\text{for } \lambda \geq 1 \text{ and } |\xi'| \geq 1 ,$$

corresponding to an expansion of the corresponding symbol g'

(1.2.41) $$g'(x',\xi',\xi_n,\eta_n) = F_{x_n\to\xi_n} \overline{F}_{y_n\to\xi_n} \tilde{g}'(x',x_n,y_n,\xi')$$

in homogeneous terms

(1.2.42) $$g'_{d-1-\ell}(x',\lambda\xi',\lambda\xi_n,\lambda\eta_n) = \lambda^{d-1-\ell} g'_{d-1-\ell}(x,\xi',\xi_n,\eta_n)$$
$$\text{for } \lambda \geq 1 \text{ and } |\xi'| \geq 1 .$$

The expansion holds in the sense that $\tilde{g}' - \Sigma_{\ell<M}\, \tilde{g}'_{d-1-\ell}$ satisfies (1.2.38) with d replaced by $d-M$, for any $M \in \mathbb{N}$. The symbol-kernel and symbol of G itself are

$$\tilde{g}(x',x_n,y_n,\xi') = \sum_{0\le j<r} \tilde{k}_j(x',x_n,\xi')D^j_{y_n}\,\delta(y_n) + \tilde{g}'(x',x_n,y_n,\xi')$$

(1.2.43)

$$g(x',\xi',\xi_n,\eta_n) = \sum_{0\le j<r} k_j(x',\xi)\eta_n^j + g'(x',\xi',\xi_n,\eta_n) .$$

The <u>principal</u> symbol-kernel and symbol are $\tilde{g}^0 = \tilde{g}_{d-1}$ resp. $g^0 = g_{d-1}$.

We define the <u>boundary symbol operator</u> $g(x',\xi',D_n)$ from $\tilde{g}$ by

$$(1.2.44)\quad g(x',\xi',D_n)u(x_n) = \sum_{0\le j<r} \tilde{k}_j(x',x_n,\xi')\gamma_j u + \int_0^\infty \tilde{g}'(x',x_n,y_n,\xi')u(y_n)dy_n ,$$

also called $\mathrm{OPG}_n(g)$; then G (also called $\mathrm{OPG}(g)$ or $\mathrm{OPG}(\tilde{g})$) can be viewed as

$$(1.2.45)\qquad G = \mathrm{OP}'(g(x',\xi',D_n)) .$$

A s.g.o. G of order d and class r defines continuous operators

$$G:\ H^s(\overline{\mathbb{R}}^n_+) \to H^{s-d}(\overline{\mathbb{R}}^n_+) \qquad \text{for } s > r - \tfrac{1}{2} ,$$

(1.2.46)

$$G:\ H^{(s,t)}(\overline{\mathbb{R}}^n_+) \to H^{(m,s-d-m+t)}(\overline{\mathbb{R}}^n_+) \qquad \text{for } s > r - \tfrac{1}{2} ,$$

$$m \ge 0 \text{ and } t \in \mathbb{R} ,$$

and maps $C^\infty_{(0)}(\overline{\mathbb{R}}^n_+)$ into $C^\infty(\overline{\mathbb{R}}^n_+)$ and, if $r = 0$, $\mathring{\mathcal{E}}'(\overline{\mathbb{R}}^n_+)$ into $\mathring{\mathcal{D}}'(\overline{\mathbb{R}}^n_+)$.

It is important to observe that s.g.o.s of class 0 have <u>adjoints</u> of the same kind: Using the OP' notation for symbols depending on (x',y') we have that

$$G = \mathrm{OPG}(\tilde{g}(x',x_n,y_n,\xi')) \quad \text{implies}$$

(1.2.47)

$$G^* = \mathrm{OPG}(\overline{\tilde{g}}(y',y_n,x_n,\xi')) \quad \text{(complex conjugate)},$$

when G is of class 0. S.g.o.s of class $r > 0$ do not have adjoints within the present calculus.

By an expansion of $\tilde{g}'$ in Laguerre series with respect to x_n and y_n (as explained e.g. in [Grubb 17], see also Section 2.2), we can write G on the form

$$(1.2.48)\qquad G = \sum_{m=1}^{\infty} K_m T_m ,$$

where the K_m and T_m are sequences of Poisson operators of order 0 resp. trace operators of order d and class r such that $K_m T_m$ is rapidly decreasing (with respect to all symbol norms, as in (1.2.38)). This shows that any singular Green operator can be written as the product of a Poisson operator (the row $\{K_m\}_{m<M}$) and a trace operator (the column $\{T_m\}_{m<M}$) plus a small s.g.o., a point of view that is fruitful e.g. in the study of inverses of

boundary problems. Moreover, the K_m and T_m can be written on a form generated from ps.d.o.s as in (1.2.28) resp. (1.2.17, 20), using the extension of [Seeley 5].

The special s.g.o. $L(P,Q)$ is generated from ps.d.o.s in a somewhat different way also. Here $L(P,Q)$ can be broken up in simpler terms

$$L(P,Q) = \sum_{0\leq j<d'} K_j \gamma_j + G^+(P)G^-(Q) \;, \tag{1.2.49}$$

where the K_j are Poisson operators derived from P and the "differential operator part" of Q (cf. (1.2.10), d' is the order of Q and the contribution vanishes when $d' \leq 0$), and $G^+(P)$ and $G^-(Q)$ are special s.g.o.s of class 0, defined by the formulas

$$\begin{aligned} G^+(P) &= r^+Pe^-J \\ G^-(Q) &= Jr^-Qe^+ = G^+(Q^*)^* \;, \end{aligned} \tag{1.2.50}$$

where J is the reflection operator

$$J\colon u(x',x_n) \rightsquigarrow u(x',-x_n) \;, \tag{1.2.51}$$

see [Grubb 17].

To the above operators defined by Fourier integral formulas, one must add the negligible operators of each type, defined as operators of the form (1.2.17), (1.2.34), (1.2.36) with S_j, T', K, K_j and G' replaced by integral operators with C^∞ kernels (up to the boundary) over the respective domains. They are of class r, when they contain trace operators γ_j for $j \leq r-1$.

The various operator classes defined above are invariant under coordinate changes in $\overline{\mathbb{R}}^n_+$ preserving the boundary $\{x_n=0\}$; this holds both for the polyhomogeneous classes and the $S_{1,0}$ classes. This is stated in [Boutet 3], with an indication of how to conclude the invariance for $S_{1,0}$ Poisson operators once it is shown for $S_{1,0}$ ps.d.o.s having the transmission property. [Rempel-Schulze 1] proves the invariance under coordinate changes in x' alone, where the rules for ps.d.o.s in x' apply. A complete proof, with formulas for the symbols of the transformed operators, is given below in Section 2.4 (see also Theorem 2.2.12 3°), including parameter-dependence.

Having defined the ingredients in A (1.2.14) we shall now look at composition rules. When A' is another such system, going from $C^\infty_{(0)}(\overline{\mathbb{R}}^n_+)^{N'} \times C^\infty_0(\mathbb{R}^{n-1})^{M'}$ to $C^\infty(\overline{\mathbb{R}}^n_+)^{N''} \times C^\infty(\mathbb{R}^{n-1})^{M''}$, and one of the operators is proper, the composition equals

$$(1.2.52)\qquad A'' = \begin{pmatrix} P_\Omega + G & K \\ T & S \end{pmatrix} \begin{pmatrix} P'_\Omega + G' & K' \\ T' & S' \end{pmatrix}$$

$$= \begin{pmatrix} (PP')_\Omega - L + P_\Omega G' + GP'_\Omega + GG' + KT' & P_\Omega K' + GK' + KS' \\ TP'_\Omega + TG' + ST' & TK' + SS' \end{pmatrix} .$$

One can now show that A'' again has the structure of a Green operator, which really amounts to showing 14 different composition rules (cf. [Boutet 3], [Rempel-Schulze 1] and [Grubb 17]); these will also be taken up in detail in Chapter 2.

It is also of interest to see whether A has an adjoint within the Green operator calculus. In view of the preceding information, it is true when G and T are of class 0 and P is of order ≤ 0 (it can also hold in certain positive order cases, cf. Lemma 1.3.1 later). There is a technical device worth mentioning here, that can transform the system A into one that does have an adjoint, namely, there exists a family of ps.d.o.s Λ^m_- such that $\Lambda^m_{-,\Omega}$ maps $H^{m_+}(\overline{\mathbb{R}}^n_+)$ homeomorphically onto $H^{m_-}(\overline{\mathbb{R}}^n_+)$, where $m_\pm = \max\{\pm m, 0\}$ (see [Boutet 3], [Rempel-Schulze 1], and Chapter 3 below). If P is of order $d > 0$ and T and G are of class $r > 0$ we can compose A to the right with

$$(1.2.52')\qquad \begin{pmatrix} \Lambda^{-m}_{-,\Omega} & 0 \\ 0 & I \end{pmatrix} ,$$

where $m = \max\{r,d\}$, which reduces A to a system of class 0. This is useful e.g. in the study of index problems.

For the study of invertible elements in the "algebra" we need to define the concept of ellipticity. This really consists of two conditions. One is that P is elliptic on $\overline{\mathbb{R}}^n_+$, i.e., the principal symbol $p^0(x,\xi)$ is an invertible function (or square matrix), when $|\xi| \ge c > 0$. The other condition is that the (x_n-independent) principal boundary symbol operator for A,

$$(1.2.53)\qquad a^0(x',\xi',D_n) = \begin{pmatrix} p^0(x',0,\xi',D_n)_\Omega + g^0(x',\xi',D_n) & k^0(x',\xi',D_n) \\ t^0(x',\xi',D_n) & s^0(x',\xi') \end{pmatrix}$$

defines a bijection from $\mathscr{S}(\overline{\mathbb{R}}_+)^N \times \mathbb{C}^M$ to $\mathscr{S}(\overline{\mathbb{R}}_+)^N \times \mathbb{C}^{M'}$, for all x', all $|\xi'| \ge c > 0$. It is an important point in the theory, that this hypothesis suffices to assure that the inverse operator $(a^0(x',\xi',D_n))^{-1}$ is again the principal boundary symbol operator $c^0(x',\xi',D_n)$ for a Green system C^0 belonging to the theory, which is not at all obvious from the mere bijectiveness. When the ellipticity holds, it is possible to construct a Green operator C (with the same principal boundary symbol operator as C^0) which is a parametrix of A , in the sense that

$$\text{(1.2.54)} \qquad AC - I \quad \text{and} \quad CA - I \quad \text{are negligible.}$$

(Note here that since A and C are not necessarily of class 0 , the negligible operators in (1.2.54) need not be of class 0.)

For a complete description, there remains to define the operators as acting in bundles over manifolds. The details can be left out here, since we do it carefully for parameter-dependent operators later on (in Section 2.4), so let us just mention that when E is a vector bundle of dimension N over the compact set $\overline{\Omega} \subset \Sigma$ with boundary Γ (described by local coordinates and trivializations in the Appendix), and P is a ps.d.o. in $\tilde{E}$ over Σ (where $E = \tilde{E}|_{\overline{\Omega}}$), having the transmission property at Γ , then Green operators considered in connection with P are of the form

$$\text{(1.2.55)} \qquad A = \begin{pmatrix} P_\Omega + G & K \\ T & S \end{pmatrix} : \begin{matrix} C^\infty(E) \\ \times \\ C^\infty(F) \end{matrix} \rightarrow \begin{matrix} C^\infty(E) \\ \times \\ C^\infty(F') \end{matrix}$$

where $\dim F = M$ and $\dim F' = M'$. Here P_Ω is defined by (1.2.5), (A.64). The terms T, K and S are often given as block matrices with different orders for different entries (fitting together as in Douglis - Nirenberg elliptic systems). The continuity statements (1.2.8), (1.2.27), (1.2.33) and (1.2.46) carry over to E and the respective bundles over Γ , when we replace $H^s_{loc}(\overline{\mathbb{R}}^n_+)$ and $H^s_{comp}(\overline{\mathbb{R}}^n_+)$ by $H^s(E)$, replace the boundary spaces similarly, and replace the $H^{(s,t)}$ spaces by $H^{(s,t)}(E_{\Sigma'_+})$ as in (A.69), inserting cut-off functions supported near Γ in the formulas containing the $H^{(s,t)}$-spaces.

When P is elliptic, it can be shown that $p^0(x',\xi',D_n)_\Omega$ is a Fredholm operator in $\mathscr{S}(\overline{\mathbb{R}}_+)^N$ (and between suitable Sobolev spaces over $\overline{\mathbb{R}}_+$, the nullspace and range complement being the same as for $\mathscr{S}(\overline{\mathbb{R}}_+)^N$), with an index depending continously on (x',ξ') . A necessary condition for the ellipticity of A is that $M' - M = \text{index } p^0$. When ellipticity holds, A itself is a Fredholm operator, both as an operator between the C^∞ spaces in (1.2.55) and as an operator between suitable Sobolev spaces; the nullspace and range complement being the same C^∞ spaces as in the situation (1.2.55). More details are given for a particular case in Section 1.4, and for the parameter-dependent case in Chapter 3. The index of A is studied in [Boutet 3] and in [Rempel-Schulze 1]. In the present work, our analysis of the trace of the heat operator will lead to a new formula for the index of normal boundary problems (see Section 4.3), which involves slightly fewer symbol data than the general formula of S. Rempel (see [Rempel-Schulze 1] or [Rempel 1]).

1.2.2 Remark. Throughout this work we use a convention for polyhomogeneous symbols $s \sim \Sigma_{\ell \geq 0} s_{d-\ell}$, where the individual terms $s_{d-\ell}$ are only assumed to be homogeneous for $|\xi| \geq 1$ (in the ps.d.o. case) resp. for $|\xi'| \geq 1$ (in the other cases); on the other hand, they are C^∞ for $|\xi| \leq 1$ resp. $|\xi'| \leq 1$ and belong to the respective symbol classes themselves. Occasionally, we also need to refer to the associated strictly homogeneous symbols, $s^h_{d-\ell}$, defined by extension by homogeneity; these are generally irregular at $\xi = 0$ resp. $\xi' = 0$, whereas the smooth terms $s_{d-\ell}$ have the advantage of being directly used in the operator definitions.

This convention is perhaps not the most usual for ps.d.o.s, but it is very useful in the study of boundary problems (as introduced in [Boutet 3]), where ξ' and ξ_n (resp. ξ', ξ_n and η_n) enter together in an intricate way. This is even more so, when we include the parameter μ later in this book, considering symbols that are homogeneous in (ξ',ξ_n,μ) (or in (ξ',ξ_n,η_n,μ)) for $|\xi'| \geq 1$ but with a complicated behavior near $\xi' = 0$ that is best described by estimates.

For simplicity, we assume in the rest of this chapter that all symbols are polyhomogeneous, although some of the results are valid also for $S_{1,0}$ symbol classes. (For, as we recall, expansions like (1.2.9) do not require polyhomogeneity. For precise statements, see Definition 2.2.7 (applied to μ-independent symbols), and Definition 2.3.13, or the previous works on the parameter-independent theory [Boutet 3], [Rempel-Schulze 1] or [Grubb 17].)

It is assumed throughout that the ps.d.o.s (on the n-dimensional manifold) have the transmission property (at the boundary).

1.3 Green's formula.

In the present calculus, the ps.d.o.s of positive order satisfy a Green's formula quite similar to the one in the differential operator case. We first observe

Lemma 1.3.1. *Let* P *be a ps.d.o. in* $\tilde{E}$, *having the transmission property at* Γ . *Then* P *can be written*

$$P = A + P' , \tag{1.3.1}$$

where A *is a differential operator of order* d *of the form*

$$A = \sum_{\ell=0}^{d} S_\ell(x,D')D_n^\ell \tag{1.3.2}$$

with tangential differential operators S_ℓ *of order* $d-\ell$ *supported near* Γ , *and* P' *is a ps.d.o. of order* d *satisfying*

$$((P')_\Omega u,v)_\Omega = (u,(P'^*)_\Omega v)_\Omega \quad \textit{for} \quad u,v \in C^\infty_{(0)}(E) . \tag{1.3.3}$$

(1.3.3) *holds for* P *itself, when* $d \leq 0$.

Proof: Consider the situation in local coordinates, where Ω , Σ and Γ are replaced by $\mathbb{R}^n_+$, $\mathbb{R}^n$ and $\mathbb{R}^{n-1}$, and the vector bundles are trivial. Let P be the operator with symbol $p(x,\xi)$ (a negligible correction will always satisfy (1.3.3)). The last statement is obvious, so we assume $d > 0$. We expand $p(x,\xi)$ in two ways: the Taylor expansion at $x_n = 0$

$$p(x,\xi) = \sum_{0\leq j<d} \frac{1}{j!} x_n^j \partial^j_{x_n} p(x',0,\xi) + x_n^d r_d(x,\xi) , \tag{1.3.4}$$

and the expansion of each $\partial^j_{x_n} p$ due to the transmission property

$$\partial^j_{x_n} p(x',0,\xi) = \sum_{0\leq \ell\leq d} s_{\ell,j}(x',\xi')\xi_n^\ell + p_j(x',\xi) \tag{1.3.5}$$

(with polynomials $s_{\ell,j}$ in ξ' of degree $d-\ell$). Then we can write (taking $\eta(x_n) \in C_0^\infty(\mathbb{R})$, equal to 1 near 0),

$$p(x,\xi) = \sum_{0\leq \ell\leq d} \sum_{0\leq j<d} \frac{1}{j!} \eta(x_n)x_n^j s_{\ell,j}(x',\xi')\xi_n^\ell + p'(x,\xi) ; \tag{1.3.6}$$

here the sum over ℓ is the symbol of the differential operator A , and $P' = OP(p')$ satisfies (1.3.3) since it is a sum of terms that either have a factor x_n^d or have a symbol that is bounded in ξ_n (the straightforward argument is given in more detail in [Grubb 17, Chapter 3]).

Coordinate transformations in x' alone preserve these properties, so the decomposition carries over to the manifold situation, by a partition of unity and local coordinates. □

The operator A indicated in the proof is not the only possible choice, see Remark 1.3.3. Lemma 1.3.1 and Proposition 1.3.2 are valid for $S_{1,0}$ as well as polyhomogeneous symbols.

1.3.2 Proposition. *Let* P *be as in Lemma 1.3.1. The following Green's formula holds for* u *and* $v \in C^\infty_{(0)}(E)$

$$(P_\Omega u,v)_\Omega - (u,P^*_\Omega v)_\Omega = (\mathcal{A}\rho u,\rho v)_\Gamma \ , \tag{1.3.7}$$

where $\mathcal{A}$ *is a (uniquely determined) matrix (the Green's matrix)*

$$\mathcal{A} = (\mathcal{A}_{jk})_{j,k=0,\dots,d-1}$$

of differential operators $\mathcal{A}_{jk}$ *in* $E|_\Gamma$ *of orders* $d-j-k-1$, *with*

$$\mathcal{A}_{jk}(x',D') = i\ S_{j+k+1}(x',0,D') + \text{lower order terms} \ , \tag{1.3.8}$$

and ρ *denotes the Cauchy boundary operator*

$$\rho u = \{\gamma_0 u,\dots,\gamma_{d-1}u\} \ . \tag{1.3.9}$$

Proof: In view of Lemma 1.3.1, one has that

$$(P_\Omega u,v) - (u,P^*_\Omega v) = (Au,v) - (u,A^*v) \ , \tag{1.3.10}$$

where the latter satisfies the usual Green's formula, that can be derived from the formula

$$(D_n u,v)_\Omega - (u,D_n v)_\Omega = i(\gamma_0 u,\gamma_0 v)_\Gamma$$

(details in the present terminology are given e.g. in [Grubb 2]). The entries in $\mathcal{A}$ are uniquely determined since ρ is surjective from $C^\infty_{(0)}(E)$ to $\Pi_{j=0}^{d-1}\, C^\infty_{(0)}(E_\Gamma)$. □

The trace operator ρ and the formulas extend to Sobolev spaces as described e.g. in [Lions-Magenes 1]. For instance, when $\overline{\Omega}$ is compact, ρ is continuous

$$\rho: H^d(E) \to \prod_{j=0}^{d-1} H^{d-j-\frac{1}{2}}(E_\Gamma) ,$$

and (1.3.7) is valid for u and $v \in H^d(E)$.

We observe that $\mathcal{A}$ has a skew-triangular character

$$(1.3.11)\quad \mathcal{A} = \mathcal{A}^0 + \mathcal{A}' = i\begin{pmatrix} S_1^0 & \cdots & S_{d-1}^0 & S_d^0 \\ S_2^0 & \cdots & S_d^0 & 0 \\ \vdots & \cdot^{\cdot^{\cdot}} & \vdots & \vdots \\ S_d^0 & \cdots & 0 & 0 \end{pmatrix}(x',0,D') + \begin{pmatrix} \text{lower} & & 0 \\ \text{order} & \cdot^{\cdot^{\cdot}} & \vdots \\ & & \vdots \\ 0 & \cdots\cdots & 0 \end{pmatrix} .$$

It is sometimes advantageous to write

$$(1.3.12)\qquad \mathcal{A} = I^\times \mathcal{A}^\times \quad \text{and} \quad \mathcal{A}^0 = I^\times \mathcal{A}^{0\times}$$

where $I^\times$ is the "skew-unit matrix"

$$(1.3.13)\qquad I^\times = (\delta_{j,d-j-k-1})_{0\leq j,k<d} ,$$

and $\mathcal{A}^\times$ and $\mathcal{A}^{0\times}$ are ordinary triangular matrices, e.g.

$$(1.3.14)\qquad \mathcal{A}^{0\times} = i\begin{pmatrix} S_d^0 & 0 & \cdots & 0 \\ S_{d-1}^0 & S_d^0 & \cdots & 0 \\ \vdots & \vdots & \ddots & \vdots \\ S_1^0 & S_2^0 & \cdots & S_d^0 \end{pmatrix} .$$

The terms in the diagonal of $\mathcal{A}^\times$ and $\mathcal{A}^{0\times}$, resp. second diagonal of $\mathcal{A}$ and $\mathcal{A}^0$, equal $i\, S_d(x',0,D') = i\, s_d(x')$, where $s_d(x')$ is the coefficient of ξ_n^d in the expansion (1.2.10) of $p(x,\xi)$ and $p^0(x,\xi)$; it is a zero order differential operator, hence a multiplication operator (a morphism in E_Γ). The operator $\mathcal{A}$ (and also $\mathcal{A}^0$) is <u>invertible</u> if and only if $s_d(x')$ is a <u>bijective</u> morphism for each $x' \in \Gamma$; and that holds exactly when Γ is <u>non-characteristic</u> for P , for the latter means that $p^0(x',0,0,\xi_n)$ is invertible for $\xi_n \neq 0$, and by (1.2.10) and the homogeneity of p^0 for $|\xi| \geq 1$,

$$(1.3.15)\qquad p^0(x',0,0,\xi_n) = s_d(x')\xi_n^d \quad \text{for } |\xi_n| \geq 1 .$$

<u>1.3.3 Remark.</u> When P is decomposed as in Lemma 1.3.1 in two different ways,

$$P = A + P' = A_1 + P_1' ,$$

the difference $A' = A - A_1$ must be such that

$$(A'u,v)_\Omega - (u,A'^*v)_\Omega = 0 \quad \text{for all} \quad u,v \in C^\infty_{(0)}(E) ,$$

i.e. the corresponding matrix $\mathcal{A}_{A'}$ in Green's formula must be zero. So A is uniquely determined up to a summand with Green's matrix zero. Such summands are the differential operators of the form

$$A' = \sum_{0\le \ell\le d} x_n^\ell \, S'_\ell(x,D')D_n^\ell ,$$

the S'_ℓ having smooth coefficients. ([Melrose 1] calls these differential operators totally characteristic.)

The following observation for the noncharacteristic case will be useful later on.

1.3.4 Lemma. *Let* P *be a ps.d.o. of order* $d > 0$, *having the transmission property at* Γ. *When* Γ *is noncharacteristic for* P *i.e.,* $s_d(x')$ *is invertible, then the principal boundary symbol operator* $p^0(x',0,\xi',D_n)$ *has the (hypoellipticity) property for* $|\xi'| \ge 1$:

When $u \in L^2(\mathbb{R}_+)^N$ *and* $p^0_\Omega u \in L^2(\mathbb{R}_+)^N$, *then* $u \in H^d(\overline{\mathbb{R}}_+)^N$.

Proof: Let u and $p^0 u \in L^2(\mathbb{R}_+)^N$. In view of Lemma 1.3.1 and (1.3.15) we can write

$$p^0_\Omega u = s_d(x')D_n^d u + \sum_{0\le j<d} s_j^0(x',\xi')D_n^j u + p'^0_\Omega u ,$$

where $p'^0_\Omega u \in L^2(\mathbb{R}_+)^N$. It follows that

$$a(D_n)u \equiv D_n^d u + \sum_{0\le j<d} s_d^{-1}s_j^0 D_n^j u + cu \in L^2(\mathbb{R}_+)^N ,$$

for any constant c; we can choose c so large that $|a(\xi_n)| = |\xi_n^d + \Sigma_{0\le j<d}\, s_d^{-1}s_j^0\xi_n^j + c| \ge c_1\langle\xi_n\rangle^d$ for some $c_1 > 0$. The problem is then reduced to the well known differential operator case; for completeness we give a full proof. It is seen by successive integration that u is in H^d on bounded intervals $I =]0,t[$: $D_n(D_n^{d-1}u + \Sigma_{1\le j<d}s_d^{-1}s_j^0 D_n^{j-1}u) \in L^2(I)^N$ implies $D_n^{d-1}u + \Sigma s_d^{-1}s_j^0 D_n^{j-1}u \in L^2(I)^N$, etc. Thus for $\chi \in C^\infty_{(0)}(\overline{\mathbb{R}}_+)$ with $\chi = 1$ near 0, $\chi u \in H^d(\overline{\mathbb{R}}_+)^N$, so also $a(D_n)[(1-\chi)u] = a(D_n)u - a(D_n)\chi u \in L^2(\mathbb{R}_+)^N$. Extending $(1-\chi)u$ by zero on $\mathbb{R}_-$, we have that $\|\langle\xi_n\rangle^d F[(1-\chi)u]\|_{L^2} \le c_1^{-1}\|a(\xi_n)F[(1-\chi)u]\|_{L^2} < \infty$, so altogether $u \in H^d(\overline{\mathbb{R}}_+)^N$. $\square$

1.4 Realizations and normal boundary conditions.

Let P be a ps.d.o. of order $d > 0$ in $\tilde{E}$ having the transmission property at Γ (we do not repeat this in the following), and assume moreover that P is elliptic. For simplicity of formulations we take $\bar{\Omega}$ compact from now on. As it is usual for differential operators, one can define the maximal and minimal realizations of P in $L^2(E)$ as the operators acting like P_Ω and determined by

$$D(P_{max}) = \{u \in L^2(E) \mid P_\Omega u \in L^2(E)\} \ , \qquad (1.4.1)$$

$$P_{min} = \text{closure of } P_\Omega|_{C_0^\infty(\Omega,E)} \quad \text{in } L^2(E)$$

(that P_{max} is closed follows from the continuity of $P: L^2(\tilde{E}) \to H^{-d}_{loc}(\tilde{E})$); hence $P_{min} \subset P_{max}$). In view of the ellipticity of P, the graph norm and the H^d-norm are equivalent on $C_0^\infty(\Omega,E)$, so

$$D(P_{min}) = \overset{\circ}{H}{}^d(E) \ . \qquad (1.4.2)$$

Moreover, one has the usual identity

$$P_{max} = (P^*_{min})^* \ , \quad \text{as operators in } L^2(E) \ , \qquad (1.4.3)$$

where P^* is the formal adjoint of P on $\tilde{E}$. To see this, note that the domain of the adjoint S of $(P^*)_{min}$ consists of those functions $u \in L^2(E)$ for which there exists $f \in L^2(E)$ so that

$$(u,(P^*)_\Omega\varphi)_\Omega = (f,\varphi)_\Omega \quad \text{for all } \varphi \in C_0^\infty(\Omega,E) \ ,$$

and then $f = Su$. But when this holds,

$$(u,(P^*)_\Omega\varphi)_\Omega = (u,r^+P^*\varphi)_\Omega = \langle e^+u,\overline{P^*\varphi}\rangle_\Sigma = \langle r^+Pe^+u,\bar{\varphi}\rangle_\Omega \ ,$$

so $u \in D(P_{max})$ with $f = P_{max}u$. Conversely, when $r^+Pe^+u = f \in L^2(E)$, it is seen that $u \in D(S)$ with $Su = f$, so altogether $S = P_{max}$.

Note that $H^d(E) \subset D(P_{max})$ (the inclusion is in general strict).

Realizations of P would in the differential operator case be defined as the operators $\tilde{P}$ acting like P and with domains lying between $D(P_{min})$ and $D(P_{max})$. Here one would in particular be interested in the realizations $\tilde{P}$ with

$$\overset{\circ}{H}{}^d(E) \subset D(\tilde{P}) \subset H^d(E) \ ,$$

which includes the realizations defined by boundary conditions satisfying the Shapiro-Lopatinski-condition (elliptic boundary conditions).

The present problems require a wider concept of realization. For one thing, it will be allowed to add a singular Green term to P, which changes the action of the operator. Another thing is that we include trace operators with non-local terms, which may very well violate the usual inclusion of $D(P_{min})$ in the domain of all realizations.

Before we give the precise definition, we shall explain the notation we use for trace operators. By a system of trace operators (or a trace operator) T associated with the order d, we understand a (column) vector of trace operators $T = \{T_0, T_1, \dots, T_{d-1}\}$ (void if $d \leq 0$), where each T_k is of order k and class $\leq d$, going from E to F_k; here $\{F_0, F_1, \dots, F_{d-1}\}$ is a given system of vector bundles over γ, with $\dim F_k = M_k \geq 0$. We denote

$$(1.4.4) \qquad F = F_0 \oplus \dots \oplus F_{d-1} \quad \text{and} \quad M = \sum_{0 \leq k < d} M_k .$$

For notational convenience, we have here included all orders k between 0 and $d-1$, but this encompasses those systems of trace operators that pick out certain orders, since we have allowed bundles F_k with dimension 0; when M_k is zero, T_k is trivial and could be omitted. (When P is scalar, it is customary to consider only scalar trace operators, i.e. $M_k = 1$ or 0, and one could label the nontrivial trace operators by a subset J of the set $\{0,1,\dots,d-1\}$. The statements will sometimes be specified for this notation.)

When T is as above, it maps $H^d(E)$ continuously into

$$(1.4.5) \qquad H_F^d = \prod_{0 \leq k < d} H^{d-k-\frac{1}{2}}(F_k) .$$

Note also that since each T_k is of class $\leq d$, it may be written

$$T_k = \sum_{0 \leq j < d} S_{kj}\gamma_j + T_k' ,$$

where the S_{kj} are ps.d.o.s from E_Γ to F_k of order $k-j$, and T_k' is of class 0, so T may be written in short form (cf. (1.3.9))

$$(1.4.6) \qquad T = S\rho + T' ,$$

where $S = (S_{kj})_{\substack{0 \leq k < d \\ 0 \leq j < d}}$, and T' is of class 0.

1.4.1 Definition. *Let* P *be a ps.d.o. of order* $d \geq 0$ *in* $\tilde{E}$, *having the transmission property at* Γ ; *let* G *be a singular Green operator in* E *of class* $\leq d$ *and order* d, *and let* $T = \{T_0,\ldots,T_{d-1}\}$ *be a system of trace operators* T_k *from* E *to* F_k, *associated with the order* d. *The operator* $B = (P+G)_T$ *in* $L^2(E)$, *acting like* $P_\Omega + G$ *and with domain*

$$D((P+G)_T) = \{u \in H^d(E) \mid Tu = 0\} \tag{1.4.7}$$

is called the H^d*-realization (or just realization) of* P *defined by the system* $\{P_\Omega+G,T\}$. *The realization is called elliptic, when the system* $\{P_\Omega+G,T\}$ *is elliptic.*

The reasons for excluding trace operators of class $> d$ will be discussed in Section 1.5. (The orders can in principle always be adapted by composition with ps.d.o.s over Γ.)

Observe that even when $G = 0$, one may have that

$$D(P_{min}) \not\subset D(P_T) , \tag{1.4.8}$$

namely whenever T contains integrals over the interior of Ω (as in Example 1.1.1 above). Here the operator P_T does act like P_Ω, but the adjoint will in general not act like P^*_Ω, in view of (1.4.8),

$$(P_T)^* \not\subset (P^*)_{max} . \tag{1.4.9}$$

For the class of normal boundary problems to be introduced further below, $(P_T)^*$ will however be of the form $(P^*+G')_{T'}$ for suitable G' and T', when considered on $H^d(E)$. See also Example 1.4.5 below.

It will be useful to make some observations on elliptic realizations here. For one thing, the ellipticity hypothesis, which states that the boundary symbol operator (defined in local coordinates)

$$\begin{pmatrix} p^0(x',0,\xi',D_n)_\Omega + g^0(x',\xi',D_n) \\ t^0(x',\xi',D_n) \end{pmatrix} : \mathscr{S}(\overline{\mathbb{R}}_+)^N \to \begin{matrix} \mathscr{S}(\overline{\mathbb{R}}_+)^N \\ \times \\ \mathbb{C}^M \end{matrix} \tag{1.4.10}$$

is bijective for all x', all $|\xi'| \geq 1$, is equivalent with the property:

$$\begin{pmatrix} p^0_\Omega + g^0 \\ t^0 \end{pmatrix} : H^d(\overline{\mathbb{R}}_+)^N \to \begin{matrix} L^2(\mathbb{R}_+)^N \\ \times \\ \mathbb{C}^M \end{matrix} \tag{1.4.11}$$

is bijective for all x', all $|\xi'| \geq 1$. This hinges on the fact that $p^0_\Omega + g^0 : H^d(\overline{\mathbb{R}}_+)^N \to L^2(\mathbb{R}_+)^N$ has nullspace (kernel) and co-range (co-kernel) in $\mathcal{S}(\overline{\mathbb{R}}_+)^N$, coinciding with the nullspace resp. co-range of $p^0_\Omega + g^0 : \mathcal{S}(\overline{\mathbb{R}}_+)^N \to \mathcal{S}(\overline{\mathbb{R}}_+)^N$. In the present case, ellipticity means that $p^0_\Omega + g^0$ is surjective, the kernel has dimension M and t^0 provides a bijection of the kernel onto $\mathbb{C}^M$; this can be expressed equally well by the bijectiveness of (1.4.10) or of (1.4.11).

Let us also introduce the boundary symbol realization $b^0(x',\xi',D_n)$, defined for each x',ξ' as the operator in $L^2(\mathbb{R}_+)^N$ acting like $p^0(x',0,\xi',D_n)_\Omega + g^0(x',\xi',D_n)$ and with domain

(1.4.11') $$D(b^0(x',\xi',D_n)) = \{u \in H^d(\overline{\mathbb{R}}_+)^N \mid t^0(x',\xi',D_n)u = 0\} .$$

Ellipticity implies that b^0 is bijective from $D(b^0)$ to $L^2(\mathbb{R}_+)^N$; the converse holds when t^0 has a right inverse, as in Lemma 1.6.6 below.

Now the ellipticity implies a Fredholm solvability of the problem

(1.4.12) $$\begin{aligned}(P_\Omega+G)u &= f ,\\ Tu &= \varphi ,\end{aligned}$$

where f and φ are given in $L^2(E)$ resp. H^d_F (cf. (1.4.5)), and u is sought in $H^d(E)$. Basically, this follows from the fact that

(1.4.13) $$A = \begin{pmatrix} P_\Omega + G \\ T \end{pmatrix} \text{ has a parametrix } C = (R \quad K)$$

within the calculus. A good analysis of the solution operator can be given by use of the fact there exists a family of ("order reducing") elliptic ps.d.o.s Λ^m_- in $\tilde{E}$ such that $\Lambda^m_{-,\Omega}$ maps $H^{m_+}(E)$ homeomorphically onto $H^{m_-}(E)$ (where $m_\pm = \max\{\pm m, 0\}$), cf. Section 1.2 and Remark 3.2.15. There is also a pseudo-differential homeomorphism Λ_Γ of H^d_F onto $L^2(F)$. Composing A to the right with $\Lambda^{-d}_{-,\Omega}$ and to the left with $\begin{pmatrix} I & 0 \\ 0 & \Lambda_\Gamma \end{pmatrix}$, we reduce it to a system

(1.4.14) $$\tilde{A} = \begin{pmatrix} P_\Omega + G \\ \Lambda_\Gamma T \end{pmatrix} \Lambda^{-d}_{-,\Omega} = \begin{pmatrix} Q_\Omega + \tilde{G} \\ \tilde{T} \end{pmatrix} : L^2(E) \to \begin{matrix} L^2(E) \\ \times \\ L^2(F) \end{matrix} ,$$

where $Q = P\Lambda^{-d}_-$ and $\tilde{G}$ have order 0, $\tilde{T}$ has order $-\frac{1}{2}$, and $\tilde{G}$ and $\tilde{T}$ have class 0. The ellipticity holds again for the system $\tilde{A}$, so $\tilde{A}$ has a parametrix

$\widetilde{C} = (\widetilde{R} \;\; \widetilde{K})$, with

(1.4.15)
$$\widetilde{A}\,\widetilde{C} = I + S = \begin{pmatrix} I & 0 \\ 0 & I \end{pmatrix} + \begin{pmatrix} S_{11} & S_{12} \\ S_{21} & S_{22} \end{pmatrix} \quad \text{in} \quad \begin{matrix} L^2(E) \\ \times \\ L^2(F) \end{matrix} \;,$$
$$\widetilde{C}\,\widetilde{A} = I + S' \quad \text{in} \quad L^2(E) \;,$$

S and S' being negligible. The latter are of class 0, so they are in fact integral operators with C^∞ kernels. They are compact operators, which implies that $\widetilde{A}$ has closed range and a finite dimensional nullspace and co-range. The nullspace $Z(\widetilde{A})$ is contained in $Z(I+S')$, so is in $C^\infty(E)$, and the co-range $R(\widetilde{A})^\perp = Z(\widetilde{A}^*)$ is of the same kind, since

$$\widetilde{C}^* \widetilde{A}^* = (\widetilde{A}\,\widetilde{C})^* = I + S^*$$

where S^* has C^∞ kernel. Returning to A by use of Λ_Γ^{-1} and $(\Lambda_{-,\Omega}^{-d})^{-1}$, one gets a parametrix $C = (R \quad K)$ for the problem (1.4.12). In particular, there is a solution for any $\{f,\varphi\} \in L^2(E) \times H_F^d$ satisfying a finite set of orthogonality relations (with smooth g_ℓ and ψ_ℓ)

(1.4.16)
$$(f,g_\ell)_{L^2(E)} + (\varphi,\psi_\ell)_{L^2(F)} = 0 \;, \quad \ell = 1,\dots,\ell_0 \;,$$

and the solution is unique modulo a finite dimensional C^∞ subspace of $H^d(E)$.

All this is quite well known, and satisfactory for the nonhomogeneous problem (1.4.12). In particular, the problem

(1.4.17)
$$(P_\Omega + G)u = f$$
$$Tu = 0$$

is solved by Rf (cf. (1.4.13)) in a parametrix sense, in that $(P_\Omega+G)R-I$ and TR are smoothing operators. But R is not of much help when we consider the realization B , for R need not map into $D(B)$ (where Tu must equal zero). On the other hand, the abovementioned Fredholm properties imply that there exists a more abstractly defined operator $R_1 : L^2(E) \to D(B)$ such that $BR_1 - I = K_1$ and $R_1 B - I = K_2$ have finite dimensional C^∞ range. We shall now show that one can in fact find a parametrix R_0 within the calculus, having all these properties (by a modification of [Grubb-Geymonat 1, Th. 5.3]).

1.4.2 Proposition. *Let* $B = (P+G)_T$ *be an elliptic realization of order* $d \geq 0$, *as introduced in Definition 1.4.1. There exists a parametrix* R_0 *with the properties:*

(i) $R_0 = (P^{-1})_\Omega + G_0$, *where* P^{-1} *is a parametrix of* P *on* $\tilde{E}$ *and* G_0 *is a singular Green operator of order* $-d$ *and class 0;*

(ii) R_0 *maps* $L^2(E)$ *into* $D(B)$ *;*

(iii) $BR_0 = I + S$, *where* S *is an integral operator with* C^∞ *kernel and finite rank. Also* R_0B-I *on* $D(B)$ *has finite dimensional* C^∞ *range.*

Proof: The point is to modify the parametrix $\tilde{C}$ of $\tilde{A}$ given above. Since we know from the above analysis, that $\tilde{A}$ has a smooth nullspace Z and its closed range has a smooth orthogonal complement Z' , we can choose $\tilde{C}'$ as the operator acting like the inverse of $\tilde{A}$ from $(L^2(E)\times L^2(F)) \ominus Z'$ to $L^2(E) \ominus Z$, and mapping Z' into 0. Then

$$\begin{aligned} \tilde{C}'\tilde{A} &= I - pr_Z \quad \text{on} \quad L^2(E) , \\ \tilde{A}\tilde{C}' &= I - pr_{Z'} \quad \text{on} \quad L^2(E)\times L^2(F) , \end{aligned} \tag{1.4.18}$$

where pr_X denotes orthogonal projection onto X ; since Z and Z' are smooth finite dimensional, pr_Z and $pr_{Z'}$ are operators with C^∞ kernel (of the form $\Sigma_{j\le j_0} u_j(x)\bar{u}_j(y)$). Now (cf.(1.4.15))

$$\tilde{C}' - \tilde{C} = (\tilde{C}\tilde{A} - S')(\tilde{C}' - \tilde{C}) = \tilde{C}(I-pr_{Z'}) - \tilde{C}(I+S) - S'(\tilde{C}' - \tilde{C})$$

maps $L^2(E)\times L^2(F)$ into $C^\infty(E)$; and its adjoint

$$(\tilde{C}' - \tilde{C})^* = (\tilde{C}^*\tilde{A}^* - S^*)(\tilde{C}'^* - \tilde{C}^*) = \tilde{C}^*(I-pr_Z) - \tilde{C}^*(I+S'^*) - S^*(\tilde{C}' - \tilde{C})^*$$

maps $L^2(E)$ into $C^\infty(E)\times C^\infty(F)$, so $\tilde{C}' - \tilde{C}$ is an integral operator with C^∞ kernel, and hence $\tilde{C}'$ belongs to the Green operator calculus. In view of the structure of $\tilde{C}$, we have that

$$\tilde{C}' = (\tilde{R}' \quad \tilde{K}'), \quad \text{with } \tilde{R}' = (Q^{-1})_\Omega + \tilde{G}' ,$$

Q^{-1} being a parametrix of Q , $\tilde{G}'$ a s.g.o. of order and class 0, and $\tilde{K}'$ a Poisson operator. $\tilde{R}'$ is a parametrix of the realization $\tilde{B} = (Q+\tilde{G})_{\tilde{T}}$.

Let us write $pr_{Z'}$ as

$$pr_{Z'} = \begin{pmatrix} \tilde{S}_{11} & \tilde{S}_{12} \\ \tilde{S}_{21} & \tilde{S}_{22} \end{pmatrix} \quad \text{in} \quad \begin{matrix} L^2(E) \\ \times \\ L^2(F) \end{matrix} ,$$

then each of the entries is smoothing, with finite rank. Let $Z'' = R(\tilde{S}_{21}^*) = Z(\tilde{S}_{21})^\perp$, it is a finite dimensional C^∞ subspace of $L^2(E)$. Let $pr_{Z''}$

be the orthogonal projection onto Z'' in $L^2(E)$. Then we have

$$\begin{pmatrix} P_\Omega + G \\ \Lambda_\Gamma T \end{pmatrix} \Lambda^{-d}_{-,\Omega} (\tilde{R}' \quad \tilde{K}') \begin{pmatrix} I - pr_{Z''} \\ 0 \end{pmatrix} = \tilde{A}\,\tilde{C}' \begin{pmatrix} I - pr_{Z''} \\ 0 \end{pmatrix}$$

$$= \begin{pmatrix} I - \tilde{S}_{11} & -\tilde{S}_{12} \\ -\tilde{S}_{21} & I - \tilde{S}_{22} \end{pmatrix} \begin{pmatrix} I - pr_{Z''} \\ 0 \end{pmatrix} = \begin{pmatrix} (I-\tilde{S}_{11})(I-pr_{Z''}) \\ 0 \end{pmatrix},$$

which shows that the operator

$$R_0 = \Lambda^{-d}_{-,\Omega}\, \tilde{R}'(I-pr_{Z''}) \tag{1.4.19}$$

satisfies $\Lambda_\Gamma T R_0 = 0$ (and hence $TR_0 = 0$), and $(P_\Omega+G)R_0 = I + S'''$, with an integral operator S''' with C^∞ kernel and finite rank. Thus R_0 maps into $D(B)$, and $BR_0 - I = S'''$. By the rules of calculus, R_0 satisfies (i), and $R_0B - I$ maps $D(B)$ into $C^\infty(E)$. By comparison with R_1 mentioned before the proposition

$$R_0B-I = (R_0-R_1)B + R_1B-I = ((R_1B-K_2)R_0-R_1(BR_0-S'''))B + K_2 = -K_2R_0B + R_1S'''B + K_2$$

has finite rank. This ends the proof. □

We now specify an interesting class of boundary conditions, the normal boundary conditions.

<u>1.4.3 Definition.</u> *Let* $T = \{T_0,\ldots,T_{d-1}\}$ *be a system of trace operators* T_k *from* E *to* F_k, *associated with the order* d. *Then* T *is said to be* <u>*normal*</u>, *when each* T_k *is of the form (void if* $M_k = 0$*)*

$$T_k = \sum_{0\le j\le k} S_{kj}\,\gamma_j + T'_k \,, \tag{1.4.20}$$

with S_{kk} *being a* <u>*surjective morphism*</u> *from* E_Γ *to* F_k, S_{kj} *a ps.d.o. from* E *to* F_k *of order* $k-j$ *for* $j < k$ *and* T'_k *a trace operator from* E *to* F_k *of order* k *and class 0. (In particular* $M_k = \dim F_k$ *is* $\le N$ *for each* k.*)* *The boundary condition* $Tu = \varphi$ *is then also called normal.*

With the notation of Definition 1.4.1, the system $\{P_\Omega+G,T\}$, *the boundary value problem*

$$(P_\Omega+G)u = f$$
$$Tu = \varphi\,,$$

and the realization $B = (P+G)_T$, *are said to be* <u>*normal*</u>, *when* Γ *is noncharacteristic for* P *and* T *is normal.*

It will be proved further below that normal realizations have dense domains in $L^2(E)$, so that they have well-defined adjoints. When $T = \gamma_0 Q_\Omega$ for a ps.d.o. Q of order $k \geq 0$, normality means that Γ is noncharacteristic for Q.

Observe that when T is written on the short form $T = S\rho + T'$ as in (1.4.6), the normality means precisely that S is triangular

$$S = \begin{pmatrix} S_{00} & 0 & & 0 \\ S_{01} & S_{11} & \cdots & 0 \\ \vdots & \vdots & \ddots & \vdots \\ S_{0,d-1} & S_{1,d-1} & \cdots & S_{d-1,d-1} \end{pmatrix} \tag{1.4.21}$$

with surjective morphisms $S_{kk}(x')$ in the diagonal. The surjectiveness of the diagonal elements S_{kk} implies surjectiveness of the trace operator $S\rho$ from $C^\infty(E)$ to $C^\infty(F)$ (cf. [Grubb 2] or Lemma 1.6.1 below), and we show in Section 1.6 that the full trace operator $T = S\rho + T'$ is surjective.

1.4.4 Remark. The definition can be written in a (perhaps) simpler way, when P is scalar (E is a trivial one-dimensional bundle). Here the M_k are necessarily 0 or 1; when $M_k = 0$ we can omit the T_k, and when $M_k = 1$, $S_{kk}(x')$ is an invertible function so a multiplication by $1/S_{kk}$ reduces T_k to the form

$$T_k = \gamma_k + \sum_{j<k} S_{kj}\gamma_j + T_k' \quad , \text{ for each } k . \tag{1.4.22}$$

Sometimes one uses a formulation here where T is written as $\{T_k\}_{k\in J}$, J being the subset of indices $k \in J_\rho \equiv \{0,1,\dots,d-1\}$ for which F_k is nonzero. (One could also write $T = \{T_{m_j}\}_{j=1,\dots,j_0}$ with mutually distinct m_j, but this gives other notational complications.) This formulation can also be used in general when

$$\begin{aligned} F_k &= E_\Gamma \quad \text{for } k \in J , \\ F_k &= 0 \quad \text{for } k \in J_\rho \setminus J , \end{aligned} \tag{1.4.23}$$

for some index set $J \subset J_\rho = \{0,1,\dots,d-1\}$.

1.4.5 Example. Consider the system in Example 1.1.1, where we denote $-\Delta = P$, and take $G' = 0$ for simplicity. Each of the boundary conditions (1.1.3) and (1.1.4) is normal (one can take $\{F_0,F_1\} = \{E_\Gamma,0\}$ resp. $\{0,E_\Gamma\}$, adding an empty first order, resp. zero order boundary condition). Let us consider the adjoint of the realization P_{T_0} defined by the boundary condition (1.1.3). For $u \in D(P_{T_0})$ and $v \in H^d(E)$ we have, in view of Green's formula for $-\Delta$ (where

$\mathcal{A} = i\begin{pmatrix} 0 & 1 \\ 1 & 0 \end{pmatrix}$, $\mathcal{A}^\times = i\begin{pmatrix} 1 & 0 \\ 0 & 1 \end{pmatrix}$),

$$\begin{aligned}(P_{T_0}u,v)_\Omega = (-\Delta u,v)_\Omega &= i(\gamma_1 u,\gamma_0 v)_\Gamma + i(\gamma_0 u,\gamma_1 v)_\Gamma + (u,-\Delta v)_\Omega \\ &= i(\gamma_1 u,\gamma_0 v)_\Gamma + i(-T_0' u,\gamma_1 v)_\Gamma + (u,-\Delta v)_\Omega \\ &= i(\gamma_1 u,\gamma_0 v)_\Gamma + (u,-\Delta v+iT_0'^*\gamma_1 v)_\Omega .\end{aligned}$$

It follows that $(P_{T_0})^*$ <u>contains</u> the realization

(1.4.24) $$(P_{T_0})^* \supset (-\Delta+iT_0'^*\gamma_1)_{\gamma_0} ,$$

an operator acting like $-\Delta$ plus the singular Green operator $i\,T_0'^*\,\gamma_1$, and with domain defined by a <u>local</u> boundary condition $\gamma_0 v = 0$. As we shall see later, there is equality in (1.4.24), when P_{T_0} is elliptic. When $T_0' \neq 0$, this is an example where (1.4.8)-(1.4.9) hold. The operator P is local and the boundary condition $T_0 u = 0$ nonlocal, whereas this situation is reversed for the adjoint $(P_{T_0})^*$.

Normal boundary conditions are of interest here, not only because they define convenient realizations (that behave nicely under composition and formation of adjoints, cf. Theorems 1.4.6 and 1.6.9 later), but also because normality is practically necessary for the parameter-ellipticity that we need to assume for the resolvent construction (see the discussion in Section 1.5).

<u>1.4.6 Theorem</u>. *Let* $B_1 = (P_1+G_1)_{T_1}$ *and* $B_2 = (P_2+G_2)_{T_2}$ *be realizations of ps.d.o.s* P_1 *and* P_2 *on* $\tilde{E}$ *of order* $d_1 \geq 0$ *resp.* $d_2 \geq 0$, *as defined in Definition 1.4.1, and denote the composition* $B_1 \circ B_2$ *by* C ; *i.e.* C *is the operator acting like*

(1.4.25) $$Cu = (P_{1,\Omega}+G_1)(P_{2,\Omega}+G_2)$$

with domain

(1.4.26) $$D(C) = \{u \in H^{d_2}(E) \mid T_2 u = 0 , (P_{2,\Omega}+G_2)u \in H^{d_1}(E) , T_1(P_{2,\Omega}+G_2)u = 0\} .$$

Then C *is an extension of the realization* $B_3 = (P_3+G_3)_{T_3}$, *where*

$$(1.4.27)\qquad \begin{aligned} P_3 &= P_1 \circ P_2\ , \quad \textit{of order}\ \ d_3 = d_1 + d_2\ ,\\ G_3 &= -L(P_1,P_2) + P_{1,\Omega}G_2 + G_1P_{2,\Omega} + G_1G_2\\ T_3 &= \begin{pmatrix} T_2 \\ T_1(P_{2,\Omega}+G_2) \end{pmatrix}\ . \end{aligned}$$

When $(P_2+G_2)_{T_2}$ *is elliptic,* C *equals* $(P_3+G_3)_{T_3}$. *When both* $(P_1+G_1)_{T_1}$ *and* $(P_2+G_2)_{T_2}$ *are elliptic, then so is* $(P_3+G_3)_{T_3}$. *When* $(P_1+G_1)_{T_1}$ *and* $(P_2+G_2)_{T_2}$ *are normal in the sense of Definition 1.4.3, then so is* $(P_3+G_3)_{T_3}$.

Proof: It is obvious from the rules of calculus that C is an extension of $(P_3+G_3)_{T_3}$ defined by (1.4.27); the latter acts like $P_{3,\Omega}+G_3$ and has the domain

$$D((P_3+G_3)_{T_3}) = \{u \in H^{d_1+d_2}(E) \mid T_3u = 0\}\ .$$

Moreover, when B_2 is elliptic, $B_2u \in H^{d_1}(E)$ implies $u \in H^{d_1+d_2}(E)$, so here $D(C) = D((P_3+G_3)_{T_3})$.

When B_1 and B_2 are elliptic, then so is B_3, for the bijectiveness of the boundary symbol operators

$$\begin{pmatrix} p^0_{1,\Omega} + g^0_1 \\ t^0_1 \end{pmatrix} \quad \text{and} \quad \begin{pmatrix} p^0_{2,\Omega} + g^0_2 \\ t^0_2 \end{pmatrix}$$

implies the bijectiveness of the composed operator

$$\begin{pmatrix} p^0_{1,\Omega} + g^0_1 & 0 \\ 0 & I \\ t^0_1 & 0 \end{pmatrix}\begin{pmatrix} p^0_{2,\Omega} + g^0_2 \\ t^0_2 \end{pmatrix} = \begin{pmatrix} (p^0_{1,\Omega} + g^0_1)(p^0_{2,\Omega} + g^0_2) \\ t^0_2 \\ t^0_1(p^0_{2,\Omega} + g^0_2) \end{pmatrix}\ .$$

Now assume that B_1 and B_2 are normal realizations. Clearly Γ is non-characteristic for P_3, and the normality of T_3 is seen as follows: There are given the trace operators $T_i = \{T_{i,0},\dots,T_{i,d_i-1}\}$ going from E to $F_i = \oplus_{0\le k<d_i} F_{i,k}$ for $i = 1,2$, and we define $T_3 = \{T_{3,0},\dots,T_{3,d_3-1}\}$ going from E to $F_3 = \oplus_{0\le k<d_3} F_{3,k}$ by

$$(1.4.28)\quad \begin{aligned} &F_{3,k} = F_{2,k} \quad \text{and} \quad T_{3,k} = T_{2,k} && \text{for} \quad 0 \le k < d_2 \\ &F_{3,k} = F_{1,k-d_2} \quad \text{and} \quad T_{3,k} = T_{1,k-d_2}(P_{2,\Omega} + G_2) && \text{for} \quad d_2 \le k < d_3 . \end{aligned}$$

The normality condition is obviously satisfied by the $T_{3,k}$ with $k < d_2$. For $k \geq d_2$, we use that

$$P_{2,\Omega} = S_{d_2}(x)D_n^{d_2} + Q ,$$

where Q is a sum of terms that either contain a factor x_n in front or have symbols that are $O(\langle\xi_n\rangle^{d_2-1})$, and

$$G_{2,\Omega} = \sum_{0 \le j < d_2} K_j\gamma_j + G_2' ,$$

with G_2' of class 0. Writing

$$T_{1,\ell} = S_{\ell\ell}(x')\gamma_\ell + \text{terms of class} \ < \ell ,$$

we then have (since $\gamma_0(x_n u) = 0$ and $\gamma_j(x_n u) = \gamma_0 D_n^j(x_n u) = j\gamma_{j-1}u$ for $j > 0$),

$$\begin{aligned} T_{3,\ell+d_2} &= T_{1,\ell}(P_{2,\Omega}+G_2) \\ &= S_{\ell\ell}(x')\gamma_\ell S_{d_2}(x)D_n^{d_2} + \text{terms of class} \ < \ell + d_2 \\ &= S_{\ell\ell}(x')S_{d_2}(x',0)\gamma_{\ell+d_2} + \text{terms of class} \ < \ell + d_2 . \end{aligned}$$

Here the coefficient $S_{\ell\ell}(x')S_{d_2}(x',0)$ is a surjective morphism from E_Γ to $F_{1,\ell}$ since $S_{\ell\ell}$ is one, and $S_{d_2}(x',0)$ is bijective in E_Γ. Then all the terms $T_{3,k}$ in (1.4.28) satisfy the condition for normality. $\square$

The normal boundary conditions and realizations are studied in further detail in Sections 1.6 and 1.7, and the motivation for their use in resolvent problems is given in Section 1.5.

1.5 Parameter-ellipticity and parabolicity.

It is well known that a boundary value problem does not have to be normal in order to have a nice solution operator (or parametrix); ellipticity suffices for this. However, it turns out that in the study of the resolvent $R_\lambda = (B-\lambda I)^{-1}$ of a realization $B = (P+G)_T$, normality of T plays a crucial rôle. We shall now introduce the concept of parameter-ellipticity for systems $\{P_\Omega+G-\lambda I,T\}$ (for λ on a ray in $\mathbb{C}$), with a discussion of the restrictions that it puts on P,G and T.

Concerning examples of systems $\{P_\Omega+G,T\}$ for which $\{P_\Omega+G-\lambda I,T\}$ has the parameter-ellipticity property, let us mention first of all, that one can show that the Dirichlet problem for a strongly elliptic ps.d.o. P (with the transmission property, as always here) belongs to this class, just as in the classical case of differential operators, see Theorem 1.7.2 later on. More generally, there is a large class of realizations that we call m-coercive (the realizations satisfying the "Gårding inequality", see Section 1.7), which have the parameter-ellipticity property; in fact, one has the following hierarchy for normal boundary value problems:

(1.5.1) $(P+G)_T$ is m-coercive

$\Downarrow$

$\begin{pmatrix} P_\Omega+G-\lambda I \\ T \end{pmatrix}$ is parameter-elliptic on all rays $\lambda = e^{i\theta}r$ with $\theta \in [\pi/2, 3\pi/2]$ (parabolicity)

$\Downarrow$

$\begin{pmatrix} P_\Omega+G-\lambda I \\ T \end{pmatrix}$ is parameter-elliptic on a ray.

An interesting observation here is that whereas the condition for parameter-ellipticity (or the condition for parabolicity) is a bijectiveness condition on certain principal symbols, hence is stable under "small" or lower order perturbations, the condition for m-coerciveness is not generally stable in this way (even perturbations of the Dirichlet problem may fail to be m-coercive, when we allow pseudo-differential elements). Much more is said about this in Section 1.7, to which the interested reader is referred.

For differential operators, the property of parameter-ellipticity was formulated explicitly in [Agmon 2,3] and in [Agranovich-Vishik 1], where it was called, respectively, the condition for having a "ray of minimal growth" and "ellipticity with a parameter". The condition was called "Agmon's condition" in [Seeley 3,4]. "Agmon's condition" for differential operators implies normality, but this was not always observed: [Agmon 2,3] assumes normality on beforehand, whereas [Agranovich-

Vishik 1] state that they are free from this assumption; and [Seeley 3] later carries out a general resolvent construction without assuming normality, remarking however at the end of the paper that normality is a consequence of the "Agmon condition" assumed throughout (so that a certain projection P_0 that complicated the presentation, was trivial). The observation is repeated in [Seeley 4] and [Seeley 6], where it is credited to T. Burak.

The reason that "Agmon's condition" for differential operators implies normality is that the condition includes (in a natural way) the invertibility of a certain limit boundary symbol operator for $\xi' \to 0$, where only the top normal order terms in the trace operator survive. In the pseudo-differential operator case, one can generalize "Agmon's condition" in various ways. There is a weak generalization (consisting of conditions (I) and (II) in Definition 1.5.5 below), which does not require normality, nor the existence of a limit operator for $\xi' \to 0$; this allows interesting non-standard estimates in constant coefficient cases, as analyzed e.g. in [Rempel-Schulze 3] and [Frank-Wendt 1,2]. However, we do not think that these phenomena are completely established in the variable-coefficient cases, where there can be a lack of control of lower order terms (we return to this problem in Remarks 1.5.16, 1.7.17, 2.1.19, 3.2.14 and 3.2.16) so we settle in this book for a definition of parameter-ellipticity that moreover requires the existence of an invertible limit operator for $\xi' \to 0$ (condition (III) in Definition 1.5.5 below). This requirement implies (in the scalar case it is) normality. (Our first announcements [Grubb 13, 14] were deficient on this point; they were corrected in [Grubb 15].)

The following explanation of the concepts gives a motivation for the general symbol classes and parameter-dependent operators studied systematically in Chapters 2 and 3. Some technicalities are only sketchily described, since they are taken up again in detail later.

Consider first the case of an operator $P = OP(p(x,\xi))$ of order $d > 0$ on a boundaryless manifold (or on $\mathbb{R}^n$). In the differential operator case, parameter-ellipticity of $P-\lambda$ (for λ on the ray $\lambda = e^{i\theta}r$, $r \geq 0$, for some fixed θ), simply means that the principal symbol $p^0(x,\xi)$ avoids the values $\lambda = e^{i\theta}r$ (has no eigenvalues on the ray $\lambda = e^{i\theta}r$, in the matrix case). The same definition is adequate for ps.d.o.s of order $d > 0$ (see below). Let us see how the property is used in the differential operator case: We write

$$-\lambda = \omega\mu^d, \quad \text{where } \mu = |\lambda|^{1/d} \text{ and } \omega = \exp(i(\theta-\pi)). \tag{1.5.2}$$

Then $p^0(x,\xi)-\lambda = p^0(x,\xi)+\omega\mu^d$ is a homogeneous polynomial of degree d in (ξ,μ) with C^∞ coefficients in x , so μ can in principle be considered as another cotangent variable. This is a very fruitful point of view, for the parameter-ellipticity now simply means that

$$(1.5.3) \qquad \det(p^0(x,\xi)+\omega\mu^d) \neq 0 \quad \text{for} \quad (\xi,\mu) \in \overline{\mathbb{R}}^{n-1}_+ \setminus 0 \ ,$$

i.e., the symbol $p(x,\xi)+\omega\mu^d$ is elliptic in the usual sense, for each x , when considered as a function of the n+1 cotangent variables (ξ,μ) . (It is an inessential restriction that μ runs on $\overline{\mathbb{R}}_+$, as long μ is used mainly as a parameter. When d is even, one can let $\mu \in \mathbb{R}$.) The well known techniques for usual elliptic operators can then be applied; for example there is a straightforward construction of a parametrix symbol; and, in the even order case, there are standard techniques to pass from informations on the operator $P+\omega D^d_{n+1}$ to $P+\omega\mu^d$, cf. [Agmon 2].

In the case of pseudo-differential operators, there are certain difficulties with this point of view, primarily because the symbol $p^0(x,\xi)+\omega\mu^d$ is not in general a standard ps.d.o. symbol in the (ξ,μ)-variables when $p^0(x,\xi)$ is a standard ps.d.o. symbol in the ξ-variable. For, in this case, the ξ-derivatives satisfy for $|\alpha| \neq 0$

$$(1.5.4) \qquad |D^\alpha_\xi(p^0(x,\xi)+\omega\mu^d)| = |D^\alpha_\xi p^0(x,\xi)| \leq c(x)\langle\xi\rangle^{d-|\alpha|} \ ,$$

where

$$(1.5.5) \qquad \langle\xi\rangle^{d-|\alpha|} \text{ is } O(\langle(\xi,\mu)\rangle^{d-|\alpha|}) \text{ on } \overline{\mathbb{R}}^{n+1}_+, \ \underline{\text{if and only if}} \ |\alpha| \leq d \ .$$

(As a matrix norm, we usually take the operator norm in euclidean space. We use from now on the abbreviations $\langle\xi,\mu\rangle = (1+|\xi|^2+|\mu|^2)^{\frac{1}{2}}$ and $|\xi,\mu| = (|\xi|^2+|\mu|^2)^{\frac{1}{2}}$, as defined in (A.1).)

1.5.1 Remark Let P be a pseudo-differential operator on $\mathbb{R}^n$ of order $d > 0$; add an extra variable t , and consider the "heat operator" $\partial_t + P$ operating on functions of $x \in \mathbb{R}^n$ and $t \in \mathbb{R}$. Then in the differential operator case, $\partial_t + P$ can be considered as a pseudo-differential operator with anisotropic symbol $i\tau + p(x,\xi)$, allowing studies of parametrices (in $S_{\rho,\delta}$ classes of ps.d.o.s) under suitable hypotheses on P . In the case where P is a genuine ps.d.o., with a principal symbol that is not polynomial in ξ , the symbol $i\tau + p(x,\xi)$ does not fall in the usual anisotropic classes, as considered in [Fabes-Rivière 1], [Lascar 1], [de Gosson 1], [Rempel-Schulze 1, Section 4.3.6] and elsewhere, which means that the heat equation for ps.d.o.s escapes treatment in the standard parabolic theory. Therefore a different kind of symbol calculus is necessary, and even more so in the case where boundary conditions are included.

The calculus we shall describe seems to have some relation to the calculus of Eskin and Čan Zui Ho (the announcement [Čan Zui Ho - Eskin 1] is all that has been available to us), except that they take ps.d.o.s in a larger sense - with very few estimates on ξ-derivatives, and no asymptotic expansions.

Other, related calculi have been studied in [Rempel-Schulze 3,4] (for ps.d.o.s not necessarily having the transmission property) and [Frank-Wendt 1,2] (for ps.d.o.s with rational symbols). See also [Vishik-Eskin 2].

In Chapter 2, we study systematically the class $S^{d,\nu}_{1,0}$ of symbols $p(x,\xi,\mu)$ of order d depending on a parameter $\mu \geq 0$ (or $\mu \in \mathbb{R}$) and satisfying

$$(1.5.6) \quad |D^\beta_x D^\alpha_\xi D^j_\mu p(x,\xi,\mu)| \leq c(x)\,(\langle\xi\rangle^{\nu-|\alpha|} + \langle\xi,\mu\rangle^{\nu-|\alpha|})\langle\xi,\mu\rangle^{d-\nu-j} ,$$

where the usual estimates in (ξ,μ) are satisfied for $|\alpha| \leq \nu$ but the control in terms of $\langle\xi,\mu\rangle$ does not improve for larger $|\alpha|$. ν is called the <u>regularity number</u>, and can have any real value; in some sense it measures how many "good" derivatives p has. When $p(x,\xi)$ is a usual symbol of order $d \geq 0$, the symbol $p(x,\xi) + \omega\mu^d$ is precisely in this class, of regularity $\nu = d$. (Differential operators are assigned the regularity $\nu = +\infty$.)

The subclass $S^{d,\nu}$ of <u>polyhomogeneous</u> symbols of order d and regularity ν consists of the symbols that have an asymptotic expansion

$$(1.5.7) \qquad p(x,\xi,\mu) \sim \sum_{\ell\in\mathbb{N}} p_{d-\ell}(x,\xi,\mu)$$

where each $p_{d-\ell}(x,\xi,\mu)$ has the homogeneity property

$$(1.5.8) \quad p_{d-\ell}(x,t\xi,t\mu) = t^{d-\ell} p_{d-\ell}(x,\xi,\mu) \quad \text{for } t \geq 1 , \quad |\xi| \geq 1 ,$$

and (1.5.7) holds in the sense that $p - \Sigma_{\ell<M} p_{d-\ell} \in S^{d-M,\nu-M}_{1,0}$ for all M. When $p(x,\xi)$ is a usual polyhomogeneous symbol of order $d \geq 0$, then $p(x,\xi) + \omega\mu^d$ is in $S^{d,d}$. As usual, we also denote $p_d(x,\xi,\mu)$ by $p^0(x,\xi,\mu)$, the principal symbol, it is uniquely determined for $|\xi| \geq 1$.

Note that the homogeneity, although valid with respect to the variables (ξ,μ) , is only assumed to hold for $|\xi| \geq 1$, which means that in the noncompact region

$$(1.5.9) \qquad V_1 = \{(\xi,\mu) \mid \mu \geq 0 , \quad |\xi| \leq 1\}$$

the only control over $p_{d-\ell}$ is furnished by estimates such as (1.5.6). This is precisely the problematic thing for the symbols $p(x,\xi) + \omega\mu^d$, when p is truly pseudo-differential. Consider $p^0(x,\xi)$. When it is a homogeneous function of ξ that is not a polynomial, it has an irregularity at $\xi = 0$, that can be handled in two ways: Either we modify $p^0(x,\xi)$ for small ξ to be a C^∞-function, loosing the homogeneity there, or we keep the strict homogeneity, but then we must deal with the fact that the function has a singularity at $\xi = 0$. Denote the strictly homogeneous version of p^0 by p^h. For $d = 0$, p^h is bounded near $\xi = 0$, and for $d > 0$ it is continuous; more precisely, the derivatives of p^h up to order ℓ, where

$$\ell = \text{largest integer} < d , \tag{1.5.10}$$

satisfy a Hölder condition of order $d - \ell$. (Note that the regularity $\nu = d$ of p^0 indicates precisely the degree of Hölder continuity.)

Both points of view will be used in the present text. The first point of view is the most commonly used in studies of ps.d.o.s; the ambiguity of p^0 near $\xi = 0$ is reflected in negligible error terms that occur everywhere in the theory anyway; and it is only because of the parameter μ that the harmless compact region $|\xi| \leq 1$ leads to the noncompact region V_1. The second point of view, the consideration of p^h, enters when we need certain precise informations for $\mu \to \infty$ (observe that when ξ is fixed and $\mu \to \infty$, the ray $\{(t\xi,t\mu) \mid t \geq 0\}$ approaches the μ-axis $\{\xi = 0\}$).

1.5.2 Definition. *The parameter-dependent,* N×N-*matrix formed symbol* $p(x,\xi,\mu) \in S^{d,\nu}$ *is said to be elliptic, when its principal symbol* p^0 *can be defined such that there are positive continuous functions* $c(x)$ *and* $c_0(x)$ *for which* $p^0(x,\xi,\mu)$ *is invertible, satisfying*

$$|p^0(x,\xi,\mu)^{-1}| \leq c(x)\langle\xi,\mu\rangle^{-d} , \tag{1.5.11}$$

for all $(\xi,\mu) \in \overline{\mathbb{R}}^{n+1}_+$ *with* $|\xi,\mu| \geq c_0(x)$.

For the original pseudo-differential operators that do not depend on μ, we can now formulate the definitions of parameter-elliptic and parabolic symbols.

<u>1.5.3 Definition.</u> *Let* $p(x,\xi) \in S^d$ *with* $d > 0$; *and let* $P = OP(p)$.

1^0 *Let* $\theta \in \mathbb{R}$. $P-\lambda$ *and* $p(x,\xi)-\lambda$ *are said to be* <u>*parameter-elliptic on the ray*</u> $\lambda = e^{i\theta}r$, *if the symbol*

$$\bar{p}(x,\xi,\mu) = p(x,\xi) + e^{i(\theta-\pi)}\mu^d , \tag{1.5.12}$$

that lies in $S^{d,d}$, *is elliptic in the sense of Definition 1.5.2, at each* x .

2^0 $P+\partial_t$, $P-\lambda$ *and* $p(x,\xi)-\lambda$ *are said to be* <u>*parabolic*</u>, *if* $p(x,\xi)-\lambda$ *is parameter-elliptic on the rays* $\lambda = e^{i\theta}r$ *for all* $\theta \in [\pi/2, 3\pi/2]$.

When $d > 0$, the ellipticity definition may in fact also be formulated in terms of the associated strictly homogeneous symbol, which is determined by

$$p^h(x,\xi) = |\xi|^d \, p^0(x,\xi/|\xi|) \quad \text{for} \quad \xi \neq 0 ; \tag{1.5.13}$$

note that $p^h(x,\xi)$ extends to a continuous function taking the value 0 at $\xi = 0$, when $d > 0$. We then have

<u>1.5.4 Lemma.</u> *Let* $p(x,\xi) \in S^d$ *with* $d > 0$, *and let* $|\omega| = 1$. *The following statements* (a), (b) *and* (c) *are equivalent:*

(a) $p(x,\xi)+\omega\mu^d$ *is elliptic in the sense of Definition 1.5.2 (i.e.,* $p(x,\xi)-\lambda$ *is parameter-elliptic on the ray* $\lambda = -\omega r$, $r \geq 0$).

(b) *For* $|\xi| = 1$, $p^0(x,\xi)$ *has no eigenvalues on the ray* $\lambda = -\omega r$, $r \geq 0$, *for each* x .

(c) $p^h(x,\xi)+\omega\mu^d$ *is bijective for all* $(\xi,\mu) \in \overline{\mathbb{R}}^{n+1}_+ \setminus 0$, *for each* x .

<u>Proof:</u> (a) obviously implies (b). (b) implies that $p^h(x,\xi)+\omega\mu^d$ is bijective for all $\xi \neq 0$, all $\mu \geq 0$, and here the values $\xi = 0$, $\mu > 0$, are included simply because $p^h(x,0) = 0$, so (b) implies (c). Now let (c) hold; then since $p^h(x,\xi)+\omega\mu^d$ is a continuous function of x and of $(\xi,\mu) \in \overline{\mathbb{R}}^{n+1}_+ \setminus 0$, and is homogeneous in (ξ,μ) of degree d , there is a continuous function $c_1(x)$ so that

$$|(p^h(x,\xi)+\omega\mu^d)^{-1}| \leq c_1(x)|\xi,\mu|^{-d} \quad \text{for} \quad (\xi,\mu) \in \overline{\mathbb{R}}^{n+1}_+ \setminus 0 . \tag{1.5.14}$$

We can choose the definition of the smooth function $p^0(x,\xi)$ for $|\xi| \le 1$ such that $|p^h(x,\xi)-p^0(x,\xi)| \le 1$ for all (x,ξ). Then

$$(1.5.15) \qquad (p^h+\omega\mu^d)^{-1}(p^0+\omega\mu^d) = I - (p^h+\omega\mu^d)^{-1}(p^h-p^0) ,$$

where

$$|(p^h+\omega\mu^d)^{-1}(p^h-p^0)| \le c_1(x)|\xi,\mu|^{-d} .$$

Let $c_0(x) = (\frac{1}{2}\, c_1(x))^{-d}$, then for $|\xi,\mu| \ge c_0(x)$,

$$|(p^h+\omega\mu^d)^{-1}(p^h-p^0)| \le \tfrac{1}{2} ,$$

so that (1.5.15) is invertible (by a Neumann series argument). Moreover, the inverse satisfies, for $|\xi,\mu| \ge c_0(x)$

$$|(p^0+\omega\mu^d)^{-1}| = |[I-(p^h+\omega\mu^d)^{-1}(p^h-p^0)]^{-1}(p^h+\omega\mu^d)^{-1}| \le 2c_1(x)|\xi,\mu|^{-d} \le c'(x)\langle\xi,\mu\rangle^{-d}$$

where we have used that $|\xi,\mu|^{-1} \le c_2(x)\langle\xi,\mu\rangle^{-1}$ for $|\xi,\mu| \ge c_0(x)$. This shows that (c) implies (a). □

The ellipticity, parameter-ellipticity, resp. parabolicity, is said to hold <u>uniformly</u> when $c(x)$ and $c_0(x)$ in the estimates can be replaced by positive constants; this is automatically satisfied when x runs in a compact set.

It is well known, already for differential operators, that <u>strong ellipticity</u> of P (positive definiteness of $p^0(x,\xi)+p^0(x,\xi)^*$ for $|\xi| \ge 1$) <u>implies parabolicity</u> of $P+\partial_t$, but that the converse holds only in the scalar case $(N=1)$. Parameter-ellipticity on a ray is of course weaker than strong ellipticity, for any $N \ge 1$.

Note that when P has the transmission property, <u>parabolicity implies that P is of even order</u>, because for odd d, σ is an eigenvalue of $p^0(x',0,0,\xi_n)$ if and only if $-\sigma$ is an eigenvalue of $p^0(x',0,0,-\xi_n)$ (for $|\xi_n| \ge 1$).

We shall now consider the same questions for boundary value problems. Here we recall that the ellipticity definition in Section 1.2 required invertibility of the boundary symbol operator for all $|\xi'| \ge c > 0$, so it is natural to expect that parameter-ellipticity amounts to the invertibility of a parameter-dependent boundary symbol operator for $|\xi',\mu| \ge c > 0$. This is also correct in the differential operator case, but in the pseudo-differential case the mere invertibility will not suffice; some extra control is needed at $\xi' = 0$, because of the lack of homogeneity of the C^∞ symbols on the region

(1.5.16) $$V_1' = \{(\xi',\mu) \mid |\xi'| \leq 1 \ , \ \mu \geq 0\} \ .$$

Let us consider an elliptic Green operator

(1.5.17) $$A = \begin{pmatrix} P_\Omega + G \\ T \end{pmatrix} \ , \quad \text{with} \quad a^0(x',\xi',D_n) = \begin{pmatrix} p_\Omega^0 + g^0 \\ t^0 \end{pmatrix}$$

as in (1.2.53), and the associated realization B, defined as in Definition 1.4.1 when the hypotheses there are satisfied. The resolvent of B

(1.5.18) $$R_\lambda = (B-\lambda)^{-1}$$

is the solution operator for the problem

$$(P_\Omega+G-\lambda)u = f \ ,$$
$$Tu = 0 \ ,$$

and it will be studied along with the inhomogeneous problem

(1.5.19) $$A_\lambda u \equiv \begin{pmatrix} P_\Omega+G-\lambda I \\ T \end{pmatrix} u = \begin{pmatrix} f \\ \varphi \end{pmatrix} \ ;$$

in fact, when A_λ is invertible, its inverse equals

(1.5.20) $$\begin{pmatrix} P_\Omega+G-\lambda I \\ T \end{pmatrix}^{-1} = (R_\lambda \quad K_\lambda) \ ,$$

where R_λ is the resolvent, and K_λ is a (parameter-dependent) Poisson operator. Together with $a^0(x',\xi',D_n)$ we consider $a^h(x',\xi',D_n)$

(1.5.21) $$a^h(x',\xi',D_n) = \begin{pmatrix} p_\Omega^h + g^h \\ t^h \end{pmatrix} \ ,$$

the operator defined for $\xi' \neq 0$ by the <u>strictly homogeneous</u> symbols $p^h(x',0,\xi',\xi_n)$, $g^h(x',\xi',\xi_n,\eta_n)$ and $t^h(x',\xi',\xi_n)$, coinciding with p^0, g^0 and t^0 (respectively) for $|\xi'| \geq 1$.

1.5.5 Definition. *Let* A *be a Green operator* (1.5.17), *with* P *of order* $d > 0$, *having the transmission property, with* G *of order* d, *and with* $T = \{T_0, T_1, \ldots, T_{d-1}\}$, T_k *of order* k *(all symbols being polyhomogeneous).*

1^0 *Let* $\theta \in \mathbb{R}$, *and let* $\omega = \exp i(\theta - \pi)$. *We say that* A_λ (1.5.19) *is* *parameter-elliptic on the ray* $\lambda = e^{i\theta} r$ $(r \geq 0)$, *when* (I)-(III) *hold:*

(I) $p^0(x,\xi)+\omega\mu^d$ *is bijective for* $|\xi| = 1$, $\mu \geq 0$ *and all* x *(i.e.,* $P-\lambda$ *is parameter-elliptic on the ray* $\lambda = e^{i\theta} r$ *).*

(II) *The parameter-dependent boundary symbol operator*

$$(1.5.22) \qquad \bar{a}^0(x',\xi',\mu,D_n) = \begin{pmatrix} p^0(x',0,\xi',D_n)_\Omega + g^0(x',\xi',D_n) + \omega\mu^d \\ t^0(x',\xi',D_n) \end{pmatrix}$$

is bijective from $\mathscr{S}(\overline{\mathbb{R}}_+)^N$ *to* $\mathscr{S}(\overline{\mathbb{R}}_+)^N \times \mathbb{C}^M$ *for* $|\xi'| = 1$, $\mu \geq 0$ *and all* x'.

(III) *The strictly homogeneous boundary symbol operators* $p^h(x',0,\xi',D_n)$, $g^h(x',\xi',D_n)$ *and* $t^h(x',\xi',D_n)$ *have limits for* $\xi' \to 0$ *in the respective principal symbol norms (to be explained further below), and the limiting parameter-dependent boundary symbol operator*

$$\bar{a}^h(x',0,\mu,D_n) = \begin{pmatrix} p^h(x',0,0,D_n)_\Omega + g^h(x',0,D_n) + \omega\mu^d \\ t^h(x',0,D_n) \end{pmatrix}$$

is bijective from $\mathscr{S}(\overline{\mathbb{R}}_+)^N$ *to* $\mathscr{S}(\overline{\mathbb{R}}_+)^N \times \mathbb{C}^M$ *for all* $\mu > 0$ *and all* x'.

When (I)-(III) *hold, we also say that the* μ*-dependent system* $\{P_\Omega + G + \omega\mu^d, T\}$, *and its set of symbols, are* *parameter-elliptic.*

2^0 *The system* $\{\partial_t + P_\Omega + G, T\}$, *and the system* A_λ, *and its symbol, are said to be* *parabolic, when* A_λ *is parameter-elliptic on each of the rays* $\lambda = e^{i\theta} r$ *with* $\theta \in [\pi/2, 3\pi/2]$.

For differential boundary problems, part 1^0 of the definition gives precisely the parameter-elliptic systems considered in [Agmon 2,3], [Agranovich-Vishik 1] and [Seeley 3, 4, 6]; and part 2^0 gives the standard parabolic systems (references to classical works on these are given in Section 4.1).

Hypotheses (II) and (III) together could also be formulated as a single hypothesis (since $\bar{a}^0 = \bar{a}^h$ for $|\xi'| \geq 1$):

(II+III) *For all* x' , *the strictly homogeneous parameter-dependent boundary symbol operator*

$$(1.5.23)\qquad \bar{a}^h(x',\xi',\mu,D_n) = \begin{pmatrix} p^h(x',0,\xi',D_n)_\Omega + g^h(x',\xi',D_n) + \omega\mu^d \\ t^h(x',\xi',D_n) \end{pmatrix}$$

is continuous (with respect to principal symbol norms) in $(\xi',\mu) \in \overline{\mathbb{R}}^n_+ \setminus 0$, *and is bijective from* $\mathcal{S}(\overline{\mathbb{R}}_+)^N$ *to* $\mathcal{S}(\overline{\mathbb{R}}_+)^N \times \mathbb{C}^M$ *for all* $(\xi',\mu) \in \overline{\mathbb{R}}^n_+ \setminus 0$.

We prefer to keep (II) and (III) apart, because there exist boundary value problems where (II) but not (III) is satisfied. Moreover, let us consider the following strengthened version of (II):

(II') *The parameter-dependent boundary symbol operator* $\bar{a}^0(x',\xi',\mu,D_n)$ (1.5.22) *is bijective from* $\mathcal{S}(\overline{\mathbb{R}}_+)^N$ *to* $\mathcal{S}(\overline{\mathbb{R}}_+)^N \times \mathbb{C}^M$ *for all* x' *and all* $(\xi',\mu) \in \overline{\mathbb{R}}^n_+$ *with* $|\xi',\mu| \geq c_0(x')$, *for some continuous function* $c_0(x') \geq 0$.

Even here, there exist pseudo-differential boundary problems satisfying (II') but not (III), see Example 1.5.13 below. This is quite different from the situation for differential operator problems, where $\bar{a}^h$ itself is C^∞ in $(\xi',\mu) \in \overline{\mathbb{R}}^n_+$ (because of the polynomial structure of the symbols), so that one naturally takes $\bar{a}^0 \equiv \bar{a}^h$. Then (II') implies (II+III).

For ps.d.o. problems, the invertibility of a given smooth principal boundary symbol operator a^0 as in (II'), does not imply (III). But it can be shown that (II'+III) is equivalent with (II+III), see Proposition 3.1.4 later.

Let us now explain the meaning of the convergence requirement in (III). (All this is taken up again in a more systematic fashion in Chapter 2, in the framework of parameter-dependent symbol classes.)

Let T be a system of trace operators T_k of orders $k=0,1,\dots,d-1$, going from E to bundles F_k over Γ . (The orders of a given system can be modified to lie between 0 and $d-1$, by left composition with invertible ps.d.o.s over the boundary, so all trace operators are in principle included; however, such a modification may interfere with the validity of parameter-ellipticity.)

Let r be such that the entries in T are of class $\leq r$. Then each T_k is of the form

$$(1.5.24) \qquad T_k = \sum_{0\leq j<r} S_{kj}\gamma_j + T'_k \quad ,$$

where T'_k is of class 0 . In particular,

$$(1.5.25) \qquad \begin{aligned} t_k^0(x',\xi) &= \sum_{0\leq j<r} s_{kj}^0(x',\xi')\xi_n^j + t_k'^0(x',\xi) \\ t_k^h(x',\xi) &= \sum_{0\leq j<r} s_{kj}^h(x',\xi')\xi_n^j + t_k'^h(x',\xi) \ . \end{aligned}$$

The convergence for $\xi' \to 0$ is simply taken in the sense that

$$(1.5.26) \qquad s_{kj}^h(x',\xi') \to s_{kj}^h(x',0) \quad \text{for each } x' \quad ,$$

$$(1.5.27) \qquad t_k'^h(x',\xi',\xi_n) \to t_k'^h(x',0,\xi_n) \ \text{ in } L^2_{\xi_n} \quad \text{for each } x' \quad ,$$

the latter corresponds to the principal estimate (the one with all indices zero) in (1.2.19). Now observe that each $s_{kj}^h(x',\xi')$ is homogeneous in ξ' of degree $k-j$. Then (1.5.26) cannot hold unless $k-j > 0$ or, in case $k-j = 0$, $s_{kj}^h(x',\xi')$ is constant in ξ' . In other words, the T_k must be of the form (of class $\leq k+1$)

$$(1.5.28) \qquad T_k = s_{kk}(x')\gamma_k + \sum_{0\leq j<k} S_{kj}\gamma_j + T'_k \ .$$

For the $t_k'^h(x',\xi',\xi_n)$ we observe that by the homogeneity,

$$(1.5.29) \qquad \|t_k'^h(x',\xi',\xi_n)\|_{L^2_{\xi_n}} \leq c_k(x')|\xi'|^{k+\frac{1}{2}} \quad ,$$

so they converge to zero for $\xi' \to 0$, since $k \geq 0$. Note also that the $s_{kj}^h(x',\xi')$ with $k > j$ go to zero for $\xi' \to 0$, since they are homogeneous of degree $k-j$. Thus the limit symbol is simply

$$(1.5.30) \qquad t_k^h(x',0,\xi_n) = s_{kk}(x')\xi_n^k \ .$$

The G term is similarly analyzed. It will in general be assumed to be of order d, and has the form

$$G = \sum_{0\leq j<r} K_j\gamma_j + G' \quad ,$$

where G' is of class 0, and the K_j are Poisson operators of order $d-j$. Now the principal homogeneous symbols satisfy

$$(1.5.31) \qquad \|k_j^h(x',\xi',\xi_n)\|_{L^2_{\xi_n}} \leq c_j(x')|\xi'|^{d-j-\frac{1}{2}} \quad ,$$

since k_j^h is homogeneous in (ξ',ξ_n) of degree $d-j-1$, and

(1.5.32) $$\| g'^h(x',\xi',\xi_n,\eta_n)\|_{L^2_{\xi_n,\eta_n}} \le c(x')|\xi'|^d ,$$

since g'^h is homogeneous in (ξ',ξ_n,η_n) of degree $d-1$. The convergence for $\xi' \to 0$ is precisely taken in the sense that the k_j^h converge in $L^2_{\xi_n}$-norm and g'^h converges in $L^2_{\xi_n,\eta_n}$ norm (in accordance with (1.2.19) and (1.2.38)). We see that g'^h is unproblematic, since $d > 0$ so that

$$g'^h(x',\xi',\xi_n,\eta_n) \to 0 \quad \text{for} \quad \xi' \to 0 ,$$

and that the k_j^h with $j < d$ behave similarly, whereas the k_j^h with $j \ge d$ blow up for $\xi' \to 0$ (unless they are zero). It follows that G must necessarily be of the form

$$G = \sum_{0\le j<d} K_j\gamma_j + G' ,$$

i.e. <u>of class</u> $\le d$. Altogether we have found

<u>1.5.6 Lemma.</u> *Let* $\begin{pmatrix} P_\Omega+G \\ T \end{pmatrix}$ *be a Green operator with* P *and* G *of order* $d > 0$, *and* $T = \{T_0,T_1,\dots,T_{d-1}\}$ *with* T_k *of order* k. *The associated strictly homogeneous symbols* g^h *and* t_k^h *have limits for* $\xi' \to 0$ *if and only if* G *and* T_k *are of the form, respectively,*

(1.5.33) $$G = \sum_{0\le j<d} K_j\gamma_j + G' ,$$
$$T_k = s_{kk}(x')\gamma_k + \sum_{0\le j<k} S_{kj}\gamma_j + T_k' ,$$

with G' *and the* T_k' *of class 0. In the affirmative case, the limits of the strictly homogeneous symbols are determined by*

(1.5.34) $$g^h(x',\xi',\xi_n,\eta_n) \to 0 , \quad t_k^h(x',\xi',\xi_n) \to s_{kk}(x')\xi_n^k , \text{ for } \xi' \to 0 .$$

Also p_Ω^h has a limit for $\xi' \to 0$ in a related sense: Write

$$p^h(x',0,\xi) = \sum_{0\le \ell\le d} s_\ell^h(x',\xi')\xi_n^\ell + p'^h(x',0,\xi) ,$$

where p'^h is $O(|\xi_n|^{-1})$, then

$$s_d^h(x',\xi') = s_d(x') \quad \text{for all } \xi' ,$$

(1.5.35) $$|s_\ell^h(x',\xi')| \le c_\ell(x')|\xi'|^{d-\ell} \to 0 \quad \text{for } \xi' \to 0 , \text{ when } \ell < d ,$$

$$\| p'^h(x,0,\xi',\xi_n)\|_{L^2_{\xi_n}} \le c'(x')|\xi'|^{d+\frac{1}{2}} \to 0 \quad \text{for } \xi' \to 0 ,$$

so we have that

(1.5.36) $$p^h(x',0,\xi) \to s_d(x')\xi_n^d \qquad \text{for } \xi' \to 0$$

in this sense. When all the symbols converge in the described way, the operator family $\bar{a}^h(x',\xi',\mu,D_n)\colon H^d(\bar{\mathbb{R}}_+)^N \to L^2(\mathbb{R}_+)^N \times \mathbb{C}^M$ depends continuously on (ξ',μ) and we see that the limit for $\xi' \to 0$ is the <u>differential operator</u>

(1.5.37) $$\bar{a}^h(x',0,\mu,D_n) = \begin{pmatrix} s_d(x')D_n^d + \omega\mu^d \\ s_{00}(x')\gamma_0 \\ \cdot \\ \cdot \\ \cdot \\ s_{d-1,d-1}(x')\gamma_{d-1} \end{pmatrix} : H^d(\bar{\mathbb{R}}_+)^N \to \begin{matrix} L^2(\mathbb{R}_+)^N \\ \times\, \mathbb{C}^{M_0} \\ \cdot \\ \cdot \\ \times\, \mathbb{C}^{M_{d-1}} \end{matrix} ,$$

it also goes from $\mathscr{S}(\bar{\mathbb{R}}_+)^N$ to $\mathscr{S}(\bar{\mathbb{R}}_+)^N \times \prod_{0\le k<d} \mathbb{C}^{M_k}$.

Condition (III) in Definition 1.5.5 requires bijectiveness of this operator, so we see immediately that a necessary condition for this is <u>surjectiveness</u> of each of the matrices $s_{kk}(x')$, i.e. <u>normality of</u> T . We have shown

<u>1.5.7 Lemma</u>. *In order for the system* $\begin{pmatrix} P_\Omega + G - \lambda \\ T \end{pmatrix}$ *to be parameter-elliptic on a ray, it is necessary that* T *be normal.*

<u>1.5.8 Remark</u>. The normality requirement is partly caused by our assumption that the strictly homogeneous symbols have a limit for $\xi' \to 0$. One can possibly make do with less, e.g. some uniformity assumption on $s_k(x',\xi')$ for $\xi' \to 0$, allowing a ps.d.o. of order 0 as coefficient of γ_k in T_k . When $M_k = N$, parameter-ellipticity will then force S_{kk} to be an elliptic ps.d.o. of order 0, and it can practically be eliminated by composition with a parametrix, which makes the trace operator normal. When $M_k < N$, one must have a surjectively elliptic coefficient S_{kk} , and the theory becomes difficult because S_{kk} is only of "regularity 0" in the sense we discussed for p above. The inclusion of such terms seems to be of marginal interest, and the theory will be complicated enough without them.

In the scalar case, the structure of (1.5.37) is fairly simple, because s_d and the nontrivial s_{jj} are then just invertible functions. In this case we can show that when (I) and (II) hold, the normality is also sufficient for the validity of (III). For then (as shown below), any normal boundary problem for $s_d(x')D_n^d + \omega\mu^d$ with the correct number of nontrivial boundary conditions (the number of roots of $s_d(x')\xi_n^d + \omega\mu^d$ in $\mathbb{C}_+$) is uniquely solvable. When $N > 1$, the cases where $s_d(x')$ is diagonalizable and the boundary space dimensions M_k are either 0 or N, can be reduced to the preceding case. When general dimensions M_k are allowed, not all normal boundary operators (1.5.37) with the correct boundary space dimension ΣM_k are wellposed, so the mentioned arguments do not work in general.

1.5.9 Proposition. *Let* $d \in \mathbb{N}_+$.

1^0 *Let* $\lambda' \in \mathbb{C}\setminus\overline{\mathbb{R}}_+$ *if* d *is even and* $\lambda' \in \mathbb{C}\setminus\mathbb{R}$ *if* d *is odd, and denote by* M *the number of roots of* $\tau^d-\lambda'$ *in* $\mathbb{C}_+$. *Let* J *be a subset of* $\{0,1,\dots,d-1\}$ *with* M *entries. The operator*

$$(1.5.38)\qquad a'(D_t) = \begin{pmatrix} D_t^d - \lambda' \\ \\ \{\gamma_j\}_{j\in J} \end{pmatrix} : H^d(\overline{\mathbb{R}}_+) \to \begin{matrix} L^2(\mathbb{R}_+) \\ \times \\ \mathbb{C}^M \end{matrix}$$

is bijective.

2^0 *Let* $\omega \in \mathbb{C}$ *with* $|\omega| = 1$, *and consider* $A_\mu = \{P_\Omega+G+\omega\mu^d, T\}$ *(for* $N\geq 1$*), with* P *and* G *of order* d *and class* $\leq d$, *and* $T = \{T_j\}_{j\in J}$ *having entries of order* j *and class* $\leq j+1$, *going from* E_Γ *to* F_j *with* $\dim F_j = N$, *for each* $j \in J$; *here* J *is a subset of* $\{0,1,\dots,d-1\}$. *Assume that* A_μ *satisfies* (I) *and* (II) *of Definition 1.5.5, and that* T *is normal. If* $N = 1$, *or if* $N > 1$ *and* $s_d(x')$ $(= p^0(x',0,0,1))$ *is diagonalizable, then* A_μ *satisfies* (III).

Proof: 1^0 The considerations here are related to the study of completely elliptic boundary problems in [Hörmander 9]. The d roots of $\tau^d-\lambda'$ have the form

$$(1.5.39)\qquad \begin{aligned} &\tau_\ell = e^{i(\sigma+2\pi\ell/d)}r, \qquad \ell=0,1,\dots,d-1 \\ &\text{with } r = |\lambda'|^{1/d} \quad \text{and} \quad \sigma = (\arg\lambda')/d \end{aligned}$$

($\arg\lambda' \in \,]0,2\pi[$). When d is even, the first $d/2$ roots lie in $\mathbb{C}_+$, and when d is odd, the first $(d+1)/2$ or $(d-1)/2$ of the roots lie in $\mathbb{C}_+$ (so $M = d/2$ resp. $(d\pm1)/2$). Let us write $J = \{m_0,m_1,\dots,m_{M-1}\}$. Since the differential operator $D_t^d - \lambda'$ is surjective from $H^d(\mathbb{R})$ to $L^2(\mathbb{R})$, it suffices to show that the problem

$$(1.5.40)\qquad \begin{aligned} &(D_t^d - \lambda')u = 0 \quad \text{on } \mathbb{R}_+ ,\\ &\gamma_{m_k} u = \varphi_k \quad \text{at } t = 0 , \text{ for } k = 0,\dots,M-1 , \end{aligned}$$

has a unique solution in $H^d(\overline{\mathbb{R}}_+)$ for each vector $\{\varphi_0,\dots,\varphi_{M-1}\} \in \mathbb{C}^M$. The full set of functions satisfying $(D_t^d - \lambda')u = 0$ on $\mathbb{R}$ is the span of the exponential solutions $\{\exp(i\tau_\ell t)\}_{0\leq \ell<d}$, and the subset of these lying in H^d on $\overline{\mathbb{R}}_+$ is spanned by $\{\exp(i\tau_\ell t)\}_{0\leq\ell<M}$. Inserting $u = \Sigma_{0\leq\ell<M}\, c_\ell \exp(i\tau_\ell t)$ in (1.5.40), we reduce the problem to solving the system of equations

$$(1.5.41)\qquad \sum_{0\leq\ell<M} c_\ell (i\tau_\ell)^{m_k} = \varphi_k , \quad 0 \leq k < M ,$$

and this is uniquely solvable when the matrix

$$(1.5.42)\qquad X = \begin{pmatrix} \tau_0^{m_0} & \tau_0^{m_1} & \cdots & \tau_0^{m_{M-1}} \\ \tau_1^{m_0} & \tau_1^{m_1} & \cdots & \tau_1^{m_{M-1}} \\ \vdots & \vdots & & \vdots \\ \tau_{M-1}^{m_0} & \tau_{M-1}^{m_1} & & \tau_{M-1}^{m_{M-1}} \end{pmatrix}$$

is invertible. In view of (1.5.39),

$$\tau_\ell^{m_k} = r^{m_k} \exp(i\sigma m_k + 2\pi m_k \ell/d) ,$$

so the entries in the k'th column have the common factor $r^{m_k}\exp(i\sigma m_k)$, and

$$(1.5.43)\qquad \det X = [\Pi_{0\leq j<M}\, r^{m_j} \exp(i\sigma m_j)] \det([\exp(i2\pi m_k/d)]^\ell)_{0\leq\ell,k<M} .$$

The last determinant is a Vandermonde determinant $\det((\lambda_k)^\ell)_{0\leq\ell,k<M}$, where the complex numbers $\lambda_k = \exp(i2\pi m_k/d)$ are nonzero and mutually distinct, since the m_k are distinct integers between 0 and $d-1$. Thus $\det X \neq 0$, and the unique solvability is proved.

2° Consider first the scalar case $N = 1$, and fix x'. Let $\mu > 0$. When (I) holds, the polynomial $s_d(x')\xi_n^d + \omega\mu^d$ can be written as $s_d(\xi_n^d - \lambda')$, with λ' as under 1°. When (II) holds, we have for each $\xi' \neq 0$ that the boundary space dimension $M_0 + \dots + M_{d-1}$ (where the M_j are either 1 or 0) must equal the index of $p^h(x',0,\xi',D_n)_\Omega + \omega\mu^d$ (since the singular Green operator $g^0(x',\xi',D_n)$ is compact from $H^d(\overline{\mathbb{R}}_+)$ to $L^2(\mathbb{R}_+)$); and since the operator family $p^h + \omega\mu^d$ depends continuously on $(\xi',\mu) \in \overline{\mathbb{R}}^n_+ \setminus 0$, the index equals the index of $p^h(x',0,0,D_n)_\Omega + \omega\mu^d = s_d(D_n^d - \lambda')$; this index is precisely M described above. The limit boundary symbol operator (1.5.37) reduces to the form (1.5.38), when we divide out the invertible coefficients and remove the zero-dimensional (trivial) trace operators, so 1° shows that the limit boundary symbol operator is bijective.

When $N > 1$, and $s_d(x')$ is diagonalizable (i.e., there is, in each local coordinate system at $\partial\Omega$, a diagonal matrix $c(x')$ and a regular matrix $b(x')$ so that $s_d(x') = b(x')c(x')b(x')^{-1}$), we reduce the original problem, by replacing $P_\Omega+G+\omega\mu^d$ locally by $b(x')(P_\Omega+G+\omega\mu^d)b(x')^{-1}$, to a case where the limit operator on $\mathbb{R}_+$ is on diagonal form. Since the M_k equal N or 0, the limit trace operator can be reduced to a system of standard normal derivatives $\{\gamma_j\}_{j\in J}$. In this way, the limit boundary symbol operator decomposes into a set of operators belonging to the scalar case, where 1° can be applied. □

1.5.10 Remark. Also in general, the dimension of the boundary bundle $M = \Sigma_{0<k<d} M_k$ is determined (for parameter-elliptic systems) by the requirement that (1.5.37) be bijective, so the standard theory for such systems implies that M equals the number of roots in $\mathbb{C}_+$ of $\det(s_d(x')\xi_n^d + \omega\mu^d)$. (We take Ω connected, cf. A.5, so that this number is independent of x'.)

In the parabolic case, d must be even, as observed after Lemma 1.5.4. Then the roots of $\det(s_d(x')\xi_n^d + \omega\mu^d)$ come in pairs $\pm\tau$, so there are exactly $Nd/2$ roots in $\mathbb{C}_+$ and $Nd/2$ roots in $\mathbb{C}_-$, and $M = Nd/2$. We have altogether:

(1.5.44) parabolicity implies d even and $M \equiv \sum_{0\le k<d} M_k = Nd/2$.

On the other hand, parameter-ellipticity on a single ray $\lambda = -\omega\mu^d$ does not require that d or Nd should be even. For example, the first order scalar ps.d.o. with symbol $\overline{\lambda}^1_+$ (cf. (3.1.48) later; it is a modification of $|\xi'| + i\xi_n$) defines a parameter-elliptic symbol $\overline{\lambda}^1_+ + \mu$, for which the Green system $\{\overline{\lambda}^1_{+,\Omega} + \mu, \gamma_0\}$ satisfies (I)-(III) in Definition 1.5.5. Also $\overline{\lambda}^1_- + \mu$ is parameter-elliptic, but here the operator $\overline{\lambda}^1_{-,\Omega} + \mu$ itself (with empty boundary condition) satisfies (I)-(III). In these examples, one can replace μ by $e^{i\theta}\mu$ for any $|\theta| < \pi/2$.

Here are some further examples of parameter-elliptic systems:

1.5.11 Example. The system

(1.5.45) $$\begin{pmatrix} -\Delta - \lambda \\ \gamma_0 \end{pmatrix}$$

associated with the Dirichlet problem for the Laplace operator, is a very simple example of a differential operator system, that is parameter elliptic on all the rays $\lambda = e^{i\theta}r$ with $\theta \in]0,2\pi[$; in particular, it is parabolic. Here (I-III) are easily verified by computation (it suffices to show injectiveness, for dimensional reasons), or one can appeal to the 1-coerciveness of $(-\Delta)_{\gamma_0}$. To get a pseudo-differential example, we can perturb (1.5.45) a little bit, replacing the parameter-independent operator by, say,

$$A = \begin{pmatrix} -\Delta + K_0\gamma_0 + K_1\gamma_1 + G' \\ \\ \gamma_0 + T_0' \end{pmatrix} , \tag{1.5.46}$$

where K_0 and K_1 are Poisson operators of orders 2 resp. 1, G' is a singular Green operator of order 2 and class 0 , and T_0' is a trace operator of order 0 and class 0 . If the principal symbol norms of K_0, K_1, G' and T_0' are sufficiently small, A_λ will again be parameter-elliptic on all the rays $\lambda = e^{i\theta}r$ with $\theta \in]0,2\pi[$. For, (II+III) for (1.5.46) can also be characterized as the bijectiveness of $\bar{a}^h(x',\xi',\mu,D_n)$: $H^2(\overline{\mathbb{R}}_+) \to L^2(\mathbb{R}_+)\times\mathbb{C}$ for all $(\xi',\mu) \in \overline{\mathbb{R}}^n_+$ with $|\xi',\mu| = 1$. Since $\overline{\mathbb{R}}^n_+ \cap S^{n-1}$ is compact, there is an upper bound on the norm of the inverse $(\bar{a}^h)^{-1}$ for $(\xi',\mu) \in \overline{\mathbb{R}}^n_+ \cap S^{n-1}$. A perturbation $\bar{a}^h(x',\xi',\mu,D_n)+s^h(x',\xi',D_n) = \bar{a}^h(I + (\bar{a}^h)^{-1}s^h)$ will then be bijective, e.g. if the operator $(\bar{a}^h)^{-1}s^h$ in $H^2(\overline{\mathbb{R}}_+)$ has norm $\leq \delta < 1$, which is assured by taking s^h small enough for $|\xi'| \leq 1$.

<u>1.5.12 Example.</u> One can also consider perturbations of the system defining the Neumann problem,

$$A' = \begin{pmatrix} -\Delta + K_0\gamma_0 + K_1\gamma_1 + G' \\ \\ \gamma_1 + S_0\gamma_0 + T_1' \end{pmatrix} \tag{1.5.47}$$

with K_0, K_1, G' as described above, S_0 a ps.d.o. of order 1 on the boundary, and T_1' a trace operator of order 1 and class 0 . Again the unperturbed operator (with K_0, K_1, G', S_0 and T_1' being zero) has the parameter-ellipticity property on all rays $\lambda = e^{i\theta}r$, $\theta \in]0,2\pi[$, so A_λ' is parameter-elliptic when the perturbing operators have a sufficiently small principal symbol norm.

In these examples, one can replace $-\Delta$ by more general strongly elliptic ps.d.o.s, as accounted for in Section 1.7.

Of course, many more examples can be given on the basis of the m-coercive cases in general (the discussion of these in Section 1.7 is included for that purpose), and here one finds in some cases (e.g. the Dirichlet case) that the m-coerciveness can be violated under small perturbations that preserve the parameter-ellipticity.

The examples given at the end of Remark 1.5.10 are parameter-elliptic but not coercive (in the sense of satisfying a Gårding inequality).

1.5.13 Example. We shall now give an example where (I) and (II), and even (II') are valid, but (III) (as well as the desired resolvent estimates) does not hold. Consider

$$A_\lambda = \begin{pmatrix} -\Delta - \lambda \\ T \end{pmatrix}, \quad \text{with } T = (1-\Delta_\Gamma)^{\frac{1}{2}} \gamma_1 R \,, \tag{1.5.48}$$

where R is the solution operator for the Dirichlet problem

$$R\colon u \rightsquigarrow v \,, \quad \text{where } -\Delta v = u \text{ on } \Omega \,, \quad \gamma_0 v = 0 \text{ on } \Gamma \,; \tag{1.5.49}$$

the bijective operator $(1-\Delta_\Gamma)^{\frac{1}{2}} = OP'(\langle\xi'\rangle)$ has been included in order to give the trace operator nonnegative order. When $\lambda \in \overline{\mathbb{R}_-}$, the problem

$$\begin{aligned} (-\Delta - \lambda)u &= f \quad \text{in } \Omega \,, \\ Tu &= \varphi \quad \text{at } \Gamma \,, \end{aligned} \tag{1.5.50}$$

can by insertion of $u = -\Delta v$ be transformed to the problem

$$\begin{aligned} \Delta^2 v + \lambda \Delta v &= f && \text{in } \Omega \,, \\ \gamma_0 v &= 0 && \text{at } \Gamma \,, \\ \gamma_1 v &= (1-\Delta_\Gamma)^{-\frac{1}{2}} \varphi && \text{at } \Gamma \,, \end{aligned} \tag{1.5.51}$$

which is simply a Dirichlet problem for the strongly elliptic fourth order operator $\Delta^2 + \lambda\Delta$. Since $\Delta^2 + \lambda\Delta \geq \Delta^2$ for $\lambda \in \overline{\mathbb{R}_-}$, there is a unique solution to (1.5.51) for all smooth data.

On the symbol level, we give $-\Delta$ the principal boundary symbol $\sigma(\xi')^2 + D_n^{\,2}$, where $\sigma(\xi')$ is a positive C^∞ function on $\mathbb{R}^{n-1}$ coinciding with $|\xi'|$ for $|\xi'| \geq 1$. Then the boundary symbol operator $r^0(x',\xi',D_n)$ for R depends smoothly on $\xi' \in \mathbb{R}^{n-1}$, and the boundary symbolic problem corresponding to (1.5.51),

$$\begin{aligned} (\sigma(\xi')^2 + D_n^2)^2 v + \mu^2(\sigma(\xi')^2 + D_n^2)v &= f && \text{on } \mathbb{R}_+ \,, \\ \gamma_0 v &= 0 && \text{at } x_n = 0 \,, \\ \gamma_1 v &= \sigma(\xi')^{-1}\varphi && \text{at } x_n = 0 \,, \end{aligned} \tag{1.5.52}$$

is uniquely solvable for all $(\xi',\mu) \in \mathbb{R}^n$. This gives us a smooth boundary symbol operator $\overline{a}^0(x',\xi',\mu,D_n)$ associated with (1.5.48), which is bijective for all $(\xi',\mu) \in \mathbb{R}^n$. Thus (I) and (II') (and hence (II)) are satisfied.

However, the trace operator T is of order 0 and class 0, so the symbol norm $\|t^h\|$ is $O(|\xi'|^{\frac{1}{2}})$ (cf. (1.5.29)); it goes to 0 for $\xi' \to 0$, and the limit operator

$$\bar{a}^h(x',0,\mu,D_n) = \begin{pmatrix} D_n^2 + \mu^2 \\ \\ 0 \end{pmatrix} : \mathscr{S}(\overline{\mathbb{R}}_+) \to \begin{matrix} \mathscr{S}(\overline{\mathbb{R}}_+) \\ \times \\ \mathbb{C} \end{matrix} \tag{1.5.53}$$

is not bijective. Thus (III) is not satisfied.

Moreover, a closer analysis of the resolvent R_λ for (1.5.48) will show that it does not satisfy the estimates we can deduce when (I-III) hold, namely, its norm as an operator in $L^2(\Omega)$ is merely $O(\langle\lambda\rangle^{-3/4})$, where the operators satisfying (I-III) will have resolvents that are $O(\langle\lambda\rangle^{-1})$ for $|\lambda| \to \infty$. (More comments are given in Remark 3.2.16 and Section 4.7.)

The example was pointed out to us by Denise Huet (in 1981), who also told us about the relation to singular perturbation theory: Set $\mu^{-1} = \varepsilon$ (i.e., $-\lambda = \varepsilon^{-2}$), and multiply the first line in (1.5.51) by ε^2, then one arrives at a type of problem

$$\begin{aligned} \varepsilon^2\Delta^2 v - \Delta v &= g \;, \\ \gamma_0 v &= \varphi_0 \;, \\ \gamma_1 v &= \varphi_1 \;, \end{aligned} \tag{1.5.54}$$

that has been widely studied in singular perturbation theory, where the point is to analyze the relation between v and the solution w of the following limit problem for $\varepsilon \to 0$:

$$\begin{aligned} -\Delta w &= g \;, \\ \gamma_0 w &= \varphi_0 \;. \end{aligned} \tag{1.5.55}$$

At first sight, our resolvent analysis did not seem to be of any help in the study of (1.5.54-55), because (1.5.48) fails to satisfy (III) in the definition of parameter-ellipticity. However, when the work on the present book was nearing the end, we found out that with some further reductions, (1.5.54) can indeed be transformed to a parameter-elliptic resolvent problem, see the complete discussion in Section 4.7.

There do exist non-normal realizations for which the resolvent is $O(\langle\lambda\rangle^{-1})$, see e.g. Remark 1.7.17 later.

We shall now introduce a further notion, that plays a great rôle in the precise calculus.

Our hypothesis (III) requires a certain continuity of the strictly homogeneous boundary symbol operators for $\xi' \to 0$. For the fine analysis we shall make, it will actually be necessary to measure the amount of continuity (and it will be possible to distinguish between various types of discontinuity) by introduction of the so-called regularity number ν for the boundary symbol operators. The concept is analogous to the concept we mentioned for $p(x,\xi)$ earlier, where the regularity number ν indicated the number of "good" estimates in terms of a calculus where (ξ,μ) is considered as the cotangent variable; - another description of ν was the amount of Hölder continuity of the strictly homogeneous symbol for $\xi \to 0$ (when $\mu > 0$) - all this generalizes to boundary symbols.

The full machinery will be introduced in Chapter 2 (where also negative regularity numbers will be included), and we here just list some practical definitions.

1.5.14 Definition. *Let* d *be a positive integer, and let* $k \geq 0$.

1^0 *A singular Green operator* G *of order* d *and class* 0 *is of regularity* $\nu(G) = d$.

2^0 *A singular Green operator* G *of order* d *and class* $r \in [1,d]$ *is of regularity* $\nu(G) = d-r+\frac{1}{2}$.

3^0 *A trace operator* T *of order* k *and class* 0 *is of regularity* $\nu(T) = k + \frac{1}{2}$.

4^0 *A trace operator of order* k *and class* $k+1$ *containing only standard terms*

$$T = \sum_{j\leq k} S_{kj}\gamma_j$$

is of regularity $\nu(T)$, *where* $\nu(T)$ *is the lowest value* $k-j$ *for which* S_{kj} *is not a differential operator* ($\nu = +\infty$ *if all* S_{kj} *are differential operators*).

5^0 *A general trace operator of order* k *and class* $k+1$,

$$T = \sum_{j\leq k} S_{kj}\gamma_j + T'$$

is of regularity $\nu(T) = \min\{\nu(\sum_{j\leq k} S_{kj}\gamma_j), \nu(T')\}$.

The motivation for 1^0 is (1.5.32), and 2^0 is motivated by (1.5.31), where the lowest value of $d-j-\frac{1}{2}$ for $j \leq r-1$ is $d-r+\frac{1}{2}$. The motivation for 3^0 is

(1.5.29), and 4^0 is motivated by the degree of homogeneity of the $s^h_{kj}(x',\xi')$; 5^0 then takes the weakest regularity number from the two preceding cases.

In particular, we have for G and T considered in Definition 1.5.5 that

$$\nu(G) = \min\{d, d-r+\tfrac{1}{2}\} \tag{1.5.56}$$

when G is of order d and class $r \leq d$, and that

$$\nu(T_k) = \min\{1, k+\tfrac{1}{2}\} \tag{1.5.57}$$

(or possibly better, if more than the first of the coefficients S_{kj} are differential operators). The smallest occurring regularity is here $\frac{1}{2}$, and the largest is d, unless there are only differential operators involved ($G = 0$ in particular), so that the regularity is $+\infty$.

In Chapter 2, the regularity numbers will be defined more generally for parameter-dependent symbols, and we shall see how Definition 1.5.14 fits together with that (see in particular Proposition 2.3.14 ff.).

<u>1.5.15 Example</u>. In the Dirichlet-type problem (1.5.46), $-\Delta$ has regularity $+\infty$ (and if it is replaced by a second order ps.d.o. P, one gets regularity 2), G' has regularity 2, γ_0 and γ_1 have regularity $+\infty$, K_0 has regularity 3/2, and K_1 and T_0' contribute with regularity $\frac{1}{2}$. As shown in Section 1.6 later (see Example 1.6.13), selfadjointness of the corresponding realization (with K_0 eliminated)requires $K_1 = iT_0'^*$ and $G' = G'^*$, here the system is of the form

$$A = \begin{pmatrix} -\Delta + iT_0'^*\gamma_1 + G' \\ \\ \gamma_0 + T_0' \end{pmatrix} . \tag{1.5.58}$$

So when $T_0' \neq 0$, we are in a (relatively difficult) case where the regularity of the system is only $\frac{1}{2}$.

For the Neumann-type problem (1.5.47) one gets selfadjointness precisely when the system, with K_1 eliminated, is of the form (cf. Example 1.6.15):

$$A' = \begin{pmatrix} -\Delta + iT_1'^*\gamma_0 + G' \\ \\ \gamma_1 + S_0\gamma_0 + T_1' \end{pmatrix} \tag{1.5.59}$$

where $G' = G'^*$ and $S_0 = -S_0^*$. Here G' again contributes with regularity 2, T_1' and $T_1'^*$ are of regularity 3/2 and S_0 is of regularity 1, so the regularity of the full system is 1, at worst.

Observe that the selfadjoint Neumann-type cases have better regularity than the Dirichlet-type cases.

Let us also mention the fact, encountered in Section 1.7, that the m-coerciveness of Neumann-type problems is stable under small perturbations, whereas it is not so for Dirichlet type problems (cf. (1.7.61) ff.).

The works of [Nambu 1] and [Triggiani 1] are only concerned with Neumann-type problems, and it would be interesting to know whether this reflects some natural preference in the applications they aim for. However, the Dirichlet-type problems (and other problems with regularity $\frac{1}{2}$) come up with full importance in our applications, as in Section 4.7 and in [Grubb-Solonnikov 1].

1.5.16 Remark. The study of boundary problems for $-\Delta - \lambda$ as in Example 1.5.13 is carried further in the articles of S. Rempel and B.W. Schulze on complex powers of ps.d.o.s without the transmission property [Rempel-Schulze 3, 4]. On one hand, they analyze the λ-dependence of the resolvent in certain cases where R_λ exists on rays but does not have minimal growth; on the other hand they exhibit examples where R_λ does have rays of minimal growth, but the boundary condition is not normal. The example presented in Remark 1.7.17 is one of theirs.

The papers of Rempel and Schulze are of interest, because they include ps.d.o.s that do not have the transmission property, which gives complications in the form of so-called Mellin terms. However, for the case of ps.d.o.s having the transmission property (the Boutet de Monvel calculus), the unusual examples are strongly linked with the case of constant coefficient operators [Rempel-Schulze 3]. For variable coefficient operators, [Rempel-Schulze 4] introduces symbol classes with a hypothesis on first ξ'-derivatives that is principally very similar to what we call regularity 1 (as defined in [Grubb 11, 13, 15] and in this book); and when this is combined with boundedness hypotheses for inverse principal symbols, one arrives at a symbol class where the strictly homogeneous symbols have invertible limits for $\xi' \to 0$ (the condition on ξ'-derivatives implies a Lipschitz continuity at $\xi' = 0$); in particular this restricts the set-up to normal boundary conditions when applied to differential boundary problems or Boutet de Monvel type problems. This is contrary to the impression given in [Rempel-Schulze 3, 4], where they make a point of being more general in this respect. The works [Grubb 11-15] and some other material of the author's was available to them, but apparently, the importance of the regularity concept in variable coefficient cases was not realized.

In the same line of thought, let us note that the general hypotheses of [Rempel-Schulze 4] do not allow regularity $\frac{1}{2}$ (thereby excluding some Dirichlet-type boundary conditions); and that they have a requirement on $D_{\xi_j} D_{\xi_n}$-derivatives of the interior pseudo-differential operator P, that is part of what we call regularity 2, and which excludes the important example $P = (-\Delta)^{\frac{1}{2}}$ of a ps.d.o. not having the transmission property.

1.6 Adjoints.

We here continue the analysis of normal boundary problems. It is shown that the full trace operator is surjective and that the realization is densely defined; the adjoint is seen to be another normal realization, and the self-adjoint realizations are characterized. Basic examples are discussed. This study is not necessary for the resolvent construction in itself, but is important for applications, in particular it prepares for the variational theory in Section 1.7. The formulas inevitably look complicated as soon as one deals with matrices, or operators of order > 2 ; we here build on the methods developed in [Grubb 2].

1.6.1 Lemma. *Let* A *be a differential operator in* $\tilde{E}$ *of order* d , *with* Γ *noncharacteristic for* A . *Let* $T = S\rho$ *be a normal trace operator with* S *going from* $E_\Gamma^d = \oplus_{0\le k<d} E_\Gamma$ *to* $F = \oplus_{0\le k<d} F_k$, *and let* $Z = \oplus_{0\le k<d} Z_k$ *be the vector bundle formed of the kernels* Z_k *of the morphisms* $S_{kk}\colon E_\Gamma \to F_k$ *(so* Z *is the kernel of* $S_{\text{diag}}\colon E_\Gamma^d \to F$, *the diagonal part of* S). *For each* k , *let* S'_{kk} *be a morphism in* E_Γ *projecting* E_Γ *onto* Z_k , *and let* S' *be the* $d\times d$ *diagonal matrix with diagonal elements* $S'_{00},\dots,S'_{d-1,d-1}$. *Let* $\tilde{S}$ *be the* $d\times d$ *matrix of ps.d.o. blocks* $\tilde{S}_{kj} = \begin{pmatrix} S_{kj} \\ S'_{kj}\end{pmatrix}$, *going from* E_Γ^d *to* $\oplus_{0\le k<d}(F_k\oplus Z_k)$, *denoted by* $\begin{pmatrix} S \\ S'\end{pmatrix}$ *for short (here* k,j *run in* $\{0,1,\dots,d-1\}$ *)* .

Then $\tilde{S} = \begin{pmatrix} S \\ S'\end{pmatrix}$ *is a bijection of* $\Pi_{0\le j<d}\, H^{s-j}(E_\Gamma)$ *onto* $\Pi_{0\le k<d}[H^{s-k}(F_k)\times H^{s-k}(Z_k)]$ *for any* $s\in\mathbb{R}$, *and the inverse* $\tilde{C}$ *is again a triangular matrix of ps.d.o.s,* $\tilde{C} = (\tilde{C}_{jk})_{0\le j,k<d}$, *where the* $\tilde{C}_{jk}$ *go from* $H^{s-k}(F_k)\times H^{s-k}(Z_k)$ to $H^{s-j}(E_\Gamma)$, *are equal to 0 for* $j<k$ *and of order* $j-k$ *for* $j\ge k$, *the* $\tilde{C}_{kk}$ *being homeomorphisms from* $F_k\oplus Z_k$ *to* E_Γ . *We here write* $\tilde{C}_{jk} = (C_{jk}\quad C'_{jk})$ *where*

$$C_{jk}\colon H^{s-k}(F_k)\to H^{s-j}(E_\Gamma) ,$$

$$C'_{jk}\colon H^{s-k}(Z_k)\to H^{s-j}(E_\Gamma) ,$$

and define the matrices $C = (C_{jk})_{0\le j,k<d}$ *and* $C' = (C'_{jk})_{0\le j,k<d}$, *writing* $\tilde{C} = (C\quad C')$ *for short, so that*

$$(1.6.1)\qquad \begin{aligned} (C\quad C')\begin{pmatrix} S \\ S'\end{pmatrix} &= I \quad \text{in} \quad \prod_{0\le k<d} H^{s-k}(E_\Gamma) ,\\ \begin{pmatrix} S \\ S'\end{pmatrix}(C\quad C') &= \begin{pmatrix} I & 0 \\ 0 & I\end{pmatrix} \quad \text{in} \quad \Pi H^{s-k}(F_k)\times \Pi H^{s-k}(Z_k) . \end{aligned}$$

Then finally, Green's formula may be written (cf. (1.3.13))

$$(1.6.2)\qquad (Au,v)_\Omega - (u,A'v)_\Omega = (\mathcal{A}\rho u,\rho v)_\Gamma = (S\rho u, I^\times S''\rho v)_\Gamma + (S'\rho u, I^\times S'''\rho v)_\Gamma\ ,$$

where the operators S'' *and* S''' *(continuous for any* $s\in\mathbb{R}$*),*

$$(1.6.3)\qquad \begin{aligned} S'' &= I^\times C^*\mathcal{A}^*\rho:\ \Pi H^{s-k}(E_\Gamma)\to \Pi H^{s-k}(F_{d-k-1})\\ S''' &= I^\times C'^*\mathcal{A}^*\rho:\ \Pi H^{s-k}(E_\Gamma)\to \Pi H^{s-k}(Z_{d-k-1})\end{aligned}$$

are again triangular systems of ps.d.o.s with morphisms in the diagonal; here C^* *and* C'^* *are surjective.*

When Γ *is noncharacteristic for* A , $S''\rho$ *and* $S'''\rho$ *are normal trace operators.*

<u>Proof</u>: Write $\tilde S = \tilde S_{diag} + \tilde S_{sub}$; here $\tilde S_{diag}$ is a diagonal matrix with homeomorphisms

$$\tilde S_{kk} = \begin{pmatrix} S_{kk}\\ S'_{kk}\end{pmatrix} : E_\Gamma \to \begin{matrix} F_k\\ \oplus\\ Z_k\end{matrix}$$

in the diagonal, and $\tilde S_{sub}$ is lower triangular, with zeroes in and above the diagonal. It is then easy to invert $\tilde S$: Observe e.g. that

$$U = (\tilde S_{diag})^{-1}\tilde S$$

is a triangular matrix of ps.d.o.s in E_Γ^d , with identities in the diagonal, so that

$$U = I + V$$

with V lower triangular and nilpotent. Then

$$(1+V)^{-1} = 1 - V + V^2 - \dots + (-V)^{d-1}\ ,$$

so that

$$(1.6.4)\qquad \tilde C = \tilde S^{-1} = U^{-1}(\tilde S_{diag})^{-1} = (1 - V + V^2 - \dots + (-V)^{d-1})\tilde S_{diag}^{-1}\ .$$

Since the iterates of V are lower triangular, with zeroes in the diagonal, $\tilde C$ is lower triangular and has $\tilde C_{diag} = \tilde S_{diag}^{-1}$. $\tilde C$ is now split into two matrices C and C' as explained in the lemma, and we insert the first identity of (1.6.1) in Green's formula for A , which gives

$$\begin{aligned}(\mathcal{A}\rho u,\rho v) &= (\mathcal{A}(CS+C'S')\rho u,\rho v)_\Gamma\\ &= (S\rho u, C^*\mathcal{A}^*\rho v)_\Gamma + (S'\rho u, C'^*\mathcal{A}^*\rho v)_\Gamma = (S\rho u, I^\times S''\rho v) + (S'\rho u, I^\times S'''\rho v)_\Gamma\end{aligned}$$

(since $I^{\times}I^{\times} = I$). Here S'' and S''' are lower triangular, since e.g. (cf. (1.3.12))

$$S'' = I^{\times}C^{*}\mathcal{A}^{\times *}I^{\times} ,$$

where $C^{*}\mathcal{A}^{\times *} = (\mathcal{A}^{\times}C)^{*}$ is upper triangular. That

$$(1.6.5) \qquad \begin{aligned} C^{*}&: \Pi_j H^{-s+j}(E_\Gamma) \to \Pi_k H^{-s+k}(F_k) \text{ and} \\ C'^{*}&: \Pi_j H^{-s+j}(E_\Gamma) \to \Pi_k H^{-s+k}(Z_k) \end{aligned}$$

are both surjective, follows from the fact that $\tilde{C}^{*} = \begin{pmatrix} C^{*} \\ C'^{*} \end{pmatrix}$ is surjective.

It remains to show that $S''\rho$ and $S'''\rho$ are normal in the non-characteristic case, but this follows readily from the fact that the diagonal elements in $\mathcal{A}^{\times}$ are then homeomorphisms in E_Γ : Since

$$(1.6.6) \qquad \begin{pmatrix} S_{kk} \\ S'_{kk} \end{pmatrix} (C_{kk} \; C'_{kk}) = \begin{pmatrix} I & 0 \\ 0 & I \end{pmatrix} \quad \text{in} \quad F_k \oplus Z_k ,$$

the C_{kk} and C'_{kk} are injective morphisms from F_k resp. Z_k to E_Γ ; then so are the diagonal elements in $\mathcal{A}^{\times}C$ resp. $\mathcal{A}^{\times}C'$; and the diagonal elements in $(\mathcal{A}^{\times}C)^{*}$ resp. $(\mathcal{A}^{\times}C')^{*}$ are then surjective from E_Γ to F_k resp. Z_k , so that finally the diagonal elements in S'' resp. S''' are surjective from E_Γ to F_{d-k-1} resp. Z_{d-k-1} . ▯

In view of Lemma 1.3.1 and Proposition 1.3.2, one has the corollary for ps.d.o.s having the transmission property at Γ (as we always assume):

<u>1.6.2 Corollary</u>. *Let* P *be a ps.d.o. in* $\tilde{E}$ *of order* d , *having the transmission property at* Γ . *Let* $T = S\rho$ *be as in Lemma 1.6.1, then the conclusions of Lemma 1.6.1 are valid, with* A *and* A^{*} *replaced by* P_Ω *and* P^{*}_Ω *(taking as* $\mathcal{A}$ *the Green's matrix satisfying Proposition 1.3.2). In particular, one has a formula*

$$(1.6.7) \qquad (P_\Omega u, v)_\Omega - (u, P^{*}_\Omega v)_\Omega = (S\rho u, I^{\times}S''\rho v)_\Gamma + (S'\rho u, I^{\times}S'''\rho v)_\Gamma$$

with normal trace operators $S'\rho$, $S''\rho$ *and* $S'''\rho$, *when* Γ *is noncharacteristic for* P .

Observe that we have in particular obtained that normality of $T = S\rho$ (which requires surjectiveness of the <u>diagonal</u> elements) <u>implies surjectiveness</u>

of the full matrix S . Since the Cauchy boundary operator ρ is surjective (from $C^\infty(E)$ to $C^\infty(E_\Gamma)^d$ or from $H^d(E)$ to $\Pi_{0\le k<d} H^{d-k-\frac{1}{2}}(E_\Gamma)$), it follows that the trace operator $T = S\rho$ is surjective

$$T = S\rho : H^d(E) \to \prod_{0\le k<d} H^{d-k-\frac{1}{2}}(F_k) \ .$$

1.6.3. Remark. The lemma and its corollary look a little simpler in the scalar case and the case (1.4.23), where they show the following, with notation as in Remark 1.4.4. When T is a normal trace operator $T = \{T_k\}_{k\in J}$, with $J \subset J_\rho = \{0,1,\dots,d-1\}$ and $T_k = \gamma_k + \Sigma_{j<k} S_{kj}\gamma_j$ for $k \in J$, then T is surjective from $H^d(E)$ to $\Pi_{k\in J} H^{d-k-\frac{1}{2}}(E_\Gamma)$. Now set

$$J' = J_\rho \setminus J$$
$$J'' = \{k \in J_\rho \mid d-k-1 \in J\}$$
$$J''' = \{k \in J_\rho \mid d-k-1 \notin J\}$$

and let $S'\rho = \{\gamma_k\}_{k\in J'}$. There exist auxiliary normal (and surjective) trace operators $S''\rho$ and $S'''\rho$ from $H^d(E)$ to $\Pi_{k\in J''} H^{d-k-\frac{1}{2}}(E_\Gamma)$ resp. $\Pi_{k\in J'''} H^{d-k-\frac{1}{2}}(E_\Gamma)$, such that

(1.6.8) $$(P_\Omega u,v)_\Omega - (u,P^*_\Omega v)_\Omega = (S\rho u,\dot{I}^\times S''\rho u)_\Gamma + (S'\rho u,\dot{I}^\times S'''\rho u)_\Gamma \ ,$$

where $\dot{I}^\times$ indicates a reflection of the index set (replacing $\{k\}_{k\in J}$ by $\{d-k-1\}_{k\in J}$).

In particular, if T is of the form $T = \{\gamma_k\}_{k\in J}$, then S and S' are the projections

(1.6.9) $$S = pr_J \quad \text{and} \quad S' = pr_{J'}$$

mapping a vector $\{\varphi_0,\varphi_1,\dots,\varphi_{d-1}\}$ into $\{\varphi_k\}_{k\in J}$ resp. $\{\varphi_k\}_{k\in J'}$; and the operators C and C' (satisfying (1.6.1)) are the injections

(1.6.10) $$C = i_J \quad \text{and} \quad C' = i_{J'} \ ,$$

supplying $\{\varphi_k\}_{k\in J}$ with zeroes on the places indexed by $k \in J'$, resp. supplying $\{\varphi_k\}_{k\in J'}$ with zeroes on the places indexed by $k \in J$. Their adjoints are again projections

$$C^* = pr_J \quad \text{and} \quad C'^* = pr_{J'} \ ,$$

so altogether,

$$(1.6.11)\qquad \begin{aligned} S'' &= \dot{I}^\times \mathrm{pr}_J \mathcal{A}^* \rho = \mathrm{pr}_{J''} I^\times \mathcal{A}^{\times *} I^\times \rho \ , \\ S''' &= \dot{I}^\times \mathrm{pr}_{J'} \mathcal{A}^* \rho = \mathrm{pr}_{J'''} I^\times \mathcal{A}^{\times *} I^\times \rho \ , \end{aligned}$$

in this case. (When T is of a more general form $S\rho$, S'' and S''' will also contain terms derived from S .)

We now turn to the more general normal trace operators $T = S\rho + T'$ considered in connection with pseudo-differential operators, and we shall show that also these are surjective. This is obtained by use of the following lemma, where $H^{(0,d)}(E_{\Sigma_+'})$ stands for the space of L^2 functions on $\Sigma_+' = \Gamma \times [0,1[$, whose x' derivatives up to order d are in L^2 (cf. also (A.23), (A.69)ff.).

1.6.4 Lemma. *The Cauchy trace operator* $\rho = \{\gamma_k\}_{0\le k<d} : H^d(E) \to \Pi_{0\le k<d} H^{d-k-\frac{1}{2}}(E_\Gamma)$ *has a family of continuous right inverses* K_δ *for* $\delta > 0$ *(Poisson operators), that map into functions supported in* Σ_+' *and whose norms, as operators from* $\Pi_{0\le k<d} H^{s-k-\frac{1}{2}}(E_\Gamma)$ *to* $H^{(0,s)}(E_{\Sigma_+'})$ *are* $O(\delta)$ *for* $\delta \to 0$, *any* s .

Proof: We first consider the situation for $\Omega = \mathbb{R}^n_+$, $N = 1$. As is well known, ρ defines an isomorphism between $H^d(\overline{\mathbb{R}}^n_+) \ominus \overset{\circ}{H}{}^d(\overline{\mathbb{R}}^n_+)$ and $\Pi H^{d-k-\frac{1}{2}}(\mathbb{R}^{n-1})$, so one cannot expect to obtain a small norm with respect to the metric of $H^d(\overline{\mathbb{R}}^n_+)$. But the situation is different with $H^{(0,d)}(\overline{\mathbb{R}}^n_+)$. We construct the right inverse as in [Hörmander 1, Theorem 2.5.7]. Let $\zeta \in C_0^\infty(\mathbb{R})$, with $\zeta = 1$ on a neighborhood of 0 and $\operatorname{supp} \zeta \subset]-1,1[$. For any $\delta > 0$, the system of operators

$$K^{(\delta)} = \{K_0^{(\delta)}, \dots, K_{d-1}^{(\delta)}\}$$

defined by

$$(1.6.12)\qquad (K_j^{(\delta)}\varphi)(x',x_n) = (2\pi)^{1-n} \int_{\mathbb{R}^{n-1}} e^{ix'\cdot\xi'} \zeta(\delta^{-2}\langle\xi'\rangle x_n) \frac{1}{j!} x_n^j \hat{\varphi}(\xi') d\xi'$$

furnishes a right inverse of ρ , and they are Poisson operators, cf. Section 1.2 or 2.3]. Here we have for $\varphi \in \mathcal{S}(\mathbb{R}^{n-1})$, any $s \in \mathbb{R}$,

$$(1.6.13)\qquad \begin{aligned} \|K_j^{(\delta)}\varphi\|^2_{H^{(0,s)}(\overline{\mathbb{R}}^n_+)} &= c_j \int_{\mathbb{R}^{n-1}} \int_0^\infty \langle\xi'\rangle^{2s} |\zeta(\delta^{-2}\langle\xi'\rangle x_n) x_n^j \hat{\varphi}(\xi')|^2 dx_n \, d\xi' \\ &= c_j \, \delta^{4j+2} \int_{\mathbb{R}^{n-1}} \langle\xi'\rangle^{2s-2j-1} |\hat{\varphi}(\xi)|^2 \int_0^\infty |\zeta(t)t^j|^2 dt \, d\xi' \\ &= c_j' \, \delta^{4j+2} \|\varphi\|^2_{H^{s-j-\frac{1}{2}}(\mathbb{R}^{n-1})} \ . \end{aligned}$$

(One can similarly show that

$$\|D^k_{x_n} K_j^{(\delta)}\varphi\|^2_{H^{(0,s)}} = c_{j,k}\ \delta^{4j-4k+2}\|\varphi\|^2_{s+k-j-\frac{1}{2}}\ .) \tag{1.6.14}$$

The desired result is obtained by taking $K_\delta = \zeta(x_n)K^{(\delta)}$. The construction is carried over to the manifold situation by use of local coordinates. $\square$

<u>1.6.5 Proposition.</u> *Let* $T = S\rho + T'$ *be a normal trace operator as defined in Definition 1.4.3. Then* T *is surjective from* $H^d(E)$ *to* $\Pi_{0\leq k<d}\ H^{d-k-\frac{1}{2}}(F_k)$ *(and from* $C^\infty(E)$ *to* $\Pi C^\infty(F_k)$*), with a continuous right inverse* K *(a Poisson operator). For any* $\varepsilon > 0$ *and any* $s \in \mathbb{R}$, K *can be chosen such that it maps into functions supported in* Σ'_+ *and its norm as an operator from* $\Pi_{0\leq k<d}\ H^{s-k-\frac{1}{2}}(F_k)$ *to* $H^{(0,s)}(E_{\Sigma'_+})$ *is smaller than* ε .

<u>Proof:</u> Consider first the case where $S = I$. We search for a solution u of the equation

$$\rho u + T'u = \varphi \tag{1.6.15}$$

on the form $u = K_\delta\psi$, where K_δ is as above. Inserting $K_\delta\psi$, we find the equation

$$\psi + T'K_\delta\psi = \varphi\ .$$

Since T' is of class 0, it is continuous from $H^{(0,d)}(E_{\Sigma'_+})$ to $\Pi H^{d-k-\frac{1}{2}}(E_\Gamma)$, so by Lemma 1.6.4, the norm of $T'K_\delta$, as an operator in $\Pi H^{d-k-\frac{1}{2}}(E_\Gamma)$, goes to 0 for $\delta \to 0$. Let δ_0 be such that this norm is $\leq \frac{1}{2}$ for $\delta \leq \delta_0$. Then $S_\delta = I + T'K_\delta$ has an inverse (a convergent Neumann series)

$$S_\delta^{-1} = (I+T'K_\delta)^{-1} = \sum_{m=0}^{\infty} (-T'K_\delta)^m \tag{1.6.16}$$

with norm ≤ 2 . Now $S_\delta = I + T'K_\delta$ is a ps.d.o. in E^d_Γ , and its invertibility implies that it is in fact elliptic (as a system $(S_{jk})_{0\leq j,k<d}$ where S_{jk} is of order $j-k$), so also S_δ^{-1} is an elliptic ps.d.o. system in E^d_Γ . Finally, (1.6.15) has the solution

$$u = K_\delta\ S_\delta^{-1}\ \varphi\ , \tag{1.6.17}$$

where $K = K_\delta\ S_\delta^{-1}$ is a Poisson operator, and we can take δ so small that the norm from $\Pi H^{s-k-\frac{1}{2}}(E_\Gamma)$ to $H^{(0,s)}(E_{\Sigma'_+})$ is $\leq \varepsilon$.

Now let S be general, so we have to solve the equation

$$S\rho u + T'u = \varphi \,. \tag{1.6.18}$$

Here we use Lemma 1.6.1 and consider the larger system

$$\begin{aligned} S\rho u + T'u &= \varphi \\ S'\rho u &= 0 \,, \end{aligned} \tag{1.6.19}$$

which by composition with $(C \quad C')$ leads to the equation

$$\rho u + C\,T'u = C\varphi \,,$$

CT' being again of class 0. By the first part of the proof, this has a solution

$$u = K_\delta\, S_\delta^{-1} C\varphi = K\varphi \,,$$

where K can be estimated as asserted. □

Before applying these results, we note that they have variants on the one-dimensional level.

1.6.6 Lemma.

1^0 *The Cauchy trace operator* $\rho\colon H^d(\overline{\mathbb{R}}_+)^N \to \Pi_{0\le j<d}\,\mathbb{C}^N$ *has a family of continuous right inverses* k_δ *for* $\delta > 0$ *(Poisson operators), mapping into functions supported in* $[0,1[$ *and with the property that the norm of* k_δ *as an operator from* $\Pi_{0\le k<d}\,\mathbb{C}^N$ *to* $L^2(\mathbb{R}_+)^N$ *goes to 0 for* $\delta \to 0$.

2^0 *When* $t^0(x',\xi',D_n)\colon H^d(\overline{\mathbb{R}}_+)^N \to \Pi_{0\le j<d}\,\mathbb{C}^{M_j}$ *is the principal boundary symbol operator associated with a normal trace operator* T , *then it has a continuous right inverse* $k^0(x',\xi',D_n)$ *(a Poisson operator); here* k^0 *can for any* $\varepsilon > 0$ *be taken with norm* $\le \varepsilon$ *as an operator from* $\Pi_{0\le j<d}\,\mathbb{C}^{M_j}$ *to* $L^2(\mathbb{R}_+)^N$, *for* x' *in a compact set and* $|\xi'| = 1$.

Proof: One can take $k_\delta = \zeta(x_n)k^{(\delta)}$, where $k^{(\delta)}\varphi = \Sigma_{0\le j<d}\, k_j^{(\delta)}\varphi_j$, with $k_j^{(\delta)}\varphi_j = \zeta(\delta^{-2}x_n)\frac{1}{j!}x_n^j\varphi_j$ for $\varphi_j \in \mathbb{C}^N$; the proof then goes practically as above. □

We recall that the Poisson operators on the one-dimensional level always map into $\mathcal{S}(\overline{\mathbb{R}}_+)^N$.

<u>1.6.7 Corollary.</u> *Let* P *be elliptic. When* $B = (P+G)_T$ *is defined by a normal trace operator* T , *then* $\begin{pmatrix} P_\Omega + G \\ T \end{pmatrix}$ *is elliptic if and only if the principal boundary realization* $b^0(x',\xi',D_n)$ *(cf.* (1.4.11')*) defines a bijection of* $D(b^0(x',\xi',D_n))$ *onto* $L^2(\mathbb{R}_+)^N$ *for all* x' , *all* $|\xi'| \geq 1$.

<u>Proof:</u> As noted in Section 1.2, ellipticity holds precisely when

$$\begin{pmatrix} p^0(x',0,\xi',D_n) + g^0(x',\xi',D_n) \\ t^0(x',\xi',D_n) \end{pmatrix} : H^d(\overline{\mathbb{R}}_+)^N \to \begin{matrix} L^2(\mathbb{R}_+)^N \\ \times \\ \mathbb{C}^M \end{matrix} \tag{1.6.20}$$

is bijective (for $|\xi'| \geq 1$). By the surjectiveness of t^0 , the solvability question for the problem

$$\begin{aligned} (p^0_\Omega + g^0)u &= f , \quad f \in L^2(\mathbb{R}_+)^N , \\ t^0 u &= \varphi , \quad \varphi \in \mathbb{C}^M = \prod_{0 \leq k < d} \mathbb{C}^{M_k} , \end{aligned} \tag{1.6.21}$$

can be reduced to the solvability question for

$$\begin{aligned} (p^0_\Omega + g^0)u &= f , \quad f \in L^2(\mathbb{R}_+)^N , \\ t^0 u &= 0 , \end{aligned}$$

so that unique solvability of (1.6.21) with $u \in H^d(\overline{\mathbb{R}}_+)^N$ holds if and only if b^0 is bijective. □

We can also use Lemma 1.6.4 to show that normal realizations have dense domains. In fact, we can show how the domains are homeomorphic to domains defined by normal boundary conditions <u>without</u> the non-local term T' ; and since $C_0^\infty(\Omega)$ is a dense subset here, the denseness follows immediately. The elimination of T' can be convenient for special purposes (and is used in Sections 1.7 and 4.4), but in the functional calculus itself, where we define functions of a realization B, one cannot simply replace B by another realization.

<u>1.6.8 Lemma.</u> *Let* $T = S\rho + T'$ *(going from* E *to* $\oplus_{0 \leq k < d} F_k$*) be a normal trace operator associated with the order* d .

1^0 *Let* C *be a right inverse of* S *according to Lemma 1.6.1. There is a* $\delta_0 > 0$ *such that for* $\delta \in]0,\delta_0]$, *the operator*

$$\Lambda = I + K_\delta C T' \tag{1.6.22}$$

(cf. Lemma 1.6.4), is a homeomorphism in $H^s(E)$ *for any* $s \geq 0$, *with inverse of the form*

$$\Lambda^{-1} = I + K_\delta Q_\delta CT' \ , \tag{1.6.22'}$$

for a δ-dependent ps.d.o. Q_δ *on* $\oplus_{0<k<d} E_\Gamma$ *that is bijective and elliptic in* $\Pi_{0\leq k<d} H^{s-k}(E_\Gamma)$. *Here one has the identity*

$$T = S\rho\Lambda \ , \tag{1.6.23}$$

and in particular, Λ *defines a bijection*

$$\Lambda\colon \{u \in H^d(E) \mid Tu=0\} \xrightarrow{\sim} \{v \in H^d(E) \mid S\rho v=0\} \ . \tag{1.6.23'}$$

2^0 *The set* $\{u \in H^d(E) \mid Tu=0\}$ *is dense in* $L^2(E)$.

Proof: Consider Λ defined by (1.6.22). By Lemma 1.6.4, K_δ maps $\Pi_{0\leq k<d} H^{-k-\frac{1}{2}}(E_\Gamma)$ into $L^2(E)$ with a norm that is $O(\delta)$, so the norm of $K_\delta CT'$ in $L^2(E)$ is $O(\delta)$. Similarly, the norm of the operator $T'K_\delta C$ in $\Pi H^{-k-\frac{1}{2}}(E_\Gamma)$ is $O(\delta)$. Taking δ_0 so small that these norms are $\leq \frac{1}{2}$ for $\delta \in]0,\delta_0]$, we can invert Λ by a Neumann series

$$\Lambda^{-1} = I + \sum_{m=1}^{\infty} (-K_\delta CT')^m \ ,$$

converging in operator norm in $L^2(E)$. In particular,

$$\sum_{m=1}^{\infty} (-K_\delta CT')^m = -K_\delta CT' - K_\delta \sum_{m=1}^{\infty} (-CT'K_\delta)^m CT'$$
$$= -K_\delta(1 + CT'K_\delta)^{-1}CT' \ ;$$

here the inverse $(I+CT'K_\delta)^{-1}$ of the ps.d.o. $I+CT'K_\delta$ in $\Pi H^{-\frac{1}{2}-k}(E_\Gamma)$ is an elliptic ps.d.o., since invertibility in the Sobolev space implies ellipticity. Thus Λ^{-1} has the form stated in (1.6.22'). Now we see that when $v = \Lambda u$ for $u \in H^d(E)$, then since $\rho K_\delta = I$ and $SC = I$ (cf. (1.6.1)),

$$S\rho v = S(\rho u + \rho K_\delta CT'u) = (S\rho + T')u \ ,$$

which shows (1.6.23). In particular, $S\rho v = 0$ if and only if $S\rho u + T'u = 0$, which completes the proof of 1^0.

Now 2^0 is an immediate consequence, for $C_0^\infty(\Omega,E)$ is contained in $\{v \in H^d(E) \mid S\rho v=0\}$, and is dense in $L^2(E)$, which by Λ^{-1} carries over to the denseness of $\Lambda^{-1}C_0^\infty(\Omega,E) \subset \{u \in H^d(E) \mid Tu=0\}$ in $L^2(E)$. □

The lemma shows that normal realizations are densely defined. When B is a normal realization, we can then analyze its <u>adjoint</u> B^* (as $L^2(E)$-operators). The <u>formal adjoint</u> B' will be defined as the operator acting like B^* and with domain

$$D(B') = D(B^*) \cap H^d(E) \,. \tag{1.6.24}$$

<u>1.6.9 Theorem</u>. *Let P be a ps.d.o. of order $d \geq 0$ in $\tilde{E}$, having the transmission property at Γ, and assume Γ is noncharacteristic for P. Let G be a s.g.o. in E of order d and class $\leq d$, let T be a normal trace operator (going from E to $\oplus_{0\leq k<d} F_k$), and let $B = (P+G)_T$ be the realization, acting like $P_\Omega + G$ and with domain*

$$D(B) = \{u \in H^d(E) \mid Tu = 0\} \,.$$

Write

$$T = S\rho + T' \,, \qquad G = K\rho + G' \,,$$

where T' and G' are of class 0, and define S', C, C', S'' and S''' as in Lemma 1.6.1. Then the formal adjoint B' of B is the realization $(P^*+\tilde{G})_{\tilde{T}}$, where $\tilde{G}$ is the s.g.o. of order d and class $\leq d$

$$\tilde{G} = -T'^*I^\times S''\rho + G'^* - T'^*C^*K^* = -T'^*C^*\mathcal{A}^*\rho + G'^* - T'^*C^*K^* \,, \tag{1.6.25}$$

and $\tilde{T}$ is the normal trace operator going from E to $\oplus_{0\leq k<d} Z_{d-k-1}$, defined by

$$\tilde{T} = S'''\rho + I^\times C'^*K^* = I^\times C'^*\mathcal{A}^*\rho + I^\times C'^*K^* \,. \tag{1.6.26}$$

When B is elliptic, B^* equals B' and is likewise elliptic.

<u>Proof</u>: Let $v \in H^d(E)$; we shall determine when it belongs to $D(B^*)$. Using the Green's formula (1.6.7) adapted specially to $S\rho$, we have for $u \in D(B)$,

$$\begin{aligned}
((P_\Omega+G)u,v)_\Omega &= (P_\Omega u,v)_\Omega + (K\rho u,v)_\Omega + (G'u,v)_\Omega \\
&= (u,P^*_\Omega v)_\Omega + (S\rho u, I^\times S''\rho v)_\Gamma + (S'\rho u, I^\times S'''\rho v)_\Gamma + ((CS+C'S')\rho u, K^*v)_\Gamma + (u,G'^*v)_\Omega \\
&= (u,P^*_\Omega v+G'^*v)_\Omega - (T'u, I^\times S''\rho v + C^*K^*v)_\Gamma + (S'\rho u\,,\, I^\times S'''\rho v + C'^*K^*v)_\Gamma \\
&= (u,(P^*_\Omega+G'^*-T'^*I^\times S''\rho - T'^*C^*K^*)v)_\Omega + (S'\rho u,(I^\times S'''\rho + C'^*K^*)v)_\Gamma \\
&= (u,(P^*_\Omega+\tilde{G})v)_\Omega + (S'\rho u, I^\times \tilde{T}v)_\Gamma
\end{aligned} \tag{1.6.27}$$

cf. (1.6.1). We clearly have that when $\widetilde{T}v = 0$, then $v \in D(B^*)$ and $B^*v = P^*_\Omega + \widetilde{G}$, so $B^* \supset (P^*+\widetilde{G})_{\widetilde{T}}$. Conversely, we must show that when $v \in D(B^*) \cap H^d(E)$, then $\widetilde{T}v = 0$, which is seen as follows.

Observe first that when $v \in D(B^*) \cap H^d(E)$, then the last line in (1.6.27) is a continuous functional on u in the $L^2(E)$-norm, so there is a constant $c(v)$ for which

$$(1.6.28) \qquad |(S'\rho u, I^\times \widetilde{T}v)_\Gamma| \leq c(v)\|u\|_0 \ , \quad \text{for } u \in D(B) \ .$$

We only have to consider the case where $\widetilde{T}$ is not void, i.e. $\Sigma_k \dim Z_k > 0$. Assume that there is a v such that $\widetilde{T}v = \psi \neq 0$; we shall show that this leads to a contradiction. Let $\varphi \in \Pi_{0 \leq k < d} C^\infty(Z_k)$, with

$$(1.6.29) \qquad (\varphi, I^\times \psi) \neq 0 \ .$$

Applying Proposition 1.6.5 to the trace operator $\begin{pmatrix} T \\ S'\rho \end{pmatrix}$, we can for each $\varepsilon > 0$ find $u_\varepsilon \in H^d(E)$, supported in Σ'_+ and with

$$Tu_\varepsilon = 0 \ , \qquad S'\rho u_\varepsilon = \varphi \ ,$$

and $\|u_\varepsilon\|_{(0,d)} \leq \varepsilon$. Now $u_\varepsilon \in D(B)$, so by (1.6.28)

$$|(S'\rho u_\varepsilon, I^\times \widetilde{T}v)| = |(\varphi, I^\times \psi)| \leq c(v)\|u_\varepsilon\|_0 \leq c(v)\|u_\varepsilon\|_{(0,d)} \to 0 \quad \text{for } \varepsilon \to 0 \ ;$$

this contradicts (1.6.29). Then $\widetilde{T}v$ must equal 0, and we have shown that $B' = (P^*+\widetilde{G})_{\widetilde{T}}$. Clearly, $\widetilde{T}$ is normal.

Consider finally the elliptic case. We first show the ellipticity of the adjoint on the boundary symbol level. Recall that in the elliptic case, the associated boundary symbol operator $\begin{pmatrix} p^0_\Omega + g^0 \\ t^0 \end{pmatrix}$ is bijective from $H^d(\overline{\mathbb{R}}_+)^N$ to $L^2(\mathbb{R}_+)^N \times \mathbb{C}^M$ with inverse $(r^0 \ \ k^0)$, and the boundary symbol realization $b^0 = (p^0_\Omega + g^0)_{t^0}$ is bijective from $D(b^0)$ to $L^2(\mathbb{R}_+)^N$ with inverse r^0 (at each (x',ξ') with $|\xi'| \geq 1$). The preceding arguments give that b^0 is densely defined and that b^{0*} is an extension of the boundary symbol realization b'^0 corresponding to the system

$$(1.6.30) \qquad \begin{pmatrix} (p^{*0})_\Omega + \widetilde{g}^0 \\ \widetilde{t}^0 \end{pmatrix}$$

associated with (1.6.25)-(1.6.26), in such a way that

$$(1.6.31) \qquad D(b^{0*}) \cap H^d(\overline{\mathbb{R}}_+)^N = D(b'^0) \ .$$

Now r^0 is a bounded operator in $L^2(\mathbb{R}_+)^N$, of the form $(p^0)^{-1}_\Omega + g^0_1$ where $(p^0)^{-1}$ is a ps.d.o. of order $-d$ and g^0_1 is a s.g.o. of order $-d$ and class 0. Then

$$(r^0)^* = ((p^0)^{-1}_\Omega + \tilde{g}^0)^* = (p^{0*})^{-1}_\Omega + g^{0*}_1 ,$$

which has a similar structure as r^0; so in particular it maps $L^2(\mathbb{R}_+)^N$ into $H^d(\overline{\mathbb{R}}_+)^N$. Then since b^0 is a bijection of the dense domain $D(b^0)$ onto $L^2(\mathbb{R}_+)^N$, it follows from an elementary result on Hilbert space operators that the adjoint b^{0*} is the inverse of the adjoint r^{0*}. In particular, the domain of b^{0*} is contained in $H^d(\overline{\mathbb{R}}_+)^N$, so in fact

$$b^{0*} = b'^0 = ((p^{0*})_\Omega + \tilde{g}^0)_{\tilde{t}^0} .$$

This implies that the problem

$$b'^0 u = f$$

has a unique solution $u \in H^d(\overline{\mathbb{R}}_+)^N$ for any $f \in L^2(\mathbb{R}_+)^N$, and hence $\{P^* + \tilde{G}, \tilde{T}\}$ and B' are elliptic, in view of Corollary 1.6.7.

It remains to show that $B' = B^*$ in the elliptic case. We here use that B has a "good" parametrix R_0 as stated in Proposition 1.4.2. Since R_0 is a bounded operator in $L^2(E)$, one has by standard arguments

$$R_0^* B^* \subset (BR_0)^* , \quad \text{which equals} \quad I + \mathcal{S}^* .$$

Then if $f \in D(B^*)$, one finds that

$$f = R_0^* B^* f - \mathcal{S}^* f \in H^d(E) ,$$

since $R_0^* = (P^{-1})^*_\Omega + G_0^*$ maps $L^2(E)$ into $H^d(E)$ and $\mathcal{S}^*$ is smoothing. Thus $B^* = B'$, in view of (1.6.24). ▯

1.6.9' Example. As a special exercise, let us find the formal adjoint of the "minimal operator" B_0 associated with $P_\Omega + G$, i.e. the realization defined from the system $\{P_\Omega + G, \rho\}$. Here when $G = K\rho + G'$, we have in fact that $B_0 = (P+G')_\rho$, since $\rho u = 0$ on the domain. The trace operator is ρ (in particular $T' = 0$), so that $S = I = C$ in the terminology of Lemma 1.6.1, and S' and C' are zero (or void), for the range of S' is zerodimensional. The formulas (1.6.25) and (1.6.26) give that $\tilde{G} = G'^*$, and $\tilde{T}$ is void. Thus the formal adjoint $(B_0)'$ equals the realization acting like $P^*_\Omega + G'^*$ and with domain $H^d(E)$, i.e. the "maximal operator" for $P^*_\Omega + G'^*$.

It follows in particular that when $B = (P+G)_T$ is a normal realization for which $D(B) \supset \mathring{H}^d(E)$ (i.e., $B \supset B_0$), then the formal adjoint B' acts like $P^* + G'^*$ (since $B' \subset (B_0)'$); where G'^* is of class 0.

An obvious corollary of the theorem is that it suffices for formal self-adjointness of B, that

$$\widetilde{G} = G \quad \text{and} \quad \widetilde{T} = \Psi T \tag{1.6.32}$$

for some homeomorphism Ψ, where $\widetilde{G}$ and $\widetilde{T}$ are defined by (1.6.25) and (1.6.26).

However, the full condition (1.6.32) is not necessary for selfadjointness, since G is in general not completely determined from a realization $B = (P+G)_T$.

In fact, since $Tu = 0$ for $u \in D(B)$, one has that

$$(P+G)_T = (P+G+K_1T)_T$$

for any Poisson operator K_1. So in order to characterize the selfadjoint realizations, we first investigate the extent to which a realization $B = (P+G)_T$ determines G and T. Here, when $T = S\rho + T'$ and $G = K\rho + G'$, an insertion of $Tu = 0$ and $\rho u = CS\rho u + C'S'\rho u$ gives that Gu can be replaced by

$$G_1u = KC'S'\rho u + (G'-KCT')u = K_1S'\rho u + G_1'u \;, \quad \text{with} \quad K_1 = KC' \text{ and } G_1' = G'-KCT' \;,$$

so we can always assume that G is given on the form $G = KS'\rho + G'$, to make the analysis more precise.

1.6.10 Theorem. *Let* $B = (P+G)_T$ *and* $\overline{B} = (\overline{P}+\overline{G})_{\overline{T}}$ *be normal realizations of a pseudo-differential operator* P *of order* $d \geq 0$, *the trace operators* $T = S\rho + T'$ *resp.* $\overline{T} = \overline{S}\rho + \overline{T}'$ *mapping into bundles* $F = \oplus_{0 \leq k < d} F_k$ *resp.* $\overline{F} = \oplus_{0 \leq k < d} \overline{F}_k$. *Then we have, with the notation of Lemmas 1.6.1 and 1.6.8 applied to the two operator systems:*

1^0 $D(B) = D(\overline{B})$ *holds if and only if:* $Z_k = \overline{Z}_k$ *for* $k = 0,\dots,d-1$, *and there exists a triangular homeomorphism* $\Psi = (\Psi_{jk})_{0 \leq j,k < d}$

$$\Psi : \prod_{0 \leq k < d} H^{s-k}(F_k) \to \prod_{0 \leq k < d} H^{s-k}(\overline{F}_k) \,, \tag{1.6.33}$$

with bijective morphisms $\Psi_{kk} : F_k \to \overline{F}_k$ *in the diagonal and ps.d.o.s* Ψ_{jk} *of order* $j-k$ *below it, such that*

$$\overline{T} = \Psi T \,, \quad i.e. \,, \quad \overline{S} = \Psi S \quad \text{and} \quad \overline{T}' = \Psi T' \,. \tag{1.6.34}$$

2^0 *Let* $D(B) = D(\overline{B})$. *Since* $Z_k = \overline{Z}_k$ *for all* k *according to* 1^0, *the operator* S' *defined in Lemma 1.6.1 supplies* S *as well as* $\overline{S}$ *to homeomorphisms in* $\prod_{0 \leq k < d} H^{s-k}(E_\Gamma)$. *Assume (as we may) that the singular Green operators* G *and* $\overline{G}$ *have been reduced to the form*

(1.6.34') $$G = KS'\rho + G' \quad \textit{resp.} \quad \overline{G} = \overline{K}S'\rho + \overline{G}' \ ,$$

by insertion of the homogeneous boundary conditions; K *and* $\overline{K}$ *go from* $\Pi_{0\leq k<d} H^{d-k-\frac{1}{2}}(Z_k)$ *to* $H^d(E)$. *Then* $B = \overline{B}$ *if and only if*

(1.6.35) $$K = \overline{K} \ , \qquad G' = \overline{G}' \ .$$

Proof: In 1°, the mentioned properties of course imply $D(B) = D(\overline{B})$. Conversely, assume that $D(B) = D(\overline{B})$, i.e.,

(1.6.36) $$Tu = 0 \Leftrightarrow \overline{T}u = 0 \ , \quad \text{for} \quad u \in H^d(E) \ .$$

Let Λ be a bijection in $H^s(E)$ for $s \geq 0$, as defined in Lemma 1.6.8, such that $T = S\rho\Lambda$; note that

$$\Lambda - I = K_\delta CT' \quad \text{and} \quad \Lambda^{-1} - I = K_\delta Q_\delta CT'$$

are singular Green operators of order and class 0 . For $v = \Lambda u \in H^d(E)$, we then have

$$S\rho v = 0 \Leftrightarrow (\overline{S}\rho + \overline{T}')\Lambda^{-1}v = 0$$

$$\Leftrightarrow \overline{S}\rho v + T'' \quad = 0 \ ,$$

where $T'' = \overline{S}\rho(\Lambda^{-1}-I) + \overline{T}'\Lambda^{-1}$ is of class 0 .

Since $S\rho v = 0$ is satisfied by all $v \in C_0^\infty(\Omega,E)$, and such v also have $\overline{S}\rho v = 0$, it follows that $T''v = 0$ for $v \in C_0^\infty(\Omega,E)$, and hence $T'' \equiv 0$ since it is of class 0 . This proves the identity

$$\overline{S}\rho = \overline{S}\rho\Lambda^{-1} + \overline{T}'\Lambda^{-1} \ ,$$

which implies

$$\overline{S}\rho\Lambda \ = \overline{S}\rho + \overline{T}' = \overline{T} \ .$$

So we now get from (1.6.36), taking $v = \Lambda u \in H^d(E)$

(1.6.37) $$S\rho v = 0 \Leftrightarrow \overline{S}\rho v = 0 \ , \quad \text{for} \quad v \in H^d(E) \ .$$

Since these trace operators are without "interior term", the techniques of [Grubb 2] can be applied.

Let us denote

$$Z^t(S) = \{\varphi \in \prod_{0\leq k<d} H^{t-k-\frac{1}{2}}(E_\Gamma) \mid S\varphi = 0\} \ ,$$

$$R^t(S) = S \prod_{0\leq k<d} H^{t-k-\frac{1}{2}}(E_\Gamma) \ ,$$

with analogous definitions for other operators. Recall that S is triangular, and is the sum of a diagonal part and a (nilpotent) subtriangular part

$$S = S_{diag} + S_{sub} ,$$

where S_{diag} has morphisms S_{kk} in the diagonal. The same holds for S', C and C' defined in Lemma 1.6.1 as well as for $\overline{S}$ and its associated operators $\overline{S}'$, $\overline{C}$ and $\overline{C}'$. Since ρ is surjective from $H^d(E)$ to $\Pi_{0\leq k<d} H^{d-k-\frac{1}{2}}(E_\Gamma)$, (1.6.37) is equivalent with the statement

$$Z^d(S) = Z^d(\overline{S}) . \tag{1.6.38}$$

Since $SC = I$, one has that

$$Z^d(S) = R^d(I - CS) ,$$

so (1.6.38) implies

$$\overline{S}(I - CS) = 0 . \tag{1.6.39}$$

All the matrices here are triangular, and we find, taking the diagonal part, that

$$\overline{S}_{diag}(I - C_{diag}\, S_{diag}) = 0 .$$

Then since $S_{diag}\, C_{diag} = I$ (the diagonal part of the equation $SC = I$), we find, using the above steps backwards, that

$$Z^d(\overline{S}_{diag}) \supset R^d(I - C_{diag}\, S_{diag}) = Z^d(S_{diag}) .$$

The argument also works for S and $\overline{S}$ interchanged, so altogether,

$$Z^d(S_{diag}) = Z^d(\overline{S}_{diag}) .$$

Since S_{diag} is a morphism, we may here identify

$$Z^d(S_{diag}) = \{\varphi \in \prod_{0\leq k<d} H^{d-k-\frac{1}{2}}(E_\Gamma) \mid S_{diag}\varphi = 0\} = \prod_{0\leq k<d} H^{d-k-\frac{1}{2}}(Z_k) ,$$

with a similar statement for $\overline{S}$, so it follows that

$$Z_k = \overline{Z}_k \quad \text{for} \quad k = 0,1,\dots,d-1 .$$

Then also $F_k \simeq \overline{F}_k$ for each k.

Finally, (1.6.39) shows that

$$\overline{S} = \overline{S}CS = \Psi S ,$$

where $\Psi = \overline{S}C$ is a surjective triangular ps.d.o. from $\Pi H^{t-k-\frac{1}{2}}(F_k)$ to

$\Pi H^{t-k-\frac{1}{2}}(\overline{F}_k)$ with morphisms in the diagonal. Since $F_k \simeq \overline{F}_k$, it follows that Ψ is bijective, with bijective morphisms in the diagonal. (This is shown in [Grubb 2, Lemma 1.13] by induction in p for the submatrices $(\Psi_{jk})_{j,k\leq p}$.) We now also have

$$\overline{T} = \overline{S}\rho\Lambda = \Psi S\rho\Lambda = \Psi T ,$$

so the proof of 1^0 is complete.

Now consider 2^0. Since S' defined in Lemma 1.6.1 is a supplement of S as well as $\overline{S}$, both G and $\overline{G}$ can be replaced by the reduced forms (1.6.34').

Clearly, (1.6.35) implies $B = \overline{B}$. For the converse direction, we again use Λ to transform the domain of B. Let $B = \overline{B}$, then

$$(1.6.40) \quad (P_\Omega+G)\Lambda v = (P_\Omega+\overline{G})\Lambda v \quad \text{for all } v \text{ with } S\rho v = 0 ,$$

and hence, for these v,

$$(1.6.41) \quad 0 = (G-\overline{G})\Lambda v = [(K-\overline{K})S'\rho + G'-\overline{G}']\Lambda v = (K-\overline{K})S'\rho + G''v ,$$

$$\text{where } G'' = (K-\overline{K})S'\rho(\Lambda-I) + (G'-\overline{G}')\Lambda \text{ is of class } 0 .$$

Since $\rho v = 0$ when $v \in C_0^\infty(\Omega, E)$, the identity (1.6.41) is satisfied for all $v \in C_0^\infty(\Omega,E)$, so we can conclude that $G'' \equiv 0$. Hence also

$$(1.6.42) \quad (K-\overline{K})S'\rho v = 0 \quad \text{for all } v \in H^d(E) \text{ with } S\rho v = 0 .$$

Here $S'\rho v$ runs through the full space $\Pi_{0<k<d} H^{d-k-\frac{1}{2}}(Z_k)$, so it follows that $K = \overline{K}$. Now $G'' = 0$ implies $(G'-\overline{G}')\Lambda = 0$, and hence $G' = \overline{G}'$. □

As an application of Theorems 1.6.9 and 1.6.10, we can characterize the formally selfadjoint realizations $B = (P+G)_T$. We here assume that G is given on reduced form as in (1.6.34')

$$G = KS'\rho + G' .$$

Then K must be replaced by KS' in the formulas (1.6.25) and (1.6.26) for the terms in the adjoint $B' = (P^*+G)_T$, which leads to the simpler formulas

$$(1.6.43) \quad \begin{aligned} \widetilde{T} &= S'''\rho + I^\times C'^*S'^*K^* = S'''\rho + I^\times K^* \equiv I^\times C'^*\mathcal{A}^*\rho + I^\times K^* , \\ \widetilde{G} &= -T'^*I^\times S''\rho + G'^* - T'^*C^*S'^*K^* = -T'^*I^\times S''\rho + G'^* \equiv -T'^*C^*\mathcal{A}^*\rho + G'^* , \end{aligned}$$

where we have used that $S'C' = I$ and $S'C = 0$, cf. (1.6.1). We assume from now on that Γ is noncharacteristic for P , so that $\mathcal{A}$ is invertible.

Now if $D(B) = D(B')$, then the above theorem tells us that Z_k equals the kernel $\overline{Z}_k$ of

$$(S''')_{kk} = (I^\times C'^*\mathcal{A}^*)_{kk} = iC'^*_{d-k-1,d-k-1}\, s_d(x') \quad \text{for each } k$$

(recall (1.3.11) ff.), which is a vector bundle morphism of E_Γ onto Z_{d-k-1} ; hence $\overline{Z}_k \simeq F_{d-k-1}$, and consequently $Z_k \simeq F_{d-k-1}$ for $k=0,1,\dots,d-1$. Moreover, there is a triangular pseudo-differential homeomorphism Ψ (with vector bundle morphisms in the diagonal) so that

$$\text{(1.6.44)} \qquad \tilde{T} = \Psi T \ ; \quad \text{in particular} \quad I^\times C'^*\mathcal{A}^* = \Psi S \quad \text{and} \quad I^\times K^* = \Psi T' \ .$$

By use of the identity $SC = I$ we find the explicit formula for Ψ :

$$\text{(1.6.45)} \qquad \Psi = I^\times C'^*\mathcal{A}^*C \quad ,$$

so this expression is seen to be a homeomorphism!

In particular, the diagonal part provides an explicit vector bundle homeomorphism of F_k onto Z_{d-k-1} :

$$\text{(1.6.46)} \qquad \Psi_{kk} \equiv C'^*_{d-k-1,d-k-1}\, s_d^*\, C_{kk} : F_k \overset{\sim}{\to} Z_{d-k-1} \ , \quad \text{for } k=0,\dots,d-1 \ .$$

In order to apply part 2^0 of the above theorem, we reduce $\tilde{G}$ to the form (1.6.34') (by insertion of $I = CS + C'S'$ and the equation $S\rho u = -T'u$, which holds on $D(B')$ when it equals $D(B)$)

$$\begin{aligned} \tilde{G}u &= -T'^*C^*\mathcal{A}^*(CS+C'S')\rho u + G'^*u \\ &= -T'^*C^*\mathcal{A}^*C'S'\rho u + G'^*u + T'^*C^*\mathcal{A}^*CT'u \ . \end{aligned}$$

Then part 2^0 of the above theorem tells us that when $P = P^*$, then $B = B'$ implies

$$\text{(1.6.48)} \qquad \begin{aligned} K &= -T'^*C^*\mathcal{A}^*C' \ , \\ G' &= G'^* + T'^*C^*\mathcal{A}^*CT' \ , \end{aligned}$$

and conversely, when (1.6.44) and (1.6.48) hold, $B = B'$. Noting in particular (cf. (1.3.7)) that

$$\text{(1.6.48')} \qquad P = P^* \quad \text{implies} \quad \mathcal{A}^* = -\mathcal{A} \ ,$$

we see that the first line in (1.6.48) can also be written

$$K = T'^*C^*\mathcal{A}C' ,$$

which is equivalent with the last statement in (1.6.44), in view of (1.6.45). Altogether, we have found:

1.6.11 Theorem. *Let* P *be a ps.d.o. of order* $d \geq 0$, *with* Γ *noncharacteristic for* P , *and let* $B = (P+G)_T$ *be a normal realization, with* $T = S\rho + T'$, *and with* G *given on the reduced form*

$$G = KS'\rho + G' . \tag{1.6.49}$$

Then the formal adjoint B' *equals* $(P^*+\tilde{G})_{\tilde{T}}$, *where*

$$\begin{aligned} \tilde{T} &= S'''\rho + I^\times K^* = I^\times C'^*\mathcal{A}^*\rho + I^\times K^* \\ \tilde{G} &= -T'^*I^\times S''\rho + G'^* = -T'^*C^*\mathcal{A}^*\rho + G'^* . \end{aligned} \tag{1.6.50}$$

Moreover, one has:

1^0 $D(B) = D(B')$ *holds if and only if*

$$\tilde{T} = \Psi T , \quad \textit{in particular} \quad S''' = \Psi S \quad \textit{and} \quad I^\times K^* = \Psi T' , \tag{1.6.51}$$

with a homeomorphism Ψ *of* $\Pi_{0\leq k<d} H^{s-k}(F_k)$ *onto* $\Pi_{0\leq k<d} H^{s-k}(Z_{d-k-1})$. *In the affirmative case,* Ψ *is determined by the formula*

$$\Psi = I^\times C'^*\mathcal{A}^*C , \tag{1.6.52}$$

and the diagonal elements in Ψ *define vector bundle homeomorphims of* F_k *onto* Z_{d-k-1} , *cf.* (1.6.46).

2^0 *Let* $P = P^*$, *so in particular,* $\mathcal{A}^* = -\mathcal{A}$. *Then* $B = B'$ *holds if and only if the conditions in* 1^0 *are satisfied along with*

$$G' = G'^* + T'^*C^*\mathcal{A}^*CT' . \tag{1.6.53}$$

As a consequence, one then also has

$$K = T'^*\Psi^*I^\times = T'^*C^*\mathcal{A}C' . \tag{1.6.54}$$

When $\{P_\Omega+G,T\}$ *is elliptic, these conditions imply that* B *is selfadjoint.*

Let us see what the results look like in the case where the F_k equal E_Γ or 0 (in particular the scalar case), as in Remark 1.6.3. Here

$$F_k = E_\Gamma \quad \text{for} \quad k \in J\ , \quad F_k = 0 \quad \text{for} \quad k \in J'$$
$$Z_k = 0 \quad \text{for} \quad k \in J\ , \quad Z_k = E_\Gamma \quad \text{for} \quad k \in J'\ ,$$

so the property $F_k \simeq Z_{d-k-1}$ implies that J' must be the "reflected" set of J

(1.6.55) $$J' = \{0 \le k < d \mid d-k-1 \in J\} = \dot{I}^\times J\ .$$

In particular, d must be even, $d = 2m$.

The special cases where J equals $J_0 = \{0,1,\dots,m-1\}$ resp. $J_1 = \{m,m+1,\dots,2m-1\}$ represent generalizations of the Dirichlet resp. Neumann condition; we shall analyze these in detail in the next examples. We denote

(1.6.56)
$$\gamma u = \{\gamma_0 u,\dots,\gamma_{m-1}u\} = \{\gamma_j u\}_{j\in J_0}\ , \quad \text{ranging in } E^0 = \Pi_{j\in J_0} E_\Gamma\ ,$$
$$\nu u = \{\gamma_m u,\dots,\gamma_{2m-1},u\} = \{\gamma_j u\}_{j\in J_1}, \quad \text{ranging in } E^1 = \Pi_{j\in J_1} E_\Gamma\ ;$$

and we split $\mathcal{A}$ in blocks corresponding to the two index sets J_0 and J_1

(1.6.57) $$\mathcal{A} = \begin{pmatrix} \mathcal{A}^{00} & \mathcal{A}^{01} \\ \mathcal{A}^{10} & 0 \end{pmatrix},$$

so that Green's formula takes the form

(1.6.58) $$(P_\Omega u,v)_\Omega - (u,P^*_\Omega v)_\Omega = (\mathcal{A}^{00}\gamma u + \mathcal{A}^{01}\nu u,\gamma v) + (\mathcal{A}^{10}\gamma u,\nu v)_\Gamma\ ,$$

here $\mathcal{A}^{01}$ and $\mathcal{A}^{10}$ are invertible. γ and ν are called the Dirichlet resp. the Neumann trace operator.

<u>1.6.12 Example</u>. Let $d = 2m$. <u>Dirichlet-type boundary conditions</u> are conditions where $J = J_0$; here $T = S\gamma + T'$ where S is a bijective operator in $\Pi_{j\in J_0} H^{s-j}(E_\Gamma)$ (in view of the triangular form and the surjectiveness of the diagonal part), so by composition with S^{-1} one can reduce the condition $Tu = 0$ to the form

(1.6.59) $$T^0 u \equiv \gamma u + T'^0 u = 0\ ,$$

where T'^0 is of class 0. We denote (as in (1.4.5))

(1.6.60) $$H^s_{E^0} = \prod_{0\le k\le m} H^{s-k-\frac12}(E_\Gamma) \quad \text{and} \quad H^s_{E^1} = \prod_{m\le k<2m} H^{s-k-\frac12}(E_\Gamma)\ .$$

Note that in this case

(1.6.61)
$$S = \mathrm{pr}_{J_0}\ , \qquad S' = \mathrm{pr}_{J_1}\ ,$$
$$C = i_{J_0} \qquad C' = i_{J_1}\ ,$$

and hence

$$(1.6.62)\qquad \begin{aligned} S'' &= I^{\times}\mathrm{pr}_{J_0}\mathcal{A}^*\rho = \dot{I}^{\times}\mathcal{A}^{00*}\gamma + \dot{I}^{\times}\mathcal{A}^{10*}\nu \\ S''' &= I^{\times}\mathrm{pr}_{J_1}\mathcal{A}^*\rho = \dot{I}^{\times}\mathcal{A}^{01*}\gamma \ , \end{aligned}$$

where $\dot{I}^{\times}$ indicates a reflection of the index set as in Remark 1.6.3. When $B = (P+G)_{T^0}$, G is on reduced form when γ is eliminated from it, i.e.,

$$(1.6.63)\qquad G = K^1\nu + G' \ .$$

Then the formal adjoint B' equals $(P^*+\tilde{G})_{\tilde{T}}$, with

$$\tilde{T} = \dot{I}^{\times}\mathcal{A}^{01*}\gamma + \dot{I}^{\times}K^1 = \dot{I}^{\times}\mathcal{A}^{01*}(\gamma + (\mathcal{A}^{01*})^{-1}K^{1*})$$

$$\tilde{G} = -T'^{0*}(\mathcal{A}^{00*}\gamma + \mathcal{A}^{10*}\nu) + G'^* \ .$$

In particular, $D(B) = D(B')$ holds if and only if

$$(1.6.64)\qquad T'^0 = (\mathcal{A}^{01*})^{-1}K^{1*} \ ,$$

and in this case, $\tilde{G}$ satisfies, for $u \in D(B)$,

$$\tilde{G}u = -T'^{0*}\mathcal{A}^{10*}\nu u + G'^*u + T'^{0*}\mathcal{A}^{00*}T'^0u \ ,$$

which shows how the reduced form looks.

Finally, when $P = P^*$, one has that $B = B'$ if and only if (1.6.64) holds and

$$(1.6.65)\qquad G' = G'^* + T'^{0*}\mathcal{A}^{00*}T'^0 \ ;$$

here one can use that

$$(1.6.65')\qquad P = P^* \text{ implies } \mathcal{A}^{01} = -\mathcal{A}^{10*} \text{ and } \mathcal{A}^{00*} = -\mathcal{A}^{00} \ .$$

Note that <u>when</u> $T'^0 = 0$ in the selfadjoint case, the singular Green operator (when on reduced form) is <u>necessarily of class</u> 0 . Note also that we have for the particular case where G and T'^0 are zero:

$$(1.6.66)\qquad (P_\gamma)' = (P^*)_\gamma \ .$$

<u>1.6.13 Example</u>. In the special case where $P = -\Delta$, the terms in Green's formula are particularly simple:

$$\mathcal{A} = i\begin{pmatrix} 0 & I \\ I & 0 \end{pmatrix} ,$$

(1.6.67)
$$(-\Delta u,v)_\Omega - (u,-\Delta v)_\Omega = i(\gamma_1 u,\gamma_0 v) + i(\gamma_0 u,\gamma_1 v)_\Gamma ,$$

(recall that $\gamma_1 u = -i\partial_{x_n} u|_\Gamma$) . Consider a Dirichlet-type realization $B = (-\Delta+G)_{T_0}$ with G on reduced form,

(1.6.68)
$$T_0 = \gamma_0 + T_0' , \qquad G = K_1\gamma_1 + G' .$$

In this case, selfadjointness holds if and only if

(1.6.69)
$$K_1 = i\, T_0'^* , \qquad G' = G'^* .$$

Note in particular that a realization $(-\Delta)_{T_0}$ without singular Green term can be selfadjoint only when $T_0' = 0$. And, a realization $(-\Delta+G)_{\gamma_0}$ can only be selfadjoint when G is of class 0 (G on reduced form).

1.6.14 Example. Let $d = 2m$. Neumann-type boundary conditions are conditions where (1.4.23) holds with $J = \{m,\dots,2m-1\} = J_1$. Here $Tu = 0$ may be reduced to the form

(1.6.70)
$$T^1 u \equiv \nu u + S^{10}\gamma u + T'^1 u = 0 ,$$

and T^1 is continuous from $H^{2m}(E)$ to $H^{2m}_{E^1}$ (cf. (1.6.56)). Now the general notation of Lemma 1.6.1 specializes to the present case as follows:

$$S\rho u = \nu u + S^{10}\gamma u , \quad S'\rho u = \gamma u ,$$

where S and S' are written together in block notation as

(1.6.71)
$$\tilde{S} = \begin{pmatrix} I & 0 \\ S^{10} & I \end{pmatrix} = \begin{pmatrix} S' \\ S \end{pmatrix} ,$$

which has the inverse

(1.6.72)
$$\tilde{C} = \begin{pmatrix} I & 0 \\ -S^{10} & I \end{pmatrix} = (C' \quad C) , \text{ with } C = \begin{pmatrix} 0 \\ I \end{pmatrix} , \quad C' = \begin{pmatrix} I \\ -S^{10} \end{pmatrix} .$$

Then

(1.6.73)
$$C^* = (0 \quad I) \text{ and } C'^* = (I \quad -S^{10*})$$

so that

$$C^*\mathcal{A}^*\rho u = (0 \quad I)\begin{pmatrix} \mathcal{A}^{00*} & \mathcal{A}^{10*} \\ \mathcal{A}^{01*} & 0 \end{pmatrix}\begin{pmatrix} \gamma u \\ \nu u \end{pmatrix} = \mathcal{A}^{01*}\gamma u ,$$

(1.6.74)

$$C'^*\mathcal{A}^*\rho u = (I \quad -S^{10*})\begin{pmatrix} \mathcal{A}^{00*}\gamma u + \mathcal{A}^{10*}\nu u \\ \mathcal{A}^{01*}\gamma u \end{pmatrix} = \mathcal{A}^{10*}\nu u + (\mathcal{A}^{00*} - S^{10*}\mathcal{A}^{01*})\gamma u .$$

Consider a realization $B = (P+G)_{T^1}$ with T^1 of the form (1.6.70) and G on reduced form

(1.6.75) $$G = K^0\gamma + G' .$$

Then the formal adjoint equals $(P^*+\tilde{G})_{\tilde{T}}$ with

(1.6.76)

$$\tilde{T} = \dot{I}^\times[\mathcal{A}^{10*}\nu u + (\mathcal{A}^{00*} - S^{10*}\mathcal{A}^{01*})\gamma u] + \dot{I}^\times K^{0*}$$

$$\tilde{G} = -T'^{1*}\mathcal{A}^{01*}\gamma u + G'^* .$$

We see that $D(B) = D(B')$ holds if and only if

$$\tilde{T} = \dot{I}^\times\mathcal{A}^{10*}T^1 ,$$

which can be specified into the following conditions

(1.6.77)

$$T'^1 = (\mathcal{A}^{10*})^{-1} K^{0*}$$

$$S^{10} = (\mathcal{A}^{10*})^{-1}(\mathcal{A}^{00*} - S^{10*}\mathcal{A}^{01*}) .$$

Furthermore, if $P = P^*$, then $B = B'$ if and only if (1.6.77) holds and

(1.6.78) $$G' = G'^* .$$

Again, the term of class 0 in T^1 is directly linked with the boundary term $K^0\gamma$ in G; in formally selfadjoint cases one of them cannot be nonzero without the other one being so.

1.6.15 Example. In the special case where $P = -\Delta$, and $B = (-\Delta+G)_T$ is a Neumann-type realization with G on reduced form:

(1.6.79) $$T = \gamma_1 + S_0\gamma_0 + T'_1 , \qquad G = K_0\gamma_0 + G' ,$$

the criteria for formal selfadjointness are:

(1.6.80)

$$K_0 = iT'^*_1 ,$$

$$S_0 = -S^*_0 ,$$

$$G' = G'^* .$$

1.6.16 Example. Let us also include a simple example of a system acting on vector valued functions, where the terms in the trace operator maps into lower dimensional bundles. Let $E = \bar{\Omega}\times\mathbb{C}^N$, $N > 1$, and let Q be an orthogonal projection of $\mathbb{C}^N$ onto a subspace V of dimension $k \geq 1$, with $k < N$. Let $Q^\perp$ be the projection onto $V^\perp$. Then

$$(1.6.81)\qquad A = \begin{pmatrix} -\Delta \\ Q\gamma_0 \\ Q^\perp\gamma_1 \end{pmatrix} : C^\infty(E) \to \begin{matrix} C^\infty(E) \\ \times \\ C^\infty(QE_\Gamma) \\ \times \\ C^\infty(Q^\perp E_\Gamma) \end{matrix}$$

defines a selfadjoint elliptic realization (in fact it breaks up into a Dirichlet problem for the sections in $QE = \bar{\Omega}\times V$ and a Neumann problem for the sections in $Q^\perp E = \bar{\Omega}\times V^\perp$). The ingredients in Lemma 1.6.1 are here:

$$(1.6.82)\qquad \begin{aligned} &F = QE_\Gamma \oplus Q^\perp E_\Gamma = (\Gamma\times V)\oplus(\Gamma\times V^\perp)\ , \quad Z = Q^\perp E_\Gamma \oplus QE_\Gamma = (\Gamma\times V^\perp)\oplus(\Gamma\times V)\ , \\ &S = \begin{pmatrix} Q & 0 \\ 0 & Q^\perp \end{pmatrix} : E_\Gamma \oplus E_\Gamma \to F\ , \qquad S' = \begin{pmatrix} Q^\perp & 0 \\ 0 & Q \end{pmatrix} : E_\Gamma \oplus E_\Gamma \to Z\ , \\ &C = S^* = \begin{pmatrix} i_V & 0 \\ 0 & i_{V^\perp} \end{pmatrix} : F \to E_\Gamma \oplus E_\Gamma\ , \quad C' = S'^* = \begin{pmatrix} i_{V^\perp} & 0 \\ 0 & i_V \end{pmatrix} : Z \to E_\Gamma \oplus E_\Gamma \end{aligned}$$

In view of (1.6.67), Green's formula (1.6.2) has the form

$$(1.6.83)\qquad (-\Delta u,v)_{L^2(\Omega)^N} - (u,-\Delta v)_{L^2(\Omega)^N} =$$

$$i\left(\begin{pmatrix} Q\gamma_0 u \\ Q^\perp\gamma_1 u \end{pmatrix}, \begin{pmatrix} Q\gamma_1 v \\ Q^\perp\gamma_0 v \end{pmatrix}\right)_\Gamma + i\left(\begin{pmatrix} Q^\perp\gamma_0 u \\ Q\gamma_1 u \end{pmatrix}, \begin{pmatrix} Q^\perp\gamma_1 v \\ Q\gamma_0 v \end{pmatrix}\right)_\Gamma .$$

Let us add a term of class 0 to the trace operator in (1.6.81), replacing it by

$$T = \begin{pmatrix} Q(\gamma_0 + T_0') \\ Q^\perp(\gamma_1 + T_1') \end{pmatrix}$$

where T_0' and T_1' are of class 0. (The second line could also include a term $Q^\perp S_0\gamma_0$, which of course complicates the calculations.) Let us also add a singular Green term to $-\Delta$, taken on reduced form (cf. (1.6.49) and (1.6.82))

$$G = K_0 Q^\perp\gamma_0 + K_1 Q\gamma_1 + G'\ ,$$

with G' of class 0. So we study the operator

$$(1.6.84) \qquad A = \begin{pmatrix} -\Delta+K_0Q^{\perp}\gamma_0 + K_1Q\gamma_1 + G' \\ Q(\gamma_0+T_0') \\ Q^{\perp}(\gamma_1+T_1') \end{pmatrix}$$

where K_0 and K_1 are Poisson operators of orders 2 resp. 1 from $\Gamma\times V$ resp. $\Gamma\times V^{\perp}$ to E, G' is a s.g.o. of class 0 and order 2, and T_0' and T_1' are trace operators of class 0 and orders 0 resp. 1. Let B be the corresponding realization. Then $B' = (-\Delta+\tilde{G})_{\tilde{T}}$, where

$$\tilde{T} = I^{\times}\begin{pmatrix} Q^{\perp} & 0 \\ 0 & Q \end{pmatrix}(-i)\begin{pmatrix} 0 & I \\ I & 0 \end{pmatrix}\rho + I^{\times}\begin{pmatrix} Q^{\perp}K_0^* \\ Q\,K_1^* \end{pmatrix} = -i\begin{pmatrix} Q\gamma_0 + i\,K_1^* \\ Q^{\perp}\gamma_1 + i\,K_0^* \end{pmatrix}$$

(1.6.85)

$$\tilde{G} = -(T_0'^*Q \quad T_1'^*Q^{\perp})\begin{pmatrix} Q & 0 \\ 0 & Q^{\perp} \end{pmatrix}(-i)\begin{pmatrix} 0 & I \\ I & 0 \end{pmatrix}\rho + G'^* = iT_1'^*Q^{\perp}\gamma_0 + i\,T_0'^*Q\gamma_1 + G'^* .$$

So, $D(B) = D(B')$ if and only if

$$(1.6.86) \qquad K_0 = i\,T_1'^*Q^{\perp}, \qquad K_1 = i\,T_0'^*Q ;$$

and since $\Delta = \Delta^*$, one has $B = B'$ precisely when (1.6.86) holds together with

$$(1.6.87) \qquad G' = G'^* .$$

Similar calculations, carried out for more general elliptic operators, will give formulas of the same kind, but with modifications stemming from $\mathcal{a}^{00}$ and S^{01} as in (1.6.65) and (1.6.77).

For other boundary conditions, the calculation of specific examples seems hardly more enlightening than the general formulas.

1.7 Semiboundedness and coerciveness.

A special class of boundary problems satisfying the parabolicity hypothesis introduced in Section 1.5 is the class of systems $\{P_\Omega+G,T\}$ defining lower bounded, "variational" realizations (with the property of m-coerciveness described in detail below). These are of interest, on one hand because there are well-established results in functional calculus for them (e.g. the theory of contraction semigroups generated by such operators, with applications to the heat equation); on the other hand they provide a good source of examples of boundary value problems that enter into our theory.

We recall that the ps.d.o. P of order d is called strongly elliptic on M (where M is a set in Σ), when the principal symbol $p^0(x,\xi)$ satisfies the (matrix-) inequality, with $c(x) > 0$,

$$\text{(1.7.1)}\qquad \operatorname{Re} p^0(x,\xi) \equiv \tfrac{1}{2}[p^0(x,\xi)+p^0(x,\xi)^*] \geq c(x)|\xi|^d I \quad \text{for } x\in M\ ,\ |\xi|\geq 1\ .$$

(One can also include operators for which the inequality (1.7.1) is obtained after multiplication by a (matrix) function of x.) Let us furthermore recall that when P satisfies (1.7.1) on a compact manifold Σ without boundary, it satisfies the Gårding inequality (for any $\varepsilon > 0$)

$$\text{(1.7.2)}\qquad \operatorname{Re}(Pu,u) \geq c\|u\|^2_{d/2} - c_1\|u\|^2_{d/2-\varepsilon} \quad \text{for } u \in H^{d/2}(\Sigma)\ (\text{or } H^{d/2}(\tilde{E}))\ ,$$

with $c > 0$.

Note that strong ellipticity of P requires that $(p^0(x,\xi)v,v)_{\mathbb{C}^N}$, for each x and each $v \in \mathbb{C}^N$, takes its values in a proper cone when $\xi \in \mathbb{R}^n$, $|\xi| \geq 1$, since

$$\text{(1.7.3)}\qquad |\operatorname{Im}(p^0(x,\xi)v,v)| \leq c'(x)|\xi|^d|v|^2 \leq c''(x)\operatorname{Re}(p^0(x,\xi)v,v)\ ,$$

when (1.7.1) holds. In particular, when P furthermore has the transmission property at $x_n = 0$, d must be even, since

$$p^0(x',0,0,\xi_n) = (-1)^d\, p^0(x,0,0,-\xi_n) \neq 0$$

for $|\xi_n| \geq 1$.

Now let P be strongly elliptic of even order $d = 2m$ on Σ. We assume throughout, that P has the transmission property at Γ. It is of interest to discuss which realizations B of P in $L^2(E)$ that satisfy lower boundedness inequalities like (1.7.2). We say that B is m-coercive, when there exist constants $C > 0$, $c > 0$ and $c_1 \in \mathbf{R}$ so that

(1.7.4) $$C\|u\|_m^2 \geq \operatorname{Re}(Bu,u) \geq c\|u\|_m^2 - c_1\|u\|_0^2 \quad \text{for} \quad u \in D(B) .$$

(Also here, one can include the realizations satisfying (1.7.4) after composition with a morphism; but we restrict the attention to systems satisfying (1.7.4) to fix the ideas.) One can call the systems $\begin{pmatrix} P_\Omega + G \\ T \end{pmatrix}$, for which $(P+G)_T$ is m-coercive, the strongly elliptic systems.

It is well-known in the differential operator case, that the Dirichlet realization of a strongly elliptic operator of order $2m$ is m-coercive [Gårding 1]. Larger classes of m-coercive elliptic realizations were introduced by S. Agmon and others in the fifties, and the complete set of m-coercive realizations defined by normal boundary conditions was thoroughly discussed in [Grubb 1-4].

The corresponding analysis has not been carried out for pseudo-differential boundary problems, as far as we know, and we shall therefore present some basic steps here, before we go on to the parameter-elliptic problems. The reader who is mainly interested in the parameter-elliptic theory can bypass this detailed analysis.

Realizations satisfying (1.7.4) can be expected to be linked with variational theory, and a standard ingredient here is the so-called Lax-Milgram lemma (that has its roots in the Friedrichs construction, and was also used by Vishik, cf. [Lions-Magenes 1, Chap. 2.10]). We formulate a version of the lemma with an outline of its proof, that will be used in the following.

1.7.1 Lemma. *Let* V *and* H *be Hilbert spaces, such that* V *is identified with a dense subspace of* H , *the respective norms satisfying (with* $c > 0$*)*

(1.7.5) $$\|u\|_V \geq c\|u\|_H \quad \text{for} \quad u \in V .$$

Let $s(u,v)$ *be a sesquilinear form on* V *that is continuous on* V, *and* V*-coercive:*

(1.7.6) $$\operatorname{Re} s(u,u) \geq c_0\|u\|_V^2 - k\|u\|_H^2 \quad \text{for} \quad u \in V ,$$

with $c_0 > 0$. *Let* S *be the operator in* H *defined from* s *and* V *as follows*

(1.7.7) $$D(S) = \{u \in V \mid \exists f \in H \text{ so that } s(u,v) = (f,v)_H \text{ for all } v \in V\}$$
$$Su = f ,$$

and let $s^*(u,v)$ *be the adjoint form on* V

(1.7.8) $$s^*(u,v) = \overline{s(v,u)} \quad \text{for} \quad u,v \in V .$$

Then S *is a closed, densely defined operator in* H , *its adjoint* S^* *is* *precisely the analogous operator defined from* s^* *and* V , *and both* S *and* S^* *have their numerical range and spectrum lying in a sector* C *of the complex plane (with* α *and* $\gamma \in \mathbb{R}$, $\beta \geq 0$ *)*

$$C = \{z \in \mathbb{C} \mid \operatorname{Re} z \geq \alpha \, , \, |\operatorname{Im} z| \leq \beta(\operatorname{Re} z + \gamma)\} \, . \tag{1.7.9}$$

Proof: The operator S is well defined, since V is dense in H . Assume first $k = 0$. The dense injection $V \subset H$ and the identification of H with H^* induce an injection $H \subset V^*$ (the space of continuous, conjugate linear functionals on V), such that

$$\langle w,\overline{v}\rangle_{V^*,V} = (w,v)_H \quad \text{when} \quad w \in H \, , \, v \in V \, ,$$

here $\langle w,\overline{v}\rangle$ denotes the value of the functional $w \in V^*$ on $v \in V$, cf. (A.12).

The sesquilinear form $s(u,v)$ induces a continuous operator $\mathcal{S}$ from V to V^* by the requirement

$$s(u,v) = \langle \mathcal{S}u,\overline{v}\rangle_{V^*,V} \quad \text{for all} \quad u \text{ and } v \in V \, ,$$

and the adjoint form $s^*(u,v)$ induces in this way precisely the adjoint $\mathcal{S}^*$ (going from V to V^*). It can be derived from the inequality (1.7.6) (with $k = 0$) that $\mathcal{S}$ and $\mathcal{S}^*$ are homeomorphisms of V onto V^* . With the injection $H \subset V^*$ defined as above, S is now simply the restriction of $\mathcal{S}$ to the space

$$D(S) = \{u \in V \mid \mathcal{S}u \in H\} \, , \tag{1.7.10}$$

and it follows that S is a bijection of $D(S)$ onto H . Define analogously S' by restriction of $\mathcal{S}^*$; then it follows by use of (1.7.5) that S^{-1} and $(S')^{-1}$ are bounded operators in H , and by (1.7.8) that they are adjoints of one another. Since $(S')^{-1}$ is injective, $R(S^{-1}) = D(S)$ is dense in H , so S^* is well defined, and it is seen that S^* equals S' . Note that S satisfies

$$\operatorname{Re}(Su,u) \geq c_0\|u\|_V^2 \geq c_0c^2\|u\|_H^2 \, ,$$

and 0 is in the resolvent set of S ; then the spectrum is contained in the halfplane $\{z \in \mathbb{C} \mid \operatorname{Re} z \geq c_0c^2\}$. The same holds for S^* .

When $k \neq 0$, we replace s by

$$s_k(u,v) = s(u,v) + (ku,v)_H \, ,$$

defining an operator S_k by the above procedure, and it is seen that $S_k =$

$S+kI$; in this case the spectrum and the numerical range of S and S^* lie in $\{z \in \mathbb{C} \mid \operatorname{Re} z \geq c_0 c^2 - k\}$. To show that they lie in a sector C , one uses that $|\operatorname{Im} s_k(u,u)| \leq \beta \operatorname{Re} s_k(u,u)$ for some β in view of (1.7.6) and the continuity of s on $V \times V$; then the above arguments can be applied to rotations $e^{\pm i\theta} S_k$ for a suitable $\theta > 0$. $\square$

When the situation of the lemma holds, we say for short that S is the variational operator associated with s on V in H , and operators generated in this way are called variational. One can show that $D(S)$ is dense even in V , so that a given variational operator S in H determines the form s and the space V it is associated with, by closure of the form $(Su,v)_H$ (the V-norm of $u \in D(S)$ being equivalent with $\operatorname{Re}(Su,u) + k\|u\|_H^2$ for a suitably large fixed k).

In the following theorem, the lemma will be applied with V equal to $\mathring{H}^m(E)$, which, as we recall, is the closure of $C_0^\infty(\Omega,E)$ in $H^m(\widetilde{E})$, and identifies with the space of functions $f \in H^m(E)$ for which $\gamma_0 f = \dots = \gamma_{m-1} f = 0$. The theorem extends the classical result of [Gårding 1] for differential operators.

1.7.2 Theorem. *Let* P *be strongly elliptic in* $\widetilde{E}$ *of even order* $d = 2m > 0$.

1^0 *The Dirichlet realization* P_γ *(hence also the system* $\{P_\Omega,\gamma\}$ *) is elliptic.*

2^0 *The Dirichlet boundary symbol realization* $b^0(x',\xi,D_n) = p^0(x',0,\xi',D_n)_\gamma$ *is* m*-coercive and positive, satisfying*

$$\operatorname{Re}(p^0_\Omega u,u) \geq c(x',\xi')\|u\|_m^2 \quad \textit{for all} \quad u \in \mathring{H}^m(\overline{\mathbb{R}}_+)^N \tag{1.7.11}$$

with $c(x',\xi') > 0$, *for any* x' , *any* $|\xi'| \geq 1$. *Moreover,* b^0 *is the variational operator associated with the sesquilinear form*

$$s(u,v) = \langle p^0_\Omega u,\overline{v}\rangle_{H^{-m},\mathring{H}^m} \tag{1.7.12}$$

on $\mathring{H}^m(\overline{\mathbb{R}}_+)^N$ *in* $L^2(\mathbb{R}_+)^N$, *the adjoint being the analogous operator for* p^{0*} .

3^0 *The realization* P_γ *(with domain* $H^{2m}(E) \cap \mathring{H}^m(E)$ *) is* m*-coercive:*

$$C\|u\|_m^2 \geq \operatorname{Re}(P_\Omega u,u) \geq c\|u\|_m^2 - c_1\|u\|_0^2 \quad \textit{for} \quad u \in D(P_\gamma) , \tag{1.7.13}$$

and it is the variational operator associated with the sesquiliniar form

$$b(u,v) = \langle P_\Omega u,\overline{v}\rangle_{H^{-m}(E),\mathring{H}^m(E)} \tag{1.7.14}$$

on $\mathring{H}^m(E)$ *in* $L^2(E)$, *the adjoint being the analogous operator for* P^* . P_γ *and* P^*_γ *have their numerical range and spectrum in a sector* C *as in* (1.7.9).

Proof: We begin the proof on the boundary symbol level. Here we first note that the strong ellipticity of P implies by Fourier transformation, since P is of order $2m$,

$$C\|u\|_m^2 \geq \mathrm{Re}(p^0u,u) \geq c\|u\|_m^2 \quad \text{for} \quad u \in \mathscr{S}(\mathbb{R})^N$$

with $C \geq c > 0$, and hence this holds in particular for $u \in C_0^\infty(\mathbb{R}_+)^N$. Thus $\mathrm{Re}(p^0u,u)$ is a norm equivalent with the H^m-norm, and in particular the completion $\mathring{H}^m(\overline{\mathbb{R}}_+)^N$ of $C_0^\infty(\mathbb{R}_+)^N$ in the H^m-norm is identical with the completion V of $C_0^\infty(\mathbb{R}_+)^N$ in the norm $\mathrm{Re}(p^0u,u)$. Hence the sesquilinear form

(1.7.15) $$s(u,v) = (p_\Omega^0 u,v) \quad \text{for} \quad u,v \in C_0^\infty(\mathbb{R}_+)^N$$

extends by continuity to a sesquilinear form $s(u,v)$ on $V = \mathring{H}^m(\overline{\mathbb{R}}_+)^N$, that is continuous and satisfies

(1.7.16) $$\mathrm{Re}\, s(u,u) \geq c\|u\|_m^2 \quad \text{for} \quad u \in V .$$

We now apply Lemma 1.7.1 and the concepts introduced there. In the present case where $V = \mathring{H}^m(\overline{\mathbb{R}}_+)^N$ and $H = L^2(\mathbb{R}_+)^N$, V^* is customarily identified with $H^{-m}(\overline{\mathbb{R}}_+)^N$ $(\subset \mathscr{D}'(\mathbb{R}_+)^N)$, whereby the duality $\langle w,\bar{v}\rangle_{V^*,V}$ coincides with the distribution duality when $v \in C_0^\infty(\mathbb{R}_+)^N$. Moreover, $p_\Omega^0 = r^+p^0e^+$ maps $\mathring{H}^m(\overline{\mathbb{R}}_+)^N$ continuously into $H^{-m}(\overline{\mathbb{R}}_+)^N$ with an adjoint $(p_\Omega^0)^*$ (from $\mathring{H}^m(\overline{\mathbb{R}}_+)^N$ to $H^{-m}(\overline{\mathbb{R}}_+)^N$) satisfying

$$(p_\Omega^0)^* = (p^{0*})_\Omega \equiv r^+p^{0*}e^+ \quad \text{on} \quad \mathring{H}^m(\overline{\mathbb{R}}_+)^N .$$

This implies that the extended sesquilinear form s on $V = \mathring{H}^m$ is in fact described by

(1.7.17) $$s(u,v) = \langle p_\Omega^0 u,v\rangle_{H^{-m},\mathring{H}^m} ,$$

which shows that S equals p_Ω^0 (as an operator from $\mathring{H}^m$ to H^{-m}). It follows in particular, that

(1.7.18) $$Su = p_\Omega^0 u \quad \text{for} \quad u \in D(S) .$$

Now recall the hypoellipticity property (Lemma 1.3.4), which shows that $D(S) \subset H^{2m}(\overline{\mathbb{R}}_+)^N$, so altogether $D(S) \subset H^{2m}(\overline{\mathbb{R}}_+)^N \cap \mathring{H}^m(\overline{\mathbb{R}}_+)^N$. On the other hand, if $u \in H^{2m}(\overline{\mathbb{R}}_+)^N \cap \mathring{H}^m(\overline{\mathbb{R}}_+)^N$, then $f = p^0_\Omega u \in L^2(\mathbb{R}_+)^N$, and

$$s(u,v) = \langle p^0_\Omega u, \bar{v}\rangle_{H^{-m},\mathring{H}^m} = (p^0_\Omega u, v)_{L^2} = (f,v)_{L^2}$$

for all $v \in V$, so $u \in D(S)$. We have then shown that

$$D(S) = H^{2m}(\overline{\mathbb{R}}_+)^N \cap \mathring{H}^m(\overline{\mathbb{R}}_+)^N . \tag{1.7.19}$$

(1.7.18) and (1.7.19) together show that S is the realization of p^0 on $\mathbb{R}_+$ defined by the (clearly normal) boundary condition $\gamma u = 0$, i.e.,

$$S = (p^0)_\gamma .$$

Since S is a bijection of $D(S)$ onto $L^2(\mathbb{R}_+)^N$, the realization is elliptic, in view of Corollary 1.6.7. Altogether, this shows 1^0 and 2^0.

Once the ellipticity is established, the rest is easy in view of Theorem 1.6.9. Like in the preceding analysis, we can extend the sesquilinear form

$$b(u,v) = (Pu,v) \quad \text{for } u,v \in C_0^\infty(\Omega,E)^N$$

by continuity to a sesquilinear form $b(u,v)$ on $\mathring{H}^m(E)^N$, and in view of (1.7.2), b is $\mathring{H}^m(E)^N$-coercive. Then b on $\mathring{H}^m(E)^N$ induces a variational operator B. This operator is an extension of the realization P_γ, since $D(P_\gamma) \subset V$ and

$$(P_\gamma u,v)_{L^2} = (P_\Omega u,v)_{L^2} = \langle P_\Omega u,\bar{v}\rangle_{H^{-m},\mathring{H}^m} = b(u,v) \quad \text{for } u \in D(P_\gamma),\ v \in V .$$

Since $(P_\Omega u,v) = (u,(P^*)_\Omega v)$ for u and $v \in C_0^\infty(\Omega,E)^N$, it follows by extension by continuity that the adjoint form b^* equals

$$b^*(u,v) = \langle (P^*)_\Omega u,\bar{v}\rangle_{H^{-m},\mathring{H}^m} .$$

The adjoint B^* of B is the variational operator defined from b^* on $\mathring{H}^m(E)^N$ in $L^2(E)^N$, so it satisfies analogously with B,

$$B^* \supset (P^*)_\gamma . \tag{1.7.20}$$

By 1^0 and 2^0 above, P_γ is elliptic, and (1.6.66) jointly with Theorem 1.6.9 show that

$$(P_\gamma)^* = (P^*)_\gamma .$$

Then (1.7.20) implies

$$B \subset P_\gamma$$

so we have altogether that $B = P_\gamma$. The general properties of P_γ and $(P^*)_\gamma$ now follow from Lemma 1.7.1. □

In particular, if P is formally selfadjoint and strongly elliptic, then P_γ is selfadjoint lower bounded, and it equals the Friedrichs extension of $P|_{C_0^\infty(\Omega,E)}$ (as the variational construction shows).

Note also that part 2^0 of the theorem implies:

1.7.3 Corollary. *When* P *is strongly elliptic in* $\widetilde{E}$ *of order* $2m$, *then* p_Ω^0 *is surjective from* $H^d(\overline{\mathbb{R}}_+)^N$ *to* $L^2(\mathbb{R}_+)^N$ *(and from* $\mathscr{S}(\overline{\mathbb{R}}_+)^N$ *to* $\mathscr{S}(\overline{\mathbb{R}}_+)^N$ *), with nullspace dimension equal to* mN *; in particular (cf.* (A.94)*)*

$$\text{index } p_\Omega^0 = mN . \tag{1.7.21}$$

With the above results as basis, one can analyze many more realizations of P defined by normal boundary conditions. First of all, the realizations P_T without a G and without "interior" terms in T can be treated by a slight extension of the theory of [Grubb 1-4]. We shall go through this rapidly, giving a variant that uses the right inverse of S according to Lemma 1.6.1 (rather than [Grubb 2, Section 1.3]). More general cases are treated later.

Let P be of even order $2m > 0$, and write Green's formula on the form

$$(P_\Omega u,v)_\Omega - (u,P_\Omega^* v)_\Omega = (\mathcal{A}^{00}\gamma u + \mathcal{A}^{01}\nu u,\gamma v)_\Gamma + (\mathcal{A}^{10}\gamma u,\nu v)_\Gamma , \tag{1.7.22}$$

that was explained in (1.6.56)-(1.6.58). Writing $P = A + P'$ as in Lemma 1.3.1 and using the "halfways" Green's formula for A ,

$$(Au,v)_\Omega = a(u,v) + (\mathcal{A}^{01}\nu u,\gamma v)$$

for a suitable integro-differential sesquilinear form $a(u,v)$ on $H^m(E) \times H^m(E)$ (cf. [Grubb 2]), we also have

$$(P_\Omega u,v) = a(u,v) + p'(u,v) + (\mathcal{A}^{01}\nu u,\gamma v) \tag{1.7.23}$$

where

$$p'(u,v) = (P'_\Omega u,v) = (u,(P'^*)_\Omega v)$$

satisfies

$$|p'(u,v)| \leq c\|u\|_m \|v\|_m \quad \text{for} \quad u,v \in H^{2m}(E) \ ;$$

this follows e.g. by interpolation from the inequalities

$$|(P'_\Omega u,v)| \leq c_1\|u\|_{2m}\|v\|_0 \ , \quad |(u,P'^*_\Omega v)| \leq c_2\|u\|_0\|v\|_{2m}$$

(or one can show it directly from the structure of P'). Altogether

(1.7.24) $$(P_\Omega u,v) = p(u,v) + (\mathcal{A}^{01}\nu u,\gamma v) \quad \text{for} \quad u,v \in H^{2m}(E) \ ,$$

where $p(u,v) = a(u,v) + p'(u,v)$ satisfies

(1.7.25) $$|p(u,v)| \leq c\|u\|_m\|v\|_m \ .$$

Consider a normal trace operator without "interior" term

(1.7.26) $$T = S\rho : H^{2m}(E) \to \prod_{0\leq k<2m} H^{2m-k-\frac{1}{2}}(F_k) \ .$$

It is advantageous to split S in blocks indexed by $J_0 = \{0,1,\dots,m-1\}$ and $J_1 = \{m,\dots,2m-1\}$

(1.7.27) $$S = \begin{pmatrix} S^{00} & 0 \\ S^{10} & S^{11} \end{pmatrix}$$

where $S^{\delta\varepsilon}$ goes from $\prod_{j\in J_\varepsilon} H^{2m-j-\frac{1}{2}}(E_\Gamma)$ to $\prod_{k\in J_\delta} H^{2m-k-\frac{1}{2}}(F_k)$. The other matrices occurring in Lemma 1.6.1 are split similarly

(1.7.28) $$S' = \begin{pmatrix} S'^{00} & 0 \\ 0 & S'^{11} \end{pmatrix} , \quad C = \begin{pmatrix} C^{00} & 0 \\ C^{10} & C^{11} \end{pmatrix} , \quad C' = \begin{pmatrix} C'^{00} & 0 \\ C'^{10} & C'^{11} \end{pmatrix}$$

which corresponds to a splitting of $\tilde{S}$ and $\tilde{C}$:

(1.7.29) $$\tilde{S} = \begin{pmatrix} \tilde{S}^{00} & 0 \\ \tilde{S}^{10} & \tilde{S}^{11} \end{pmatrix} = \begin{pmatrix} \begin{pmatrix} S^{00} \\ S'^{00} \end{pmatrix} & \begin{pmatrix} 0 \\ 0 \end{pmatrix} \\ \begin{pmatrix} S^{10} \\ 0 \end{pmatrix} & \begin{pmatrix} S^{11} \\ S'^{11} \end{pmatrix} \end{pmatrix}$$

etcetera. Because of the triangular nature of the operators, the inversion formulas (1.6.1) imply in particular

$$(1.7.30)\qquad C^{00}S^{00} + C'^{00}S'^{00} = 1 \quad \text{on } \oplus_{J_0} E_\Gamma$$

$$\begin{pmatrix} S^{00}C^{00} & S^{00}C'^{00} \\ S'^{00}C^{00} & S'^{00}C'^{00} \end{pmatrix} = \begin{pmatrix} I & 0 \\ 0 & I \end{pmatrix} \quad \text{on } \begin{matrix} \oplus_{J_0} F_k \\ \oplus_{J_0} Z_k \end{matrix} ,$$

with analogous formulas with 0 replaced by 1; the S-operators being surjective and the C-operators injective. Now the boundary condition may be written

$$(1.7.31)\qquad \begin{aligned} S^{00}\gamma u &= 0 , \\ S^{10}\gamma u + S^{11}\nu u &= 0 . \end{aligned}$$

In view of (1.7.30), the condition (1.7.31) may equivalently be written as:

$$(1.7.32)\qquad \begin{aligned} &\text{(i)} \quad \gamma u = C'^{00}\varphi_0 \quad \text{and} \\ &\text{(ii)} \quad \nu u = -C^{11}S^{10}\gamma u + C'^{11}\varphi_1 \\ &\text{for some } \{\varphi_0,\varphi_1\} \in \Pi_{J_0} H^{2m-k-\frac{1}{2}}(Z_k) \times \Pi_{J_1} H^{2m-k-\frac{1}{2}}(Z_k) , \end{aligned}$$

for one can take $\varphi_0 = S'^{00}\gamma u$ and $\varphi_1 = S'^{11}\nu u$; (1.7.32) gives a parametrization of the space of boundary values. We first study weak semiboundedness, see (1.7.33) below, which is necessary for m-coerciveness, or for P_T being variational or even just semibounded in L^2-norm.

1.7.4 Theorem. *Let* P *be a ps.d.o. on* $\widetilde{E}$ *of even order* $2m > 0$, *with* Γ *noncharacteristic for* P . *Let* P_T *be the realization defined by a normal trace operator of the form* $T = S\rho$. *Then* P_T *is weakly semibounded, i.e. there exists a* $\theta \in \mathbb{R}$ *and a constant* c *such that*

$$(1.7.33)\qquad \operatorname{Re} e^{i\theta}(P_\Omega u,u) \le c\|u\|_m^2 \quad \textit{for all } u \in D(P_T) ;$$

if and only if

$$(1.7.34)\qquad C'^{00*}\mathcal{A}^{01}C'^{11} = 0 .$$

In the affirmative case, one has in fact

$$(1.7.35)\qquad \begin{aligned} &(\mathcal{A}^{01}\nu u,\gamma u) = -(\mathcal{A}^{01}C^{11}S^{10}\gamma u,\gamma v) \quad \textit{and} \\ &|(P_\Omega u,v)| \le c'\|u\|_m\|v\|_m \quad \textit{for all } u,v \in D(P_T) . \end{aligned}$$

Proof: The only term in the right hand side of (1.7.23), that is not always H^m-bounded, is $(\mathcal{A}^{01}\nu u,\gamma u)$. Inserting (1.7.32), we may write it as

$$(1.7.36)\qquad (\mathcal{A}^{01}\nu u,\gamma v)_\Gamma = (\mathcal{A}^{01}(-C'^{11}S^{10}\gamma u + C'^{11}\varphi_1),\gamma v)_\Gamma$$
$$= (\mathcal{A}^{01}C'^{11}\varphi_1, C'^{00}\varphi_0)_\Gamma - (\mathcal{A}^{01}C'^{11}S^{10}\gamma u,\gamma v)_\Gamma \ .$$

The last term here is again H^m-bounded, so (1.7.33) holds if and only if

$$(1.7.37)\qquad \mathrm{Re}\ e^{i\theta}(C'^{00*}\mathcal{A}^{01}C'^{11}\varphi_1,\varphi_0) \le c_1\|u\|_m^2 \quad \text{for } u \in D(P_T) \text{ with (1.7.32).}$$

Addition of $w \in C_0^\infty(\Omega,E)$ to u does not change the left hand side; then since

$$\inf\{\|u+w\|_m \mid w \in C_0^\infty(\Omega,E)\} \simeq \|\gamma u\|^2_{\Pi_{J_0} H^{m-k-\frac12}(E_\Gamma)}$$

(cf. [Lions and Magenes 1]), (1.7.37) holds if and only if, for some constant c_2 ,

$$(1.7.38)\quad \mathrm{Re}\ e^{i\theta}(C'^{00*}\mathcal{A}^{01}C'^{11}\varphi_1,\varphi_0) \le c_2\|\gamma u\|_{\Pi H^{m-k-\frac12}} \quad \text{for } u \in D(P_T) \text{ with (1.7.32).}$$

This holds of course if $C'^{00*}\mathcal{A}^{01}C'^{11} = 0$, and then (1.7.35) is also valid in view of (1.7.36). Conversely, (1.7.38) implies, since $\gamma u = S^{00}\varphi_0$,

$$(1.7.39)\qquad \mathrm{Re}\ e^{i\theta}(C'^{00*}\mathcal{A}^{01}C'^{11}\varphi_1,\varphi_0) \le c_3\|\varphi_0\|^2_{\Pi H^{m-k-\frac12}(Z_k)} \ ,$$

where φ_1 and φ_0 run freely in $\Pi_{J_1} H^{2m-k-\frac12}(Z_k)$ resp. $\Pi_{J_0} H^{2m-k-\frac12}(Z_k)$. This implies (1.7.34). □

Throughout the proof it is used that the various matrix operators over Γ have the appropriate continuity properties, which we do not always recall explicitly.

Note in particular that the Dirichlet condition corresponds to the case where C'^{00} is zero-dimensional, and the Neumann-type conditions correspond to the case where C'^{11} is zero-dimensional, so in these cases (1.7.34) is trivially satisfied, and the realizations are weakly semibounded (as we know more or less already).

As observed for the differential operator case in earlier works, the condition (1.7.34) is not always stable under lower order perturbations of the problem. It holds stably, when F satisfies (1.4.23) and J is one of the sets $\{0,1,\dots,p-1,m,m+1,m+p-1\}$ (cf. e.g. [Fujiwara-Shimakura 1]), but otherwise it requires exact identity between certain operators, that can be violated by perturbation of low order terms.

The link between weakly semibounded boundary problems and sesquilinear forms can now be elaborated as in [Grubb 2]. The discussion of m-coerciveness will be based on a decomposition that was derived, in the differential operator case, from an abstract theorem in [Grubb 4]; the more direct proof that we give here was explained in [Grubb 3]. We now assume that P is strongly elliptic, and we set

$$P^r = \frac{1}{2}(P+P^*) ,$$

it is formally selfadjoint and strongly elliptic. By application of Theorem 1.7.2 to P^r we have (after adding a constant to P if necessary) that P^r_γ is selfadjoint positive, satisfying

$$(1.7.40)\qquad (P^r_\Omega u,v) \geq c\|v\|^2_m \quad \text{for } u \in D(P^r_\gamma) ,$$

with $c > 0$. Now we decompose $H^{2m}(E)$ as follows

$$H^{2m}(E) = D(P^r_\gamma) \dotplus Z^{2m}(P^r_\Omega) ,$$

where $Z^{2m}(P^r_\Omega)$ is the space of $z \in H^{2m}(E)$ with $P^r_\Omega z = 0$; the decomposition is described by

$$(1.7.41)\qquad \begin{gathered} u = u^r_\gamma + u^r_\zeta , \\ u^r_\gamma = (P^r_\gamma)^{-1} P^r_\Omega u , \quad u^r_\zeta = u - u^r_\gamma . \end{gathered}$$

Observe that since

$$(1.7.42)\qquad \begin{gathered} P^r_\Omega u^r_\zeta = 0 , \quad \text{and} \\ \gamma u^r_\gamma = 0 \quad \text{so that } \gamma u^r_\zeta = \gamma u , \end{gathered}$$

u^r_ζ is likewise determined by

$$(1.7.43)\qquad u^r_\zeta = K^r \gamma u ,$$

where K^r is the Poisson operator $K^r : \varphi \mapsto z$ solving

$$(1.7.44)\qquad \begin{gathered} P^r_\Omega z = 0 \\ \gamma z = \varphi . \end{gathered}$$

Let us now also define the ps.d.o. Q^r over Γ by

$$(1.7.45)\qquad Q^r = \nu K^r ,$$

then Q^r maps γz into νz when $z \in Z^{2m}(P^r_\Omega)$. (Q^r is an <u>elliptic</u> ps.d.o. if and only if the Neumann problem for P^r is elliptic.)

<u>1.7.5 Lemma</u>. *For any* $u \in H^{2m}(E)$ *one has*

$$\mathrm{Re}(P_\Omega u,u)_\Omega = (P^r_\Omega u^r_\gamma, u^r_\gamma)_\Omega + \mathrm{Re}(\mathcal{A}^{01}\nu u,\gamma u)_\Gamma + (\mathcal{B}\gamma u,\gamma u) \tag{1.7.46}$$

where $\mathcal{B}$ *is the ps.d.o. system on* Γ

$$\mathcal{B} = \tfrac{1}{2}[\mathcal{A}^{00*} + (\mathcal{A}^{10*} - \mathcal{A}^{01})Q^r] \ . \tag{1.7.47}$$

When P *is selfadjoint,* $\mathcal{B} = -\frac{1}{2}\mathcal{A}^{00} - \mathcal{A}^{01}Q^r$, *cf.* (1.6.65').

<u>Proof</u>: Let $u \in H^{2m}(E)$ and set $v = u^r_\gamma$, $z = u^r_\zeta$. By Green's formula (1.7.22)

$$(v,P_\Omega v)_\Omega = (P^*_\Omega v,v)_\Omega$$

$$(z,P_\Omega v)_\Omega = (P^*_\Omega z,v)_\Omega + (\gamma z,\mathcal{A}^{01}\nu v)_\Gamma$$

$$\begin{aligned}(z,P_\Omega z)_\Omega &= (P^*_\Omega z,z)_\Omega + (\gamma z,\mathcal{A}^{00}\gamma z + \mathcal{A}^{01}\nu z) + (\nu z,\mathcal{A}^{10}\gamma z)\\ &= (P^*_\Omega z,z)_\Omega + ([\mathcal{A}^{00*} + \mathcal{A}^{10*}Q^r]\gamma z,\gamma z) + (\gamma z,\mathcal{A}^{01}Q^r\gamma z) \ .\end{aligned}$$

Then

$$\begin{aligned}\mathrm{Re}(P_\Omega u,u) &= \tfrac{1}{2}(P_\Omega u,u) + \tfrac{1}{2}(u,P_\Omega u)\\
&= \tfrac{1}{2}[(P_\Omega v,v) + (P_\Omega v,z) + (P_\Omega z,v) + (P_\Omega z,z)]\\
&\qquad + \tfrac{1}{2}[(v,P_\Omega v) + (v,P_\Omega z) + (z,P_\Omega v) + (z,P_\Omega z)]\\
&= (P^r_\Omega v,v) + (v,P^r_\Omega z) + (P^r_\Omega z,v) + (P^r_\Omega z,z) + \tfrac{1}{2}(\mathcal{A}^{01}\nu v,\gamma z) + \tfrac{1}{2}(\gamma z,\mathcal{A}^{01}\nu v)\\
&\qquad + \tfrac{1}{2}([\mathcal{A}^{00*} + \mathcal{A}^{10*}Q^r]\gamma z,\gamma z) + \tfrac{1}{2}(\gamma z,\mathcal{A}^{01}Q^r\gamma z)\\
&= (P^r_\Omega v,v) + \mathrm{Re}(\mathcal{A}^{01}(\nu u - Q^r\gamma u),\gamma z) + \tfrac{1}{2}([\mathcal{A}^{00*} + \mathcal{A}^{10*}Q^r]\gamma z,\gamma z) + \tfrac{1}{2}(\gamma z,\mathcal{A}^{01}Q^r\gamma z)\end{aligned}$$

since $P^r_\Omega z = 0$, $\nu v = \nu(u - u^r_\zeta) = \nu u - Q^r\gamma u$, and $\gamma z = \gamma u$; this gives (1.7.46). □

<u>1.7.6 Example</u>. In the case $P = -\Delta$, P^r equals P, and the formula (1.7.46) becomes

$$\mathrm{Re}(P_\Omega u,u) = (P_\Omega u^r_\gamma, u^r_\gamma) + \mathrm{Re}(i\gamma_1 u,\gamma_0 u) + (-iQ^r\gamma_0 u,\gamma_0 u) \tag{1.7.48}$$

where we recall that $i\gamma_1 u = \partial u/\partial n$, and iQ^r is the ps.d.o. sending γz into

$\partial z/\partial n$ when $\Delta z = 0$. In this case, $-iQ^r$ is a positive ps.d.o. on Γ with principal symbol $|\xi'|$.

Since P^r_γ satisfies (1.7.40), it depends on the boundary term in (1.7.46) whether a given realization P_T is m-coercive. When P_T is weakly semibounded, we find by insertion of (1.7.32) in (1.7.46), in view of (1.7.35),

(1.7.49)
$$\begin{aligned}\operatorname{Re}(P_\Omega u,u)_\Omega &= (P^r_\Omega u^r_\gamma, u^r_\gamma)_\Omega + ((\mathcal{B} - \mathcal{A}^{01}C^{11}S^{10})\gamma u, \gamma u)_\Gamma \\ &= (P^r_\Omega u^r_\gamma, u^r_\gamma)_\Omega + (C'^{00*}(\mathcal{B} - \mathcal{A}^{01}C^{11}S^{10})C'^{00}\varphi_0, \varphi_0)\ ,\end{aligned}$$

when $u \in D(P_T)$ and $\gamma u = C'^{00}\varphi_0$.

1.7.7 Lemma. *For any* $\varepsilon \in [0,\frac{1}{2}[$ *there are positive constants* c_ε *and* C_ε *so that for* $u \in H^{m-\varepsilon}(E)$ *(with the notation* (1.6.60)*)*,

(1.7.50)
$$c_\varepsilon \|u\|^2_{m-\varepsilon} \le \|u^r_\gamma\|^2_{m-\varepsilon} + \|\gamma u\|^2_{H^{m-\varepsilon}_{E^0}} \le C_\varepsilon \|u\|^2_{m-\varepsilon}\ .$$

Proof: Recall from (1.7.43) that

$$u^r_\gamma = u - K^r\gamma u$$

for $u \in H^{2m}(E)$, this definition extends to $H^m(E)$ (as does the formula $u^r_\gamma = (P^r_\gamma)^{-1}P_\Omega$ in view of the proof of Theorem 1.7.2); and here

$$\|K^r\gamma u\|_s \le c_s \|\gamma u\|_{H^s_{E^0}}$$

for any s. This implies the left hand inequality in (1.7.50):

$$\begin{aligned}\|u\|_{m-\varepsilon} = \|u^r_\gamma + K^r\gamma u\|_{m-\varepsilon} &\le \|u^r_\gamma\|_{m-\varepsilon} + \|K^r\gamma u\|_{m-\varepsilon} \\ &\le \|u^r_\gamma\|_{m-\varepsilon} + c_{m-\varepsilon}\|\gamma u\|_{H^{m-\varepsilon}_{E^0}}\ .\end{aligned}$$

The right hand inequality now follows by using that γ is continuous from $H^s(E)$ to $H^s_{E^0}$ for $s > m - \frac{1}{2}$,

$$\begin{aligned}\|u^r_\gamma\|_{m-\varepsilon} + \|\gamma u\|_{H^{m-\varepsilon}_{E^0}} &= \|u - K^r\gamma u\|_{m-\varepsilon} + \|\gamma u\|_{H^{m-\varepsilon}_{E^0}} \\ &\le \|u\|_{m-\varepsilon} + c_1\|\gamma u\|_{H^{m-\varepsilon}_{E^0}} \le c_2\|u\|_{m-\varepsilon}\ .\end{aligned}$$
□

1.7.8 Theorem. *Let* P *be a strongly elliptic ps.d.o. of order* $2m > 0$, *and let* P_T *be the realization defined by a normal trace operator* T *of the form* $T = S\rho$. *It is necessary and sufficient for m-coerciveness of* P_T *that the following two conditions hold:*

(i) P_T *is weakly semibounded, i.e.,*

$$C'^{00*}\mathcal{A}^{01}C'^{11} = 0 . \tag{1.7.51}$$

(ii) *The ps.d.o. over* Γ *(cf. (1.7.47))*

$$\mathcal{L} = C'^{00*}(\mathcal{B} - \mathcal{A}^{01}C^{11}S^{10})C'^{00} , \tag{1.7.52}$$

acting in $\Pi_{k\in J_0} H^{m-k-\frac{1}{2}}(Z_k)$, *is strongly elliptic, i.e.,*

$$\sigma^0(\mathcal{L}) + \sigma^0(\mathcal{L})^* \quad \textit{is positive definite}$$

at each (x',ξ') *with* $|\xi'| \geq 1$.

When furthermore $M \equiv \dim \oplus_{0\leq k<2m} F_k$ *equals* mN , *then* (i) *and* (ii) *imply that* P_T *is elliptic.*

Proof: Clearly, weak semiboundedness is necessary for m-coerciveness, so we may assume (1.7.51) in view of Theorem 1.7.4. Then, as shown in (1.7.49), one has for $u \in D(P_T)$,

$$\mathrm{Re}(P_\Omega u,u) = (P^r_\Omega u^r_\gamma, P^r_\Omega u^r_\gamma) + \mathrm{Re}((\mathcal{B} - \mathcal{A}^{01}C^{11}S^{10})\gamma u,\gamma u) . \tag{1.7.53}$$

Since $\gamma u = S^{00}\varphi_0$ and $\varphi_0 = C^{00}\gamma u$, where φ_0 runs through $\Pi_{J_0} H^{2m-k-\frac{1}{2}}(Z_k)$ when u runs through $D(P_T)$, one has in view of the continuity of the operators S^{00} and C^{00} that there exist $\varepsilon \in]0,\frac{1}{2}[$, $c > 0$ and $c' \in \mathbb{R}$ so that

$$\mathrm{Re}((\mathcal{B} - \mathcal{A}^{01}C^{11}S^{10})\gamma u,\gamma u) \geq c\|\gamma u\|^2_{H^m_{E^0}} - c'\|\gamma u\|^2_{H^{m-\varepsilon}_{E^0}} \tag{1.7.54}$$

holds for $u \in D(P_T)$, <u>if and only if</u>

$$\mathrm{Re}(\mathcal{L}\varphi_0,\varphi_0) \geq c_1\|\varphi_0\|^2_{\{m-k-\frac{1}{2}\}} - c_1'\|\varphi_0\|^2_{\{m-\varepsilon-k-\frac{1}{2}\}} \tag{1.7.55}$$

holds for $\varphi_0 \in \Pi_{J_0} H^{m-k-\frac{1}{2}}(Z_k)$, with $c_1 > 0$ and $c_1' \in \mathbb{R}$; here $\|\varphi_0\|_{\{m-\varepsilon-k-\frac{1}{2}\}}$ is the norm in $\Pi_{k\in J_0} H^{m-\varepsilon-k-\frac{1}{2}}(Z_k)$. Here (1.7.55) holds precisely when $\mathcal{L}$ is strongly elliptic. In view of (1.7.40) we find by use of Lemma 1.7.7 that when (1.7.54) holds, then

(1.7.56) $$\operatorname{Re}(P_\Omega u,u) \geq c''\|u\|_m^2 - c_1''\|u\|_{m-\varepsilon}^2 \quad \text{for } u \in D(P_T) ,$$

i.e., P_T is strongly elliptic. This shows that (ii) implies m-coerciveness.

Conversely, assume (1.7.56). We can substitute u by $u - w_j$ ($j \in \mathbb{N}$), where $w_j \in C_0^\infty(\Omega,E)$ and $w_j \to u_\gamma^r$ in $\mathring{H}^m(E)$; then since $(u-w_j)_\gamma^r = u_\gamma^r - w_j$ and

$$\|u_\gamma^r - w_j\|_m \to 0 \quad , \quad \|u_\gamma^r - w_j\|_{m-\varepsilon} \to 0 \ ,$$
$$(P_\Omega^r(u_\gamma^r - w_j), u_\gamma^r - w_j) \to 0$$

for $j \to \infty$, and $\gamma(u-w_j) = \gamma u$, it follows that (1.7.54) holds (with the same constants $c = c''$, $c' = c_1''$). Then also (1.7.55) holds, and the necessity of (ii) is proved.

Finally, consider the case where $M = mN$. This means that $\oplus_{k<2m} F_k$ and $\oplus_{k<2m} Z_k$ have the same dimension. The above calculations, carried out on the boundary symbol level, show that $b^0 = (p^0)_{t^0}$ is positive and hence

$$\begin{pmatrix} p_\Omega^0 \\ t^0 \end{pmatrix} : H^{2m}(\overline{\mathbb{R}}_+)^N \to \begin{matrix} L^2(\mathbb{R}_+)^N \\ \times \\ \mathbb{C}^M \end{matrix}$$

is injective; then in view of Corollary 1.7.3 it must be bijective, and P_T is elliptic (cf. Corollary 1.6.7). $\square$

1.7.9 Remark. One can furthermore show, exactly as in [Grubb 1 and 2], that when $M = mN$, then weak semiboundedness of P_T implies weak semiboundedness of the formal adjoint $(P^*)_{\tilde{T}}$ (which likewise has no "interior" terms), and in fact weak semiboundedness here is equivalent with the property

(1.7.57) $$\gamma D(P_T) = \gamma D((P^*)_{\tilde{T}}) ,$$

and with the equation

(1.7.58) $$S^{00}(\mathcal{A}^{01*})^{-1}S^{11*} = 0 .$$

Moreover, in this case, a consideration of b^0 and b^{0*} shows that m-coerciveness of P_T will imply m-coerciveness of $(P^*)_{\tilde{T}}$, which is then elliptic and equal to the true adjoint of P_T ; in particular, both operators are variational. The associated sesquilinear forms have domain equal to the closed subspace of $H^m(E)$ defined by

(1.7.59) $$V_{S^{00}} = \{u \in H^m(E) \mid S^{00}\gamma u = 0\} .$$

(The sesquilinear forms can be realized as expressions $a_T(u,v)+p'(u,v)$, where p' is as in (1.7.23) and a_T is an integro-differential form, specially adapted to the boundary condition as in [Grubb 2], when S is differential.)

The proof of Theorem 1.7.8 for differential operators was originally based on a more abstract analysis of the realizations $\widetilde{P}$ of P_Ω , acting like P_Ω and with

(1.7.60) $$D(P_{min}) \subset D(\widetilde{P}) \subset D(P_{max}) ,$$

cf. e.g. [Grubb 4]; the theory has its roots in works of [Krein 1], [Birman 1] and [Vishik 1].

We now turn to general realizations $B = (P+G)_T$ where "interior" terms are allowed. Here B need no longer act like P_Ω , and its domain need not satisfy inclusions as in (1.7.60) (recall e.g. Example 1.4.5), so the abstract theory of [Grubb 4] cannot be used. However, it is possible to discuss m-coerciveness and semiboundedness by direct use of the boundary operator caculus.

Let us first consider weak semiboundedness, i.e. the property that for some θ and $c \in \mathbb{R}$,

(1.7.61) $$\text{Re } e^{i\theta}(Bu,u)_\Omega \leq c\|u\|_m^2 \quad \text{when} \quad u \in D(B) ,$$

necessary for the other semiboundedness or variationality properties, we are interested in. It is not hard to show that Neumann-type realizations are weakly semibounded. But we now meet the surprising fact that Dirichlet-type realizations need not be weakly semibounded; there is a necessary and sufficient condition on the interior terms for this to hold, and the condition is exact, i.e. not stable under small perturbations. Let us first consider the Neumann- and Dirichlet-type conditions, and afterwards give the more technical discussion of the general case. We write G and T as in Examples 1.6.12-15

(1.7.62) $$G = K^1\nu + K^0\gamma + G' ,$$

with a Dirichlet-type trace operator

(1.7.63) $$T^0 = \gamma + T'^0 ,$$

or a Neumann-type trace operator

(1.7.64) $$T^1 = \nu + S^0\gamma + T'^1 .$$

By (1.7.24), one has in general

$$(1.7.65)\qquad (Bu,v)_\Omega = p(u,v) + (Gu,v)_\Omega + (\mathcal{A}^{01}\nu u,\gamma v)_\Gamma$$
$$= (K^1\nu u,v)_\Omega + (\mathcal{A}^{01}\nu u,\gamma v)_\Gamma + \text{m-bounded terms}$$

(we say that $f(u,v)$ is m-bounded when $|f(u,v)| \le c\|u\|_m\|v\|_m$). Now in the Neumann-type problems where $T^1u = 0$,

$$(K^1\nu u,v)_\Omega + (\mathcal{A}^{01}\nu u,\gamma v)_\Gamma = -(K^1(S^0\gamma u + T'^1u)\ ,v)_\Omega - (\mathcal{A}^{01}(S^0\gamma u + T'^1u)\ ,\gamma v)_\Gamma$$
$$= -(S^0\gamma u + T'^1u\ , K^{1*}v + \mathcal{A}^{01*}\gamma v)_\Gamma$$

which is m-bounded, since all the operators are defined on $H^m(E)$, and the orders fit together. (E.g., T'^1: $H^m(E) \to \Pi_{k\in J_1} H^{m-k-\frac{1}{2}}(E_\Gamma)$ and K^{1*}: $H^m(E) \to \Pi_{k\in J_1} H^{-m+k+\frac{1}{2}}(E_\Gamma)$ are continuous.) So in this case, weak semiboundedness holds without restrictions, and one even has m-boundedness:

$$(1.7.66)\qquad |(Bu,v)| \le c\|u\|_m\|v\|_m \quad \text{for } u,v \in D(B)\ .$$

Now consider a Dirichlet-type condition. Here, when $u \in D(B)$ (so that $\gamma u = -T'^0u$),

$$(1.7.67)\qquad (K^1\nu u,u) + (\mathcal{A}^{01}\nu u,\gamma u) = (\nu u,T''u)$$

where T'' is the trace operator of class 0

$$(1.7.68)\qquad T'' = K^{1*} - \mathcal{A}^{01*}T'^0\ .$$

Since ν is not defined as a continuous operator on $H^m(E)$, there is something wrong, when $T''u \neq 0$. Of course, $T'' = 0$ <u>implies</u> weak semiboundedness (in fact, (1.7.66) will hold). We shall now show that $T'' = 0$ is also <u>necessary</u> for weak semiboundedness. This requires a deeper analysis of the surjectiveness properties of our trace operators. For the present case, we need the construction of a sequence u_ε in $D(B)$ where $(\nu u_\varepsilon\ , T''u_\varepsilon)$ is not dominated by $\|u_\varepsilon\|_m^2$ if $T'' \neq 0$. A tool for constructing this is furnished by the following lemma, which is formulated such that it is also useful in the general case.

<u>1.7.10. Lemma</u>. *Let* T *be a normal trace operator associated with the order* $2m$, *and write it in blocks corresponding to the index sets* $J_0 = \{0,1,\dots,m-1\}$ *and* $J_1 = \{m,m+1,\dots,2m-1\}$

$$T = \begin{pmatrix} S^{00} & 0 \\ S^{10} & S^{11} \end{pmatrix}\begin{pmatrix} \gamma \\ \nu \end{pmatrix} + \begin{pmatrix} T'^{0} \\ T'^{1} \end{pmatrix} , \tag{1.7.69}$$

Let S' *be a morphism supplementing* S *as in Lemma 1.6.1, written in blocks*

$$S' = \begin{pmatrix} S'^{00} & 0 \\ 0 & S'^{11} \end{pmatrix} , \tag{1.7.70}$$

so that the full system $\{T, S'\rho\}$ *is normal. Let* $\varphi \in \Pi_{k\in J_1} H^{2m-k-\frac{1}{2}}(Z_k)$, *with* $\varphi \neq 0$. *There is a family of functions* $u_\varepsilon \in H^{2m}(E)$ *for* $\varepsilon \in]0,\varepsilon_0]$ *such that for each* ε ,

$$Tu_\varepsilon = 0 \quad , \quad S'^{00}\gamma u_\varepsilon = 0 \quad , \quad S'^{11}\nu u_\varepsilon = \varphi , \quad \textit{and} \tag{1.7.71}$$

$$\|u_\varepsilon\|_m \le c(\varphi)\varepsilon \quad , \tag{1.7.72}$$

where $c(\varphi)$ *depends only on* φ . *(In fact,* $c(\varphi) \le \text{const.}\|\varphi\|_{\Pi_{k\in J_1} H^{m-k-\frac{1}{2}}(Z_k)}$ *.)*

<u>Proof</u>: The right inverse K_δ of ρ constructed in Lemma 1.6.4 can be written

$$\begin{gathered} K_\delta = (K_\delta^0 \;\; K_\delta^1) \quad , \quad \text{where} \\ \gamma K_\delta^0 = I \; , \quad \gamma K_\delta^1 = 0 \; , \quad \nu K_\delta^0 = 0 \; , \quad \nu K_\delta^0 = I \; ; \end{gathered} \tag{1.7.73}$$

and it is seen from the estimates (1.6.14), applied with $k = 0,1,\dots,m$ and $s = m-k$, that

$$\|K_\delta^1\psi_1\|_{H^m(E)} \le \delta \,\text{const.}\|\psi_1\|_{\Pi_{j\in J_1} H^{m-j-\frac{1}{2}}(E_\Gamma)} \tag{1.7.74}$$

for all $\psi_1 \in \Pi_{J_1} H^{m-j-\frac{1}{2}}(E_\Gamma)$. Now we search for the solution of (1.7.71) on the form $u_\varepsilon = (K_\delta^0 \;\; K_\varepsilon^1)\begin{pmatrix} \psi_0 \\ \psi_1 \end{pmatrix}$, which gives

$$\begin{pmatrix} \begin{pmatrix} S^{00}\gamma + T'^{0} \\ S'^{00}\gamma \end{pmatrix} \\ \\ \begin{pmatrix} S^{11}\nu + S^{10}\gamma + T'^{1} \\ S'^{11}\nu \end{pmatrix} \end{pmatrix} (K_\delta^0\psi_0 + K_\varepsilon^1\psi_1) = \begin{pmatrix} \begin{pmatrix} 0 \\ 0 \end{pmatrix} \\ \\ \begin{pmatrix} 0 \\ \varphi \end{pmatrix} \end{pmatrix} \tag{1.7.75}$$

and hence, in view of (1.7.73) (recall also (1.7.29))

$$(1.7.76)\qquad \begin{aligned} &\tilde{S}^{00}\psi_0 + \begin{pmatrix} T'^0 \\ 0 \end{pmatrix} K^0_\delta \psi_0 = -\begin{pmatrix} T'^0 \\ 0 \end{pmatrix} K^1_\varepsilon \psi_1 \ , \\ &\tilde{S}^{11}\psi_1 + \begin{pmatrix} S^{10}\psi_0 \\ 0 \end{pmatrix} + \begin{pmatrix} T'^1 \\ 0 \end{pmatrix} (K^0_\delta \psi_0 + K^1_\varepsilon \psi_1) = \begin{pmatrix} 0 \\ \varphi \end{pmatrix} . \end{aligned}$$

The first line can be solved for ψ_0 (as in Proposition 1.6.5) by taking δ so small that the norm of

$$\approx{S} = (\tilde{S}^{00})^{-1} \begin{pmatrix} T'^0 \\ 0 \end{pmatrix} K^0_\delta$$

in $H^s_{E^0}$ is $\leq \frac{1}{2}$ for $s \in [0,2m]$ (it is seen from (1.6.13) that the tangential order s does not really interfere with the estimates, see also (1.6.60)); then

$$(1.7.77)\qquad \psi_0 = -[1 + \approx{S}]^{-1} (\tilde{S}^{00})^{-1} \begin{pmatrix} T'^0 \\ 0 \end{pmatrix} K^1_\varepsilon \psi_1 = \tilde{T}\, K^1_\varepsilon \psi_1 \ ,$$

with a trace operator $\tilde{T}$ of class 0. δ is kept fixed in the following, and we insert (1.7.77) in the second line of (1.7.76). This leads to an equation

$$\tilde{S}^{11}\psi_1 + \approx{T}\, K^1_\varepsilon \psi_1 = \begin{pmatrix} 0 \\ \varphi \end{pmatrix} .$$

Now we can choose ε_0 so small that the norm of $(\tilde{S}^{11})^{-1}\, \approx{T}\, K^1_\varepsilon$ is $\leq \frac{1}{2}$ for $\varepsilon \leq \varepsilon_0$, and invert $[I + (\tilde{S}^{11})^{-1}\, \approx{T}\, K^1_\varepsilon]$; then we find

$$(1.7.78)\qquad \psi_1 = [I + (\tilde{S}^{11})^{-1}\, \approx{T}\, K^1_\varepsilon]^{-1} (\tilde{S}^{11})^{-1} \begin{pmatrix} 0 \\ \varphi \end{pmatrix} ,$$

where the norm of the operator $[\ ]^{-1}$ is ≤ 2 for $\varepsilon \leq \varepsilon_0$. With ψ_1 and ψ_0 determined by (1.7.78) and (1.7.77), we now have a solution of (1.7.71)

$$u_\varepsilon = K^0_\delta \psi_0 + K^1_\varepsilon \psi_1 = (I + K^0_\delta \tilde{T}) K^1_\varepsilon \psi_1 \ ,$$

and since $K^0_\delta \tilde{T}$ is a singular Green operator of order and class 0 (hence bounded in $H^m(E)$), it follows in view of (1.7.74) that also (1.7.72) is satisfied. The last statement is seen from an inspection of the proof. □

In the special case (1.7.63), the lemma means that for any $\varphi \in H^m_{E^1}$ (1.6.60), there is a family of functions u_ε such that

$$(1.7.79)\qquad T^0 u_\varepsilon = 0 \ , \quad \nu u_\varepsilon = \varphi \ , \quad \text{and} \quad \|u_\varepsilon\|_m \leq c\varepsilon \quad \text{for} \quad \varepsilon \in \,]0,\varepsilon_0] \ .$$

<u>1.7.11 Proposition</u>. *Let the ps.d.o.* P *be of order* 2m *on* $\tilde{E}$. *Let* $B = (P+G)_T$ *be the realization of* P *determined by a Dirichlet-type boundary condition* $T^0 \equiv \gamma u + T'^0 u = 0$, *with* G *written as in* (1.7.62). *Then* B *is weakly semi-bounded if and only if*

$$K^{1*} = \mathcal{a}^{01*}\, T'^0 , \tag{1.7.80}$$

and in the affirmative case, B *is* m*-bounded (satisfies* (1.7.66)*).*

<u>Proof</u>: Let $T'' = K^1{*} - \mathcal{a}^{01*}\, T'^0$. That $T'' = 0$ implies weak semiboundedness and (1.7.66) was accounted for above. Now we shall show that the inequality

$$\operatorname{Re} e^{i\theta}(Bu,u) \leq c\|u\|_m^2 \quad \text{for } u \in D(B),$$

for some θ and c, implies $T'' = 0$. To fix the ideas, we can take $\theta = 0$. Assume that $T'' \neq 0$, then for some $v \in L^2(E)$, $T''v \neq 0$. Since the set of $u \in H^{2m}(E)$, for which $T^0u = 0$ and $\nu u = 0$, is dense in $L^2(E)$ (Lemma 1.6.8), we can find such a u for which $T''u \neq 0$; we denote $T''u = \varphi$. Now take a sequence u_ε satisfying (1.7.79), and let $v_\varepsilon = \varepsilon u + u_{a\varepsilon}$, where $a \in \,]0,1]$ will be fixed later. In view of (1.7.65) and (1.7.67) we have (for $\varepsilon \leq \varepsilon_0$)

$$\operatorname{Re}(Bv_\varepsilon,v_\varepsilon) = \operatorname{Re}(\nu v_\varepsilon, T''v_\varepsilon) + \mathcal{O}(\|v_\varepsilon\|_m^2) ,$$

where

$$\operatorname{Re}(\nu v_\varepsilon, T''v_\varepsilon) = \operatorname{Re}(\varphi\, , \varepsilon\varphi + T''u_{a\varepsilon}) = \varepsilon\|\varphi\|_{L^2}^2 + \operatorname{Re}(\varphi\, , T''u_{a\varepsilon}) .$$

Since T'' is continuous on $H^m(E)$, <u>we can choose</u> a <u>so small</u> that

$$\begin{aligned} |\operatorname{Re}(\varphi\, , T''u_{a\varepsilon})| &\leq \|\varphi\|_{H^m_{E^1}} \|T''u_{a\varepsilon}\|_{(H^m_{E^1})^*} \leq c_1\|\varphi\|_{H^m_{E^1}} \|u_{a\varepsilon}\|_m \\ &\leq c_2 a\varepsilon c(\varphi)\|\varphi\|_{H^m_{E^1}} \leq \tfrac{\varepsilon}{2}\|\varphi\|_{L^2}^2 , \end{aligned}$$

whereby

$$\operatorname{Re}(\nu v_\varepsilon, T''v_\varepsilon) \geq \tfrac{\varepsilon}{2}\|\varphi\|_{L^2}^2 .$$

On the other hand,

$$\|v_\varepsilon\|_m^2 = \|u_{a\varepsilon} + \varepsilon u\|_m^2 \leq \varepsilon^2(c^2c(\varphi)^2 + \|u\|_m)^2 \leq c_3\varepsilon^2 .$$

The inequality

$$\varepsilon\|\varphi\|_{L^2}^2 \leq \varepsilon^2 c_1$$

cannot hold for $\varepsilon \to 0$ with $\varphi \neq 0$, so the hypothesis $T'' \neq 0$ is contradicted.

□

Note that (1.7.80) is part of the condition for selfadjointness, cf. (1.6.64).

Now consider the general case. The result has the same flavor as in the preceding cases, but is more technical to state and to prove. With the formulation (1.7.69) we write the condition $Tu = 0$ on the form

$$S^{00}\gamma u + T'^{0}u = 0 \,, \qquad (1.7.81)$$
$$S^{11}\nu u + S^{10}\gamma u + T'^{1}u = 0 \,.$$

We define S'^{00} , C^{00} , etc. by Lemma 1.6.1 and (1.7.28) ff.

1.7.12 Lemma. *The function* $u \in H^{2m}(E)$ *satisfies* (1.7.81) *if and only if it satisfies*

$$\gamma u = C'^{00}\varphi_0 - C^{00}T'^{0}u \qquad (1.7.82)$$
$$\nu u = C'^{11}\varphi_1 - C^{11}S^{10}\gamma u - C^{11}T'^{1}u$$

for some $\varphi_0 \in \Pi_{k\in J_0} H^{2m-k-\frac{1}{2}}(Z_k)$ *and* $\varphi_1 \in \Pi_{k\in J_1} H^{2m-k-\frac{1}{2}}(Z_k)$; *in fact one then has*

$$\varphi_0 = S'^{00}\gamma u \ \text{ and } \ \varphi_1 = S'^{11}\nu u \,. \qquad (1.7.83)$$

For any given pair $\{\varphi_0, \varphi_1\}$ *there exists* u *satisfying* (1.7.82).

Proof: If u satisfies (1.7.81) and we define φ_0 and φ_1 by (1.7.83), then (1.7.82) is obtained by multiplication on the left by C^{00} resp. C^{11} , using that $C^{\delta\delta}S^{\delta\delta} = I - C'^{\delta\delta}S'^{\delta\delta}$ for $\delta = 0,1$. If u satisfies (1.7.82) then (1.7.83) follows by composition with S'^{00} resp. S'^{11} , and (1.7.81) follows by composition with S^{00} resp. S^{11} (cf. (1.7.30)). The last statement follows from the fact that the trace operator $\{T, S'^{00}\gamma, S'^{11}\nu\}$ is normal and hence surjective. $\square$

1.7.13 Theorem. *Let* $B = (P+G)_T$ *be a normal realization on* E *of a pseudo-differential operator* P *of order* $2m$ *on* $\tilde{E}$, *and write* G *and* T *as in* (1.7.62) *and* (1.7.69). *Then* B *is weakly semibounded (satisfies* (1.7.61) *for some* c *and* θ *) if and only if the following equations hold*

$$C'^{00*}\mathcal{A}^{01}C'^{11} = 0 \,, \qquad (1.7.84)$$

$$C'^{11*}(K^{1*} - \mathcal{A}^{01*}C^{00}T'^{0}) = 0 \,. \qquad (1.7.85)$$

When they hold, B *is in fact* m*-bounded (satisfies* (1.7.66)*).*

In particular, symmetric realizations have these properties.

Proof: In view of (1.7.65) and Lemma 1.7.12, we have

$$(Bu,v) = (K^1\nu u,v) + (\mathcal{Q}^{01}\nu u,\gamma v) + \text{m-bounded terms}$$

$$= (C'^{11}S'^{11}\nu u - C^{11}S^{10}\gamma u - C^{11}T'^{1}u, K^{1*}v + \mathcal{Q}^{01*}\gamma v) + \text{m-bounded terms}$$

$$= (C'^{11}S'^{11}\nu u, K^{1*}v + \mathcal{Q}^{01*}\gamma v) + \text{m-bounded terms}.$$

Consider the inequality (1.7.61) with $\theta = 0$, to fix the ideas. It clearly holds if and only if for some c_1,

(1.7.86) $$\operatorname{Re} f(u,u) \leq c_1\|u\|_m^2 \quad \text{for} \quad u \in D(B),$$

where

(1.7.87) $$f(u,v) = (C'^{11}S'^{11}\nu u, K^{1*}v + \mathcal{Q}^{01*}\gamma v).$$

Now when $u,u' \in D(B)$, we have from Lemma 1.7.12

$$f(u,u') = (C'^{11}\varphi_1, K^{1*}u' + \mathcal{Q}^{01*}(C'^{00}\varphi_0' - C^{00}T'^{0}u'))$$

$$= (\varphi_1, C'^{11*}\mathcal{Q}^{01*}C'^{00}\varphi_0' + T''u'),$$

where

$$T'' = C'^{11*}(K^{1*} - \mathcal{Q}^{01*}C^{00}T'^{0}) \; ; \quad \text{it is of class } 0 .$$

If (1.7.84) and (1.7.85) hold, then $f(u,u') = 0$ so that B is weakly semi-bounded and satisfies (1.7.66). Conversely assume (1.7.86). To see that T'' must be the zero operator, suppose that $T''v \neq 0$ for some $v \in L^2(E)$. Since the set of $u \in H^{2m}(E)$, for which Tu, $S'^{00}\gamma u$ and $S'^{11}u$ are zero, is dense in $L^2(E)$ (Lemma 1.6.8), there is a u with these properties such that $T''u = \varphi \neq 0$. Now take a sequence u_ε according to Lemma 1.7.10, and set $v_\varepsilon = \varepsilon u + u_{a\varepsilon}$, with $a \leq 1$ to be chosen later. The proof now proceeds just as in Proposition 1.7.11: On one hand,

$$\operatorname{Re} f(v_\varepsilon,v_\varepsilon) = \operatorname{Re}(\varphi, T''v_\varepsilon) = \operatorname{Re}(\varphi, \varepsilon\varphi + T''u_{a\varepsilon}) \geq \varepsilon\|\varphi\|^2 - |(\varphi, T''u_{a\varepsilon})| ,$$

where we can take a so small that $|(\varphi,T''u_{a\varepsilon})| \leq \frac{\varepsilon}{2}\|\varphi\|^2$, so that

$$\operatorname{Re} f(v_\varepsilon,v_\varepsilon) \geq \frac{\varepsilon}{2}\|\varphi\|^2 .$$

On the other hand

$$\|v_\varepsilon\|_m^2 = \|\varepsilon u + u_{a\varepsilon}\|_m^2 \leq \varepsilon^2 \cdot c ,$$

and then

$$\operatorname{Re} f(v_\varepsilon,v_\varepsilon) \leq c_1\|v_\varepsilon\|_m^2 \leq c_2\,\varepsilon^2 .$$

This cannot hold for $\varepsilon \to 0$, and it follows that T" must equal 0.

Thus, in the case where (1.7.86) holds,

$$f(u,u) = (\varphi_1, C'^{11*}\mathcal{A}^{01*}C'^{00}\varphi_0) \quad \text{for } u \in D(B) ,$$

where $\varphi_1 = S'^{11}\nu u$ and $\varphi_0 = S'^{00}\gamma u$. We shall now show that (1.7.86) also implies that

$$\mathcal{C} \equiv C'^{11*}\mathcal{A}^{01*}C'^{00} = 0$$

(or, equivalently, $\mathcal{C}^* = 0$; this is (1.7.84)). If $\mathcal{C}$ is not zero, there exists a smooth φ_0 with $\mathcal{C}\varphi_0 \neq 0$, and there exists a $u \in D(B)$ with

$$Tu = 0 , \quad S'^{11}\nu u = 0 \quad \text{and} \quad S'^{00}\gamma u = \varphi_0 ,$$

since this system of trace operators is normal. Construct u_ε according to Lemma 1.7.10 with $\varphi = \mathcal{C}\varphi_0$. Now take $v_\varepsilon = \varepsilon u + u_\varepsilon$. Again we have that

$$\|v_\varepsilon\|_m \leq \text{const. } \varepsilon^2 ,$$

whereas

$$\begin{aligned} \operatorname{Re} f(v_\varepsilon, v_\varepsilon) &= \operatorname{Re}(S'^{11}\nu v_\varepsilon, \mathcal{C}S'^{00}\gamma v_\varepsilon) = \operatorname{Re}(\mathcal{C}\varphi_0, \varepsilon\mathcal{C}\varphi_0) \\ &= \varepsilon\|\mathcal{C}\varphi_0\|^2 . \end{aligned}$$

For $\varepsilon \to 0$ we get a contradiction to (1.7.86), which shows that $\mathcal{C}$ must equal zero, and the proof is complete. ▯

Finally, let us consider the question of m-coerciveness for realizations $B = (P+G)_T$ satisfying the properties listed in Theorem 1.7.13. By use of Lemma 1.7.5 one finds for $u \in D(B)$, in view of Lemma 1.7.12 and Theorem 1.7.13:

$$\begin{aligned} (1.7.88) \qquad \operatorname{Re}(Bu,u) &= \operatorname{Re}(P_\Omega u,u) + \operatorname{Re}(Gu,u) \\ &= (P^r_\Omega u^r_\gamma, u^r_\gamma) + \operatorname{Re}(\mathcal{A}^{01}\nu u, \gamma u) + (\mathcal{B}\gamma u, \gamma u) + \operatorname{Re}(Gu,u) \\ &= (P^r_\Omega u^r_\gamma, u^r_\gamma) + \operatorname{Re}(\nu u, \mathcal{A}^{01*}\gamma u + K^{1*}u) + (\mathcal{B}\gamma u,\gamma u) + \operatorname{Re}(K^0\gamma u, u) + \operatorname{Re}(G'u,u) \\ &= (P^r_\Omega u^r_\gamma, u^r_\gamma) + \operatorname{Re}(-C^{11}S^{10}\gamma u - C^{11}T'^1 u, \mathcal{A}^{01*}\gamma u + K^{1*}u) + (\mathcal{B}\gamma u,\gamma u) + \\ &\qquad + \operatorname{Re}(K^0\gamma u,u) + \operatorname{Re}(G'u,u) \\ &\equiv s(u,u) , \end{aligned}$$

where we used that $f(u,u)$ defined by (1.7.87) is zero. There is of course no hope to write this as an integro-differential form like in the differential operator case; let us note that $s(u,u)$, or rather, the associated sesquilinear form

$$(1.7.89) \qquad s(u,v) \equiv \frac{1}{4}\Big[s(u+v,u+v) - s(u-v,u-v) + is(u+iv,u+iv) - is(u-iv,u-iv)\Big] ,$$

has the general structure

$$(1.7.90) \qquad s(u,v) = (P^r_\Omega u^r_\gamma , v^r_\gamma) + (G''u , v) + (K'\gamma u , v) + (\mathcal{E}\gamma u ,\gamma v) ,$$

where G'' is a s.g.o. of order $2m$ and class 0, and K' and $\mathcal{E}$ are Poisson resp. ps.d.o.s of the appropriate orders, so that $s(u,v)$ is continuous on $H^m(E)$. In view of (1.7.41-43), and (1.7.82), $s(u,u)$ may for $u \in D(B)$ be further reduced to the form

$$(1.7.91) \qquad s(u,u) = (P^r_\Omega u^r_\gamma , u^r_\gamma) + (G''u , u) + (K_0' \varphi_0 , u) + (\mathcal{E}_0 \varphi_0 , \varphi_0)$$

$$\text{where } \varphi_0 = S'^{00} \gamma u .$$

In this way, the m-coerciveness problem for $(P+G)_T$ is reduced to an m-coerciveness problem for a continuous sesquilinear form on a subset of $H^m(E)$. In the differential operator case (as considered in [Grubb 1 , 4], the s.g.o. term with G'' and the "mixed" term with K_0' do not appear, and m-coerciveness holds if and only if $\mathcal{E}_0$ is strongly elliptic. For the present case one finds at least that it suffices for m-coerciveness that $G'' \geq 0$, K_0 is suitably small and $\mathcal{E}_0$ is strongly elliptic. We shall not give a complete analysis here, but restrict ourselves to some indications of what more can be done.

The first thing we observe is that for the question of m-coerciveness of $s(u,v)$ on $D(B)$ it is only the "Dirichlet" part of the boundary condition, namely the condition

$$(1.7.92) \qquad S^{00}\gamma u + T_0' u = 0 ,$$

that matters. For one has:

1.7.14 Lemma. *The closure of* $D(B)$ *in* $H^m(E)$ *equals the set*

$$(1.7.93) \qquad V = \{u \in H^m(E) \mid S^{00}\gamma u + T_0' u = 0\} .$$

In particular, $s(u,v)$ *satisfies the inequality*

(1.7.94) $$s(u,u) \geq c\|u\|_m^2 - c_0\|u\|_0^2$$

for $u \in D(B)$, *if and only if it does so for* $u \in V$.

Proof. Clearly, $D(B) \subset V$, so we have to prove that $D(B)$ is dense in V in the $H^m(E)$-topology.

Let $v \in V$. First take a sequence of functions $w_j \in C^\infty(E)$ such that $w_j \to v$ in $H^m(E)$; then by continuity, $S^{00}\gamma w_j + T_0' w_j \to 0$ in $\Pi_{0\leq k<m} H^{m-k-\frac{1}{2}}(F_k)$. Next, we construct $x_j \in C^\infty(E)$ such that $x_j \to 0$ in $H^m(E)$ and

$$S^{00}\gamma x_j + T_0' x_j = S^{00}\gamma w_j + T_0' w_j \quad \text{for each } j .$$

This is found by setting $x_j = K_\delta^0 \varphi_j$ (cf. (1.7.73)) and using the existence of C^{00} so that $S^{00}C^{00} = I$; here φ_j must solve

$$S^{00}(\varphi_j + C^{00}T_0'K_\delta^0\varphi_j) = S^{00}(\gamma w_j + C^{00}T_0' w_j) ,$$

which is obtained by taking

$$\varphi_j = (1 + C^{00}T_0'K_\delta^0)^{-1} (\gamma w_j + C^{00}T_0' w_j) ,$$

that is well defined for δ sufficiently small. By continuity, $\varphi_j \to 0$ in $\Pi H^{m-k-\frac{1}{2}}(E_\Gamma)$ and hence $x_j \to 0$ in $H^m(E)$. So now we have a sequence of smooth functions $y_j = w_j - x_j \to v$ in $H^m(E)$, with $S^{00}\gamma y_j + T_0' y_j = 0$. Here

$$S^{11}\nu y_j + S^{10}\gamma y_j + T_1' y_j = \rho_j$$

exists for each j . By a variant of Lemma 1.7.10 we can construct $z_j \in H^m(E)$ such that

$$S^{10}\gamma z_j + T_0' z_j = 0$$

$$S^{11}\nu z_j + S^{10}\gamma z_j + T_1' z_j = \rho_j$$

for each j , and $\|z_j\|_m \to 0$ for $j \to \infty$. Then finally $z_j - y_j$ lies in $D(B)$ and approximates v in $H^m(E)$. - The last statement in the lemma is an easy consequence. □

The next observation is that we can simplify the sesquilinear form problem by transformation to a more local condition by use of Lemma 1.6.8.

Let C^{00} be the right inverse of S^{00} defined above and set

$$(1.7.95) \qquad \Lambda = I + K_\delta^0 C^{00} T_0' ,$$

then by Lemma 1.6.8, there is a $\delta_0 > 0$ so that for $\delta \in]0,\delta_0]$, Λ is a bijection in $H^s(E)$ <u>for all</u> $s \geq 0$, with an inverse of the form $\Lambda^{-1} = I + K_\delta^0 Q_\delta C^{00} T_0'$ (Q_δ being a ps.d.o. in $\Pi_{0 \leq k < m} H^{-k}(E_\Gamma)$) , such that

$$(1.7.96) \qquad T_0 \equiv S^{00}\gamma + T_0' = S^{00}\gamma\Lambda ;$$

and Λ defines in particular a bijection of V (1.7.93) onto

$$(1.7.97) \qquad V_1 = \{v \in H^m(E) \mid S^{00}\gamma v = 0\} .$$

In this way, the validity of (1.7.94) for $u \in V$ is equivalent with the validity of an inequality

$$(1.7.98) \qquad s_1(v,v) \geq c_1\|v\|_m^2 - c_2\|v\|_0^2 \qquad \text{for } v \in V_1 ,$$

where s_1 is the sesquilinear form

$$(1.7.99) \qquad s_1(u,v) \equiv s(\Lambda u , \Lambda v) \qquad \text{for } u,v \in H^m(E) ;$$

note that it is again of the kind (1.7.90).

The advantage of reducing the problem to one where V is replaced by V_1 (1.7.98), is that now

$$\mathring{H}^m(E) \subset V_1 \subset H^m(E) ,$$

so that we can write

$$V_1 = \mathring{H}^m(E) \dotplus W ,$$

where $W \simeq \gamma V_1$ represents the boundary values (in fact K^r (1.7.43) defines a bijection of γV_1 onto W). Here m-coerciveness of s_1 requires m-coerciveness of its restrictions to $\mathring{H}^m(E)$ and to W ; the former represents m-coerciveness of the Dirichlet problem for an associated operator $P_\Omega + G'''$, and the latter can be reduced to strong ellipticity of a certain pseudo-differential system over Γ . Conversely, we expect that m-coerciveness of these two restrictions will imply m-coerciveness of the full form on V_1 by methods like those in [Grubb 5',4], under reasonable hypotheses on s_1 .

The detailed analysis will be left to a treatment elsewhere, and we here just end by two particular examples.

1.7.15 Example. Consider a formally selfadjoint Dirichlet-type realization, as described in Example 1.6.12. (Note that $P^r = P$ here; for precision we keep the notation of (1.7.41)ff. anyway.) For $u \in D(B)$ we have, using Lemma 1.7.5, formulas (1.6.64)-(1.6.65') and the fact that $\gamma u = -T'^0 u$,

$$
\begin{aligned}
\operatorname{Re}(Bu,u) &= \operatorname{Re}(P_\Omega u,u) + \operatorname{Re}(Gu,u) \\
&= (P_\Omega u^r_\gamma, u^r_\gamma) - \operatorname{Re}(\mathcal{A}^{01}\nu u, T'^0 u) + (\mathcal{B}T'^0 u, T'^0 u) \\
&\qquad + \operatorname{Re}(K^1 \nu u,u) + \operatorname{Re}(G'u,u) \\
&= (P_\Omega u^r_\gamma, u^r_\gamma) + (G''u,u) ,
\end{aligned} \tag{1.7.100}
$$

where

$$G'' = T'^{0*}(-\tfrac{1}{2}\mathcal{A}^{00} - \mathcal{A}^{01}Q^r)T'^0 + \tfrac{1}{2}(G'+G'^*) = -T'^{0*}(\mathcal{A}^{00} + \mathcal{A}^{01}Q^r)T'^0 + G' .$$

When $G'' \geq 0$,

$$\operatorname{Re}(Bu,u) \geq (P_\Omega u^r_\gamma, u^r_\gamma) \geq c\|u^r_\gamma\|^2_m , \tag{1.7.101}$$

this gives at least positivity of the realization. As for m-coerciveness, we observe that for $\varepsilon \in [0,\frac{1}{2}[$, one has by (1.7.43), when $u \in D(B)$,

$$\|u^r_\gamma\|_{m-\varepsilon} = \|u-K^r\gamma u\|_{m-\varepsilon} = \|u+K^rT'^0u\|_{m-\varepsilon} \simeq \|u\|_{m-\varepsilon}$$

$$\text{when } I + K^rT'^0 \text{ is invertible;}$$

here the invertibility holds at least when T'^0 has a sufficiently small norm, or when $K^rT'^0$ has a lower bound > -1 . In those cases, the inequality (1.7.101) implies m-coerciveness. Then one can furthermore allow a small negative lower bound for G'' .

In the case $P = -\Delta$, $\mathcal{A}^{00} = 0$, and $-\mathcal{A}^{01}Q^r = -iQ^r$ is a positive selfadjoint ps.d.o. with principal symbol $|\xi'|$, so the first term in G'' is ≥ 0 . Then it suffices that $\operatorname{Re} G' \geq 0$ or has a suitably small negative lower bound.

By use of (1.7.95-97), one gets another criterion. Here the space

$$V = \{u \in H^m(E) \mid \gamma u + T'_0 u = 0\}$$

can be replaced by $V_1 = \mathring{H}^m(E)$ by insertion of $u = \Lambda^{-1}v = (I - K^0_\delta T_1)v$ for suitable δ and T_1 . Noting that $u^r_\gamma = (1 - K^r\gamma)u$ we then find for the form $\operatorname{Re}(Bu,u)$, when $\gamma v = 0$,

$$
\begin{aligned}
\operatorname{Re}(Bu,u) &= \operatorname{Re}(B\Lambda^{-1}v,\Lambda^{-1}v) \\
&= (P_\Omega(I-K^r\gamma)\Lambda^{-1}v,(I-K^r\gamma)\Lambda^{-1}v) + (G''\Lambda^{-1}v,\Lambda^{-1}v)
\end{aligned} \tag{1.7.102}
$$

$$= (P_\Omega(v+(K^r-K_\delta^0)T_1v),v + (K^r-K_\delta^0)T_1v) + (G''\Lambda^{-1}v,\Lambda^{-1}v)$$

$$= (P_\Omega v,v) + (G'''v,v) \equiv s_1(v,v)$$

where G''' is the s.g.o. of class 0 and order $2m$

$$G''' = (\Lambda^{-1})^*G''\Lambda^{-1} + T_1^*(K^r-K_\delta^0)^*P_\Omega(K^r-K_\delta^0)T_1 + 2\,\mathrm{Re}[P_\Omega(K^r-K_\delta^0)T_1] .$$

Here m-coerciveness on $\mathring{H}^m(E)$ is obtained e.g. if $G''' \geq 0$, but also a moderate negativity can be allowed, since P_Ω itself is m-coercive on $\mathring{H}^m(E)$. (We expect that the m-coerciveness of $P_\Omega + G'''$ on $\mathring{H}^m(E)$ can be characterized by a positivity estimate of a model sesquilinear form at the boundary.)

1.7.16 Example. Let $P = -\Delta$ and consider a formally self-adjoint Neumann-type realization, as described in Example 1.6.15. Here we have for $u \in D(B)$, since $\gamma_1 u = -S_0\gamma_0 u - T'^1 u$,

$$\begin{aligned} \mathrm{Re}(Bu,u) &= (P_\Omega u_\gamma^r,u_\gamma^r) - \mathrm{Re}(i(S_0\gamma_0 u + T_1'u),\gamma_0 u) \\ &\quad + (\mathcal{B}\gamma_0 u,\gamma_0 u) + \mathrm{Re}(K_0\gamma_0 u,u) + \mathrm{Re}(G'u,u) \\ &= (P_\Omega u_\gamma^r,u_\gamma^r) + ((iS_0-iQ^r)\gamma_0 u,\gamma_0 u) - 2\mathrm{Re}(iT_1'u,\gamma_0 u) + (G'u,u) , \end{aligned} \tag{1.7.103}$$

by use of (1.6.80) and Lemma 1.7.5, as in the preceding example. Note that iS_0-iQ^r is self-adjoint with $-iQ^r$ positive (with symbol $|\xi'|$). The form is to be considered on $H^1(E)$ (in view of Lemma 1.7.14), so we do not need a transformation as in Lemma 1.7.15. Now 1-coerciveness is assured e.g. if $G' \geq 0$, iS_0 has principal symbol $> -|\xi'|$, and T_1' is small, for we can then obtain that

$$\mathrm{Re}(Bu,u) \geq c_1\|u_\gamma^r\|_1^2 - c_\varepsilon\|u^r\|_{1-\varepsilon}^2 + c_1'\|\gamma_0 u\|_{\frac{1}{2}}^2 - c_\varepsilon'\|\gamma_0 u\|_{\frac{1}{2}-\varepsilon}^2$$

with positive constants c_1 and c_1'; and this gives 1-coerciveness by Lemma 1.7.7. Also a small negative lower bound on G' can be allowed. Sharper criteria can probably be obtained if $u = u_\gamma^r + K^r\gamma_0 u$ is inserted in the last terms in (1.7.103).

1.7.17 Remark. It is seen from the preceding discussion how the normality of the boundary condition is of great help in the systematic discussion of coerciveness problems. Let us however also remark, that there do exist elliptic boundary problems that are not normal but define m-coercive realizations anyway. An example is the following:

$$\begin{aligned} -\Delta u &= f && \text{on } \mathbb{R}_+^n , \\ D_{x_1}\gamma_1 u + (1-\Delta_{x'})\gamma_0 u &= \varphi && \text{at } x_n = 0 , \end{aligned} \tag{1.7.104}$$

which was considered in [Rempel-Schulze 3, 4] in connection with resolvent studies. Since the coefficient of γ_1 is not invertible, the boundary condition is not normal, and it remains non-normal if multiplied by $(1-\Delta_{x'})^{-\frac{1}{2}}$. However, the realization B of $-\Delta$ with domain

(1.7.105) $$D(B) = \{u \in H^2(\overline{\mathbb{R}}^n_+) \mid D_{x_1}\gamma_1 u + (1-\Delta_{x'})\gamma_0 u = 0\}$$

is 1-coercive, for one finds by writing the boundary condition on the form

(1.7.106) $$\gamma_0 u = -(1-\Delta_{x'})^{-1} D_{x_1}\gamma_1 u \left[\equiv -S\gamma_1 u \right],$$

and applying Green's formula, that

(1.7.107) $$\begin{aligned} \operatorname{Re}(Bu,u) &= \operatorname{Re}\left[\sum_{j=1}^{n} \|D_j u\|_0^2 + i\int_{\mathbb{R}^{n-1}} \gamma_1 u\, \overline{S\gamma_1 u}\, dx' \right] \\ &= \sum_{j=1}^{n} \|D_j u\|_0^2 = \|u\|_1^2 - \|u\|_0^2 \quad \text{for } u \in D(B), \end{aligned}$$

since S is a constant coefficient ps.d.o. on $\mathbb{R}^{n-1}$ with real symbol. It is not hard to check that the problem is elliptic; and an argument as in Theorem 1.6.9 shows that the adjoint B^* is the analogous operator defined by the boundary condition

(1.7.108) $$\gamma_0 u = +S\gamma_1 u .$$

Here B^* is likewise elliptic and 1-coercive.

In particular, the inequalities for B and B^* imply that $(B-\lambda)^{-1}$ exists for $\operatorname{Re}\lambda \le -1$ and satisfies

(1.7.109) $$\|(B-\lambda)^{-1}f\|_1 + (|\operatorname{Re}\lambda| - 1)\,\|(B-\lambda)^{-1}f\|_0 \le \|f\|_0 ,$$

so the norm of the resolvent in $L^2(\mathbb{R}^n_+)$ is $O(|\operatorname{Re}\lambda|^{-1})$ for $\operatorname{Re}\lambda \to -\infty$. In particular, the norm is $O(\langle\lambda\rangle^{-1})$ on each ray $\{\lambda = re^{i\theta} \mid r \in [r_0,+\infty[\}$ with argument $\theta \in \,]\pi/2, 3\pi/2[$, as shown in [Rempel-Schulze 3,4]. (The considerations extend to various other non-normal boundary conditions on the form $S_1\gamma_1 u + S_0\gamma_0 u = 0$.)

In the systematic calculus of variable coefficient operators, this type of example has too low "regularity" to enter in our general theory, and only partial results are obtained; see the discussion in Remark 3.2.16. Actually, the systematic variable coefficient calculus of [Rempel-Schulze 4] contains requirements that essentially imply regularity 1 (in a less apparent way), and which are not verified by this example, so it is not covered by their general theory either, as one might think; it is linked with the constant coefficient case. (See also our Remark 1.5.16.)

CHAPTER 2

THE CALCULUS OF PARAMETER-DEPENDENT OPERATORS

2.1 Parameter-dependent pseudo-differential operators.

The next two chapters contain the full machinery of the parameter-dependent calculus, which is developed in order to give a convenient framework for the study of resolvents and evolution problems, and other problems with an unbounded parameter. To demonstrate the usefulness of the systematic point of view, we have inserted Examples 2.1.18 and 2.2.13, where some "hand" calculations in connection with the simple pseudo-differential operator

$$P + \mu^2 = (D^4_{x_1} + D^4_{x_2}) \,/\, (D^2_{x_1} + D^2_{x_2}) + \mu^2$$

are attempted.

The standard (parameter-independent) calculus is contained in the theory as a special case, so the present chapters can also be used as a complete and self-contained presentation of that calculus (if somewhat heavy because of the precautions due to the parameter). As in [Grubb 17], we systematically include a study of the kernels of the boundary symbolic operators (whereas [Boutet 3] and [Rempel-Schulze 1] place more emphasis on complex function techniques), using the symbol-kernels whenever they make the understanding easier.

Here is an overview of the contents of Chapter 2:

Section 2.1 presents the setting of parameter-dependent "interior" pseudo-differential operators, defined so that the class includes resolvents $(P+\mu^d)^{-1}$ as considered in Section 1.5. Related symbol classes occur e.g. in [Robert 1], [Mohamed 1, 2], [Iwasaki 1] (including non-elliptic operators); in [Vishik-Eskin 2], [Čan Zui Ho-Eskin 1], [Eskin 1], [Drin 1, 2, 3], [Eidelman-Drin 1, 2], [Giga 1], [Rempel-Schulze 4] (considering cases with essentially a principal estimate and a control over lower order terms); and in [Grubb 6] (with finitely many estimates of type $S_{\rho,\delta}$, $\rho = 1-\delta > \frac{1}{2}$) . The symbol spaces are classified according to the so-called regularity number ν , indicating how close the behavior of the parameter

μ is to that of a usual cotangent variable; this is discussed carefully, in preparation for some more complicated considerations for boundary operators.

Section 2.2 begins with an account of certain function spaces on $\mathbb{R}$ defined by asymptotic properties, that are basic for the Boutet de Monvel theory; - in particular the relation between $\mathscr{S}(\overline{\mathbb{R}}_+)$ and its Fourier transform H^+, and the representation of its functions by Laguerre series expansions. After this, the transmission property is introduced for our ps.d.o. symbols, and the asymptotic function spaces are used in a crucial analysis of the symbol classes, showing how they behave under projections (e.g. onto H^+, and in Laguerre series expansions) with respect to the conormal variable ξ_n.

Now the boundary symbol classes are introduced in Section 2.3 to match the above ps.d.o. symbols and their projections. There are many types of operators and many kinds of symbol seminorms to define, and the whole rest of the calculus depends on a good choice. (For example, the present choice gives better results than a calculus based on Laguerre series norms which we used in an earlier version.) Negligible symbols (those of order $-\infty$) are introduced - here the new complication is that there is one class for each regularity number. We also show how the parameter-independent symbols fit into the calculus.

In Section 2.4 we define the operators associated with the symbols; both the OP_n version with operators on $\overline{\mathbb{R}}_+$, parametrized by (x',y',ξ',μ), and the full operators OP on $\overline{\mathbb{R}}^n_+$ (here OP is supplied with a suffix T, K or G, according to whether it is a trace operator, Poisson operator or singular Green operator). Classes of negligible operators are introduced, allowing us to establish rules for reductions of symbols and passage to adjoints, and to define operators acting in vector bundles over manifolds. There is given a complete proof that our operator classes are invariant under diffeomorphisms preserving the boundary (this fills a gap also in the standard theory of pseudo-differential boundary problems).

In Section 2.5 we stablish estimates of the operators in parametrized Sobolev spaces.

Sections 2.6 and 2.7 are concerned with composition rules. The point is to show that the composition of two operators again belongs to our calculus, and there is a large number of different compositions to handle here. In most cases, it is found that the regularity numbers behave as for ps.d.o.s, namely the regularity ν'' of a composition of two operators with regularities ν resp. ν' satisfies

$$\nu'' = m(\nu,\nu') \equiv \min\{\nu,\nu',\nu+\nu'\} \quad .$$

But there is one case where there can be a loss of $\frac{1}{2}$, namely $\nu''=m(\nu-\frac{1}{2},\nu'-\frac{1}{2})$ for the singular Green operator term $L(P_\mu,P'_\mu)$ arising from composition of two truncated ps.d.o.s, in the general parameter-dependent case.

With a view to the resolvent construction we show however, that ν'' equals $m(\nu,\nu')$ in some cases where one factor is μ-independent; this is important for the applications.

Section 2.6 treats the composition of operators OP_n in the case of x_n-independent symbols; whereas Section 2.7 provides the extra technical considerations needed for the x_n-dependent case, and treats the compositions with respect to the full x-variable.

Finally, we give in Section 2.8 a brief description of some classes of strictly homogeneous, generally nonsmooth symbols, that are needed in connection with the main spaces of smooth symbols; and it is shown how the regularity number ν is connected with a Hölder continuity property.

All this prepares for the study of elliptic symbols and invertibility, that is taken up in Chapter 3.

Let us now define the relevant pseudo-differential symbols, motivated by the considerations in Section 1.5. We consider symbols $p(X,\xi,\mu)$ depending on $X \in \Xi$ open $\subset \mathbb{R}^{n'}$, $\xi \in \mathbb{R}^n$ and $\mu \in \overline{\mathbb{R}}_+$, where n and n' are positive integers. In the following, d and ν are real numbers. We use the notation (A.1), (A.2) for $\langle\xi\rangle$, $\langle\xi,\mu\rangle$, $\rho(\xi,\mu)$ and $|\xi,\mu|$; and we often replace $\langle\xi,\mu\rangle$ by a function $\kappa(\xi,\mu)$, where

$$(2.1.1)\quad \kappa(\xi,\mu) = |\xi,\mu| \text{ for } |\xi,\mu| \geq 1 \text{ and is } C^\infty \text{ and } > 0 \text{ on } \overline{\mathbb{R}}^{n+1}_+ .$$

(For a fixed μ_0, $\rho(\xi,\mu_0) \simeq 1$ and $\kappa(\xi,\mu_0) \simeq \langle\xi\rangle$.) Note that

$$(2.1.1')\quad \begin{aligned}(\rho(\xi,\mu)^\nu+1)\langle\xi,\mu\rangle^d &= \langle\xi\rangle^\nu\langle\xi,\mu\rangle^{d-\nu}+\langle\xi,\mu\rangle^d \leq \langle\xi\rangle^{\nu+1}\langle\xi,\mu\rangle^{d-\nu}+\langle\xi,\mu\rangle^{d+1}\\ &= (\rho(\xi,\mu)^{\nu+1}+1)\langle\xi,\mu\rangle^{d+1} .\end{aligned}$$

2.1.1 Definition.

1^0 *The space* $S^{d,\nu}_{1,0}(\Xi,\overline{\mathbb{R}}^{n+1}_+)$ *of pseudo-differential symbols of order* d *and regularity* ν *consists of the functions* $p(X,\xi,\mu) \in C^\infty(\Xi\times\mathbb{R}^n\times\overline{\mathbb{R}}_+)$ *satisfying, for any indices* $\beta \in \mathbb{N}^{n'}$, $\alpha \in \mathbb{N}^n$ *and* $j \in \mathbb{N}$,

$$(2.1.2)\quad \begin{aligned}|D^\beta_X D^\alpha_\xi D^j_\mu p(X,\xi,\mu)| &\leq c(X)(\langle\xi\rangle^{\nu-|\alpha|}+\langle\xi,\mu\rangle^{\nu-|\alpha|})\langle\xi,\mu\rangle^{d-\nu-j}\\ &\equiv c(X)(\rho(\xi,\mu)^{\nu-|\alpha|}+1)\langle\xi,\mu\rangle^{d-|\alpha|-j}\end{aligned}$$

for all X,ξ,μ, *with a continuous function* $c(X)$ *depending on the indices.*

2^0 *The space* $S^{d,\nu}(\Xi,\overline{\mathbb{R}}^{n+1}_+)$ *of polyhomogeneous pseudo-differential symbols of order* d *and regularity* ν *consists of the functions* $p \in S^{d,\nu}_{1,0}(\Xi,\overline{\mathbb{R}}^{n+1}_+)$

that furthermore have asymptotic expansions

$$(2.1.3)\qquad p(X,\xi,\mu) \sim \sum_{\ell\in\mathbb{N}} p_{d-\ell}(X,\xi,\mu)$$

in the sense that $p - \Sigma_{\ell<M} p_{d-\ell} \in S^{d-M,\nu-M}_{1,0}(\Xi,\overline{\mathbb{R}}^{n+1}_+)$ *and each term* $p_{d-\ell}$ *is homogeneous in* (ξ,μ)

$$(2.1.4)\qquad p_{d-\ell}(X,t\xi,t\mu) = t^{d-\ell} p_{d-\ell}(X,\xi,\mu) \quad \textit{for} \quad t \geq 1,\ |\xi| \geq 1 .$$

We often denote $p_d = p^0$, the principal part; and there is a certain freedom of choice of p^0 for $|\xi| \leq 1$.

Occasionally, there is a need for the analogous definition where $\overline{\mathbb{R}}^{n+1}_+$ is replaced by a relatively open conical subset V, and c depends on the ray; we then write $S^{d,\nu}_{1,0}(\Xi,V)$ resp. $S^{d,\nu}(\Xi,V)$. In the following, we sometimes formulate the statements for such general V. The indication of $\overline{\mathbb{R}}^{n+1}_+$ (or V) and Ξ are often omitted altogether, when this causes no confusion.

One can also define parameter-dependent spaces $S^{d,\nu}_{\rho,\delta}(\Xi,V)$ with ρ and $1-\delta \in]0,1]$; this is not included here since we do not need the concept (see however [Grubb 6], where more delicate resolvent estimates used such spaces).

<u>2.1.2 Definition.</u> *The* $N\times N$*-matrix formed symbol* $p(X,\xi,\mu) \in S^{d,\nu}(\Xi,V) \otimes L(\mathbb{C}^N,\mathbb{C}^N)$ *is said to be elliptic at* X, *when* p^0 *can be chosen such that for certain positive constants* $c(X)$ *and* $c_0(X)$, $p^0(X,\xi,\mu)$ *is invertible with*

$$(2.1.5)\qquad |p^0(X,\xi,\mu)^{-1}| \leq c(X)\langle\xi,\mu\rangle^{-d}$$

for $|\xi,\mu| \geq c_0(X)$. *When* Ξ' *is a subset of* Ξ, p *is said to be elliptic on* Ξ', *when* p^0 *can be chosen so that* (2.1.5) *is valid for all* $X \in \Xi'$.

<u>2.1.3 Remark.</u> For $N\times M$-matrix formed symbols, one can similarly introduce the notions "surjectively elliptic" and "injectively elliptic" where p^0 is surjective resp. injective, with a right resp. left inverse satisfying estimates (2.1.5).

The symbol spaces are Fréchet spaces in the following way: For each compact $K \subset \Xi$, take $c(X)$ as small as possible in (2.1.2); then $\max_{X\in K} c(X)$ is a seminorm on $S^{d,\nu}_{1,0}(\Xi,\overline{\mathbb{R}}^{n+1}_+)$; we denote it $\|p\|_{\alpha,\beta,j,K}$. With the topology defined by the system of all such seminorms, $S^{d,\nu}_{1,0}(\Xi,\overline{\mathbb{R}}^{n+1}_+)$ is a Fréchet space. For $S^{d,\nu}(\Xi,\overline{\mathbb{R}}^{n+1}_+)$, one gets a Fréchet space when furthermore the seminorms of the lower order parts $p - \Sigma_{\ell<M} p_{d-\ell}$ are adjoined. For the spaces $S^{d,\nu}_{1,0}(\Xi,V)$ and $S^{d,\nu}(\Xi,V)$, $c(X)$ is replaced by $c(X,\omega)$ depending furthermore on the direction $\omega = (\xi,\mu)/|\xi,\mu|$, and K is replaced by $K\times V'$, where $V' \cap S^{n-1}$ is compact in $V \cap S^{n-1}$.

2.1.4 Remark. In the applications, X will usually be taken equal to x, running in an open subset Σ of $\mathbb{R}^n$, or $X = (x,y) \in \Sigma\times\Sigma$; and μ will be treated as a parameter. One can also define spaces with μ running on $\mathbb{R}$ instead of $\overline{\mathbb{R}}_+$, and let X contain a space variable s for which μ is the cotangent variable (this permits a time-dependence for the operator $P + \partial_t$ in the heat equation). These extensions are not very deep, so we omit them for simplicity in the following.

Observe that when $p \in S^{d,\nu}_{1,0}$, then $D^\alpha_\xi p \in S^{d-|\alpha|,\nu-|\alpha|}_{1,0}$ and $D^j_\mu p \in S^{d-j,\nu}_{1,0}$, and when $p \in S^{d,\nu}$ then $p_{d-\ell} \in S^{d-\ell,\nu-\ell}$. One has obviously (cf. 2.1.1'),

$$S^{d-N-N',\nu-N} \subset S^{d-N,\nu-N} \subset S^{d,\nu} \subset S^{d,\nu-N''} \tag{2.1.6}$$

when N and $N' \in \mathbb{N}$ and $N'' \in \overline{\mathbb{R}}_+$. Similar observations hold for the $S^{d,\nu}_{1,0}$-classes, with arbitrary N,N' and $N'' \in \overline{\mathbb{R}}_+$. We denote

$$\bigcap_{N\in\mathbb{R}} S^{d,N}_{1,0} = S^{d,\infty}_{1,0} \quad \text{and} \quad \bigcap_{N\in\mathbb{R}} S^{d,N} = S^{d,\infty}\ ; \tag{2.1.7}$$

the spaces of symbols with regularity $+\infty$; they behave like standard symbols in $n+1$ cotangent variables. A variant of the usual reconstruction lemma for ps.d.o.s holds:

When there are given symbols $p_{d-\ell} \in S^{d-\ell,\nu-\ell}$ for $\ell \in \mathbb{N}$, then there exists a symbol $p \in S^{d,\nu}$ such that $p - \Sigma_{\ell<M} p_{d-\ell} \in S^{d-M,\nu-M}$ for all $M \in \mathbb{N}$. We say that $p \sim \sum_{\ell\in\mathbb{N}} p_{d-\ell}$ also in this situation (the series can be rearranged in a series of homogeneous terms). The same statements hold with S replaced by $S_{1,0}$ (except for what was said about a rearrangement). (Cf. Remark 2.3.8'.)

By considering the functions $p(X,\xi)$ in $S^d_{1,0}(\Xi,\mathbb{R}^n)$ or $S^d(\Xi,\mathbb{R}^n)$ (see (1.2.2) ff.) as constant in μ, one has the identifications

$$\begin{aligned} &S^d_{1,0}(\Xi,\mathbb{R}^n) \subset S^{d,d}_{1,0}(\Xi,\overline{\mathbb{R}}^{n+1}_+)\ ,\\ &S^d(\Xi,\mathbb{R}^n) \subset S^{d,d}(\Xi,\overline{\mathbb{R}}^{n+1}_+)\ , \end{aligned} \tag{2.1.8}$$

(cf. (2.1.2), first line). In the converse direction, when $p(X,\xi,\mu) \in S^{d,\nu}_{1,0}(\Xi,\overline{\mathbb{R}}^{n+1}_+)$ is considered for a fixed μ, it defines a function in $S^d_{1,0}(\Xi,\mathbb{R}^n)$, and also when $p(X,\xi,\mu) \in S^{d,\nu}(\Xi,\overline{\mathbb{R}}^{n+1}_+)$ is considered for a fixed μ, it defines a function in $S^d_{1,0}(\Xi,\mathbb{R}^n)$.

In the following, we concentrate the efforts on the spaces $S^{d,\nu}(\Xi,\overline{\mathbb{R}}^{n+1}_+)$. Almost all results have a counterpart in the $S^{d,\nu}_{1,0}$ spaces, with a similar or easier proof.

The $S^{d,\nu}$ spaces have slightly complicated multiplication properties because of the presence of ν . The general rule is

<u>2.1.5 Proposition.</u> *For* d, d', ν *and* $\nu' \in \mathbb{R}$ *one has*

$$S^{d,\nu}(\Xi,V)\cdot S^{d',\nu'}(\Xi,V) \subset S^{d+d',m(\nu,\nu')}(\Xi,V) \quad , \tag{2.1.9}$$

where

$$m(\nu,\nu') = \min\{\nu,\nu',\ \nu+\nu'\} \quad . \tag{2.1.10}$$

<u>Proof.</u> Note first that for $\rho \in]0,1]$,

$$(\rho^{\nu}+1)(\rho^{\nu'}+1) = \rho^{\nu} + \rho^{\nu'} + \rho^{\nu+\nu'}+1 \leq 3\rho^{m(\nu,\nu')} + 1 \quad . \tag{2.1.11}$$

Then when $p \in S^{d,\nu}$ and $p' \in S^{d',\nu'}$, one has by use of the Leibniz formula,

$$\begin{aligned} |D^{\beta}_{X}D^{\alpha}_{\xi}D^{j}_{\mu}(pp')| &= |D^{\beta}_{X} \sum_{\gamma\leq\alpha,k\leq j} c_{\gamma,k}(D^{\alpha-\gamma}_{\xi}D^{j-k}_{\mu}p)(D^{\gamma}_{\xi}D^{k}p')| \\ &\leq c(X) \sum_{\gamma\leq\alpha,k\leq j} (\rho(\xi,\mu)^{\nu-|\alpha-\gamma|}+1)(\rho(\xi,\mu)^{\nu'-|\gamma|}+1)\langle\xi,\mu\rangle^{d+d'-|\alpha|-j} \\ &\leq c'(X) \sum_{\gamma\leq\alpha,k\leq j} (\rho^{m(\nu-|\alpha-\gamma|,\nu'-|\gamma|)}+1)\langle\xi,\mu\rangle^{d+d'-|\alpha|-j} \quad . \end{aligned}$$

Since

$$\min_{\gamma\leq\alpha} \min\{\nu-|\alpha-\gamma|,\nu'-|\gamma|,\nu+\nu'-|\alpha|\} = \min\{\nu,\nu',\nu+\nu'\} - |\alpha| \quad ,$$

it follows that

$$|D^{\beta}_{X}D^{\alpha}_{\xi}D^{j}_{\mu}(pp')| \leq c''(X)(\rho^{m(\nu,\nu')-|\alpha|}+1)\langle\xi,\mu\rangle^{d+d'-|\alpha|-j} \quad .$$

The homogeneous products $p_{d-\ell}p'_{d'-\ell'}$ satisfy similar estimates, with regularity number $m(\nu-\ell,\nu'-\ell') \geq m(\nu,\nu')-\ell-\ell'$ and degree $d+d'-\ell-\ell'$. Arranging $p'' = pp'$ as an asymptotic series of homogeneous terms $p''_{d+d'-\ell}$ $(\ell \in \mathbb{N})$, one can write $p'' - \sum_{\ell<M} p''_{d+d'-\ell}$ as a product $(p - \sum_{\ell<M} p_{d-\ell})(p' - \sum_{\ell<M} p'_{d'-\ell'})$ plus a finite number of terms $p_{d-j}p'_{d'-k}$ with $j+k \geq M$; this satisfies the required estimates with regularity number $m(\nu,\nu') - M$ and degree $d+d' - M$. $\square$

Note that the function $m(\nu,\nu')$ equals $\nu+\nu'$ only when ν and ν' are ≤ 0; the regularity can generally not be improved by multiplication with an element of large positive regularity. There are a few exceptions to this rule, occuring when one factor is independent of μ, which we list since it is of importance in the theory to obtain as high a regularity as possible.

2.1.6 Lemma. *Let* $p \in S^d(\Xi,\mathbb{R}^n)$ *and* $q \in S^{d',\nu}(\Xi,\overline{\mathbb{R}}^{n+1}_+)$. *Then*

$$p(X,\xi)q(X,\xi,\mu) \in S^{d+d',\min\{d+\nu,d\}}(\Xi,\overline{\mathbb{R}}^{n+1}_+) \quad . \tag{2.1.12}$$

If, moreover, $d \in \mathbb{N}$ *and* p *is polynomial in* ξ , *then*

$$p(X,\xi)q(X,\xi,\mu) \in S^{d+d',d+\nu}(\Xi,\overline{\mathbb{R}}^{n+1}_+) \quad . \tag{2.1.13}$$

Proof: One has here that

$$\begin{aligned} |D^\beta_X D^\alpha_\xi D^j_\mu(pq)| &= |D^\beta_X \sum_{\gamma\leq\alpha} c_\gamma (D^{\alpha-\gamma}_\xi p)(D^\gamma_\xi D^j_\mu q)| \\ &\leq c(X) \sum_{\gamma\leq\alpha} \langle\xi\rangle^{d-|\alpha-\gamma|}(\rho^{\nu-|\gamma|}+1)\langle\xi,\mu\rangle^{d'-|\gamma|-j} \\ &= c(X) \sum_{\gamma\leq\alpha} (\langle\xi\rangle^{d+\nu-|\alpha|}\langle\xi,\mu\rangle^{d'-\nu-j} + \langle\xi\rangle^{d-|\alpha-\gamma|}\langle\xi,\mu\rangle^{d'-|\gamma|-j}) \\ &\leq c'(X)(\langle\xi\rangle^{d+\nu-|\alpha|}\langle\xi,\mu\rangle^{d'-\nu-j} + \langle\xi\rangle^{d-|\alpha|}\langle\xi,\mu\rangle^{d'-j}) \ . \end{aligned} \tag{2.1.14}$$

The weakest regularity is here $\min\{d+\nu,d\}$. For the lower order parts, $p - \Sigma_{\ell<M} p_{d-\ell} \in S^{d-M}$ and $q - \Sigma_{\ell<N} q_{d'-\ell} \in S^{d'-N,\nu-N}$, so the derivatives of the products are estimated with weakest regularity number $\min\{d-M+\nu-N,d-M\} \geq \min\{d+\nu,d\} - M-N$. With similar estimates for the homogeneous terms, we have the ingredients to conclude (2.1.12).

Now if p is polynomial in ξ , $|\alpha-\gamma|\leq d$ in the nonzero terms, so we get

$$\begin{aligned} |D^\beta_X D^\alpha_\xi D^j_\mu(pq)| &\leq c(X) \sum_{|\alpha-\gamma|\leq d} \langle\xi\rangle^{d-|\alpha-\gamma|}(\rho^{\nu-|\gamma|}+1)\langle\xi,\mu\rangle^{d'-|\gamma|-j} \\ &\leq c(X)(\langle\xi\rangle^{d+\nu-|\alpha|}\langle\xi,\mu\rangle^{d'-\nu-j} + \langle\xi,\mu\rangle^{d+d'-|\alpha|-j}) \end{aligned} \tag{2.1.15}$$

since $\langle\xi\rangle^{d-|\alpha-\gamma|} \leq \langle\xi,\mu\rangle^{d-|\alpha-\gamma|}$ here; this and similar estimates for lower order terms and parts shows (2.1.13). □

We show these simple rules in detail, since they are heavily used and generalized later on.

2.1.7 Remark. It is sometimes fruitful to observe that when $p_{d-\ell}(X,\xi,\mu)$ is a C^∞ function of $(X,\xi,\mu) \in \Xi \times \overline{\mathbb{R}}^{n+1}_+$ that is homogeneous of degree $d-\ell$ in (ξ,μ) for $|\xi| \geq 1$, then the validity of the inequalities (for all α,β and j)

$$(2.1.16)\quad |D^\beta_X D^\alpha_\xi D^j_\mu p_{d-\ell}(X,\xi,\mu)| \leq c(X)(\langle\mu\rangle^{d-\nu-j} + \langle\mu\rangle^{d-\ell-|\alpha|-j}) \quad \text{for } |\xi| \leq 1 ,$$

in the region $\{(\xi,\mu) \mid |\xi| \leq 1, \mu \geq 0\}$, suffice to assure that $p_{d-\ell}$ is in $S^{d,\nu}(\Xi, \overline{\mathbb{R}}^{n+1}_+)$ (by extension by homogeneity from $\{|\xi|=1\}$). In other words, for such homogeneous functions, it is their magnitude in μ on the set $\{|\xi|\leq 1, \mu \geq 0\}$ that determines which space they belong to. Let us also observe that

$$(2.1.17)\qquad (\rho(\xi,\mu)^\nu + 1)\langle\xi,\mu\rangle^d \simeq \begin{cases} \langle\xi,\mu\rangle^d & \text{when } \nu \geq 0 , \\ \langle\xi\rangle^\nu\langle\xi,\mu\rangle^{d-\nu} & \text{when } \nu \leq 0 . \end{cases}$$

Another interesting aspect of the polyhomogeneous symbols p is the study of the strictly homogeneous symbol associated with p^0, in particular its behavior for $\xi \to 0$ (μ fixed > 0). (Of course, the behavior for $\mu \to \infty$ ($\xi \neq 0$) or for $\xi \to 0$ ($\mu \neq 0$) of a function homogeneous in $(\xi,\mu) \in \overline{\mathbb{R}}^{n+1}_+ \setminus 0$ are two sides of the same thing.)

2.1.8 Definition. *Let* $p \in S^{d,\nu}(\Xi,V)$. *Then* $p^h(X,\xi,\mu)$ *is defined as the function that coincides with* $p^0(X,\xi,\mu)$ *for* $|\xi| \geq 1$ *and is strictly homogeneous, i.e.*

$$(2.1.18)\qquad p^h(X,\xi,\mu) = |\xi|^d p^0(X,\xi/|\xi|,\mu/|\xi|) , \quad \textit{for all } (\xi,\mu) \textit{ with } \xi \neq 0 .$$

p^h *is extended by continuity to* $\xi = 0$ *if possible.*

2.1.9 Lemma.

1^0 *Let* $p \in S^{d,\nu}(\Xi,V)$. *The associated strictly homogeneous function* p^h *satisfies the estimates, for all indices,*

$$(2.1.19)\quad |D^\beta_X D^\alpha_\xi D^j_\mu p^h(X,\xi,\mu)| \leq c(X)(|\xi|^{\nu-|\alpha|}|\xi,\mu|^{d-\nu-j} + |\xi,\mu|^{d-|\alpha|-j})$$

for $\xi \neq 0$.

2^0 *Let* $p \in S^{d,\nu}(\Xi,\overline{\mathbb{R}}^{n+1}_+)$ *with* $\nu \geq 0$. *Then*

$$(2.1.20)\qquad |p^0(X,\xi,\mu) - p^h(X,\xi,\mu)| \leq c(X)\langle\xi,\mu\rangle^{d-\nu}$$

for all X *and all* (ξ,μ) *with* $|\xi,\mu| \geq 1$.

Proof: Similarly to (2.1.17), one has

$$(2.1.21) \qquad (|\xi|^{\nu} + |\xi,\mu|^{\nu})|\xi,\mu|^{d-\nu} \simeq \begin{cases} |\xi,\mu|^{d} & \text{when } \nu \geq 0 \\ |\xi|^{\nu}|\xi,\mu|^{d-\nu} & \text{when } \nu \leq 0 , \end{cases}$$

since $|\xi| \leq |\xi,\mu|$. Now if $\nu \geq 0$, the observation (2.1.17) implies

$$\begin{aligned} |p^h(X,\xi,\mu)| &= ||\xi|^d p^h(X,\xi/|\xi|,\mu/|\xi|)| \\ &\leq c(X)|\xi|^d(1+1+\mu^2/|\xi|^2)^{d/2} = c(X)(2|\xi|^2+\mu^2)^{d/2} \\ &\leq c'(X)|\xi,\mu|^d , \end{aligned}$$

and if $\nu \leq 0$ we have

$$\begin{aligned} |p^h(X,\xi,\mu)| &= ||\xi|^d p^h(X,\xi/|\xi|,\mu/|\xi|)| \\ &\leq c(X)|\xi|^d\, 1^{\nu}(1+1+\mu^2/|\xi|^2)^{(d-\nu)/2} \leq c'(X)|\xi|^{\nu}|\xi,\mu|^{d-\nu} . \end{aligned}$$

With similar arguments applied to the derivatives, we find (2.1.19). This shows 1°.

For 2°, assume that $\nu \geq 0$ and $V = \overline{\mathbb{R}}^{n+1}_+$. When $\nu = 0$, the estimate is obvious in view of (2.1.19). Then consider the case $\nu \in]0,1]$, and set $p' = p^0 - p^h$. Since p' vanishes for $|\xi| \geq 1$, we have e.g. for $\xi_1 > 0$,

$$p'(X,\xi_1,\xi_2,\dots,\xi_n,\mu) = -\int_{\xi_1}^{1} \partial_{\xi_1} p'(X,t,\xi_2,\dots,\xi_n,\mu)dt .$$

Inserting the estimates for $\partial_{\xi_1} p^0$ and $\partial_{\xi_1} p^h$, we find that for $\xi_1 > 0$, $|\xi,\mu| \geq 2$, $|\xi| \leq 1$,

$$\begin{aligned} |p'(X,\xi_1,\dots,\xi_n,\mu)| &\leq c(X)\Big(\langle\mu\rangle^{d-\nu} + \int_{\xi_1}^{1} (t+|\xi_2,\dots,\xi_n|)^{\nu-1}(t+|\xi_2,\dots,\xi_n,\mu|)^{d-\nu}dt\Big) \\ &\leq c'(X)\langle\mu\rangle^{d-\nu} \qquad (\text{since } \nu-1 > -1) , \end{aligned}$$

implying (2.1.20) for $\xi_1 > 0$. Other halfspaces ($\xi_1<0$, or $\pm\xi_n>0$) are treated similarly. When $\nu=\ell+\tau$, $\ell \in \mathbb{N}$ and $\tau\in]0,1]$, one expresses p' by a derivative of order $\ell+1$. □

The arguments extend to functions with values in a Banach space.

The space of functions $p^h(X,\xi,\mu) \in C^{\infty}(\Xi \times (\mathbb{R}^n \setminus 0) \times \overline{\mathbb{R}}_+)$ that are homogeneous in (ξ,μ) of degree d on the whole set $(\mathbb{R}^n \setminus 0) \times \overline{\mathbb{R}}_+$ and satisfy (2.1.19) for all indices, is called the space of strictly homogeneous ps.d.o. symbols of order d and regularity ν ; it will be denoted $S^{d,\nu}_{hom}(\Xi, \overline{\mathbb{R}}^{n+1}_+)$. The functions here are extended by continuity to $\xi = 0$ $(\mu > 0)$ whenever possible. (For a conical subset $V \subset \overline{\mathbb{R}}^{n+1}_+$, $S^{d,\nu}_{hom}(\Xi,V)$ is defined such that the estimates (2.1.19) hold on $\Xi \times V$, with c depending also on the direction $\omega = (\xi,\mu)/|\xi,\mu|$.)

When $\nu \geq 1$, the estimate (2.1.19) for first order derivatives, combined with Taylor's formula, shows that

$$p^h(X,\xi,\mu) - p^h(X,\xi^*,\mu^*) \leq c(X)(|\xi-\xi^*| + |\mu-\mu^*|) \tag{2.1.22}$$

for (ξ,μ) and (ξ^*,μ^*) in a neighborhood of the point $(0,1)$, ξ and $\xi^* \neq 0$. This Lipschitz condition, together with the continuity of $p^h(X,\xi,\mu)$ on the unit half-sphere $\{(\xi,\mu) \mid |\xi,\mu| = 1, \mu \geq 0\}$ minus the point $(0,1)$, implies that p^h is continuous on the full unit half-sphere. By homogeneity, p^h is then in fact continuous for all $(\xi,\mu) \in \overline{\mathbb{R}}^{n+1}_+ \setminus 0$. Similarly, when $\nu \geq \ell+1$ for some $\ell \in \mathbb{N}$, p^h has ℓ continuous derivatives for $(\xi,\mu) \in \overline{\mathbb{R}}^{n+1}_+ \setminus 0$.

Actually, one can establish continuity at $\xi = 0$ when ν is merely > 0. For our purposes, this is of interest mainly for the boundary symbol classes introduced later on, so we give a proof that prepares for the case of such symbols also. The main ingredient is the following elementary lemma.

<u>2.1.10 Lemma</u>. *Let* U *be a Banach space with norm* $\|\cdot\|$, *let* $n \geq 2$, *and let* $\sigma \in [0,1[$. *Let* f *be a function from* $\dot{B}_1 = \{\xi \mid 0 < |\xi| \leq 1\}$ *to* U *such that* $f \in C^1(\dot{B}_1,U)$ *(has continuous strong derivatives), and*

$$\|\partial_j f(\xi)\| \leq c|\xi|^{-\sigma} \quad \textit{for } \xi \in \dot{B}_1 . \tag{2.1.23}$$

Then f *extends to a continuous function* f *on* $B_1 = \{\xi \mid |\xi| \leq 1\}$, *with the Hölder property*

$$\|f(\xi)-f(\eta)\| \leq c_{\sigma,n}\, c\, |\xi-\eta|^{1-\sigma} \quad \textit{for } \xi,\eta \in B_1 , \tag{2.1.24}$$

where $c_{\sigma,n}$ *depends on* σ *and* n *only.*

<u>Proof</u>: Let $\xi \in \dot{B}_1$. For $a > 0$ such that $a\xi \in \dot{B}_1$,

$$\begin{aligned} f(a\xi) &= f(\xi) + \int_1^a \frac{\partial}{\partial t} f(t\xi)dt \\ &= f(\xi) + \int_1^a \sum_{j=1}^n \partial_j f(t\xi)\xi_j dt , \end{aligned} \tag{2.1.25}$$

where the integrand satisfies, in view of (2.1.23),

$$\Big|\sum_{j=1}^n \partial_j f(t\xi)\xi_j\Big| \leq \sqrt{n}\, c\, t^{-\sigma}|\xi|^{1-\sigma} .$$

Since $-\sigma > -1$, $f(a\xi)$ converges for $a \to 0$ to a value we call $f(0)$, it is defined by (2.1.25) for $a = 0$. Moreover, we have for any a,

$$|f(a\xi) - f(\xi)| \le \sqrt{n}\, c|\xi|^{1-\sigma} \int_1^a t^{-\sigma} dt$$

$$= \frac{\sqrt{n}\, c}{1-\sigma} |\xi|^{1-\sigma} |a^{1-\sigma} - 1| \le \frac{\sqrt{n}\, c}{1-\sigma} |\xi a - \xi|^{1-\sigma}$$

(using the inequality $|a^{1-\sigma} - b^{1-\sigma}| \le |a-b|^{1-\sigma}$).

On the other hand if $\eta \in \dot{B}_1$ with $|\eta| = |\xi|$, then we can estimate $|f(\xi)-f(\eta)|$ by the integral of $|\partial_t f(\varphi(t))|$ where $\varphi(t)$ describes a curve from η to ξ in the sphere $\{\zeta \mid |\zeta| = |\xi|\}$. Since the curve can be chosen as an arc with length $\le \pi|\xi-\eta|$, we get, by (2.1.23),

$$|f(\xi)-f(\eta)| \le \pi|\xi-\eta|\, \sqrt{n}\, c\, |\xi|^{-\sigma}$$

$$= \pi\sqrt{n}\, c\, |\xi-\eta|^{1-\sigma} |\xi-\eta|^{\sigma} |\xi|^{-\sigma} \le 2^{\sigma}\sqrt{n}\, \pi\, c|\xi-\eta|^{1-\sigma},$$

since $|\xi-\eta|^{\sigma} \le (2|\xi|)^{\sigma}$.

So we see that (2.1.24) holds when ξ and η are on the same ray and when they are on the same sphere. Note also that the latter estimate implies that $f(0)$ equals $\lim_{a\to 0} f(a\eta)$ for any η.

Finally, when ξ and η are two arbitrary points in $\dot{B}_1$, with $|\eta| \ge |\xi|$, say, then we set $\zeta = \eta|\xi|/|\eta|$, so that $\eta = a\zeta$ with $a = |\eta|/|\xi| \ge 1$. The three points ξ, η and ζ define a triangle where the angle at ζ is $\ge \pi/2$, so

$$|\eta-\zeta| \le |\xi-\eta| \quad \text{and} \quad |\xi-\zeta| \le |\xi-\eta|,$$

whereby the above estimates imply

$$|f(\xi)-f(\eta)| \le |f(\xi) - f(\zeta)| + |f(\zeta) - f(\eta)|$$

$$\le \sqrt{n}\, c(2^{\sigma}\pi|\xi-\zeta|^{1-\sigma} + \frac{1}{1-\sigma} |\zeta-\eta|^{1-\sigma}) \le c_{n,\sigma}\, c|\xi-\eta|^{1-\sigma},$$

completing the proof of (2.1.24) in general; in particular, f is Hölder continuous. □

2.1.10' Definition. *Let* U *be a Banach space, and let* $\nu \in \overline{\mathbb{R}}_+$. *The space* $C^{\nu-}(\Xi \times \overline{\mathbb{R}}^{n+1}_+, U)$ *is defined as the set of functions* f *from* $(X,\xi,\mu) \in \Xi \times \overline{\mathbb{R}}^{n+1}_+$ *to* U *with the following properties:*

Let ℓ *be the largest integer strictly less than* ν. *Then* f *has continuous (strong) derivatives of order* $\le \ell$ *on* $\Xi \times (\overline{\mathbb{R}}^{n+1} \setminus 0)$, *and the derivatives of order* ℓ *are Hölder continuous on* $\Xi \times (\overline{\mathbb{R}}^{n+1}_+ \setminus 0)$ *with exponent* $\nu-\ell$:

$$\|D^\alpha f(X,\xi,\mu) - D^\alpha f(X^*,\xi^*,\mu^*)\|_U \le c(|X-X^*|^{\nu-\ell} + |\xi-\xi^*|^{\nu-\ell} + |\mu-\mu^*|^{\nu-\ell}), \tag{2.1.26}$$

$$\text{on compact subsets of } \Xi \times (\overline{\mathbb{R}}^{n+1}_+ \setminus 0), \quad \text{for } |\alpha| = \ell .$$

Moreover, if ν *is integer, the derivatives of order* ν *exist for* $\xi \neq 0$ *and are bounded on compact subsets of* $\Xi \times (\overline{\mathbb{R}}^{n+1}_+ \setminus 0)$ *(this covers the case* $\nu=0$*).*

2.1.11 Proposition.

1^0 *For* $\nu \ge 0$, *one has*

$$S^{d,\nu}_{hom}(\Xi, \overline{\mathbb{R}}^{n+1}_+) \otimes L(\mathbb{C}^N,\mathbb{C}^N) \subset C^{\nu-}(\Xi \times \overline{\mathbb{R}}^{n+1}_+, L(\mathbb{C}^N,\mathbb{C}^N)) . \tag{2.1.27}$$

Thus, if $p \in S^{d,\nu}(\Xi, \overline{\mathbb{R}}^{n+1}_+) \otimes L(\mathbb{C}^N,\mathbb{C}^N)$ *with* $\nu = \ell+\tau$ *where* $\ell \in \mathbb{N}$ *and* $\tau \in]0,1]$, *and* p^h *is the associated strictly homogeneous symbol, then* p^h *and its derivatives up to order* ℓ *are Hölder continuous with exponent* τ *for* $(\xi,\mu) \neq 0$, *and satisfy in particular (for any* $a > 0$*)*

$$|D^\alpha_\xi p^h(X,\xi,\mu) - D^\alpha_\xi p^h(X,0,\mu)| \le c(a,X)|\xi|^\tau |\xi,\mu|^{d-|\alpha|-\tau} \tag{2.1.28}$$

$$\text{for } |\xi| \le a\mu , \quad |\alpha| = \ell .$$

2^0 *If* $p \in S^{d,\nu}(\Xi, \overline{\mathbb{R}}^{n+1}_+)$ *with* $\nu > 0$, *then* p *is elliptic on* Ξ *if and only if* $p^h(X,\xi,\mu)$ *is bijective for all* X , *all* $(\xi,\mu) \in \overline{\mathbb{R}}^{n+1}_+ \setminus 0$.

Similar statements hold for $S^{d,\nu}(\Xi,V)$, *when* V *is a cone in* $\overline{\mathbb{R}}^{n+1}_+$.

Proof: For $\nu = 0$, the inclusion (2.1.27) is a consequence of the definition of $S^{d,\nu}_{hom}$, and so are the statements on boundedness of derivatives when ν is integer. For $\nu = \ell+\tau$ ($\ell \in \mathbb{N}$ and $\tau \in]0,1]$) , the Hölder continuity properties follow immediately from an application of Lemma 2.1.10 to the derivatives up to order ℓ .

Now since $p \in S^{d,\nu}$ implies $p^h \in S^{d,\nu}_{hom}$ by Lemma 2.1.9, we get the next statements on p^h by a specification of the preceding arguments; in particular, (2.1.28) follows from the Hölder estimate at $\xi = 0$, $\mu = 1$, and the homogeneity in (ξ,μ) .

Finally, we show the statement in 2^0 as follows: When p is elliptic and p^0 is a principal symbol satisfying (2.1.5), we have for $\xi \neq 0$

$$|p^h(X,\xi,\mu)^{-1}| = |\xi|^{-d}|p^0(X,\xi/|\xi|,\mu/|\xi|)|^{-1}$$
$$\le c(X)|\xi|^{-d}(1+1+\mu^2/|\xi|^2)^{-d/2} = c(X)(2|\xi|^2+\mu^2)^{-d/2} .$$

In particular, $(p^h)^{-1}$ is bounded for $\xi \to 0$ when $\mu > 0$. Since $\nu > 0$, p^h is continuous at $\xi = 0$ (for $\mu > 0$); then $p^h(X,0,\mu)$ must be bijective.

Conversely, let $p^h(X,\xi,\mu)$ be bijective for all (X,ξ,μ) with $(\xi,\mu) \neq (0,0)$. Then by homogeneity, and compactness of the unit half-sphere, p^h satisfies the estimate

$$|p^h(X,\xi,\mu)^{-1}| \leq c(X)|\xi,\mu|^{-d} \quad \text{for } (\xi,\mu) \neq 0 .$$

When p^0 is a principal symbol of p, we now have

$$p^0(X,\xi,\mu) = p^h(p^h)^{-1}(p^h+p^0-p^h) = p^h(I + (p^h)^{-1}(p^0-p^h)) ,$$

where (2.1.20) implies that

$$|(p^h)^{-1}(p^0-p^h)| \begin{cases} \leq c_1(X)\langle\xi,\mu\rangle^{-\nu} & \text{for } |\xi,\mu| \geq 1 , \\ = 0 & \text{for } |\xi| \geq 1 . \end{cases}$$

Define $c_0(X) \geq 1$ such that $\mu \geq c_0(X)$ implies $c_1(X)\langle\mu\rangle^{-\nu} \leq \frac{1}{2}$ (possible since $\nu > 0$), then $I + (p^h)^{-1}(p^0-p^h)$ is invertible for all (X,ξ,μ) with $\mu \geq c_0(X)$, by the Neumann series

$$(I + (p^h)^{-1}(p^0-p^h))^{-1} = \sum_{k\geq 0} [-(p^h)^{-1}(p^0-p^h)]^k \quad ;$$

note that the norm is less than 2. It follows that $p^0(X,\xi,\mu)$ is invertible both on the set $\{|\xi| \leq 1 , \ \mu \geq c_0(X)\}$ and on the set $\{|\xi| \geq 1\}$, satisfying

$$|p^0(X,\xi,\mu)^{-1}| \leq c_2(X)\langle\xi,\mu\rangle^{-d}$$

there, so p is elliptic as defined in Definition 2.1.2. □

2.1.11' Remark. It is of interest to know whether every strictly homogeneous symbol $p^h \in S^{d,\nu}_{\text{hom}}$ has an associated smooth symbol $p^0 \in S^{d,\nu}$ such that $p^h = p^0$ for $|\xi| \geq 1$. This is easily obtained if $\nu \leq 0$, for then we can simply take $p^0(X,\xi,\mu) = \zeta(|\xi|)p^h(X,\xi,\mu)$, where $\zeta(t) \in C^\infty(\mathbb{R})$ equals 1 for $|t| \geq 1$ and 0 for $|t| \leq 1/2$. For $\nu > 0$, this does not work, since ζ itself is only in $S^{0,0}$. But here one can do the following: Let N be the smallest integer $\geq \nu$ and let $p^h_N(X,\xi,\mu) = \Sigma_{|\alpha|<N}\partial^\alpha_\xi p^h(X,0,\mu)\xi^\alpha/\alpha!$ (the Taylor polynomial at $\xi = 0$); then set

$$\text{(2.1.28')} \quad \begin{aligned} p^0(X,\xi,\mu) &= \zeta(|\xi,\mu|)[\zeta(|\xi|)p^h(X,\xi,\mu) + (1-\zeta(|\xi|))p^h_N(X,\xi,\mu)] \\ &= \zeta(|\xi,\mu|)[p^h_N(X,\xi,\mu) + \zeta(|\xi|)(p^h(X,\xi,\mu) - p^h_N(X,\xi,\mu))] . \end{aligned}$$

It is clearly smooth and satisfies $p^0 = p^h$ for $|\xi| \geq 1$; and the inequalities (2.1.16) hold for $|\xi| \leq 1$, since $p^h - p^h_N$ on the set $\{\frac{1}{2}\leq|\xi|\leq 1, \ \mu \geq \frac{1}{2}\}$ has estimates like a symbol in $S^{d-N,\nu-N}$ in view of Taylor's formula (A.6), with $\nu-N \leq 0$ so that the factor $\zeta(|\xi|)$ does not change the regularity. See also Remark 2.8.4.

The discussion of strictly homogeneous symbols is taken up again in Section 2.8, where we introduce the corresponding symbol spaces for the boundary operator.

We now turn to the calculus of the associated pseudo-differential operators. Again we spend some detail on fairly elementary arguments, because they will be used again in the more complicated boundary operator calculus, with reference to the presentation here.

In the following, we specialize Ξ to be either $\Sigma \subset \mathbb{R}^n$ (with points x) or $\Sigma\times\Sigma$ (with points (x,y)). One could of course also let Ξ contain some parameters. The μ-dependent symbols $p(x,\xi,\mu)$ and $p(x,y,\xi,\mu)$ in $S^{d,\nu}_{1,0}$ define μ-dependent pseudo-differential operators $P_\mu = OP(p(x,\xi,\mu))$ or $OP(p(x,y,\xi,\mu))$ by the formulas

$$(P_\mu u)(x) = (2\pi)^{-n}\int_{\mathbb{R}^n} e^{ix\cdot\xi}p(x,\xi,\mu)\hat{u}(\xi)d\xi \quad \text{for} \quad u \in \mathscr{S}(\mathbb{R}^n)\ , \quad \text{resp.}$$

(2.1.29)

$$(P_\mu u)(x) = (2\pi)^{-n}\int_{\mathbb{R}^{2n}} e^{i(x-y)\cdot\xi}p(x,y,\xi,\mu)u(y)dyd\xi, \text{ for } \quad u \in C_0^\infty(\Sigma)\ .$$

The second formula applies to $p(x,\xi,\mu)$ when it is considered as a constant function of y . The first formula determines $p(x,\xi,\mu)$ uniquely from the values of $(P_\mu u)(x)$ for all $u \in \mathscr{S}(\mathbb{R}^n)$, all μ , whereas $p(x,y,\xi,\mu)$ in the second formula is not determined from p . (For example, if $a(x) \in C_0^\infty(\Sigma)$, $a \neq 0$, then $p(x,y,\xi,\mu) = \xi_1 a(y) - a(x)\xi_1 - D_{x_1}a$ defines the zero operator.) The second formula is understood in the sense of oscillatory integrals [Hörmander 6] (since the symbol is in $S^d_{1,0}$ for each fixed μ) .

When p in (2.1.29) depends only on (x,ξ,μ) or on (y,ξ,μ) , or, say, on (x',y_n,ξ,μ) , we say that the operator (and its symbol) is "on x-form" , "on y-form" , on "(x',y_n)-form" etc. The class of operators defined from symbols in $S^{d,\nu}_{1,0}(\Xi,\overline{\mathbb{R}}^{n+1}_+)$ is often denoted OP $S^{d,\nu}_{1,0}(\Xi,\overline{\mathbb{R}}^{n+1}_+)$, etc.

When the definition (2.1.29) is applied with respect to the x'-coordinate only, the operator is called $OP'(p)$, and when it is applied for the x_n-coordinate only, the operator is called $OP_n(p)$, or $p(x,y,\xi',\mu,D_n)$. Note that when p is independent of x_n and y ,

$$(2.1.30)\quad \|OP_n(p)u\|_{H^{s,\mu}(\mathbb{R})} = (2\pi)^{-\frac{1}{2}}\|\langle\xi,\mu\rangle^s p(x',\xi)\hat{u}(\xi_n)\|_{L^2(\mathbb{R})}$$
$$\leq \sup_{\xi_n}|\langle\xi,\mu\rangle^{-d}p(x',\xi)|\|u\|_{H^{s+d,\mu}(\mathbb{R})}\ .$$

We shall now prove parameter-dependent versions of the well known statements in Proposition 1.2.1 .

<u>2.1.12 Proposition.</u> *Let* $P_\mu = OP(p(x,y,\xi,\mu))$, *where* $p \in S^{d,\nu}_{1,0}(\mathbb{R}^n \times \mathbb{R}^n, \overline{\mathbb{R}}^{n+1}_+)$ *and vanishes for* (x,y) *outside a compact set. For any* $s \in \mathbb{R}$ *there are constants* c_s *and* c'_s *so that when* $u \in \mathcal{S}(\mathbb{R}^n)$ *and* $\mu \geq 0$

$$(2.1.31)\quad \|P_\mu u\|_{H^{s-d,\mu}(\mathbb{R}^n)} \leq c_s \|(\rho(\xi,\mu)^\nu + 1)\langle\xi,\mu\rangle^s \hat{u}(\xi)\|_0$$
$$\leq c'_s(\langle\mu\rangle^{-\nu}+1)\|u\|_{H^{s,\mu}(\mathbb{R}^n)} \quad .$$

<u>Proof</u>: Let u and $v \in \mathcal{S}(\mathbb{R}^n)$. Then

$$|(P_\mu u,v)| = |\int e^{i(x-y)\cdot\xi} p(x,y,\xi,\mu)u(y)\overline{v}(x)dyd\xi dx|$$
$$= c_1 |\int e^{i(x\xi - y\xi + y\theta - x\eta)} p(x,y,\xi,\mu)\hat{u}(\theta)\overline{\hat{v}}(\eta)d\theta d\eta dyd\xi dx|$$
$$= c_2 |\int \hat{p}(\eta-\xi,\xi-\theta,\xi,\mu)\hat{u}(\theta)\overline{\hat{v}}(\eta)d\theta d\eta d\xi| \quad ,$$

where $\hat{p}(\zeta,\sigma,\xi,\mu)$ denotes the Fourier transform of p in x and y . In view of the estimates (2.1.2) for $\beta \in \mathbb{N}^{2n}$ ($|\alpha|$ and $M=0$) and the compact support of p , one has for any $\mathbb{N}$,

$$(2.1.32)\quad |\hat{p}(\zeta,\sigma,\xi,\mu)| \leq c_N \langle\zeta\rangle^{-N}\langle\sigma\rangle^{-N} \begin{cases} \langle\xi,\mu\rangle^d & \text{if } \nu \geq 0 , \\ \langle\xi\rangle^\nu\langle\xi,\mu\rangle^{d-\nu} & \text{if } \nu < 0 . \end{cases}$$

By a well-known inequality (A.17), one has for each s

$$(2.1.32')\quad \begin{aligned} \langle\xi,\mu\rangle^d &= \langle\xi,\mu\rangle^{d-s}\langle\xi,\mu\rangle^s \leq c_3\langle\eta,\mu\rangle^{d-s}\langle\xi-\eta\rangle^{|d-s|}\langle\theta,\mu\rangle^s\langle\xi-\theta\rangle^{|s|} \quad , \\ \langle\xi\rangle^\nu\langle\xi,\mu\rangle^{d-\nu} &= \langle\xi,\mu\rangle^{d-s}\langle\xi\rangle^\nu \langle\xi,\mu\rangle^{s-\nu} \\ &\leq c_4\langle\eta,\mu\rangle^{d-s}\langle\xi-\eta\rangle^{|d-s|}\langle\theta\rangle^\nu\langle\theta,\mu\rangle^{s-\nu}\langle\xi-\theta\rangle^{|\nu|+|s-\nu|} \quad . \end{aligned}$$

Insertion of these inequalities in the integral and application of the Cauchy-Schwarz inequality, with

$$(2.1.33)\qquad N > n + \max\{|d-s|,\ |s|,\ |\nu|+|s-\nu|\}$$

gives

$$|(P_\mu u,v)| \leq c_5 |||\hat{u}(\theta)|||\ \|\langle\eta,\mu\rangle^{d-s}\hat{v}(\eta)\|_0 \quad ,$$

where $|||\hat{u}||| = \|\langle\theta,\mu\rangle^s\hat{u}(\theta)\|_0$ when $\nu \geq 0$ and $|||\hat{u}||| = \|\langle\theta\rangle^\nu\langle\theta,\mu\rangle^{s-\nu}\hat{u}(\theta)\|_0$ when $\nu < 0$. This shows that

$$\|P_\mu u\|_{s-d,\mu} = \sup_{v\neq 0} \frac{|(P_\mu u,v)|}{\|v\|_{d-s,\mu}}$$

$$\leq c_5 |||\hat{u}||| \leq c_6 \|(\rho(\theta,\mu)^\nu+1)\langle\theta,\mu\rangle^s \hat{u}(\theta)\|_0 \quad ,$$

cf. (2.1.17). The last inequality follows from the fact that for all ξ,μ ,

(2.1.34) $$\langle\mu\rangle^{-1} \leq \rho(\xi,\mu) = \langle\xi\rangle/\langle\xi,\mu\rangle \leq 1 \quad . \qquad \square$$

2.1.13 Remark. It is seen from the proof that c_s and c'_s can be estimated by a finite number of seminorms on p, namely those involving the (x,y)-derivatives of order $\leq 2N$ where N is any integer satisfying (2.1.33).

For fixed μ we have from the standard theory that $P = OP(p(x,y,\xi,\mu))$ maps $C_0^\infty(\Sigma)$ into $C^\infty(\Sigma)$, with an adjoint P^* going from $\mathcal{E}'(\Sigma)$ to $\mathcal{D}'(\Sigma)$, in the sense that

(2.1.35) $$\langle P^*u,\overline{\varphi}\rangle = \langle u,\overline{P\varphi}\rangle \qquad \text{for} \quad \varphi \in C_0^\infty(\Sigma) \ ,$$

and it seen by inspection of the formula that when $u \in C_0^\infty(\Sigma)$, then $P^*u = OP(p')u$, where

(2.1.36) $$p'(x,y,\xi,\mu) = \overline{p}(y,x,\xi,\mu) \ .$$

This is a well known way to extend $OP(p')$ (and then also $OP(p)$) to distribution spaces.

Consider in particular the symbols satisfying

(2.1.37) $$p(x,y,\xi,\mu) \in \bigcap_{N\in\mathbb{N}} S_{1,0}^{d-N,\nu-N}(\Sigma\times\Sigma,\ \overline{\mathbb{R}}_+^{n+1}) \ .$$

It is seen from the first inequality in (2.1.2) and the inequality (A.17) that they satisfy

(2.1.38) $$|D_{x,y}^\beta D_\xi^\alpha D_\mu^j p(x,y,\xi,\mu)| \leq c(x,y)\langle\xi\rangle^{-N'}\langle\mu\rangle^{d-\nu-j}$$

for any α,β,j and N' , and this actually characterizes the symbols satisfying (2.1.37). (The explicit estimates are taken up again in (2.3.43)-(2.3.44).) Note that the estimates (2.1.38) involve only the difference $\nu' = \nu-d$. We define the space of negligible ps.d.o. symbols of regularity ν , denoted $S^{-\infty,\nu-\infty}(\Sigma\times\Sigma,\ \overline{\mathbb{R}}_+^{n+1})$, by

$$S^{-\infty,\nu-\infty} = \bigcap_{N\in\mathbb{N}} S_{1,0}^{-N,\nu-N} ,$$

(2.1.39)

also equal to $\bigcap_{N\in\mathbb{N}} S^{-N,\nu-N}$ and $\bigcap_{N\in\mathbb{N}} S^{d-N,\nu-d-N}$, any d .

The corresponding operators are called the <u>negligible ps.d.o.s of regularity</u> ν. In view of the considerations around (1.2.4), a negligible ps.d.o. of regularity μ is an integral operator with C^∞ kernel $r(x,y,\mu)$, and estimates as in (2.1.38) imply

$$|D^\beta_{x,y} D^j_\mu r(x,y,\mu)| \le c'(x,y)\langle\mu\rangle^{-\nu-j} \tag{2.1.40}$$

for all indices (one can also appeal to the continuity statements (2.1.31) for the operator and its adjoint). Conversely, if $r(x,y,\mu)$ is a C^∞ function on $\Sigma\times\Sigma\times\overline{\mathbb{R}}_+$ satisfying (2.1.40), it is the kernel of the operator with symbol

$$p(x,y,\xi,\mu) = \exp[i(y-x)\cdot\xi]h(\xi)r(x,y,\mu) , \tag{2.1.41}$$

where $h(\xi) \in C_0^\infty(\mathbb{R}^n)$ with $(2\pi)^{-n}\int h(\xi)d\xi = 1$; this symbol is in $S^{-\infty,\nu-\infty}$.

<u>2.1.14 Definition.</u> *We say that* $p \sim p'$ *(modulo regularity* ν*), and* $P_\mu \sim P'_\mu$ *(mod. reg.* ν*), when* $p-p'$ *resp.* $P_\mu - P'_\mu$ *is negligible of regularity* ν .

Observe in particular that when $p \in S^{d,\nu}_{1,0}(\Sigma\times\Sigma, \overline{\mathbb{R}}^{n+1}_+)$, and $\chi(x,y) \in C^\infty(\Sigma\times\Sigma)$ is zero on a neighbourhood of the diagonal $x=y$ in $\Sigma\times\Sigma$, then

$$R_\mu = OP(\chi(x,y)p(x,y,\xi,\mu)) \tag{2.1.42}$$

is negligible of regularity $\nu-d$, for χ can be replaced by $|x-y|^{2N}\chi_N$ with $\chi_N \in C^\infty(\Sigma\times\Sigma)$, for any $N \in \mathbb{N}$, so that

$$\begin{aligned} R_\mu u &= (2\pi)^{-n} \int e^{i(x-y)\cdot\xi}|x-y|^{2N}\chi_N\, p\, u(y)dyd\xi \\ &= (2\pi)^{-n} \int\Big[(-\Delta_\xi)^N e^{i(x-y)\cdot\xi}\Big]\chi_N p\, u\, dyd\xi \\ &= (2\pi)^{-n} \int e^{i(x-y)\cdot\xi}\chi_N(+\Delta_\xi)^N\, p\, u\, dyd\xi \end{aligned} \tag{2.1.43}$$

with a symbol in $S^{d-2N,\nu-2N}$, any N. This allows one to write

$$P_\mu = OP(p(x,y,\xi,\mu)) = OP((1-\chi)p) + OP(\chi p) = P_{1,\mu} + R_\mu \tag{2.1.44}$$

where R_μ is negligible of regularity ν-d and $P_{1,\mu}$ is proper (as in Proposition 1.2.1 3^0); in fact, $P_{1,\mu}$ can be taken with kernel support in any neighbourhood of the diagonal $\{x=y\}$ in $\Sigma\times\Sigma$.

A proper ps.d.o. maps $C_0^\infty(\Sigma)$ into $C_0^\infty(\Sigma)$ and therefore extends to a mapping from $\mathcal{D}'(\Sigma)$ into $\mathcal{D}'(\Sigma)$ (by use of the formal adjoint). One can show that a proper ps.d.o. $P_{1,\mu}$ on $\mathbb{R}^n$ can always be represented on the form

$$P_{1,\mu} = OP(p_1(x,\xi,\mu))$$

(where p_1 is uniquely determined from $P_{1,\mu}$, as noted after (2.1.29)), and that p_1 is determined by the formula

$$e^{ix\cdot\xi}p_1(x,\xi,\mu) = P_{1,\mu}(\exp(iy\cdot\xi))(x) \ . \tag{2.1.45}$$

The proof that (2.1.45) actually defines a symbol of $P_{1,\mu}$ on x-form, lying in the right symbol class, is based on an analysis that is rather similar to the proof of part 1^0 of the following theorem; and we shall leave out further explanation.

The parameter-dependent version of the usual composition rules goes as follows. (Our formulation lies close to that of [Hörmander 6], so we shall not be very detailed.)

2.1.15 Theorem. *Let* d, ν, d' *and* $\nu' \in \mathbb{R}$.

1^0 *Let* $p(x,y,\xi,\mu) \in S^{d,\nu}(\Sigma\times\Sigma,\ \overline{\mathbb{R}}_+^{n+1})$. *Then there are symbols* $p_1(x,\xi,\mu)$ *and* $p_2(y,\xi,\mu) \in S^{d,\nu}(\Sigma,\ \overline{\mathbb{R}}_+^{n+1})$ *such that*

(2.1.46) (i) $OP(p(x,y,\xi,\mu)) \sim OP(p_1(x,\xi,\mu)) \sim OP(p_2(y,\xi,\mu))$, *with*

(ii) $p_1(x,\xi,\mu) \sim \sum_{\alpha\in\mathbb{N}^n} \frac{1}{\alpha!} D_\xi^\alpha\partial_y^\alpha p(x,y,\xi,\mu)\big|_{y=x}$,

(iii) $p_2(y,\xi,\mu) \sim \sum_{\alpha\in\mathbb{N}^n} \frac{1}{\alpha!} \overline{D}_\xi^\alpha\partial_x^\alpha p(x,y,\xi,\mu)\big|_{x=y}$,

modulo regularity ν-d *(the series are rearranged according to homogeneity).*

2^0 *Let* $p(x,\xi,\mu) \in S^{d,\nu}(\Sigma,\ \overline{\mathbb{R}}_+^{n+1})$. *Then there is a symbol* $p_1(x,\xi,\mu) \in S^{d,\nu}(\Sigma,\ \overline{\mathbb{R}}_+^{n+1})$, *such that*

(2.1.47) (i) $OP(p(x,\xi,\mu))^* \sim OP(p_1(x,\xi,\mu))$, *with*

(ii) $p_1(x,\xi,\mu) \sim \sum_{\alpha\in\mathbb{N}} \frac{1}{\alpha!} D_\xi^\alpha\partial_x^\alpha\overline{p}(x,\xi,\mu)$,

modulo regularity ν-d .

3^0 *Let* $p \in S^{d,\nu}(\Sigma,\overline{\mathbb{R}}^{n+1}_+)$ *and* $p' \in S^{d',\nu'}(\Sigma,\overline{\mathbb{R}}^{n+1}_+)$, *with* p *or* p' *vanishing for* x *outside a compact set in* Σ. *There is a symbol* $p'' \in S^{d+d',m(\nu,\nu')}(\Sigma,\overline{\mathbb{R}}^{n+1}_+)$, *such that*

(2.1.48) (i) $OP(p(x,\xi,\mu))OP(p'(x,\xi,\mu)) \sim OP(p''(x,\xi,\mu))$, *with*

$$\text{(ii)}\quad p''(x,\xi,\mu) \sim \sum_{\alpha\in\mathbb{N}^n} \frac{1}{\alpha!} D_\xi^\alpha p(x,\xi,\mu)\partial_x^\alpha p'(x,\xi,\mu) \quad ,$$

modulo regularity $m(\nu,\nu')-d-d'$.

Proof (indications): The principal step is the proof of 1^0. Here one inserts in (2.1.29) a Taylor expansion (A.6) of $p(x,y,\xi,\mu)$ in y at $y=x$, which gives, by integration by parts like in (2.1.43),

$$OP(p) = (2\pi)^{-n}\int e^{i(x-y)\cdot\xi} \sum_{|\alpha|<N} \frac{1}{\alpha!} D_\xi^\alpha \partial_y^\alpha p(x,x,\xi,\mu)u(y)dyd\xi + R_N u \; ,$$

where the symbol in the first term is part of the series (2.1.46 ii), and

$$R_N u = (2\pi)^{-n}\int e^{i(x-y)\cdot\xi} \sum_{|\alpha|=N} \frac{N}{\alpha!} D_\xi^\alpha \int_0^1 (1-h)^{N-1}\partial_y^\alpha p(x,x+hy,\xi,\mu)udhdyd\xi$$

$$= \int r_N(x,y,\mu)u(y)dy \; ,$$

with r_N continuous and $O(\langle\mu\rangle^{d-\nu})$ for sufficiently large N. More precisely, the symbol estimates imply that for $N > \nu + |d-\nu|+n+k$,

$$\text{(2.1.49)}\qquad |D_{x,y}^\beta D_\mu^j r_N(x,y,\mu)| \le c(x,y)\langle\mu\rangle^{d-\nu-j} \qquad \text{for } |\beta| \le k \;,\; \text{all } j \;.$$

One can now associate a symbol $p_1(x,\xi,\mu)$ to the series in (2.1.46 ii), lying in $S^{d,\nu}$. Finally one shows the first equivalence in (2.1.46 i) by using that one has

$$OP(p) - OP(p_1) = R_N - OP\Big(p_1 - \sum_{|\alpha|<N} \frac{1}{\alpha!} D_\xi^\alpha \partial_y^\alpha p(x,x,\xi,\mu)\Big)$$

for any N. For each k, we can take N so large that the two terms on the right have kernels satisfying estimates as in (2.1.49) for $|\beta| \le k$. This shows that $OP(p)-OP(p_1)$ is negligible. The proof of the remaining part of (2.1.46 i), and (iii), is based in a similar way on inserting Taylor expansions with terms $\frac{1}{\alpha!}(x-y)^\alpha\, \partial_x^\alpha p(y,y,\xi,\mu)$. This shows 1^0.

2^0 is now a simple corollary, when we note that P_μ^* has the symbol on y-form $\overline{p}(y,\xi,\mu)$ (recall (2.1.36)), which is carried over to x-form by use of (2.1.46 ii).

Also 3^0 can be obtained as a corollary: One transforms p' to a y-form $p_1(y,\xi,\mu)$ by use of (2.1.46 iii), then $P'' \sim OP(p(x,\xi,\mu)p_1(y,\xi,\mu))$, where the symbol $p(x,\xi,\mu)p_1(y,\xi,\mu)$ is transformed to x-form by use of (2.1.46 ii), and reduced to the actual form (2.1.48 ii) by use of an identity for binomial coefficients. There is also a direct proof by insertion of Taylor expansions at $\theta=\xi$ for $p(x,\xi,\mu)$ in the composed formula:

$$P_\mu P'_\mu u = (2\pi)^{-2n}\int e^{i(x-z)\xi+i(z-y)\theta}p(x,\xi,\mu)p'(x,\theta,\mu)u(y)dyd\theta dzd\xi$$

$$(2.1.50)\qquad = (2\pi)^{-2n}\int e^{i(x-z)\xi+i(z-y)\theta}\sum_{|\alpha|<N}\frac{(\xi-\theta)^\alpha}{\alpha!}\partial_\xi^\alpha p(x,\theta,\mu)p'(z,\theta,\mu)udyd\theta dzd\xi + \int r_N(x,y,\mu)u(y)dy \quad ;$$

here the first integral gives the terms in the series in (2.1.48 ii) and the second integral is shown to have a continuous kernel for large N, satisfying more and more of the required estimates for kernels of negligible operators, the larger N is. □

The rules are likewise valid for the $S^{d,\nu}_{1,0}$ spaces.
For the composition described in (2.1.48 ii) we denote

$$p''(x,\xi,\mu) = p(x,\xi,\mu)\circ p'(x,\xi,\mu) \quad ,$$

this expression is also defined for symbols with (ξ,μ) in a cone V.

We shall now show that elliptic symbols (cf. Definition 2.1.2) of regularity ≥ 0 have inverses with respect to this composition product.

<u>2.1.16 Theorem.</u> *Let* d *and* $\nu\in\mathbb{R}$ *with* $\nu\geq 0$.

1^0 *Let* $p(x,\xi,\mu)$ *be an* $N\times N$*-matrix formed symbol in* $S^{d,\nu}(\Sigma,V)\otimes L(\mathbb{C}^N,\mathbb{C}^N)$, *and assume that* p *is elliptic. There exists a symbol* $q\in S^{-d,\nu}(\Sigma,V)\otimes L(\mathbb{C}^N,\mathbb{C}^N)$, *such that*

$$(2.1.51)\qquad p\circ q\sim q\circ p\sim I \quad (\textit{mod.reg. } \nu)\ .$$

q *is uniquely determined (mod.reg.* $\nu+d$) *and is called a parametrix symbol for* p.

2^0 *Let* $p\in S^{d,\nu}(\Sigma,\overline{\mathbb{R}}^{n+1}_+)\otimes L(\mathbb{C}^N,\mathbb{C}^N)$ *be elliptic, let* q *be a parametrix symbol for* p *and let* P_μ *and* Q_μ *be proper ps.d.o.s in* Σ *with symbol* p *resp.* q. *Then*

(2.1.52) $$P_\mu Q_\mu - I \sim Q_\mu P_\mu - I \sim 0 \qquad (\textit{mod.reg. } \nu)\ ,$$

Q_μ *is called a parametrix of* P_μ .

Proof: It is necessary for (2.1.51) that q^0 equals $(p^0)^{-1}$ for large $|\xi|$, since all other terms give contributions of order ≤ -1 . Let $\zeta(x,\xi,\mu)$ be a C^∞ function equal to 0 for $|\xi,\mu| \leq c_0(x)$ and equal to 1 for $|\xi,\mu| \geq 2c_0(x)$. Let

$$q^0(x,\xi,\mu) = \zeta(x,\xi,\mu)p^0(x,\xi,\mu)^{-1} \qquad (\text{for } (\xi,\mu) \in V)\ .$$

Then q^0 is homogeneous of degree $-d$ in (ξ,μ) for $|\xi| \geq 1$, $|\xi,\mu| \geq 2c_0(x)$, and it is $O(\langle\xi,\mu\rangle^{-d})$. By use of the Leibniz formula for $|\xi,\mu| \geq 2c_0(x)$:

(2.1.53) $$0 = D^\alpha(p^0q^0) = \sum_{\beta\leq\alpha} c_{\alpha,\beta} D^{\alpha-\beta}p^0\, D^\beta q^0 \qquad \text{for} \quad \alpha \neq 0\ ,$$

implying

(2.1.54) $$D^\alpha q^0 = -q^0 \sum_{\beta<\alpha} c_{\alpha,\beta}\, D^{\alpha-\beta}p^0\, D^\beta q^0 \qquad \text{for} \quad \alpha \neq 0\ ,$$

one finds successively, in view of Proposition 2.1.5,

$$D_{\xi_i} q^0 = -q^0(D_{\xi_i} p^0)q^0 = O(\langle\xi,\mu\rangle^{-d}(\rho^{\nu-1} + 1))$$
$$\vdots$$
$$\begin{aligned} D_\xi^\alpha q^0 &= -q^0 \sum_{\beta<\alpha} c_{\alpha,\beta}(D^{\alpha-\beta}p^0)D_\xi^\beta q^0 \\ &= O(\langle\xi,\mu\rangle^{-d} \sum_{\beta<\alpha} (\rho^{\nu-|\alpha-\beta|} + 1)(\rho^{\nu-|\beta|} + 1)) \\ &= O(\langle\xi,\mu\rangle^{-d}(\rho^{\nu-|\alpha|} + 1))\ , \quad \underline{\text{since}} \quad \nu \geq 0\ . \end{aligned}$$

The x and μ-derivatives are similarly investigated, and one finds altogether that $q^0 \in S^{-d,\nu}$.

Now by Theorem 2.1.15 3^0,

$$\begin{aligned} p \circ q^0 &\sim p^0q^0 + (p-p^0)q^0 + \sum_{\alpha\neq 0} \frac{1}{\alpha!} D_\xi^\alpha p\, \partial_x^\alpha q^0 \\ &\sim I - r(x,\xi,\mu) \qquad (\text{mod.reg. } \nu)\ , \end{aligned}$$

where $r(x,\xi,\mu) \in S^{-1,\nu-1}(\Omega,V)$ (here we have used again that $\nu \geq 0$) . Defining $s(x,\xi,\mu)$ as a symbol in $S^{0,\nu}$ satisfying

$$s(x,\xi,\mu) \sim \sum_{k=0}^{\infty} r(x,\xi,\mu)^{\circ k} \qquad (\text{mod.reg. } \nu)$$

where $r^{\circ k}$ stands for the composition of k factors r (it lies in $S^{-k,\nu-k}$), we find that

$$p \circ q^0 \circ s \sim I \qquad (\text{mod.reg. } \nu),$$

so any $q \sim q^0 \circ s$ is a right inverse of p with respect to the symbol composition. A left inverse q' is found similarly, and a standard argument (as in Theorem 3.2.7 below) shows that $q \sim q'$. The uniqueness of the lower order part of q is seen by comparison of homogeneous terms.

Statement 2^0 in the proposition now follows from Theorem 2.1.15 3^0. □

When $\nu < 0$, the arguments go badly, since much regularity is lost in compositions; anyway, the ellipticity definition is somewhat unnatural in that case. Further comments are given in Remark 2.1.19.

For the definition of operators on manifolds we have to investigate the behavior under coordinate changes. It is found that the new classes $S^{d,\nu}$ are coordinate invariant; the proof is a straightforward modification of the usual proof (in the form presented e.g. in [Boutet 4]).

2.1.17 Lemma. *Let* d *and* $\nu \in \mathbb{R}$. *Let* Σ *and* $\underline{\Sigma}$ *be open subsets of* $\mathbb{R}^n$, *and* $\kappa: x \mapsto \underline{x}$ *a* C^∞ *diffeomorphism of* Σ *onto* $\underline{\Sigma}$. *Let* $p(x,y,\xi,\mu) \in S^{d,\nu}_{1,0}(\Sigma\times\Sigma,\overline{\mathbb{R}}^{n+1}_+)$ *defining the parametrized family of ps.d.o.s* $P_\mu = OP(p)$. *Then the operator family* $\underline{P}_\mu$ *defined by*

$$\underline{P}_\mu u = P_\mu(u \circ \kappa) \circ \kappa^{-1} \tag{2.1.55}$$

is a parametrized ps.d.o. family with a symbol $\underline{p} \in S^{d,\nu}_{1,0}(\underline{\Sigma}\times\underline{\Sigma},\overline{\mathbb{R}}^{n+1}_+)$ *satisfying*

$$\underline{p}(\underline{x},\underline{y},\underline{\xi},\mu) = p(x,y,{}^tM(x,y)\underline{\xi},\mu) \cdot |\det {}^tM(x,y)|\,|\det \kappa'(y)^{-1}|, \tag{2.1.56}$$

for (x,y) *in a neighbourhood* Ξ *of the diagonal* $\{x=y\}$ *in* $\Sigma\times\Sigma$; *here* κ' *is the Jacobian matrix* $(\partial\kappa_i/\partial x_j)$ *and* M *is the matrix defined (on* Ξ*) by*

$$\underline{x} - \underline{y} = M(x,y)(x-y). \tag{2.1.57}$$

In particular, $M(x,x) = \kappa'(x)$, *and* $\underline{p}(\underline{x},\underline{x},\underline{\xi},\mu) = p(x,x,{}^t\kappa'(x)\underline{\xi},\mu)$. *If* p *is polyhomogeneous, then so is* $\underline{p}$. *Then if* p *is on* x*-form, the* x*-form* $\underline{p}_1$ *of* $\underline{p}$ *satisfies*

$$\underline{p}_1^0(\underline{x},\underline{\xi},\mu) = p^0(x,{}^t\kappa'(x)\underline{\xi},\mu). \tag{2.1.58}$$

Proof: By Taylor's formula for each κ_i , we can take

$$M(x,y) = \int_0^1 \kappa'(x+t(y-x))dt \quad ; \tag{2.1.58'}$$

here $M(x,x) = \kappa'(x)$ is invertible, so $M(x,y)$ is invertible for (x,y) in a neighborhood Ξ of the diagonal $x=y$. Modulo a $(\nu-d)$-negligible changement of p (clearly invariant under diffeomorphisms, cf. (2.1.40)), we can assume that p is supported in Ξ . Then we have for $u \in C_0^\infty(\underline{\Sigma})$, setting $\xi = {}^tM(x,y)\underline{\xi}$,

$$\begin{aligned}(\underline{P}_\mu u)(\kappa(x)) &= (2\pi)^{-n}\int e^{i(x-y)\cdot\xi}p(x,y,\xi,\mu)u(\kappa(y))dyd\xi \\ &= (2\pi)^{-n}\int e^{i(x-y)\cdot{}^tM\underline{\xi}}\, p(x,y,\xi,\mu)u(\underline{y})|\det \kappa'(y)^{-1}|\ |\det{}^tM(x,y)|d\underline{y}d\underline{\xi} \\ &= (2\pi)^{-n}\int e^{i(\underline{x}-\underline{y})\cdot\underline{\xi}}\,\underline{p}(\underline{x},\underline{y},\underline{\xi},\mu)u(\underline{y})d\underline{\eta}d\underline{\xi} \quad ,\end{aligned}$$

with $\underline{p}$ defined by (2.1.56). Clearly, $\underline{p}$ is a symbol in $S^{d,\nu}_{1,0}$ as asserted. When $x=y$, $\det \kappa'(y)^{-1}$ and $\det{}^tM(x,y)$ cancel out.

The formula (2.1.56) shows that polyhomogeneity is preserved. When p is on x-form, one finds the x-form $\underline{p}_1$ of the symbol of $\underline{P}_\mu$ by application of Theorem 2.1.15 1°; this gives in particular (2.1.58). □

It is sometimes of interest to use a more precise formula for the x-form $\underline{p}_1$, given by [Hörmander 2] (or see [Hörmander 8, Theorem 18.1.17]):

$$\underline{p}_1(\underline{x},\underline{\xi},\mu) \sim p(x,{}^t\kappa'(x)\underline{\xi},\mu) + \sum_{|\alpha|\geq 1} D_\xi^\alpha p(x,{}^t\kappa'(x)\underline{\xi},\mu)\varphi_\alpha(x,\underline{\xi}) \ , \tag{2.1.59}$$

where the φ_α are certain polynomials in $\underline{\xi}$ of degree $\leq |\alpha|/2$, with C^∞ coefficients in x . The formula is valid for $S_{1,0}$ symbols, and gives in the polyhomogeneous case an expansion in homogeneous terms, by rearrangement according to homogeneity degree.

Lemma 2.1.17 is used when one wants to define the ps.d.o.s on manifolds, and (2.1.58) implies that the principal symbol has an invariant meaning as a function on the cotangent bundle.

The main difference between the above arguments and the standard arguments in the ps.d.o. calculus is that we had to keep track of negligible terms with different regularities ν , instead of just operators with C^∞ kernels. This becomes more complicated when boundary operators are included.

2.1.18 Example. We shall discuss a simple example of a parameter-dependent ps. d.o. with finite regularity. Let $n = 2$ and write (x_1, x_2) as (x, y) and (ξ_1, ξ_2) as (ξ, η) ; and let ζ be an excision function ($1 - \zeta \in C_0^\infty(\mathbb{R}^2)$ and $\zeta = 0$ near 0). The following symbol defines a ps.d.o. of order 2

$$p(\xi,\eta) = \frac{\xi^4+\eta^4}{\xi^2+\eta^2}\zeta(\xi,\eta) \ , \tag{2.1.60}$$

which is essentially the quotient between two elliptic differential operators

$$P \sim (D_x^4 + D_y^4)/(D_x^2+D_y^2) \ ;$$

such examples occur e.g. in manipulations with matrices (as for example in [Grubb 5]).

When regarded as μ-dependent, P has regularity 2 but not more, for we can write

$$p^h(\xi,\eta) = \eta^2 - \xi^2 + 2\xi^4/(\xi^2+\eta^2) \ , \tag{2.1.61}$$

where

$$\partial_\eta^3[\xi^4/(\xi^2+\eta^2)] = \xi^4\eta(8\xi^2-4\eta^2)/(\xi^2+\eta^2)^4$$

is only $O(|\xi,\eta|^{-1})$ for $|\xi,\eta| \to 0$ (compare Lemma 2.1.9 1^0). Now let

$$q(\xi,\eta,\mu) = (p^h(\xi,\eta)+\mu^2)^{-1}\zeta(\xi,\eta) = \frac{\xi^2+\eta^2}{\xi^4+\eta^4+\mu^2(\xi^2+\eta^2)}\zeta(\xi,\eta) \ . \tag{2.1.62}$$

Since $p(\xi,\eta) + \mu^2$ is elliptic of order 2 in view of Proposition 2.1.11, q is elliptic of order -2 , and has regularity 2, by Theorem 2.1.16. The example will be followed up later on, in Example 2.2.13.

2.1.19 Remark. We shall give some comments on what happens when the hypotheses of Theorem 2.1.16 are violated. On one hand, p may be of negative regularity, still satisfying the ellipticity requirement, and on the other hand, p may be of nonnegative regularity having an inverse p^{-1} that does not satisfy the inequality in Definition 2.1.2. In constant coefficient cases, one can easily operate with such symbols anyway; but in variable coefficient cases, the lower order terms arizing in symbol compositions may give trouble.

Let us consider the following two examples:

$$\begin{aligned} p_1(x,\xi,\mu) &= \mu^4/a(x,\xi) + \zeta(\xi/|\xi,\mu|)\varphi(x)(a(x,\xi)+\mu^2), \quad \varphi > 0 \ , \quad |\xi,\mu| \geq 1 \ , \\ p_2(x,\xi,\mu) &= a(x,\xi)(a(x,\xi)+\mu^2)+\zeta(\xi/|\xi,\mu|)\varphi(x)(a(x,\xi)+\mu^2)^2, \quad |\xi,\mu| \geq 1 \ , \end{aligned} \tag{2.1.63}$$

where $a(x,\xi)$ is homogeneous of degree 2 in $|\xi|$ for $|\xi| \geq 1$ and satisfies (with $0 < c_1 \leq c_2$)

$$c_1\langle\xi\rangle^2 \leq a(x,\xi) \leq c_2\langle\xi\rangle^2 .$$

Here $p_1 \in S^{2,-2}$ (negative regularity), but its inverse p_1^{-1} lies in $S^{-2,2}$ and is $O(\langle\xi,\mu\rangle^{-2})$; on the other hand, $p_2 \in S^{4,2}$ (positive regularity), but p_2^{-1} is in $S^{-4,-2}$ and is only $O(\langle\xi\rangle^{-2}\langle\xi,\mu\rangle^{-2})$. The composition rules (Proposition 2.1.5 and Theorem 2.1.15) give that

$$r_1(x,\xi) = I - p_1\circ p_1^{-1} \sim -\sum_{|\alpha|\geq 1} \frac{1}{\alpha!} D_\xi^\alpha p_1 \partial_x^\alpha p_1^{-1} \in \bigcup_{|\alpha|\geq 1} S^{2-|\alpha|,-2-|\alpha|} . S^{-2,2} \subset S^{-1,-3} ,$$

$$r_2(x,\xi) = I - p_2\circ p_2^{-1} \sim -\sum_{|\alpha|\geq 1} \frac{1}{\alpha!} D_\xi^\alpha p_2 \partial_x^\alpha p_2^{-1} \in \bigcup_{|\alpha|\geq 1} S^{4-|\alpha|,\, 2-|\alpha|} . S^{-4,-2} \subset S^{-1,-2} .$$

One can now formally construct right parametrix symbols q_1 resp. q_2 for p_1 and p_2 by iteration as in Theorem 2.1.16, but the lower order terms here will get worse and worse in their μ-dependence: Since $r_1^{\circ k} \in S^{-k,-3k}$ and $r_2^{\circ k} \in S^{-k,-2k}$, for $k = 1,2,\ldots$, we find (in a quite formal sense)

(2.1.64)
$$q_1 \sim p_1^{-1} \circ \sum_{k=0}^{\infty} r_1^{\circ k} \in S^{-2,2} \cup \bigcup_{k=1}^{\infty} S^{-2-k,-3k} ,$$

$$q_2 \sim p_2^{-1} \circ \sum_{k=0}^{\infty} r_2^{\circ k} \in S^{-4,-2} \cup \bigcup_{k=1}^{\infty} S^{-4-k,-2k} .$$

The symbol classes are not contained in a single space $S^{d,\nu}$. Note that one has e.g.

$$p_2^{-1}\circ r_2^{\circ k} \quad \text{is} \quad O(\langle\xi\rangle^{-2k}\langle\xi,\mu\rangle^{-4+k}) \quad \text{for} \quad k \geq 1 ,$$

which shows an increase in the power of μ for increasing k .

The lack of control for increasing k stems from the fact that r_1 and r_2 lie in $S^{-1,\nu''}$ with $\nu'' < -1$. There are some mildly irregular cases where one can get $\nu'' \geq -1$ in either the right or the left parametrix construction. More specifically, let $p \in S^{d,\nu}$ with $p^{-1} \in S^{-d,\nu'}$, then

(2.1.65)
$$r = I - p\circ p^{-1} \in S^{-1,m(\nu-1,\nu')}$$

$$r' = I - p^{-1}\circ p \in S^{-1,m(\nu'-1,\nu)} ,$$

by a precise application of (2.1.48). Here, if $m(\nu-1,\nu') = -1+\delta$ with $\delta \in [0,1]$, then $r^{\circ k} \in S^{-k,-k+\delta}$ for all k , so that there is a right parametrix symbol

$$q \sim p^{-1} \circ \sum_{k=0}^{\infty} r^{\circ k} \in S^{-d,\nu'} \cdot S^{0,\delta} \subset S^{-d,m(\nu',\delta)} \quad .$$

Similarly, if $m(\nu'-1,\nu) = -1+\delta'$ with $\delta' \in [0,1]$, then there is a left parametri[x] symbol q' with

$$q' \sim \sum_{k=0}^{\infty} (r')^{\circ k} \circ p^{-1} \in S^{-d,m(\nu',\delta')} \quad .$$

Both properties $m(\nu-1,\nu') \geq -1$ and $m(\nu'-1,\nu) \geq -1$ are satisfied only when the assumptions of Theorem 2.1.16 hold. For an example where one of the properties holds take

(2.1.66) $\quad p_3(x,\xi,\mu) = b(x,\xi)(b(x,\xi)+\mu) + \zeta(\xi/|\xi,\mu|)(b(x,\xi)+\mu)^2 \ , \ b > 0 \ , \ |\xi,\mu| \geq 1 \ ,$

where $b(x,\xi)$ is elliptic of order 1 (and invertible), and $(b(x,\xi) + \mu)^{-1}$ is $O(\langle\xi,\mu\rangle^{-1})$. Here $p_3 \in S^{2,1}$, and $p_3^{-1} \in S^{-2,-1}$, so that $r_3 = I - p_3 \circ p_3^{-1}$ is in $S^{-1,-1}$ and there is a right parametrix symbol

(2.1.67) $$q_3 \sim p_3^{-1} \circ \sum_{k=0}^{\infty} r_3^{\circ k} \in S^{-2,-1} \quad .$$

But $r_3' = I - p_3^{-1} \circ p_3$ is in $S^{-1,-2}$, and the left parametrix symbol is a formal series with the same kind of bad μ-behavior as for q_1 and q_2 further above.

In the cases where δ or $\delta' \geq 0$, there are uniform μ-estimates of the lower order parts of the right resp. left parametrix. Let us however observe, that one needs strictly positive regularity of the parametrix symbol (be it one-sided or two-sided) in order for its principal part to have a better behavior in μ than the other parts of the symbol.

All these observations generalize to the boundary symbol calculus introduced later on; they are taken up in Remark 3.2.16.

It is very likely that the irregular symbols can be handled within the more general symbol calculi that have been developed for non-elliptic operators, as described e.g. in [Hörmander 8]. At any rate, the difficulty really lies in the boundary symbol calculus. We hope to return to this problem in future works.

2.2 The transmission property.

The transmission property (at $x_n=0$) for symbols in $S^d_{1,0}(\mathbb{R}^n,\mathbb{R}^n)$ was presented in Section 1.2 (see (1.2.6)-(1.2.7)), and it was mentioned without further explanation that the property implies a certain asymptotic expansion (1.2.9) of the symbols in powers of ξ_n. Since we shall now define the transmission property for parameter-dependent symbols, we insert a full explanation of the properties of functions of one variable ξ_n (denoted t below), that are needed here. (It is consistent with the Wiener-Hopf calculus in [Boutet 3].)

For each integer $d \in \mathbb{Z}$ one defines the space H_d as the space of C^∞ functions $f(t)$ on $\mathbb{R}$ with the asymptotic property: There exist complex numbers $s_d, s_{d-1}, \dots$ such that for all indices k, ℓ and $N \in \mathbb{N}$,

$$\partial_t^\ell[t^k f(t) - \sum_{d-N\leq j\leq d} s_j t^{j+k}] \text{ is } O(|t|^{d-N-1+k-\ell}) \quad \text{for } |t| \to \infty . \tag{2.2.1}$$

The s_j are uniquely determined from f. It is not hard to show that if we make the changes of variable, where σ is a parameter > 0,

$$t = \tau^{-1} \quad \text{resp.} \quad t = \frac{\sigma}{i}\frac{1-z}{1+z} \quad (\text{i.e., } z = \frac{\sigma - it}{\sigma + it} = \frac{2\sigma}{\sigma + it} - 1) \tag{2.2.2}$$

and define the functions

$$k(\tau) = \tau^d f(\tau^{-1}) , \quad g(z) = (1+z)^d f(\tfrac{\sigma}{i}\tfrac{1-z}{1+z}) , \tag{2.2.3}$$

then $f \in H_d$ is equivalent with each of the statements

$$f \in C^\infty(\mathbb{R}) \text{ and } k \in C^\infty(\mathbb{R}) \quad (\text{including } \tau = 0) , \tag{2.2.4}$$

$$g(z) \text{ is } C^\infty \text{ on the circle } \{|z| = 1\} \quad (\text{including } z = -1) , \tag{2.2.5}$$

in fact the coefficients s_{d-j} are certain constants times the Taylor coefficients of $k(\tau)$ at $\tau = 0$. We denote

$$H = \bigcup_{d\in\mathbb{Z}} H_d , \tag{2.2.6}$$

and observe also the decomposition in a direct sum

$$H = H_{-1} \dotplus \mathbb{C}[t] , \tag{2.2.7}$$

where $\mathbb{C}[t]$ is the space of polynomials in t. The corresponding projection of H onto H_{-1} is denoted h_{-1}, and $(I-h_{-1})f$ is called the polynomial part of f. Occasionally, we also use the projector h_0 of H onto H_0, that removes $\Sigma_{1\leq j\leq d} s_j t^j$ from f. The spaces H_d have Fréchet topologies (defined by families of seminorms in relation to (2.2.1), (2.2.4) or (2.2.5)), and H is an inductive limit of such spaces.

There is a fourth description of H that is likewise interesting. Let $f(t) \in H_{-1}$, let $g(z) = (1+z)^{-1} f(-i\sigma(1-z)/(1+z))$ (as in (2.2.3)) and consider its Fourier series development (for $z = e^{i\theta}$), with the convention

$$\text{(2.2.8)} \quad \begin{aligned} &g(z) = (2\sigma)^{-\frac{1}{2}} \sum_{k\in\mathbb{Z}} b_k z^k , \quad \text{decomposed in} \\ &g^+(z) = (2\sigma)^{-\frac{1}{2}} \sum_{k\geq 0} b_k z^k \quad \text{and} \quad g^-(z) = (2\sigma)^{-\frac{1}{2}} \sum_{k<0} b_k z^k . \end{aligned}$$

Note that $g \in C^\infty(\{|z|=1\})$ means precisely that the sequence $(b_k)_{k\in\mathbb{Z}}$ is <u>rapidly decreasing</u> for $|k| \to \infty$, i.e. the sequences $(k^N b_k)_{k\in\mathbb{Z}}$ are bounded for all N. The space of sequences (b_k) will be denoted $s(\mathbb{Z})$, and we shall also use some of the following norms

$$\text{(2.2.9)} \quad \begin{aligned} &\|(b_k)_{k\in\mathbb{Z}}\|_{\ell^p_N} = \Big(\sum_{k\in\mathbb{Z}} |(1+|k|)^N b_k|^p\Big)^{1/p} , \quad 1 \leq p < \infty , \\ &\|(b_k)_{k\in\mathbb{Z}}\|_{\ell^\infty_N} = \sup_k (1+|k|)^N |b_k| , \end{aligned}$$

and corresponding Banach spaces $\ell^p_N(\mathbb{Z})$, $1 \leq p \leq \infty$. Also the analogous notation with $\mathbb{Z}$ replaced by $\mathbb{N}$ will be used. For $f(t) = (1+z)g(z)$, this leads to an expansion of f in terms of the functions $(\sigma-it)^k/(\sigma+it)^{k+1}$ that are <u>orthogonal in</u> $L^2(\mathbb{R})$. Moreover, we have that

$$\text{(2.2.10)} \quad \frac{1}{a}(1+z)z^k = a\frac{(\sigma-it)^k}{(\sigma+it)^{k+1}} = F_{x\to t}[\varphi_k(x,\sigma)] \equiv \hat{\varphi}_k(t,\sigma) , \qquad a = (2\sigma)^{\frac{1}{2}} ,$$

where φ_k is the (variant of a) <u>Laguerre function</u> defined as follows:

$$\text{(2.2.11)} \quad \begin{aligned} &\text{When } k \geq 0 , \quad \varphi_k(x,\sigma) = \begin{cases} (2\sigma)^{\frac{1}{2}}(\sigma-\partial_x)^k (x^k e^{-x\sigma}/k!) & \text{for } x \geq 0 , \\ 0 & \text{for } x \leq 0 . \end{cases} \\ &\text{When } k < 0 , \quad \varphi_k(x,\sigma) = \varphi_{-k-1}(-x,\sigma) . \end{aligned}$$

The functions $(\varphi_k)_{k\in\mathbb{Z}}$ form a complete orthonormal system in $L^2(\mathbb{R})$, and the functions with $k \geq 0$ resp. $k < 0$ span $L^2(\mathbb{R}_+)$ resp. $L^2(\mathbb{R}_-)$ (we often write φ_k for $r^+\varphi_k$ when $k \geq 0$, and φ_k for $r^-\varphi_k$ when $k < 0$). The φ_k

with $k \geq 0$ are the eigenfunctions of the (unconventional) Laguerre operator

(2.2.12) $$L_{\sigma,+} = -\sigma^{-1}(\sigma+\partial_x)x(\sigma-\partial_x) = -\sigma^{-1}\partial_x x\partial_x + \sigma x + 1$$

in $L^2(\mathbb{R}_+)$, with simple eigenvalues $2(k+1)$, and the φ_k with $k < 0$ are the eigenfunctions for $L_{\sigma,-}$ defined by the same expression on $\mathbb{R}_-$.

(We use the notation $\hat{L}_{\sigma,\pm}$ for the Fourier transformed versions of $L_{\sigma,\pm}$,

(2.2.13) $$\hat{L}_{\sigma,\pm}f = h^{\pm}_{-1}[-i\sigma^{-1}(\sigma+it)\partial_t(\sigma-it)f(t)] = h^{\pm}_{-1}[(i\sigma^{-1}t\partial_t t - i\sigma\partial_t + 1)f] ;$$

they map H^+ resp. H^-_{-1} into H^+ resp. H^-_{-1} , and define unbounded selfadjoint operators in $L^{\pm} = F(e^{\pm}L^2(\mathbb{R}_{\pm}))$; the maps and spaces are explained further below.)

The property of the Laguerre system that is of particular interest here is the fact that rapidly decreasing coefficient series correspond to functions in $\mathcal{S}(\overline{\mathbb{R}}_+)$.

<u>2.2.1 Lemma</u>. *Let* $u \in L^2(\mathbb{R}_+)$, *expanded in the Laguerre system* $(\varphi_k)_{k\in\mathbb{N}}$, *by*

$$u(x) = \sum_{k\in\mathbb{N}} b_k\varphi_k(x,\sigma) .$$

Then $u \in \mathcal{S}(\overline{\mathbb{R}}_+)$ *if and only if* $(b_k)_{k\in\mathbb{N}}$ *is rapidly decreasing. More precisely, one has the identity*

(2.2.14) $$\|u\|_{L^2(\mathbb{R}_+)} = \|(b_k(u))_{k\in\mathbb{N}}\|_{\ell^2_0} ,$$

and there are the estimates of $(b_k)_{k\in\mathbb{N}}$ *in terms of* u:

(2.2.15) $$\|(b_k(u))_{k\in\mathbb{N}}\|_{\ell^2_N} = 2^{-N}\|(b_k(L^N_{\sigma,+}u))\|_{\ell^2_0}$$

$$= 2^{-N}\|L^N_{\sigma,+}u\|_{L^2} \leq c_N \max_{j+\ell\leq N} \sigma^{j-\ell}\|x^{j+\ell}\partial_x^{2\ell}u\|_{L^2} ;$$

and the estimates of u *in terms of* $(b_k)_{k\in\mathbb{N}}$:

(2.2.16) $$\|xu\|_{L^2} = \|(b_k(xu))\|_{\ell^2_0} \leq c\sigma^{-1}\|(b_k(u))\|_{\ell^2_1} ,$$

(2.2.17) $$\|\partial_x u\|_{L^2} = \|(b_k(\partial_x u))\|_{\ell^2_0} \leq c_\varepsilon \sigma\|(b_k(u))\|_{\ell^2_{1+\varepsilon}} ,$$

(2.2.18) $$\|x\partial_x u\|_{L^2} = \|(b_k(x\partial_x u))\|_{\ell^2_0} \leq c'\|(b_k(u))\|_{\ell^2_1} ,$$

$$(2.2.19)\quad \| \partial_x (x\partial_x u)\|_{L^2} = \| (b_k(\partial_x x \partial_x u))\|_{\ell^2_0} \leq c''\sigma \| (b_k(u))\|_{\ell^2_1} \quad ,$$

for any $\varepsilon > 0$, *as well as higher order estimates*

$$(2.2.20)\quad \| x^i \partial_x^j (x\partial_x)^k (\partial_x x \partial_x)^\ell u\|_{L^2} \leq c''' \sigma^{-i+j+\ell} \| (b_k(u))\|_{\ell^2_m} \quad ,$$

with $m = i+(1+\varepsilon)j+k+\ell$.

Proof: The identity (2.2.14) follows from the orthonormality and completeness of the system φ_k in $L^2(\mathbb{R}_+)$. (2.2.15) is then obvious from the eigenvalue property of the φ_k ; in fact when $u \in \mathcal{S}(\overline{\mathbb{R}}_+)$, the ℓ^2_N-norms of $(b_k)_{k\in\mathbb{N}}$ are estimated by

$$\begin{aligned}
\| (b_k)_{k\in\mathbb{N}} \|^2_{\ell^2_N} &= \sum_{k\in\mathbb{N}} |(1+k)^N b_k|^2 \\
&= 2^{-2N} \| \sum_{k\in\mathbb{N}} b_k \, L^N_{\sigma,+} \varphi_k \|^2 \\
&= 2^{-2N} \| L^N_{\sigma,+} u \|^2_{L^2} = 2^{-2N} \| (-\sigma^{-1}\partial_x x \partial_x + \sigma x + 1)^N u \|^2_{L^2} \\
&\leq c_N \max_{j+\ell\leq N} \sigma^{2(j-\ell)} \| x^{j+\ell} \partial_x^{2\ell} u \|^2_{L^2} \quad ,
\end{aligned}$$

since $x\partial_x = \partial_x x - I$. For the estimates (2.2.16)-(2.2.19) we calculate the formulas:

$$\partial_t \frac{(\sigma-it)^k}{(\sigma+it)^{k+1}} = \frac{-ik}{\sigma} \frac{(\sigma-it)^{k-1}}{(\sigma+it)^k} + \frac{-i(2k+1)}{2\sigma} \frac{(\sigma-it)^k}{(\sigma+it)^{k+1}} + \frac{-i(k+1)}{2\sigma} \frac{(\sigma-it)^{k+1}}{(\sigma+it)^{k+2}}$$

and

$$(2.2.21)\quad it \frac{(\sigma-it)^k}{(\sigma+it)^{k+1}} = - \frac{(\sigma-it)^k}{(\sigma+it)^{k+1}} + \frac{(\sigma-it)^k}{(\sigma+it)^k}$$

where, for $k > 0$,

$$(2.2.22)\quad \frac{(\sigma-it)^k}{(\sigma+it)^k} = \sigma \frac{(\sigma-it)^{k-1}}{(\sigma+it)^k} - it \frac{(\sigma-it)^{k-1}}{(\sigma+it)^k} = \ldots = 2\sigma \sum_{0\leq j\leq k-1} (-1)^{k-1-j} \frac{(\sigma-it)^{j-1}}{(\sigma+it)^j} + (\ldots$$

and, moreover,

$$\partial_t \left(t \frac{(\sigma-it)^k}{(\sigma+it)^{k+1}}\right) = - \frac{k}{2} \frac{(\sigma-it)^{k-1}}{(\sigma+it)^k} + \frac{1}{2} \frac{(\sigma-it)^k}{(\sigma+it)^{k+1}} + \frac{k+1}{2} \frac{(\sigma-it)^{k+1}}{(\sigma+it)^{k+2}} \quad .$$

This gives the formulas for φ_k and its Fourier transform $\hat{\varphi}_k$ (cf. (2.2.10))

$$\partial_t\hat{\varphi}_k(t,\sigma) = \frac{-i}{2\sigma}(k\hat{\varphi}_{k-1} + (2k+1)\hat{\varphi}_k + (k+1)\hat{\varphi}_{k+1}), \quad \text{for } k \in \mathbb{Z},$$

$$it\hat{\varphi}_k(t,\sigma) = -\sigma\hat{\varphi}_k + 2\sigma \sum_{0\le j\le k-1} (-1)^{k-1-j}\hat{\varphi}_j + (-1)^k(2\sigma)^{\frac{1}{2}} \quad \text{for } k \ge 0$$

$$\partial_t(t\hat{\varphi}_k(t,\sigma)) = -\frac{k}{2}\hat{\varphi}_{k-1} + \frac{1}{2}\hat{\varphi}_k + \frac{k+1}{2}\hat{\varphi}_{k+1} \quad \text{for } k \in \mathbb{Z},$$

(2.2.23)

$$x\varphi_k(x,\sigma) = \frac{1}{2\sigma}(k\varphi_{k-1} + (2k+1)\varphi_k + (k+1)\varphi_{k+1}) \quad \text{for } k \in \mathbb{Z},$$

$$\partial_x\varphi_k(x,\sigma) = -\sigma\varphi_k + 2\sigma \sum_{0\le j\le k-1} (-1)^{k-1-j}\varphi_j + (-1)^k(2\sigma)^{\frac{1}{2}}\delta \quad \text{for } k \ge 0,$$

$$x\partial_x\varphi_k(x,\sigma) = \frac{k}{2}\varphi_{k-1} - \frac{1}{2}\varphi_k - \frac{k+1}{2}\varphi_{k+1} \quad \text{for } k \in \mathbb{Z}.$$

In particular, one has for the restriction to $\mathbb{R}_+$

(2.2.24) $$r^+\partial_x\varphi_k = -\sigma\varphi_k + 2\sigma \sum_{0\le j\le k-1} (-1)^{k-1-j}\varphi_j, \quad k \ge 0.$$

Using these formulas, we find

$$\|xu\|^2_{L^2} = \|\sum_{k\in\mathbb{N}} b_k \frac{1}{2\sigma}(k\varphi_{k-1} + (2k+1)\varphi_k + (k+1)\varphi_{k+1})\|^2_{L^2}$$
$$\le c\sigma^{-2}\Sigma|(1+k)b_k|^2 = c\sigma^{-2}\|b_k\|^2_{\ell^2_1},$$

$$\|\partial_x u\|^2_{L^2} = \sigma^2\|\sum_{k\in\mathbb{N}} b_k(-\varphi_k + 2\sum_{0\le j<k}(-1)^{k-1-j}\varphi_j)\|^2_{L^2}$$
$$= \sigma^2\|\sum_{k\in\mathbb{N}}(-b_k + 2b_{k+1} - 2b_{k+2} + \dots)\varphi_k\|^2_{L^2}$$
$$\le 4\sigma^2 \sum_{k\in\mathbb{N}}(\sum_{j\ge k}|b_j|)^2$$
$$\le c'\sigma^2 \sum_{k\in\mathbb{N}}(\sum_{j\ge k}(1+j)^{2N}|b_j|^2)(\sum_{\ell\ge k}(1+\ell)^{-2N})$$
$$\le c''\sigma^2 \sum_{j\in\mathbb{N}}(1+j)^{2N}|b_j|^2 \sum_{k\in\mathbb{N}}(1+k)^{-2N+1}$$
$$\le c'''\sigma^2 \sum_{j\in\mathbb{N}}(1+j)^{2N}|b_j|^2, \quad \text{when } N > 1,$$

it suffices to take $N = 1+\varepsilon$, any $\varepsilon > 0$. We also have

$$\|x\partial_x u\|^2_{L^2} = \|\sum_{k\in\mathbb{N}} b_k(\frac{k}{2}\varphi_{k-1} - \frac{1}{2}\varphi_k - \frac{k+1}{2}\varphi_{k+1})\|^2_{L^2} \le c'\|(b_k)_{k\in\mathbb{N}}\|^2_{\ell^2_1}$$

and finally

$$\|\partial_x x \partial_x u\|^2_{L^2} = \|(-\sigma L_{\sigma,+}u + \sigma^2 xu - \sigma u)\|^2_{L^2} \leq c\sigma^2 \|(b_k)_{k\in\mathbb{N}}\|^2_{\ell^2_1}$$

by the preceding cases. The higher order estimates follow from the fact that in the above proofs we can replace ℓ^2_0 and ℓ^2_1 (resp. $\ell^2_{1+\varepsilon}$) by ℓ^2_N and ℓ^2_{N+1} (resp. $\ell^2_{N+1+\varepsilon}$) . □

By the help of (2.2.20) one can estimate any expression $\|x^k \partial^\ell_x u\|_{L^2}$ in terms of $\|(b_k)\|_{\ell^2_m}$, where the lowest possible value of m is determined by different formulas for $k \geq \ell$, $k \geq \ell/2$ and $k < \ell/2$.

We now return to the decomposition (2.2.8), which has the counterpart for $f(t)$, in view of (2.2.10) ,

$$f(t) = f^+(t) + f^-(t) \,, \quad \text{where}$$

(2.2.25)

$$f^+(t) = \sum_{k\geq 0} b_k \hat{\varphi}_k(t,\sigma) \quad \text{and} \quad f^-(t) = \sum_{k<0} b_k \hat{\varphi}_k(t,\sigma) \quad ;$$

the sequences $(b_k)_{k\geq 0}$ and $(b_k)_{k<0}$ can be any rapidly decreasing sequences. We denote the corresponding decomposition of the whole space H_{-1} by

(2.2.26) $$H_{-1} = H^+_{-1} \dotplus H^-_{-1} \,,$$

with projections denoted h^+_{-1} resp. h^-_{-1} ; note that they are orthogonal projections with respect to $L^2(\mathbb{R})$-norm (since the $\hat{\varphi}_k$ are mutually orthogonal), so that

(2.2.27) $$\|f^+\|_{L^2} \leq \|f\|_{L^2} \quad ; \quad \|f^-\|_{L^2} \leq \|f\|_{L^2} \quad \text{for } f \in H^-_{-1} \quad .$$

Observe that

(2.2.28) $$\overline{\hat{\varphi}_k} = \hat{\varphi}_{-k-1} \quad \text{for all } k \,,$$

which shows that

(2.2.29) $$f \in H^+_{-1} \Leftrightarrow \overline{f} \in H^-_{-1} \quad .$$

By Lemma 2.2.1 we now see that H^+_{-1} is precisely the space of Fourier transform of functions $e^+u(x)$, where $u \in \mathscr{S}(\overline{\mathbb{R}}_+)$,

(2.2.30) $$H_{-1}^{+} = F(e^{+}\mathcal{S}(\overline{\mathbb{R}}_{+})) \ .$$

Similarly (cf. (2.2.29))

(2.2.31) $$H_{-1}^{-} = F(e^{-}\mathcal{S}(\overline{\mathbb{R}}_{-})) \ .$$

The above analysis is concerned with H_{-1} ; for the complete description of H one has to adjoin $\mathbb{C}[t]$ (cf. (2.2.7)), and it is customary to define (with a slight asymmetry)

(2.2.32) $$H^{+} = H_{-1}^{+} \ , \qquad H^{-} = H_{-1}^{-} \dotplus \mathbb{C}[t] \ .$$

Note that $\mathbb{C}[t]$ is the space of Fourier transforms of the "polynomials" $\Sigma c_k \delta^{(k)}$ where $\delta^{(k)} = D_x^k \delta$; we call the latter space $\mathbb{C}[\delta']$. Then the analysis can be summed up in the following statement.

2.2.2 Proposition.

1^0 *The space* $H = \cup_{d\in\mathbb{Z}} H_d$ *admits a decomposition in a direct sum*

(2.2.33) $$H = H^{+} \dotplus H^{-} \ ,$$

with projections denoted h^{+} *and* h^{-} ; *moreover (cf. also* (2.2.26)ff.*)*

$$H^{-} = H_{-1}^{-} \dotplus \mathbb{C}[t] \ , \qquad H^{+} = H_{-1}^{+} \ .$$

The decompositions are defined in such a way that the space H , *by inverse Fourier transformation, is mapped onto the space*

(2.2.34) $$\dot{\mathcal{S}}(\mathbb{R}) \equiv e^{+}\mathcal{S}(\overline{\mathbb{R}}_{+}) \dotplus e^{-}\mathcal{S}(\overline{\mathbb{R}}_{-}) \dotplus \mathbb{C}[\delta'] \ ,$$
$$\text{where } F^{-1}H^{+} = e^{+}\mathcal{S}(\overline{\mathbb{R}}_{+}) \text{ and } F^{-1}H^{-} = e^{-}\mathcal{S}(\overline{\mathbb{R}}_{-}) \dotplus \mathbb{C}[\delta'] \ ,$$

the projectors h^{+} *and* h^{-} *carrying over to the projectors* $e^{+}r^{+}$ *and* $I-e^{+}r^{+}$; *here* $(I-e^{+}r^{+})v = e^{-}r^{-}v$ *when* v *is a* *function* *in* $\dot{\mathcal{S}}(\mathbb{R})$.

2^0 *The spaces* $\mathcal{S}(\overline{\mathbb{R}}_{+})$ *and* H^{+} *are described as the spaces of functions*

(2.2.35) $$u(x) = \sum_{k\in\mathbb{N}} b_k \varphi_k(x,\sigma) \quad \textit{resp.} \quad f(t) = \sum_{k\in\mathbb{N}} b_k \hat{\varphi}_k(t,\sigma) \ ,$$

expanded in the orthonormal Laguerre system $(\varphi_k)_{k\in\mathbb{N}}$ *on* $\mathbb{R}_{+}$ *(cf.* (2.2.11)*)* , *resp. the Fourier transformed Laguerre system* $(\hat{\varphi}_k)_{k\in\mathbb{N}}$ *in* $L^2(\mathbb{R})$ *(cf.* (2.2.10)*)*, *with rapidly decreasing coefficient series* $(b_k)_{k\in\mathbb{N}}$. *There are similar statements for* $\mathcal{S}(\overline{\mathbb{R}_-})$ *and* H^-_{-1} *using* $(\varphi_k)_{k<0}$ *on* $\mathbb{R}_-$ *resp.* $(\hat\varphi_k)_{k<0}$ *in* $L^2(\mathbb{R})$ *(the latter is the same as* $(\overline{\hat\varphi}_\ell)_{\ell\geq 0}$ *).*

2.2.3 Remark. The decomposition $f = h^+f + h^-f$ can also be described by writing f as a sum of two terms h^+f and h^-f that have C^∞ extensions to $\overline{\mathbb{C}_-}$ resp. $\overline{\mathbb{C}_+}$, that are holomorphic in $\mathbb{C}_-$ resp. $\mathbb{C}_+$, cf. (A.4); with h^+f satisfying estimates (2.2.1) with $d^+ = -1$ in $\overline{\mathbb{C}_-}$ and h^-f satisfying estimates (2.2.1) with $d^- = \max\{d,-1\}$ in $\overline{\mathbb{C}_+}$. One can show this by using that in the coordinate change (2.2.2), $t \in \mathbb{C}_-$ corresponds to $|z| < 1$ and $t \in \mathbb{C}_+$ corresponds to $|z| > 1$; and g^+ resp. g^- have holomorphic extensions in $|z| < 1$ resp. $|z| > 1$. Also the Paley-Wiener theorem (linking the analyticity of h^+f in $\mathbb{C}_-$ with the fact that $\operatorname{supp} F^{-1}f \subset \overline{\mathbb{R}}_+$) could be used. h^+f and h^-f can then be defined from f by Cauchy integrals; this point of view has a prominent rôle in [Boutet 3]. However, the analytic extensions seem less convenient for estimates where one cannot play on homogeneity.

The (Fréchet) topology on the space H^+ is defined by either of the following three systems of norms (where $f = Fe^+u$, with expansions (2.2.35)):

$$\|(b_k)_{k\in\mathbb{N}}\|_{\ell^2_N}, \quad N \in \mathbb{N} \qquad \text{(cf. (2.2.9))},$$

$$\text{(2.2.36)} \qquad \|x^j D_x^m u(x)\|_{L^2(\mathbb{R}_+)}$$

$$= (2\pi)^{-\frac12}\|h^+(D_t^j[t^m f(t)])\|_{L^2(\mathbb{R})}, \quad j,m \in \mathbb{N}.$$

The equivalence of the first and second norm systems was shown in Lemma 2.2.1, and for the second and third norm we simply use the Plancherel-Parseval theorem. The application of h^+ here just removes polynomial terms; it corresponds to removing those singularities at zero that arize from differentiating e^+u.

The mapping that assigns the coefficients s_j in (2.2.1) to a function $f \in H$ will now be investigated. As mentioned before, they are linked with the Taylor coefficients of $k(\tau)$ at $\tau = 0$, cf. (2.2.3). When we consider $F^{-1}f \in \dot{\mathcal{S}}(\mathbb{R})$, we see that the coefficients s_j with $j \geq 0$ appear here as coefficients in a "polynomial" $\Sigma c_k \delta^{(k)}$. For the coefficients with $j < 0$, we begin with an analysis of the case where $f \in H^+$.

Let $f \in H^+$, having the asymptotic expansion

(2.2.37) $$f(t) \sim s_{-1}t^{-1} + s_{-2}t^{-2} + \ldots \quad ,$$

and let $u \in \mathscr{S}(\overline{\mathbb{R}}_+)$ be such that $Fe^+u = f$. Green's formula (recall $\gamma_j u = D_x^j u(0)$),

(2.2.38) $$(D_x^N u,v)_{\mathbb{R}_+} - (u, D_x^N v)_{\mathbb{R}_+} = i \sum_{0\leq k\leq N-1} \gamma_{N-k-1}u\cdot\gamma_k v \quad \text{for } u,v \in \mathscr{S}(\overline{\mathbb{R}}_+) ,$$

can be written on distribution form when we insert $v = r^+\varphi$ for some $\varphi \in \mathscr{S}(\mathbb{R})$ and observe that

$$(D_x^N u, v)_{\mathbb{R}_+} = (e^+D_x^N u, \varphi)_{\mathbb{R}} = \langle e^+D_x^N u, \overline{\varphi}\rangle_{\mathbb{R}}$$

$$(u, D_x^N v)_{\mathbb{R}_+} = \langle e^+u, \overline{D_x^N\varphi}\rangle_{\mathbb{R}} = \langle D_x^N e^+u, \overline{\varphi}\rangle_{\mathbb{R}} ,$$

$$\gamma_j u\cdot\overline{\gamma_k v} = \langle(\gamma_j u)\delta, \overline{D_x^k\varphi}\rangle_{\mathbb{R}} = \langle(\gamma_j u)D_x^k\delta, \overline{\varphi}\rangle_{\mathbb{R}} ,$$

whereby (2.2.38) becomes

(2.2.39) $$D_x^N e^+u = -i \sum_{0\leq k\leq N-1} (\gamma_{N-k-1}u)D_x^k\delta + e^+D_x^N u , \quad \text{for any } N \in \mathbb{N} .$$

This gives by Fourier transformation

(2.2.40) $$t^N f(t) = -i \sum_{0\leq k<N} (\gamma_{N-k-1}u)t^k + g_N(t) \quad ,$$

where $g_N(t) \in H^+ \subset H_{-1}$. By the uniqueness of the coefficients in (2.2.37) one may conclude

(2.2.41) $$\gamma_j u = i\, s_{-1-j} \qquad \text{for all } j \geq 0 .$$

Another interpretation of the coefficients can be given by use of the "plus-integral" $\int^+ f(t)dt$, which is defined on H by linear extension of the two cases: $\int^+ f(t)dt = \int_{\mathbb{R}} f(t)dt$ when $f \in L^1(\mathbb{R})$ $(f \in H_{-2})$; and $\int^+ f(t)dt = \int_C f(t)dt$ when f is meromorphic in $\mathbb{C}_+$, C denoting a contour around the poles in $\mathbb{C}_+$. In particular,

(2.2.42) $$\int^+ f(t)dt = 0 \quad \text{if} \quad f \in H^- \quad \text{or} \quad f \in H^+ \cap H_{-2}$$

(in the latter case, the integration can be carried over to a contour in $\mathbb{C}_-$) . Then when $f \in H^+$ and has the expansion (2.2.37), we can write, for any $k \in \mathbb{N}$,

(2.2.43) $$t^k f(t) = \sum_{-k\le j\le -1} s_j t^{j+k} + s_{-1-k}\frac{1}{t-i} + g_k(t) \quad \text{with} \quad g_k(t) \in H^+ \cap H_{-2} ,$$

and conclude that

(2.2.44) $$\frac{1}{2\pi i}\int^+ t^k f(t)dt = s_{-1-k} = -i\gamma_k u \quad .$$

Here s_{-1-k} can be estimated by use of the standard trace estimates

(2.2.45) $$|s_{-1-k}|^2 = |\gamma_k u|^2 = -\int_0^\infty \partial_x[u^{(k)}\overline{u}^{(k)}]dx$$

$$\le 2\|D_x^k u\|_{L^2(\mathbb{R}_+)} \|D_x^{k+1}u\|_{L^2(\mathbb{R}_+)}$$

$$\le \frac{1}{\pi}\|h^+(t^k f)\|_{L^2(\mathbb{R})} \|h^+(t^{k+1}f)\|_{L^2(\mathbb{R})} , \quad \text{when} \quad f \in H^+ .$$

Consider now the more general case where $f \in H_{-1}$. Let $f^+ = h^+ f$ and $f^- = h^- f$; the functions all have expansions

$$f(t) \sim \sum_{k\in\mathbb{N}} s_{-1-k}t^{-1-k} , \quad f^\pm(t) \sim \sum_{k\in\mathbb{N}} s^\pm_{-1-k}t^{-1-k} ,$$

and by the uniqueness of the coefficients, one must have

$$s_{-1-k} = s^+_{-1-k} + s^-_{-1-k} \quad .$$

Let $F^{-1}f = v$, then $v \in e^+\mathscr{S}(\overline{\mathbb{R}}_+) \dotplus e^-\mathscr{S}(\overline{\mathbb{R}}_-)$ and by (2.2.41),

(2.2.46) $$\begin{aligned} s^+_{-1-k} &= -i\ \gamma^+_k v \equiv -i \lim_{x\to 0+} D^k_x v(x) \\ s^-_{-1-k} &= \ i\ \gamma^-_k v \equiv \ i \lim_{x\to 0-} D^k_x v(x) , \end{aligned}$$

The latter formula follows by application of the first formula to $\overline{f}(t)$. It follows in particular that

(2.2.47) $$s_{-1-k} = -i(\gamma^+_k v - \gamma^-_k v)$$

so s_{-1-k} is determined from the jump of $D^k_x v$ at $x = 0$. Moreover, we can estimate s_{-1-k} by use of (2.2.45):

(2.2.48) $$|s_{-1-k}|^2 \le 2(|s^+_{-1-k}|^2 + |s^-_{-1-k}|^2)$$

$$\le \frac{2}{\pi}(\|h^+(t^k f)\| \|h^+(t^{k+1}f)\| + \|h^+(t^k\overline{f})\| \|h^+(t^{k+1}\overline{f})\|)$$

$$\le \frac{4}{\pi}\|h_{-1}(t^k f)\| \|h_{-1}(t^{k+1}f)\| \qquad \text{for} \quad k \in \mathbb{N} ,$$

since $\|h^+g\| \leq \|h_{-1}g\| = \|h_{-1}\bar{g}\|$ in general. The estimate (2.2.48) for the coefficients with index < 0 extends to all $f \in H$, since the polynomial part is removed by the projections h^+ and h_{-1}.

By a combination of the asymptotic expansion (2.2.1) and the Laguerre series expansion of functions in H_{-1}, one has the general expansions with uniquely determined coefficients

(2.2.49) $$f(t) = \sum_{0\leq j\leq d} s_j t^j + \sum_{k\in\mathbb{Z}} b_k \hat{\varphi}_k(t,\sigma) \quad ,$$

where the last sum equals $h_{-1}f$, and

$$h^+f(t) = \sum_{k\geq 0} b_k \hat{\varphi}_k(t,\sigma)$$

(2.2.50) $$h^-f(t) = \sum_{0\leq j\leq d} s_j t^j + \sum_{k<0} b_k \hat{\varphi}_k(t,\sigma)$$

$$= \sum_{0\leq j\leq d} s_j t^j + \sum_{\ell>0} b_\ell \bar{\hat{\varphi}}_\ell(t,\sigma) \quad .$$

Note that

(2.2.51) $$\sum_{k\in\mathbb{Z}} |b_k|^2 = (2\pi)^{-1} \|h_{-1}f\|^2_{L^2(\mathbb{R})} \quad .$$

The two sums in the expression (2.2.49) for f are really quite different in nature; and a multiplication of $f(t)$ by t^k gives a shift in the decomposition that is not particularly simple; see the formula for $t\hat{\varphi}_k$ in (2.2.23). However, there is one other decomposition that we do need on some occasions. Define the functions, for $k \in \mathbb{Z}$,

(2.2.52) $$\hat{\psi}_k(t,\sigma) = \frac{(\sigma-it)^k}{(\sigma+it)^k} \quad [= (2\sigma)^{-\frac{1}{2}}(\sigma+it)\hat{\varphi}_k(t,\sigma)] \ ,$$

and note that

(2.2.53) $$\hat{\varphi}_k(t,\sigma) = (2\sigma)^{\frac{1}{2}} \frac{(\sigma-it)^k}{(\sigma+it)^{k+1}} \quad \frac{(\sigma-it+\sigma+it)}{2\sigma}$$

$$= (2\sigma)^{-\frac{1}{2}} (\hat{\psi}_{k+1}(t,\sigma) + \hat{\psi}_k(t,\sigma)) \ .$$

Inserting this in (2.2.49), we find the expansion

(2.2.54) $$f(t) = \sum_{1\leq j\leq d} s_j t^j + \sum_{k\in\mathbb{Z}} a_k \hat{\psi}_k(t,\sigma) \quad ,$$

where the s_j for $j \geq 1$ are the same as in (2.2.49), and the other coefficients are determined by the formulas

$$a_k = (2\sigma)^{-\frac{1}{2}}(b_k + b_{k-1}) \quad ,$$
$$a_0 = (2\sigma)^{-\frac{1}{2}}(b_0 + b_{-1}) + s_0 \ . \tag{2.2.55}$$

The system $\hat{\psi}_k$ is a complete orthogonal system in the weighted L^2 space over $\mathbb{R}$ with weight $(\sigma^2+t^2)^{-1}$. Their inverse Fourier transforms are the distributions (cf. (2.2.23))

$$\psi_k(x,\sigma) = (2\sigma)^{\frac{1}{2}} \sum_{0\leq j\leq k-1} (-1)^{k-1-j}\varphi_j + (-1)^k\delta \qquad \text{for } k \geq 0 \ ,$$
$$\psi_k(x,\sigma) = \psi_{-k}(-x,\sigma) \qquad \text{for } k \leq 0 \ . \tag{2.2.56}$$

We now turn to the general analysis of the pseudo-differential symbols introduced in Section 2.1, using the above calculus with respect to the variable x_n. In the following, Ω' denotes a subset of $\mathbb{R}^{n-1}$ and we set

$$\Sigma = \Omega' \times \mathbb{R} \ , \qquad \Omega = \Omega' \times \mathbb{R}_+ \ . \tag{2.2.57}$$

(One can replace Ω' by an open set $\Xi \subset \mathbb{R}^{n'}$ for some other n'.)

<u>2.2.4 Definition.</u> *Let* $d \in \mathbb{Z}$ *and* $\nu \in \mathbb{R}$. *A symbol* $p(x,y,\xi,\mu) \in S^{d,\nu}(\Sigma\times\Sigma,\overline{\mathbb{R}}^{n+1}_+)$ *is said to have the transmission property at* $x_n = a$, *when the homogeneous terms in* p *have the following symmetry property*

$$D^\beta_{x,y} D^\alpha_\xi D^j_\mu p_{d-\ell}(x,y,(0,\xi_n),0) = (-1)^{d-\ell-|\alpha|-j} D^\beta_{x,y} D^\alpha_\xi D^j_\mu p_{d-\ell}(x,y,(0,-\xi_n),0) \tag{2.2.58}$$

for $x_n = y_n = a$ *and* $|\xi_n| \geq 1$, *all* $\alpha \in \mathbb{N}^n$, $\beta \in \mathbb{N}^{2n}$, j *and* $\ell \in \mathbb{N}$. *For* $a = 0$, p *is simply said to have the transmission property, and the space of such symbols is denoted* $S^{d,\nu}_{tr}(\Sigma\times\Sigma,\overline{\mathbb{R}}^{n+1}_+)$. *For symbols independent of* y *(or* x*), the space is denoted* $S^{d,\nu}_{tr}(\Sigma,\overline{\mathbb{R}}^{n+1}_+)$.

The operator P_μ defined for each μ from the symbol $p(x,y,\xi,\mu)$ on

$\Omega = \Omega' \times \mathbb{R}$ by formula (2.1.29) gives by restriction to $\Omega = \Omega' \times \mathbb{R}_+$ the operator

$$P_{\mu,\Omega} = r_\Omega P_\mu e_\Omega \qquad (= r^+ P_\mu e^+)$$

(cf. (A.30,31,64)). The transmission property at $x_n = 0$ assures that $P_{\mu,\Omega}$ sends $C^\infty_{(0)}(\Omega' \times \overline{\mathbb{R}}_+)$ into $C^\infty(\Omega' \times \overline{\mathbb{R}}_+)$, and it implies certain continuity properties in relation to the parametrized Sobolev spaces $H^{s,\mu}(\Omega' \times \overline{\mathbb{R}}_+)$ and $H^{(s,t),\mu}(\Omega' \times \overline{\mathbb{R}}_+)$. This will be investigated in Section 2.5.

The following analysis of the symbols with the transmission property is crucial for all that follows. (Cf. (A.1)-(A.2) for notation.)

2.2.5 Theorem. *Let* $p \in S^{d,\nu}_{tr}(\Sigma \times \Sigma, \overline{\mathbb{R}}^{n+1}_+)$. *For all integers* $k \le d$ *there exist (uniquely determined) functions* $s_k(x',y',\xi',\mu) \in C^\infty(\Omega' \times \Omega' \times \overline{\mathbb{R}}^n_+)$, *polynomial in* (ξ',μ) *of degree* $d-k$, *such that* $p \sim \sum_{k \le d} s_k \xi_n^k$ *at* $x_n = 0$ *in the sense that for any* $m \in \mathbb{N}$,

$$\begin{aligned} &|\xi_n^m p(x',0,y',0,\xi',\xi_n,\mu) - \sum_{-m \le k \le d} s_k(x',y',\xi',\mu)\xi_n^{k+m}| \\ &\le c(x',y')(\rho(\xi',\mu)^{\nu+m} + 1)\langle \xi',\mu\rangle^{d+1+m}\langle \xi,\mu\rangle^{-1} . \end{aligned} \tag{2.2.59}$$

There are analogous expansions of the derived functions $D^\beta_{x,y} D^\alpha_\xi D^j_\mu [p - \Sigma_{\ell<M} p_{d-\ell}] \in S^{d-|\alpha|-j-M,\nu-|\alpha|-M}$, *in such a way that the coefficients* s_k *in the derived expressions fit together with the original* s_k *under termwise differentiation, e.g.*

$$\begin{aligned} &|D^\beta_{x',y'} D^\alpha_\xi D^j_\mu [\xi_n^m p_{d-\ell} - \sum_{-m \le k \le d-\ell-\alpha_n-j} s_{d-\ell,k}\, \xi_n^{k+m}]| \\ &\le c(x',y')(\rho(\xi',\mu)^{\nu+m-\ell-|\alpha|} + 1)\langle \xi',\mu\rangle^{d+1+m-\ell-|\alpha|-j}\langle \xi,\mu\rangle^{-1} , \end{aligned} \tag{2.2.60}$$

with similar estimates for the functions $p - \Sigma_{\ell<M} p_{d-\ell}$ *and the* (x_n,y_n)*-derivatives.*

Conversely, if $p \in S^{d,\nu}$ *and has expansions in this way, then it has the transmission property.*

Proof: The last statement is easy to prove, for when p satisfies (2.2.59), then

$$p(x',0,y',0,0,\xi_n,0) - s_d(x',y',0,0)\xi_n^d \text{ is } O(|\xi_n|^{d-1}) \text{ for } \xi_n \to \pm\infty$$

(as is seen by taking $m = 0$ if $d \ge 0$ and $m = -d$ if $d < 0$) ; since $p - p_d$ is also $O(|\xi_n|^{d-1})$ at $\xi' = \mu = 0$, this gives (2.2.58) for α,β, j and ℓ

equal to zero. Similarly, (2.2.60) ff. imply that $D^{\beta}_{x,y}D^{\alpha}_{\xi}D^{j}_{\mu}p_{d-\ell}$ has an asymptotic expansion in ξ_n at $x_n = y_n = 0$ beginning with a term $s(x',y')\xi_n^{d-\ell-|\alpha|-j}$ (the rest being of lower order in ξ_n), so that (2.2.58) holds in general.

Now consider the main statement. Let $p \in S^{d,\nu}$ have the transmission property. The first and most important step in the proof is to show (2.2.59) for p_d in the (x',y')-independent case (the estimates in the (x',y')-dependent case are easily seen to be locally uniform), so let us take $p = p_d = p_d(\xi,\mu)$ satisfying Definition 2.2.4.

Since multiplication by ξ_n^m maps $S^{d,\nu}$ into $S^{d+m,\nu+m}$ (cf. Lemma 2.1.6), and obviously preserves the transmission property, it suffices to consider the case $m = 0$. Observe that when $d \leq -1$, the sum over k vanishes, and

(2.2.61)
$$\begin{aligned} |p(x,y,\xi,\mu)| &\leq c(x,y)(\rho(\xi,\mu)^{\nu} + 1)\langle\xi,\mu\rangle^{d} \\ &\leq c'(x,y)(\rho(\xi',\mu)^{\nu} + 1)\langle\xi',\mu\rangle^{d+1}\langle\xi,\mu\rangle^{-1} , \end{aligned}$$

since $\langle\xi,\mu\rangle^{-1} \leq \langle\xi',\mu\rangle^{-1}$ and

(2.2.62)
$$\rho(\xi',\mu) \leq \rho(\xi,\mu) \leq 1 .$$

It remains to consider the case $d \geq 0$. Denote $(\xi',\mu) = \eta$, with $\xi_i = \eta_i$ for $i < n$ and $\eta_n = \mu$.

Let us write, with a slight abuse of notation,

$$p(\eta,\xi_n) = p(\xi',\xi_n,\mu) ,$$

it satisfies

(2.2.63)
$$p(s\eta,s\xi_n) = s^d p(\eta,\xi_n) \quad \text{for } s \geq 1 , \ |\eta',\xi_n| \geq 1 .$$

Introduce

(2.2.64)
$$k(\eta,\tau) = \tau^d p(\eta,\tau^{-1}) \quad \text{for } \tau \neq 0 ;$$

here in view of (2.2.63),

$$k(\eta,\tau) = (\pm 1)^d \, p(\pm\tau\eta,\pm 1) \quad \text{for } \pm\tau \in \,]0,1] .$$

Now (2.2.58) implies that for all η,

$$\lim_{\tau\to 0+} k(\eta,\tau) = \lim_{\tau\to 0-} k(\eta,\tau) = p(0,1)$$

and since this holds locally uniformly in η, k extends to a continuous function of $\eta \in \overline{\mathbb{R}}^n_+$, $\tau \in \mathbb{R}$. Moreover, we have

$$\partial_\tau k(\eta,\tau) = (\pm 1)^{d-1} \sum_{i=1}^{n} \eta_i \partial_{\eta_i} p\,(\pm\tau\eta, \pm 1) \quad \text{for} \quad \pm\tau \in \,]0,1]$$

respectively, so we find successively

$$\partial_\tau^j k(\eta,\tau) = (\pm 1)^{d-j} \sum_{|\alpha|=j} \frac{j!}{\alpha!} \eta^\alpha \partial_\eta^\alpha p\,(\pm\tau\eta, \pm 1) \quad \text{for} \quad \pm\tau \in \,]0,1]$$

for any $j \in \mathbb{N}$. Applications of (2.2.58) show that all τ-derivatives of k are continuous in $(\eta,\tau) \in \overline{\mathbb{R}}^n_+ \times \mathbb{R}$, with

$$\partial_\tau^j(\eta,0) = \sum_{|\alpha|=j} \frac{j!}{\alpha!} \eta^\alpha \partial_\eta^\alpha p(0,1) \,. \tag{2.2.67}$$

Since by Taylor's formula

$$k(\eta,\tau) = \sum_{0\le j\le d} \frac{\tau^j}{j!} \partial_\tau^j k(\eta,0) + \mathcal{O}(\tau^{d+1})$$

(locally uniformly in η), it follows (cf. (2.2.64)) that

$$p(\eta,\xi_n) = \sum_{0\le j\le d} \frac{1}{j!} \partial_\tau^j k(\eta,0) \xi_n^{d-j} + \mathcal{O}(\langle\xi_n\rangle^{-1}) \quad , \tag{2.2.68}$$

locally uniformly in η. We now set (cf. (2.2.67))

$$s_k(\eta) = \frac{1}{(d-k)!} \partial_\tau^{d-k} k(\eta,0) = \sum_{|\alpha|=d-k} \frac{\eta^\alpha}{\alpha!} \partial_\eta^\alpha p(0,1) \,, \tag{2.2.69}$$

so we have that

$$p(\eta,\xi_n) = \sum_{0\le k\le d} s_k(\eta) \xi_n^k + \mathcal{O}(\langle\xi_n\rangle^{-1}) \tag{2.2.70}$$

locally uniformly in η. - This shows a "pointwise" version of (2.2.59), where the uniqueness of the coefficients s_k is already apparent; their consistency with expansion coefficients for the derived functions follows from this uniqueness.

It remains to estimate the remainder in (2.2.70) more precisely. We denote it

$$p'(\eta,\xi_n) = p(\eta,\xi_n) - \sum_{0\le k\le d} s_k(\eta)\xi_n^k \quad , \tag{2.2.71}$$

it is clearly homogeneous in (η,ξ_n) of degree d for $|\eta',\xi_n| \ge 1$ (recall that s_k is polynomial), and it belongs to $S^{d,\nu}$. Define also

$$k'(\eta,\tau) = \tau^d p'(\eta,\tau^{-1}) \,. \tag{2.2.72}$$

When we consider this function for $|\tau| \le 1$, we are in the range where $|\xi_n| = |\tau^{-1}| \ge 1$, and we find just as in the treatment of $k(\eta,\tau)$ above that k' is C^∞ in (η,τ) also at $\tau = 0$. For this function, the first d terms in the Taylor expansion in τ are zero, so

$$(2.2.73) \qquad k'(\eta,\tau) \quad \text{is} \quad O(\tau^{d+1}) \quad \text{for } \tau \to 0 ,$$

locally uniformly in η. We now estimate p' as follows:
The region $|\xi_n| \ge \langle\eta\rangle$. Write $\xi_n = \langle\eta\rangle\tau^{-1}$, where $|\tau| \le 1$. Then since we are in the range where p' is truly homogeneous (of degree d),

$$p'(\eta,\xi_n) = \langle\eta\rangle^d p'(\eta/\langle\eta\rangle, \tau^{-1}) = \langle\eta\rangle^d \tau^{-d} k'(\eta/\langle\eta\rangle, \tau) .$$

Here $\eta/\langle\eta\rangle$ runs in a compact set, so (2.2.73) allows the conclusion

$$(2.2.74) \quad |p'(\eta,\xi_n)| \le c\langle\eta\rangle^d|\tau| = c\langle\eta\rangle^{d+1}|\xi_n^{-1}| \le c'\langle\eta\rangle^{d+1}\langle\xi\rangle^{-1}$$

(since $|\xi_n| \ge \frac{1}{2}(|\xi_n| + \langle\eta\rangle) \sim \frac{1}{2}\langle\xi\rangle$ when $|\xi_n| \ge \langle\eta\rangle$).
The region $|\xi_n| \le \langle\eta\rangle$. Here we use the estimates valid for the class $S^{d,\nu}$ (recall that $\eta = (\xi',\mu)$). If $\nu \ge 0$,

$$(2.2.75) \quad \begin{aligned} |p'(\eta,\xi_n)| &\le c\langle\eta,\xi_n\rangle^d \le c\langle\eta,\xi_n\rangle^{-1} \sup_{|\xi_n|\le\langle\eta\rangle} \langle\eta,\xi_n\rangle^{d+1} \\ &\le c'\langle\eta\rangle^{d+1}\langle\eta,\xi_n\rangle^{-1} . \end{aligned}$$

If $\nu \le 0$,

$$(2.2.76) \quad |p'(\eta,\xi_n)| \le c\langle\xi',\xi_n\rangle^{\nu}\langle\eta,\xi_n\rangle^{d-\nu} \le c'\langle\xi'\rangle^{\nu}\langle\eta\rangle^{d-\nu+1}\langle\eta,\xi_n\rangle^{-1} .$$

Collecting (2.2.74), (2.2.75) and (2.2.76), we have altogether

$$|p'(\eta,\xi_n)| \le c(\langle\xi'\rangle^{\nu} + \langle\xi',\eta\rangle^{\nu})\langle\xi',\mu\rangle^{d-\nu+1}\langle\xi,\mu\rangle^{-1} ,$$

completing the proof of the estimate (2.2.59) in this particular case.

The corresponding estimates of derivatives are shown by applying the above arguments directly to the resulting homogeneous symbols. Finally, the estimates of the symbol associated with a full asymptotic series is obtained by applying the above considerations to symbols of order ≥ 0 and using that the remainders satisfy the estimates in a trivial way, as in (2.2.61). □

Observe that the estimates (2.2.59)-(2.2.60) for fixed (x',y',ξ',μ) express precisely that p is in H_d as a function of ξ_n, cf. (2.2.1). In fact, the function estimated in (2.2.59) is the h_{-1}-projection of $\xi_n^m p$, (cf. (2.2.7)),

$$\xi_n^m p - \sum_{-m\leq k\leq d} s_k \xi_n^{k+m} = h_{-1}[\xi_n^m p] \ , \tag{2.2.77}$$

(also denoted $h_{-1,\xi_n}[\xi_n^m p]$) . The precise estimates on the dependence of the parameters play an important role.

2.2.6 Example. Let $p(\xi_n,\mu) = (1+i\xi_n+\mu)^{-1}$, then clearly $p \in S_{tr}^{-1,\infty}(\mathbb{R}, \overline{\mathbb{R}}_+^2)$. Observe here that even though $\xi_n p(\xi_n,\mu)$ is $O(\langle\mu\rangle^{-1})$, its h_{-1}-projection is not, for

$$h_{-1}[\xi_n p(\xi_n,\mu)] = -ih_{-1}[1 - \frac{1+\mu}{1+i\xi_n+\mu}] = i\,\frac{1+\mu}{1+i\xi_n+\mu} \ ,$$

which is only $O(\langle\mu\rangle^0)$. (Note that moreover, $h_{-1}[\xi_n p] \in H^+$ and equals $h^+[\xi_n p]$.) This demonstrates one of the fundamental difficulties in the resolvent calculus for boundary problems, namely that estimates in the spectral parameter μ are generally not preserved under composition with differential operator symbols followed by application of projectors h_{-1} (or h^+) .

For the more general symbol spaces $S_{1,0}^{d,\nu}$, we can now define the transmission property consistently with Definition 2.2.4 by the requirement that expansions as in Theorem 2.2.5 should hold. The resulting symbol class is called $S_{1,0,tr}^{d,\nu}(\Sigma\times\Sigma, \overline{\mathbb{R}}_+^{n+1})$.

2.2.7 Definition. *A symbol* $p \in S_{1,0}^{d,\nu}(\Sigma\times\Sigma, \overline{\mathbb{R}}_+^{n+1})$ *is said to have the transmission property at* $x_n = a$, *when there exist* C^∞ *functions* $s_{k,\alpha,\beta,j}(x',y',\xi',\mu)$ *such that for any indices* α,β,j *and* m,

$$\begin{aligned} &|\xi_n^m D_{x,y}^\beta D_\xi^\alpha D_\mu^j p(x,y,\xi,\mu) - \sum_{-m\leq k\leq d-|\alpha|-j} s_{k,\alpha,\beta,j}(x',y',\xi',\mu)\xi_n^{k+m}| \\ &\leq c(x',y')(\rho(\xi',\mu)^{\nu+m-|\alpha|} + 1)\langle\xi',\mu\rangle^{d+1+m-|\alpha|-j}\langle\xi,\mu\rangle^{-1} \ , \end{aligned} \tag{2.2.78}$$

for $x_n = y_n = a$. *In other words,* p *and its derivatives are in* H *as functions of* ξ_n *(for* $x_n=y_n=a$*)*, *with the estimates* (2.2.78) *for the* h_{-1}*-projections (the expressions in* $|\ |$*) of* $\xi_n^m p$ *and the derived functions.*

One observes that since the $s_{k,\alpha,\beta,j}$ in the various expressions must fit together under termwise differentiation, they must be zero after a finite number of differentiations in ξ' and μ', hence are necessarily polynomial in (ξ',μ) (are symbols of differential operators).

Note that when $p \in S^{d,\nu}_{1,0,tr}(\Sigma\times\Sigma,\ \overline{\mathbb{R}}^{n+1}_{+})$, then for each fixed (x',y',ξ',μ) , the operator $p(x',y',\xi',\mu,D_n) = OP_n(p(x',0,y',0,\xi',\xi_n,\mu))$ on $\mathbb{R}$ has the kernel

$$(2.2.79)\qquad K_p(x',x_n,y',y_n,\xi',\mu) = \tilde{p}(x',0,y',0,\xi',z,\mu)\big|_{z=x_n-y_n}\ ;\ \text{where}\ \ \tilde{p} = F^{-1}_{\xi_n\to z}\, p\ ,$$

here $\tilde{p}$ lies in $\dot{\mathscr{S}}(\mathbb{R})$ as a distribution in z , for each $(x',y',\xi',\mu) \in \Omega'\times\Omega'\times\overline{\mathbb{R}}^{n}_{+}$.

We shall also study $L^2_{\xi_n}$-norms and Laguerre expansions associated with p . In the following, we use the functions

$$(2.2.80)\qquad \begin{aligned} &\rho(\xi',\mu) = \langle\xi'\rangle/\langle\xi',\mu\rangle\ ,\\ &\kappa(\xi',\mu) = |\xi',\mu|\ \ \text{for}\ \ |\xi',\mu| \geq 1\ \ \text{and is}\ C^\infty\ \text{and}\ > 0\ \text{on}\ \ \overline{\mathbb{R}}^{n}_{+}\ ; \end{aligned}$$

note that $\kappa(\xi',\mu) \sim \langle\xi',\mu\rangle$ but is more homogeneous. Observe also that

$$(2.2.81)\qquad \|\langle\xi,\mu\rangle^{-1}\|^2_{L^2_{\xi_n}} = \int_{\mathbb{R}} (\langle\xi',\mu\rangle^2 + \xi_n^2)^{-1}d\xi_n = \tfrac{\pi}{2}\langle\xi',\mu\rangle^{-1}\ .$$

<u>2.2.8 Theorem</u>. *Let* $p \in S^{d,\nu}_{tr}(\Sigma\times\Sigma,\ \overline{\mathbb{R}}^{n+1}_{+})$, *so that* $p(x',0,y',0,\xi,\mu)$ *and its terms are in* H_d *as functions of* ξ_n , *according to Theorem 2.2.5. One has for all indices* $\beta \in \mathbb{N}^{2n}$, $\alpha \in \mathbb{N}^n$, j,m *and* $M \in \mathbb{N}$,

$$(2.2.82)\qquad \begin{aligned} &\|h^+_{\xi_n} D^\beta_{x,y} D^\alpha_\xi \xi_n^m [p(x',0,y',0,\xi,\mu) - \sum_{\ell<M} p_{d-\ell}(x',0,y',0,\xi,\mu)]\|_{L^2_{\xi_n}}\\ &\leq \|h_{-1,\xi_n} D^\beta_{x,y} D^\alpha_\xi \xi_n^m [p - \sum_{\ell<M} p_{d-\ell}]\|_{L^2_{\xi_n}}\\ &\leq c(x',y')(\rho(\xi',\mu)^{\nu-|\alpha|+m-M} + 1)\langle\xi',\mu\rangle^{d+\frac12-|\alpha|-j+m-M}\ . \end{aligned}$$

Moreover, p *has the expansions, for* $x_n = y_n = 0$,

$$(2.2.83)\qquad p = \sum_{0\leq j\leq d} s_j\xi_n^j + \sum_{k\in\mathbb{Z}} b_k(x',y',\xi',\mu)\hat{\varphi}_k(\xi_n,\kappa(\xi',\mu))$$

$$(2.2.84)\qquad p = \sum_{1\leq j\leq d} s_j\xi_n^j + \sum_{k\in\mathbb{Z}} a_k(x',y',\xi',\mu)\hat{\psi}_k(\xi_n,\kappa(\xi',\mu))\ ,$$

here the s_j *are the polynomials determined in Theorem 2.2.5;* $(b_k)_{k\in\mathbb{Z}}$ *is a sequence of functions in* $S^{d+\frac12,\nu}(\Omega'\times\Omega',\ \overline{\mathbb{R}}^{n}_{+})$ *with the properties: when* b_k *is expanded in a series of homogeneous terms* $b_{k,d+\frac12-\ell}$, *then for all indices* $\alpha \in \mathbb{N}^{n-1}$, $\beta \in \mathbb{N}^{2n-2}$,j,M *and* $N \in \mathbb{N}$,

$$(2.2.85)\quad \|(D^{\beta}_{x',y'}D^{\alpha}_{\xi'}D^{j}_{\mu}[b_k - \sum_{\ell<M} b_{k,d+\frac12-\ell}])_{k\in\mathbb{Z}}\|_{\ell^2_N} \le c(x',y')(\rho^{\nu-|\alpha|-j-M-N} + 1)\kappa^{d+\frac12-|\alpha|-j-M} \;;$$

and $(a_k)_{k\in\mathbb{Z}}$ *is a sequence in* $S^{d,\nu}(\Omega'\times\Omega',\overline{\mathbb{R}}^n_+)$, *satisfying the analogous estimates with* $d+\frac12$ *replaced by* d. *Similar results hold for* $S^{d,\nu}_{1,0,tr}(\Sigma\times\Sigma,\overline{\mathbb{R}}^{n+1}_+)$.

Proof: The estimates (2.2.82) follow straightforwardly from the estimates in Theorem 2.2.5, by use of (2.2.27) (cf. also (2.2.77)). Now $h_{-1}p$ can be expanded in the orthonormal system $(\hat\varphi_k(\xi_n,\kappa))_{k\in\mathbb{Z}}$

$$(2.2.86)\quad \begin{aligned} h_{-1}p &= \sum_{k\in\mathbb{Z}} b_k(x',y',\xi',\mu)\hat\varphi_k(\xi_n,\kappa(\xi',\mu)) ,\\ b_k(x',y',\xi',\mu) &= \int_{\mathbb{R}} (h_{-1}p)(x',y',\xi,\mu)\overline{\hat\varphi_k}(\xi_n,\kappa(\xi',\mu))d\xi_n \end{aligned}$$

and here we can apply Lemma 2.2.1 (the easy estimate (2.2.15)) to the inverse Fourier transforms of $h^+p = \sum_{k>0} b_k\hat\varphi_k$ and $h^-_{-1}p = \overline{h^+\overline{p}} = \sum_{k<0} b_k\hat\varphi_k$ to derive from (2.2.82) that the estimates (2.2.85) are satisfied for all N and M and all β, with α and j equal to zero (for the (x',y')-derivatives and the consideration of lower order parts carry directly over to p). For the study of derivatives in ξ' and μ we need to calculate (cf. (2.2.10))

$$(2.2.87)\quad \begin{aligned} \partial_{\xi_j}\hat\varphi(\xi_n,\kappa) &= \frac{1}{2\kappa}(k\hat\varphi_{k-1} - \hat\varphi_k - (k+1)\hat\varphi_{k+1})\partial_{\xi_j}\kappa,\\ \partial_{\mu}\hat\varphi(\xi_n,\kappa) &= \frac{1}{2\kappa}(k\hat\varphi_{k-1} - \hat\varphi_k - (k+1)\hat\varphi_{k+1})\partial_{\mu}\kappa , \end{aligned}$$

where

$$\partial_{\xi_j}\kappa = \frac{\xi_j}{|\xi',\mu|} \quad \text{for } |\xi',\mu| \ge 1 , \quad j < n ,$$

$$\partial_{\mu}\kappa = \frac{\mu}{|\xi',\mu|} \quad \text{for } |\xi',\mu| \ge 1 .$$

In view of the already proved estimates we then have e.g.

$$\begin{aligned} (2\pi)^{\frac12}\|(\partial_{\xi_j}b_k)_{k\in\mathbb{Z}}\|_{\ell^2_0} &= \|\sum_{k\in\mathbb{Z}}(\partial_{\xi_j}b_k)\hat\varphi_k\|_{L^2}\\ &= \|\partial_{\xi_j}(\Sigma\, b_k\hat\varphi_k) - \Sigma b_k(\partial_{\xi_j}\hat\varphi_k)\|_{L^2}\\ &\le c_1(\rho^{\nu-1}+1)\kappa^{d-\frac12} + c_2\kappa^{-1}(\rho^{\nu-1}+1)\kappa^{d+\frac12}\\ &\le c_3(\rho^{\nu-1}+1)\kappa^{d-\frac12} \end{aligned}$$

and

$$(2\pi)^{\frac{1}{2}} \| (\partial_\mu b_k)_{k\in\mathbb{Z}} \|_{\ell_0^2} = \| \Sigma(\partial_\mu b_k)\hat{\varphi}_k \|_{L^2}$$

$$= \| \partial_\mu (\Sigma b_k \hat{\varphi}_k) - \Sigma b_k (\partial_\mu \hat{\varphi}_k) \|_{L^2}$$

$$\leq c_1'(\rho^\nu + 1)\kappa^{d-\frac{1}{2}} + c_2'\kappa^{-1}(\rho^{\nu-1}+1)\kappa^{d+\frac{1}{2}}$$

$$\leq c_3'(\rho^{\nu-1}+1)\kappa^{d-\frac{1}{2}} \quad ,$$

which shows (2.2.85) for β, M and N equal to 0 and $|\alpha|$ or $j = 1$. This gives the general pattern for the ℓ_N^2-norms: there is a loss of regularity each time $|\alpha|$ and j are augmented; and one finds the general estimates by successive applications of the preceding arguments.

Note however, that for each fixed k, the μ-derivatives are better behaved:

$$|\partial_\mu b_k| = |\int_{\mathbb{R}} \partial_\mu (h_{-1}p)\bar{\hat{\varphi}}_k d\xi_n + \int_{\mathbb{R}} (h_{-1}p)\partial_\mu \bar{\hat{\varphi}}_k d\xi_n|$$

$$\leq c(\| \partial_\mu (h_{-1}p) \|_{L^2} + k\kappa^{-1} \| (h_{-1}p) \|_{L^2})$$

$$\leq c_k(\rho^\nu + 1)\kappa^{d-\frac{1}{2}} \quad ,$$

and similarly for mixed derivatives and lower order parts, so that each b_k is in $S^{d+\frac{1}{2},\nu}(\Omega'\times\Omega', \overline{\mathbb{R}}^n_+)$. The assertions on the functions $a_k(x',y',\xi',\mu)$ follow immediately from the formulas (2.2.55) (with $\sigma = \kappa(\xi',\mu)$). □

2.2.9 Remark. For fixed μ_0, $\kappa(\xi',\mu_0) \simeq \langle\xi'\rangle$ and

$$(\rho(\xi',\mu_0)^\nu + 1)\kappa(\xi',\mu_0)^d \simeq \langle\xi'\rangle^d \quad .$$

It is then easy to conclude by use of the second half of Lemma 2.2.1 that the expansion (2.2.83) together with the estimates (2.2.85) are sufficient for the original transmission property (as well as necessary), in the μ-independent case.

But in the μ-dependent case, the estimates (2.2.85) do not imply that the symbol belongs to our $S^{d,\nu}$-space ; there is too much loss of regularity, both in applications of ∂_μ, $\partial_{\xi_n}\xi_n$ and ξ_n to $h_{-1}p$; for the latter, see (2.2.18) and (2.2.16).

2.2.10 Remark. It will sometimes be of interest to have estimates like (2.2.85) for noninteger N. This is easily obtained by interpolation between neighboring integer cases, when the two exponents on ρ are both ≥ 0 or both ≤ 0, by use of (2.1.17): For $\theta \in]0,1[$, one has by Hölder's inequality, since $N+\theta = N(1-\theta)+(N+1)\theta$,

$$\begin{aligned}
(2.2.87) \qquad \|(b_k)_{k\in\mathbb{Z}}\|^2_{\ell^2_{N+\theta}} &= \sum_{k\in\mathbb{Z}} (1+|k|)^{2(N+\theta)}|b_k|^2 \\
&= \sum_{k\in\mathbb{Z}} (1+|k|)^{2N(1-\theta)}|b_k|^{2(1-\theta)}(1+|k|)^{2(N+1)\theta}|b_k|^{2\theta} \\
&\leq \Big(\sum_{k\in\mathbb{Z}} (1+|k|)^{2N}|b_k|^2\Big)^{1-\theta}\Big(\sum_{k\in\mathbb{Z}} (1+|k|)^{2(N+1)}|b_k|^2\Big)^{\theta} \\
&= \|(b_k)_{k\in\mathbb{Z}}\|^{2(1-\theta)}_{\ell^2_N}\ \|(b_k)_{k\in\mathbb{Z}}\|^{2\theta}_{\ell^2_{N+1}} \qquad ;
\end{aligned}$$

here in view of (2.1.17),

$$(2.2.88) \qquad \begin{aligned}
(\rho^{\nu-N}+1)^{1-\theta}(\rho^{\nu-N-1}+1)^{\theta} &\simeq \text{const.} \\
&\simeq (\rho^{\nu-N-\theta}+1)\ , \quad \text{when} \quad \nu-N \geq 1\ , \\
(\rho^{\nu-N}+1)^{1-\theta}(\rho^{\nu-N-1}+1)^{\theta} & \\
&\simeq (\langle\xi'\rangle^{\nu-N}\langle\xi',\mu\rangle^{N-\nu})^{1-\theta}(\langle\xi'\rangle^{\nu-N-1}\langle\xi',\mu\rangle^{N+1-\nu})^{\theta} \\
&= \langle\xi'\rangle^{\nu-N-\theta}\langle\xi',\mu\rangle^{N+\theta-\nu} \\
&\simeq (\rho^{\nu-N-\theta}+1)\ , \quad \text{when} \quad \nu-N \leq 0\ .
\end{aligned}$$

Thus the inequalities

$$(2.2.89) \qquad \|(b_k)_{k\in\mathbb{Z}}\|_{\ell^2_N} \leq c(X)(\rho^{\nu-N}+1)\kappa^{d+\frac{1}{2}}$$

extend to all values of $N \in \overline{\mathbb{R}}_+$, when ν is <u>integer or</u> ≤ 0. However, if ν is a positive non-integer, interpolation gives a weaker result for $\nu-N \in]0,1[$. For example, if $\nu = \frac{1}{2}$ and (2.2.89) is known to hold for integer N, the ℓ^2_ε norm for small $\varepsilon > 0$ is only estimated by

$$\begin{aligned}
(2.2.90) \qquad \|(b_k)_{k\in\mathbb{Z}}\|_{\ell^2_\varepsilon} &\leq \|(b_k)\|^{1-\varepsilon}_{\ell^2_0}\ \|(b_k)\|^{\varepsilon}_{\ell^2_1} \\
&\leq c(X)(\rho^{\frac{1}{2}}+1)^{1-\varepsilon}(\rho^{-\frac{1}{2}}+1)^{\varepsilon}\ \kappa^{d+\frac{1}{2}} \\
&\simeq c(X)\langle\xi'\rangle^{-\varepsilon/2}\ \kappa^{d+\frac{1}{2}+\varepsilon/2} \simeq c(X)(\rho^{-\varepsilon/2}+1)\kappa^{d+\frac{1}{2}}\ ,
\end{aligned}$$

by (2.1.17); and this is a much weaker estimate than the estimate with $\rho^{\frac{1}{2}-\varepsilon}$

that would be desired in analogy with (2.2.89). We cannot quite avoid cases with regularity $\frac{1}{2}$ (Dirichlet-type boundary conditions give such cases); then for certain important Laguerre series estimates we invoke other tricks (in Section 2.6).

2.2.11 Remark. The Laguerre expansions have a limited importance in the description of the calculus, and as we have seen, the estimates formulated in terms of these are not as efficient as the estimates (2.2.82), for the parameter dependent classes. However, the Laguerre estimates play a rôle in the ellipticity discussion (Chapter 3). At present we note that when the symbol of p is written on the form (2.2.84), the bounded part $p' = \Sigma_{k\in\mathbb{Z}} a_k\hat{\psi}_k$ defines a "convolution" operator

$$p'(D_n)v = F^{-1}\sum_{k\in\mathbb{Z}} a_k\hat{\psi}_k \sum_{m\in\mathbb{Z}} v_m\hat{\varphi}_m \quad (\text{where } v = \sum_{m\in\mathbb{Z}} v_m\varphi_m)$$

$$= F^{-1}\sum_{k,m\in\mathbb{Z}} a_k v_m \hat{\varphi}_{k+m} = \sum_{\ell,m\in\mathbb{Z}} a_{\ell-m} v_m \varphi_\ell ,$$

since $\hat{\psi}_k\hat{\varphi}_m = \hat{\varphi}_{k+m}$, cf. (2.2.10) and (2.2.52). The restriction to $\mathbb{R}_+$ is then a Toeplitz operator: When $u \in \mathscr{S}(\overline{\mathbb{R}}_+)$ and is written $u = \Sigma_{m\in\mathbb{N}} u_m\varphi_m$ (with $u_m = (u,\varphi_m)$) then

(2.2.91)
$$p'(D_n)_\Omega u = r^+ F^{-1}\sum_{k\in\mathbb{Z}} a_k\hat{\psi}_k \sum_{m\in\mathbb{N}} u_m\hat{\varphi}_m$$

$$= r^+ F^{-1}\sum_{\ell\in\mathbb{Z},m\in\mathbb{N}} a_{\ell-m} u_m\hat{\varphi}_\ell$$

$$= \sum_{\ell,m\in\mathbb{N}} a_{\ell-m} u_m \varphi_\ell ,$$

so p'_Ω is the operator with coefficient matrix $(a_{\ell-m})_{\ell,m\in\mathbb{N}}$ in the Laguerre system. Also the differential operator part can be viewed as a Toeplitz operator, since e.g., by (2.2.23),

$$r^+\partial_x u = \sum_{m\in\mathbb{N}} (-\kappa u_m\varphi_m + 2\kappa \sum_{0\leq j<m} (-1)^{m-1-j} u_m\varphi_j)$$

$$= \sum_{\ell,m\in\mathbb{N}} c_{\ell-m}u_m\varphi_\ell \quad \text{with}$$

(2.2.92)
$$c_j = 2\kappa(-1)^{j-1} \text{ for } j > 0 , \quad c_0 = -\kappa , \quad c_j = 0 \text{ for } j < 0 .$$

The transmission property is preserved under various transformation rules established in Section 2.1, as follows.

2.2.12 Theorem. *Let* $\Sigma = \Omega' \times \mathbb{R}$, Ω' *open* $\subset \mathbb{R}^{n-1}$.

1^0 *When* p *and* p' *are as in Theorem 2.1.15 and furthermore have the transmission property, then the resulting symbols listed in the theorem have the transmission property. The statement holds also for* $S_{1,0}$ *symbol spaces.*

2^0 *When* p *is as in Theorem 2.1.16 and has the transmission property, then the parametrix symbol* q *likewise has the transmission property.*

3^0 *Let* $\Sigma = \Omega' \times \mathbb{R}$ *and* $\underline{\Sigma} = \underline{\Omega}' \times \mathbb{R}$, *and let* $\kappa: x \mapsto \underline{x}$ *be a diffeomorphism of* Σ *onto* $\underline{\Sigma}$ *mapping* $\Omega' \times \{0\}$ *onto* $\underline{\Omega}' \times \{0\}$ *(in short: preserving the set* $\{x_n = 0\}$ *). Let* p, $\underline{p}$ *and* $\underline{p}_1$ *be as in Lemma 2.1.17. When* p *has the transmission property, then so does* $\underline{p}$ *(for small* $\underline{x}-\underline{y}$ *), and so does* $\underline{p}_1$; *i.e., the transmission property is invariant under such diffeomorphisms.*

Proof: A basic ingredient in these statements is, that when p has an asymptotic expansion in a series of terms of decreasing order, where the individual terms have the transmission property, then p has the transmission property. This follows rather immediately from Definition 2.2.4 in the polyhomogeneous case, when the series is rearranged according to homogeneity.

In the $S_{1,0}$ case, where one has to verify estimates of the type (2.2.78), one uses that for each α,β,j,m, only finitely many terms in the expansion of p contribute to the polynomial part, and they do so in a controlled way. More precisely, let $p \sim \Sigma_{\ell \geq 0} p_{d-\ell}$, where $p_{d-\ell}$ belongs to $S_{1,0}^{d-\ell,\nu-\ell}$ and has the transmission property (2.2.78), expressed briefly as follows:

$$p_{d-\ell} \sim \sum_{-\infty < k \leq d-\ell} s_{d-\ell,k}(x',y',\xi',\mu)\xi_n^k ,$$

for each $\ell \geq 0$ (recall that the coefficients $s_{d-\ell,k}$ are polynomials in (ξ',μ)). Then if we define s_{d-j} as the finite sum

$$s_{d-j} = \sum_{0 \leq \ell \leq j} s_{d-\ell,d-j} , \quad \text{for each } j \in \mathbb{N} ,$$

we have, for $d+m \geq 0$,

$$\xi_n^m p - \sum_{-m \leq k \leq d} s_k \xi_n^{k+m} = \sum_{0 \leq \ell \leq d+m} \xi_n^m \big(p_{d-\ell} - \sum_{-m \leq k \leq d-\ell} s_{d-\ell,k}\xi_n^k\big) + \xi_n^m \big(p - \sum_{0 \leq \ell \leq d+m} p_{d-\ell}\big) ,$$

where the terms in the first sum over $0 \leq \ell \leq d+m$ are $O((\rho(\xi',\mu)^{\nu-\ell+m}+1)\langle\xi',\mu\rangle^{d-\ell+1+m}\langle\xi,\mu\rangle^{-1})$, and the second sum is of order -1 and regularity ν-d-1 in view of Lemma 2.1.6, so that, by (2.2.61),

$$|\xi_n^m(p - \sum_{0\le\ell\le d+m} p_{d-\ell})| \le c(\rho(\xi',\mu)^{\nu-d-1}+1)\langle\xi,\mu\rangle^{-1} .$$

We find altogether (recall (2.1.1')) that $p \in H_d$, with

$$(2.2.93)\qquad |h_{-1}\xi_n^m p| = |\xi_n^m p - \sum_{-m\le k\le d} s_k\xi_n^{k+m}| \le c(\rho(\xi',\mu)^{\nu+m}+1)\langle\xi',\mu\rangle^{d+1+m}\langle\xi,\mu\rangle^{-1} .$$

The derivatives are treated in a similar way, and it follows that p has the transmission property.

Now 1° follows from the explicit form of the composition rules, and 2° follows from the formula for q.

For the treatment of 3°, we make some observations on the structure of the diffeomorphism $\kappa = (\kappa_1,\dots,\kappa_n) : x \sim \underline{x}$ and the induced transformation $\xi = {}^tM\underline{\xi}$. The assumption that κ preserves the set $\{x_n=0\}$ implies that $\kappa_n(x',0) = 0$ for all x'. Then, by Taylor's formula,

$$(2.2.94)\qquad \begin{aligned} \underline{x}_n &= \kappa_n(x',x_n) = \kappa_n(x',0) + x_n\int_0^1 \partial_n\kappa_n(x',hx_n)\,dh \\ &= C_1(x)x_n \end{aligned}$$

for a C^∞ function $C_1(x) \neq 0$. Since moreover $\partial_j\kappa_n(x',0) = 0$ for $j=1,\dots,n-1$, the matrix $M(x,y)$ defined in (2.1.57), (2.1.58'), can be written in blocks

$$(2.2.95)\qquad M(x,y) = \begin{pmatrix} A(x,y) & B(x,y) \\ B'(x,y) & C(x,y) \end{pmatrix} ,$$

$$\text{with } A = (M_{jk})_{1\le j,k\le n-1} ,\qquad B = (M_{jn})_{1\le j\le n-1} ,$$

$$B' = (M_{nk})_{1\le k\le n-1} ,\qquad C = M_{nn} ,$$

having the properties:

(2.2.96)

(i) $A_0(x') \equiv A(x',0,x',0) = (\partial_k\kappa_j(x',0))_{j,k\le n-1}$, invertible matrix;

(ii) $B_0(x') \equiv B(x',0,x',0) = (\partial_n\kappa_j(x',0))_{j\le n-1}$;

(iii) $B'(x',0,y',0) = 0$ for all $x',y' \in \Omega'$;

(iv) $C_0(x') \equiv C_1(x',0) \equiv C(x',0,x',0) = \partial_n\kappa_n(x',0)$, invertible function.

The invertibility of A_0 follows from the invertibility of $M(x,x)$ together with (iii), (iv). By continuity, one also has invertibility of $A(x,y)$ and $C(x,y)$ when $|x'-y'|$, x_n and y_n are close to zero. Observe that (cf. (2.1.57))

$$(2.2.97)\qquad \underline{x}' - \underline{y}' = A(x,y)(x'-y') + B(x,y)(x_n-y_n) .$$

We can modify p so that the invertibility holds on $\operatorname{supp} p$. The relation

$$\xi = {}^t M(x,y)\underline{\xi} ,$$

used in Lemma 2.1.17, can now be written

$$(2.2.98) \qquad \xi = \begin{pmatrix} \xi' \\ \xi_n \end{pmatrix} = \begin{pmatrix} {}^tA(x,y)\underline{\xi}' + {}^tB'(x,y)\underline{\xi}_n \\ {}^tB(x,y)\underline{\xi}' + C(x,y)\underline{\xi}_n \end{pmatrix} ,$$

and in particular:

$$(2.2.99) \qquad \begin{aligned} \xi' &= {}^tA\underline{\xi}' \quad \text{for } (x,y) = (x',0,y',0) , \\ \xi_n &= {}^tB\underline{\xi}' + C\underline{\xi}_n . \end{aligned}$$

When (2.2.99) holds, and $x'-y'$ is small,

$$(2.2.100) \qquad \underline{\xi} = \begin{pmatrix} \underline{\xi}' \\ \underline{\xi}_n \end{pmatrix} = \begin{pmatrix} {}^tA^{-1}\xi' \\ -\frac{1}{C}\,{}^tB\;{}^tA^{-1}\xi' + \frac{1}{C}\,\xi_n \end{pmatrix} .$$

Moreover, when we set $\underline{v}(\underline{\xi}) = v(\xi)$, we have for the corresponding derivatives:

$$(2.2.101) \qquad \begin{aligned} D_{\underline{\xi}'}\,\underline{v}(\underline{\xi}) &= \Big(\sum_{k=1}^{n} \frac{\partial \xi_k}{\partial \underline{\xi}_j} D_{\xi_k} v(\xi)\Big)_{1\le j\le n-1} = AD_{\xi'}v(\xi) + BD_{\xi_n}v(\xi) \\ D_{\underline{\xi}_n}\underline{v}(\underline{\xi}) &= CD_{\xi_n}v(\xi) . \end{aligned}$$

Observe also that when (2.2.99) holds, then there are positive functions $c_1(x',y')$ and $c_2(x',y')$ defined for $x'-y'$ close to zero, so that

$$(2.2.102) \qquad \begin{aligned} c_1(x',y')\langle\xi'\rangle &\le \langle\underline{\xi}'\rangle \le c_1(x',y')^{-1}\langle\xi'\rangle , \\ c_2(x',y')\langle\xi',\mu\rangle &\le \langle\underline{\xi}',\mu\rangle \le c_2(x',y')^{-1}\langle\xi',\mu\rangle , \end{aligned}$$

besides the usual equivalences $\langle\xi\rangle \sim \langle\underline{\xi}\rangle$ and $\langle\xi,\mu\rangle \sim \langle\underline{\xi},\mu\rangle$.

For the polyhomogeneous symbols, we now use that when $\underline{x}_n = \underline{y}_n = 0$ in the formula (2.1.56) (corresponding to $x_n = y_n = 0$), then the point $(\underline{\xi}',\underline{\xi}_n) = (0,\underline{\xi}_n)$ corresponds to the point $(\xi',\xi_n) = (0,C\underline{\xi}_n)$ (cf. (2.2.99)) when $x'-y'$ is close to 0, so that the symmetry property in (2.2.58) for p carries over to the transformed symbol $\underline{p}$, for small $x-y$. Derivatives behave similarly. To show the transmission property for the $\underline{x}$-form $\underline{p}_1$, we can appeal to point 1° in this theorem; note also that (2.1.59) shows more explicitly how the coordinate change (for p on x-form) gives rise to polynomial factors φ_α that do not violate the transmission property.

For $S_{1,0}$ symbols, we must show the validity of estimates as in (2.2.78) for the transformed symbol. Here we note that when p satisfies the estimates (2.2.78), one has in particular, when ξ and $\underline{\xi}$ are related by (2.2.99),

$$|\xi_n^m p(x',0,y',0,\xi,\mu) - \sum_{-m\le k\le d} s_k(x',y',\xi',\mu)\xi_n^{k+m}|$$

$$\equiv |({}^t B\underline{\xi}' + C\underline{\xi}_n)^m p(\dots, {}^t M\underline{\xi},\mu) - \sum_{-m\le k\le d} s_k(\dots, {}^t A\underline{\xi}',\mu)({}^t B\underline{\xi}' + C\underline{\xi}_n)^{k+m}|$$

$$\le c(x',y')(\rho(\xi',\mu)^{\nu+m} + 1)\langle\xi',\mu\rangle^{d+m+1}\langle\xi,\mu\rangle^{-1}$$

$$\le c'(x',y')(\rho(\underline{\xi}',\mu)^{\nu+m} + 1)\langle\underline{\xi}',\mu\rangle^{d+m+1}\langle\underline{\xi},\mu\rangle^{-1} ,$$

in view of (2.2.102). For small $x'-y'$ (where $C \neq 0$), one can use these estimates successively for $m=0,1,\dots$ to determine expansion coefficients $s'_k(x',y',\underline{\xi}',\mu)$ such that

$$(2.2.103)\quad |\underline{\xi}_n^m p(x',0,y',0,{}^t M\underline{\xi},\mu) - \sum_{-m\le k\le d} s'_k(x',y',\underline{\xi}',\mu)\underline{\xi}_n^{m+k}|$$
$$\le c(x',y')(\rho(\underline{\xi}',\mu)^{\nu+m} + 1)\langle\underline{\xi}',\mu\rangle^{d+m+1}\langle\underline{\xi},\mu\rangle^{-1}$$

for all m; here the s'_k are polynomials in $(\underline{\xi}',\mu)$, since the s_k are polynomials in (ξ',μ). In particular, we see that $p(x',0,y',0,{}^t M\underline{\xi},\mu)$ is in H_d as a function of $\underline{\xi}_n$, and the procedure we have just described shows how to find $h_{-1,\underline{\xi}_n}[\underline{\xi}_n^m p(x',0,y',0,{}^t M\underline{\xi},\mu)]$ for each m.

As for derivatives of p, we note that $D_{\underline{\xi}_n} p$ is directly related to $D_{\xi_n} p$ by (2.2.101), whereas $D_{\underline{\xi}_j} p$ for $j < n$ is a sum of derivatives of p, but again the terms can be regrouped to furnish the desired expansions as in (2.2.78). The derivatives in x,y and μ are easily included, so it is found altogether that when $p(x,y,\xi,\mu)$ has the transmission property at $x_n = 0$, then so does $p(x,y,{}^t M(x,y)\underline{\xi},\mu)$, as a function of $(x,y,\underline{\xi},\mu)$.

In the complete transformation formula (2.1.56), one must furthermore consider derivatives in $\underline{x}$ and $\underline{y}$ (including differentiations through ${}^t M(x,y)\underline{\xi}$), and the determinant factors must be taken into account; but all this can be handled in the same way as above. It follows that $\underline{p}$, defined by (2.1.56), has the transmission property (as a function of $(\underline{x},\underline{y},\underline{\xi},\mu)$). The symbol on $\underline{x}$-form likewise has the transmission property (in view of 1° above, or by use of the explicit form (2.1.59)).

□

The invariance of the transmission property for parameter-independent polyhomogeneous symbols is shown in [Boutet 1] and in [Hörmander 8, Section 18.2], whereas the $S_{1,0}$ class has not, to our knowledge, been discussed in detail before.

2.2.13 Example. The symbols p and q in Example 2.1.18 have the transmission property at $y = 0$ (in fact at $y = a$ for any a), since the symmetry conditions in (2.2.58) are obviously verified. Thus q can be decomposed

(2.2.104) $$q(\xi,\eta,\mu) = q^+(\xi,\eta,\mu) + q^-(\xi,\eta,\mu) \ ,$$

where q^+ and q^- are in H^+ resp. H^- as functions of η . Theorem 2.2.8 shows e.g. that

(2.2.105) $$\| D_\xi^k D_\eta^\ell q^+ \|_{L^2_\eta} \leq c(\rho^{2-k-\ell} + 1)\kappa^{-2-k-\ell} \ , \quad \rho = \frac{\langle\xi\rangle}{\langle\xi,\mu\rangle} \ , \quad \kappa \sim \langle\xi,\mu\rangle \ ,$$

and in particular the associated strictly homogeneous symbol satisfies (cf. Lemma 2.1.9)

(2.2.106) $$\| D_\xi^k D_\eta^\ell (q^+)^h \|_{L^2_\eta} \leq c|\xi|^{2-k-\ell} |\xi,\mu|^{-4}$$

when $k + \ell \geq 2$. This would be quite cumbersome to deduce by explicit calculations, even in this simple case. To show what the calculations amount to, let us sketch a few steps. We factor the denominator polynomial f in q^h (cf. (2.1.62)) considered as a function of η ,

(2.2.107) $$\begin{aligned} f_{\xi,\mu}(\eta) &= \eta^4 + \mu^2\eta^2 + (\mu^2\xi^2+\xi^4) = (\eta^2+\lambda_1)(\eta^2+\lambda_2) \\ &= (\lambda_1^{\frac12}+i\eta)(\lambda_1^{\frac12}-i\eta)(\lambda_2^{\frac12}+i\eta)(\lambda_2^{\frac12}-i\eta) \ , \end{aligned}$$

where

(2.2.108) $$\begin{aligned} \lambda_1 &= \tfrac12\mu^2 + \tfrac12(\mu^4-4(\mu^2\xi^2+\xi^4))^{\frac12} \\ \lambda_2 &= \tfrac12\mu^2 - \tfrac12(\mu^4-4(\mu^2\xi^2+\xi^4))^{\frac12} \ ; \end{aligned}$$

here we take $\lambda_1^{\frac12}$ and $\lambda_2^{\frac12}$ with positive real part (when possible). Note that $|\lambda_j^{\frac12}|$ is $O(|\xi,\mu|)$; but as for a lower bound, $\lambda_2^{\frac12}$ behaves for $\mu \to \infty$ like

(2.2.109) $$|\mu(1 - [1-4\xi^2(\mu^2+\xi^2)/\mu^4]^{\frac12})^{\frac12}| \sim |\mu\xi(\mu^2+\xi^2)^{\frac12}/\mu^2| = |\xi||\xi,\mu||\mu|^{-1} \ .$$

Now

$$q^h(\xi,\eta,\mu) = \frac{\xi^2+\eta^2}{(\eta^2+\lambda_1)(\eta^2+\lambda_2)} = (\xi^2+\eta^2)\,\frac{1}{\lambda_2-\lambda_1}\left[\frac{1}{\eta^2+\lambda_1} - \frac{1}{\eta^2+\lambda_2}\right] \ ,$$

where

$$\frac{1}{\eta^2+\lambda_j} = \frac{1}{\lambda_j^{\frac12}+i\eta}\,\frac{1}{\lambda_j^{\frac12}-i\eta} = \frac{1}{2\lambda_j^{\frac12}}\left[\frac{1}{\lambda_j^{\frac12}+i\eta} + \frac{1}{\lambda_j^{\frac12}-i\eta}\right] \ .$$

The contributions to q^+ are of the form

$$\frac{c\xi^2}{\lambda_2 - \lambda_1} \frac{1}{\lambda_j^{\frac{1}{2}}} \frac{1}{\lambda_j^{\frac{1}{2}} + i\eta} \quad \text{and} \quad \frac{c\lambda_j}{\lambda_2 - \lambda_1} \frac{1}{\lambda_j^{\frac{1}{2}}} \frac{1}{\lambda_j^{\frac{1}{2}} + i\eta}$$

for $j = 1,2$, where the behavior in terms of ξ and μ can be analyzed using (2.2.109) and other estimates. Our general theorem shows that the terms add up to a function with the behavior (2.2.105-106).

2.3 Parameter-dependent boundary symbols.

Our choice of boundary symbol classes will be governed by the requirement that if P_μ is a parameter-dependent ps.d.o. on $\mathbb{R}^n$ having the transmission property (Definitions 2.2.4 and 2.2.7), then the operators defined by

$$
\begin{aligned}
&T_\mu u = \gamma_0 P_{\mu,\Omega} u \quad , \\
(2.3.1) \qquad &K_\mu v = r^+ P_\mu (v(x')\otimes\delta(x_n)) \quad , \\
&G^+(P_\mu) u = r^+ P_\mu e^- J u \qquad \text{(recall (A.31-32))} \ ,
\end{aligned}
$$

for $u \in C^\infty_{(0)}(\overline{\mathbb{R}}^n_+)$ and $v \in C^\infty_0(\mathbb{R}^{n-1})$, should be, respectively, a (parametrized) trace operator, Poisson operator and singular Green operator, belonging to the calculus. (Similar observations were made for the non-parametrized case in Section 1.2.)

2.3.1 Example. Let $p(\xi_n) \in H_d$; it is then a symbol in $S^d(\mathbb{R},\mathbb{R})$ having the transmission property (at any x_n, since it is constant in x_n). Denote $h^+p = p^+$ and $h^-p = p^-$, and recall the notation $\tilde p = F^{-1}_{\xi_n\to x_n} p$. Then for $u \in \mathcal{S}(\overline{\mathbb{R}}_+)$ with $f = \widehat{e^+u}$,

$$(2.3.2) \qquad \gamma_0 p(D_n)_\Omega u = \gamma_0 r^+ F^{-1}(p(\xi_n)\widehat{e^+u}(\xi_n)) = \gamma_0 r^+ F^{-1}(h^+(pf))$$

$$= \frac{1}{2\pi}\int^+ h^+(p^+f + p^-f)d\xi_n = \frac{1}{2\pi}\int^+ h^+(p^-f)d\xi_n = \frac{1}{2\pi}\int^+ p^- f d\xi_n \quad ,$$

where we have used (2.2.44) and (2.2.42) ($p^+f \in H^+ \cap H_{-2}$ and $h^-(p^-f) \in H^-$). Trace operators on $\mathbb{R}_+$ are generally of the form

$$(2.3.3) \qquad t(D_n)u = \frac{1}{2\pi}\int^+ t(\xi_n)\widehat{e^+u}(\xi_n)d\xi_n \ , \quad \text{where } t(\xi_n) \in H^- \ .$$

For $v \in \mathbb{C}$, we have, by approximating δ by a sequence of smooth functions w_ε supported in $[-2\varepsilon,-\varepsilon]$, such that $\hat w_\varepsilon \in H^-$, $\hat w_\varepsilon \to 1$,

$$(2.3.4) \quad r^+p(D_n)(v\otimes\delta) = \lim_{\varepsilon\to 0} r^+p(D_n)(vw_\varepsilon(x_n)) = \lim_{\varepsilon\to 0} r^+ \frac{1}{2\pi}\int e^{ix_n\xi_n} p(\xi_n) v\hat w_\varepsilon(\xi_n)d\xi_n$$

$$= \lim_{\varepsilon\to 0} r^+ \frac{1}{2\pi}\int e^{ix_n\xi_n}(p^+ + p^-)\hat w_\varepsilon d\xi_n v = \lim_{\varepsilon\to 0} r^+ \frac{1}{2\pi}\int e^{ix_n\xi_n} p^+\hat w_\varepsilon d\xi_n \, v$$

$$= r^+ F^{-1}_{\xi_n\to x_n} p^+ v = \tilde p^+(x_n)\cdot v \quad ,$$

since $F^{-1}(p^-\hat w_\varepsilon)$ is supported in $\overline{\mathbb{R}}_-$, as F^{-1} of a function in H^- . Poisson operators on $\mathbb{R}_+$ are generally of the form

$$k(D_n)v = \tilde{k}(x_n)\cdot v \ , \tag{2.3.5}$$

where $\tilde{k}(x_n) = F^{-1}_{\xi_n\to x_n} k(\xi_n)$ with $k \in H^+$. Finally, we have (reading the integrals as Fourier transforms and using that the distribution kernel of p equals $\tilde{p}(x_n-y_n)$ where $\tilde{p} \in \dot{\mathscr{S}}(\mathbb{R})$)

$$\begin{aligned} r^+p(D_n)e^-Ju &= r^+ \frac{1}{2\pi}\int e^{ix_n\xi_n}p(\xi_n)\int_{-\infty}^0 e^{-iy_n\xi_n}u(-y_n)dy_n\, d\xi_n \\ &= r^+ \frac{1}{2\pi}\int e^{ix_n\xi_n}p(\xi_n)\int_0^{\infty} e^{iy_n\xi_n}u(y_n)dy_n\, d\xi_n \\ &= r^+\int_0^\infty \tilde{p}(x_n+y_n)u(y_n)dy_n = r^+\int_0^\infty \tilde{p}^+(x_n+y_n)u(y_n)dy_n \end{aligned} \tag{2.3.6}$$

since $\tilde{p}(z) = \tilde{p}^+(z)$ for $z > 0$. Here $\tilde{p}^+(z) \in \mathscr{S}(\overline{\mathbb{R}}_+)$, so r^+pe^-J is an integral operator with kernel $\tilde{p}^+(x_n+y_n) \in \mathscr{S}(\overline{\mathbb{R}}^2_{++})$. General singular Green operators of class 0 on $\mathbb{R}_+$ are simply integral operators

$$Gu = \int_0^\infty \tilde{g}(x_n,y_n)u(y_n)dy_n \tag{2.3.7}$$

with kernel $\tilde{g} \in \mathscr{S}(\overline{\mathbb{R}}^2_{++})$. (These phenomena are also considered in [Grubb 17].)

For a good calculus associated with the ps.d.o.s with symbols p in $S^{d,\nu}_{tr}$, the trace, Poisson and Green symbols must encompass the functions derived from p with respect to ξ_n as in the above example. Thus it can be expected that the symbols satisfy estimates like (2.2.82), taking the regularity number ν into account. Actually, the estimates will be a little weaker; more comments are made on this in Remark 2.3.6 below.

In the following, we use the convention

$$N_\pm = \max\{\pm N, 0\} \ , \tag{2.3.8}$$

and ρ and κ are considered as functions of (ξ',μ) (as in (2.2.80)).

<u>2.3.2 Definition.</u> *Let* d *and* $\nu \in \mathbb{R}$, *let* $r \in \mathbb{N}$, *and let* Ξ *be open* $\subset \mathbb{R}^{n'}$. *Let* K *stand for either of the spaces* H_{r-1} , H^-_{r-1} *or* H^+ .

1^0 *The space* $S^{d,\nu}_{1,0}(\Xi,\overline{\mathbb{R}}^n_+,K)$ *consists of the functions* $f(X,\xi',\xi_n,\mu) \in C^\infty(\Xi\times\overline{\mathbb{R}}^{n+1}_+)$, *lying in* K *with respect to* ξ_n , *such that when* f *is written on the form*

(2.3.9) $$f(X,\xi',\xi_n,\mu) = \sum_{0\le j\le r-1} s_j(X,\xi',\mu)\xi_n^j + f'(X,\xi,\mu)$$

with $f' = h_{-1,\xi_n} f$, *then* $s_j(X,\xi',\mu) \in S^{d-j,\nu}_{1,0}(\Xi,\overline{\mathbb{R}}^n_+)$ *(Definition* 2.1.1*) and* f' *satisfies, for all indices* $\beta \in \mathbb{N}^{n'}$, $\alpha \in \mathbb{N}^{n-1}$, j , k *and* $k' \in \mathbb{N}$,

(2.3.10) $$\| D^\beta_X D^\alpha_{\xi'} D^j_\mu h_{-1}(D^k_{\xi_n} \xi_n^{k'} f')\|_{L^2_{\xi_n}}$$

$$\le c(X)(\rho^{\nu-[k-k']_+-|\alpha|} + 1)\kappa^{d+\frac{1}{2}-k+k'-|\alpha|-j} , \textit{ all } \xi',\mu ,$$

(with $\rho(\xi',\mu)$ *and* $\kappa(\xi',\mu)$ *defined in* (2.2.80)*).*

2^0 *The space* $S^{d,\nu}(\Xi,\overline{\mathbb{R}}^n_+,K)$ *of polyhomogeneous symbols (in* $S^{d,\nu}_{1,0}$ *) consists of those symbols* $f \in S^{d,\nu}_{1,0}(\Xi,\overline{\mathbb{R}}^n_+,K)$ *that furthermore have asymptotic expansions*

(2.3.11) $$f \sim \sum_{\ell\in\mathbb{N}} f_{d-\ell} ,$$

where $f - \Sigma_{\ell<M} f_{d-\ell} \in S^{d-M,\nu-M}_{1,0}(\Xi,\overline{\mathbb{R}}^n_+,K)$ *for any* $M \in \mathbb{N}$, *and the symbols* $f_{d-\ell}$ *are homogeneous of degree* $d-\ell$ *in* (ξ,μ) *on the set where* $|\xi'| \ge 1$ *:*

(2.3.12) $$f_{d-\ell}(X,t\xi,t\mu) = t^{d-\ell} f_{d-\ell}(X,\xi,\mu) \textit{ for } |\xi'| \ge 1 , \quad t \ge 1 .$$

The principal symbol f_d is also denoted f^0 .

The system of <u>symbol seminorms</u> on $S^{d,\nu}_{1,0}(\Xi, \overline{\mathbb{R}}^n_+,K)$ consists of the symbol seminorms of the coefficients s_j in the spaces $S^{d-j,\nu}_{1,0}(\Xi, \overline{\mathbb{R}}^n_+)$ for $0\le j\le r-1$, together with the symbol seminorms defined from (2.3.10) over compact sets $K\subset\Xi$:

(2.3.12') $$|||h_{-1}f|||_{\alpha,\beta,j,k,k',K} = \sup_{X\in K} c_0(X) ,$$

where $c_0(X)$ is the infimum of the possible values of $c(X)$ in (2.3.10). For the polyhomogenoeus symbol spaces $S^{d,\nu}(\Xi, \overline{\mathbb{R}}^n_+,K)$, we adjoin to these seminorms the seminorms of the lower order parts $f - \Sigma_{\ell<M} f_{d-\ell}$ for all $M \in \mathbb{N}$. In this way the symbol spaces have Fréchet topologies. The set Ξ can be taken on other forms, e.g. as the closure of an open set. - By the <u>first symbol seminorms</u> of $f \in S^{d,\nu}_{1,0}(\Xi, \overline{\mathbb{R}}^n_+,K)$ we understand the system of norms

(2.3.13) $$|||s_j|||_{0,0,0,K} , \; j=0,\dots,r-1 , \text{ and } |||h_{-1}f|||_{0,0,0,0,0,K} ,$$

where K runs through the compact subsets of Ξ . In the polyhomogeneous case, one here inserts the principal parts.

2.3.3 Definition.

1^0 $S^{d,\nu}_{1,0}(\Xi,\overline{\mathbb{R}}^n_+,H^-_{r-1})$ *and* $S^{d,\nu}(\Xi,\overline{\mathbb{R}}^n_+,H^-_{r-1})$ *are called, respectively, the space of* $S_{1,0}$ *or polyhomogeneous (parameter-dependent) trace symbols of degree* d , *class* r *and regularity* ν , *and of order* d .

2^0 $S^{d,\nu}_{1,0}(\Xi,\overline{\mathbb{R}}^n_+,H^+)$ *and* $S^{d,\nu}(\Xi,\overline{\mathbb{R}}^n_+,H^+)$ *are called, respectively, the space of* $S_{1,0}$ *or polyhomogeneous (parameter-dependent) Poisson symbols of degree* d *and regularity* ν , *and of order* $d+1$.

The order convention is consistent with that of the usual trace and Poisson operators, as in Section 1.2.

In the following, we mostly restrict the attention to polyhomogeneous symbols, to avoid repetitions (a good deal of the calculus extends readily to $S_{1,0}$-symbols).

2.3.4 Remark. The space $\overline{\mathbb{R}}^n_+$ can in the definitions, and much of the calculus, be replaced by an open cone V ; the spaces are then denoted $S^{d,\nu}_{1,0}(\Xi,V,K)$ and $S^{d,\nu}(\Xi,V,K)$. One can also let μ run through $\mathbb{R}$, and let Ξ contain an associated space variable, cf. Remark 2.1.4. The indications Ξ , $\overline{\mathbb{R}}^n_+$ (or Ξ , V) are often omitted in the following, when this causes no confusion.

We have as an immediate consequence of Theorems 2.2.5 and 2.2.8:

2.3.5 Corollary. *Let* $d\in\mathbb{Z}$, *let* $\Omega'\subset\mathbb{R}^{n-1}$ *and let* $\Sigma=\Omega'\times\mathbb{R}$. *When* $p(x,y,\xi,\mu)\in S^{d,\nu}_{tr}(\Sigma\times\Sigma,\overline{\mathbb{R}}^{n+1}_+)$, *then, with* $r=[d+1]_+$,

$$p(x',0,y',0,\xi,\mu)\in S^{d,\nu}(\Omega'\times\Omega',\overline{\mathbb{R}}^n_+,H^-_{r-1})$$

$$h^+_{\xi_n}p(x',0,y',0,\xi,\mu)\in S^{d,\nu}(\Omega'\times\Omega',\overline{\mathbb{R}}^n_+,H^+) \tag{2.3.14}$$

$$h^-_{\xi_n}p(x',0,y',0,\xi,\mu)\in S^{d,\nu}(\Omega'\times\Omega',\overline{\mathbb{R}}^n_+,H^-_{r-1})\ ,$$

and there are related statements for the lower order parts and derived symbols.

2.3.6 Remark. One may observe that the estimates in (2.2.82) are slightly better than what is required in (2.3.10) (the regularity increases by multiplication with ξ_n^m). But this cannot be upheld when we include terms like $s_j(X,\xi',\mu)\xi_n^j$ with non-polynomial coefficients s_j ; here a multiplication by ξ_n^m does not increase the regularity. (Recall that ps.d.o. symbols in $S_{tr}^{d,\nu}$ are more special than the functions f ; they have polynomials as coefficients s_j .) The ps.d.o. coefficients s_j in f arize naturally through the composition rules, also in the parameter-independent case.

In order to define the appropriate symbol spaces for singular Green operators, we insert a complement to the presentation of the spaces H_d, H^+ and H_d^- in Section 2.2, concerning tensor products.

It is well known (cf. e.g. [Treves 1]) that the space $\mathscr{S}(\overline{\mathbb{R}}_+)$ is nuclear, so that the completed tensor products of $\mathscr{S}(\overline{\mathbb{R}}_+)$ with itself in various topologies coincide, in fact with

$$\mathscr{S}(\overline{\mathbb{R}}_+)\hat{\otimes}\mathscr{S}(\overline{\mathbb{R}}_+) = \mathscr{S}(\overline{\mathbb{R}}^2_{++}) ,$$

the restriction of $\mathscr{S}(\mathbb{R}^2)$ to $\overline{\mathbb{R}}^2_{++}$. The (nuclear Fréchet) topology on $\mathscr{S}(\overline{\mathbb{R}}^2_{++})$ is described e.g. by the system of seminorms

$$\|x_n^k D_{x_n}^{k'} y_n^m D_{y_n}^{m'} \tilde{g}(x_n,y_n)\|_{L(\mathbb{R}^2_{++})} \quad \text{for } \tilde{g}(x_n,y_n) \in \mathscr{S}(\overline{\mathbb{R}}^2_{++}) . \tag{2.3.15}$$

By Fourier transformation in x_n and co-Fourier transformation in y_n of $e^{++}\mathscr{S}(\overline{\mathbb{R}}^2_{++})$, we obtain (in view of (2.2.30) and (2.2.29)) the completed tensor product

$$\begin{aligned} &F_{x_n\to\xi_n} \overline{F}_{y_n\to\eta_n}(e^{++}\mathscr{S}(\overline{\mathbb{R}}^2_{++})) \\ &= F_{x_n\to\xi_n} \overline{F}_{y_n\to\eta_n}(e^{+}\mathscr{S}(\overline{\mathbb{R}}_+)\hat{\otimes} e^{+}\mathscr{S}(\overline{\mathbb{R}}_+)) = H^+\hat{\otimes} H^-_{-1} , \end{aligned} \tag{2.3.16}$$

here the sesqui-Fourier transform acts as a homeomorphism. In particular the (semi)norm (2.3.15) on $\tilde{g}(x_n,y_n)$ carries over to the (semi)norm

$$\frac{1}{2\pi}\|h^+_{\xi_n} h^-_{-1,\eta_n}(D^k_{\xi_n}\xi_n^{k'} D^m_{\eta_n}\eta_n^{m'} g(\xi_n,\eta_n))\|_{L^2(\mathbb{R}^2)} \quad \text{for } g\in H^+\hat{\otimes} H^-_{-1} , \tag{2.3.17}$$

where

$$g(\xi_n,\eta_n) = (2\pi)^{-2}\int_{\mathbb{R}^2} e^{-ix_n\xi_n+iy_n\eta_n}\tilde{g}(x_n,y_n)\,dx_n\,dy_n ; \tag{2.3.18}$$

and the system of seminorms (2.3.17) defines the (nuclear Fréchet) topology on $H^+ \hat{\otimes} H^-_{-1}$.

We can also set this in relation to the orthonormal double sequence $(\varphi_\ell(x_n,\sigma)\varphi_m(y_n,\sigma))_{\ell,m\in\mathbb{N}}$ formed of the Laguerre functions; it is a complete orthonormal basis for $L^2(\mathbb{R}^2_{++})$, and it lies in $\mathscr{S}(\overline{\mathbb{R}}^2_{++})$. Here, when $\tilde{g}$ and g are expanded in double Laguerre series:

$$(2.3.19)\qquad \begin{aligned} \tilde{g}(x_n,y_n) &= \sum_{\ell,m\in\mathbb{N}} c_{\ell m}\, \varphi_\ell(x_n,\sigma)\varphi_m(y_n,\sigma) , \\ g(\xi_n,\eta_n) &= \sum_{\ell,m\in\mathbb{N}} c_{\ell m}\, \hat{\varphi}_\ell(\xi_n,\sigma)\overline{\hat{\varphi}_m(\eta_n,\sigma)} , \end{aligned}$$

one has that

$$(2.3.20)\qquad \tilde{g} \in \mathscr{S}(\overline{\mathbb{R}}^2_{++}) \;\Leftrightarrow\; g \in H^+ \hat{\otimes} H^-_{-1} \;\Leftrightarrow\; (c_{\ell m})_{\ell,m\in\mathbb{N}} \in \mathfrak{s}(\mathbb{N}\times\mathbb{N}) ,$$

where $\mathfrak{s}(\mathbb{N}\times\mathbb{N})$ is the space of rapidly decreasing sequences indexed by $(\ell,m) \in \mathbb{N}\times\mathbb{N}$. The system of (semi) norms where N and N' run through $\mathbb{N}$,

$$(2.3.21)\qquad \|(c_{\ell m})_{\ell,m\in\mathbb{N}}\|_{\ell^2_{N,N'}} \equiv \Big(\sum_{\ell,m\in\mathbb{N}} |(1+\ell)^N(1+m)^{N'} c_{\ell m}|^2\Big)^{\frac{1}{2}} ,$$

defines the topology, equivalently with the systems (2.3.15) and (2.3.17).

By use of the direct sum decomposition

$$H^-_{r-1} = H^-_{-1} \dotplus \mathbb{C}_{r-1}[t] , \quad \text{for } r \geq 0 ,$$

where $\mathbb{C}_{r-1}[t]$ denotes the (r-dimensional) space of polynomials of degree $< r$, we can likewise define $H^+ \hat{\otimes} H^-_{r-1}$, with elements

$$f(\xi_n,\eta_n) = \sum_{0\leq j\leq r-1} s_j \eta_n^j + f'(\xi_n,\eta_n) ,$$

here the s_j are constants, and $f' \in H^+ \hat{\otimes} H^-_{-1}$.

These are the spaces we shall use most; it is not harder (but requires more notation) to define also the completed tensor product $H_{-1} \hat{\otimes} H_{-1}$ (as the Fourier transform of $[e^+ \mathscr{S}(\overline{\mathbb{R}}_+) \dotplus e^- \mathscr{S}(\overline{\mathbb{R}}_-)] \hat{\otimes} [e^+ \mathscr{S}(\overline{\mathbb{R}}_+) \dotplus e^- \mathscr{S}(\overline{\mathbb{R}}_-)]$, that consists of sums of restrictions of $\mathscr{S}(\mathbb{R}^2)$ to the four quadrants), and $H_{-1} \hat{\otimes} H_{r-1}$ (that is obtained by adjoining $H_{-1} \otimes \mathbb{C}_{r-1}[\eta_n]$ to the preceding space). Such spaces permit a unified point of view on boundary operators, see Section 3.2.

2.3.7 Definition. *Let* d *and* $\nu \in \mathbb{R}$, *let* $r \in \mathbb{N}$, *and let* Ξ *be open* $\subset \mathbb{R}^{n'}$. *Let* K *stand for either of the spaces* $H^+ \hat{\otimes} H^-_{r-1}$ *or* $H_{-1} \hat{\otimes} H_{r-1}$.

1^0 *The space* $S^{d,\nu}_{1,0}(\Xi,\overline{\mathbb{R}}^n_+,K)$ *consists of the functions* $f(X,\xi',\xi_n,\eta_n,\mu) \in C^\infty(\Xi\times\overline{\mathbb{R}}^{n+2}_+)$, *lying in* K *with respect to* (ξ_n,η_n), *such that when* f *is written on the form*

$$(2.3.22) \qquad f(X,\xi',\xi_n,\eta_n,\mu) = \sum_{0\le j<r} k_j(X,\xi',\xi_n,\mu)\eta_n^j + f'(X,\xi',\xi_n,\eta_n,\mu)$$

with $f' = h_{-1,\eta_n} f$, *then* $k_j \in S^{d-j,\nu}(\Xi,\overline{\mathbb{R}}^n_+,H^+)$ *for each* j, *and* f' *satisfies the estimates, for all* $\beta \in \mathbb{N}^{n'}$, $\alpha \in \mathbb{N}^{n-1}$, j, k, k', m *and* $m' \in \mathbb{N}$,

$$(2.3.23) \qquad \| D^\beta_X D^\alpha_{\xi'} D^j_\mu h_{-1,\xi_n} h_{-1,\eta_n}(D^k_{\xi_n}\xi_n^{k'} D^m_{\eta_n}\eta_n^{m'} f')\|_{L^2(\mathbb{R}^2)}$$
$$\le c(X)(\rho^{\nu-[k-k']_+-[m-m']_+-|\alpha|}+1)\kappa^{d+1-k+k'-m+m'-|\alpha|-j} .$$

2^0 *The space* $S^{d,\nu}(\Xi,\overline{\mathbb{R}}^n_+,K)$ *consists of those symbols* $f \in S^{d,\nu}_{1,0}(\Xi,\overline{\mathbb{R}}^n_+,K)$ *that furthermore have asymptotic expansions*

$$f \sim \sum_{\ell\in\mathbb{N}} f_{d-\ell} ,$$

where $f - \Sigma_{\ell<M} f_{d-\ell} \in S^{d-M,\nu-M}_{1,0}(\Xi,\overline{\mathbb{R}}^n_+,K)$ *for any* $M \in \mathbb{N}$, *and the symbols* $f_{d-\ell}$ *are homogeneous of degree* $d-\ell$ *in* (ξ',ξ_n,η_n,η) *on the set where* $|\xi'|\ge 1$ *:*

$$(2.3.24) \qquad f_{d-\ell}(X,t\xi',t\xi_n,t\eta_n,t\mu) = t^{d-\ell} f_{d-\ell}(X,\xi',\xi_n,\eta_n,\mu) \text{ for } t \text{ and } |\xi'|\ge 1 .$$

As usual, we also denote the principal symbol f_d by f^0. The system of <u>symbol seminorms</u> on $S^{d,\nu}_{1,0}$ consists of the symbol seminorms of the k_j and the seminorms defined from (2.3.23) (in the same way as after Definition 2.3.2). The <u>first symbol seminorms</u> are the seminorms without derivatives and powers of ξ_n, η_n, and without lower order terms in the polyhomogeneous case.

<u>2.3.8 Definition.</u> $S^{d,\nu}_{1,0}(\Xi,\overline{\mathbb{R}}^n_+,H^+\hat{\otimes}H^-_{r-1})$ *and* $S^{d,\nu}(\Xi,\overline{\mathbb{R}}^n_+,H^+\hat{\otimes}H^-_{r-1})$ *are called, respectively, the space of* $S_{1,0}$ *or polyhomogeneous (parameter-dependent)* <u>*singular Green symbols*</u> *of degree* d, *class* r *and regularity* ν, *and of order* $d+1$.

The connection between this choice of symbol spaces and the requirement that (2.3.6) defines an operator in our calculus, is analyzed in Theorem 2.6.6 later on.

Remark 2.3.4 applies also to these symbol spaces.

To the symbols of class 0 we associate the symbol-kernels by inverse Fourier and co-Fourier transformation

$$
\begin{aligned}
&\tilde t(X,x_n,\xi',\mu) = \frac{1}{2\pi}\int e^{-ix_n\xi_n}\, t(X,\xi',\xi_n,\mu)\,d\xi_n\\
(2.3.25)\quad &\tilde k(X,x_n,\xi',\mu) = \frac{1}{2\pi}\int e^{ix_n\xi_n}\, k(X,\xi',\xi_n,\mu)\,d\xi_n\\
&\tilde g(X,x_n,y_n,\xi',\mu) = (2\pi)^{-2}\int e^{ix_n\xi_n-iy_n\eta_n}\, g(X,\xi',\xi_n,\eta_n,\mu)\,d\xi_n\,d\eta_n\,,
\end{aligned}
$$

here $\tilde t$ and $\tilde k$ run through the space $S^{d,\nu}_{1,0}(\Xi,\overline{\mathbb{R}}^n_+,\mathscr{S}(\overline{\mathbb{R}}_+))$ resp. $S^{d,\nu}(\Xi,\overline{\mathbb{R}}^n_+,\mathscr{S}(\overline{\mathbb{R}}_+))$ of functions satisfying

$$
(2.3.26)\qquad \|D^\beta_X D^\alpha_{\xi'} D^j_\mu x_n^k D^{k'}_{x_n}\tilde t(X,x_n,\xi',\mu)\|_{L^2(\mathbb{R}_+)} \le c(X)(\rho^{\nu-[k-k']_+-|\alpha|}+1)\kappa^{d+\frac12-k+k'-|\alpha|-j}\,,
$$

with related estimates for $\tilde t - \Sigma_{\ell<M}\tilde t_{d-\ell}$ in the polyhomogeneous case; and the terms $\tilde t_{d-\ell} = F^{-1}_{\xi_n\to x_n} t_{d-\ell}$ are quasi-homogeneous:

$$
(2.3.27)\qquad \tilde t_{d-\ell}(X,\tfrac{1}{\lambda}x_n,\lambda\xi',\lambda\mu) = \lambda^{d-\ell+1}\,\tilde t_{d-\ell}(X,x_n,\xi',\mu) \quad\text{for } \lambda \text{ and } |\xi'|\ge 1\,.
$$

$\tilde g$ runs through the space $S^{d,\nu}_{1,0}(\Xi,\overline{\mathbb{R}}^n_+,\mathscr{S}(\overline{\mathbb{R}}^2_{++}))$ resp. $S^{d,\nu}(\Xi,\overline{\mathbb{R}}^n_+,\mathscr{S}(\overline{\mathbb{R}}^2_{++}))$ of functions satisfying

$$
(2.3.28)\qquad \|D^\beta_X D^\alpha_{\xi'} D^j_\mu x_n^k D^{k'}_{x_n} y_n^m D^{m'}_{y_n}\tilde g(X,x_n,y_n,\xi',\mu)\|_{L^2(\mathbb{R}^2_{++})} \le c(X)(\rho^{\nu-[k-k']_+-[m-m']_+-|\alpha|}+1)\kappa^{d+1-k+k'-m+m'-|\alpha|-j}\,,
$$

with related estimates for $\tilde g - \Sigma_{\ell<M}\tilde g_{d-\ell}$ in the polyhomogeneous case; here the terms $\tilde g_{d-\ell} = F^{-1}_{\xi_n\to x_n}\overline{F}^{-1}_{\eta_n\to y_n} g_{d-\ell}$ are quasi-homogeneous:

$$
(2.3.29)\qquad \tilde g_{d-\ell}(X,\tfrac{1}{\lambda}x_n,\tfrac{1}{\lambda}y_n,\lambda\xi',\lambda\mu) = \lambda^{d-\ell+2}\tilde g_{d-\ell}(X,x_n,y_n,\xi',\mu) \quad\text{for } \lambda \text{ and } |\xi'|\ge 1\,.
$$

Working with symbol-kernels is often more manageable than working with the symbols themselves. (Remark 2.3.4 applies to these spaces.)

Finally, one can also use the Laguerre expansions for symbols of class 0

$$k(X,\xi',\xi_n,\mu) = \sum_{m\in\mathbb{N}} a_m(X,\xi',\mu)\hat{\varphi}_m(\xi_n,\kappa)$$

(2.3.30) $$t(X,\xi',\xi_n,\mu) = \sum_{m\in\mathbb{N}} b_m(X,\xi',\mu)\overline{\hat{\varphi}_m}(\xi_n,\kappa)$$

$$g(X,\xi',\eta_n,\mu) = \sum_{k,m\in\mathbb{N}} c_{km}(X,\xi',\mu)\hat{\varphi}_k(\xi_n,\kappa)\overline{\hat{\varphi}_m}(\eta_n,\kappa)$$

where the coefficients are in $S_{1,0}^{d+\frac{1}{2},\nu}(\Xi,\overline{\mathbb{R}}_+^n)$ resp. $S^{d+\frac{1}{2},\nu}(\Xi,\overline{\mathbb{R}}_+^n)$ in the first two cases and in $S_{1,0}^{d+1,\nu}(\Xi,\overline{\mathbb{R}}_+^n)$ resp. $S^{d+1,\nu}(\Xi,\overline{\mathbb{R}}_+^n)$ in the third case. One finds as in the proof of Theorem 2.2.8 that they satisfy estimates:

(2.3.31) $$\|(D_X^\beta D_{\xi'}^\alpha D_\mu^j a_m(X,\xi',\mu))_{m\in\mathbb{N}}\|_{\ell_N^2} \le c(X)(\rho^{\nu-|\alpha|-j-N}+1)\kappa^{d+\frac{1}{2}-|\alpha|-j}$$

for the coefficients of k, similar estimates for the coefficients $(b_m)_{m\in\mathbb{N}}$ of t, and (cf. (2.3.21))

(2.3.32) $$\|(D_X^\beta D_{\xi'}^\alpha D_\mu^j c_{km}(X,\xi',\mu))_{k,m\in\mathbb{N}}\|_{\ell_{N,N'}^2} \le c(X)(\rho^{\nu-|\alpha|-j-N-N'}+1)\kappa^{d+1-|\alpha|-j}$$

for the coefficients of g.

<u>2.3.8' Remark</u>. For each of the symbol spaces, we have introduced above, one has the "reconstruction principle", that when $(f_{d-\ell})_{\ell\in\mathbb{N}}$ is a sequence of symbols in $S_{1,0}^{d-\ell,\nu-\ell}$, then there <u>exists</u> a symbol $f\in S_{1,0}^{d,\nu}$ (polyhomogeneous when the $f_{d-\ell}$ are so), such that $f-\Sigma_{\ell<M} f_{d-\ell}$ is in $S_{1,0}^{d-M,\nu-M}$ for any $M\in\mathbb{N}$. Here f can be constructed by a variant of the usual reconstruction procedure for ps.d.o. symbols, introduced originally in [Hörmander 2], also explained in [Hörmander 8, Section 18.1].

Consider e.g. the case of a sequence of Poisson symbol-kernels $\tilde{k}_{d-\ell}$ lying in $S_{1,0}^{d-\ell,\nu-\ell}(\Xi,\overline{\mathbb{R}}_+^n,\mathscr{S}(\overline{\mathbb{R}}_+))$. Let Ξ' be open, with $\overline{\Xi'}$ compact $\subset\Xi$. Take $\varphi\in C_0^\infty(\mathbb{R}^{n-1})$ equal to 1 in a neighborhood of 0, and let $\chi_\ell(\xi') = 1-\varphi(\varepsilon_\ell\xi')$, where ε_ℓ is a sequence of positive numbers converging to zero so rapidly that

$$\|D_X^\beta D_{\xi'}^\alpha D_\mu^j x_n^k D_{x_n}^{k'}[\chi_\ell(\xi')\tilde{k}_{d-\ell}(X,x_n,\xi',\mu)]\|_{L_{x_n}^2} \le 2^{-\ell}\langle\xi'\rangle^{\nu-\ell-[k-k']_+-|\alpha|+1}\langle\xi',\mu\rangle^{d+\frac{1}{2}+[k-k']_--j} \quad \text{for } \ell\ge\nu,\ |\alpha|+|\beta|+j+k+k'\le\ell .$$

This can be obtained on the basis of the symbol estimates, where we use (2.1.17) in view of the negative regularity. We can now set

$$(2.3.32') \quad \tilde{k}(X,x_n,\xi',\mu) = \sum_{0\le \ell<\nu} \tilde{k}_{d-\ell}(X,x_n,\xi',\mu) + \sum_{\ell\ge\nu} \chi_\ell(\xi')\tilde{k}_{d-\ell}(X,x_n,\xi',\mu) ,$$

where the series converges, because it is locally finite in ξ'. To see that the sum is in $S^{d,\nu}_{1,0}(\Xi', \overline{\mathbb{R}}^n_+, \mathcal{S}(\overline{\mathbb{R}}_+))$, it suffices to show that the sum over $\ell \ge \nu$ is an element of $S^{d,\nu}_{1,0}$, and we here use that for each α,β,j,k,k' we can take $N \ge |\alpha|+|\beta|+j+k+k'$ and estimate

$$\begin{aligned} \| D^\beta_X D^\alpha_{\xi'} D^j_\mu x^k_n D^{k'}_{x_n} \sum_{\ell\ge\nu} \chi_\ell \tilde{k}_{d-\ell}\|_{L^2_{x_n}} &\le \| D^\beta_X D^\alpha_{\xi'} D^j_\mu x^k_n D^{k'}_{x_n} \sum_{\nu\le\ell\le N} \chi_\ell \tilde{k}_{d-\ell}\| \\ &\quad + \| D^\beta_X D^\alpha_{\xi'} D^j_\mu x^k_n D^{k'}_{x_n} \sum_{\ell>N} \chi_\ell \tilde{k}_{d-\ell}\| \\ &\le (C + \sum_{\ell>N} 2^{-\ell})\langle\xi'\rangle^{\nu - [k-k']_+ - |\alpha|} \langle\xi',\mu\rangle^{d+\frac12+[k-k']_- - j} , \end{aligned}$$

for $X \in \Xi'$, $(\xi',\mu) \in \overline{\mathbb{R}}^n_+$ (cf. also (2.1.1')). It is verified in a similar way that $\tilde{k} - \Sigma_{\ell<M}\tilde{k}_{d-\ell}$ is in $S^{d-M,\nu-M}_{1,0}$ for each M. A symbol having these properties on all of Ξ is obtained by a partition of unity.

There are similar proofs for the other symbol classes.

It is obvious from the definitions, that the following mappings are continuous

$$(2.3.33) \qquad S^{d,\nu}_{1,0}(K) \ni f \mapsto D^\beta_X D^\alpha_{\xi'} D^j_\mu f \in S^{d-|\alpha|-j,\nu-|\alpha|}(K) .$$

Also mappings $\xi^k_n D^m_{\xi_n}$ have to be considered, and here it is less obvious how they behave, because of the projections h_{-1} and the different conditions on the polynomial part and the h_{-1} part of the symbols. The following lemma is important for the usefulness of our symbol classes, especially with respect to composition rules.

2.3.9 Lemma. *Let* d *and* $\nu \in \mathbb{R}$, *and* $r \in \mathbb{N}$. *The following mappings are continuous from* $S^{d,\nu}_{1,0}(\Xi,\overline{\mathbb{R}}^n_+, H_{r-1})$

$$(2.3.34) \qquad 1^0 \quad f \mapsto h_{-1}f \in S^{d,\nu}_{1,0}(H_{-1}) ,$$

$$2^0 \quad f \mapsto h^+f \in S^{d,\nu}_{1,0}(H^+) ,$$

3^0 $f \sim h^-_{-1}f \in S^{d,\nu}_{1,0}(H^-_{-1})$,

4^0 $f \sim s_j(X,\xi',\mu) \in S^{d-j,\nu}_{1,0}(\Xi,\overline{\mathbb{R}}^n_+)$ *for* $-\infty < j < r$, *cf.* (2.3.38),

5^0 $f \sim h^-f \in S^{d,\nu}_{1,0}(H^-_{r-1})$,

6^0 $f \sim h_0 f \in S^{d,\nu}_{1,0}(H_0)$,

7^0 $f \sim \xi_n^{m'} f \in S^{d+m',\nu}_{1,0}(H_{r+m'-1})$,

8^0 $f \sim D^m_{\xi_n} f \in S^{d-m,\nu-m}_{1,0}(H_{[r-m]_+-1})$

9^0 $f \sim D^m_{\xi_n}\xi_n^{m'} f \in S^{d',\nu'}_{1,0}(H_{r'-1})$, *with*

$$d' = d-m+m' \ , \ r' = [r-m+m']_+ \ , \ \nu' = \nu - [m-m']_+ \ .$$

Proof: Statement 1^0 follows from the definition, and statements 2^0 and 3^0 are obvious consequences, since h^+ and h^-_{-1} are (complementing) L^2-orthogonal projections in H_{-1} (recall (2.2.27)):

$$\|h^+g\|_0 = \|h^+h_{-1}g\|_0 \le \|h_{-1}g\|_0 \tag{2.3.35}$$
$$\|h^-_{-1}g\|_0 = \|h^-_{-1}h_{-1}g\|_0 = \|h_{-1}g\|_0$$

for $g \in H$, and this applies to all the functions derived from f . Statement 4^0 for $j \ge 0$ follows from the definition (s_j is the coefficient in the expansion (2.3.9)) and then 5^0 follows, since $h^-f = \Sigma_{0\le j<r}\, s_j\xi_n^j + h^-_{-1}f$, and 6^0 follows, since $h_0 f = h_{-1}f + s_0$.

It remains to show 4^0 for $j < 0$, and 7^0-9^0. If f is polynomial

$$f = \sum_{0\le j<r} s_j\xi_n^j \ ,$$

4^0 is trivially verified (since $s_j = 0$ for $j < 0$), and 7^0 is obvious, for

$$D^m_{\xi_m}\xi_n^{m'} f = \sum_{[m'-m]_+\le j<r} s_j c_{j,m,m'}\xi_n^{j-m+m'} \tag{2.3.36}$$

with constants $c_{j,m,m'}$, and $s_j c_{j,m,m'}$ has order $d-j = d'-(j-m+m')$ and regularity $\nu \ge \nu' = \nu - [m-m']_+$ by hypothesis. We can then assume that $f \in S^{d,\nu}_{1,0}(H_{-1})$. Here the coefficients s_{-1-k} are estimated by use of (2.2.48):

$$(2.3.37) \qquad |s_{-1-k}(X,\xi',\mu)|^2 \le 4\pi^{-1}\|h_{-1}(\xi_n^k f(X,\xi,\mu))\|_0 \|h_{-1}(\xi_n^{k+1} f(X,\xi,\mu))\|_0$$

$$\le c(X)[(\rho^\nu+1)\kappa^{d+\frac{1}{2}+k}][(\rho^\nu+1)\kappa^{d+\frac{3}{2}+k}]$$

$$= [c'(X)(\rho^\nu+1)\kappa^{d+1+k}]^2 .$$

With similar estimates of derivatives and lower order parts, one gets all the estimates required for the statement $s_{-1-k} \in S^{d+1+k,\nu}(\Sigma,\overline{\mathbb{R}}^n_+)$, and 4^0 is shown.

Now consider 7^0. Here

$$(2.3.38) \qquad \xi_n^{m'} f = \sum_{-m'\le j<r} s_j \xi_n^{j+m'} + h_{-1}(\xi_n^{m'} f) ,$$

where $s_j \in S^{d-j,\nu}_{1,0}(\Sigma,\overline{\mathbb{R}}^n_+) = S^{d+m'-j-m',\nu}_{1,0}(\Sigma,\overline{\mathbb{R}}^n_+)$ and for the h_{-1}-part we observe that

$$(2.3.39) \qquad h_{-1}D^k_{\xi_n}\xi_n^{k'}h_{-1}\xi_n^{m'}f = h_{-1}D^k_{\xi_n}\xi_n^{k'+m'}f$$

for any k, k' and $m' \in \mathbb{N}$ (for both expressions equal $D^k_{\xi_n}\sum_{j<k-k'-m'} s_j\xi_n^j$). Then the desired estimates for $h_{-1}(\xi_n^{m'}f)$ follows from (2.3.10), e.g.

$$\|h_{-1}(D^k_{\xi_n}\xi_n^{k'}h_{-1}(\xi_n^{m'}f))\| = \|h_{-1}(D^k_{\xi_n}\xi_n^{k'+m'}f)\|$$

$$\le c(X)(\rho^{\nu-[k-k'-m']_+-|\alpha|}+1)\kappa^{d+\frac{1}{2}-k+k'+m'-|\alpha|-j}$$

$$\le c(X)(\rho^{\nu-[k-k']_+-|\alpha|}+1)\kappa^{d+\frac{1}{2}-k+k'+m'-|\alpha|-j} .$$

With similar estimates for derivatives in X, ξ', μ it is seen that 7^0 holds.

The proof of 8^0 follows straightforwardly from the given estimates on f. The general case $D^m_{\xi_n}\xi_n^{m'}f$ follows the pattern of 7^0 when $m' \ge m$ and 8^0 when $m \ge m'$, which leads to (2.3.34) 9^0. □

There is a similar lemma for the s.g.o. symbols.

2.3.10 Lemma. *Let* d *and* $\nu \in \mathbb{R}$, *and* $r \in \mathbb{N}$. *The following mappings are continuous from* $S^{d,\nu}_{1,0}(\Xi,\overline{\mathbb{R}}^n_+, H_{-1}\hat{\otimes}H_{r-1})$:

(2.3.40) 1^0 $f \sim h_{-1,\eta_n} f \in S^{d,\nu}_{1,0}(H_{-1} \hat{\otimes} H_{-1})$,

2^0 $f \sim h^+_{\xi_n} f \in S^{d,\nu}_{1,0}(H^+ \hat{\otimes} H_{r-1})$,

3^0 $f \sim h^-_{-1,\eta_n} f \in S^{d,\nu}_{1,0}(H_{-1} \hat{\otimes} H^-_{-1})$,

4^0 $f \sim k_j(X,\xi,\mu) \in S^{d-j,\nu}_{1,0}(H_{-1})$, $-\infty < j < r$, *cf.* (2.3.41),

5^0 $f \sim h^-_{\eta_n} f \in S^{d,\nu}_{1,0}(H_{-1} \hat{\otimes} H^-_{r-1})$,

6^0 $f \sim h_{-1,\xi_n} D^k_{\xi_n} \xi^{k'}_n D^m_{\eta_n} \eta^{m'}_n f \in S^{d',\nu'}_{1,0}(H_{-1} \hat{\otimes} H^-_{r'-1})$ *with*

$$d' = d-k+k'-m+m' \ , \ r' = [r-m+m']_+ , \ \nu' = \nu-[k-k']_+-[m-m']_+ \ .$$

Proof: The statements 1^0 - 3^0 follow easily from the definition. As for 4^0, k_j is defined for $j \geq 0$ in Definition 2.3.7 (and 4^0 obviously holds in this case); and in general it stands for the coefficient with respect to the expansion in powers of η_n

$$f(X,\xi',\xi_n,\eta_n,\mu) \sim \sum_{-\infty<j<r} k_j(X,\xi',\xi_n,\mu)\eta_n^j \ . \tag{2.3.41}$$

For $j = -1-m$, $m \in \mathbb{N}$, we use the estimates (2.2.48) as in Lemma 2.3.9, setting $f' = h_{-1,\eta_n} f$:

(2.3.42)
$$\begin{aligned} &\int_{\mathbb{R}} |k_{-1-m}(X,\xi',\xi_n,\mu)|^2 d\xi_n \\ &\leq 4\pi^{-1} \int_{\mathbb{R}} \|h_{-1,\eta_n}(\eta_n^m f')\|_{L^2_{\eta_n}} \|h_{-1,\eta_n}(\eta_n^{m+1} f')\|_{L^2_{\eta_n}} d\xi_n \\ &\leq 4\pi^{-1} \|h_{-1,\eta_n}(\eta_n^m f')\|_{L^2_{\xi_n,\eta_n}} \|h_{-1,\eta_n}(\eta_n^{m+1} f')\|_{L^2_{\xi_n,\eta_n}} \\ &\leq c(X)(\rho^\nu+1)^2 \kappa^{d+1+m} \kappa^{d+1+m+1} \\ &= [c'(X)(\rho^\nu+1)\kappa^{d+\frac{1}{2}+1+m}]^2 \ . \end{aligned}$$

With similar estimates for derivatives and lower order parts, and for the functions obtained by applying $h_{-1,\xi_n} D^\ell_{\xi_n} \xi^{\ell'}_n$ (with use of (2.3.39)), we find that $k_{-1-m} \in S^{d+1+m,\nu}(H_{-1})$, showing 4^0 also for $j < 0$. The proofs of 5^0 and 6^0 are now completed much as in Lemma 2.3.9. ▯

One has again the inclusions listed in (2.1.6); and the notation (2.1.7) will be used. Moreover, one can define the spaces $S^{-\infty,\nu-\infty}$ of negligible symbols (or symbol-kernels) of regularity ν, as in (2.1.39). Because of the special rôle of ξ_n-derivatives and factors, the structure is a little complicated. Note that one has, for a and $\sigma \in \mathbb{R}$ and $N \geq \sigma$, by the inequality (A.17)

$$(2.3.43) \qquad (\rho^{\sigma-N}+1)\kappa^{a-N} = \langle\xi'\rangle^{\sigma-N}\langle\xi',\mu\rangle^{a-\sigma} + \langle\xi',\mu\rangle^{a-N} \leq 2\langle\xi'\rangle^{\sigma-N}\langle\xi',\mu\rangle^{a-\sigma}$$
$$\leq c\langle\xi'\rangle^{\sigma-N+|a-\sigma|}\langle\mu\rangle^{a-\sigma} \quad ,$$

and in the converse direction

$$(2.3.44) \qquad \langle\xi'\rangle^{-M}\langle\mu\rangle^{a-\sigma} \leq c\langle\xi'\rangle^{-M+|a-\sigma|}\langle\xi',\mu\rangle^{a-\sigma}$$

for any M.

2.3.11 Lemma.

1^0 *Let* K *equal* H^+, H^-_{-1} *or* H_{-1}. *Then* $f(X,\xi,\mu) \in S^{-\infty,\nu-\infty}(\Xi,\overline{\mathbb{R}}^n_+,K)$ *if and only if* f *is in* K *with respect to* ξ_n *and satisfies, for all indices* $\alpha, \beta, j, k, k', M$,

$$(2.3.45) \qquad \| D^\beta_X D^\alpha_{\xi'} D^j_\mu h_{-1} D^k_{\xi_n} \xi_n^{k'} f \|_{L^2_{\xi_n}} \leq c(X)\langle\xi'\rangle^{-M}\langle\mu\rangle^{\frac{1}{2}-\nu+[k-k']_- - j} \; .$$

2^0 *Let* K *equal* $H^+\hat{\otimes} H^-_{-1}$ *or* $H_{-1}\hat{\otimes} H_{-1}$. *Then* $f(X,\xi',\xi_n,\eta_n,\mu) \in S^{-\infty,\nu-\infty}(\Xi,\overline{\mathbb{R}}^n_+,K)$ *if and only if* f *is in* K *with respect to* (ξ_n,η_n) *and satisfies, for all indices* $\alpha, \beta, j, k, k', m, m', M$,

$$(2.3.46) \qquad \| D^\beta_X D^\alpha_{\xi'} D^j_\mu h_{-1,\xi_n} h_{-1,\eta_n} D^k_{\xi_n} \xi_n^k D^m_{\eta_n} \eta_n^{m'} f \|_{L^2_{\xi_n,\eta_n}}$$
$$\leq c(X)\langle\xi'\rangle^{-M}\langle\mu\rangle^{1-\nu+[k-k']_- + [m-m']_- - j} \; .$$

There are similar statements for negligible symbol-kernels.

Proof: By definition, $S^{-\infty,\nu-\infty} = \cap S^{-N,\nu-N}$. Consider case 1^0, and let $f \in S^{-\infty,\nu-\infty}(K)$. For each index set α, β, j, k, k' we find by taking N large and applying (2.3.43)

$$\|D_X^\beta D_{\xi'}^\alpha D_\mu^j h_{-1} D_{\xi_n}^k \xi_n^{k'} f\|_{L^2_{\xi_n}} \le c(X)\langle\xi'\rangle^{\nu-[k-k']_+-|\alpha|-N}\langle\xi',\mu\rangle^{\frac12-\nu+[k-k']_--j}$$
$$\le c'(x)\langle\xi'\rangle^{-M}\langle\mu\rangle^{\frac12-\nu+[k-k']_--j}$$

where M increases with increasing N. Conversely, if f satisfies the estimates (2.3.45), we can for each index set $\alpha, \beta, j, k, k', N$ take M so large that an application of (2.3.44) gives (2.3.10) with ν and d replaced by $\nu - N$ and $-N$, which shows that $f \in S^{-\infty,\nu-\infty}(K)$. The proof of 2^0 is similar. ▯

2.3.12 Remark. With the order conventions, we have made earlier, we have that the negligible Poisson symbols of order d and regularity ν constitute the space $S^{-\infty,\nu-d+1-\infty}(\Xi,\overline{\mathbb{R}}^n_+,H^+)$. The negligible trace symbols of order d, class r and regularity ν constitute the space $S^{-\infty,\nu-d-\infty}(\Xi,\overline{\mathbb{R}}^n_+,H^-_{r-1})$; they are of the form (2.3.9) with $s_j \in S^{-\infty,\nu-d+j-\infty}(\Xi,\overline{\mathbb{R}}^n_+)$ and $f' \in S^{-\infty,\nu-d-\infty}(\Xi,\overline{\mathbb{R}}^n_+,H^-_{-1})$. The negligible singular Green symbols of order d, class r and regularity ν constitute the space $S^{-\infty,\nu-d+1-\infty}(\Xi,\overline{\mathbb{R}}^n_+,H^+\hat{\otimes}H^-_{r-1})$; here the coefficients k_j in (2.3.21) are in $S^{-\infty,\nu-d+j+1-\infty}(\Xi,\overline{\mathbb{R}}^n_+,H^+)$.

The definition of the original trace, Poisson and singular Green operators was given in Section 1.2 (mainly by use of the symbol-kernel terminology which does not require explanation of the H, H^+, H^- spaces). They are consistent with our new definitions, in that the original operators are the μ-independent elements of the new classes. (When μ is fixed, $\rho^{\nu-N}+1$ can be disregarded, and κ is equivalent with $\langle\xi'\rangle$.) To clarify the relations we list the definitions of the parameter-independent symbols in complete form (cf. also Section 1.2, [Boutet 3], [Rempel-Schulze 1], [Grubb 17]).

2.3.13 Definition. *The original symbol classes (without a special parameter μ) of order $d \in \mathbb{R}$ and class $r \in \mathbb{N}$ are defined as follows.*

1^0 *The space* $S^{d-1}_{1,0}(\Xi,\mathbb{R}^{n-1},H^+)$ *of* $S_{1,0}$ *Poisson symbols of degree* $d-1$, *and order* d, *consists of the* C^∞ *function* $k(X,\xi',\xi_n)$ *that lie in* H^+ *with respect to* ξ_n, *satisfying, for all indices,*

$$(2.3.47) \qquad \| D_X^\beta D_{\xi'}^\alpha h_{-1}[D_{\xi_n}^m \xi_n^{m'} k(X,\xi',\xi_n)]\|_{L^2_{\xi_n}(\mathbb{R})} \le c(x')\langle\xi'\rangle^{d-\frac12-m+m'-|\alpha|} .$$

The space $S^{d-1}(\Xi,\mathbb{R}^{n-1},H^+)$ *of polyhomogeneous Poisson symbols of degree* $d-1$ *and order* d *consists of the functions* $k\in S^{d-1}_{1,0}(\Xi,\mathbb{R}^{n-1},H^+)$ *that have asymptotic expansions* $k\sim\Sigma_{\ell\in\mathbb{N}}k_{d-1-\ell}$ *in functions* $k_{d-1-\ell}$ *that are homogeneous in* ξ *of degree* $d-1-\ell$ *for* $|\xi'|\ge 1$, *such that* $k-\Sigma_{\ell<M}k_{d-1-\ell}\in S^{d-1-M}_{1,0}$ *for all* $M\in\mathbb{N}$.

2^0 *The spaces* $S^d_{1,0}(\Xi,\mathbb{R}^{n-1},H^-_{-1})$ *and* $S^d(\Xi,\mathbb{R}^{n-1},H^-_{-1})$ *of* $S_{1,0}$ *resp. polyhomogeneous trace symbols of degree* d *and class* 0, *and order* d, *are defined as the spaces* S^d *in* 1^0 *with* H^+ *replaced by* H^-_{-1}. *The spaces* $S^d_{1,0}(\Xi,\mathbb{R}^{n-1},H^-_{r-1})$ *and* $S^d(\Xi,\mathbb{R}^{n-1},H^-_{r-1})$ *of* $S_{1,0}$ *resp. polyhomogeneous trace symbols of degree* d *and class* r, *and order* d, *consist of the functions*

$$(2.3.48) \qquad t(X,\xi) = \sum_{0\le j<r} s_j(X,\xi')\xi_n^j + t'(X,\xi)$$

where t' *is as just defined, and* $s_j\in S^{d-j}_{1,0}(\Xi,\mathbb{R}^{n-1})$ *resp.* $S^{d-j}(\Xi,\mathbb{R}^{n-1})$.

3^0 *Inverse Fourier transformation* $F^{-1}_{\xi_n\to x_n}$ *of the spaces* $S^d_{1,0}(\Xi,\mathbb{R}^{n-1},H^+)$ *resp.* $S^d_{1,0}(\Xi,\mathbb{R}^{n-1},H^+)$, *and inverse co-Fourier transformation* $\overline{F}^{-1}_{\xi_n\to x_n}$ *of the spaces* $S^d_{1,0}(\Xi,\mathbb{R}^{n-1},H^-_{-1})$ *resp.* $S^d(\Xi,\mathbb{R}^{n-1},H^-_{-1})$, *give the spaces* $S^d_{1,0}(\Xi,\mathbb{R}^{n-1},\mathscr{S}(\overline{\mathbb{R}}_+))$ *resp.* $S^d(\Xi,\mathbb{R}^{n-1},\mathscr{S}(\overline{\mathbb{R}}_+))$ *of* $S_{1,0}$ *resp. polyhomogeneous symbol-kernels (for (Poisson operators of order* $d+1$ *and trace operators of order* d *and class* 0*); they are described by* (1.2.19), (1.2.21-23).

4^0 *The space* $S^{d-1}_{1,0}(\Xi,\mathbb{R}^{n-1},H^+\hat\otimes H^-_{r-1})$ *of* $S_{1,0}$ *singular Green symbols of degree* $d-1$ *and class* r, *and order* d, *consists of the* C^∞ *functions* $g(X,\xi',\xi_n,\eta_n)$ *that lie in* $H^+\hat\otimes H^-_{r-1}$ *with respect to* (ξ_n,η_n), *such that*

$$(2.3.49) \qquad g(X,\xi',\xi_n,\eta_n) = \sum_{0\le j<r} k_j(X,\xi',\xi_n)\eta_n^j + g'(X,\xi',\xi_n,\eta_n)$$

with $k_j\in S^{d-1-j}_{1,0}(\Xi,\mathbb{R}^{n-1},H^+)$ *and* g' *(of class* 0*) satisfying, for all indices,*

$$(2.3.50) \qquad \| D_X^\beta D_{\xi'}^{\alpha'} h_{-1,\xi_n} h_{-1,\eta_n}[D_{\xi_n}^k \xi_n^{k'} D_{\eta_n}^m \eta_n^{m'} g'(X,\xi',\xi_n,\eta_n)]\|_{L^2_{\xi_n,\eta_n}(\mathbb{R}^2)} \le c(X)\langle\xi'\rangle^{d-k+k'-m+m'-|\alpha|} .$$

The space $S^{d-1}(\Xi,\mathbb{R}^{n-1},H^+\hat{\otimes}H^-_{r-1})$ *of polyhomogeneous singular Green symbols of degree* $d-1$ *and class* r, *and order* d, *consist of the function* $g \in S^{d-1}_{1,0}(\Xi,\mathbb{R}^{n-1},H^+\hat{\otimes}H^-_{r-1})$ *for which* k_j *(in* (2.3.49)*) is polyhomogeneous, and* g' *has an asymptotic expansion* $g' \sim \Sigma_{\ell\in\mathbb{N}} g'_{d-1-\ell}$ *in functions that are homogeneous in* (ξ',ξ_n,η_n) *of degree* $d-1-\ell$ *for* $|\xi'| \geq 1$, *such that* $g' - \Sigma_{\ell<M} g'_{d-1-\ell} \in S^{d-1-M}_{1,0}$ *for all* $M \in \mathbb{N}$.

5^0 *Inverse sesqui-Fourier transformation* $F^{-1}_{\xi_n\to x_n}\overline{F}^{-1}_{\eta_n\to y_n}$ *of the spaces* $S^{d-1}_{1,0}(\Xi,\mathbb{R}^{n-1},H^+\hat{\otimes}H^-_{-1})$ *resp.* $S^{d-1}(\Xi,\mathbb{R}^{n-1},H^+\hat{\otimes}H^-_{-1})$ *gives the spaces* $S^{d-1}_{1,0}(\Xi,\mathbb{R}^{n-1},\mathcal{S}(\overline{\mathbb{R}}^2_{++}))$ *resp.* $S^{d-1}(\Xi,\mathbb{R}^{n-1},\mathcal{S}(\overline{\mathbb{R}}^2_{++}))$ *of* $S_{1,0}$ *resp. polyhomogeneous singular Green symbol-kernels* $\tilde{g}(X,x_n,y_n,\xi')$ *of order* d *and class* 0*; they are described by* (1.2.38-40).

From these estimates it is easily seen how the μ-independent symbols fit into the μ-dependent set-up.

2.3.14 Proposition. *When the symbols listed in Definition 2.3.13 are considered as functions of* μ *that are constant in* μ, *one has the following continuous injections of the symbols spaces into parameter-dependent symbol spaces:*

$$(2.3.51)\qquad S^{d-1}(\Xi,\mathbb{R}^{n-1},H^+) \subset S^{d-1,d-\frac{1}{2}}(\Xi,\overline{\mathbb{R}}^n_+,H^+)\,,$$

$$(2.3.52)\qquad S^{d}(\Xi,\mathbb{R}^{n-1},H^-_{-1}) \subset S^{d,d+\frac{1}{2}}(\Xi,\overline{\mathbb{R}}^n_+,H^-_{-1})\,,$$

$$(2.3.53)\qquad S^{d}(\Xi,\mathbb{R}^{n-1},H^-_{r-1}) \subset S^{d,d-r+1}(\Xi,\overline{\mathbb{R}}^n_+,H^-_{r-1}) \quad \textit{when } r>0\,,$$

$$(2.3.54)\qquad S^{d-1}(\Xi,\mathbb{R}^{n-1},H^+\hat{\otimes}H^-_{-1}) \subset S^{d-1,d}(\Xi,\overline{\mathbb{R}}^n_+,H^+\hat{\otimes}H^-_{-1})\,,$$

$$(2.3.55)\qquad S^{d-1}(\Xi,\mathbb{R}^{n-1},H^+\hat{\otimes}H^-_{r-1}) \subset S^{d-1,d-r+\frac{1}{2}}(\Xi,\overline{\mathbb{R}}^n_+,H^+\hat{\otimes}H^-_{r-1}) \quad \textit{when } r>0\,.$$

Similar inclusions hold for the $S_{1,0}$ *spaces. In the converse direction, one has that the elements in each* $S^{d,\nu}$ *space define elements in the corresponding* $S^d_{1,0}$ *space, when considered for fixed* μ.

Proof. (2.3.51) is concerned with Poisson symbols of order d. Consider here (2.3.47) and write

$$m - m' = [m-m']_+ - [m-m']_- , \tag{2.3.56}$$

cf. (2.3.8). When $[m-m']_+ + |\alpha| \leq d - \frac{1}{2}$, one has à fortiori that $d - \frac{1}{2} - m + m' - |\alpha| \geq 0$, so that

$$\begin{aligned} \langle\xi'\rangle^{d-\frac{1}{2}-m+m'-|\alpha|} &\leq \langle\xi',\mu\rangle^{d-\frac{1}{2}-m+m'-|\alpha|} \\ &\leq (\rho^{d-\frac{1}{2}-[m-m']_+-|\alpha|}+1)\langle\xi',\mu\rangle^{d-\frac{1}{2}-m+m'-|\alpha|} . \end{aligned} \tag{2.3.57}$$

When $[m-m']_+ + |\alpha| \geq d - \frac{1}{2}$, we have since $[m-m']_- \geq 0$,

$$\begin{aligned} \langle\xi'\rangle^{d-\frac{1}{2}-m+m'-|\alpha|} &= \langle\xi'\rangle^{d-\frac{1}{2}-[m-m']_+-|\alpha|}\langle\xi'\rangle^{[m-m']_-} \\ &\leq \langle\xi'\rangle^{d-\frac{1}{2}-[m-m']_+-|\alpha|}\langle\xi',\mu\rangle^{[m-m']_-} \\ &\leq (\rho^{d-\frac{1}{2}-[m-m']_+-|\alpha|}+1)\langle\xi',\mu\rangle^{d-\frac{1}{2}-m+m'-|\alpha|} , \end{aligned} \tag{2.3.58}$$

cf. also (2.1.17). This shows that the symbols satisfy (2.3.10) with $\nu = d - \frac{1}{2}$ (note that the symbols vanish when μ-derivatives are taken).

The proof of (2.3.52) for trace symbols of order d and class 0 is completely similar.

In (2.3.53) for trace symbols of order d and class $r > 0$ we note that the part of class 0 contributes with a regularity $d + \frac{1}{2}$, but now the coefficients s_j in (2.3.48) also enter. Here s_j is of order $d - j$ and therefore generally contributes with regularity $d - j$ (as in (2.1.8)); the lowest regularity is obtained for $j = r - 1$, and this shows (2.3.53), since $-r + 1 < \frac{1}{2}$.

(2.3.54) is shown by inequalities like (2.3.57) and (2.3.58), only with $d - \frac{1}{2}$ replaced by d and $m - m'$ replaced by

$$k - k' + m - m' = [k-k']_+ + [m-m']_+ - [k-k']_- - [m-m']_- . \tag{2.3.59}$$

For (2.3.55) we observe that the part of class 0 contributes with regularity d, whereas the terms k_j in (2.3.49) are of order $d - j$ and hence contribute with regularity $d - j - \frac{1}{2}$; the lowest regularity is here $d - r + \frac{1}{2}$ since $r \geq 1$. The same arguments apply to the $S_{1,0}$ spaces. The last statement is obvious. ∎

Differential trace symbols are of course assigned regularity $+\infty$:

$$\sum_{0\le j<r} s_j(X,\xi')\xi_n^j \in S^{d,\infty}(\Xi,\overline{\mathbb{R}}^n_+,H^-_{r-1}) , \tag{2.3.60}$$

when the s_j are symbols of differential operators, i.e. polynomial in ξ' .

Note also that when $p \in S^d_{tr}(\Xi\times\mathbb{R},\mathbb{R}^n)$, then

$$t(X,\xi) = h^-p(X,\xi) \in S^{d,d+\frac{1}{2}}(\Xi,\overline{\mathbb{R}}^n_+,H^-_d) , \tag{2.3.61}$$

with a better regularity than (2.3.53) gives when $d \ge 0$, this comes from the fact that the s_j coefficients are here polynomial, cf. Theorem 2.2.5.

The inclusions in (2.3.51)-(2.3.55) and (2.3.60) show that our general regularity concept is consistent with the concept introduced in Definition 1.5.14.

Along with the various polyhomogeneous symbol classes, we shall also consider the associated strictly homogeneous symbols , see Section 2.8.

2.4 Operators and kernels.

For each fixed μ, the symbols we have defined lie in the original $S^d_{1,0}$-spaces and hence define operators according to the rules given in Section 1.2. The formulation was given in terms of symbol-kernels; we now also write the formulas with symbols, using oscillatory integrals and the plus-integral (explained in Section 2.2 around formula (2.2.42)). The space called Ξ in Section 2.3 will now be taken to be Ω' or $\Omega'\times\Omega'$, where Ω' is an open subset of $\mathbb{R}^{n-1}$. Pseudo-differential symbols will be defined for $x\in\Sigma$ or $(x,y)\in\Sigma\times\Sigma$ where $\Sigma=\Omega'\times\mathbb{R}$. We set $\Omega=\Omega'\times\mathbb{R}_+$. The following abbreviations will be used:

$$(2.4.1)\quad \begin{array}{l} V' \text{ stands for } (\Omega',\overline{\mathbb{R}}^n_+) \text{ and } V \text{ stands for } (\Omega'\times\mathbb{R},\overline{\mathbb{R}}^{n+1}_+) \\ W' \text{ stands for } (\Omega'\times\Omega',\overline{\mathbb{R}}^n_+) \text{ and } W \text{ stands for } (\Omega'\times\mathbb{R}\times\Omega'\times\mathbb{R},\overline{\mathbb{R}}^{n+1}_+)\ . \end{array}$$

We first list the conventions for operators acting with respect to the x_n-variable only (boundary symbol operators):

$$(2.4.2)\quad [p(x,y,\xi',\mu,D_n)u](x_n)=OP_n(p(x,y,\xi,\mu))u$$
$$=\frac{1}{2\pi}\int e^{i(x_n-y_n)\xi_n}\,p(x,y,\xi,\mu)u(y_n)dy_nd\xi_n\quad ;$$

$$(2.4.3)\quad [p(x,y,\xi',\mu,D_n)_\Omega u](x_n)=r^+p(x,y,\xi',\mu,D_n)e^+u\quad ;$$

$$(2.4.4)\quad [k(x',y',\xi',\mu,D_n)v](x_n)=\frac{1}{2\pi}\int e^{ix_n\xi_n}\,k(x',y',\xi,\mu)vd\xi_n=\tilde{k}(x,y',\xi',\mu)\cdot v$$

$$(2.4.5)\quad t(x',y',\xi',\mu,D_n)u=\frac{1}{2\pi}\int^+ t(x',y',\xi,\mu)\widehat{e^+u}(\xi_n)d\xi_n$$
$$=\sum_{0\le j<r}s_j(x',y',\xi',\mu)\gamma_ju+\int_0^\infty\tilde{t}'(x',y',y_n,\xi',\mu)u(y_n)dy_n\quad ;$$

$$(2.4.6)\quad [g(x',y',\xi',\mu,D_n)u](x_n)=(2\pi)^{-2}\int e^{ix_n\xi_n}\int^+ g(x',y',\xi,\eta_n,\mu)\widehat{e^+u}(\eta_n)d\eta_nd\xi_n$$
$$=\sum_{0\le j<r}\tilde{k}_j(x,y',\xi',\mu)\gamma_ju+\int_0^\infty\tilde{g}'(x',x_n,y',y_n,\xi',\mu)u(y_n)dy_n\quad ;$$

they will also sometimes be denoted $OP_n(p)$, $OPK_n(k)$, $OPT_n(t)$ resp. $OPG_n(g)$. (It is assumed that t and g are decomposed as in (2.3.9) resp. (2.3.22).) Here $u\in\mathscr{S}(\overline{\mathbb{R}}_+)$ and $v\in\mathbb{C}$ (or they can be vector valued); $t(D_n)$ maps into $\mathbb{C}$, and $k(D_n)$ and $g(D_n)$ map into $\mathscr{S}(\overline{\mathbb{R}}_+)$ (or N-tuples of such spaces). Indexations are dropped whenever convenient.

Now the operators in the full space variables are defined by applying the usual ps.d.o. definition in the x'-variable:

$$\text{(2.4.7)}\qquad OP'(a(x',y',\xi',\mu,D_n))w = (2\pi)^{1-n}\int_{\mathbb{R}^{2n-2}} e^{i(x'-y')\cdot\xi'}\, a(x',y',\xi',\mu,D_n)w\, dy'd\xi'$$

where $w = u(y',y_n)$ or $v(y')$, according to which type of operator is being considered.

For y-independent symbols, the formulas can be abbreviated by use of the Fourier transform (as in (2.1.29)).

The resulting operators are also denoted

$$\text{(2.4.8)}\qquad \begin{aligned} P_{\mu,\Omega} &= OP'\ OP_n(p)_\Omega = OP(p)_\Omega, && \text{parameter-dependent ps.d.o.,}\\ K_\mu &= OP'\ OPK_n(k) = OPK(k), && \text{parameter-dependent Poisson operator,}\\ T_\mu &= OP'\ OPT_n(t) = OPT(t), && \text{parameter-dependent trace operator,}\\ G_\mu &= OP'\ OPG_n(g) = OPG(g), && \text{param.-dep. singular Green operator,}\end{aligned}$$

and we shall furthermore need parameter-dependent ps.d.o.s in x',

$$\text{(2.4.9)}\qquad S_\mu = OP'(s(x',y',\xi',\mu)) .$$

Composition of operators (and the symbol compositions this induces) will be indicated by $\circ$, (or $\circ_n$ for compositions with respect to x_n only, if precision is needed). In Definition 2.4.5 below we add certain negligible operators.

For each fixed μ, the operators have the continuity properties listed in Section 1.2; we show the more general μ-dependent versions in Section 2.5.

A system of such operators

$$\text{(2.4.10)}\qquad A_\mu = \begin{pmatrix} P_{\mu,\Omega} + G_\mu & K_\mu \\ T_\mu & S_\mu \end{pmatrix} : \begin{matrix} C^\infty_{(0)}(\Omega'\times\overline{\mathbb{R}}_+)^N \\ \times \\ C^\infty_0(\Omega')^M \end{matrix} \to \begin{matrix} C^\infty(\Omega'\times\overline{\mathbb{R}}_+)^{N'} \\ \times \\ C^\infty(\Omega')^{M'} \end{matrix}$$

is called a parameter-dependent Green operator (it is assumed that P_μ has the transmission property). Here $P_{\mu,\Omega}$ is often called "the pseudo-differential part", and the other entries, or the full matrix

$$(2.4.11)\qquad H_\mu = \begin{pmatrix} G_\mu & K_\mu \\ T_\mu & S_\mu \end{pmatrix} ,$$

are called "the singular part". In this way H_μ gets the label "singular Green operator", which is not really misleading; in fact it can sometimes be fruitful to view K_μ, T_μ and S_μ as special cases of s.g.o.s with finite rank (by an imbedding procedure that we take up again in Section 3.2 when we need it). The word parameter-dependent (or parametrized) is often omitted in the following.

By consideration of scalar products over $\mathbb{R}^n_+$ resp. $\mathbb{R}^{n-1}$ one verifies immediately the following rules for adjoints:

2.4.1 Proposition.

1° $K_\mu = OPK(k(x',y',\xi,\mu))$ *is a Poisson operator with symbol in* $S^{d,\nu}_{1,0}(W',H^+)$ *if and only if its adjoint* K^*_μ *is a trace operator of class* 0, *with symbol in* $S^{d,\nu}_{1,0}(W',H^-_{-1})$; *and here*

$$(2.4.12)\qquad \begin{aligned} &(K_\mu)^* = T_\mu = OPT(t(x',y',\xi,\mu)) \quad \text{with} \\ &t(x',y',\xi,\mu) = \bar{k}(y',x',\xi,\mu) . \end{aligned}$$

For the symbol-kernels, $\tilde{t}(x',y',x_n,\xi',\mu) = \bar{\tilde{k}}(y',x',x_n,\xi',\mu)$.

2° $G_\mu = OPG(g(x',y',\xi,\eta_n,\mu))$ *is a singular Green operator of class* 0 *with symbol in* $S^{d,\nu}_{1,0}(W', H^+ \hat{\otimes} H^-_{-1})$, *if and only if its adjoint* G^*_μ *is a singular Green operator of class* 0 , *with symbol in* $S^{d,\nu}_{1,0}(W', H^+ \hat{\otimes} H^-_{-1})$; *and here*

$$(2.4.13)\qquad \begin{aligned} &G^*_\mu = OPG(g_1(x',y',\xi,\eta_n,\mu)) , \quad \text{with} \\ &g_1(x',y',\xi',\xi_n,\eta_n,\mu) = \bar{g}(y',x',\xi',\eta_n,\xi_n,\mu) . \end{aligned}$$

For the symbol-kernels, $\tilde{g}_1(x,y,\xi',\mu) = \bar{\tilde{g}}(y,x,\xi',\mu)$.

Reductions to x'-form or y'-form take place very similarly to Theorem 2.1.15, we return to this in Theorem 2.4.6 below. In preparation for that, we consider negligible operators. First we observe the following simple rules for composition with the multiplication operator x_n^N .

2.4.2 Lemma. *When* k,t *and* g *are Poisson, trace resp. singular Green symbols, and* M *and* $N \in \mathbb{N}$, *then*

$$x_n^M \circ OPK(k) = OPK(\overline{D}_{\xi_n}^M k)$$

(2.4.14) $$OPT(t) \circ x_n^N = OPT(D_{\xi_n}^N t)$$

$$x_n^M \circ OPG(g) \circ x_n^N = OPG(\overline{D}_{\xi_n}^M D_{\eta_n}^N g) .$$

In particular, there are continuous maps

$$x_n^M : S^{d,\nu}(W', \mathscr{S}(\overline{\mathbb{R}}_+)) \to S^{d-M,\nu-M}(W', \mathscr{S}(\overline{\mathbb{R}}_+)) ,$$

(2.4.15)

$$x_n^M y_n^N : S^{d,\nu}(W', \mathscr{S}(\overline{\mathbb{R}}^2_{++})) \to S^{d-M-N,\nu-M-N}(W', \mathscr{S}(\overline{\mathbb{R}}^2_{++})) .$$

Proof: Follows easily from the definitions (note e.g. that when $\tilde{k}$ is the symbol-kernel associated with the symbol k, then $x_n^M \tilde{k}$ is the symbol-kernel associated with $(i\partial_{\xi_n})^M k = \overline{D}_{\xi_n}^M k)$. The last statement also follows from the definitions (or see Lemmas 2.3.9 - 10). □

The negligible symbols (of the various classes) give rise to integral operators with kernels that are smooth, but have a particular behavior with respect to the x_n-direction, in relation to the μ-dependence. We see from Lemma 2.4.2 that a multiplication by a smooth function $\zeta(x_n)$ supported away from $x_n = 0$ (so that $\zeta(x_n)$ is $O(x_n^M)$ for all x_n) turns the operator into an operator with a negligible symbol. We shall now investigate these negligible operators. It is seen that they retain some global properties with respect to $x_n \in \overline{\mathbb{R}}_+$, and for a convenient calculus on manifolds we shall later have to adjoin negligible operators with a more "loose" behavior for x_n at a distance from 0.

2.4.3 Lemma.

1^0 K_μ *is a Poisson operator defined from a negligible symbol of regularity* $\nu+1$

$$k(x',y',\xi,\mu) \in S^{-\infty,\nu+1-\infty}(W', H^+)$$

if and only if K_μ *is an integral operator from* Ω' *to* $\Omega' \times \overline{\mathbb{R}}_+$ *with a kernel* $r_K(x,y',\mu)$ *satisfying*

$$\| D^\beta_{x',y'} x_n^m D_{x_n}^{m'} D_\mu^j r_K(x,y',\mu) \|_{L^2_{x_n}(\mathbb{R}_+)} \le c(x',y') \langle\mu\rangle^{-\nu-\frac{1}{2}+[m-m']_- -j} \quad (2.4.16)$$

for all indices, $(x',y') \in \Omega' \times \Omega'$ *and* $\mu \geq 0$.

2^0 T_μ *is a trace operator of class* r *defined from a negligible symbol of regularity* ν

$$t(x',y',\xi,\mu) \in S^{-\infty,\nu-\infty}(W', H^-_{r-1})$$

if and only if T_μ *is of the form*

$$(2.4.17) \qquad T_\mu = \sum_{0\le j<r} S_{j,\mu}\gamma_j + T'_\mu \ ,$$

where the $S_{j,\mu}$ *are ps.d.o.s on* Ω', *negligible of regularity* $\nu+j$, *and* T'_μ *is an integral operator from* $\Omega'\times\overline{\mathbb{R}}_+$ *to* Ω' *with a kernel* $r_{T'}(x',y,\mu)$ *satisfying*

$$(2.4.18) \quad \| D^\beta_{x',y'} y_n^m D^{m'}_{y_n} D^j_\mu r_{T'}(x',y,\mu) \|_{L^2_{y_n}(\mathbb{R}_+)} \le c(x',y')\langle\mu\rangle^{-\nu+\frac12+[m-m']_- -j}$$

for all indices, all $(x',y') \in \Omega'\times\Omega'$ *and* $\mu \ge 0$.

3^0 G_μ *is a s.g.o. of class* r *defined from a negligible symbol of regularity* $\nu+1$

$$g(x',y',\xi,\eta_n,\mu) \in S^{-\infty,\nu+1-\infty}(W', H^+ \hat{\otimes} H^-_{r-1})$$

if and only if G_μ *is of the form*

$$(2.4.19) \qquad G_\mu = \sum_{0\le j<r} K_{j,\mu}\gamma_j + G'_\mu \ ,$$

where the $K_{j,\mu}$ *are Poisson operators defined from negligible symbols of regularity* $\nu+1+j$, *and* G' *is an integral operator on* $\Omega'\times\overline{\mathbb{R}}_+$ *with a kernel* $r_{G'}(x,y,\mu)$ *satisfying*

$$(2.4.20) \quad \| D^\beta_{x',y'} x_n^k D^{k'}_{x_n} y_n^m D^{m'}_{y_n} D^j_\mu r_{G'}(x,y,\mu) \|_{L^2_{x_n,y_n}(\mathbb{R}^2_{++})} \le c(x',y')\langle\mu\rangle^{-\nu+[k-k']_+ +[m-m']_- -j}$$

for all indices, all $(x',y') \in \Omega'\times\Omega'$ *and* $\mu \ge 0$.

<u>Proof</u>: 1^0 Let $k \in S^{-\infty,\nu+1-\infty}(W',H^+)$. Then for $v \in C_0^\infty(\Omega') \subset C_0^\infty(\mathbb{R}^{n-1})$,

$$\mathrm{OPK}(k)v = (2\pi)^{-n+1} \int_{\mathbb{R}^{2n-2}} e^{i(x'-y')\xi'} \tilde{k}(x',x_n,y',\xi',\mu) v(y')\,dy'd\xi' \ ,$$

so that OPK(k) is the operator with kernel

$$(2.4.21) \quad r_K(x,y',\mu) = (2\pi)^{-n+1} \int_{\mathbb{R}^{n-1}} e^{i(x'-y')\xi'} \tilde{k}(x',x_n,y',\xi',\mu)\,d\xi'$$

$$= F^{-1}_{\xi'\to z'} \tilde{k}(x',x_n,y',\xi',\mu)\big|_{z'=x'-y'}$$

where $\tilde{K}$ satisfies estimates corresponding to (2.3.45) (with ν replaced by $\nu+1$). For the $L^2_{x_n}$-estimates of r_K and its derivatives it is convenient to consider the expressions, for $v \in L^2(\mathbb{R}_+)$,

$$(D^\beta_{x',y'} x_n^m D^{m'}_{x_n} D^j_\mu r_K \, , \, v(x_n))_{L^2(\mathbb{R}_+)}$$
$$= F^{-1}_{\xi'\to z'} \Big[\int_0^\infty D^\beta_{x',y'} x_n^m D^{m'}_{x_n} \tilde{K}(x',x_n,y',\xi',\mu)\bar{v}(x_n)dx_n \Big]_{z'=x'-y'} .$$

Here one can use the standard estimate for Fourier transforms in $\mathbb{R}^{n-1}$

$$\text{(2.4.22)} \qquad \sup_{x'} |D^\theta_{x'} u(x')| \leq c\|(\xi')^\theta \hat{u}(\xi')\|_{L^1(\mathbb{R})} \leq c_1 \sup_{\xi'} |\langle\xi'\rangle^{n+|\theta|}\hat{u}(\xi')|$$

together with the Cauchy-Schwarz inequality in x_n, to show that the estimates of $\tilde{K}$ imply the estimates (2.4.16).

Conversely, when r_K is given, satisfying the estimates (2.4.16), we define the symbol-kernel $\tilde{K}$ by

$$\text{(2.4.23)} \qquad \tilde{K}(x',x_n,y',\xi',\mu) = r_K(x,y',\mu)e^{i(y'-x')\xi'}h(\xi') ,$$

where $h \in C_0^\infty(\mathbb{R}^{n-1})$, is supported in $\{|\xi'| \leq 1\}$, and has $\int h(\xi')d\xi' = (2\pi)^{n-1}$. Then $\tilde{K}$ is a negligible Poisson symbol-kernel of regularity $\nu+1$, (cf. Lemma 2.3.11), and the Poisson operator, it defines, is the operator with kernel $r_K(x,y',\mu)$.

The proofs of 2° and 3° follow the same lines (in 3°, one uses scalar products over $\mathbb{R}^2_{++}$). ▯

The L^2-estimates (2.4.16) imply the following slightly weaker sup-norm estimates

$$\text{(2.4.24)} \qquad \sup_{x_n} |D^\beta_{x',y'} x_n^m D^{m'}_{x_n} D^j_\mu r_K(x,y',\mu)| \leq c(x',y')\langle\mu\rangle^{-\nu+[m-m']_- -j}$$

by use of the standard inequality (cf. (2. 2.45))

$$\text{(2.4.25)} \qquad |v(x_n)|^2 \leq 2\|v\|_{L^2(\mathbb{R}_+)} \|D_{x_n} v\|_{L^2(\mathbb{R}_+)} .$$

In the other direction, if r_K satisfies (2.4.24) and we define $\tilde{K}$ by (2.4.23), we can only conclude that

$$\text{(2.4.26)} \quad \|D^\beta_{x',y'} x_n^m D^{m'}_{x_n} D^\alpha_{\xi'} D^j_\mu \tilde{K}\|_{L^2(\mathbb{R}_+)} \leq c(x',y')\langle\xi'\rangle^{-M}\langle\mu\rangle^{-\nu+[m-m']_- -j} ,$$

with regularity $\frac{1}{2}$ weaker than in the class $S^{-\infty,\nu+1-\infty}$. The L^2-norms are preferable to sup-norms, because of the sharp correlation they give between estimates of $\tilde{k}$ and r_K.

Observe however, that at positive distances from the boundary $\{x_n = 0\}$ one gets much better estimates. Let

$$\zeta(x_n) \in C^\infty(\mathbb{R}) , \quad \zeta = 0 \text{ near } x_n = 0 , \quad \zeta = 1 \text{ for } |x_n| \geq a > 0 . \tag{2.4.27}$$

Then when $\tilde{k} \in S^{-\infty,\nu+1-\infty}(W', \mathscr{S}(\overline{\mathbb{R}}_+))$, we find, writing (for large N)

$$\zeta(x_n)\tilde{k} = \frac{\zeta(x_n)}{x_n^N} x_n^N \tilde{k} , \qquad D_{x_n}(\zeta\tilde{k}) = \frac{D\zeta}{x_n^N} x_n^N \tilde{k} + \frac{\zeta}{x_n^N} x_n^N D\tilde{k} ,$$

etc., that r_K satisfies estimates

$$|D^\beta_{x',y'} x_n^m D_{x_n}^{m'} D_\mu^j \zeta r_K(x,y',\mu)| \leq c(x',y')\langle\mu\rangle^{-\nu-\frac{1}{2}-j} \tag{2.4.28}$$

for all indices β, m, m', j.

Similar observations hold for the other kinds of operators. This motivates the following definitions.

<u>2.4.4 Definition</u>. *Let* $0 < b < a \leq \infty$ *and let* $I_a = [0,a[$, $I_b = [0,b[$.

1^0 *A (parameter-dependent) negligible Poisson operator* K_μ *on* $\Omega' \times I_a$ *of regularity* $\nu+1$ *is an integral operator from* Ω' *to* $\Omega' \times I_a$ *with kernel* $r_K(x,y',\mu)$ *satisfying the estimates*

$$\| D^\beta_{x',y'} x_n^m D_{x_n}^{m'} D_\mu^j r_K(x,y',\mu) \|_{L^2_{x_n}(I_b)} \leq c(x',y')\langle\mu\rangle^{-\nu-\frac{1}{2}+[m-m']_- -j} \tag{2.4.29}$$

and, for any $\varphi \in C_0^\infty(\Omega' \times]0,a[\times \Omega')$,

$$\sup_{x,y'} |\varphi(x,y') D^\theta_{x,y'} D_\mu^j r_K(x,y',\mu)| \leq c\langle\mu\rangle^{-\nu-\frac{1}{2}-j} , \tag{2.4.30}$$

for all indices. When (2.4.30) *moreover holds for any* $\varphi \in C^\infty_{(0)}(\Omega' \times I_a \times \Omega')$, K_μ *is said to be* <u>*uniformly*</u> *negligible of regularity* $\nu+1$.

2^0 *A (parameter-dependent) negligible trace operator* T_μ *on* $\Omega' \times I_a$ *of class* r *and regularity* ν *is an operator*

$$T_\mu = \sum_{0 \leq j < r} S_{j,\mu} \gamma_j + T'_\mu ,$$

where the $S_{j,\mu}$ *are negligible ps.d.o.s on* Ω' *of regularity* $\nu+j$, *and*

T'_μ *is an integral operator from* $\Omega'\times I_a$ *to* Ω' *with kernel* $r_{T'}(x',y,\mu)$ *satisfying the estimates*

$$(2.4.31)\quad \| D^\beta_{x',y'} y_n^m D^{m'}_{y_n} D^j_\mu r_{T'}(x',y,\mu)\|_{L^2_{y_n}(I_b)} \le c(x',y')\langle\mu\rangle^{-\nu+\frac12+[m-m']_- -j}$$

and, for any $\varphi \in C_0^\infty(\Omega'\times\Omega'\times]0,a[)$,

$$(2.4.32)\qquad \sup_{x',y} |\varphi(x',y) D^\theta_{x',y} D^j_\mu r_{T'}(x',y,\mu)| \le c\langle\mu\rangle^{-\nu+\frac12-j} ,$$

for all indices. When (2.4.32) *moreover holds for any* $\varphi \in C^\infty_{(0)}(\Omega'\times\Omega'\times I_a)$, T_μ *is said to be* *uniformly* *negligible of regularity* ν .

3^0 *A (parameter-dependent) negligible singular Green operator* G_μ *on* $\Omega'\times I_a$ *of class* r *and regularity* $\nu+1$ *is an operator*

$$G_\mu = \sum_{0\le j<r} K_{j,\mu}\gamma_j + G'_\mu ,$$

where the $K_{j,\mu}$ *are negligible Poisson operators on* $\Omega'\times I_a$ *of regularity* $\nu+1+j$, *and* G'_μ *is an integral operator on* $\Omega'\times I_a$ *with kernel* $r_{G'}(x,y,\mu)$ *satisfying the estimates*

$$(2.4.33)\quad \| D^\beta_{x',y'} x_n^k D^{k'}_{x_n} y_n^m D^{m'}_{y_n} r_{G'}(x,y,\mu)\|_{L^2_{x_n,y_n}(I_b\times I_b)}$$
$$\le c(x',y')\langle\mu\rangle^{-\nu+[k-k']_- +[m-m']_- -j} ,$$

and, for any $\varphi \in C_0^\infty(\Omega'\times]0,a[\times\Omega'\times]0,a[)$,

$$(2.4.34)\qquad \sup_{x,y} |\varphi(x,y) D^\theta_{x,y} D^j_\mu r_{G'}(x,y,\mu)| \le c\langle\mu\rangle^{-\nu-j} ,$$

for all indices. When (2.4.34) *moreover holds for any* $\varphi \in C^\infty_{(0)}(\Omega'\times I_a\times\Omega'\times I_a)$, G_μ *is said to be* *uniformly* *negligible of regularity* $\nu+1$.

Note in particular, that the space of uniformly negligible singular Green operators of class 0 and regularity $\nu+1$ coincides with the space of integral operators on $\Omega'\times\overline{\mathbb{R}}_+$ with C^∞ kernels satisfying (2.1.40), i.e. the operators P_Ω (for $\Omega = \Omega'\times\mathbb{R}_+$) where P is a negligible ps.d.o. of regularity ν on $\Omega'\times\mathbb{R}$.

The complete operator classes consist of the respective operators defined from symbols, with negligible operators added to them. (It does not suffice, as one might expect from the ps.d.o. case, to adjoin the operators defined from symbols on (x',y')-form. However, it would be possible to formulate the general operators in a calculus where the Poisson, trace and Green symbols depend moreover on x_n, resp. y_n, resp. (x_n,y_n). Such operators are taken up in Remark 2.4.9.)

2.4.5 Definition. *Let* d *and* $\nu \in \mathbb{R}$, *and* $r \in \mathbb{N}$.

1^0 *A (parameter-dependent) Poisson operator on* $\Omega' \times \overline{\mathbb{R}}_+$ *of order* d *and regularity* ν *is the sum of an operator* $\mathrm{OPK}(k(x',y',\xi,\mu))$ *with* $k \in S^{d-1,\nu}(W',H^+)$ *and a negligible Poisson operator on* $\Omega' \times \overline{\mathbb{R}}_+$ *of regularity* $\nu+1-d$.

2^0 *A (parameter-dependent) trace operator on* $\Omega' \times \overline{\mathbb{R}}_+$ *of order* d, *class* r *and regularity* ν *is the sum of an operator* $\mathrm{OPT}(t(x',y',\xi,\mu))$ *with* $t \in S^{d,\nu}(W',H^-_{r-1})$ *and a negligible trace operator on* $\Omega' \times \overline{\mathbb{R}}_+$ *of regularity* $\nu-d$.

3^0 *A (parameter-dependent) singular Green operator on* $\Omega' \times \overline{\mathbb{R}}_+$ *of order* d, *class* r *and regularity* ν *is the sum of an operator* $\mathrm{OPG}(g(x',y',\xi,\eta_n,\mu))$ *with* $g \in S^{d-1,\nu}(W',H^+ \hat{\otimes} H^-_{r-1})$ *and a negligible singular Green operator of class* r *and regularity* $\nu+1-d$.

The terminology is also used for operators going from Σ_1 to Σ_2 (for Σ_1 and Σ_2 relatively open in Ω' or $\Omega' \times \overline{\mathbb{R}}_+$, respectively), defined from one of the above operators by first injecting $C^\infty_{(0)}(\Sigma_1)$ into $C^\infty_{(0)}(\Omega')$ resp. $C^\infty_{(0)}(\Omega' \times \overline{\mathbb{R}}_+)$, and afterwards restricting to Σ_2.

The definitions extend in an obvious way to matrix formed operators; here M×N-matrix formed symbols are indicated by adjoining $\otimes L(\mathbb{C}^N,\mathbb{C}^M)$ to the symbol spaces.

We use these conventions also on the boundary symbol operator level.

After the concept of negligible symbols and operators has been clarified, it is easy to formulate the rules for reduction of symbols and passage to adjoints.

2.4.6 Theorem. *Let* d *and* $\nu \in \mathbb{R}$, *and* $r \in \mathbb{N}$. *The sums in the following are for* $\alpha \in \mathbb{N}^{n-1}$.

1^0 *Let* $K = H^+$, H^-_{r-1} *or* $H^+ \hat{\otimes} H^-_{r-1}$; *and let* $a(x',y',\eta,\mu) \in S^{d,\nu}(\Omega' \times \Omega', \overline{\mathbb{R}}^n_+, K)$; *here* η *stands for* (ξ',ξ_n) *or* (ξ',ξ_n,η_n), *respectively.*

Let $a(x',y',\xi',\mu,D_n)$ *be the associated operator on* $\mathbb{R}_+$ *(defined as* $OPK_n(a)$, $OPT_n(a)$ *or* $OPG_n(a)$, *respectively). Then there are symbols* $a_1(x',\eta,\mu)$ *and* $a_2(y',\eta,\mu)$ *in* $S^{d,\nu}(\Omega',\overline{\mathbb{R}}^n_+,K)$ *such that*

$$\text{(i)}\quad OP'(a(x',y',\xi',\mu,D_n)) \sim OP'(a_1(x',\xi',\mu,D_n)) \sim OP'(a_2(y',\xi',\mu,D_n)) ,$$

$$(2.4.35)\quad \text{(ii)}\quad a_1(x',\eta,\mu) \sim \sum_\alpha \frac{1}{\alpha!} D^\alpha_{\xi'} \partial^\alpha_{y'} a(x',y',\eta,\mu)|_{y'=x'}$$

$$\text{(iii)}\quad a_2(y',\eta,\mu) \sim \sum_\alpha \frac{1}{\alpha!} \overline{D}^\alpha_{\xi'} \partial^\alpha_{x'} a(x',y',\eta,\mu)|_{x'=y'}$$

modulo regularity $\nu-d$.

2^0 *Let* $k(x',\xi,\mu) \in S^{d,\nu}(\Omega',\overline{\mathbb{R}}^n_+,H^+)$. *Then there is a symbol* $t(x',\xi,\mu) \in S^{d,\nu}(\Omega',\overline{\mathbb{R}}^n_+,H^-_{-1})$ *such that*

$$(2.4.36)\quad \text{(i)}\quad OPK(k)^* \sim OPT(t) , \quad \textit{with}$$

$$\text{(ii)}\quad t(x',\xi,\mu) \sim \sum_\alpha \frac{1}{\alpha!} D^\alpha_{\xi'} \partial^\alpha_{x'} \overline{k}(x',\xi,\mu)$$

modulo regularity $\nu-d$. *Similarly, when* $t(x',\xi,\mu) \in S^{d,\nu}(\Omega',\overline{\mathbb{R}}^n_+,H^-_{-1})$, *there is a symbol* $k(x',\xi,\mu) \in S^{d,\nu}(\Omega',\overline{\mathbb{R}}^n_+,H^+)$ *such that*

$$(2.4.37)\quad \text{(i)}\quad OPT(t)^* \sim OPK(k) , \quad \textit{with}$$

$$\text{(ii)}\quad k(x',\xi,\mu) \sim \sum_\alpha \frac{1}{\alpha!} D^\alpha_{\xi'} \partial^\alpha_{x'} \overline{t}(x',\xi,\mu) ,$$

modulo regularity $\nu-d$.

3^0 *Let* $g(x',\xi,\eta_n,\mu) \in S^{d,\nu}(\Omega',\overline{\mathbb{R}}^n_+,H^+ \hat{\otimes} H^-_{-1})$. *Then there is a symbol* $g_1(x',\xi,\eta_n,\mu) \in S^{d,\nu}(\Omega',\overline{\mathbb{R}}^n_+,H^+ \hat{\otimes} H^-_{-1})$ *such that*

$$(2.4.38)\quad \text{(i)}\quad OPG(g)^* \sim OPG(g_1) , \quad \textit{with}$$

$$\text{(ii)}\quad g_1(x',\xi',\xi_n,\eta_n,\mu) \sim \sum_\alpha \frac{1}{\alpha!} D^\alpha_{\xi'} \partial^\alpha_{x'} \overline{g}(x',\xi',\theta_n,\zeta_n,\mu)|_{\substack{\theta_n=\eta_n \\ \zeta_n=\xi_n}}$$

modulo regularity $\nu-d$.

In 2^0 *and* 3^0, *the complex conjugate is replaced by the adjoint in case of matrices. - There are similar rules for the symbol-kernels.*

<u>Proof</u>: One uses a modification of the proof of Theorem 2.1.15 (applied with respect to the x' variables), taking the rules in Proposition 2.4.1 into account, replacing sup norms by L^2-norms in the appropriate places, and applying the "reconstruction principle" described in Remark 2.3.8'. □

Observe the following immediate consequence of point 1^0 in the theorem:

2.4.7 Corollary. *Notation as in Theorem* 2.4.6. *Let* φ *and* ψ *be* C^∞ *functions on* $\mathbb{R}^{n-1}$ *such that* $\varphi\psi = 0$. *Then if* $a(x',y',\eta,\mu) \in S^{d,\nu}(\Omega'\times\Omega',\overline{\mathbb{R}}^n_+,K)$ *the operator* $\varphi\,OP'(a(x',y',\xi',\mu,D_n))\psi$ *is negligible of regularity* ν-d.

Proof: We observe that $\varphi = 0$ on any open set where $\psi \neq 0$, and vice versa. The considered operator has the symbol $\varphi(x')a(x',y',\eta,\mu)\psi(y')$, which by reduction to x'-form as in (2.4.35)(ii) gives zero (since $\partial^\alpha_{y'}\psi$ vanishes on the open set where $\varphi \neq 0$) . Then the operator is negligible of regularity ν-d . ◻

Concerning cut-off functions in the normal variable, one has:

2.4.8 Lemma. *Let* K_μ , T_μ *and* G_μ *be Poisson, trace and singular Green operators on* $\Omega'\times\overline{\mathbb{R}}_+$ *of order* d , *class* r *and regularity* ν . *Let* $\varphi(x_n)$ *and* $\psi(x_n)$ *be as in* (2.4.27). *Then one has for the operators composed with multiplication by* φ *or* ψ :

1^0 $\varphi\circ K_\mu$ *is uniformly negligible of regularity* $\nu+1-d$.

2^0 $T_\mu\circ\psi$ *is of class* 0 *and is uniformly negligible of regularity* $\nu-d$.

3^0 $\varphi\circ G_\mu$ *is negligible of regularity* $\nu+1-d$.

4^0 $G_\mu\circ\psi$ *is of class* 0 *and negligible of regularity* $\nu+1-d$.

5^0 $\varphi\circ G_\mu\circ\psi$ *is of class* 0 *and uniformly negligible of regularity* $\nu+1-d$.

Proof: Let $K_\mu = OPK(k) + R_\mu$, where R_μ is negligible of regularity $\nu+1-d$. Since we may write, for any $N \in \mathbb{N}$,

$$\varphi(x_n)\tilde{k} = \frac{\varphi(x_n)}{x_n^N}\, x_n^N\tilde{k} , \tag{2.4.39}$$

where $x_n^N\tilde{k} \in S^{d-1-N,\nu-N}$, the contribution from $\varphi\circ OPK(k)$ is also negligible of regularity $\nu+1-d$. Since $\varphi = \zeta\varphi$ for some ζ as in (2.4.27), with a smaller a , the argument following (2.4.27) show that K_μ is indeed uniformly negligible of regularity $\nu+1-d$.

The proofs of 2^0 and 5^0 are similar (note that the terms with γ_j are cancelled). In case 3^0 and 4^0 we just get negligible terms, since either the y_n-derivatives or the x_n-derivatives are not so well controlled. ◻

2.4.9 Remark. It can sometimes be advantageous to work in a framework where one allows x_n-dependent Poisson symbols, y_n-dependent trace symbols and (x_n,y_n)-dependent s.g.o. symbols, and one defines (by oscillatory integrals)

$$(K_\mu v)(x) = (2\pi)^{-n}\int e^{i(x'-y')\cdot\xi' + ix_n\xi_n} k(x,y',\xi,\mu)v(y')dy'd\xi \ ,$$

$$(2.4.40)\quad (T_\mu u)(x') = (2\pi)^{-n}\int e^{i(x'-y')\cdot\xi' - iy_n\xi_n} t(x',y,\xi,\mu)u(y)dyd\xi \ ,$$

$$(G_\mu u)(x) = (2\pi)^{-n-1}\int e^{i(x'-y')\cdot\xi'+ix_n\xi_n-iy_n\eta_n} g(x,y,\xi,\eta_n,\mu)u(y)dyd\xi d\eta_n \ ,$$

where $k \in S^{d-1,\nu}_{1,0}(\Omega'\times\overline{\mathbb{R}}_+\times\Omega', \overline{\mathbb{R}}^n_+,H^+)$, $t \in S^{d,\nu}_{1,0}(\Omega'\times\Omega'\times\overline{\mathbb{R}}_+, \overline{\mathbb{R}}^n_+,H^-_{r-1})$, and $g \in S^{d-1,\nu}_{1,0}(\Omega'\times\overline{\mathbb{R}}_+\times\Omega'\times\overline{\mathbb{R}}_+, \overline{\mathbb{R}}^n_+,H^+\hat{\otimes} H^-_{r-1})$. To show that these operators are Poisson, trace resp. singular Green operators of order d and regularity ν, in the same sense as Definition 2.4.5, one uses a reduction to eliminate x_n, y_n, resp. (x_n,y_n) from the symbol, by insertion of Taylor expansions in x_n, y_n, resp. (x_n,y_n), and estimates of remainders somewhat like in the proof of Theorem 2.1.15 1°. We show the details for K_μ, taking $\Omega' = \mathbb{R}^{n-1}$ for simplicity. Insert

$$k(x,y',\xi,\mu) = \sum_{j<M}\frac{1}{j!}x_n^j\partial^j_{x_n}k(x',0,y',\xi,\mu) + x_n^M k_{(M)}(x',x_n,y',\xi,\mu)$$

in the above definition of $K_\mu v$; then we have in view of (2.4.14):

$$K_\mu = \mathrm{OPK}\Big(\sum_{j<M}\frac{1}{j!}\overline{D}^j_{\xi_n}\partial^j_{x_n}k(x',0,y',\xi,\mu)\Big) + K_{\mu,(M)} \ ,$$

where $K_{\mu,(M)}$ satisfies (formally by integration by parts)

$$K_{\mu,(M)}v = (2\pi)^{-n}\int e^{i(x'-y')\cdot\xi'+ix_n\xi_n} x_n^M k_{(M)}(x,y',\xi,\mu)v(y')dy'd\xi$$

$$= (2\pi)^{-n}\int e^{i(x'-y')\cdot\xi'+ix_n\xi_n} \overline{D}^M_{\xi_n} k_{(M)}(x,y',\xi,\mu)v(y')dy'd\xi \ .$$

Since $k_{(M)}$ is in $S^{d-1,\nu}_{1,0}(\overline{\mathbb{R}}^{2n-1}_+, \overline{\mathbb{R}}^n_+,H^+)$, the derivative $\overline{D}^M_{\xi_n}k_{(M)}$ is in $S^{d-1-M,\nu-M}$, so $K_{\mu,(M)}$ is an operator with kernel

$$K_{(M)}(x,y,\mu) = (2\pi)^{-n}\int e^{i(x'-y')\cdot\xi'+ix_n\xi_n} k_{(M)}(x,y',\xi,\mu)d\xi$$

satisfying (in view of (2.3.43))

$$\|x_n^m D^{m'}_{x_n} D_{x',y'} D^j_\mu K_{(M)}\|_{L^2_{x_n}([0,a])} \le c(x',y')\langle\mu\rangle^{d-\frac12-\nu+[m-m']_- - j} \ , \quad \text{any } a > 0 \ ,$$

for M large in comparison with m,m', $|\alpha|$ and j. Let $k_1(x',y',\xi,\mu)$ be a symbol with the asymptotic expansion

$$(2.4.41)\qquad k_1(x',y',\xi,\mu) \sim \sum_{j\in\mathbb{N}} \frac{1}{j!}\overline{D}^j_{\xi_n}\partial^j_{x_n} k(x',0,y',\xi,\mu)\ .$$

Then since

$$K_\mu - OPK(k_1) = K_{\mu,(M)} - OPK([k_1 - \sum_{j<M}\frac{1}{j!}\overline{D}^j_{\xi_n}\partial^j_{x_n}k])$$

for any M, where the symbol in [ ] is in $S^{d-1-M,\nu-M}_{1,0}$, the kernel of $K_\mu - OPK(k_1)$ satisfies all the estimates (2.4.16) required for the kernel of a negligible Poisson operator of regularity $\nu-d+1$. Thus K_μ is a Poisson operator, with symbol (2.4.41).

With suitable modifications of the above proof, one likewise finds that T_μ and G_μ are (parametrized) trace, resp. singular Green operators, with symbols, respectively,

$$(2.4.42)\qquad \begin{aligned} t_1(x',y',\xi,\mu) &\sim \sum_{j\in\mathbb{N}} \frac{1}{j!} D^j_{\xi_n}\partial^j_{y_n} t(x',y',0,\xi,\mu)\\ g_1(x',y',\xi,\mu) &\sim \sum_{j,k\in\mathbb{N}} \frac{1}{j!k!}\overline{D}^j_{\xi_n} D^k_{\eta_n}\partial^j_{x_n}\partial^k_{y_n} g(x',0,y',0,\xi,\eta_n,\mu)\ .\end{aligned}$$

The symbols can of course be further reduced as in Theorem 2.4.6. Note also that when k, t or g is polyhomogeneous, then so is k_1, t_1 resp. g_1.

We finally consider coordinate changes. One finds by a straightforward application of the proof of Lemma 2.1.17 that coordinate changes in x' alone preserve the symbol and operator classes. However, for a truly invariant definition of the operators on manifolds, the classes should also be independent of the choice of normal coordinate, which means that one should have invariance under coordinate changes $\kappa\colon \Omega'\times\overline{\mathbb{R}}_+ \to \underline{\Omega}'\times\overline{\mathbb{R}}_+$ sending $\Omega'\times\{0\}$ into $\underline{\Omega}'\times\{0\}$ - or, if we restrict the attention to the case $\Omega' = \underline{\Omega}' = \mathbb{R}^{n-1}$ (which is no essential limitation), invariance under coordinate changes in $\overline{\mathbb{R}}^n_+$ preserving the set $\{x_n=0\}$. The literature on the parameter-independent case does not give much information on this; so we shall give a proof that covers both parameter-dependent and parameter-independent, $S_{1,0}$ and polyhomogeneous cases. (The author was greatly helped by L. Hörmander in working this out, and we also owe Remark 2.4.12 to him.)

We use the notation and formulas established in the proof of Theorem 2.2.12 3° see (2.2.94-102). The diffeomorphism $\kappa\colon x \mapsto \underline{x}$ preserves the positivity of x_n. (This would have been natural to assume also in Theorem 2.2.12, but is not strictly necessary there, since the order d is integer; see [Hörmander 8, Section 18.2] for a treatment of general orders.) Since $\{x_n > 0\}$ is mapped into $\{\underline{x}_n > 0\}$, the function $C_1(x)$ in (2.2.94) is > 0. The induced diffeomorphism in $\partial\overline{\mathbb{R}}^n_+ = \mathbb{R}^{n-1}$ will be called λ,

$$\lambda(x') = (\kappa_1(x',0), \ldots, \kappa_{n-1}(x',0)) \,. \tag{2.4.43}$$

When K, T and G are Poisson, trace and singular Green operators in $\overline{\mathbb{R}}^n_+$, we define the transformed operators $\underline{K}$, $\underline{T}$ and $\underline{G}$ by, respectively,

$$\begin{aligned} &\text{(i)} && (\underline{K}v)\circ\kappa = K(v\circ\lambda) \,, \\ (2.4.44)\quad &\text{(ii)} && (\underline{T}u)\circ\lambda = T(u\circ\kappa) \,, \\ &\text{(iii)} && (\underline{G}u)\circ\kappa = G(u\circ\kappa) \,, \end{aligned}$$

when v and u are functions on $\mathbb{R}^{n-1}$ resp. $\overline{\mathbb{R}}^n_+$.

Consider first the negligible operators.

2.4.10 Proposition. *The spaces of negligible Poisson, trace or singular Green operators in* $\overline{\mathbb{R}}^n_+$ *of a given regularity are invariant under coordinate changes preserving the boundary* $\{x_n=0\}$.

Proof: When u is a function of x, we set

$$\underline{u}(\underline{x}) = u(\kappa^{-1}(\underline{x})) = u(x) \,, \qquad \text{where} \quad \underline{x} = \kappa(x) \,. \tag{2.4.45}$$

By (2.2.94) applied to κ and its inverse,

$$\underline{x}_n = C_1(x)x_n \qquad \text{and} \qquad x_n = C_2(x)\underline{x}_n \tag{2.4.46}$$

for certain positive C^∞ functions C_1 and C_2; then we furthermore have for $j < n$

$$\begin{aligned} \partial_{\underline{x}_j}\underline{u}(\underline{x}) &= \sum_{1\le k\le n} \frac{\partial x_k}{\partial \underline{x}_j}\partial_{x_k}u(x) \\ &= \sum_{1\le k\le n-1} \frac{\partial x_k}{\partial \underline{x}_j}\partial_{x_k}u(x) + \partial_{\underline{x}_j}C_2(x)\underline{x}_n\partial_{x_n}u(x) \\ &= \sum_{1\le k\le n-1} b_{kj}(x)\partial_{x_k}u(x) + c_j(x)x_n\partial_{x_n}u(x) \end{aligned} \tag{2.4.47}$$

with C^∞ functions b_{kj} and c_j . Moreover,

$$\partial_{\underline{x}_n}\underline{u}(\underline{x}) = \sum_{1\le k\le n} \frac{\partial x_k}{\partial \underline{x}_n}\partial_{x_k}u(x) = \sum_{1\le k\le n} b_{kn}(x)\partial_{x_k}u(x) \,. \tag{2.4.48}$$

Let G_μ be an integral operator with kernel $r_G(x,y,\mu)$. Then the transformed operator $\underline{G}_\mu$ defined by (2.4.44 iii) satisfies

$$(\underline{G}_{\mu}u)(\underline{x}) = \int_{\mathbb{R}^n_+} r_G(x,y,\mu)u(y)dy$$

$$= \int_{\mathbb{R}^n_+} r_G(\kappa^{-1}(\underline{x}),\kappa^{-1}(\underline{y}),\mu)\underline{u}(\underline{y})\,|\det \kappa'(y)^{-1}|\,d\underline{y}$$

so the kernel of $\underline{G}_{\mu}$ is

(2.4.49) $r_{\underline{G}}(\underline{x},\underline{y},\mu) = r_G(x,y,\mu)\,|\det \kappa'(y)^{-1}|$ where $\underline{x} = \kappa(x)$, $\underline{y} = \kappa(y)$.

Now if G_{μ} is negligible of regularity $\nu+1$, so that r_G satisfies the estimates (2.4.33), we have in view of (2.4.46), for a suitable $b_1(x',y') > 0$,

$$\|r_{\underline{G}}(\underline{x},\underline{y},\mu)\|_{L^2_{\underline{x}_n,\underline{y}_n}([0,b_1]\times[0,b_1])}$$

$$\leq c(x',y')\|r_G(x,y,\mu)\|_{L^2_{x_n,y_n}([0,b]\times[0,b])}$$

$$\leq c'(x',y')\langle\mu\rangle^{-\nu} .$$

Moreover, we have for the derivatives, by (2.4.47-48):

$$\|\underline{x}_n^k\partial_{\underline{x}_n}^{k'} r_{\underline{G}}\|_{L^2_{\underline{x}_n,\underline{y}_n}} = \|C_1(x)^k x_n^k(\sum_{1\leq j\leq n} b_{jn}\partial_{x_j})^{k'} r_G\|_{L^2_{\underline{x}_n,\underline{y}_n}} \leq c(x',y')\langle\mu\rangle^{-\nu+[k-k']_-}$$

(the worst contribution comes from $x_n^k\partial_{x_n}^{k'}$) and, for $j < n$,

$$\|\partial_{\underline{x}_j} r_{\underline{G}}\| = \|\sum_{1\leq k\leq n-1}\frac{\partial x_k}{\partial \underline{x}_j}\partial_{x_k} r_G + c_j x_n\partial_{x_n} r_G\| \leq c(x',y')\langle\mu\rangle^{-\nu} ,$$

with similar estimates for the $\underline{y}$-derivatives. Application of D^j_{μ} of course improve the estimates by a factor $c\langle\mu\rangle^{-j}$. This shows the general pattern of how the derivatives behave, and it is found altogether that $r_{\underline{G}}(\underline{x},\underline{y},\mu)$ satisfies the estimates (2.4.33), so that $\underline{G}_{\mu}$ is a negligible singular Green operator of regularity $\nu+1$. (At a distance from the boundary, we have the simpler estimates for uniformly negligible operators.)

The other types of symbols and operators are treated similarly. □

Now let us treat the general operators.

2.4.11 Theorem. *The spaces of parameter-dependent polyhomogeneous, resp.* $S_{1,0}$, *Poisson, trace and singular Green operators of a given order, class and regularity, are invariant under diffeomorphisms of* $\overline{\mathbb{R}}^n_+$ *preserving the boundary* $\{x_n=0\}$. *Formulas for symbols of the transformed operators (using the notation of Lemma 2.1.17, Theorem 2.2.12 3° and* (2.4.44)*), are given in* (2.4.53, 57-59') *for Poisson operators, in* (2.4.60-61) *for trace operators, and in* (2.4.63-66) *for singular Green operators.*

In particular, the spaces of standard (non-parametrized) polyhomogeneous resp. $S_{1,0}$ *Poisson, trace and singular Green operators of a given order and class are invariant under diffeomorphisms of* $\overline{\mathbb{R}}^n_+$ *preserving* $\{x_n=0\}$.

In the polyhomogeneous cases, the principal symbols multiplied by $\delta(x_n)$ *are invariant under such diffeomorphisms.*

Proof: Let us go directly to the parameter-dependent case, which requires the most careful study of estimates (the μ-independent case is covered by this, but can also be explained more directly, see Remark 2.4.12 below). We first consider a Poisson operator; trace operators are then easily included, whereas singular Green operators require further developments.

Let $k(x',y',\xi,\mu) \in S^{d-1,\nu}_{1,0}(\mathbb{R}^{n-1}\times\mathbb{R}^{n-1},\overline{\mathbb{R}}^n_+,H^+)$, defining the operator K_μ by the formula

$$(2.4.50)\qquad (K_\mu u)(x) = (2\pi)^{-n}\int_{\mathbb{R}^n} e^{i(x'-y')\cdot\xi' + ix_n\xi_n}\varphi(x_n)k(x',y',\xi,\mu)u(y')\,dy'd\xi$$

(with appropriate intepretations of the integral). We have inserted a cut-off function $\varphi(x_n) \in C_0^\infty(\mathbb{R})$ equal to 1 on a neighborhood of 0 ; it suffices to analyze operators with such a factor, in view of the preceding proposition. Consider $\underline{K}_\mu$ defined from K_μ by (2.4.44 i). We want to write $\underline{K}_\mu$ as $\mathrm{OPK}(k_1)$ plus a negligible term, for a suitable Poisson symbol k_1 . We shall use the transformation rule

$$(2.4.51)\qquad \begin{aligned} \xi' &= {}^tA(x,y',0)\underline{\xi}' \\ \xi_n &= {}^tB(x,y',0)\underline{\xi}' + C_1(x)\underline{\xi}_n \end{aligned}$$

where A, B and C_1 were defined in (2.2.95) and (2.2.94); recall that $A(x,y',0)$ is invertible for (x,y') in a neighborhood of the set $\{x'=y',\ x_n=0\}$, and $C_1 > 0$. We assume (as we may in view of Proposition 2.4.10) that $\varphi(x_n)k(x',y',\xi,\mu)$ is supported in this neighborhood.

Then we have, writing $\underline{x} = \kappa(x)$, $\underline{y}' = \lambda(y')$, $\underline{v}(\underline{y}') = v(y')$, and using (2.4.51), (2.2.97) and (2.2.94):

$$\begin{aligned}(2.4.52)\qquad (\underline{K}_\mu v)(\kappa(x)) &= (2\pi)^{-n}\int_{\mathbb{R}^n} e^{i(x'-y')\cdot\xi'+ix_n\xi_n}\varphi(x_n)k(x',y',\xi,\mu)v(y')dy'd\xi \\ &= (2\pi)^{-n}\int_{\mathbb{R}^n} e^{iA(x'-y')\cdot\underline{\xi}'+iBx_n\cdot\underline{\xi}'+iC_1x_n\underline{\xi}_n}\varphi(x_n)k(x',y',\xi,\mu)v(y')dy'd\xi\end{aligned}$$

$$= (2\pi)^{-n}\int_{\mathbb{R}^n} e^{i(\underline{x}'-\underline{y}')\cdot\underline{\xi}'+i\underline{x}_n\underline{\xi}_n}\varphi(x_n)k(x',y',\xi,\mu)|\det\lambda'(y')^{-1}||\det A|C_1\underline{v}(\underline{y}')d\underline{y}'d\underline{\xi}$$

$$= (2\pi)^{-n}\int_{\mathbb{R}^n} e^{i(\underline{x}'-\underline{y}')\cdot\underline{\xi}'+i\underline{x}_n\underline{\xi}_n}\underline{k}(\underline{x},\underline{y}',\underline{\xi},\mu)\underline{v}(\underline{y}')d\underline{y}'d\underline{\xi}\ ,$$

where we have set

$$\text{(2.4.53)}\quad \underline{k}(\underline{x},\underline{y}',\underline{\xi},\mu) = \varphi(x_n)k(x',y',{}^tA(x,y',0)\underline{\xi}',{}^tB(x,y',0)\underline{\xi}' + C_1(x)\underline{\xi}_n,\mu)\cdot$$

$$\cdot|\det\lambda'(y')^{-1}A(x,y',0)|C_1(x),\ \text{with}\ \ \underline{x}=\kappa(x),\ \underline{y}' = \lambda(y')\ .$$

This gives $\underline{K}_\mu$ on the form considered in Remark 2.4.9 (with a symbol depending on $\underline{x}_n$), so to show that $\underline{K}_\mu$ is a Poisson operator of order d and regularity ν, we just have to show that $\underline{k}$ is in $S^{d-1,\nu}_{1,0}(\overline{\mathbb{R}}^{2n-1}_+,\overline{\mathbb{R}}^n_+,H^+)$.

Note first that $\underline{k}$ is in H^+ as a function of $\underline{\xi}_n$, for each $x,y',\underline{\xi}',\mu$, since ${}^tB\underline{\xi}'$ is real and $C_1(x)$ is positive. Moreover, homogeneity of k in (ξ,μ) for $|\xi'|\geq 1$ will clearly imply homogeneity of $\underline{k}$ in $(\underline{\xi},\mu)$ for $|\underline{\xi}'|\geq c(x',y')>0$ (cf. (2.2.102)), so polyhomogeneity carries over, as soon as we have obtained the preservation of the estimates required for the space $S^{d-1,\nu}_{1,0}$. An important point is now the rôle of the h_{-1}-projection in the new coordinate $\underline{\xi}_n$. Fortunately, the description of its effect is not much different from the description of the h_{-1}-projection in the proof of the invariance of the transmission property in Theorem 2.2.12 3°. We have for each $m\in\mathbb{N}$ the decomposition of $\xi_n^m k$

$$\text{(2.4.54)}\quad \xi_n^m k(x',y',\xi,\mu) = \sum_{-m\leq j<0} s_j(x',y',\xi',\mu)\xi_n^{j+m} + h_{-1}(\xi_n^m k)\ ,$$

where $s_j\in S^{d-1-j,\nu}_{1,0}(\mathbb{R}^{2n-2},\overline{\mathbb{R}}^n_+)$ and $h_{-1}(\xi_n^m k)\in S^{d-1+m,\nu}_{1,0}(\mathbb{R}^{2n-2},\overline{\mathbb{R}}^n_+,H^+)$, cf. (2.3.9) ff. and Lemma 2.3.9 4°, 7°, 1°. Insertion of (2.4.51) gives, for each m (recall also (2.4.53) and (2.2.102)), that the $L^2_{\xi_n}$-norm satisfies

$$\|({}^tB\underline{\xi}'+C_1\underline{\xi}_n)^m\underline{k}(\underline{x},\underline{y}',\underline{\xi},\mu) - \sum_{-m\leq j<0} s_j(x',y',{}^tA\underline{\xi}',\mu)({}^tB\underline{\xi}'+C_1\underline{\xi}_n)^{j+m}\varphi|\det\lambda'^{-1}A|C_1$$

$$\leq c(x,y')(\rho(\xi',\mu)^\nu+1)\langle\xi',\mu\rangle^{d-\frac12+m}$$

$$\leq c'(\underline{x},\underline{y}')(\rho(\underline{\xi}',\mu)^\nu+1)\langle\underline{\xi}',\mu\rangle^{d-\frac12+m}\ .$$

Since $C_1>0$, this can be used successively for $m=0,1,\dots$ to determine the expansion coefficients s_j' of $\underline{k}$, satisfying

$$\|\underline{\xi}_n^m\underline{k} - \sum_{-m\leq j<0} s_j'(\underline{x},\underline{y}',\underline{\xi}',\mu)\underline{\xi}_n^{j+m}\|_{L^2_{\underline{\xi}_n}} \leq c''(\underline{x},\underline{y}')(\rho(\underline{\xi}',\mu)^\nu+1)\langle\underline{\xi}',\mu\rangle^{d-\frac12+m}\ ;$$

here each s_j' is a linear combination of the s_k with $k \geq j$, with coefficients that are polynomial in $\underline{\xi}'$ and C^∞ in $\underline{x}$ and $\underline{y}'$.

Now let us consider derivatives. For the derivatives in $\underline{\xi}'$ and $\underline{\xi}_n$ we use (2.2.99-101) with C replaced by C_1. Both the application of $D_{\underline{\xi}'}$ and the multiplication by $\underline{\xi}_n$ mix the coordinates, but it turns out that the parameter-dependent estimates are not endangered by this:

$$(2.4.55)\quad \underline{\xi}_n^{m'} D^m_{\underline{\xi}_n} D^\alpha_{\underline{\xi}'}\underline{k} = \Big(-\frac{1}{C_1}\,{}^tB\,{}^tA^{-1}\xi' + \frac{1}{C_1}\xi_n\Big)^{m'}(C_1 D_{\xi_n})^m (AD_{\xi'} + BD_{\xi_n})^\alpha k$$

is a sum of terms of degree $d-1+m'-m-|\alpha|$, where the regularity exponents (the power of ρ in the estimates) for the "extreme" terms satisfy:

(2.4.56)

(i) $\xi_n^{m'} D^m_{\xi_n} D^\alpha_{\xi'} k$ has regularity exponent $\nu-[m-m']_+ - |\alpha|$,

(ii) $\xi_n^{m'} D^{m+|\alpha|}_{\xi_n} k$ has regularity exponent $\nu-[m+|\alpha|-m']_+$,

(iii) $(\xi')^\theta D^m_{\xi_n} D^\alpha_{\xi'} k$ for $|\theta| = m'$ has regularity exponent $\nu+m'-m-|\alpha|$,

(iv) $(\xi')^\theta D^{m+|\alpha|}_{\xi_n} k$ for $|\theta| = m'$ has regularity exponent $\nu+m'-m-|\alpha|$,

and all these exponents are $\geq \nu-[m-m']_+ - |\alpha|$ which is what we want for (2.4.55). In (iii) and (iv), we have used that multiplication by a polynomial in ξ' lifts the regularity exponent; this observation is quite parallel to the second statement in Lemma 2.1.6, and is formally written up in Lemma 2.6.2 3^0 later.

Combining these considerations with the analysis of h_{-1}-projections given further above, we find altogether that

$$\|h_{-1,\underline{\xi}_n}\underline{\xi}_n^{m'} D^m_{\underline{\xi}_n} D^\alpha_{\underline{\xi}'}\underline{k}\|_{L^2_{\underline{\xi}_n}} \leq c(\underline{x},\underline{y}')\big(\rho(\underline{\xi}',\mu)^{\nu-[m-m']_+-|\alpha|} + 1\big)\langle\underline{\xi}',\mu\rangle^{d-\frac{1}{2}-m+m'-|\alpha|}$$

for each set of indices.

Concerning $(\underline{x},\underline{y}')$-derivatives of $\underline{k}$, we note that these are built up from differentiations of k in (x',y') and differentiations through ξ' and ξ_n that lead to polynomial factors. First order differentiations through ξ' give rise to expressions of the form $\Sigma_{\ell,j\leq n-1} c_{\ell m}(x,y')\xi_\ell\partial_{\xi_j}k$, and first order differentiations through ξ_n give rise to expressions $\Sigma_{\ell\leq n} c_\ell(x,y')\xi_\ell\partial_{\xi_n}k$. In all cases the degree and regularity are preserved (use Lemma 2.6.2 3^0 for the expressions $\xi_\ell\partial_{\xi_j}k$ with $\ell \leq n-1$, $j \leq n$, and use Lemma 2.3.9 9^0 for the expression $\xi_n\partial_{\xi_n}k$). Higher order differentiations will then also preserve degree and regularity. - The estimates of μ-derivatives carry over directly to $\underline{k}$.

Altogether, it is found that indeed $\underline{k}$ belongs to $S^{d-1,\nu}_{1,0}(\overline{\mathbb{R}}^{2n-1}_+, \overline{\mathbb{R}}^n_+, H^+)$, so $\underline{K}_\mu$ is a Poisson operator of order d and regularity ν, in view of Remark

2.4.9. A symbol on $(\underline{x}',\underline{y}')$-form is

$$(2.4.57) \qquad \underline{k}_1(\underline{x}',\underline{y}',\underline{\xi},\mu) \sim \sum_{j\in\mathbb{N}} \frac{1}{j!} \bar{D}^j_{\underline{\xi}_n} \partial^j_{\underline{x}_n} \underline{k}(\underline{x}',0,\underline{y}',\underline{\xi},\mu)$$

and a symbol on $\underline{x}'$-form is

$$(2.4.58) \qquad \underline{k}_2(\underline{x}',\underline{\xi},\mu) \sim \sum_{j\in\mathbb{N},\alpha\in\mathbb{N}^{n-1}} \frac{1}{j!\alpha!} \bar{D}^j_{\underline{\xi}_n} \partial^j_{\underline{x}_n} D^\alpha_{\underline{\xi}'} \partial^\alpha_{\underline{y}'} \underline{k}|_{\underline{x}_n=0,\ \underline{y}'=\underline{x}'} \quad .$$

When k is polyhomogeneous, then so are $\underline{k}$, $\underline{k}_1$ and $\underline{k}_2$. Note that when k is given on x'-form, then the principal symbol of $\underline{k}_2$ is (cf. (2.2.96))

$$(2.4.59) \quad \underline{k}^0_2(\underline{x}',\underline{\xi},\mu) = k^0(x',{}^tA_0(x')\underline{\xi}',{}^tB_0(x')\,\underline{\xi}' + C_0(x')\underline{\xi}_n,\mu)\,|\det\lambda'(x')^{-1}A_0(x')|C_0(x')$$

$$= k^0(x',{}^t\kappa(x',0)\underline{\xi},\mu)\partial_n\kappa_n(x',0)$$

The resemblance with the rule for ps.d.o. symbols becomes more clear when we consider k multiplied with the distribution $\delta(x_n)$, which goes into $\delta(\underline{x}_n)/\partial_n\kappa_n(x',0)$; then we get the invariant expression

$$(2.4.59') \qquad \underline{k}^0_2(\underline{x}',\underline{\xi},\mu)\delta(\underline{x}_n) = k^0(x',{}^t\kappa(x',0)\underline{\xi},\mu)\delta(x_n) \ .$$

Now consider a trace operator T_μ of order d and regularity ν with symbol $t(x',y',\xi,\mu)\varphi(y_n)$; we shall analyze the transformed operator $\underline{T}_\mu$ defined by (2.4.44 ii). Here it is perhaps easiest to treat the terms in the decomposition

$$T_\mu = \sum_{0\le j<r} S_{j,\mu}\gamma_0 D^j_{x_n} + T'_\mu$$

separately. The space of parameter-dependent ps.d.o.s $S_{j,\mu}$ in $\mathbb{R}^{n-1}$ of order $d-j$ and regularity ν is preserved under the coordinate change $\underline{x}' = \lambda(x')$, in view of Lemma 2.1.17; the differential operator $D^j_{x_n}$ carries over to another μ-independent differential operator; and T'_μ is treated as the adjoint of the Poisson operator $(T'_\mu)^*$, so it carries over to a trace operator of class 0, order d and regularity ν. Altogether, one finds that $\underline{T}_\mu$ is a trace operator of class r, order d and regularity ν, having the symbols

$$\underline{t}(\underline{x}',\underline{y},\underline{\xi},\mu) = t(x',y',{}^tA(x',0,y)\underline{\xi}',\ {}^tB(x',0,y)\underline{\xi}' + C_1(y)\underline{\xi}_n,\mu)|\det\lambda'^{-1}A|C_1\varphi \ ,$$

$$\text{with } \underline{x}' = \lambda(x'),\ \underline{y} = \kappa(y) \ ;$$

(2.4.60)

$$\underline{t}_1(\underline{x}',\underline{y}',\underline{\xi},\mu) \sim \sum_{j\in\mathbb{N}} \frac{1}{j!} D^j_{\underline{\xi}_n} \partial^j_{\underline{y}_n} \underline{t}(\underline{x}',\underline{y}',0,\underline{\xi},\mu)$$

$$\underline{t}_2(\underline{x}',\underline{\xi},\mu) \sim \sum_{j\in\mathbb{N},\ \alpha\in\mathbb{N}^{n-1}} \frac{1}{j!\alpha!} D^j_{\underline{\xi}_n} \partial^j_{\underline{y}_n} D^\alpha_{\underline{\xi}'} \partial^\alpha_{\underline{y}'} \underline{t}|_{\underline{y}_n=0,\ \underline{y}'=\underline{x}'} \quad .$$

When t is polyhomogeneous, so are $\underline{t}$, $\underline{t}_1$ and $\underline{t}_2$. Here, if t is on x'-form we have for the principal symbols

$$\underline{t}_2^0(\underline{x}',\underline{\xi},\mu) = t^0(x',{}^t\kappa(x',0)\underline{\xi},\mu)\partial_n\kappa_n(x',0) \ ; \tag{2.4.61}$$

and the principal symbol multiplied with $\delta(x_n)$ is invariant under diffeomorphisms, in the same way as in (2.4.59').

Finally, consider a singular Green operator G_μ of order d and regularity ν , defined from a symbol $g(x',y',\xi',\xi_n,\eta_n,\mu) \in S^{d-1,\nu}_{1,0}(\mathbb{R}^{2n-2}, \overline{\mathbb{R}}^n_+, H^+ \hat{\otimes} H^-_{r-1})$. Whereas the transformation rules for K_μ and T_μ were rather close to those for ps.d.o.s, the transformation rule for G_μ is different in an essential way, since the defining formula

$$G_\mu u(x) = (2\pi)^{-n-1}\int e^{i(x'-y')\cdot\xi+ix_n\xi_n-iy_n\eta_n} g\ u\ dy d\xi d\eta_n$$

contains integrations in both ξ_n and η_n . So let us make an explicit study of $\underline{G}_\mu$, defined from G_μ by (2.4.44 iii). It turns out that $\underline{G}_\mu$ can be nicely described if we set (using (2.2.94-102))

$$\begin{aligned} \xi' &= {}^tA(x,y)\underline{\xi}' \\ \xi_n &= {}^tB(x,y)\underline{\xi}' + C_1(x)\underline{\xi}_n \\ \eta_n &= {}^tB(x,y)\underline{\xi}' + C_1(y)\underline{\eta}_n \ . \end{aligned} \tag{2.4.62}$$

We also replace g by $\varphi(x_n)\varphi(y_n)g$, assuming moreover that g vanishes for $x'-y'$ outside a neighborhood of zero (as we may in view of Proposition 2.4.10), such that $A(x,y)$ is invertible where $\varphi(x_n)\varphi(y_n)g$ does not vanish. Then we have, writing $\underline{x} = \kappa(x)$, $\underline{y} = \kappa(y)$, $\underline{u}(\underline{x}) = u(x)$, and using (2.2.97),

$$\begin{aligned} (\underline{G}_\mu\underline{u})(\underline{x}) &= (G_\mu u)(x) \\ &= (2\pi)^{-n-1}\int e^{i(x'-y')\cdot\xi'+ix_n\xi_n-iy_n\eta_n}\varphi(x_n)\varphi(y_n)g(x',y',\xi,\eta_n,\mu)u(y)dyd\xi d\eta_n \\ &= (2\pi)^{-n-1}\int e^{i(\underline{x}'-\underline{y}')\cdot\underline{\xi}'+i\underline{x}_n\underline{\xi}_n-i\underline{y}_n\underline{\eta}_n}\underline{g}(\underline{x},\underline{y},\underline{\xi},\underline{\eta}_n,\mu)\underline{u}(\underline{y})d\underline{y}d\underline{\xi}d\underline{\eta}_n \quad , \end{aligned}$$

where $\underline{g}$ is defined by

$$\begin{aligned} \underline{g}(\underline{x},\underline{y},\underline{\xi},\underline{\eta}_n,\mu) = \varphi(x_n)\varphi(y_n)g(x',y',{}^tA(x,y)\underline{\xi}',{}^tB(x,y)\underline{\xi}'+C_1(x)\underline{\xi}_n,{}^tB(x,y)\underline{\xi}' + \\ C_1(y)\underline{\eta}_n,\mu)\cdot|\det\kappa'(y)^{-1}||\det A(x,y)|C_1(x)C_1(y) \ , \\ \text{with } \underline{x} = \kappa(x) \ , \ \underline{y} = \kappa(y) \ . \end{aligned} \tag{2.4.63}$$

An analysis like in the Poisson operator case shows that $\underline{g} \in S_{1,0}^{d-1,\nu}(\overline{\mathbb{R}}_{++}^{2n}, \overline{\mathbb{R}}_{+}^{n}, H^{+} \hat{\otimes} H_{r-1}^{-})$, so $\underline{G}_{\mu}$ is a singular Green operator in view of Remark 2.4.9. A symbol on $(\underline{x}',\underline{y}')$-form for $\underline{G}_{\mu}$ is

$$(2.4.64)\quad \underline{g}_1(\underline{x}',\underline{y}',\underline{\xi},\underline{\eta}_n,\mu) \sim \sum_{j,k\in\mathbb{N}} \frac{1}{j!k!} \overline{D}_{\underline{\xi}_n}^{j} D_{\underline{\eta}_n}^{k} \partial_{\underline{x}_n}^{j} \partial_{\underline{y}_n}^{k} \underline{g}(\underline{x}',0,\underline{y}',0,\underline{\xi},\underline{\eta}_n,\mu) ,$$

and a symbol on $\underline{x}'$-form is

$$(2.4.65)\quad \underline{g}_2(\underline{x}',\underline{\xi},\underline{\eta}_n,\mu) \sim \sum_{j,k,\alpha} \frac{1}{j!k!\alpha!} D_{\underline{\xi}'}^{\alpha} \overline{D}_{\underline{\xi}_n}^{j} D_{\underline{\eta}_n}^{k} \partial_{\underline{y}'}^{\alpha} \partial_{\underline{x}_n}^{j} \partial_{\underline{y}_n}^{k} \underline{g}(\underline{x},\underline{y},\underline{\xi},\underline{\eta}_n,\mu) \Big|_{\substack{\underline{x}_n=\underline{y}_n=0 \\ \underline{y}'=\underline{x}'}}$$

When g is polyhomogeneous, then so are $\underline{g}$, $\underline{g}_1$ and $\underline{g}_2$. Then if g is on x'-form, the principal symbols satisfy (cf. (2.2.96))

$$(2.4.66)\quad \underline{g}_2^0(\underline{x}',\underline{\xi},\underline{\eta}_n,\mu) = g^0(x',{}^t\kappa'(x',0)\underline{\xi} , [{}^t\kappa'(x',0)\{\underline{\xi}',\underline{\eta}_n\}]_n,\mu)\partial_n\kappa_n(x',0) .$$

In particular, this gives an invariant expression for the principal symbol multiplied by $\delta(x_n)$. See also Remark 2.4.12. □

2.4.12 Remark. In the non-parametrized case, it is perhaps simpler to see the invariance of the operator classes by an application of the point of view of [Hörmander 8, Section 18.2].

To take a typical case, let $g(x',\xi',\xi_n,\eta_n)$ be a singular Green symbol of order d (degree $d-1$) and class 0 on $\overline{\mathbb{R}}_{+}^{n}$. The corresponding operator $G = OPG(g)$ has the distribution kernel (an inverse Fourier transform)

$$K_G(x,y) = (2\pi)^{-n-1}\int e^{i(x'-y')\cdot\xi+ix_n\xi_n-iy_n\eta_n} g(x',\xi,\eta_n)d\xi d\eta_n$$

for x and $y \in \overline{\mathbb{R}}_{+}^{n}$. In order to apply [Hörmander 8], we first show how to obtain G on the form $r^+G_1e^+$, where G_1 is an operator defined on all of $\mathbb{R}^n$ from a symbol g_1 with properties similar to those of g :

Consider $\tilde{g}(x',x_n,y_n,\xi') = F^{-1}_{\xi_n\to x_n} \overline{F}^{-1}_{\eta_n\to y_n} g$; it lies in $\mathscr{S}(\overline{\mathbb{R}}_{++}^2)$ as a function of (x_n,y_n) , and satisfies the estimates (for each index set)

$$(2.4.67)\quad \|D_{x'}^{\beta} x_n^k D_{x_n}^{k'} y_n^m D_{y_n}^{m'} D_{\xi'}^{\alpha} \tilde{g}\|_{L^2_{x_n,y_n}} \le c(x')\langle\xi'\rangle^{d-k+k'-m+m'-|\alpha|} .$$

Let $\tilde{g}_1$ be an extension of $\tilde{g}$ to all $(x_n,y_n) \in \mathbb{R}^2$, such that $\tilde{g}_1$ is in $\mathscr{S}(\mathbb{R}^2)$ as a function of (x_n,y_n) and satisfies estimates like (2.4.67) for the norms in $L^2(\mathbb{R}^2)$; $\tilde{g}_1$ can be constructed by use of the extension procedure of [Seeley 5] in x_n and y_n , as done for Poisson symbol-kernels in [Boutet 3, (3.5)]. Now let $g_1 = F_{x_n\to\xi_n} \overline{F}_{y_n\to\eta_n} \tilde{g}_1$, then the operator G_1 defined by

$$(2.4.68)\quad (G_1u)(x) = (2\pi)^{-n-1}\int e^{i(x'-y')\cdot\xi'+ix_n\xi_n-iy_n\eta_n} g_1(x',\xi,\eta_n)u(y)dyd\xi d\eta_n \quad ,$$

is such that

$$r^+G_1e^+ = G \ ,$$

and its distribution kernel

$$(2.4.69)\quad K_{G_1}(x,y) = (2\pi)^{-n-1}\int e^{i(x'-y')\cdot\xi'+ix_n\xi_n-iy_n\eta_n} g_1(x',\xi,\eta_n)d\xi d\eta_n$$

gives K_G by restriction to $(x_n,y_n) \in \overline{\mathbb{R}}^2_{++}$.

We are considering coordinate changes $\kappa\colon x \sim \underline{x}$ in $\mathbb{R}^n$ preserving the set $\{x_n=0\}$. For the kernel of an operator on $\mathbb{R}^n$, this gives rise to a coordinate change $(x,y) \sim (\underline{x},\underline{y}) = (\kappa(x),\kappa(y))$ in $X = \mathbb{R}^{2n}$ preserving the set

$$Y = \{(x,y) \mid x'=y',\ \ x_n=0,\ \ y_n=0\} \quad ;$$

and the kernel in new coordinates is described as in (2.4.49).

The estimates (2.4.67) for $\tilde{g}_1$ correspond to the estimates for g_1

$$\| D^\beta_{x'} D^k_{\xi_n} \xi_n^{k'} D^m_{\eta_n} \eta_n^{m'} D^\alpha_{\xi'} g_1(x',\xi,\eta_n)\|_{L^2_{\xi_n,\eta_n}} \le c(x')\langle\xi'\rangle^{d-k+k'-m+m'-|\alpha|}$$

(note that the use of the Seeley extension has eliminated the need to take h_{-1}-projections). These are equivalent with the system of estimates (cf. (2.4.25))

$$|D^\beta_{x'} D^k_{\xi_n} D^m_{\eta_n} D^\alpha_{\xi'} g_1| \le c(x')\langle\xi'\rangle^{d-k-m-|\alpha|-1}\Big(\frac{\langle\xi'\rangle}{\langle\xi',\xi_n\rangle}\Big)^{k'}\Big(\frac{\langle\xi'\rangle}{\langle\xi',\eta_n\rangle}\Big)^{m'}$$
$$\text{for } \alpha,\beta \in \mathbb{N}^{n-1},\ k,k',m,m' \in \mathbb{N}\ ;$$

which may in turn be replaced by the system of estimates

$$(2.4.70)\qquad |D^\beta_{x'} D^\theta_{\xi,\eta_n} g_1| \le c(x')\langle\xi'\rangle^{d-1-|\theta|}\Big(\frac{\langle\xi'\rangle}{(1+|\xi'|^2+\xi_n^2+\eta_n^2)^{\frac12}}\Big)^N$$
$$\text{for } \beta \in \mathbb{N}^{n-1},\ \theta \in \mathbb{N}^{n+1} \text{ and } N \in \mathbb{N}\ ,$$

where we have used that $(1+|\xi_n/\langle\xi'\rangle|^2)^{-1}(1+|\eta_n/\langle\xi'\rangle|^2)^{-1} \le c(1+|\xi_n/\langle\xi'\rangle|^2+|\eta_n/\langle\xi'\rangle|^2)^{-1}$, cf. (A.17). (2.4.70) shows in particular that g_1 is a ps.d.o. symbol in $S^{d-1}(\mathbb{R}^{n-1}, \mathbb{R}^{n+1})$. Then, according to [Hörmander

8, Section 18.2], the formula (2.4.69) is precisely the formula for a distribution K_{G_1} on $X = \mathbb{R}^{2n}$ that is conormal with respect to the submanifold Y defined above. The property of being conormal is invariant under coordinate changes in X preserving Y, and there are formulas for the transformation of the symbol g_1 under such coordinate changes, see [Hörmander 8, Theorems 18.2.8-9].

Now (2.4.70) moreover contains the information that g_1 is rapidly decreasing for $|\xi_n|$ and $|\eta_n| \to \infty$. Also this can be expressed in an invariant way, namely as the property that g_1 is rapidly decreasing along $N(Y) \cap N(Z)$, where $N(Y)$ and $N(Z)$ are the normal bundles in $T^*(X)$ of Y resp. Z, where Y is defined above and

$$Z = \{(x,y) \in X \mid x_n=0,\ \ y_n=0\} \ .$$

In the given coordinates, $N(Y)$ is described as the set of vectors $\{(x',0,x',0,\xi',-\xi',\xi_n,-\eta_n)\}$, and $N(Z)$ is described as the set of vectors $\{(x',0,y',0,0,0,\xi_n,-\eta_n)\}$, so $N(Y) \cap N(Z)$ is the set $\{(x',0,x',0,0,0,\xi_n,-\eta_n)\}$. Since each of these normal bundles has an invariant meaning, the property that the symbol is rapidly decreasing along $N(Y) \cap N(Z)$, is invariant, under the coordinate changes we are considering.

Altogether, the singular Green operators of class 0 and order d are, after Seeley extension, characterized by having distribution kernels that are conormal with respect to Y and are defined from symbols of degree $d-1$ that are rapidly decreasing along $N(Y) \cap N(Z)$.

For Poisson operators of order d; one can make a similar analysis of the kernels (after a Seeley extension with respect to x_n), obtaining here that their kernels are distributions on $X_1 = \mathbb{R}^n \times \mathbb{R}^{n-1}$ that are conormal with respect to $Y_1 = \{(x,y') \mid x'=y',\ x_n=0\}$, defined from symbols of degree $d-1$ that are rapidly decreasing on $N(Y_1) \cap N(Z_1)$, where $Z_1 = \{(x,y') \mid x_n=0\}$. For trace operators there is a similar statement with the rôle of $\mathbb{R}^n$ and $\mathbb{R}^{n-1}$ interchanged (and a slightly different order convention). Operators of class > 0 are easily included.

In the case of a manifold $\bar{\Omega} \subset \Sigma$ with boundary $\partial\bar{\Omega} = \Gamma$, singular Green operators on $\bar{\Omega}$ behave as described above, with $X = \Sigma\times\Sigma$ and $Y = \mathrm{diag}(\Gamma\times\Gamma)$. Denote by j the natural mapping of $T^*(\Sigma)|_\Gamma$ onto $T^*(\Gamma)$ (the adjoint of the injection $T(\Gamma) \subset T(\Sigma)|_\Gamma$) and let $j\times(-j)$ stand for the mapping $\{\xi,\eta\} \rightsquigarrow \{j(\xi),-j(\eta)\}$; then one can identify the normal bundle of Y in $T^*(X)$ with the following inverse image

$$N(Y) = (j \times (-j))^{-1}\ \mathrm{diag}(T^*(\Gamma) \times T^*(\Gamma)) \quad \text{in } T^*(X) . \tag{2.4.71}$$

For Poisson operators, one takes $X_1 = \Sigma\times\Gamma$ and $Y_1 = \mathrm{diag}(\Gamma\times\Gamma)$, then the normal bundle of Y_1 in $T^*(X_1)$ identifies with

$$N(Y_1) = (j\times(-\mathrm{id}))^{-1}\,\mathrm{diag}(T^*(\Gamma)\times T^*(\Gamma)) \quad \text{in} \quad T^*(X_1)\,. \tag{2.4.72}$$

For trace operators, one takes $X_2 = \Gamma\times\Sigma$ and $Y_2 = \mathrm{diag}(\Gamma\times\Gamma)$, so

$$N(Y_2) = (\mathrm{id}\times(-j))^{-1}\,\mathrm{diag}(T^*(\Gamma)\times T^*(\Gamma)) \quad \text{in} \quad T^*(X_2)\,. \tag{2.4.73}$$

It is explained in [Hörmander 8] around Theorem 18.2.11 how one can associate principal symbols to these operators in a completely invariant way, as elements of quotient spaces between successive symbol spaces consisting of sections in half-density bundles over the normal bundles $N(Y)$, $N(Y_1)$ resp. $N(Y_2)$.

We can now define the operators acting on sections in vector bundles over manifolds.

Here we consider the set-up defined in the Appendix, with $\tilde{E}$ being an N-dimensional vector bundle over Σ (and $\tilde{E}|_{\overline{\Omega}} = E$), and F being an M-dimensional vector bundle over Γ (here $\overline{\Omega}$ and Γ are compact).

Let there be given families of operators

$$\begin{aligned} P_\mu &: C_0^\infty(\Sigma,\tilde{E}) \to C^\infty(\tilde{E})\\ K_\mu &: C^\infty(F) \to C^\infty(E)\\ T_\mu &: C^\infty(E) \to C^\infty(F)\\ G_\mu &: C^\infty(E) \to C^\infty(E)\,, \end{aligned} \tag{2.4.74}$$

depending on the parameter $\mu \in \overline{\mathbb{R}}_+$. With reference to a specific system of trivializations $\psi_i : \tilde{E}|_{\Sigma_i} \to \Xi_i\times\mathbb{C}^N$, $\zeta_i : F|_{\Gamma_i} \to \Xi_i'\times\mathbb{C}^M$, we say that the operators are, respectively, a (parameter-dependent) ps.d.o. on $\tilde{E}$ having the transmission property at Γ, a Poisson operator on E, a trace operator on E or a singular Green operator on E, of order d, regularity ν and class r (when relevant), if the corresponding operators between open sets

$$\begin{aligned} P_\mu^{(i,j)} &: C_0^\infty(\Xi_i)^N \ni u \sim \psi_j \circ P_\mu(\psi_i^{-1}\circ u)\\ K_\mu^{(i,j)} &: C_0^\infty(\Xi_i')^M \ni v \sim \psi_j \circ K_\mu(\zeta_i^{-1}\circ v)\\ T_\mu^{(i,j)} &: C_{(0)}^\infty(\Xi_i\cap\overline{\mathbb{R}}_+^n) \ni u \sim \zeta_j \circ T_\mu(\psi_i^{-1}\circ u)\\ G_\mu^{(i,j)} &: C_{(0)}^\infty(\Xi_i\cap\overline{\mathbb{R}}_+^n) \ni u \sim \psi_i \circ G_\mu(\psi_i^{-1}\circ u) \end{aligned} \tag{2.4.75}$$

are the respective kinds of operators according to Definitions 2.2.7 and 2.4.5 ff. (As we have seen, the operator classes are independent of the choice of coordinates.)

Conversely, it is of interest to construct such operators from operators given relatively to $\mathbb{R}^n$ and its subsets.

For instance, if

(2.4.76) $$(\varphi_i)_{i=1,\dots,i_2}$$

is a system of functions $\varphi_i \in C_0^\infty(\Sigma_i)$ such that $\Sigma_{1\le i\le i_2}\varphi_i = 1$ on $\bar\Omega$; and $G_\mu^{(1)},\dots,G_\mu^{(i_2)}$ are parameter-dependent N×N-matrix formed singular Green operators of order d and regularity ν in $\Xi_1,\dots,\Xi_{i_2}$ (where only the sets $\Xi_1,\dots,\Xi_{i_1}$ intersect $\{x_n = 0\}$) , then

(2.4.77) $$G_\mu u = \sum_{i=1}^{i_2} \varphi_i\, [\psi_i^{-1}\circ G_\mu^{(i)}(\psi_i\circ(\varphi_i u))]$$

is a s.g.o. on E . When we work with a fixed normal coordinate x_n (as described in the Appendix), the principal boundary symbol operator

(2.4.78) $$g^{0(i)}(x',\xi',\mu,D_n) : \mathscr{S}(\bar{\mathbb{R}}_+)^N \to \mathscr{S}(\bar{\mathbb{R}}_+)^N ,$$

defined for $(x',\xi')\in\Xi_i'\times\mathbb{R}^{n-1}$ (modulo symbols of order $d-1$ and regularity $\nu-1$) transforms as an operator valued section in the bundle obtained by pulling E_Γ back to the cotangent bundle $T^*(\Gamma)$. G_μ then has the principal symbol obtained by superposition of the symbols stemming from each $G_\mu^{(i)}$. If we choose the φ_i with $i\le i_1$ as simple products $\varphi_i'(x')\varphi_{i,n}(x_n)$ with $\varphi_{i,n}(0)=1$, we get the principal symbol g^0 as the sum

(2.4.79) $$g^0(x',\xi',\mu,D_n) = \sum_{i=1}^{i_1} \varphi_i'(x')\Big[(\psi_{i,x'}')^{-1}\circ g^{0(i)}(\tilde\kappa_i(x',\xi'),\mu,D_n)\Big] .$$

Note also that a modification of G_μ at a distance from Γ , e.g. a replacement of G_μ by $\varphi(x_n)G_\mu\varphi(x_n)$ where $\varphi = 1$ near Γ and is supported in a neighbourhood of Γ , gives an error that is negligible of regularity $\nu-d+1$ (in view of Lemma 2.4.8, 3^0 and 4^0).

Similar considerations hold for the other types of operators.

In this way one obtains Green operators acting as follows

(2.4.80) $$A_\mu = \begin{pmatrix} P_{\mu,\Omega}+G_\mu & K_\mu \\ T_\mu & S_\mu \end{pmatrix} : \begin{matrix} C^\infty(E) \\ \times \\ C^\infty(F) \end{matrix} \to \begin{matrix} C^\infty(E) \\ \times \\ C^\infty(F') \end{matrix}$$

where F and F' are vector bundles over Γ . The Green operator is said to be of order d , regularity ν and class r , when the entries are so. Actually we shall often have to consider the slightly more general case where T_μ and K_μ are column, resp. row, vectors with entries of different orders; this is handled either by a somewhat more complicated notation, or by reducing the orders to the same order as P and G by composition with suitable ps.d.o.s over Γ .

One can of course let $P_{\mu,\Omega}$, G_μ and K_μ map into a different vector bundle E' over $\bar{\Omega}$, but we shall rarely have occasion to use this.

2.4.13 Remark. Occasionally, one needs to use the simple coordinate change $x \to \sigma x$ in $\mathbb{R}^n$, for $\sigma > 0$. Let us describe the effects of the dilation explicitly:

When $u \in \mathcal{S}(\mathbb{R}^n)$ and $\sigma > 0$, we denote

(2.4.81)
$$u_\sigma(x) = u(\sigma x) \, , \quad \text{whereby}$$
$$F(u_\sigma)(\xi) = \sigma^{-n}\,\hat{u}(\sigma^{-1}\xi) \; ;$$

the formulas extend to more general functions (or distributions) and their restrictions to $\mathbb{R}^n_+$, For pseudo-differential operators on $\mathbb{R}^n$, we then have

(2.4.82)
$$OP(p(x,\xi,\mu))(u_\sigma) = (2\pi)^{-n}\int e^{ix\cdot\xi}p(x,\xi)\hat{u}(\sigma^{-1}\xi)\sigma^{-n}d\xi = [OP(p(x,\sigma\xi,\mu))u]_\sigma \, ,$$

and for the various boundary operators (with various interpretations of the integrals)

(2.4.83)
$$\begin{aligned} OPT(t(x',\xi,\mu))(u_\sigma) &= (2\pi)^{-n}\int e^{ix'\cdot\xi'}t(x',\xi,\mu)\widehat{e^+u}(\sigma^{-1}\xi)\sigma^{-n}d\xi \\ &= [OPT(t(x',\sigma\xi,\mu))u]_\sigma \, , \\ OPK(k(x',\xi,\mu))(v_\sigma) &= (2\pi)^{-n}\int e^{ix\cdot\xi}k(x',\xi,\mu)\hat{v}(\sigma^{-1}\xi')\sigma^{-n+1}d\xi \\ &= \sigma[OPK(k(x',\sigma\xi,\mu))v]_\sigma \, , \\ OPG(g(x',\xi,\eta_n,\mu))(u_\sigma) &= (2\pi)^{-n}\int e^{ix\cdot\xi}g(x',\xi,\eta_n,\mu)\widehat{e^+u}(\sigma^{-1}\xi)\sigma^{-n}d\xi d\eta_n \\ &= \sigma[OPG(g(x',\sigma\xi,\sigma\eta_n,\mu)u]_\sigma \, , \end{aligned}$$

when $u \in \mathcal{S}(\mathbb{R}^n)$ resp. $\mathcal{S}(\bar{\mathbb{R}}^n_+)$ and $v \in \mathcal{S}(\mathbb{R}^{n-1})$. The formulas can be applied to boundary symbol operators also (when we consider ξ' as a parameter), showing that

$$(2.4.84)\quad OP_n(p(x',\xi,\mu))_\Omega[u(\sigma x_n)] = [OP_n(p(x',\xi',\sigma\xi_n,\mu))_\Omega u]_\sigma$$

$$OPT_n(t(x',\xi,\mu))[u(\sigma x_n)] = OPT_n(t(x',\xi',\sigma\xi_n,\mu))u$$

$$OPK_n(k(x',\xi,\mu))v = \sigma[OPK_n(k(x',\xi',\sigma\xi_n,\mu))v]_\sigma$$

$$OPG_n(g(x',\xi,\eta_n,\mu))[u(\sigma x_n)] = \sigma[OPG_n(g(x',\xi',\sigma\xi_n,\sigma\eta_n,\mu))u]_\sigma .$$

It is seen in particular that the formulas fit nicely with homogeneity properties of the symbols, in the sense that if t and p are homogeneous of degree d, and k and g are homogeneous of degree $d-1$ (in (ξ,μ) resp. (ξ,η_n,μ)), the right hand sides in (2.4.84) may be written as $\sigma^d[OP_n(p(x',\sigma^{-1}\xi',\xi_n,\sigma^{-1}\mu))u]_\sigma$, etc., all with the factor σ^d in front and $\sigma^{-1}\xi',\sigma^{-1}\mu$ in the appropriate place. Thus, for symbols with the homogeneity corresponding to order d, one has in all cases:

$$(2.4.85)\qquad a(x',\xi',\mu,D_n)u_\sigma = [\sigma^d a(x',\sigma^{-1}\xi',\sigma^{-1}\mu,D_n)u]_\sigma$$

for $u \in \mathscr{S}(\overline{\mathbb{R}}_+)$ or $v \in \mathbb{C}$ (here $v_\sigma = v$).

We also observe that the parameter-dependent norms satisfy, for $\mu \geq 1$ and $\sigma \geq 1$,

$$(2.4.86)\qquad \|u_\sigma\|_{s,\mu} \sim \| |\xi,\mu|^s \sigma^{-n}\hat{u}(\sigma^{-1}\xi)\|_0$$

$$= \sigma^{s-n/2}\| |\eta,\mu/\sigma|^s \hat{u}(\eta)\|_0$$

$$\sim \sigma^{s-n/2}\|u\|_{s,\mu/\sigma} ,$$

on $\mathbb{R}^n$, with related formulas for anisotropic norms.

2.5 Continuity.

In the following, we show continuity properties of the various operators in relation to the parameter-dependent Sobolev spaces introduced in the Appendix. As in Section 2.4, Ω' is open $\subset \mathbb{R}^{n-1}$, $\Omega = \Omega' \times \mathbb{R}_+$ and $\Sigma = \Omega' \times \mathbb{R}$, and we use (2.4.1) and the usual abbreviations $\rho = \langle\xi'\rangle/\langle\xi',\mu\rangle$, $\kappa \sim \langle\xi',\mu\rangle$.

2.5.1 Theorem. *Let* $K_\mu = \mathrm{OPK}(k(x',y',\xi,\mu))$, *where* $k \in S^{d-1,\nu}_{1,0}(W',H^+)$ *and vanishes for* (x',y') *outside a compact subset of* $\Omega' \times \Omega'$. *For any* s *and* $t \in \mathbb{R}$ *there are constants so that when* $v \in \mathscr{S}(\mathbb{R}^{n-1})$ *and* $\mu \geq 0$,

$$\|K_\mu v\|_{H^{(s,t),\mu}(\overline{\mathbb{R}}^n_+)} \leq c_{s,t}\|(\rho^\nu+1)\kappa^{d-\frac12+s+t}\hat{v}(\xi')\|_0 \tag{2.5.1}$$

$$\leq c'_{s,t}(\langle\mu\rangle^{-\nu}+1)\|v\|_{H^{d-\frac12+s+t,\mu}(\mathbb{R}^{n-1})},$$

and in particular,

$$\|K_\mu v\|_{H^{s-d,\mu}(\overline{\mathbb{R}}^n_+)} \leq c_s(\langle\mu\rangle^{-\nu}+1)\|v\|_{H^{s-\frac12,\mu}(\mathbb{R}^{n-1})}. \tag{2.5.2}$$

Proof: We first observe that for any $\beta \in \mathbb{N}^{2n-2}$ one has for $v \in \mathbb{C}$,

$$\|D^\beta_{x',y'}D^j_{x_n}\tilde{k}(x',x_n,y',\xi',\mu)v\|_{L^2_{x_n}(\mathbb{R}_+)} \leq c(\rho(\xi',\mu)^\nu+1)\langle\xi',\mu\rangle^{d-\frac12+j}|v|,$$

so that the function $\hat{\tilde{k}}$ obtained by Fourier transformation in x' and y' satisfies, for any N,

$$\|D^j_{x_n}\hat{\tilde{k}}(\zeta',x_n,\sigma',\xi',\mu)\|_{L^2_{x_n}(\mathbb{R}_+)} \tag{2.5.3}$$

$$\leq c_N\langle\zeta'\rangle^{-N}\langle\sigma'\rangle^{-N}(\rho^\nu+1)\langle\xi',\mu\rangle^{d-\frac12+j}.$$

Now let $v \in \mathscr{S}(\mathbb{R}^{n-1})$. Then for any $w \in C^\infty_{(0)}(\overline{\mathbb{R}}^n_+)$ we have as in Proposition 2.1.12, setting $\acute{w} = F_{x'\to\eta'}w$,

$$|(D^j_{x_n}K_\mu v,w)_{L^2(\mathbb{R}^n_+)}| \tag{2.5.4}$$

$$= c\left|\int_0^\infty\int_{\mathbb{R}^{3n-3}} e^{i(x'-y')\xi'}D^j_{x_n}\tilde{k}(x',x_n,y',\xi',\mu)v(y')\overline{w}(x)\,dy'd\xi'dx\right|$$

$$= c_1\left|\int_0^\infty\int_{\mathbb{R}^{3n-3}} D^j_{x_n}\hat{\tilde{k}}(\eta'-\xi',x_n,\xi'-\theta',\xi',\mu)\hat{v}(\theta')\overline{\acute{w}}(\eta',x_n)\,d\theta'd\eta'd\xi'dx_n\right|.$$

We here apply the Cauchy-Schwarz inequality in x_n and use the inequality (A.17) as in Proposition 2.1.12, which gives, for an arbitrary $r \in \mathbb{R}$, in view of (2.5.3)

$$|(D^j_{x_n} K_\mu v, w)|$$
$$\leq c_2 \int_{\mathbb{R}^{3n-3}} \langle \eta'-\xi'\rangle^{-N} \langle \xi'-\theta'\rangle^{-N} (\rho^\nu+1) \langle \xi',\mu\rangle^{d-\frac{1}{2}+j} |\hat{v}(\theta')| \, \|\tilde{w}(\eta',x_n)\|_{L^2_{x_n}} \, d\theta' d\eta' d\xi'$$
$$\leq c_3 |||\hat{v}(\theta')||| \; \| \langle \eta',\mu\rangle^{d-\frac{1}{2}+j-r} \tilde{w}(\eta',x_n)\|_{L^2(\mathbb{R}^n_+)} ,$$

with (cf. (2.1.17))

$$|||v(\theta')||| = \begin{cases} \| \langle \theta',\mu\rangle^r \hat{v}(\theta')\|_0 & \text{when } \nu \geq 0 , \\ \| \langle \theta'\rangle^\nu \; \langle \theta',\mu\rangle^{r-\nu} \hat{v}(\theta')\|_0 & \text{when } \nu \leq 0 . \end{cases}$$

In view of the duality between $H^{(0,t),\mu}(\overline{\mathbb{R}}^n_+)$ and $H^{(0,-t),\mu}(\overline{\mathbb{R}}^n_+)$, this implies

$$\text{(2.5.5)} \qquad \| D^j_{x_n} K_\mu v\|_{H^{(0,r-d-j+\frac{1}{2}),\mu}(\overline{\mathbb{R}}^n_+)} \leq c_4 |||\hat{v}||| \leq c_5 \| (\rho^\nu+1) \langle \xi',\mu\rangle^r \hat{v}(\xi')\|_0 .$$

When $s \leq 0$, we get from (2.5.5) with $j = 0$, in view of (A.27),

$$\| K_\mu v\|_{(s,t),\mu} \leq \| K_\mu v\|_{(0,s+t),\mu} \leq c_6 \| (\rho^\nu+1) \langle \xi',\mu\rangle^{s+t+d+j-\frac{1}{2}} \hat{v}\|_0 ,$$

showing (2.5.1) for any $s \leq 0$ and $t \in \mathbb{R}$. For s equal to a positive integer m, we use (2.5.5) for $j = 0,1,\dots,m$ and $r = t+m+d-\frac{1}{2}$, which gives (cf. (A.42))

$$\| K_\mu v\|_{(m,t),\mu} \leq c_7 \sum_{j=0}^{m} \| D^j_{x_n} K_\mu v\|_{(0,t+m-j),\mu} \leq c_8 \| v\|_{t+m+d-\frac{1}{2},\mu} ,$$

which shows (2.5.1) for integer s. The non-integer cases are included by interpolation, and the estimates with $\langle\mu\rangle^{-\nu}$ follow, in view of (2.1.34). ▯

It is seen from the proof how the constants $c_{s,t}$ etc. depend on a finite number of symbol seminorms.

A very similar proof applies to trace operators of class 0.

<u>2.5.2 Proposition.</u> *Let* $T_\mu = \mathrm{OPT}(t(x',y',\xi,\mu))$ *where* $t \in S^{d,\nu}_{1,0}(W',H^-_{-1})$ *and vanishes for* (x',y') *outside a compact subset of* $\Omega' \times \Omega'$ *. For any* $s \in \mathbb{R}$ *there are constants so that when* $u \in \mathscr{S}(\overline{\mathbb{R}}^n_+)$ *and* $\mu \geq 0$,

(2.5.6) $$\|T_\mu u\|_{H^{s,\mu}(\mathbb{R}^{n-1})} \le c_s \|(\rho^\nu+1)\kappa^{d+\frac12+s}\acute{u}(\xi',x_n)\|_{L^2},$$
$$\le c_s'(\langle\mu\rangle^{-\nu}+1)\|u\|_{H^{(0,d+\frac12+s),\mu}(\overline{\mathbb{R}}^n_+)} \quad .$$

Proof: Let $v \in \mathscr{S}(\mathbb{R}^{n-1})$. Then we have

$$|(T_\mu u,v)| = c\, \left|\int_{\mathbb{R}^{3n-3}}\int_0^\infty e^{i(x'-y')\xi'}\tilde{t}(x',y',y_n,\xi',\mu)u(y)\overline{v}(x')dyd\xi'dx'\right|$$
$$= c_1\left|\int_{\mathbb{R}^{3n-3}}\int_0^\infty \hat{\tilde{t}}(\eta'-\xi',\xi'-\theta',y_n,\xi',\mu)\acute{u}(\theta',y_n)\overline{\hat{v}}(\eta')dy_n d\theta' d\eta' d\xi'\right| ,$$

where we insert an inequality for $\hat{\tilde{t}}$ like (2.5.3) (with $j = 0$) and inequalities like (2.1.32'). After an application of the Cauchy-Schwarz inequality, we then find (as in the preceding proof)

$$|(T_\mu u,v)| \le \|(\rho^\nu+1)\kappa^{d+\frac12+s}\acute{u}(\xi',y_n)\|_0\ \|\kappa^{-d-\frac12-s}\hat{v}\|_0 \quad ,$$

which shows (2.5.6) in view of the duality between $H^{r,\mu}(\mathbb{R}^{n-1})$ and $H^{-r,\mu}(\mathbb{R}^{n-1})$ (for $r = d+\frac12+s$). ▯

The inequalities (2.5.6) show à fortiori, for $m \in \mathbb{N}$,

(2.5.7) $$\|T_\mu u\|_{s,\mu} \le c_s \sum_{0\le j\le m} \|(\rho^\nu+1)\kappa^{d+\frac12+s-j}D^j_{x_n}\acute{u}(\xi',x_n)\|_0$$
$$\le c_{s,m}(\langle\mu\rangle^{-\nu}+1)\|u\|_{H^{(m,d+s-m+\frac12),\mu}(\overline{\mathbb{R}}^n_+)} \quad ,$$

and one even has

(2.5.8) $$\|T_\mu u\|_{s,\mu} \le c_{s,t}(\langle\mu\rangle^{-\nu}+1)\|u\|_{(t,d+s-t+\frac12),\mu}$$

for $t \in \overline{\mathbb{R}}_+$ (cf. (A.28)). This inequality can be extended to be valid also for $t \in\,]-\frac12,0[$ by use of the duality between $H^{t,\kappa}(\overline{\mathbb{R}}_+)$ and $H^{-t,\kappa}(\overline{\mathbb{R}}_+)$ for $t \in\,]-\frac12,0[$, where $\tilde{t}$ is estimated in $H^{t,\kappa}(\overline{\mathbb{R}}_+)$ by interpolation. Since the extension is not necessary for our main purposes, we omit further developments in this direction.

2.5.3 Theorem. *Let* $T_\mu = OPT(t(x',y',\xi,\mu))$, *where* $t \in S^{d,\nu}_{1,0}(W',H^-_{r-1})$ *and vanishes for* (x',y') *outside a compact subset of* $\Omega' \times \Omega'$. *For any* $s \in \mathbb{R}$ *and any integer* $m \geq r$, *there are constants so that for* $u \in \mathcal{S}(\overline{\mathbb{R}}^n_+)$,

$$(2.5.9) \qquad \|T_\mu u\|_{H^{s,\mu}(\mathbb{R}^{n-1})} \leq c_{s,m} \sum_{0\leq j\leq m} \|(\rho^\nu+1)\kappa^{d+\frac{1}{2}+s-j} D^j_{x_n} \acute{u}(\xi',x_n)\|_0$$

$$\leq c'_{s,m}(\langle\mu\rangle^{-\nu}+1)\|u\|_{H^{(m,d+s-m+\frac{1}{2}),\mu}(\overline{\mathbb{R}}^n_+)} \quad .$$

One also has for $t > r - \frac{1}{2}$, $t \in \overline{\mathbb{R}}_+$,

$$(2.5.10) \qquad \|T_\mu u\|_{s,\mu} \leq c_{s,t}(\langle\mu\rangle^{-\nu}+1)\|u\|_{(t,d+s-t+\frac{1}{2}),\mu} \quad .$$

In particular,

$$(2.5.11) \qquad \|T_\mu u\|_{t-d-\frac{1}{2},\mu} \leq c_s(\langle\mu\rangle^{-\nu}+1)\|u\|_{t,\mu} \quad \textit{for } t > r - \frac{1}{2} .$$

Proof: Write

$$T_\mu = \sum_{0\leq j<r} S_{j,\mu}\gamma_j + T'_\mu \quad ,$$

where T'_μ is of class 0 and the $S_{j,\mu}$ are ps.d.o.s on Ω' with symbols $s_j(x',y',\xi',\mu) \in S^{d-j,\nu}_{1,0}$. For T'_μ we have the inequalities (2.5.9) by the preceding proposition, and for each $S_{j,\mu}\gamma_j$ we find them by a combination of the methods of Lemma A.2 and Proposition 2.1.12. The extension (2.5.10) holds since the relevant estimates for γ_j are valid for $t > r - \frac{1}{2}$. □

The corresponding estimates for singular Green operators are established quite analogously.

2.5.4 Theorem. *Let* $G = OPG(g(x',y',\xi,\eta_n,\mu))$, *where* $g \in S^{d-1,\nu}_{1,0}(W',H^+\hat{\otimes}H^-_{r-1})$ *and vanishes for* (x',y') *outside a compact subset of* $\Omega' \times \Omega'$. *For any* $s \in \overline{\mathbb{R}}_+$, $t \in \mathbb{R}$ *and integer* $m \geq r$, *there are constants so that for* $u \in \mathcal{S}(\overline{\mathbb{R}}^n_+)$,

$$(2.5.12) \qquad \|G_\mu u\|_{H^{(s,t),\mu}(\mathbb{R}^n_+)} \leq c_{s,t,m} \sum_{0\leq j\leq m} \|(\rho^\nu+1)\kappa^{d+s+t-j} D^j_{x_n} \acute{u}(\xi',x_n)\|_0$$

$$\leq c'_{s,t,m}(\langle\mu\rangle^{-\nu}+1)\|u\|_{H^{(m,d+s+t-m),\mu}(\overline{\mathbb{R}}^n_+)} \quad .$$

One also has for $s' > r - \frac{1}{2}$, $s' \in \overline{\mathbb{R}}_+$,

(2.5.13) $$\|G_\mu u\|_{(s,t),\mu} \le c_{s,t,s'}(\langle\mu\rangle^{-\nu}+1)\|u\|_{(s',d+s+t-s'),\mu} \;,$$

and in particular,

(2.5.14) $$\|G_\mu u\|_{s'-d,\mu} \le c_{s'}(\langle\mu\rangle^{-\nu}+1)\|u\|_{s',\mu} \quad \textit{for} \quad s' > r-\tfrac{1}{2} \;.$$

Proof: Write

$$G_\mu = \sum_{0\le j<r} K_{j,\mu}\gamma_j + G'_\mu \;,$$

where G'_μ is of class 0 and the $K_{j,\mu}$ are Poisson operators of order $d-j$ and regularity ν. For the term $K_{j,\mu}\gamma_j$ the result follows by a combination of the methods of Theorem 2.5.1 and Lemma A.2, and for the term G'_μ the result is shown by a proof very similar to that of Theorem 2.5.1, studying the scalar product $(G'_\mu u,v)$ for u and $v \in \mathscr{S}(\overline{\mathbb{R}}^n_+)$. □

We shall also show the continuity of $P_{\mu,\Omega}$ in these Sobolev spaces. The basic fact that we use, is that at $x_n = 0$, P_μ and each expression $D^j_{x_n}P_\mu$ is the sum of a differential operator and an operator whose symbol is $O(\langle\xi_n\rangle^{-1})$. When the symbol depends on x_n, there is moreover a general pseudo-differential term, that we can write such that it contains a factor x_n^M with a large M.

2.5.5 Theorem. *Let* $P_\mu = OP(p(x,y,\xi,\mu))$, *where* $p \in S^{d,\nu}_{1,0,tr}(W)$ *and vanishes for* (x,y) *outside a compact subset of* $\Omega\times\Omega$. *For any* $m \in \mathbb{N}$, *any* $t \in \mathbb{R}$ *there are constants so that for* $u \in \mathscr{S}(\overline{\mathbb{R}}^n_+)$,

(2.5.15) $$\|P_{\mu,\Omega}u\|_{H^{(m-d,t),\mu}(\overline{\mathbb{R}}^n_+)} \le c_{m,t}\sum_{0\le j\le m}\|(\rho^\nu+1)\kappa^{m-j+t}D^j_{x_n}\acute{u}(\xi',x_n)\|_0$$
$$\le c'_{m,t}(\langle\mu\rangle^{-\nu}+1)\|u\|_{H^{(m,t),\mu}(\overline{\mathbb{R}}^n_+)} \;.$$

One also has, for $s \in \overline{\mathbb{R}}_+$,

(2.5.16) $$\|P_{\mu,\Omega}u\|_{(s-d,t),\mu} \le c_{s,t}(\langle\mu\rangle^{-\nu}+1)\|u\|_{(s,t),\mu} \;,$$

and in particular,

(2.5.17) $$\|P_{\mu,\Omega}u\|_{s-d,\mu} \le c_s(\langle\mu\rangle^{-\nu}+1)\|u\|_{s,\mu} \quad \textit{for} \quad s \ge 0 \;.$$

Proof: We first observe that Proposition 2.1.12 has an anisotropic generalization: For all $s,t \in \mathbb{R}$

$$\|P_\mu u\|_{H^{(s-d,t),\mu}(\mathbb{R}^n)} \leq c_{s,t}\|(\rho(\xi,\mu)^\nu+1)\langle\xi,\mu\rangle^s\langle\xi',\mu\rangle^t\hat{u}(\xi)\|_0 \tag{2.5.18}$$

$$\leq c'_{s,t}(\langle\mu\rangle^{-\nu}+1)\|u\|_{H^{(s,t),\mu}(\mathbb{R}^n)} \;;$$

this is shown by a modification of the proof of Proposition 2.1.12 where one also inserts the inequality

$$1 = \langle\theta',\mu\rangle^t\langle\theta',\mu\rangle^{-t} \leq c\langle\theta',\mu\rangle^t\langle\theta'-\xi'\rangle^{|t|}\langle\xi'-\eta'\rangle^{|t|}\langle\eta',\mu\rangle^{-t}$$

in the integral before applying the Schwarz inequality. Then for $m = 0$, one has immediately, using (2.2.62),

$$\|P_{\mu,\Omega}u\|_{H^{(-d,t),\mu}(\overline{\mathbb{R}}^n_+)} \leq \|P_\mu e^+u\|_{H^{(-d,t),\mu}(\mathbb{R}^n)} \tag{2.5.19}$$

$$\leq c_1\|(\rho(\xi,\mu)^\nu+1)\langle\xi',\mu\rangle^t\,\widehat{e^+u}\|_{L^2(\mathbb{R}^n)}$$

$$\leq c_2\|(\rho(\xi',\mu)^\nu+1)\langle\xi',\mu\rangle^t\,\widehat{e^+u}\|_{L^2(\mathbb{R}^n)}$$

$$= c_3\|(\rho(\xi',\mu)^\nu+1)\langle\xi',\mu\rangle^t\,\acute{u}(\xi',x_n)\|_{L^2(\mathbb{R}^n_+)}$$

$$\leq c_4(\langle\mu\rangle^{-\nu}+1)\|u\|_{(0,t),\mu} \;;$$

this shows (2.5.15) for $m = 0$.

Now consider the case where m is large positive, with $m-d = k \geq 0$. We shall estimate $\|D^k_{x_n}P_{\mu,\Omega}u\|_{(0,t),\mu}$. For this we write

$$D^k_{x_n}P_\mu = \varphi\,OP(q(y,\xi,\mu))\varphi + R_\mu\,, \tag{2.5.20}$$

where $\varphi \in C_0^\infty(\mathbb{R}^n)$ is chosen such that $\varphi(x)$ and $\varphi(y)$ are 1 for (x,y) in the support of p, R_μ is negligible (of regularity $\nu - d$) and q is on y-form, derived from $\xi_n^k\,p(x,y,\xi,\mu)$ by use of Theorem 2.1.15 1°. In view of Lemma 2.1.6 (second statement), q is in $S^{d+k,\nu+k}_{1,0,tr}$. For R_μ, we have by use of (2.5.18), since its symbol belongs to $\cap_N S^{d-N,\nu-N}$,

$$\|R_{\mu,\Omega}u\|_{(0,t),\mu} \leq c_1\|(\rho(\xi,\mu)^{\nu-N}+1)\langle\xi,\mu\rangle^{d-N}\langle\xi',\mu\rangle^t\,\widehat{e^+u}\|_0 \leq \tag{2.5.21}$$

$$\le c_2 \| (\langle\xi\rangle^{\nu-N}\langle\xi,\mu\rangle^{d-\nu} + \langle\xi,\mu\rangle^{d-N})\langle\xi',\mu\rangle^{t}\, \widehat{e^+u}\|_0$$

$$\le c_3 \| \langle\xi\rangle^{-N'}\langle\mu\rangle^{d-\nu}\langle\xi',\mu\rangle^{t}\, \widehat{e^+u}\|_0$$

$$\le c_3 \| \langle\xi'\rangle^{-N'}\langle\mu\rangle^{d-\nu}\langle\xi',\mu\rangle^{t}\, \acute{u}(\xi',x_n)\|_{L^2(\mathbb{R}^n_+)}$$

$$\le c_4 \| \langle\xi'\rangle^{-N''}\langle\xi',\mu\rangle^{t+d-\nu}\, \acute{u}(\xi',x_n)\|_{L^2}$$

where we use (2.3.43), (2.3.44) and take N large (so also N' and N'' are large). (2.5.21) will be used with $k \ge |\nu|$.

The symbol q is split in three terms, by use of (2.2.78) and a Taylor expansion:

$$q(y,\xi,\mu) = \sum_{0\le j<N} \frac{y_n^j}{j!}\,\partial_{y_n}^j\, q(y',0,\xi',\mu) + y_n^N r_N(y,\xi,\mu) \tag{2.5.22}$$
$$= q' + q'' + y_n^N r_N ,$$

where q' is a <u>differential</u> operator symbol of order $d+k=m$ (defined from the polynomial parts of the $\partial_{y_n}^j q$) , q'' satisfies

$$|D_y^\beta q''(y,\xi,\mu)| \le c_\beta(y)(\rho(\xi',\mu)^{\nu+k} + 1)\langle\xi',\mu\rangle^{d+k+1}\langle\xi,\mu\rangle^{-1} \tag{2.5.23}$$

for any β , and $r_N \in S^{d+k,\nu+k}_{1,0,tr}$. For the differential operator term one easily finds

$$\|\varphi\, OP(q')_\Omega \varphi u\|_{H^{(0,t),\mu}(\overline{\mathbb{R}}^n_+)} \le c \sum_{0\le j\le m} \|\kappa^{m-j+t} D_{x_n}^j \acute{u}(\xi',x_n)\|_0 . \tag{2.5.24}$$

For the term q'' one shows, by another variant of the proof of Proposition 2.1.12

$$\|\varphi\, OP(q'')_\Omega \varphi u\|_{H^{(0,t),\mu}(\overline{\mathbb{R}}^n_+)} \le \|\varphi\, OP(q'')\varphi e^+u\|_{H^{(0,t),\mu}(\mathbb{R}^n)} \tag{2.5.25}$$
$$\le c\|(\rho(\xi',\mu)^{\nu+k} + 1)\langle\xi',\mu\rangle^{d+k+t}\, \acute{u}(\xi',x_n)\|_{L^2(\mathbb{R}^n_+)} ,$$

and for the last term, one finds by (2.5.18) and the fact that multiplication by x_n^N maps $H^N(\overline{\mathbb{R}}^n_+)$ into $\mathring{H}^N(\overline{\mathbb{R}}^n_+)$

(2.5.26) $$\|\varphi\ OP(y_n^N r_N)_\Omega \varphi u\|_{H^{(0,t),\mu}(\overline{\mathbb{R}}^n_+)}$$

$$\leq c\|(\rho(\xi,\mu)^{\nu+k}+1)\langle\xi,\mu\rangle^{d+k}\langle\xi',\mu\rangle^t F[e^+x_n^N\varphi u]\|_0$$

$$\leq c_1 \sum_{0\leq j<m} \|(\rho(\xi',\mu)^{\nu+k}+1)\kappa^{m-j+t} D^j_{x_n}(e^+x_n^N(\widehat{\varphi u})(\xi',x_n))\|_0$$

$$\leq c_2 \sum_{0\leq j\leq m} \|(\rho^{\nu+k}+1)\kappa^{m-j+t} D^j_{x_n}\acute{u}(\xi',x_n)\|_{L^2(\mathbb{R}^n_+)}$$

when N is large enough (it suffices to take $N\geq m$). Collecting (2.5.21, 24, 25, 26) (assuming $k\geq|\nu|$), we find altogether

$$\|D^k_{x_n} P_{\mu,\Omega}u\|_{(0,t),\mu} \leq c \sum_{0\leq j\leq m} \|(\rho^{\nu+k}+1)\kappa^{m-j+t} D^j_{x_n}\acute{u}(\xi',x_n)\|_0 \quad ,$$

which shows (2.5.15) for large m, since we already have it for $m-d=0$. The intermediate values of m can be included by interpolation, and the remaining statements are collaries. ▯

With a little more effort, one can extend (2.5.16) to $s\in\,]-\frac{1}{2},0[$, since in (2.5.19) one also has that e^+u in $H^{(s,t),\mu}(\mathbb{R}^n)$ can be identified with $u\in H^{(s,t),\mu}(\overline{\mathbb{R}}^n_+)$ when $s\in\,]-\frac{1}{2},0[$.

It is also possible to treat $P_{\mu,\Omega}$ by use of the continuity properties of the boundary operators, as in [Boutet 1] or [Rempel-Schulze 1].

The above results are formulated in terms of spaces over $\overline{\mathbb{R}}^n_+$ and one can of course deduce more general statements from these by use of partitions of unity.

In particular, we can obtain the continuity properties for operators over manifolds, by use of the rules for coordinate changes and multiplications by cut-off functions derived at the end of Section 2.4:

<u>2.5.6 Corollary</u>. *Let* A_μ *be a parameter-dependent Green operator*

$$A_\mu = \begin{pmatrix} P_{\mu,\Omega}+G_\mu & K_\mu \\ T_\mu & S_\mu \end{pmatrix} : \begin{matrix} C^\infty(E) \\ \times \\ C^\infty(F) \end{matrix} \to \begin{matrix} C^\infty(E) \\ \times \\ C^\infty(F') \end{matrix}$$

of order d , *class* r *and regularity* ν . *Then* A_μ *defines a mapping*

$$A_\mu : H^{s,\mu}(E) \times H^{s-\frac{1}{2},\mu}(F) \to H^{s-d,\mu}(E) \times H^{s-d-\frac{1}{2},\mu}(F') ,$$

for any $s \in \overline{\mathbb{R}}_+$, $s > r - \frac{1}{2}$, *whose norm is* $O(\langle\mu\rangle^{-\nu}+1)$ *for* $\mu \in \overline{\mathbb{R}}_+$.

The μ-dependence is such that when $\nu \geq 0$, the estimates are simply uniform in μ, but when $\nu < 0$, the control is weaker for $\mu \to \infty$.

Note that when $P_{\mu,\Omega} + G_\mu$ is of order $d < -n$ and class 0, then both $P_{\mu,\Omega} + G_\mu$ and its adjoint $P^*_{\mu,\Omega} + G^*_\mu$ are continuous mappings from $L^2(E)$ into $H^d(E)$ with norms that are $O(\langle\mu\rangle^{-\nu}+1)$. This implies (cf. [Agmon 4]) that $P_{\mu,\Omega} + G_\mu$ is an integral operator with a kernel $K(x,y,\mu)$ that is continuous in (x,y) for each μ, and whose sup-norm is $O(\langle\mu\rangle^{-\nu}+1)$.

The kernels can also be studied directly on the basis of the symbol estimates. We do that for important particular cases in Sections 3.3 and 4.2, with the additional structure furnished by polyhomogeneity.

It is not hard to modify the above proofs to include norm estimates for operators defined from symbols depending on x (resp. x') that are constant in x (resp. x') for large $x \in \mathbb{R}^n$ (resp. $x' \in \mathbb{R}^{n-1}$).

2.6 Composition of x_n-independent boundary symbol operators.

The study of composition rules is complicated, because there are so many different kinds of operators involved. In fact, the composition of two Green operators A_μ and A'_μ gives an operator A''_μ

$$(2.6.1)\qquad A''_\mu = A_\mu A'_\mu = \begin{pmatrix} P_{\mu,\Omega} + G_\mu & K_\mu \\ T_\mu & S_\mu \end{pmatrix} \begin{pmatrix} P'_{\mu,\Omega} + G'_\mu & K'_\mu \\ T'_\mu & S'_\mu \end{pmatrix}$$

$$= \begin{pmatrix} P''_{\mu,\Omega} + G''_\mu & K''_\mu \\ T''_\mu & S''_\mu \end{pmatrix} ,$$

which is shown to be a Green operator by showing that

(2.6.2)

(i) $P''_\mu = P_\mu P'_\mu$ is a ps.d.o. with the transmission property,

(ii) $L(P_\mu, P'_\mu) = (P_\mu P'_\mu)_\Omega - P_{\mu,\Omega} P'_{\mu,\Omega}$ is a s.g.o.,

(iii) $G'''_\mu = P_{\mu,\Omega} G'_\mu + G_\mu P'_{\mu,\Omega} + G_\mu G'_\mu + K_\mu T'_\mu$ is a s.g.o.,

(iv) $T''_\mu = T_\mu P'_{\mu,\Omega} + T_\mu G'_\mu + S_\mu T'_\mu$ is a trace operator,

(v) $K''_\mu = P_{\mu,\Omega} K'_\mu + G_\mu K'_\mu + K_\mu S'_\mu$ is a Poisson operator,

(vi) $S''_\mu = T_\mu K'_\mu + S_\mu S'_\mu$ is a ps.d.o. on Γ ,

here there are 13 new rules to show for the boundary operators, where in each case the dependence on μ (the regularity) must be investigated. The behavior with respect to the x_n-direction is of course the most interesting; for the rules with respect to the x' variables are just like those in Theorem 2.1.15 3^0.

The notation $\circ$ is used for all compositions, both in compositions of operators and in formulas for the corresponding symbols. Composition of boundary symbol operators (acting in the x_n-variable only), and the resulting symbols, are sometimes indicated by $\circ_n$ for clarity.

The analysis is carried out in three steps: First we deduce the neat results for $\circ_n$-compositions, in the case where the ps.d.o. part has symbol independent of x_n, this is done in the present section. In Section 2.7, we treat the more complicated case where x_n-dependence is allowed, and finally we show the composition rules that hold for operators acting in the full x-variable.

level, using the decomposition of t and p in polynomial and H_{-1} part.) The other rules are similarly easy (altogether, they are the first step in the original, non-parametrized calculus). In the present context, the new thing is to account for the symbol estimates contained in (2.6.5).

8^0 is a consequence of Proposition 2.1.5, and 12^0 follows in the same way from the Leibniz formula; and 3^0 and 6^0 are covered by Lemma 2.6.2 1^0. 1^0 and 4^0 follow from Lemma 2.6.2 4^0 combined with Lemma 2.3.9 2^0 resp. 5^0. 9^0 and 10^0 follow from Lemma 2.6.3 4^0 and 5^0 combined with Lemma 2.3.10 2^0 and 5^0. It remains to treat the cases containing $\int^+$ integrals, namely (2.6.5), 2^0, 5^0, 7^0 and 11^0.

Let us begin with 7^0. Here, when

$$t(X,\xi,\mu) = \sum_{0\le j<r} s_j(X,\xi',\mu)\xi_n^j + t_1(X,\xi,\mu) \quad ,$$

with t_1 of class 0, then

$$t\,k' = \sum_{0\le j<r} s_j(X,\xi',\mu)\xi_n^j k'(X,\xi,\mu) + t_1(X,\xi,\mu)k'(X,\xi,\mu) \quad ,$$

where the two terms are treated differently. For the last term we have

$$(2.6.12)\qquad \left|\frac{1}{2\pi}\int^+ t_1(X,\xi,\mu)k'(X,\xi,\mu)d\xi_n\right| = \left|\frac{1}{2\pi}\int_{\mathbb{R}} t_1k'd\xi_n\right|$$
$$\le c\|t_1\|_{L^2_{\xi_n}}\|k'\|_{L^2_{\xi_n}} \le c(X)(\rho^{\nu}+1)\kappa^{d+\frac12}(\rho^{\nu'}+1)\kappa^{d'-\frac12}$$
$$\le c'(X)(\rho^{m(\nu,\nu')}+1)\kappa^{d+d'} \quad ,$$

showing the basic estimate for this contribution. Derivatives and lower order parts are easily treated. For the sum over j, we calculate $\int^+ \xi_n^j k'\,d\xi_n$ by use of the expansions

$$(2.6.13)\qquad \xi_n^j k'(X,\xi,\mu) = \sum_{0\le m<j} s'_{-1-m}(X,\xi',\mu)\xi_n^{j-1-m} + k'_{(-j)}(X,\xi,\mu)$$

(with $k'_{(-j)} \in H^+$) , for, by (2.2.44),

$$(2.6.14)\qquad \frac{1}{2\pi}\int^+ \xi_n^j k'(X,\xi,\mu)d\xi_n = i\, s'_{-1-j}(X,\xi',\mu) \quad .$$

Lemma 2.3.9 4^0 assures that $s'_{-1-j} \in S^{d'+j,\nu'}$, and then multiplication by $s_j \in S^{d-j,\nu}$ gives a symbol in $S^{d+d',m(\nu,\nu')}(\Xi,\overline{\mathbb{R}}^n_+)$. Altogether, this shows 7^0.

The proofs of 2^0, 5^0 and 11^0 follow the same pattern, except that one has to take some more variables into account, and use the preceding rules. For instance in 11^0, one considers the product

$$g(X,\xi',\xi_n,\zeta_n,\mu)\ g'(X,\xi',\zeta_n,\eta_n,\mu)$$

$$= (\sum_{0\le j<r} k_j\zeta_n^j + g_1)(\sum_{0\le m<r'} k'_m\eta_n^m + g'_1) \quad ,$$

where the integral $\int^+\zeta_n^j k'_m(X,\xi',\zeta_n)d\zeta_n$ is treated as in (2.6.14), the integral $\int^+\zeta_n^j g'_1 d\zeta_n$ gives rise to a trace symbol, and the two integrals $\int^+ g_1 k'_m d\zeta_n$ and $\int^+ g_1 g'_1 d\zeta_n$ are estimated by the Cauchy-Schwarz inequality. The various product rules imply that the result is of regularity $m(\nu,\nu')$. The proofs work likewise for the $S^{d,\nu}_{1,0}$-spaces of symbols. ▯

The rules 1^0-12^0 are all very satisfactory in that they give a regularity that is as good as that obtained for pseudo-differential operators. However, it still remains to treat the 13'th term

(2.6.15) $$L(p,p') = (pp')_\Omega - p_\Omega \circ p'_\Omega \quad ,$$ stemming from

(2.6.16) $$L(P_\mu,P'_\mu) = (P_\mu P'_\mu)_\Omega - P_{\mu,\Omega}P'_{\mu,\Omega} \quad ,$$

and here one effectively gets lower regularity. It is known from the non-parametrized case that $L(P_\mu,P'_\mu)$ is a singular Green operator for each fixed μ , and $L(p,p')$ is a singular Green boundary symbol operator (we shall use $L(p,p')$ to denote its symbol also); the proof we give below furthermore takes μ into account. The structure can be analyzed in various ways; a very simple and direct method is given in [Grubb 17], which adapts very well to the parameter-dependent situation.

Consider operators on $\mathbb{R}^n$, and write

(2.6.17)
$$P_\mu = \sum_{0\le\ell\le d} S_{\ell,\mu}(x,D')D_n^\ell + Q_\mu \quad ,$$
$$P'_\mu = \sum_{0\le\ell\le d'} S'_{\ell,\mu}(x,D')D_n^\ell + Q'_\mu \quad ,$$

where the $S_{\ell,\mu}$ and $S'_{\ell,\mu}$ are differential operators (with symbols polynomial in (ξ',μ)) and $Q_\mu e^+$ and $Q'_\mu e^+$ map $C^\infty_{(0)}(\overline{\mathbb{R}}^n_+)$ into $L^2_{loc}(\mathbb{R}^n)$; here we use the same decomposition as in Lemma 1.3.1 and the proof of Theorem 2.5.5; Q_μ contains negligible terms, terms whose symbols are $O(\langle\xi_n\rangle^{-1})$, and terms containing a factor x_n^d . One finds the following rules: Since differential operators are local,

(2.6.18) $$L(\sum_{0\le\ell\le d} S_{\ell,\mu}D_n^\ell \, , \, P'_\mu) = 0 \quad .$$

By Green's formula (2.2.39),

$$\text{(2.6.19)}\quad L(P_\mu, \sum_{0\le \ell\le d'} S'_{\ell,\mu} D_n)u =$$

$$= r^+ \sum_{0\le\ell\le d'} P_\mu S'_{\ell,\mu}(D_n^\ell e^+u - e^+D_n^\ell u)$$

$$= -ir^+P_\mu \sum_{0\le\ell\le d'} S'_{\ell,\mu} \sum_{0\le k<\ell} (\gamma_{\ell-k-1}u \otimes D_n^k\delta)$$

$$= \sum_{0\le m\le d'-1} K_{m,\mu}\gamma_m u \quad ,$$

where $K_{m,\mu}$ is the Poisson operator of order $d+d'-m$ (as in Example 2.3.1)

$$\text{(2.6.20)}\quad K_{m,\mu}v = -i \sum_{\ell=m+1}^{d'} r^+P_\mu S'_{\ell,\mu} D_{x_n}^{\ell-1-m}(v(x') \otimes \delta(x_n)) \quad .$$

Finally, we write (cf. (A.32))

$$\text{(2.6.21)}\quad L(Q_\mu,Q'_\mu) = r^+Q_\mu Q'_\mu e^+ - r^+Q_\mu e^+r^+Q'_\mu e^+$$

$$= r^+Q_\mu(I - e^+r^+)Q'_\mu e^+$$

$$= r^+Q_\mu e^-r^-Q'_\mu e^+$$

$$= (r^+Q_\mu e^-J)(Jr^-Q'_\mu e^+)$$

$$= G^+(Q_\mu)\, G^-(Q'_\mu) \quad ,$$

where $G^+(Q_\mu)$ and $G^-(Q'_\mu)$ are the special singular Green operators of class 0 and orders d resp. d'

$$\text{(2.6.22)}\quad \begin{aligned} G^+(Q_\mu) &= r^+Q_\mu e^-J = r^+P_\mu e^-J = G^+(P_\mu) \\ G^-(Q'_\mu) &= Jr^-Q'_\mu e^+ = Jr^-P'_\mu e^+ = G^-(P'_\mu) = [G^+(P'^*_\mu)]^* \;, \end{aligned}$$

cf. [Grubb 17, Chapter 3]. Similar formulas hold on the boundary symbol level. Altogether,

$$\text{(2.6.23)}\quad L(P_\mu,P'_\mu) = \sum_{0\le m<d'} K_{m,\mu}\gamma_m + G^+(Q_\mu)\, G^-(Q'_\mu) \quad .$$

We shall now analyze the dependence on μ .

First there are the operators $K_{m,\mu}$ defined from P_μ by (2.6.20). In the product $P_\mu S'_{\ell,\mu} D_n^{\ell-1-m}$, $S'_{\ell,\mu}$ is of order $d'-\ell$ and regularity $+\infty$, so that $P_\mu S'_{\ell,\mu}$ is again of regularity ν . Composition with $D_n^{\ell-1-m}$ is governed by the rule (2.1.13) together with (2.1.48 ii), which shows that

(2.6.24) symbol of $PS'_{\ell} D_n^{\ell-1-m} \in S_{tr}^{d+d'-1-m,\nu+\ell-1-m}$.

We then need the general rule for Poisson symbols derived from ps.d.o. symbols:

2.6.4 Lemma. *Let* p *be as in Theorem 2.6.1. Then the mapping*

(2.6.25) $k(X,\xi',\mu,D_n) : v \leadsto r^+p(X,\xi',\mu,D_n)[\delta(x_n)v]\,, \quad v \in \mathbb{C}\,,$

is a Poisson operator on $\overline{\mathbb{R}}_+$, *defined from the symbol resp. symbol-kernel*

(2.6.26)
$$k(X,\xi,\mu) = h^+_{\xi_n} p(X,\xi,\mu) \in S^{d,\nu}(\Xi, \overline{\mathbb{R}}^n_+, H^+)$$
$$\tilde{k}(X,x_n,\xi',\mu) = r^+\tilde{p}(X,x_n,\xi',\mu) \in S^{d,\nu}(\Xi, \overline{\mathbb{R}}^n_+, \mathscr{S}(\overline{\mathbb{R}}_+)) \quad ,$$

here

(2.6.27) $\tilde{p}(X,z,\xi',\mu) = F^{-1}_{\xi_n \to z} p(X,\xi,\mu)$.

Proof: The necessary calculations were made in Example 2.3.1, see (2.3.4), which shows the formulas for the symbol and symbol-kernel. It follows from Corollary 2.3.5 that they lie in the symbol spaces as asserted. □

2.6.5 Proposition. *Let* $d \in \mathbb{Z}$, $d' \in \mathbb{N}$ *and* $\nu \in \mathbb{R}$, *and let* $p(X,x_n,\xi,\mu) \in S_{tr}^{d,\nu}(\Xi\times\mathbb{R}\,,\overline{\mathbb{R}}^{n+1}_+)$ *and* $s'_{\ell}(X,x_n,\xi',\mu) \in S^{d'-\ell,\infty}(\Xi\times\mathbb{R}\,,\overline{\mathbb{R}}^n_+)$ *for* $0 \le \ell \le d'$. *Then one has for the singular Green symbol resulting from the formation of* $L\Big(p(X,0,\xi',\mu,D_n),\ \sum_{0\le\ell\le d'} s'_{\ell}(X,0,\xi',\mu)D_n^{\ell}\Big)$,

(2.6.28) $L\Big(p\,,\,\Sigma\, s'_{\ell}\xi_n^{\ell}\Big) = \sum_{0\le m<d'} k_m(X,\xi,\mu)\eta_n^m \in S^{d+d'-1,\nu}(\Xi,\overline{\mathbb{R}}^n_+\,,\, H^+\hat{\otimes} H^-_{d'-1})$,

a singular Green symbol of order d+d' , *regularity* ν *and class* d' .

Proof: By (2.6.20), (2.6.24) and Lemma 2.6.4, k_m is a sum of terms of degree $d+d'-1-m$ and regularity $\nu+\ell-1-m$. Here $\ell \ge m+1$, so the regularity is $\ge \nu$. In view of the conventions for s.g.o. symbols in Definition 2.3.8, the sum in (2.6.28) is of order d+d' , regularity ν and class d' . □

Next, we consider the factors in the product (2.6.21).

2.6.6 Theorem. *Let* d *and* $\nu \in \mathbb{Z}$, *and let* p *satisfy* (2.6.3 i). *Then the singular Green operators (cf.* (A.32)*)*

$$g^+(p) = r^+p(X,0,\xi',\mu,D_n)e^-J$$
(2.6.29)
$$g^-(p) = Je^-p(X,0,\xi',\mu,D_n)e^+$$

have symbol-kernels resp. symbols (cf. (2.6.27)*)*

$$\tilde{g}^+(p)(X,x_n,y_n,\xi',\mu) = r^+\tilde{p}((X,0),x_n+y_n,\xi',\mu) \quad \text{and}$$
(2.6.30)
$$\tilde{g}^-(p)(X,x_n,y_n,\xi',\mu) = r^-\tilde{p}((X,0),-x_n-y_n,\xi',\mu) \in S^{d-1,\nu-\frac{1}{2}}(\Xi,\overline{\mathbb{R}}^n_+,\mathscr{S}(\overline{\mathbb{R}}^2_{++})) ,$$

$$g^+(p)(X,\xi,\eta_n,\mu) = F_{x_n\to\xi_n}\overline{F}_{y_n\to\eta_n}\tilde{g}^+(p) \quad \text{and}$$
(2.6.31)
$$g^-(p)(X,\xi,\eta_n,\mu) = F_{x_n\to\xi_n}\overline{F}_{y_n\to\eta_n}\tilde{g}^-(p) \in S^{d-1,\nu-\frac{1}{2}}(\Xi,\overline{\mathbb{R}}^n_+,H^+\hat{\otimes}H^-_{-1}) ,$$

of order d , *class* 0 *and regularity* $\nu-\frac{1}{2}$. *The symbols and operators vanish when* $p(D_n)$ *is a differential operator. Similar results hold for the* $S^{d,\nu}_{1,0}$ *symbol spaces.*

Proof: As shown in Example 2.3.1 (see (2.3.6)) $g^+(p)$ is the integral operator on $\mathbb{R}_+$ with kernel (where $(X,0)$ is written as X)

$$\tilde{g}^+(p)(X,x_n,y_n,\xi',\mu) = r^+\tilde{p}(X,x_n+y_n,\xi',\mu) = \tilde{p}^+(X,x_n+y_n,\xi',\mu) ,$$

(we denote $h^\pm p = p^\pm$) . The kernel is estimated by use of coordinate changes $z = x_n-y_n$, $w = x_n+y_n$,

$$(2.6.32)\quad \|\tilde{g}^+(p)(X,x_n,y_n,\xi',\mu)\|^2_{L^2_{x_n,y_n}(\mathbb{R}^2_{++})} = \int_0^\infty\int_0^\infty|\tilde{p}^+(X,x_n+y_n,\xi',\mu)|^2dx_ndy_n$$
$$= \frac{1}{2}\int_0^\infty dw\int_{-w}^{w}|(\tilde{p}^+(X,w,\xi',\mu)|^2dzdw$$
$$= \int_0^\infty w|\tilde{p}^+(X,w,\xi',\mu)|^2dw = c\int_{\mathbb{R}} p^+(X,\xi,\mu)D_{\xi_n}\overline{p^+(X,\xi,\mu)}d\xi_n$$
$$\leq c\|p^+(X,\xi,\mu)\|_{L^2_{\xi_n}(\mathbb{R})}\|D_{\xi_n}p^+(X,\xi,\mu)\|_{L^2_{\xi_n}(\mathbb{R})}$$
$$\leq c(X)(\rho^\nu+1)\kappa^{d+\frac{1}{2}}(\rho^{\nu-1}+1)\kappa^{d-\frac{1}{2}} \leq c'(X)[(\rho^{\nu-\frac{1}{2}}+1)\kappa^d]^2 ,$$

since ν is integer. (In fact, $(\rho^{\nu}+1)(\rho^{\nu-1}+1)$ is $O(1)$ when $\nu \geq 1$, and $(\rho^{\nu}+1)(\rho^{\nu-1}+1)$ is $O((\rho^{\nu-\frac{1}{2}}+1)^2)$ when $\nu \leq 0$, cf. (2.1.17).) This is the basic estimate, which easily generalizes to derivatives in X, ξ' and μ, and to lower order parts. For the symbol-kernels resulting from application of $x_n^k D_{x_n}^{k'} \, y_n^m D_{y_n}^{m'}$ we observe that for $w = x_n + y_n$,

$$x_n^k \, y_n^m \leq w^{k+m} \qquad \text{when} \quad x_n, y_n \geq 0 \ ,$$

and

$$D_{x_n}^{k'} \, D_{y_n}^{m'} \, \tilde{p}(X, x_n+y_n, \xi', \mu) = D_w^{k'+m'} \tilde{p}(X, w, \xi', \mu) \quad ,$$

so that

$$\text{(2.6.33)} \quad \| x_n^k D_{x_n}^{k'} \, y_n^m D_{y_n}^{m'} \tilde{g}^+(p) \|^2_{L^2_{x_n,y_n}(\mathbb{R}^2_{++})} \leq \int_0^\infty w^{2k+2m+1} \, | D_w^{k'+m'} \tilde{p}^+(X, w, \xi', \mu)|^2 \, dw$$

$$\leq \| w^{k+m} D_w^{k'+m'} \tilde{p}^+ \|_{L^2_w(\mathbb{R}_+)} \; \| w^{k+m+1} D_w^{k'+m'} \tilde{p}^+ \|_{L^2_w(\mathbb{R}_+)}$$

$$= c \| (D_{\xi_n}^{k+m} \xi_n^{k'+m'} p)^+ \|_{L^2_{\xi_n}(\mathbb{R})} \; \| (D_{\xi_n}^{k+m+1} \xi_n^{k'+m'} p)^+ \|_{L^2_{\xi_n}(\mathbb{R})}$$

$$\leq c(X) (\rho^{\nu-k-m+k'+m'} + 1)(\rho^{\nu-k-m-1+k'+m'} + 1) \kappa^{2d}$$

$$\leq c'(X) [(\rho^{\nu-k-m+k'+m'-\frac{1}{2}} + 1) \kappa^d]^2 \ .$$

Altogether, it is seen that $\tilde{g}^+(p)$ satisfies all the estimates required of a symbol-kernel in $S^{d,\nu-\frac{1}{2}}(\Xi, \mathbb{R}^n_+, H^+ \hat{\otimes} H^-_{-1})$. The proof for $\tilde{g}^-(p)$ is similar, based on the identities

$$\text{(2.6.34)} \quad Jr^- pe^+ u = Jr^- \frac{1}{2\pi} \int_{\mathbb{R}} e^{ix_n \xi_n} \, p(\xi_n) \int_0^\infty e^{-iy_n \xi_n} \, u(y_n) dy_n d\xi_n$$

$$= r^+ \frac{1}{2\pi} \int_{\mathbb{R}} e^{-ix_n \xi_n} \, p(\xi_n) \int_0^\infty e^{-iy_n \xi_n} \, u(y_n) dy_n d\xi_n$$

$$= r^+ \int_0^\infty \tilde{p}(X, -x_n - y_n, \xi', \mu) u(y_n) dy_n \ . \qquad \square$$

<u>2.6.7 Remark.</u> Observe that the symbols and symbol-kernels actually satisfy better estimates than required for the space $S^{d,\nu-\frac{1}{2}}$, e.g.

$$(2.6.35)\quad \| D_X^{\beta} D_{\xi'}^{\alpha} D_{\mu}^{j} x_n^{k} D_{x_n}^{k'} y_n^{m} D_{y_n}^{m'} \tilde{g}^{+}(p) \|_{L^2_{x_n,y_n}(\mathbb{R}^2_{++})}$$

$$\leq c(X)(\rho^{\nu-\frac{1}{2}-|\alpha|-k+k'-m+m'}+1)\kappa^{d-|\alpha|-k+k'-m+m'-j} \quad ,$$

where $-k+k'-m+m' \geq -[k-k']_+ - [m-m']_+$. However, this property is lost in compositions.

2.6.8 Remark. The hypothesis in the theorem, that ν be integer, is important. If ν is > 0 and not integer, one only concludes from the estimates in (2.6.32-33) that $g^+(p)$ is of regularity $\nu-\frac{1}{2}-(\nu-[\nu])/2$. (This can probably be improved by use of fractional powers of the (x_n,y_n)-operators in (2.6.35).) Since we shall not need these cases in the resolvent construction, the statements are omitted for simplicity.

The reduction of the regularity by $1/2$ appears to be unavoidable for general μ-dependent ps.d.o. symbols. However, it is worth noting that μ-independent symbols do not suffer that reduction.

2.6.9 Proposition. *When* $p \in S^d_{tr}(\Xi\times\mathbb{R}, \mathbb{R}^n)$ *then* $g^+(p)$ *and* $g^-(p)$ *belong to* $S^{d-1}(\Xi, \mathbb{R}^{n-1}, H^+\hat{\otimes} H^-_{-1})$ *and hence*

$$(2.6.36)\qquad g^+(p) \text{ and } g^-(p) \in S^{d-1,d}(\Xi, \overline{\mathbb{R}}^n_+, H^+\hat{\otimes} H^-_{-1}) \quad ,$$

when considered as μ*-dependent symbols.*

Proof: The fact that $g^+(p)$ and $g^-(p)$ are s.g.o. symbols of order d (degree $d-1$) is seen by a simpler variant of the proof of Theorem 2.6.6 above, and was shown in detail in [Grubb 17]. Now (2.6.36) follows immediately from Proposition 2.3.14, see (2.3.54). □

The term $G^+(Q_\mu)\, G^-(Q'_\mu)$ in (2.6.23) is now estimated (on the x_n-composition level) by a straightforward application of Theorem 2.6.1 11°, and we find altogether, in view of Proposition 2.6.5 and Theorem 2.6.6:

2.6.10 Theorem. *Let* p *and* p' *be as in Theorem 2.6.1, with integer regularities* ν *and* ν', *and write*

$$(2.6.37)\qquad p' = \sum_{0\leq j\leq d'} s'_j(X,\xi',\mu)\xi_n^j + h_{-1}p' \quad .$$

Then $L(p,p')$ *(defined by* (2.6.15)*) is a singular Green boundary symbol operator, of order* $d+d'$, *class* $[d']_+$ *and regularity*

$$\nu'' = m(\nu-\tfrac{1}{2},\ \nu'-\tfrac{1}{2}) \quad . \tag{2.6.38}$$

More precisely, the symbol is defined by

$$L(p,p')(X,\xi,\eta_n,\mu) = \sum_{0\le m<d'} k_m(X,\xi,\mu)\eta_n^m + g^+(p)\circ_n g^-(p') \quad , \tag{2.6.39}$$

where

$$k_m(X,\xi,\mu) = h^+(-i\sum_{\ell=m+1}^{d'} p(X,\xi,\mu)s'_\ell(X,\xi',\mu)\xi_n^{\ell-1-m}) \,, \tag{2.6.40}$$

and $g^+(p)$ *and* $g^-(p')$ *are as defined in Theorem 2.6.6. In particular,* $L(p,p')$ *depends only on* h^+p *and* h^-p' . *The results are also valid for* $S_{1,0}$ *symbol spaces.*

Note that the k_m contribute with regularity ν , which is better than ν'' in (2.6.38). Note that h^-p gives zero in (2.6.40).

Better results can sometimes be obtained, when one factor is independent of μ, cf. Proposition 2.6.9. Let us here just account for a case that will be of particular importance in the resolvent construction later on, and where there is no loss of regularity.

2.6.11 Proposition. *let* d *be integer* >0 , *let* $p\in S^d_{tr}(\Xi\times\mathbb{R},\ \mathbb{R}^n)$ *and let* $q\in S^{-d,d}_{tr}(\Xi\times\mathbb{R},\ \overline{\mathbb{R}}^{n+1}_+)$. *Then* $L(p,q)$ *is of order* 0 *and regularity* d , *i.e. lies in* $S^{-1,d}(\Xi,\ \overline{\mathbb{R}}^n_+,\ H^+\hat{\otimes}H^-_{-1})$.

Proof: To avoid repetitions, we have formulated the result for the general x_n-dependent case. We presently give the proof for the x_n-independent case; then it is extended to the x_n-dependent case just as in Theorems 2.7.6-7 below.

Since q is of order ≤ 0 ,

$$L(p,q) = g^+(p)\circ g^-(q) \quad .$$

Here $g^+(p)\in S^{d-1,d}(\Xi,\overline{\mathbb{R}}^n_+,H^+\hat{\otimes}H^-_{-1})$, as noted already in Proposition 2.6.9, but for q , Theorem 2.6.6 only gives that $g^-(q)\in S^{-d-1,d-\frac{1}{2}}(\Xi,\ \overline{\mathbb{R}}^n_+,\ H^+\hat{\otimes}H^-_{-1})$. A direct application of Theorem 2.6.1 11^0 would only give $L(p,q)$ regularity $d-\frac{1}{2}$, but now we proceed instead as in Lemma 2.1.6:

In view of the composition formula (2.6.5) 11^0 and the Cauchy-Schwarz inequality, we have since $g^+(p)$ is independent of μ ,

$$\|D^{\alpha}_{\xi'}(g^+(p)\circ g^-(q))\|_{L^2(\mathbb{R}^2)} = \|\sum_{\gamma\le\alpha} c_\gamma D^{\alpha-\gamma}_{\xi'} g^+(p)\circ D^{\gamma}_{\xi'} g^-(q)\|_{L^2(\mathbb{R}^2)}$$

$$\le c \sum_{\gamma\le\alpha} \|D^{\alpha-\gamma}_{\xi'} g^+(p)\|_{L^2(\mathbb{R}^2)} \|D^{\gamma}_{\xi'} g^-(q)\|_{L^2(\mathbb{R}^2)}$$

$$\le c(X) \sum_{\gamma\le\alpha} \langle\xi'\rangle^{d-|\alpha-\gamma|}(\rho^{d-\frac12-|\gamma|}+1)\kappa^{-d-|\gamma|}$$

$$\le c_1(X) \sum_{\gamma\le\alpha}\Big(\langle\xi'\rangle^{2d-\frac12-|\alpha|}\kappa^{-2d+\frac12} + \langle\xi'\rangle^{d-|\alpha-\gamma|}\kappa^{-d-|\gamma|}\Big)$$

$$\le c_2(X)\Big(\langle\xi'\rangle^{2d-\frac12-|\alpha|}\kappa^{-2d+\frac12} + \langle\xi'\rangle^{d-|\alpha|}\kappa^{-d}\Big)$$

$$\le c_3(X)\langle\xi'\rangle^{d-|\alpha|}\kappa^{-d} \le c_3(X)(\rho^{d-|\alpha|}+1)\kappa^{-|\alpha|} .$$

Furthermore, since

$$h^+_{\xi_n} h^-_{-1,\eta_n} D^k_{\xi_n}\xi_n^{k'} D^m_{\eta_n}\eta_n^{m'}(g^+(p)\circ g^-(q)) = (h^+_{\xi_n} D^k_{\xi_n}\xi_n^{k'} g^+(p))\circ(h^-_{-1,\eta_n} D^m_{\eta_n}\eta_n^{m'} g^-(q))$$

(cf. (2.6.5) 11^0), one has (cf. (2.6.35) , one really just needs s.g.o.-estimates),

$$\|h^+_{\xi_n} h^-_{-1,\eta_n} D^k_{\xi_n}\xi_n^{k'} D^m_{\eta_n}\eta_n^{m'}(g^+(p)\circ g^-(q))\|_{L^2(\mathbb{R}^2)}$$

$$\le c(X)\langle\xi'\rangle^{d-k+k'}(\rho^{d-\frac12-m+m'}+1)\kappa^{-d-m+m'}$$

$$= c(X)(\langle\xi'\rangle^{2d-\frac12-k+k'-m+m'}\kappa^{-2d+\frac12} + \langle\xi'\rangle^{d-k+k'}\kappa^{-d-m+m'})$$

$$\le 2c(X)\langle\xi'\rangle^{d-[k-k']_+-[m-m']_+}\kappa^{[k-k']_- + [m-m']_-} ,$$

since $2d-\frac12 > d$, $-k+k' \ge -[k-k']_+$ and $-m+m' \ge -[m-m']_+$. X-derivatives and lower order parts are likewise estimated as in Lemma 2.1.6, and by a combination of the various estimates, we find the desired result. □

The regularity obtained above follows the pattern of Lemma 2.1.6; it equals $\min\{d+\nu,d\} = \min\{2d-\frac12,d\} = d$. In fact, Lemma 2.1.6 extends to certain boundary operator compositions (in cases of "class 0"), but there are other cases where composition with a high order μ-independent operator does not improve the regularity. For example, $D^m_{x_n}\circ g$ is of regularity ν when g is of regularity ν (by the basic boundary symbol estimates) although its order is increased by m . We return to this phenomenon at the end of Section 3.3.

We end this section with some special considerations on the Laguerre series estimates, that have otherwise not been very much in focus. (The reader can bypass the details until they are needed.) Recall that, as mentioned in connection with Theorem 2.2.8 (cf. Remark 2.2.9), the ℓ^2_N and $\ell^2_{N,N'}$ norms of the Laguerre series expansions of our symbols can be derived from the $S^{d,\nu}(K)$ seminorms, but there is a loss of regularity in the converse direction (one does not get back the full informations on derived symbols such as $\partial_{\xi_n}\xi_n f$ and $\xi_n f$ when passing from ℓ^2_N norms to $S^{d,\nu}$ norms). On the other hand, the Laguerre norms have the advantage that they allow noninteger N in a very natural way, and interpolate well, which can be convenient in cases where noninteger regularity occurs. We discuss such a case below (composition rules involving $g^+(p)$ when p is of regularity 1), and then we derive the general rules for the behavior of Laguerre norms under compositions. A place where this is used is the proof of the central Theorem 3.3.10. (Let us also remark, that one can introduce norms with noninteger indexation for the $S^{d,}$ spaces by means of fractional powers of the Laguerre operator (2.2.12); this is use around Proposition 4.5.8 (with reference to [Grubb 17]), where it is needed in the non-parametrized setting, with $\sigma \sim \langle\xi'\rangle$. The purpose there is slightly different and we have chosen not to burden the whole book with this extra notation.)

Observe first the general rule, that when a singular Green symbol-kernel of class 0 is expanded in a Laguerre series

$$\tilde{g}(X,x_n,y_n,\xi',\mu) = \sum_{\ell,m\in\mathbb{N}} c_{\ell m}(\tilde{g})(X,\xi',\mu)\varphi_\ell(x_n,\kappa)\varphi_m(y_n,\kappa) ,$$

then the coefficient sequence satisfies, by a generalization of (2.2.15)

$$\text{(2.6.41)}\quad \|(c_{\ell m}(\tilde{g}))_{\ell,m\in\mathbb{N}}\|_{\ell^2_{N,N'}} \equiv \Big(\sum_{\ell,m} |(1+\ell)^N(1+m)^{N'}c_{\ell m}(\tilde{g})|^2\Big)^{\frac12}$$

$$= 2^{-N-N'}\|(c_{\ell m}(L^N_{\kappa,x_n}L^{N'}_{\kappa,y_n}\tilde{g}))_{\ell,m\in\mathbb{N}}\|_{\ell^2_{0,0}} = 2^{-N-N'}\|L^N_{\kappa,x_n}L^{N'}_{\kappa,y_n}\tilde{g}\|_{L^2_{x_n,y_n}} \quad \text{for } N,N'\in\mathbb{N}$$

(we use (2.2.12), replacing + by the variable in which the operator acts). Now let p be as in Theorem 2.6.6, with integer order d and integer regularity ν ; then $\tilde{g}^+(p)$ is of order d and regularity $\nu-\frac12$ by that theorem. When $\tilde{g}^+(p)$ is expanded in a Laguerre series $\tilde{g}^+(p) = \Sigma_{\ell,m\in\mathbb{N}} c^+_{\ell m}\varphi_\ell(x_n,\kappa)\varphi_m(y_n,\kappa)$, then in view of (2.6.41), the coefficient series satisfies

$$\text{(2.6.42)}\qquad \kappa^{-d}\|(c^+_{\ell m})_{\ell,m\in\mathbb{N}}\|_{\ell^2_{N,N'}} \le c(X)(\rho^{\nu-\frac12-N-N'}+1) ,$$

for all integer N and $N' \geq 0$ (this is a special case of (2.3.32)). When $\nu-\frac{1}{2}-N-N' \geq 0$, the right hand side can be replaced by $c(X)$, so the estimate is uniform in (ξ',μ). Now if $\nu = 1$, there is negative regularity, hence no uniform estimate, already when N or N' equals 1. In Theorem 3.3.10, we need the uniformity just for small $N,N' > 0$. However, a standard interpolation of (2.6.42) between integer cases (as described in Remark 2.2.10, extended in an obvious way to double series), does not give a uniform bound for any N or $N' > 0$, if $\nu = 1$. More generally, when ν is integer ≥ 1, there are certain cases with $\nu-N-N'$ close to zero, to which (2.6.42) does not "extend by interpolation".

What we shall explain now is how to circumvent this complication, by using that the Laguerre expansion of $\tilde{g}^+(p)$ is directly related to the expansion (2.2.84) of p. There, one can interpolate between integer values without problems, which will lead to the validity of (2.6.42) for $\tilde{g}^+(p)$ for all N and $N' \in \mathbb{R}_+$. Some related rules for other boundary operators will be considered afterwards.

2.6.12 Proposition. *Let* d *and* $\nu \in \mathbb{Z}$, *and let* p *satisfy* (2.6.3 i), *with the expansion*

(2.6.43) $$p(X,\xi,\mu) = \sum_{1\leq j\leq d} s_j(X,\xi',\mu)\xi_n^j + \sum_{k\in\mathbb{Z}} a_k(X,\xi',\mu)\hat{\psi}_k(\xi_n,\kappa) ,$$

cf. (2.2.52)ff . *Then the coefficients of* $g^\pm(p)$ *in the Laguerre series expansions*

(2.6.44) $$g^\pm(p)(X,\xi,\eta_n,\mu) = \sum_{\ell,m\in\mathbb{N}} c^\pm_{\ell m}(X,\xi',\mu)\hat{\varphi}_\ell(\xi_n,\kappa)\bar{\hat{\varphi}}_m(\eta_n,\kappa)$$

have the Hankel matrix structure

(2.6.45) $$c^+_{\ell m} = a_{\ell+m+1} \quad \textit{and} \quad c^-_{\ell m} = a_{-\ell-m-1} \quad \textit{for} \quad \ell,m \in \mathbb{N} ,$$

and the following estimates are valid for all $\alpha \in \mathbb{N}^{n-1}$, $\beta \in \mathbb{N}^{n'}$, $j \in \mathbb{N}$, N *and* $N' \in \overline{\mathbb{R}}_+$:

(2.6.46) $$\|(D_X^\beta D_{\xi'}^\alpha D_\mu^j c^\pm_{\ell m})_{\ell,m\in\mathbb{N}}\|_{\ell^2_{N,N'}} \leq c(X)(\rho^{\nu-\frac{1}{2}-|\alpha|-j-N-N'}+1)\kappa^{d-|\alpha|-j} ,$$

with similar estimates for lower order parts. Such results also hold for $S_{1,0}$ *symbols.*

Proof: The proof of (2.6.45), showing that $g^+(p(D_n))$ is a Hankel operator related to the Toeplitz operator $p(D_n)_\Omega$ (cf. Remark 2.2.11) is a discrete variant of (2.3.6):

Let $u \in \mathcal{S}(\overline{\mathbb{R}}_+)$, with the Laguerre expansion

(2.6.47) $$u(x_n) = \sum_{m\in\mathbb{N}} u_m \varphi_m(x_n,\kappa) , \qquad u_m = (u,\varphi_m) .$$

Then since $J\varphi_m(x_n,\kappa) = \varphi_m(-x_n,\kappa) = \varphi_{-1-m}(x_n,\kappa)$,

$$F(e^-Ju) = \sum_{m\in\mathbb{N}} u_m \hat{\varphi}_{-1-m}(\xi_n,\kappa) = (2\kappa)^{\frac12} \sum_{m\in\mathbb{N}} u_m \frac{(\kappa-i\xi_n)^{-m-1}}{(\kappa+i\xi_n)^{-m}}$$

(cf. Section 2.2). Now $g^+(p)$ depends only on the H_0-part of p (since r^+Qe^- vanishes when Q is a differential operator), and acts as follows:

$$g^+(p)u = r^+p(D_n)e^-Ju = r^+F^{-1} \sum_{\substack{k\in\mathbb{Z}\\ m\in\mathbb{N}}} a_k \frac{(\kappa-i\xi_n)^k}{(\kappa+i\xi_n)^k} u_m \frac{(\kappa-i\xi_n)^{-m-1}}{(\kappa+i\xi_n)^{-m}} (2\kappa)^{\frac12} .$$

$$= r^+F^{-1} \sum_{\substack{k\in\mathbb{Z}\\ m\in\mathbb{N}}} a_k u_m \hat{\varphi}_{k-m-1}(\xi_n,\kappa)$$

$$= \sum_{\substack{k\in\mathbb{Z},\, m\in\mathbb{N}\\ k-m-1\geq 0}} a_k u_m \varphi_{k-m-1}(x_n,\kappa) = \sum_{\ell,m\in\mathbb{N}} a_{\ell+m+1} (u,\varphi_m)\varphi_\ell(x_n,\kappa) .$$

This shows that $g^+(p)$ has the symbol-kernel

$$\tilde{g}^+(p) = \sum_{\ell,m\in\mathbb{N}} a_{\ell+m+1}(X,\xi',\mu)\varphi_m(x_n,\kappa)\varphi_\ell(y_n,\kappa) ,$$

which proves the first identity in (2.6.45). The second identity is shown by a variant of the arguments, or by use of the fact that

$$g^-(p(D_n)) = [g^+(p(D_n)^*)]^* . \tag{2.6.48}$$

Now the estimates for the sequence a_k obtained in Theorem 2.2.8 for integer N

$$\|(D_X^\beta D_\xi^\alpha, D_\mu^j a_k)_{k\in\mathbb{Z}}\|_{\ell^2_N} \leq c(X)(\rho^{\nu-|\alpha|-j-N}+1)\kappa^{d-|\alpha|-j} \tag{2.6.49}$$

extend by interpolation (using (2.1.17)) to all $N\in\overline{\mathbb{R}}_+$, since ν, $|\alpha|$ and j are integers, as observed in Remark 2.2.10. This implies for the $c^+_{\ell m}$:

$$\sum_{\ell,m\in\mathbb{N}} |(1+\ell)^N(1+m)^{N'}c^+_{\ell m}|^2 = \sum_{\ell,m\in\mathbb{N}} |(1+\ell)^N(1+m)^{N'} a_{\ell+m+1}|^2 \tag{2.6.50}$$

$$\leq \sum_{j\geq 1} j^{2N+2N'+1}|a_j|^2 \leq c(X)[(\rho^{\nu-N-N'-\frac12}+1)\kappa^d]^2 ,$$

giving the basic one of the estimates (2.6.46). The estimates of derivatives and lower order parts are derived similarly from the corresponding estimates of (a_k), and the proof for $c^-_{\ell m}$ is completely analogous. □

Now we can apply the following composition rule:

2.6.13 Lemma. *Let* g^i $(i=1,2)$ *be singular Green symbols of class* 0, *with Laguerre series expansions*

$$(2.6.51)\qquad g^i(X,\xi,\eta_n,\mu) = \sum_{\ell,m\in\mathbb{N}} c^i_{\ell m}(X,\xi',\mu)\hat\varphi_\ell(\xi_n,\kappa)\bar{\hat\varphi}_m(\eta_n,\kappa) \quad ,$$

where the coefficient sequences satisfy the estimates for all indices $\beta\in\mathbb{N}^{n'}$, $\alpha\in\mathbb{N}^{n-1}$, N *and* $N'\in\mathbb{N}$,

$$(2.6.52)\qquad \| (D_X^\beta D_{\xi'}^\alpha D_\mu^j c^i_{\ell m})_{\ell,m\in\mathbb{N}} \|_{\ell^2_{N,N'}} \le c(X)(\rho^{\nu^i-|\alpha|-j-N-N'}+1)\kappa^{d^i-|\alpha|-j} \quad ,$$

and related estimates for lower order parts. Then $g^1\circ g^2$ *equals* g^3, *where* g^3 *has the expansion* (2.6.51), *with*

$$(2.6.53)\qquad c^3_{\ell m} = \sum_{k\in\mathbb{N}} c^1_{\ell k} c^2_{km} \quad ,$$

satisfying the estimates (2.6.52) *with*

$$d^3 = d^1 + d^2 \quad \text{and} \quad \nu^3 = m(\nu^1,\nu^2) \quad ,$$

and moreover

$$(2.6.54)\qquad \| (D_X^\beta D_{\xi'}^\alpha D_\mu^j c^3_{\ell m})_{\ell,m\in\mathbb{N}} \|_{\ell^2_{N,N'}} \le c(X)(\rho^{m(\nu^1-N,\nu^2-N')-|\alpha|-j}+1)\kappa^{d^3-|\alpha|-j} \quad .$$

If the given estimates extend to N *and* $N'\in\overline{\mathbb{R}}_+$, *so do the estimates for* g^3.

Proof: This is derived in a straightforward way by use of the Cauchy-Schwarz inequality, e.g.

$$(2.6.55)\qquad \| (c^3_{\ell m})_{\ell,m\in\mathbb{N}} \|_{\ell^2_{N,N'}} \le \| (c^1_{\ell m}) \|_{\ell^2_{N,0}} \, \| (c^2_{\ell m}) \|_{\ell^2_{0,N'}}$$

combined with the Leibniz formula. □

In particular, we get for $L(p,p')$:

2.6.14 Corollary. *Let* p *and* p' *be as in Theorem 2.6.1, with integer regularities* ν *and* $\nu'\in\mathbb{Z}$. *Then the coefficient series* $(c_{\ell m})_{\ell,m\in\mathbb{N}}$ *in the Laguerre series expansion of* $g^+(p)\circ g^-(p')$ *satisfies estimates* (2.6.52) *(and the associated estimates for lower order parts), with* ν^i *replaced by* $m(\nu-\frac12,\nu'-\frac12)$ *and* d^i *replaced by* $d+d'$, *for all indices* $\beta\in\mathbb{N}^{n'}$, $\alpha\in\mathbb{N}^{n-1}$, $j\in\mathbb{N}$, N *and* $N'\in\overline{\mathbb{R}}_+$. *Also* (2.6.54) *is satisfied with* $\nu^1=\nu-\frac12$, $\nu^2=\nu'-\frac12$ *and* $d^3=d+d'$.

<u>Proof:</u> One combines Proposition 2.6.12 with Lemma 2.6.13 for $d^1 = d$, $d^2 = d'$, $\nu^1 = \nu - \frac{1}{2}$ and $\nu^2 = \nu' - \frac{1}{2}$. $\qquad\qquad\qquad\qquad\qquad\qquad$ □

Similar observations will be needed for compositions among the other types of operators, when symbols with regularity $\frac{1}{2}$ occur (as for instance in certain Dirichlet type boundary problems). We list the most necessary rules and estimates in the following proposition.

<u>2.6.15 Proposition.</u> *Let there be given sequences* $(a_j)_{j\in\mathbb{Z}}$, $(b_j)_{j\in\mathbb{N}}$ *and* $(c_{\ell m})_{\ell,m\in\mathbb{N}}$ *in* $s(\mathbb{Z})$, $s(\mathbb{N})$ *and* $s(\mathbb{N}\times\mathbb{N})$, *respectively, and let* p, g, t *and* k *be pseudo-differential, singular Green, trace and Poisson operators on* $\mathbb{R}_+$ *with symbols, respectively*

$$\text{(i)} \quad p(\xi,\mu) = \sum_{j\in\mathbb{Z}} a_j \hat{\psi}_j(\xi_n,\kappa) = s_0 + \sum_{j\in\mathbb{Z}} a_j'\hat{\varphi}_j(\xi_n,\kappa) \quad (\text{cf. (2.2.54-55))},$$

(2.6.56)
$$\text{(ii)} \quad g(\xi,\eta_n,\mu) = \sum_{\ell,m\in\mathbb{N}} c_{\ell m}\,\hat{\varphi}_\ell(\xi_n,\kappa)\overline{\hat{\varphi}}_m(\eta_n,\kappa),$$

$$\text{(iii)} \quad t(\xi,\mu) = \sum_{j\in\mathbb{N}} b_j\overline{\hat{\varphi}}_j(\xi_n,\kappa),$$

$$\text{(iv)} \quad k(\xi,\mu) = \sum_{j\in\mathbb{N}} b_j'\hat{\varphi}_j(\xi_n,\kappa).$$

Then one has for the symbols of the various compositions, for any N *and* N' $\in \overline{\mathbb{R}}_+$ *and* $\varepsilon \in \mathbb{R}_+$,

(2.6.57)
$$t \circ p_\Omega = \sum_{j\in\mathbb{N}} b_j''\overline{\hat{\varphi}}_j, \quad where \quad b_j'' = \sum_{\ell\in\mathbb{N}} b_\ell a_{\ell-j},$$

$$satisfying \quad \|(b_j'')\|_{\ell_N^2} \le \|(b_j)\|_{\ell_N^2}\|(a_j)\|_{\ell_N^1} \le c_\varepsilon\|(b_j)\|_{\ell_N^2}\|(a_j)\|_{\ell_{N+\frac{1}{2}+\varepsilon}^2};$$

(2.6.58)
$$p_\Omega \circ k = \sum_{j\in\mathbb{N}} c_j\hat{\varphi}_j, \quad where \quad c_j = \sum_{\ell\in\mathbb{N}} a_{j-\ell}b_\ell',$$

$$satisfying \quad \|(c_j)\|_{\ell_N^2} \le \|(a_j)\|_{\ell_N^1}\|(b_j')\|_{\ell_N^2} \le c_\varepsilon\|(a_j)\|_{\ell_{N+\frac{1}{2}+\varepsilon}^2}\|(b_j')\|_{\ell_N^2};$$

(2.6.59)
$$g \circ p_\Omega = \sum_{\ell,m\in\mathbb{N}} c_{\ell m}'\,\hat{\varphi}_\ell\overline{\hat{\varphi}}_m, \quad where \quad c_{\ell m}' = \sum_{j\in\mathbb{N}} c_{\ell j}a_{j-m},$$

$$satisfying \quad \|(c_{\ell m}')\|_{\ell_{N,N'}^2} \le \|(c_{\ell m})\|_{\ell_{N,N'}^2}\|(a_j)\|_{\ell_{N'}^1} \le c_\varepsilon\|(c_{\ell m})\|_{\ell_{N,N'}^2}\|(a_j)\|_{\ell_{N'+\frac{1}{2}+\varepsilon}^2}$$

$$(2.6.60)\quad \begin{aligned}&p_\Omega \circ g = \sum_{\ell,m\in\mathbb{N}} c''_{\ell m}\,\hat\varphi_\ell \overline{\hat\varphi}_m\ , \quad \textit{where}\quad c''_{\ell m} = \sum_{j\in\mathbb{N}} a_{\ell-j}c_{jm}\ ,\\ &\textit{satisfying}\quad \|(c''_{\ell m})\|_{\ell^2_{N,N'}} \le \|(a_j)\|_{\ell^1_N}\|(c_{\ell m})\|_{\ell^2_{N,N'}} \le c_\varepsilon \|(a_j)\|_{\ell^2_{N+\frac12+\varepsilon}}\ \|(c_{\ell m})\|_{\ell^2_{N,N'}}\ ;\end{aligned}$$

$$(2.6.61)\quad \begin{aligned}&t\circ g = \sum_{j\in\mathbb{N}} c'_j\,\overline{\hat\varphi}_j\ , \quad \textit{where}\quad c'_j = \sum_{\ell\in\mathbb{N}} b_\ell c_{\ell j}\ ,\\ &\textit{satisfying}\quad \|(c'_j)\|_{\ell^2_N} \le \|(b_j)\|_{\ell^2_0}\ \|(c_{\ell m})\|_{\ell^2_{0,N}}\ ;\end{aligned}$$

$$(2.6.62)\quad \begin{aligned}&g\circ k = \sum_{j\in\mathbb{N}} c''_j\hat\varphi_j\ , \quad \textit{where}\quad c''_j = \sum_{\ell\in\mathbb{N}} c_{j\ell}b'_\ell\ ,\\ &\textit{satisfying}\quad \|(c''_j)\|_{\ell^2_N} \le \|(c_{j\ell})\|_{\ell^2_{N,0}}\ \ \|(b'_\ell)\|_{\ell^2_0}\ ;\end{aligned}$$

$$(2.6.63)\quad \begin{aligned}&k\circ t = \sum_{\ell,m\in\mathbb{N}} b'_\ell b_m\ \hat\varphi_\ell\overline{\hat\varphi}_m\ ,\\ &\textit{satisfying}\quad \|(b'_\ell b_m)\|_{\ell^2_{N,N'}} = \|(b'_\ell)\|_{\ell^2_N}\ \|(b_m)\|_{\ell^2_{N'}}\ ;\end{aligned}$$

$$(2.6.64)\qquad \gamma_0\circ p_\Omega \quad \textit{has symbol}\quad \sum_{j<0} a'_j\,\hat\varphi_j\ .$$

<u>Proof</u>: As a typical case, consider e.g. (2.6.57). With $u\in\mathscr{S}(\overline{\mathbb{R}}_+)$ written as in (2.6.47), we have in view of (2.2.91) and (2.4.5), and the orthonormality,

$$\begin{aligned}t(D_n)p(D_n)_\Omega u &= t(D_n)\sum_{\ell,m\in\mathbb{N}} a_{\ell-m}\,u_m\varphi_\ell(x_n,\kappa)\\ &= \int_0^\infty \sum_{j,\ell,m\in\mathbb{N}} b_j\varphi_j(x_n,\kappa)a_{\ell-m}\,u_m\varphi_\ell(x_n,\kappa)dx_n\\ &= \sum_{\ell,m\in\mathbb{N}} b_\ell a_{\ell-m}\,u_m\ ,\end{aligned}$$

which shows that $t\circ p_\Omega$ has the symbol $\sum_{\ell,m} b_\ell a_{\ell-m}\overline{\hat\varphi}_m$ as asserted. The estimate is obtained by a standard convolution technique:

$$\begin{aligned}\|(b''_j)\|_{\ell^2_N} &= \Big(\sum_{j\in\mathbb{N}}(1+j)^{2N}\big|\sum_\ell b_\ell a_{\ell-j}\big|^2\Big)^{\frac12} \le \Big(\sum_{j\in\mathbb{N}}\big[\sum_\ell(1+\ell)^N|b_\ell|(1+|j-\ell|)^N|a_{\ell-j}|\big]^2\Big)^{\frac12}\\ &\le \Big(\sum_{j\in\mathbb{N}}\big[\sum_\ell(1+\ell)^{2N}|b_\ell|^2(1+|\ell-j|)^N|a_{\ell-j}|\big]\big[\sum_{\ell\in\mathbb{Z}}(1+|\ell-j|)^N|a_{\ell-j}|\big]\Big)^{\frac12}\end{aligned}$$

$$= \Big(\sum_{j\in\mathbb{N},\, \ell\in\mathbb{N}} (1+\ell)^{2N}|b_\ell|^2(1+|\ell-j|)^N|a_{\ell-j}|\Big)^{\frac12} \|(a_k)\|^{\frac12}_{\ell^1_N} \le \|(b_\ell)\|_{\ell^2_N}\|(a_k)\|_{\ell^1_N} ,$$

here we can furthermore use that for $\varepsilon > 0$,

$$(2.6.65)\quad \|(a_k)\|_{\ell^1_N} = \sum_{k\in\mathbb{Z}} (1+|k|)^{N+\frac12+\varepsilon}(1+|k|)^{-\frac12-\varepsilon}|a_k| \le \|(a_k)\|_{\ell^2_{N+\frac12+\varepsilon}} \Big(\sum_{k\in\mathbb{Z}} (1+|k|)^{-1-2\varepsilon}\Big)^{\frac12}$$

This shows (2.6.57). The proof of (2.6.59) is similar with respect to the second index on $c'_{\ell m}$; the first indexation is easily included.(2.6.58) and (2.6.60) are proved by a transposed version of these methods, and the proofs of (2.6.61)-(2.6.64) are straightforward. □

If it is more advantageous in an application, one can let the ℓ^1_N-norm fall on the other factor in (2.6.57-60).

The consequences for our general symbol classes can be explicited whenever necessary.

2.6.16 Remark. The rules are of interest e.g. when we consider μ-independent symbols with low regularity. For example, in the consideration of a trace symbol $t(X,\xi) \in S^0(\Xi, \mathbb{R}^{n-1}, H^-_{-1})$ of order and class 0 , we have by (2.3.52) that t has regularity $\frac12$ when regarded in the μ-dependent setting, and the ℓ^2_N estimates (2.3.31) for integer N do not give a good $\ell^2_{\frac12}$ estimate by interpolation (cf. Remark 2.2.10). But here we have in fact for the coefficient series of t ,

$$(2.6.66)\qquad t(X,\xi) = \sum_{k\in\mathbb{N}} b_k(X,\xi',\mu)\bar{\hat{\varphi}}_k(\xi_n,\kappa) ,$$

that

$$\|(b_k)\|_{\ell^2_0} = c\|t\|_{L^2_{\xi_n}} \le c(X)\langle\xi'\rangle^{\frac12} ,$$

and by (2.3.31)

$$\|(b_k)\|_{\ell^2_1} \le c'(X)\langle\xi'\rangle^{-\frac12}\langle\xi',\mu\rangle ,$$

which by interpolation gives

$$(2.6.67)\qquad \|(b_k)\|_{\ell^2_{\frac12}} \le c''(X)\langle\xi',\mu\rangle^{\frac12} .$$

This is much better than what (2.2.90) would give; and can be used in compositions as in Proposition 2.6.15 above. (In this special case one can even sharpen the estimate in (2.6.57) a little bit (remove an ε) by working with integral formulas instead of series, but we leave that out, to reduce technicalities.)

2.7 Compositions in general.

In preparation for the complete composition rules, we must now include those $\circ_n$-compositions, where the ps.d.o. symbols p and p' depend on x_n (and have the transmission property at $x_n = 0$ only).

One here uses a Taylor expansion

$$(2.7.1) \qquad p(X,x_n,\xi,\mu) = \sum_{j<M} \frac{x_n^j}{j!} \partial_{x_n}^j p(X,0,\xi,\mu) + x_n^M r_M(X,x_n,\xi,\mu) \quad ,$$

and the terms in the sum over $j < M$ are easy to handle since, as we have seen earlier, multiplication of the "singular" operators by x_n^j corresponds to a differentiation $D_{\xi_n}^j$ or $\bar{D}_{\xi_n}^j$ of the symbol. But also the remainder has to be treated; here r_M is of just as high order as p <u>and</u> x_n-dependent, and the point is to use the factor x_n^M.

The end results have precisely the form one would expect, and the proof we give is somewhat long, so the reader can very well skip the details in a first reading.

Actually, it is quite possible that part of the argumentation may be streamlined by passing via the Poisson, trace and singular Green operators defined from <u>symbols depending on</u> x_n, y_n <u>resp.</u> (x_n, y_n) that were briefly introduced in Remark 2.4.9. One would then have to justify certain composition formulas before application of the Taylor expansion, that provides the basic argument for Remark 2.4.9 as well as for the present section.

For the time being, we refrain from polishing the explanation in this way, and keep the original presentation from [Grubb 11], also because there is some information of independent interest in the lemmas.

We begin with the cases where one factor depends on x_n , namely the compositions

$$(2.7.2) \qquad p_\Omega \circ_n k' \; , \quad t \circ_n p'_\Omega \; , \quad p_\Omega \circ_n g' \quad \text{and} \quad g \circ_n p'_\Omega \quad ,$$

where $p(X,x_n,\xi,\mu) \in S_{tr}^{d,\nu}(\Xi \times \mathbb{R}, \overline{\mathbb{R}}_+^{n+1})$, p' is similarly in $S_{tr}^{d',\nu'}$ (of integer orders), and k', t, g and g' are as in (2.6.3). A prototype is the composition $p_\Omega \circ k'$. Assume that p vanishes for x_n outside a bounded interval $[-a,a]$ of $\mathbb{R}$ $(a > 0)$, let $\varphi \in C_0^\infty(\mathbb{R})$ with $\varphi = 1$ on $[-a,a]$, and write, for some $M \in \mathbb{N}$,

$$(2.7.3)\qquad p(X,x_n\xi,\mu) = \varphi(x_n) \sum_{j<M} \frac{x_n^j}{j!} \partial_{x_n}^j p(X,0,\xi,\mu) + x_n^M \varphi(x_n) r_M(X,x_n,\xi,\mu) .$$

We have for $k' \in S^{d'-1,\nu'}(\Xi, \overline{\mathbb{R}}^n_+, H^+)$, in view of (2.6.5) 1^0 and Lemma 2.4.2,

$$(2.7.4)\qquad \begin{aligned} k_j(D_n) &\equiv OP_n(x_n^j \partial_{x_n}^j p(X,0,\xi,\mu))\, OPK_n(k') \\ &= x_n^j \circ OPK_n(h^+(\partial_{x_n}^j p\cdot k')) \\ &= OPK_n(\overline{D}_{\xi_n}^j h^+(\partial_{x_n}^j p\cdot k')) \quad , \end{aligned}$$

where the symbol lies in $S^{d+d'-1-j,m(\nu,\nu')-j}(H^+)$, by the already established rules of calculus. Note also that $(1-\varphi(x_n))k_j(D_n)$ is uniformly negligible of regularity $m(\nu,\nu')-d-d'+1$, cf. Lemma 2.4.8, so $\varphi k_j(D_n)$ is a Poisson boundary symbol operator of order $d+d'-j$ and regularity $m(\nu,\nu')-j$ (in the extended sense of Definition 2.4.5). It remains to treat the composition

$$OP_n(x_n^M \varphi(x_n) r_M) \circ OPK_n(k') \quad ,$$

where we shall play on the fact that M can be taken to be arbitrarily large. The following two rules will be useful:

$$(2.7.5)\qquad \begin{aligned} &\text{(i)}\quad x_n^M \circ OP_n(p(X,x_n,\xi,\mu)) = \sum_{0\le j\le M} \binom{M}{j} OP_n(\overline{D}_{\xi_n}^j p)\circ x_n^{M-j} \quad , \\ &\text{(ii)}\quad \partial_{x_n}^M \circ OP_n(p(X,x_n,\xi,\mu)) = \sum_{0\le j\le M} \binom{M}{j} OP_n(\partial_{x_n}^j p)\circ \partial_{x_n}^{M-j} \quad , \end{aligned}$$

that are shown by iteration from the elementary formulas for $M = 1$:

$$\begin{aligned} x_n \int e^{ix_n\xi_n} p\,\hat{u}\,d\xi_n &= \int\Big(D_{\xi_n} e^{ix_n\xi_n}\Big) p\,\hat{u}\,d\xi_n = \int e^{ix_n\xi_n}\,\overline{D}_{\xi_n}(p\,\hat{u})d\xi_n \\ &= \int e^{ix_n\xi_n}(\overline{D}_{\xi_n} p)\hat{u}\,d\xi_n + \int e^{ix_n\xi_n} p\,\widehat{x_n u}\,d\xi_n \quad , \end{aligned}$$

$$\begin{aligned} \partial_{x_n} \int e^{ix_n\xi_n} p\,\hat{u}\,d\xi_n &= \int e^{ix_n\xi_n}(\partial_{x_n} p)\hat{u}\,d\xi_n + \int e^{ix_n\xi_n} p\, i\xi_n\,\hat{u}\,d\xi_n \\ &= \int e^{ix_n\xi_n}\,[(\partial_{x_n} p)\hat{u} + p\widehat{(\partial_{x_n} u)}]\,d\xi_n \quad . \end{aligned}$$

When M is large and we apply (2.7.5 i), we obtain $x_n^M\, p(D_n)$ as a sum of terms that are either of low order or have a factor $x_n^{M'}$ on the right, with M' large. The two kinds of terms are treated differently.

<u>2.7.1 Lemma</u>. *Let* $p \in S^{d,\nu}(\Xi\times\mathbb{R},\ \overline{\mathbb{R}}_+^{n+1})$ *(not necessarily having the transmission property), vanishing for* x_n *outside a compact interval* $I \subset \mathbb{R}$. *Let* $k' \in S^{d'-1,\nu'}(\Xi,\ \overline{\mathbb{R}}_+^n,\ H^+)$, *and let* M *be an integer* $\geq d$. *Then for each* (X,ξ',μ) , *the operator* $p(D_n)_\Omega \circ x_n^M \circ k'(D_n)$ *maps* $\mathbb{C}$ *into* $L^2(\mathbb{R}_+)$ *and is the multiplication by a function* $\tilde{r}(X,x_n,\xi',\mu)$, *that vanishes for* $x_n \notin I$ *and satisfies the estimates, for* $(X,\xi',\mu) \in \Xi\times\overline{\mathbb{R}}_+^n$,

$$(2.7.6)\qquad \| D_X^\beta D_{\xi'}^\alpha D_\mu^j x_n^N D_{x_n}^{N'} D_\mu^j \tilde{r} \|_{L^2_{x_n}(\mathbb{R}_+)} \leq c(X)(\rho^{m(\nu,\nu')-M-N+N'-|\alpha|}+1)\kappa^{d+d'-\frac{1}{2}-M-N+N'-|\alpha|-j} ,$$

when $M \geq N' + d_+$; *the other indices can vary freely (cf.* (A.4')*)*.

<u>Proof</u>: Since $p(D_n)_\Omega \circ x_n^M k'(D_n)v = p(D_n)_\Omega (x_n^M \tilde{k}'(X,x_n,\xi',\mu)v)$ for $v \in \mathbb{C}$, we have to estimate $p(D_n)_\Omega (x_n^M \tilde{k}')$ and derived expressions. A one-dimensional variant of Proposition 2.1.12 shows that (with ρ,κ defined as usual by (2.2.80))

$$\| p_\Omega (x_n^M \tilde{k}')\|_{L^2(\mathbb{R}_+)} \leq \| p(e^+ x_n^M \tilde{k}')\|_{L^2(\mathbb{R})} \leq c(X)(\rho^\nu+1)\| e^+ x_n^M \tilde{k}'\|_{H^{d,\kappa}(\mathbb{R})} ,$$

here $e^+ x_n^M \tilde{k}' \in H^d(\mathbb{R})$ since $M \geq d$. (Recall that $H^{s,\kappa}(\mathbb{R})$ is the space with norm $\|u\|_{s,\kappa} = c\|\langle\xi_n,\kappa\rangle^s \hat{u}(\xi_n)\|_0$.) When $d \geq 0$, we have by (A.25,42),

$$\begin{aligned} \| e^+ x_n^M \tilde{k}'\|_{H^{d,\kappa}(\mathbb{R})} &\simeq \sum_{0\leq j\leq d} \kappa^{d-j} \| D_{x_n}^j x_n^M \tilde{k}'\|_{L^2(\mathbb{R}_+)} \\ &\leq c(X) \sum_{0\leq j\leq d} \kappa^{d-j} (\rho^{\nu'-[M-j]_+}+1)\kappa^{d'-\frac{1}{2}-M+j} \\ &\leq c'(X)(\rho^{\nu'-M}+1)\kappa^{d+d'-\frac{1}{2}-M} , \end{aligned}$$

since $M \geq d$; and when $d \leq 0$,

$$\| e^+ x_n^M \tilde{k}'\|_{H^{d,\kappa}(\mathbb{R})} \leq \kappa^d \| x_n^M \tilde{k}'\|_0 \leq c(X)\kappa^{d+d'-\frac{1}{2}-M}(\rho^{\nu'-M}+1) .$$

Altogether,

$$(2.7.7)\qquad \| p_\Omega (x_n^M \tilde{k}')\|_{L^2(\mathbb{R}_+)} \leq c(X)(\rho^{m(\nu,\nu')-M}+1)\kappa^{d+d'-\frac{1}{2}-M} .$$

For the derived functions, we use that for each N, N', there is a formula

(2.7.8)
$$x_n^N \circ D_{x_n}^{N'} \circ p = \sum_{\substack{0 \le j \le N \\ 0 \le \ell \le N'}} p_{j,\ell} \circ x_n^j \circ D_{x_n}^{\ell} \quad , \text{ with}$$
$$p_{j,\ell} \in S^{d-N+j,\nu-N+j}(\Xi \times \mathbb{R}, \overline{\mathbb{R}}_+^{n+1}) \quad ,$$

derived from (2.7.5). Consider $(p_{j,\ell} \circ x_n^j \circ D_n^{\ell})(e^+ x_n^M \tilde{k}')$. Taking $N' \le M$, we have since $\ell \le N'$ that $D_n^{\ell} e^+ x_n^M \tilde{k}'$ is an integrable function on $\mathbb{R}$ (equal to $e^+ D_n^{\ell} x_n^M \tilde{k}'$), so

$$x_n^N D_n^{N'} p_\Omega (x_n^M \tilde{k}') = \sum_{j,\ell} r^+ p_{j,\ell,\Omega}(x_n^j D_n^{\ell} x_n^M \tilde{k}') \quad .$$

Here $x_n^j D_n^{\ell} x_n^M \tilde{k}'$ is of order $d'-M-j+\ell$ and regularity $\nu'-[M-\ell]_+-j = \nu'-M+\ell-j$ (since $\ell \le M$), so (2.7.7) applied to the $p_{j,\ell}$ shows:

$$\begin{aligned}
\|x_n^N D_n^{N'} p_\Omega (x_n^M \tilde{k}')\|_{L^2(\mathbb{R}_+)} &\le \sum_{j,\ell} \|p_{j,\ell,\Omega}(x_n^j D_n^{\ell} x_n^M \tilde{k}')\| \\
&\le c(X) \sum_{j,\ell} (\rho^{m(\nu-N+j,\nu'-M+\ell-j)}+1)\kappa^{d+d'-\frac{1}{2}-N-M+\ell} \\
&\le c'(X) \sum_{\ell} (\rho^{m(\nu,\nu')-N-M+\ell}+1)\kappa^{d+d'-\frac{1}{2}-N-M+\ell} \\
&\le c''(X)(\rho^{m(\nu,\nu')-N-M+N'}+1)\kappa^{d+d'-\frac{1}{2}-N-M+N'} \quad ,
\end{aligned}$$

since $\ell \le N' \le M$, and generally $S^{d-m,\nu-m} \subset S^{d,\nu}$ when $m \in \mathbb{N}$. Similar considerations apply to the (X,ξ',μ)-derivatives, altogether implying (2.7.6).

□

<u>2.7.2 Lemma</u>. *Let* $p \in S^{d-M,\nu-M}(\Xi \times \mathbb{R}, \overline{\mathbb{R}}_+^{n+1})$ *(not necessarily having the transmission property), vanishing for* x_n *outside a compact interval* $I \subset \mathbb{R}$. *Assume that* $M \ge d+2$. *Let* $k' \in S^{d'-1,\nu'}(\Xi, \overline{\mathbb{R}}_+^n, H^+)$. *Then for each* (X,ξ',μ), *the operator* $p(D_n)_\Omega \circ k'(D_n)$ *maps* $\mathbb{C}$ *into* $C^0(\overline{\mathbb{R}}_+)$ *and is the multiplication by a function* $\tilde{r}(X,x_n,\xi',\mu)$ *that vanishes for* $x_n \notin I$ *and satisfies the estimates, for* $(X,\xi',\mu) \in \Xi \times \overline{\mathbb{R}}_+^n$,

(2.7.9)
$$\|D_X^\beta D_{\xi'}^\alpha D_\mu^j x_n^N D_{x_n}^{N'} \tilde{r}\|_{L^2_{x_n}(\mathbb{R}_+)} \le c(X)(\rho^{m(\nu,\nu')-M+1+N'-|\alpha|}+1)\kappa^{d+d'-\frac{1}{2}-M+1+N'-|\alpha|-}$$

when $M \ge N' + 1 + [d-|\alpha|-j]_+$; *the other indices can vary freely.*

Proof: As in the preceding proof, $\tilde{r}$ equals $p(D_n)_\Omega(\tilde{k}')$. Here, since p is of order ≤ -2 , it has the continuous kernel $\tilde{p}(X,x_n,z,\xi',\mu)|_{z=x_n-y_n}$ on $\mathbb{R}$, and then since p vanishes for $x_n \notin I$, we have by the Cauchy-Schwarz inequality

$$(2.7.10)\quad \| D_X^\beta D_{\xi'}^\alpha D_\mu^j x_n^N D_{x_n}^{N'} p(D_n)_\Omega(\tilde{k}')\|_{L^2_{x_n}(\mathbb{R}_+)}$$

$$\leq c_N \sup_{x_n\in I} | D_X^\beta D_{\xi'}^\alpha D_\mu^j D_{x_n}^{N'} p(D_n)_\Omega(\tilde{k}')|$$

$$\leq c_N' \sup_{x_n} \sum_{\gamma\leq\theta} \| D_{X,\xi',\mu}^{\theta-\gamma}\tilde{p}'(X,x_n,x_n-y_n,\xi',\mu)\|_{L^2_{y_n}} \| D_{X,\xi',\mu}^\gamma \tilde{k}'\|_{L^2_{y_n}} \quad ,$$

where we have denoted $(\beta,\alpha,j) = \theta$ and $D_{x_n}^{N'}\tilde{p} = \tilde{p}'$. Here

$$\| D_X^\beta D_{\xi'}^\alpha D_\mu^j \tilde{p}(X,x_n,x_n-y_n,\xi',\mu)\|_{L^2_{y_n}} = c\| D_X^\beta D_{\xi'}^\alpha D_\mu^j p(X,x_n,\xi',\xi_n,\mu)\|_{L^2_{\xi_n}}$$

$$\leq c(X)\|(\rho(\xi,\mu)^{\nu-M-|\alpha|}+1)\langle\xi,\mu\rangle^{d-M-|\alpha|-j}\|_{L^2_{\xi_n}}$$

$$\leq c'(X)(\rho(\xi',\mu)^{\nu-M-|\alpha|+1}+1)\langle\xi',\mu\rangle^{d-M-|\alpha|-j+1} \quad ,$$

where we have used that for $s \geq 1$, any σ ,

$$(2.7.11)\quad \|(\rho(\xi,\mu)^\sigma +1)\langle\xi,\mu\rangle^{-s}\|_{L^2_{\xi_n}} \leq \|\langle\xi\rangle^{-1}(\rho(\xi,\mu)^{\sigma+1} +1)\langle\xi,\mu\rangle^{-s+1}\|_{L^2_{\xi_n}}$$

$$\leq c(\rho(\xi',\mu)^{\sigma+1} +1)\langle\xi',\mu\rangle^{-s+1} \quad ,$$

cf. (2.2.62). For the x_n-derivatives, we apply the chain rule:

$$\| D_{x_n} \tilde{p}(X,x_n,x_n-y_n,\xi',\mu)\|_{L^2_{y_n}}$$

$$\leq \| D_{x_n}\tilde{p}(X,x_n,z,\xi',\mu)\|_{L^2_z} + \| D_z\tilde{p}(X,x_n,z,\xi,\mu)\|_{L^2_z}$$

$$= c(\| D_{x_n}p\|_{L^2_{\xi_n}} + \|\xi_n p\|_{L^2_{\xi_n}}) \leq c(X)\|(\rho(\xi,\mu)^{\nu-M+1} +1)\langle\xi,\mu\rangle^{d-M+1}\|_{L^2_{\xi_n}}$$

$$\leq c'(X)(\rho(\xi',\mu)^{\nu-M+2}+1)\langle\xi',\mu\rangle^{d-M+2}$$

as above. More generally,

(2.7.12) $\| D_X^\beta D_{\xi'}^\alpha D_\mu^j D_{x_n}^{N'} \tilde{p} \|_{L^2_{y_n}} \leq c(X)(\rho(\xi',\mu)^{\nu-|\alpha|-M+1+N'}+1)\kappa^{d-|\alpha|-j-M+1+N'}$,

when $M \geq [d-|\alpha|-j]_+ + N' + 1$.

Insertion of (2.7.12) and the given informations on k' leads to (2.7.9).

□

2.7.3 Theorem. *Let* $p \in S^{d,\nu}_{tr}(\Xi\times\mathbb{R}, \overline{\mathbb{R}}^{n+1}_+)$, *vanishing for* x_n *outside a compact interval* I , *and let* $k' \in S^{d'-1,\nu'}(\Xi, \overline{\mathbb{R}}^n_+, H^+)$. *Then*

$$k''(D_n) = p(D_n)_\Omega \circ_n k'(D_n)$$

is a Poisson boundary symbol operator of order $d+d'$ *and regularity* $m(\nu,\nu')$, *defined by a symbol* $k'' = p \circ k' \in S^{d+d'-1,m(\nu,\nu')}(\Xi, \overline{\mathbb{R}}^n_+, H^+)$ *satisfying*

(2.7.13) $$k''(X,\xi,\mu) \sim \sum_{j\in\mathbb{N}} \frac{1}{j!} h^+_{\xi_n}(\overline{D}^j_{\xi_n}[\partial^j_{x_n} p(X,0,\xi,\mu)k'(X,\xi,\mu)]) .$$

The result is also valid for $S_{1,0}$ *symbol classes.*

Proof: Denote $d+d' = d''$ and $m(\nu,\nu') = \nu''$. We have already seen that $k''(D_n)$ is, for any $M \in \mathbb{N}$, of the form (where $\varphi \in C_0^\infty(\mathbb{R})$, $\varphi = 1$ on I)

(2.7.14) $$k''(D_n) = \sum_{j<M} \frac{1}{j!} OPK_n(h^+_{\xi_n}\overline{D}^j_{\xi_n}[\varphi(\partial^j_{x_n} p)k']) + x_n^M \varphi(x_n) r_M(D_n)_\Omega k'(D_n) + r'(D_n)$$

where $r'(D_n)$ is a negligible Poisson operator with the right regularity. The sum over j has symbol in $S^{d''-1,\nu''}(H^+)$, so we just have to account for $x_n^M \varphi r_M k'$. Taking some more terms in the Taylor expansion, we have, with any $M' > M$,

(2.7.15) $$x_N^M \varphi r_{M,\Omega} \circ k' = \varphi f_{M,M',\Omega} \circ k' + x_n^{M'} \varphi r_{M',\Omega} \circ k' ,$$

where $f_{M,M',\Omega} \circ k'$ has symbol kernel in $S^{d''-1-M,\nu''-M}(\mathscr{S}(\overline{\mathbb{R}}_+))$ (since it is found in a simple way from the Taylor coefficients), and $\varphi f_{M,M',\Omega} \circ k'$ likewise is defined by a symbol-kernel in $S^{d''-1-M,\nu''-M}(\mathscr{S}(\overline{\mathbb{R}}_+))$, since φ has compact support. Now write (cf. (2.7.5))

$$x_n^{M'} \circ \varphi r_{M',\Omega} \circ k' = \sum_{0\leq\ell\leq M'} \binom{M'}{\ell} OP_n[\overline{D}^\ell_{\xi_n} \varphi r_{M'}]_\Omega \circ x_n^{M'-\ell} \circ k' .$$

Set $M'' = M'/2$, assuming M' even. In the terms on the right hand side, either x_n appears in a power $M'-\ell \geq M''$, or the pseudo-differential symbol is in $S^{d-M'',\nu-M''}$. In the former case we apply Lemma 2.7.1 and in the latter case

Lemma 2.7.2. Each application leads to a multiplication operator with a kernel $\tilde{r}$ satisfying some of the estimates (2.7.6), in such a way that for $M' \to \infty$ all index sets are reached. Altogether, $x_n^M \varphi r_{M,\Omega} \circ k'$ is a multiplication operator by a function that is, by varying the choice of M' , seen to satisfy all the estimates required for a symbol kernel in $S^{d''-1-M,\nu''-M}(\mathscr{S}(\overline{\mathbb{R}}_+))$.

This completes the proof that $p(D_n)_\Omega \circ k'$ is a Poisson operator of order d'' and regularity $m(\nu,\nu')$. Since p is compactly supported in x_n , there is a precise symbol-kernel $\tilde{k}''$ defining the operator, and the preceding account shows that the associated symbol satisfies (2.7.13). The last remark is obvious from the proof. □

The terms in the series (2.7.13) can be rearranged according to homogeneity degree, in the polyhomogeneous case. (Another proof might be based on Remark 2.4.9.)

The above account gives the basic ideas in the treatment of all the products (2.7.2). In some of the other products, one also has to deal with terms of class > 0 , and in the compositions with singular Green operators, one has to keep track of more variables. Since there are no new difficulties, the reader will probably prefer to use his own imagination to fill in the details, and we just state the results here (details can also be found in the prepublication [Grubb 11 , Theorem II 4.8]).

2.7.4 Theorem. *Let* t , g *and* g' *be given as in Theorem 2.6.1, and let* $p(X,x_n,\xi,\mu) \in S_{tr}^{d,\nu}(\Xi\times\mathbb{R}, \overline{\mathbb{R}}_+^{n+1})$ *and* $p' \in S_{tr}^{d',\nu'}(\Xi\times\mathbb{R}, \overline{\mathbb{R}}_+^{n+1})$, *both vanishing for* x_n *outside a compact interval. Denote* $[r+d']_+ = r''$. *Then we have for* $\circ_n$-*compositions:*

1^0 $t(D_n) \circ_n p'(D_n)_\Omega$ *is a trace operator* $t''(D_n)$ *defined by a symbol*

$$t \circ_n p'_\Omega = t'' \in S^{d+d',m(\nu,\nu')}(\Xi, \overline{\mathbb{R}}_+^n, H^-_{r''-1}) , \quad \text{with}$$

(2.7.16)

$$t''(X,\xi,\mu) \sim \sum_{j\in\mathbb{N}} \frac{1}{j!} h^-_{\xi_n}\Big([D^j_{\xi_n} t(X,\xi,\mu)]\partial^j_{x_n} p'(X,0,\xi,\mu)\Big) .$$

2^0 $p(D_n)_\Omega \circ_n g'(D_n)$ *is a singular Green operator* $g''(D_n)$ *defined by a symbol*

$$p_\Omega \circ_n g' = g'' \in S^{d+d'-1,m(\nu,\nu')}(\Xi, \overline{\mathbb{R}}_+^n, H^+ \hat{\otimes} H^-_{r'-1}) , \quad \text{with}$$

(2.7.17)

$$g''(X,\xi,\mu) \sim \sum_{j\in\mathbb{N}} \frac{1}{j!} h^+_{\xi_n}\Big(\overline{D}^j_{\xi_n}[\partial^j_{x_n} p(X,0,\xi,\mu) g'(X,\xi,\eta_n,\mu)]\Big) .$$

3^0 $g(D_n) \circ_n p'(D_n)_\Omega$ *is a singular Green operator* $g'''(D_n)$ *defined by a symbol*

$$g \circ_n p'_\Omega = g''' \in S^{d+d'-1,m(\nu,\nu')}(\Xi,\overline{\mathbb{R}}^n_+,H^+\hat{\otimes}H^-_{r''-1}) , \quad \textit{with}$$

(2.7.18)

$$g'''(X,\xi,\mu) \sim \sum_{j\in\mathbb{N}} \frac{1}{j!} h^-_{\eta_n}\Big([D^j_{\eta_n} g(X,\xi,\eta_n,\mu)]\partial^j_{x_n} p'(X,0,\xi',\eta_n,\mu)\Big) .$$

Similar statements hold for the $S_{1,0}$ *symbol spaces.*

The treatment of the last remaining composition $L(p,p')$, in the case of x_n-dependent symbols, is well prepared by the considerations leading to Theorem 2.6.10. We first prove generalizations of Lemma 2.6.4 and Theorem 2.6.6 (of interest in themselves).

2.7.5 Theorem. *Let* $d\in\mathbb{Z}$ *and* $\nu\in\mathbb{R}$, *and let* $p(X,x_n,y_n,\xi,\mu)\in S^{d,\nu}_{tr}(\Xi\times\mathbb{R}\times\mathbb{R},\overline{\mathbb{R}}^{n+1}_+)$. *Then* $k(D_n)$ *defined by*

(2.7.19) $$k(D_n)v = r^+p(D_n)[\delta(x_n)v] \quad \textit{for} \quad v\in\mathbb{C} ,$$

is a Poisson (boundary symbol) operator of order $d+1$ *and regularity* ν , *with a symbol* $k(X,\xi,\mu)\in S^{d,\nu}(\Xi,\overline{\mathbb{R}}^n_+,H^+)$ *having the asymptotic expansion*

(2.7.20) $$k(X,\xi,\mu) \sim \sum_{j\in\mathbb{N}} \frac{1}{j!} h^+_{\xi_n}\Big(\overline{D}^j_{\xi_n}\partial^j_{x_n} p(X,0,0,\xi,\mu)\Big) .$$

Proof: By Theorem 2.1.15 and Theorem 2.2.12, $p(D_n)=p'(D_n)+p''(D_n)$, where $p'(D_n)$ has a symbol p' on y_n-form satisfying

$$p'(X,y_n,\xi,\mu) \sim \sum_{j\in\mathbb{N}} \frac{1}{j!} \overline{D}^j_{\xi_n} \partial^j_{x_n} p(X,x_n,y_n,\xi,\mu)\Big|_{x_n=y_n} ,$$

and $p''(D_n)$ is negligible of regularity $\nu-d$; p' has the transmission property. Here $p''(D_n)$ is defined by a kernel $r(X,x_n,y_n,\xi',\mu)$ satisfying

(2.7.21) $$|D^\beta_X D^\alpha_{\xi'} D^j_\mu x^m_n D^{m'}_{x_n} r| \le c(X,x_n,y_n)\langle\xi'\rangle^{-M}\langle\mu\rangle^{d-\nu-j}$$

for all indices, cf. (2.1.38). Then the operator k'' obtained by inserting p'' has the symbol-kernel $r(X,x_n,0,\xi',\mu)$ which in view of the estimates (2.7.21) is a uniformly negligible Poisson operator of regularity $\nu-d+\frac{1}{2}$, cf. Definition 2.4.4.

For k' obtained by inserting p' in (2.7.19) we have, for $v \in \mathbb{C}$,

$$r^+ OP_n(p')\delta(x_n)v = r^+ \frac{1}{2\pi}\int_{\mathbb{R}^2} e^{i(x_n-y_n)\xi_n} p'(X,y_n,\xi,\mu)v\delta(y_n)dy_n\, d\xi_n$$

$$= r^+ \frac{1}{2\pi}\int_{\mathbb{R}} e^{ix_n\xi_n} p'(X,0,\xi,\mu)d\xi_n \cdot v$$

$$= r^+ \tilde{p}'(X,x_n,\xi',\mu)v = \tilde{k}'(X,x_n,\xi',\mu)v ,$$

where $\tilde{k}' = r^+\tilde{p}' = r^+ F^{-1}_{\xi_n\to x_n} p' = r^+ F^{-1}_{\xi_n\to x_n}(h^+p')$ (recall that $p'(X,0,\xi,\mu) \in H$); the integrals can be justified as in (2.3.4). In view of Corollary 2.3.5, $k'(X,\xi,\mu)$ belongs to $S^{d,\nu}(H^+)$, so $k'(D_n)$ is the Poisson operator with symbol-kernel $\tilde{k}'$ and symbol $k' = h^+p'(X,0,\xi,\mu)$.

Altogether, k is a Poisson operator of <u>order</u> $d+1$ and regularity ν, cf. Definition 2.4.5. (The negligible elements in this class have to be of regularity $\nu - d$, which is amply satisfied by k''.) □

It follows in particular that when p and s_ℓ $(0 \le \ell \le d')$ are as in Proposition 2.6.5, then the Poisson boundary symbol operator

$$k_m(D_n)v = -i \sum_{\ell=m+1}^{d'} r^+p(D_n)\, s'_\ell\, D_{x_n}^{\ell-1-m}(\delta(x_n)v) \tag{2.7.22}$$

has a symbol $k_m(X,\xi,\mu) \in S^{d+d'-1,\nu}(\Xi, \overline{\mathbb{R}}^n_+, H^+)$, satisfying

$$k_m(X,\xi,\mu) \sim -i \sum_{\ell=m+1}^{d'} \sum_{j\in\mathbb{N}} \frac{1}{j!} h^+_{\xi_n}\left(\overline{D}^j_{\xi_n} \partial^j_{x_n} [p \circ s'_\ell \circ \xi_n^{\ell-1-m}]\right)_{x_n=0} . \tag{2.7.23}$$

<u>2.7.6 Theorem</u>. *Let* d *and* $\nu \in \mathbb{Z}$, *and let* $p(X,x_n,\xi,\mu) \in S^{d,\nu}_{tr}(\Xi\times\mathbb{R}, \overline{\mathbb{R}}^{n+1}_+)$, *vanishing for* x_n *outside a compact interval* I. *Then*

$$g^+(p(D_n)) = r^+p(D_n)e^-J \quad \textit{and} \quad g^-(p(D_n)) = Jr^-p(D_n)e^+ \tag{2.7.24}$$

are singular Green (boundary symbol) operators, of order d, *class* 0 *and regularity* $\nu - \frac{1}{2}$, *defined from symbols* $g^+(p)(X,\xi,\eta_n,\mu)$ *and* $g^-(p)(X,\xi,\eta_n,\mu)$ *in* $S^{d-1,\nu-\frac{1}{2}}(\Xi,\overline{\mathbb{R}}^n_+, H^+\hat{\otimes}H^-_{-1})$, *satisfying*

$$g^+(p)(X,\xi,\eta_n,\mu) \sim \sum_{j\in\mathbb{N}} \frac{1}{j!} \overline{D}^j_{\xi_n} g^+[\partial^j_{x_n} p(X,0,\xi,\mu)] \tag{2.7.25}$$

(2.7.26) $$g^-(p)(X,\xi,\eta_n,\mu) \sim \sum_{j\in\mathbb{N}} \frac{1}{j!} \bar{D}^j_{\xi_n} g^-[\partial^j_{x_n} p(X,0,\xi,\mu)] ,$$

where g^+ *and* g^- *of the* x_n*-independent symbols are described in Theorem 2.6.6. Similar results are valid for the* $S_{1,0}$ *symbol classes.*

Proof: Consider $g^+(p)$; the proof for $g^-(p)$ is similar. Let $\varphi \in C_0^\infty(\mathbb{R})$ with $\varphi = 1$ on I . Inserting the Taylor expansion (2.7.1), we find

$$g^+(p(D_n)) = \varphi \sum_{j<M} \frac{1}{j!} x_n^j\, g^+[OP_n(\partial^j_{x_n} p(X,0,\xi,\mu))] + g^+(\varphi x_n^M r_M) .$$

Theorem 2.6.6 and Lemma 2.4.2 imply that $\Sigma_{j<M} \frac{1}{j!} x_n^j g^+[OP_n(\partial^j_{x_n} p)]$ is a s.g.o. of order d and regularity $\nu - \frac{1}{2}$, with symbol equal to the sum for $0 \le j \le M-1$ in (2.7.25). Multiplication by φ gives another s.g.o., again with symbol-kernel in $S^{d-1,\nu-\frac{1}{2}}(\mathscr{S}(\bar{\mathbb{R}}^2_{++}))$ (the deviation from the preceding symbol-kernel being uniformly negligible of regularity $\nu - d + \frac{1}{2}$ in view of (2.6.35)). It remains to consider the term $g^+(\varphi x_n^M r_M) = g_M$.

The ps.d.o. with symbol $r'_M(X,x_n,\xi,\mu) = \varphi(x_n) r_M(X,x_n,\xi,\mu)$ has the distribution kernel

$$\tilde{r}'_M(X,x_n,x_n-y_n,\xi',\mu) = F^{-1}_{\xi_n\to z}\, r'_M(X,x_n,z,\xi',\mu)\Big]_{z=x_n-y_n} ,$$

where the general ps.d.o. properties merely assure that $\tilde{r}'_M$ is a smooth function for $x_n \neq y_n$, rapidly decreasing for $|x_n-y_n| \to \infty$ and such that high powers of $x_n - y_n$ kill the singularities at $x_n = y_n$.

The point is now that

(2.7.27) $$g^+(x_n^M \tilde{r}'_M) \text{ has the kernel } x_n^M \tilde{r}'_M(X,x_n,x_n+y_n,\xi',\mu) ,$$

considered in the quadrant $x_n > 0$, $y_n > 0$. Here the only singularity can occur at $x_n = y_n = 0$, and it can be eliminated by multiplication by $x_n + y_n$ to high powers. We then use that $0 \le x_n^M \le (x_n+y_n)^M$, where we can arrange to have M arbitrarily large.

The corresponding analysis was carried out for the μ-independent case in [Grubb 17], and all we have to do now is to insert the present μ-dependent estimates in the right places. Let us consider the basic estimate, where k , k' , m and $m' \in \mathbb{N}$ and $M \ge k'$:

(2.7.28)
$$\| x_n^k D_{x_n}^k y_n^m D_{y_n}^{m'} x_n^M \tilde{r}_M' \|_{L^2_{x_n,y_n}}$$
$$\le c \sup_{x_n \in I} \Big(\int_0^\infty | x_n^k D_{x_n}^{k'} y_n^m D_{y_n}^{m'} x_n^M \tilde{r}_M'(X,x_n,x_n+y_n,\xi',\mu)|^2 \, dy_n \Big)^{\frac12}$$
$$\le c' \sum_{0 \le j \le k'} \Big(\sup_{x_n} \int_{x_n}^\infty | w^{k+m+M-j} (D_{x_n} + D_w)^{k'-j} D_w^{m'} \tilde{r}_M'(X,x_n,w,\xi',\mu)|^2 \, dw \Big)^{\frac12}$$
$$\le c'' \sum_{\substack{j \le k' \\ \ell \le k'-j}} \sup_{x_n} \| D_{\xi_n}^{k+m+M-j} D_{x_n}^{k'-j-\ell} \xi_n^{m'+\ell} r_M'(X,x_n,\xi,\mu) \|_{L^2_{\xi_n}}$$
$$\le c(X) \sum_{\substack{j \le k' \\ \ell \le k'-j}} \| (\rho(\xi,\mu)^{\nu-k-m-M+j+m'+\ell} + 1) \langle \xi,\mu \rangle^{d-k-m-M+j+m'+\ell} \|_{L^2_{\xi_n}}$$
$$\le c'(X) \| (\rho(\xi,\mu)^{\nu-k-m-M+k'+m'} + 1) \langle \xi,\mu \rangle^{d-k-m-M+k'+m} \|_{L^2_{\xi_n}}$$
$$\le c''(X) (\rho(\xi',\mu)^{\nu-k-m-M+k'+m'+1} + 1) \kappa^{d-k-m-M+k'+m'+1}$$

since $j + \ell \le k'$; we have here used the estimates for r_M' valid according to Theorem 2.2.5 when

(2.7.29)
$$d - k - m - M + k' + m' \le -1 ,$$

and we have used (2.7.11). There are similar estimates for derivatives in (X,ξ',μ) . Since all index sets are reached when $M \to \infty$, an argument as in the proof of Theorem 2.7.3 permits the conclusion that g_M is indeed a s.g.o. with the stated regularity, and that the symbol of $g^+(p)$ satisfies (2.7.25). □

We finally conclude:

<u>2.7.7 Theorem.</u> *Let* d, d', ν *and* $\nu' \in \mathbb{Z}$, *and let* $p(X,x_n,\xi,\mu) \in S_{tr}^{d,\nu}(\Xi \times \mathbb{R}, \overline{\mathbb{R}}_+^{n+1})$ *and* $p'(X,x_n,\xi,\mu) \in S^{d',\nu'}(\Xi \times \mathbb{R}, \overline{\mathbb{R}}_+^{n+1})$, *vanishing for* x_n *outside a compact interval* I . *Then*

(2.7.30)
$$g(X,\xi',\mu,D_n) = L(p(X,x_n,\xi',\mu,D_n), p'(X,x_n,\xi',\mu,D_n)) \equiv (pp')_\Omega - p_\Omega p'_\Omega$$

is a singular Green (boundary symbol) operator defined by a symbol $g(X,\xi,\eta_n,\mu)$ $\in S^{d+d'-1,m(\nu-\frac{1}{2},\nu'-\frac{1}{2})}(\Xi,\overline{\mathbb{R}}^n_+,H^+\hat{\otimes}H^-_{r-1})$, *where* $r=[d']_+$. *The symbol* g *has an asymptotic expansion determined from the Taylor expansions of* p *and* p' *in* x_n *at* $x_n=0$, *by the formula*

(2.7.31) $$g(X,\xi,\eta_n,\mu)\sim$$

$$\sim\sum_{j,\ell,m\in\mathbb{N}}\frac{(-1)^{j+\ell}}{j!\ell!m!}D^j_{\xi_n}D^m_{\eta_n}L\Big(\partial^j_{x_n}p(X,0,\xi,\mu),D^\ell_{\eta_n}\partial^{\ell+m}_{x_n}p'(X,0,\xi',\eta_n,\mu)\Big),$$

where the terms in the series are determined by Theorem 2.6.10. The results likewise hold for $S_{1,0}$ *symbol spaces.*

Proof: p' can be written as a sum

$$p'=\sum_{0\le\ell<r}s'(X,\xi',\mu)\xi_n^\ell+p''+x_n^r p''' ,$$

where p'' is $O(\langle\xi_n\rangle^{-1})$ and $p'''\in S^{d',\nu'}$, as in (2.5.22) with $N=r=[d']_+$. Then

$$L(p,p')=L(p,\Sigma s'_\ell D^\ell_n)+L(p,p''+x^r_n p''') .$$

The first term is treated as in (2.6.19), which gives

$$L(p,\Sigma s'_\ell D^\ell_n)=\sum_{0\le m<r}k_m\gamma_m$$

where k_m is determined by (2.7.22) (which uses Theorem 2.7.5). For the second term we have, since $x^r_n p'''(D_n)e^+$ maps $L^2(\mathbb{R}_+)$ into $L^2_{loc}(\mathbb{R})$,

$$\begin{aligned}L(p,p''+x^r_n p''')&=r^+p(I-e^+r^+)(p''+x^r_n p''')e^+\\&=r^+pe^-r^-(p''+x^r_n p''')e^+=g^+(p)\circ g^-(p''+x^r_n p''') .\end{aligned}$$

Since $g^-(p''+x^r_n p''')=g^-(p')$, we find altogether

(2.7.32) $$L(p,p')=\sum_{0\le j<r}k_m\gamma_m+g^+(p)\circ g^-(p') .$$

The symbols of $g^+(p)$ and $g^-(p')$ are determined by Theorem 2.7.6, and the symbol of the composition is determined from this by formula (2.6.5) 11°. To show the formula (2.7.31), we write p and p' in Taylor expansions (2.7.1), which gives

$$L(p,p') = \sum_{j,k<M} \frac{1}{j!k!} L(x_n^j \partial_{x_n}^j p(X,0,\xi,\mu), x_n^k \partial_{x_n}^k p'(X,0,\xi,\mu)) + g_M ,$$

where g_M is of order $d+d'-M$. Note that all terms in the symbol $L(p,p')$ will be included in the sum over j , k when M goes to ∞ . Now by (2.7.5)

$$\frac{1}{j!k!} L\Big(x_n^j \, OP_n(\partial_{x_n}^j p), x_n^k \, OP_n(\partial_{x_n}^k p')\Big)$$

$$= \frac{1}{j!k!} x_n^j \circ L\Big(OP_n(\partial_{x_n}^j p), \sum_{\ell\leq k} \binom{k}{\ell} OP_n(\overline{D}_{\eta_n}^{\ell} \partial_{x_n}^k p')\Big) \circ x_n^{k-\ell} ,$$

which by Lemma 2.4.2 has the symbol

$$\frac{1}{j!k!} \sum_{\ell\leq k} \binom{k}{\ell} \overline{D}_{\xi_n}^{j} D_{\eta_n}^{k-\ell} L(\partial_{x_n}^j p , \overline{D}_{\eta_n}^{\ell} \partial_{x_n}^k p')$$

$$= \sum_{\text{finite}} \frac{(-1)^{j+\ell}}{j!(k-\ell)!\ell!} D_{\xi_n}^{j} D_{\eta_n}^{k-\ell} L(\partial_{x_n}^j p , D_{\eta_n}^{\ell} \partial_{x_n}^k p') .$$

For $M \to \infty$ we find the asymptotic series (2.7.31). □

Other formulas can be inferred from (2.7.32).

2.7.8 Remark. The result extends to symbols that are not compactly supported in x_n , as long as the corresponding operators can be composed. E.g., if $p(X,x_n,\xi,\mu)$ and $q(X,x_n,\xi,\mu)$ are ps.d.o. symbols for which $p_\Omega q_\Omega u$ is well defined for $u \in C_{(0)}^\infty(\overline{\mathbb{R}}_+)$, then one gets by truncation with real functions φ and $\varphi_1 \in C^\infty(\mathbb{R})$, equal to 1 for $x_n \leq a$ and 0 for $x_n \geq 2a$ for some $a > 0$, with $\varphi\varphi_1 = \varphi$,

$$\begin{aligned} L(p,q)u &= L(p,\varphi_1 q)u + L(p, (1-\varphi_1)q)u \\ &= L(p,\varphi_1 q)u + r^+p(I-e^+r^+)(1-\varphi_1)qe^+u \\ &= L(p,\varphi_1 q)u \end{aligned}$$

since $(I-e^+r^+)(1-\varphi_1) = 0$, and then furthermore

$$L(p,\varphi_1 q)u = L(p\varphi,\varphi_1 q)u = g^+(p\varphi)g^-(\varphi_1 q)u + \sum_{m<d'} k_m\gamma_m u ,$$

where $g^-(\varphi_1 q)$ and $g^+(p\varphi) = [g^-(\varphi p^*)]^*$ are s.g.o.s by Theorem 2.7.6 and Proposition 2.4.1, and the factors

$$k_m = -i \sum_{\ell=m+1}^{d'} r^+ p \varphi s'_\ell D_n^{\ell-1-m} \delta$$

are Poisson operators by Theorem 2.7.5. Altogether, it is found that

(2.7.33) $$L(p,q) = L(p\varphi, \varphi_1 q) \quad \text{is a s.g.o.} ,$$

with symbol determined as in Theorem 2.7.7. Similar extensions can be made for the compositions in Theorems 2.7.3 and 2.7.4, by use of Lemma 2.4.8.

The composition rules will finally be established in full generality. This is easily obtained by a combination of the above results with the techniques of Theorem 2.1.15, applied in the x'-direction.

We formulate the results for the case where Ξ is specialized to be either Ω' or $\Omega'\times\Omega'$, where Ω' is an open set in $\mathbb{R}^{n-1}$ (so the corresponding operators on $\Omega'\times\overline{\mathbb{R}}_+$ are defined by (2.4.7)). Of course one could retain some extra parameters in Ξ (taking e.g. $\Xi = \Omega'\times\Xi_1$ for some parameter set Ξ_1) ; - we shall later need this in connection with symbols depending on a complex parameter, but leave out that extra burden on the notations here.

2.7.9 Theorem. *Let* $\Xi = \Omega' \subset \mathbb{R}^{n-1}$, *let* p *and* p' *be as in Theorem 2.7.7 or Remark 2.7.8, and let* g, t, k, s *and* g', t', k', s' , *be as in Theorem 2.6.1. Let* $a(x',\xi',\mu,D_n)$ *and* A_μ *stand for the boundary symbol operator on* $\overline{\mathbb{R}}_+$, *resp. the operator* $OP'(a(x',\xi',\mu,D_n))$ *on* $\overline{\mathbb{R}}^n_+$, *defined from one of the symbols* p, g, t, k *or* s , *and let* $a'(x',\xi',\mu,D_n)$ *and* A'_μ *denote operators derived similarly from one of the primed operators. (The symbols can also be given on* (x',y')*-form, in which case one begins by reducing them to* x'*-form by Theorem 2.4.6.)*

1° *Consider one of the compositions*

$$A''_\mu = A_\mu A'_\mu = OP'(a)OP'(a')$$

listed in (2.6.2) (iii)-(vi). *Here* A''_μ *is equivalent with an operator of the form* $OP'(a'')$, *where* a'' *is described by the sum over* $\alpha \in \mathbb{N}^{n-1}$

(2.7.34) $$a''(x',\xi',\mu,D_n) \sim \sum_\alpha \frac{1}{\alpha!} D^\alpha_{\xi'} a(x',\xi',\mu,D_n) \circ \partial^\alpha_{x'} a'(x',\xi',\mu,D_n) ,$$

each term being determined by the appropriate composition rule in Theorems 2.1.15, 2.6.1, 2.7.3 and 2.7.4. The equivalences hold in the space of operators and symbols of order $d+d'$, *class* r' *resp.* $[r+d']_+$ *(in the relevant cases) and regularity* $m(\nu,\nu')$.

2^0 *Consider the singular Green operator*

$$G_\mu = L(P_\mu,P'_\mu) \equiv (P_\mu P'_\mu)_\Omega - P_{\mu,\Omega}P'_{\mu,\Omega}$$

derived from P_μ *and* P'_μ . *Here* G_μ *is equivalent with an operator of the form* $OP'(g)$, *where*

$$g(x',\xi',\mu,D_n) \sim \sum_\alpha \frac{1}{\alpha!} L(D^\alpha_\xi p(x,\xi',\mu,D_n),\partial^\alpha_x p'(x,\xi',\mu,D_n)) , \tag{2.7.35}$$

with the terms defined by Theorem 2.7.7. The equivalences hold in the space of operators and symbols of order $d+d'$, *class* $[d']_+$ *and regularity* $m(\nu-\frac{1}{2},\nu'-\frac{1}{2})$.

The results are likewise valid in $S_{1,0}$ *symbol spaces.*

Proof: 1^0 If a' were on y'-form, the resulting operator would simply have the boundary symbol operator on (x',y')-form

$$a(x',\xi',\mu,D_n) \circ_n a'(y',\xi',\mu,D_n) .$$

This can be reduced to x'-form by Theorem 2.4.6. The procedure of replacing a' by its y'-form and reducing the resulting product to x'-form gives altogether the formula (2.7.34) (just as in the ps.d.o. case). One can also show (2.7.34) directly by integrations by part and estimates, similarly to (2.1.50). The proof of 2^0 is similar. □

The regularity $m(\nu-\frac{1}{2}, \nu'-\frac{1}{2})$ for $L(P_\Omega,P'_\Omega)$ can be improved in certain cases, for instance in the situation of Proposition 2.6.11, where the proof is now completed, and extended to the full x-dependent calculus, just as above. This gives

2.7.10 Theorem. *Let* d *be integer* > 0 , *let* $p \in S^{d}_{tr}(\Omega' \times \mathbb{R}, \mathbb{R}^n)$ *and let* $q \in S^{-d,d}_{tr}(\Omega' \times \mathbb{R}, \overline{\mathbb{R}}^{n+1}_{+})$. *Then* $L(P,Q_\mu)$ *is of regularity* d *and of order and class* 0).

We finally formulate the composition results for operators on manifolds.

2.7.11 Corollary. *Let* A_μ *and* A'_μ *be parameter-dependent Green operators*

$$(2.7.36)\qquad \begin{aligned} A_\mu &= \begin{pmatrix} P_{\mu,\Omega} + G_\mu & K_\mu \\ T_\mu & S_\mu \end{pmatrix} : \begin{matrix} C^\infty(E) \\ \times \\ C^\infty(F) \end{matrix} \to \begin{matrix} C^\infty(E') \\ \times \\ C^\infty(F') \end{matrix} \\ A'_\mu &= \begin{pmatrix} P'_{\mu,\Omega} + G_\mu & K_\mu \\ T_\mu & S_\mu \end{pmatrix} : \begin{matrix} C^\infty(E') \\ \times \\ C^\infty(F') \end{matrix} \to \begin{matrix} C^\infty(E'') \\ \times \\ C^\infty(F'') \end{matrix} \end{aligned}$$

where E, E' *and* E" *are* C^∞ *vector bundles (of dimensions* N, N' *and* N" ≥ 1*) over an* n*-dimensional compact* C^∞ *manifold* $\overline{\Omega}$ *with boundary* Γ ; *and* F, F' *and* F" *are* C^∞ *vector bundles over* Γ *(of dimensions* M, M' *and* M" ≥ 0 *)* . *Assume that the ps.d.o. parts of* A_μ *and* A'_μ *have regularities* ν_1 *resp.* ν'_1 , *and the singular parts have regularities* ν_2 *resp.* ν'_2 . *Then the composition*

$$(2.7.37)\qquad A''_\mu = A_\mu \circ A'_\mu$$

is a parameter-dependent Green operator with ps.d.o. part resp. singular part of regularities ν''_1 *resp.* ν''_2 , *where*

$$(2.7.38)\qquad \begin{aligned} \nu''_1 &= m(\nu_1, \nu'_1) \\ \nu''_2 &= \min\{\nu_1 - \tfrac{1}{2},\ \nu_2,\ \nu'_1 - \tfrac{1}{2},\ \nu'_2,\ \nu_1 + \nu'_1 - 1,\ \nu_1 + \nu'_2,\ \nu_2 + \nu'_1,\ \nu_2 + \nu'_2\} . \end{aligned}$$

The principal (x_n*-independent) boundary symbol operator for* A''_μ *is the composition of the corresponding operators for* A_μ *and* A'_μ .

Proof: By partitions of unity and local coordinate systems (see Section 2.4), the study of (2.7.37) is carried over to a study of compositions of operators on $\overline{\mathbb{R}}^n_+$, where Theorem 2.7.9 can be applied. The operator A''_μ is then again a parametrized Green operator, and a consideration of the various rules for the terms in (2.6.2) shows that the resulting regularities satisfy (2.7.38). The last statement refers to the boundary symbol operators parametrized by μ and the point in the cotangent space of Γ, see (2.4.78) ff. □

The result extends to noncompact manifolds, when suitable precautions are taken concerning supports.

The formula for ν_2'' is simpler in the case where $\nu_2 \leq \nu_1 - \frac{1}{2}$ and $\nu_2' \leq \nu_1' - \frac{1}{2}$, for then

$$\nu_2'' = m(\nu_2, \nu_2') . \tag{2.7.39}$$

This will also hold if $\nu_2 \leq \nu_1$, $\nu_2' \leq \nu_1'$ and $L(p,q)$ is of regularity $m(\nu_1,\nu_1')$ (as for instance in Theorem 2.7.10).

2.8 Strictly homogeneous symbols.

In the same way as we associated a strictly homogeneous symbol $p^h(X,\xi,\mu)$ to each polyhomogeneous symbol $p(X,\xi,\mu)$ in Definition 2.1.8, we can associate strictly homogeneous symbols to the various types of symbols for boundary operators introduced in Section 2.3. Since we shall occasionally need to use such symbols and their calculus explicitly, we introduce a precise terminology.

For each of the polyhomogeneous spaces $S^{d,\nu}(\dots)$ there is an associated space of strictly homogeneous symbols, that we denote by $S^{d,\nu}_{hom}(\dots)$.

Recall from Section 2.1 that for the pseudo-differential symbols, $S^{d,\nu}_{hom}(\Xi,\overline{\mathbb{R}}^{n+1}_+)$ is defined as the space of functions $p^h(X,\xi,\mu) \in C^\infty(\Xi\times(\mathbb{R}^n\setminus 0)\times\overline{\mathbb{R}}_+)$ that are homogeneous of degree d in (ξ,μ) and satisfy the estimates (2.1.19).

We proceed similarly for the other spaces. Take K equal to H^+, H^-_{r-1} or H_{r-1}, then we define the space $S^{d,\nu}_{hom}(\Xi,\overline{\mathbb{R}}^n_+,K)$ as the space of C^∞ functions $f^h(X,\xi',\xi_n,\mu)$ on $\Xi\times(\mathbb{R}^{n-1}\setminus 0)\times\mathbb{R}\times\overline{\mathbb{R}}_+$, that are homogeneous in (ξ',ξ_n,μ) of degree d (for $\xi'\neq 0$) and satisfy, for all indices,

$$(2.8.1)\qquad \|D^\beta_X D^\alpha_{\xi'} D^j_\mu h_{-1}(D^k_{\xi_n}\xi_n^{k'}f^h)\|_{L^2_{\xi_n}}$$

$$\leq c(X)(|\xi'|^{\nu-[k-k']_+-|\alpha|}\,|\xi',\mu|^{d+\frac12-\nu+[k-k']_- -j} + |\xi',\mu|^{d+\frac12-k+k'-|\alpha|-j})\ .$$

Now when f is given in $S^{d,\nu}(\Xi,\overline{\mathbb{R}}^n_+,K)$, we define the associated strictly homogeneous symbol f^h by the formula

$$(2.8.2)\qquad f^h(X,\xi,\mu) = |\xi'|^d f^0(X,\xi/|\xi'|,\mu/|\xi'|)\quad\text{for all } (\xi,\mu) \text{ with } \xi'\neq 0\ ;$$

it is seen to belong to $S^{d,\nu}_{hom}(\Xi,\overline{\mathbb{R}}^n_+,K)$ by an argument as in the proof of Lemma 2.1.9 1°. The strictly homogeneous symbols are decomposed into a strictly homogeneous polynomial part (in ξ_n) and a strictly homogeneous part of class 0 :

$$(2.8.3)\qquad f^h = \sum_{0\leq j<r} s^h_{d-j}(X,\xi',\mu)\xi_n^j + f'^h\ ,$$

where the s^h_{d-j} lie in $S^{d-j,\nu}_{hom}(\Xi,\overline{\mathbb{R}}^n_+)$ and $f'^h\in S^{d,\nu}_{hom}(\Xi,\overline{\mathbb{R}}^n_+,K)$ is of class 0 ; when f^h is the strictly homogeneous symbol associated with a smooth symbol f, the terms s^h_{d-j} and f'^h are precisely the strictly homogeneous symbols associated with the terms s_{d-j} and f' in the decomposition of f (2.3.9). The coefficients in the Laguerre expansion of f'^h

(2.8.4) $$f'^h(X,\xi,\mu) = \sum_{m\in\mathbb{Z}} b_m^h(X,\xi',\mu)\hat{\varphi}_m(\xi_n,\kappa^h) \quad , \qquad \kappa^h = |\xi',\mu| \ ,$$

are strictly homogeneous symbols associated with the coefficients of f' , cf. (2.3.30).

Similarly, when we take K equal to $H^+ \hat{\otimes} H^-_{r-1}$ or $H_{-1} \hat{\otimes} H_{r-1}$, we define the space $S^{d,\nu}_{hom}(\Xi, \overline{\mathbb{R}}^n_+, K)$ as the space of C^∞ functions $f^h(X,\xi',\xi_n,\eta_n,\mu)$ on $\Xi\times(\mathbb{R}^{n-1}\setminus 0)\times\mathbb{R}^2\times\overline{\mathbb{R}}_+$, that are homogeneous in (ξ',ξ_n,η_n,μ) of degree d (for $\xi' \neq 0$) and satisfy, for all indices,

(2.8.5) $$\| D_X^\beta D_{\xi'}^\alpha D_\mu^j h_{-1,\xi_n} h_{-1,\eta_n}(D_{\xi_n}^k \xi_n^{k'} D_{\eta_n}^m \eta_n^{m'} f^h)\|_{L^2_{\xi_n,\eta_n}(\mathbb{R}^2)}$$

$$\leq c(X)(|\xi'|^M |\xi',\mu|^{M'} + |\xi',\mu|^{d+1-k+k'-m+m'-|\alpha|-j}) \ ,$$

where $M = \nu - [k-k']_+ - [m-m']_+ - |\alpha|$, $M' = d+1-\nu+[k-k']_- + [m-m']_- - j$.

Again, when f is given in $S^{d,\nu}(\Xi, \overline{\mathbb{R}}^n_+, K)$, then we define the associated strictly homogeneous symbol f^h from f^0 by

(2.8.6) $f^h(X,\xi,\eta_n,\mu) = |\xi'|^d f^0(X,\xi/|\xi'|,\eta_n/|\xi'|,\mu/|\xi'|)$ for all (ξ,η_n,μ) with $\xi' \neq 0$,

then $f^h \in S^{d,\nu}_{hom}$. And also here, we have a decomposition

(2.8.7) $$f^h = \sum_{0\leq j<r} k^h_{d-j}(X,\xi,\mu)\eta_n^j + f'^h \ ,$$

where $k^h_{d-j} \in S^{d-j,\nu}_{hom}(\Xi, \overline{\mathbb{R}}^n_+, K_1)$ (for $K_1 = H^+$ resp. H_{-1}) and f'^h is of class 0 . When f^h is the strictly homogeneous symbol associated with a smooth symbol $f \in S^{d,\nu}(\Xi, \overline{\mathbb{R}}^n_+, K)$, these terms are precisely the strictly homogeneous symbols associated with the terms k_{d-j} resp. f' in the decomposition (2.3.22) of f . There is also a Laguerre expansion

(2.8.8) $$f'^h = \sum_{\ell,m\in\mathbb{N}} c^h_{\ell m}(X,\xi',\mu)\hat{\varphi}_\ell(\xi_n,\kappa^h)\bar{\hat{\varphi}}_m(\eta_n,\kappa^h)$$

where the $c^h_{\ell,m}$ are the strictly homogeneous symbols associated with the Laguerre coefficients of f' , cf. (2.3.30).

The symbols are extended by continuity to $\xi' = 0$ whenever possible.

Let us also include the definition of the spaces $S^{d,\nu}_{tr,hom}(\Sigma\times\Sigma, \overline{\mathbb{R}}^{n+1}_+)$ of strictly homogeneous ps.d.o. symbols having the transmission property ; it is the subspace of $S^{d,\nu}_{hom}(\Sigma\times\Sigma,\overline{\mathbb{R}}^{n+1}_+)$ consisting of the symbols p^h satisfying (2.2.58) (with $\ell = 0$) , and it contains the strictly homogeneous symbols associated with symbols in $S^{d,\nu}_{tr}(\Sigma\times\Sigma,\overline{\mathbb{R}}^{n+1}_+)$.

It is now easy to verify that when the various reductions and product rules for the $S^{d,\nu}$-symbols are applied to the strictly homogeneous symbols for $\xi' \neq 0$, one gets mappings from $S^{d,\nu}_{hom}$-spaces to $S^{d',\nu'}_{hom}$-spaces with precisely the same rules concerning the orders and regularities as before: The mapping properties in Lemmas 2.3.9 and 2.3.10 are immediately generalized; and the rules in Section 2.6 for multiplications of symbols and compositions of operators on $\overline{\mathbb{R}}_+$ all have their counterpart in the strictly homogeneous setting; and so do the observations on Laguerre expansions in Section 2.6. One simply uses the same techniques of proof. Also the estimates of boundary symbol operators entering in Section 2.5 generalize (with strict homogeneity of the norms).

As for the inversion of symbols in the elliptic case, let us observe the following.

<u>2.8.1 Lemma</u>. *Let* $p^h(X,\xi,\mu) \in S^{d,\nu}_{hom}(\Xi,\overline{\mathbb{R}}^{n+1}_+) \otimes L(\mathbb{C}^N,\mathbb{C}^N)$. *Let* $\nu > 0$, *so that* p^h *extends by continuity to all* $(\xi,\mu) \in \overline{\mathbb{R}}^{n+1}_+\setminus 0$. *When* $p^h(X,\xi,\mu)$ *is bijective for all* $(X,\xi,\mu) \in \Xi\times(\overline{\mathbb{R}}^{n+1}_+\setminus 0)$, *then the inverse* $(p^h(X,\xi,\mu))^{-1}$ *is a symbol in* $S^{-d,\nu}_{hom}(\Xi,\overline{\mathbb{R}}^{n+1}_+)$. *Similar statements hold for symbols defined on a cone* $\Xi\times V$ *in* $\Xi\times\overline{\mathbb{R}}^{n+1}_+$.

<u>Proof:</u> One uses the Leibniz formula as in (2.1.53-54), replacing $\langle\xi,\mu\rangle$ by $|\xi,\mu|$ and ρ by $|\xi|/|\xi,\mu|$. ▯

The result is consistent with Theorem 2.1.16 in view of Proposition 2.1.11 2°; and invertible symbols of this kind are said to be <u>elliptic</u>.

The inversion of boundary symbol operators will be discussed in Chapter 3, and for that discussion we also need an analysis of the continuity properties (in ξ') of associated strictly homogeneous symbols, as well as an estimate of their deviation from the smooth symbols. One has:

<u>2.8.2 Proposition</u>. *Let* $\nu \geq 0$.

1° *When* $f \in S^{d,\nu}(\Xi,\overline{\mathbb{R}}^n_+,K)$ *for* $K = H^+$, H^-_{r-1} *or* H_{r-1} , *and* f *and the associated strictly homogeneous symbol* f^h *are decomposed as in* (2.3.9) *resp.* (2.8.3), *then*

$$(2.8.9)\qquad \begin{aligned} &|s^0_{d-j}(X,\xi',\mu) - s^h_{d-j}(X,\xi',\mu)| \le c(X)\langle\xi',\mu\rangle^{d-j-\nu} ,\\ &\|f'^0(X,\xi,\mu) - f'^h(X,\xi,\mu)\|_{L^2_{\xi_n}} \le c(X)\langle\xi',\mu\rangle^{d+\frac12-\nu} , \quad for \quad |\xi',\mu| \ge 1 . \end{aligned}$$

Moreover, the coefficients in the Laguerre expansions (2.3.30) *resp.* (2.8.4) *satisfy*

$$(2.8.10)\qquad (\sum_m |b^0_m(X,\xi',\mu) - b^h_m(X,\xi',\mu)|^2)^{\frac12} \le c(X)\langle\xi',\mu\rangle^{d+\frac12-\nu} \quad for \quad |\xi',\mu| \ge 1 .$$

2^0 *When* $f \in S^{d,\nu}(\Xi, \overline{\mathbb{R}}^n_+, K)$ *for* $K = H^+ \hat\otimes H^-_{r-1}$ *or* $H_{-1} \hat\otimes H_{r-1}$, *and* f *and the associated strictly homogeneous symbol* f^h *are decomposed as in* (2.3.22) *resp.* (2.8.7), *then*

$$(2.8.11)\qquad \begin{aligned} &\|k^0_{d-j}(X,\xi,\mu) - k^h_{d-j}(X,\xi,\mu)\|_{L^2_{\xi_n}} \le c(X)\langle\xi',\mu\rangle^{d+\frac12-j-\nu} ,\\ &\|f'^0(X,\xi,\eta_n,\mu) - f'^h(X,\xi,\eta_n,\mu)\|_{L^2_{\xi_n,\eta_n}} \le c(X)\langle\xi',\mu\rangle^{d+1-\nu} , \quad for \quad |\xi',\mu| \ge 1 . \end{aligned}$$

Moreover, the coefficients in the Laguerre expansions (2.3.30) *resp.* (2.8.8) *satisfy*

$$(2.8.12)\qquad (\sum_{\ell,m} |c^0_{\ell m}(X,\xi',\mu) - c^h_{\ell m}(X,\xi',\mu)|^2)^{\frac12} \le c(X)\langle\xi',\mu\rangle^{d+1-\nu} \quad for \quad |\xi',\mu| \ge 1 .$$

Proof: One proceeds as in the proof of Lemma 2.1.9 2^0, extending the estimates to functions valued in the spaces $L^2_{\xi_n}$ and $L^2_{\xi_n,\eta_n}$ (strong derivatives). The Laguerre series estimates are direct consequences of the L^2 estimates. □

Symbols of positive regularity have a Hölder property (recall Definition 2.1.10'):

2.8.3 Proposition. *Let* $\nu \ge 0$.

1^0 *Let* $K = H^+$, H^-_{r-1} *or* H_{r-1} . *For* $f^h \in S^{d,\nu}(\Xi, \overline{\mathbb{R}}^n_+, K)$, *decomposed as in* (2.8.4), *one has that*

$$(2.8.13)\qquad \begin{aligned} &s^h_{d-j} \in C^{\nu-}(\Xi \times \overline{\mathbb{R}}^n_+, \mathbb{C}) ,\\ &f'^h \in C^{\nu-}(\Xi \times \overline{\mathbb{R}}^n_+, L^2_{\xi_n}(\mathbb{R})) . \end{aligned}$$

In particular, if $\nu = \ell+\tau$ *with* $\ell \in \mathbb{N}$ *and* $\tau \in]0,1]$, *then* $D^\alpha_{\xi'} s^h_{d-j}$ *and* $D^\alpha_{\xi'} f'^h$ *with* $|\alpha|=\ell$ *are continuous (in symbol norm) at* $\xi'=0$ *for* $\mu > 0$, *satisfying for* $|\xi'| \le c\mu$:

$$(2.8.14)\quad \begin{aligned} &|D^\alpha_{\xi'} s^h_{d-j}(X,\xi',\mu) - D^\alpha_{\xi'} s^h_{d-j}(X,0,\mu)| \le c(X)|\xi'|^\tau |\xi',\mu|^{d-j-|\alpha|-\tau} ,\\ &\| D^\alpha_{\xi'} f'^h(X,\xi',\xi_n,\mu) - D^\alpha_{\xi'} f'^h(X,0,\xi_n,\mu)\|_{L^2_{\xi_n}} \le c(X)|\xi'|^\tau |\xi',\mu|^{d+\frac12-|\alpha|-\tau} . \end{aligned}$$

2^0 *Let* $K = H^+ \hat\otimes H^-_{r-1}$ *or* $H_{-1} \otimes H_{r-1}$. *For* $f^h \in S^{d,\nu}(\Xi, \overline{\mathbb{R}}^n_+, K)$, *decomposed as in* (2.8.7), *one has that*

$$(2.8.15)\quad \begin{aligned} &k^h_{d-j} \in C^{\nu-}(\Xi \times \overline{\mathbb{R}}^n_+, L^2_{\xi_n}(\mathbb{R})) ,\\ &f'^h \in C^{\nu-}(\Xi \times \overline{\mathbb{R}}^n_+, L^2_{\xi_n,\eta_n}(\mathbb{R}^2)) . \end{aligned}$$

In particular, if $\nu = \ell+\tau$ *with* $\ell \in \mathbb{N}$ *and* $\tau \in]0,1]$, *then* $D^\alpha_{\xi'} k^h_{d-j}$ *and* $D^\alpha_{\xi'} f'^h$ *with* $|\alpha|=\ell$ *are continuous (in symbol norm) at* $\xi' = 0$ *for* $\mu > 0$, *satisfying for* $|\xi'| \le c\mu$:

$$(2.8.16)\quad \begin{aligned} &\| D^\alpha_{\xi'} k^h_{d-j}(X,\xi',\xi_n,\mu) - D^\alpha_{\xi'} k^h_{d-j}(X,0,\xi_n,\mu)\|_{L^2_{\xi_n}} \le c(X)|\xi'|^\tau |\xi',\mu|^{d+\frac12-|\alpha|-\tau} ,\\ &\| D^\alpha_{\xi'} f'^h(X,\xi',\xi_n,\eta_n,\mu) - D^\alpha_{\xi'} f'^h(X,0,\xi_n,\eta_n,\mu)\|_{L^2_{\xi_n,\eta_n}} \le c(X)|\xi'|^\tau |\xi',\mu|^{d+1-|\alpha|-\tau} \end{aligned}$$

<u>Proof</u>: One applies Lemma 2.1.10 (with U equal to $\mathbb{C}$, $L^2_{\xi_n}$ or $L^2_{\xi_n,\eta_n}$ as needed), in the same way as in the proof of Proposition 2.1.11 1^0. □

So when $\nu>0$, the symbols are continuous at $\xi' = 0$ for $\mu > 0$, in the respective symbol norms. We then extend the symbols to be defined for all $\xi'=0$, $\mu > 0$. Here it should be noted that the value of a symbol at a point $(X,0,\mu)$ cannot be expected to be an H-function (or $H \hat\otimes H$-function) of ξ_n (or (ξ_n,η_n)) but only a polynomial in ξ_n plus an $L^2_{\xi_n}$-function (resp. a sum of terms $k(\xi_n)\eta_n + g(\xi_n,\eta_n)$ where $k \in L^2_{\xi_n}$ and $g \in L^2_{\xi_n,\eta_n}$) satisfying those estimates in (ξ_n,μ) (resp. (ξ_n,η_n,μ)) that survive for $\xi' \to 0$. But the boundary symbol operators can be defined anyway, by the formulas (2.4.2-6), and we use the same notation; and the corresponding operator families have continuity properties as derived in Section 2.5. For example, when $d \in \mathbb{Z}$ and $\nu > 0$, and a^h is a system of scalar boundary symbol operators of order d, regularity $\nu>0$ and class $r \leq d_+$ (cf. A.1):

$$a^h(X,\xi',\mu,D_n) = \begin{pmatrix} p^h(X,\xi',\mu,D_n)_\Omega + g^h(X,\xi',\mu,D_n) & k^h(X,\xi',\mu,D_n) \\ t^h(X,\xi',\mu,D_n) & s^h(X,\xi',\mu) \end{pmatrix} ,$$

then $a^h(X,\xi',\mu,D_n)$ is continuous from $H^{d_+}(\overline{\mathbb{R}}_+)\times\mathbb{C}$ to $H^{d_-}(\overline{\mathbb{R}}_+)\times\mathbb{C}$ for all $(\xi',\mu) \in \overline{\mathbb{R}}^n_+\setminus 0$ and all X, and the norm depends continuously on (X,ξ',μ).

The considerations on the strictly homogeneous symbols apply in particular to the parameter-independent symbols considered in Chapter 1; and the observations in Section 1.5 after Definition 1.5.5 come out as a special case. Note that the limiting operators for $\xi' \to 0$ do in fact have symbols in H for $\xi'=0$, for they are differential operator symbols there.

The calculus with strictly homogeneous symbols is somewhat simpler than the calculus with smooth symbols. We have placed so much emphasis on the calculus with smooth symbols, because this is what is needed for the full operator calculus, applied to general C^∞-, Sobolev- and distribution-spaces.

The upper h on strictly homogeneous symbols is included for precision, but it can be left out, when the homogeneity assumption is clear from the context.

2.8.4 Remark. Like in the pseudo-differential case (cf. Remark 2.1.11'), one can ask whether any given strictly homogeneous symbol $f^h \in S^{d,\nu}_{hom}(\Xi, \overline{\mathbb{R}}^n_+,K)$ stems from a smooth symbol $f^0 \in S^{d,\nu}(\Xi, \overline{\mathbb{R}}^n_+,K)$ such that $f^h = f^0$ for $|\xi'| \geq 1$. When $\nu \leq 0$, the question is easily answered in the affirmative, for one can then simply take

$$f^0 = \zeta(|\xi'|)f^h , \tag{2.8.17}$$

where ζ is as in Remark 2.1.11'. However, if $\nu > 0$, f^0 defined by (2.8.17) is generally only in $S^{d,0}$, since $\zeta(|\xi'|) \in S^{0,0}$ (cf. Proposition 2.1.5, or note that the derivatives of f^0 are not in general sufficiently controlled with respect to μ on the set where the derivatives of $\zeta(|\xi'|)$ are not zero).

In the case where $\nu > 0$, more caution is needed. Here the construction in Remark 2.1.11' can be directly generalized, if the Taylor coefficients of f^h in ξ' at $\xi' = 0$ are sufficiently nice functions, so that the Taylor polynomial lies in $S^{d,\nu}$. This holds for the parameter-independent normal boundary problems, where the Taylor coefficients are symbols of differential operators, but it need not hold for general, strictly homogeneous symbols. However, one can then instead make a construction using the Taylor expansion at a nearby point $\eta' \neq 0$, except in some two-dimensional cases where the symbols for $\xi' > 0$ and $\xi' < 0$ are not compatible. We give the details for Poisson symbols:

Let $\nu > 0$. Note first that a function $\tilde{k}(X,x_n,\xi',\mu) \in C^\infty(\Xi \times \overline{\mathbb{R}}_+ \times \mathbb{R}^{n-1} \times \overline{\mathbb{R}}_+)$ with the quasi-homogeneity property (cf. (2.3.27))

$$(2.8.18) \qquad \tilde{k}(X,x_n,\xi',\mu) = |\xi'|^{d+1}\tilde{k}(X,x_n|\xi'|,\xi'/|\xi'|,\mu/|\xi'|) \quad \text{for } |\xi'| \geq 1 ,$$

belongs to $S^{d,\nu}(\Xi, \overline{\mathbb{R}}^n_+, \mathscr{S}(\overline{\mathbb{R}}_+))$ if and only if it satisfies the estimates on the set $|\xi'| \leq 1$:

$$(2.8.19) \quad \|D^\beta_X D^\alpha_{\xi'} D^j_\mu x^m_n D^{m'}_{x_n} \tilde{k}\|_{L^2_{x_n}(\mathbb{R}_+)} \leq c(X)(\langle\mu\rangle^{d+\frac12-\nu+[m-m']_- -j} + \langle\mu\rangle^{d+\frac12-m+m'-|\alpha|-j})$$

(for these estimates imply the required estimates for $|\xi'| \geq 1$ by extension by homogeneity). Here the first term in the right hand side of (2.8.19) dominates when $[k-k']_+ + |\alpha| \geq \nu$, and the second term dominates when $[k-k']_+ + |\alpha| \leq \nu$. Let N be the smallest integer $\geq \nu$. Take a fixed η' with $|\eta'| \leq 1$, and consider the Taylor polynomial in ξ' at η',

$$(2.8.20) \qquad \tilde{k}_N(X,x_n,\xi',\mu;\eta') = \sum_{|\alpha|<N} \frac{(\xi'-\eta')^\alpha}{\alpha!} \partial^\alpha_{\eta'}\tilde{k}(X,x_n,\eta',\mu) ,$$

it likewise satisfies the estimates (2.8.19) for $|\xi'| \leq 1$. Moreover, one finds by use of Taylor's formula (A.6) the estimates (of $L^2(\mathbb{R}_+)$-norms in x_n):

$$(2.8.21) \quad \|D^\beta_X D^\alpha_{\xi'} D^j_\mu x^m_n D^{m'}_{x_n}[\tilde{k}-\tilde{k}_N]\| \leq c(X)\langle\mu\rangle^{d+\frac12-\nu+[m-m']_- -j} , \quad \text{for } |\xi'| \leq 1 .$$

For the associated strictly homogeneous symbol k^h, the symbol-kernel $\tilde{k}^h$ is defined from $\tilde{k}$ as the right hand side of (2.8.18) for all $\xi' \neq 0$. For $\frac12 \leq |\xi'| \leq 1$ and $\mu \geq \frac12$, $\tilde{k}^h$ satisfies an analogue of (2.8.19), and the Taylor remainder satisfies

$$(2.8.22) \quad \|D^\beta_X D^\alpha_{\xi'} D^j_\mu x^m_n D^{m'}_{x_n}[\tilde{k}^h(X,x_n,\xi',\mu) - \tilde{k}^h_N(X,x_n,\xi',\mu;\eta')]\|$$
$$\leq c(X)\langle\mu\rangle^{d+\frac12-\nu+[m-m']_- -j} \quad \text{for } \tfrac12 \leq |\xi'| \leq 1 , \quad \mu \geq \tfrac12 ;$$

this is derived from (2.8.21) as in Lemma 2.1.9 1°.

Now, assume conversely that there is given an arbitrary, strictly homogeneous symbol-kernel $\tilde{k}^h \in S^{d,\nu}_{hom}(\Xi, \overline{\mathbb{R}}^n_+, \mathscr{S}(\overline{\mathbb{R}}_+)) = F^{-1}_{\xi_n \to x_n} S^{d,\nu}_{hom}(\Xi, \overline{\mathbb{R}}^n_+, H^+)$. It satisfies the estimates (2.8.19) for $\mu \geq \frac12$, when ξ' lies in the set

(2.8.23) $$U = \{\xi' \mid \tfrac{1}{2} \leq |\xi'| \leq 1\} \ .$$

We choose a point $\eta' \in U$ and consider the Taylor polynomial $\tilde{k}^h_N$ defined as in (2.8.20). It follows from Taylor's formula that (2.8.22) holds if the segment between ξ' and η' lies in U, and further applications show that (2.8.22) is satisfied for all $\xi' \in U$ that can be connected with η' by a smooth curve running in U. When $n \geq 3$, all $\xi' \in U$ are reached in this way, so (2.8.22) holds for all $\xi' \in U$. When $n=2$, U has two components $\{-1 \leq \xi' \leq -\frac{1}{2}\}$ and $\{\frac{1}{2} \leq \xi' \leq 1\}$, and the estimates (2.8.22) may well be violated if ξ' and η' have opposite signs. In fact, the estimates (2.8.22) then hold in general if and only if they hold for one case of ξ' and η' with opposite signs.

When (2.8.22) does hold for all $\xi' \in U$, it is easy to define a smooth symbol-kernel $\tilde{k}$ in $S^{d,\nu}$ coinciding with $\tilde{k}^h$ for $|\xi'| \geq 1$, namely

(2.8.24) $$\begin{aligned}\tilde{k}(X,x_n,\xi',\mu) &= \zeta(|\xi',\mu|)[\zeta(|\xi'|)\tilde{k}^h + (1-\zeta(|\xi'|))\tilde{k}^h_N] \\ &= \zeta(|\xi',\mu|)[\tilde{k}^h_N + \zeta(|\xi'|)(\tilde{k}^h - \tilde{k}^h_N)] \ .\end{aligned}$$

Clearly it equals $\tilde{k}^h$ for $|\xi'| \geq 1$, and for $|\xi'| \leq \frac{1}{2}$ it equals $\zeta(|\xi',\mu|)$ times the Taylor polynomial, so (2.8.19) is satisfied there. For $\frac{1}{2} \leq |\xi'| \leq 1$, the estimates (2.8.22) for $\tilde{k}^h - \tilde{k}^h_N$ imply similar estimates for $\zeta(|\xi',\mu|)\zeta(|\xi'|)(\tilde{k}^h - \tilde{k}^h_N)$ (one may observe that since $\tilde{k}^h - \tilde{k}^h_N$ is estimated for $\xi' \in U$ like a symbol of order $d-N$ and regularity $\nu-N \leq 0$, the multiplication by $\zeta(|\xi'|)$ does not interfere with the regularity). Then (2.8.19) holds also for $\xi' \in U$, so $\tilde{k}$ satisfies the desired estimates.

Altogether, a strictly homogeneous symbol $k^h \in S^{d,\nu}_{hom}(\Xi, \overline{\mathbb{R}}^n_+, H^+)$ stems from a smooth symbol $k \in S^{d,\nu}(\Xi, \overline{\mathbb{R}}^n_+, H^+)$ always when $n \geq 3$; and when $n=2$ it does so if and only if the values of k^h for $\xi' > 0$ and $\xi' < 0$ are compatible in the sense that (2.8.22) holds for some $\xi' > 0$ and $\eta' < 0$. There are similar results for the other symbol classes.

This analysis clarifies some of the manipulations with our symbol classes, and it is important for a deeper understanding of their structure and possible generalizations. In the present work, we do not actually use it (e.g., the proof of Theorem 3.2.3 below shows the existence of a smooth parametrix symbol using other arguments); but it is included here for the convenience of subsequent studies.

CHAPTER 3

PARAMETRIX AND RESOLVENT CONSTRUCTIONS

3.1 Ellipticity. Auxiliary elliptic operators.

Chapter 3 is devoted to the study of parametrices (inverses) of the operators in elliptic cases.

In Section 3.1, we introduce the general concept of parameter-ellipticity. The ps.d.o. part and the "singular" (boundary operator) part are taken of regularities $\nu_1 \geq 1$ resp. $\nu_2 > 0$, and the ellipticity is described in terms of bijectiveness of certain principal symbols and boundary symbol operators. We show the existence of suitable auxiliary elliptic systems that will permit a reduction to operators of the form $I+H$, where H is a singular Green operator, in an extended sense.

The parametrix construction in Section 3.2 is carried out first on the boundary symbol level (i.e., for operators on $\mathbb{R}_+$). Here, when $A = I+H$ with H small, it is not hard to show by consideration of the Neumann series that A^{-1} exists and is of the form $I+H'$ with H' belonging to the calculus. For general H, we use the Laguerre expansions to write H as the sum of a part H_M with finite rank and a small part $H_M^\dagger$. A couple of reductions by composition with operators such as $(I+H_M^\dagger)^{-1}$ carry the situation over to a finite dimensional case, where the ellipticity means invertibility of a matrix, and our precise hypotheses assure a certain uniformity in the parameters. (It would be easier to presuppose the needed uniform estimates explicitly, as in [Rempel-Schulze 4, Sections 4.1-2], but then the applicability remains to be investigated; cf. Remark 1.5.16.) In a sense, the reduction to finite dimension here corresponds to the "reduction to the boundary" used in differential operator problems. - For the full n-dimensional operators, we now construct a parametrix within our operator classes, by use of the rules of calculus. The parametrix generally has regularity ν_1 in the pseudo-differential part and regularity $\nu_3 = \min\{\nu_1 - \frac{1}{2}, \nu_2\}$ in the boundary operator terms.

Section 3.3 presents the consequences for the resolvent construction. Because of the special observations on the regularity in some cases of compositions where one factor is parameter-independent, the resulting regularity is here improved to $\nu_3' = \min\{\nu_1,\nu_2\}$. A number of consequences are drawn; in particular the kernel is studied, and we produce our main results concerning the singular Green operator part of the resolvent.

In Section 3.4, we treat some other special cases, of interest e.g. for evolution problems; also in these cases, the general loss of $\frac{1}{2}$ regularity is avoided.

Let us now consider the boundary symbolic calculus and the ellipticity concept. We recall from Chapter 2 that ellipticity of parameter-dependent pseudo-differential symbols was defined in Definition 2.1.2; when the regularity ν is positive, it is a simple bijectiveness property (Proposition 2.1.11 2^0). Ellipticity and nonnegative regularity assure a certain uniform invertibility of the corresponding operator family, cf. Theorem 2.1.16.

In particular, the principal symbol p^0 defines an operator $p^0(D_n) = OP_n(p^0)$ on $\mathbb{R}$ that is bijective in $\mathscr{S}(\mathbb{R})$. The operator $p^0(D_n)_\Omega$ obtained by restriction to $\overline{\mathbb{R}}_+$ according to the formula (1.2.5) is generally not bijective on the restricted space $\mathscr{S}(\overline{\mathbb{R}}_+)$. But when p has the transmission property, one can show that p^0_Ω is a Fredholm operator in $\mathscr{S}(\overline{\mathbb{R}}_+)$ and in Sobolev spaces over $\overline{\mathbb{R}}_+$ (see below). To eliminate the nullspace and range complement, one adjoins here the various "singular" operators (trace, Poisson and singular Green operators, ps.d.o.s in x') , and the theory is then concerned with invertibility of a full system A_μ as in (2.4.10), where the point is to analyze the hypotheses on symbols leading to good properties of the inverse. We shall here use several versions of boundary symbol operators on $\overline{\mathbb{R}}_+$ associated with a system of symbols $p(X,x_n,\xi,\mu)$, $t(X,\xi,\mu)$, $k(X,\xi,\mu)$ $g(X,\xi,\eta_n,\mu)$ and $s(X,\xi',\mu)$, namely:

- the (x_n-independent) principal boundary symbol operator:

$$(3.1.1)\quad a^0(X,\xi',\mu,D_n) = \begin{pmatrix} p^0(X,0,\xi',\mu,D_n)_\Omega + g^0(X,\xi',\mu,D_n) & k^0(X,\xi',\mu,D_n) \\ t^0(X,\xi',\mu,D_n) & s^0(X,\xi',\mu) \end{pmatrix} ;$$

- the strictly homogeneous boundary symbol operator

$$(3.1.2)\quad a^h(X,\xi',\mu,D_n) = \begin{pmatrix} p^h(X,0,\xi',\mu,D_n)_\Omega + g^h(X,\xi',\mu,D_n) & k^h(X,\xi',\mu,D_n) \\ t^h(X,\xi',\mu,D_n) & s^h(X,\xi',\mu) \end{pmatrix} ;$$

and along with these operators also some operators taking more of the symbol structu into account:

- the boundary symbol operator

$$(3.1.3)\quad a(X,x_n,\xi',\mu,D_n) = \begin{pmatrix} p(X,x_n,\xi',\mu,D_n)_\Omega + g(X,\xi',\mu,D_n) & k(X,\xi',\mu,D_n) \\ t(X,\xi',\mu,D_n) & s(X,\xi',\mu) \end{pmatrix}$$

- the x_n-dependent principal boundary symbol operator

$$(3.1.4)\quad a^0(X,x_n,\xi',\mu,D_n) = \begin{pmatrix} p^0(X,x_n,\xi',\mu,D_n)_\Omega + g^0(X,\xi',\mu,D_n) & k^0(X,\xi',\mu,D_n) \\ t^0(X,\xi',\mu,D_n) & s^0(X,\xi',\mu) \end{pmatrix}$$

In all the definitions, X,ξ' and μ are genuine parameters, whereas in (3.1.3) and (3.1.4) x_n just indicates that the operator is defined from an x_n-dependent symbol (a standard convention for ps.d.o.s). The terms can be matrix formed; for simplicity of notation we do not always recall this in the following.

Before we define the ellipticity hypotheses in relation to these systems, we shall study p_Ω .

The Fredholm property is shown in the following lemma, that is sufficiently general to cover also the possibly less regular symbols $p^h(X,0,0,\xi_n,\mu)$ obtained for $\xi' \to 0$ in the strictly homogeneous symbol, when the regularity is ≥ 1 (cf. Proposition 2.8.3 ff.).

3.1.1 Lemma. *Let* $\kappa > 0$, *and let (cf.* (2.2.52-54)*)*

$$(3.1.5)\qquad p(\xi_n) = \sum_{k\in\mathbb{Z}} a_k\,\hat{\psi}_k(\xi_n,\kappa)$$

be a function on $\mathbb{R}$ *such that*

$$(3.1.6)\qquad \|(a_k)_{k\in\mathbb{Z}}\|_{\ell_1^2} \equiv \Big(\sum_{k\in\mathbb{Z}} (1+|k|)^2|a_k|^2\Big)^{\frac{1}{2}} < \infty\ .$$

Assume furthermore that p *is invertible, and* $p(\xi_n)^{-1} \equiv q(\xi_n)$ *satisfies*

$$(3.1.7)\qquad q(\xi_n) = \sum_{k\in\mathbb{Z}} a'_k\,\hat{\psi}_k(\xi_n,\kappa)$$

with

$$\|(a'_k)_{k\in\mathbb{Z}}\|_{\ell^2_1} < \infty \quad . \tag{3.1.8}$$

Then $p(D_n)_\Omega$ *and* $q(D_n)_\Omega$ *are Fredholm operators in* $L^2(\mathbb{R}_+)$, *with*

$$\text{index } p(D_n)_\Omega = - \text{index } q(D_n)_\Omega \ . \tag{3.1.9}$$

Moreover $g_1(D_n) = I-p(D_n)_\Omega q(D_n)_\Omega$ *and* $g_2(D_n) = I-q(D_n)_\Omega p(D_n)_\Omega$ *are Hilbert-Schmidt operators in* $L^2(\mathbb{R}_+)$, *with coefficient matrices in* $\ell^2_{\frac{1}{2},\frac{1}{2}}$ *with respect to the Laguerre system* $(\varphi_k(x_n,\kappa))_{k\in\mathbb{N}}$; *in fact their respective Laguerre coefficient systems* $(c^{(i)}_{\ell m})_{\ell,m\in\mathbb{N}}$ *(for* $i=1,2$*) satisfy*

$$\|(c^{(i)}_{\ell m})_{\ell,m\in\mathbb{N}}\|_{\ell^2_{\frac{1}{2},\frac{1}{2}}} \le \|(a_k)_{k\in\mathbb{N}}\|_{\ell^2_1} \|(a'_k)_{k\in\mathbb{N}}\|_{\ell^2_1} \ . \tag{3.1.10}$$

Proof: The ℓ^2 summability of (a_k) assures that p is a function in the weighted L^2 space with weight $(1+\xi_n^2)^{-1}$, cf. (2.2.54) ff.; in fact p is a constant plus an $L^2(\mathbb{R})$ function. Now since $pq = 1$, it is seen just as in Proposition 2.6.12 that

$$g_1(D_n) \equiv I - p(D_n)_\Omega q(D_n)_\Omega = (pq)_\Omega - p_\Omega q_\Omega$$

$$= L(p,q) = g^+(p)g^-(q) \tag{3.1.11}$$

$$g_2(D_n) \equiv I - q(D_n)_\Omega p(D_n)_\Omega = L(q,p) = g^+(q)g^-(p)$$

where $g^+(p)$, $g^-(p)$, $g^+(q)$ and $g^-(q)$ are Hankel operators with respect to the Laguerre system $(\varphi_k(x_n,\kappa))_{k\in\mathbb{N}}$, with integral operator kernels

$$\tilde{g}^+(p)(x_n,y_n) = \sum_{\ell,m\in\mathbb{N}} a_{1+\ell+m}\varphi_\ell(x_n,\kappa)\varphi_m(y_n,\kappa) \quad ,$$
$$\tilde{g}^-(p)(x_n,y_n) = \sum_{\ell,m\in\mathbb{N}} a_{-1-\ell-m}\varphi_\ell(x_n,\kappa)\varphi_m(y_n,\kappa) \quad , \tag{3.1.12}$$

and the analogous formulas for $\tilde{g}^\pm(q)$. Here

$$\|(a_{1+\ell+m})_{\ell,m\in\mathbb{N}}\|_{\ell^2_{\frac{1}{2},0}} = \Big(\sum_{\ell,m\in\mathbb{N}} (1+\ell)|a_{1+\ell+m}|^2\Big)^{\frac{1}{2}} \tag{3.1.13}$$
$$\le \Big(\sum_{\ell,m\in\mathbb{N}} (1+\ell+m)|a_{1+\ell+m}|^2\Big)^{\frac{1}{2}} = \Big(\sum_{j\in\mathbb{N}} (1+j)^2|a_{1+j}|^2\Big)^{\frac{1}{2}} \le \|(a_k)_{k\in\mathbb{Z}}\|_{\ell^2_1} \ ,$$

and the $\ell^2_{0,\frac{1}{2}}$ norm is likewise bounded by $\|(a_k)_{k\in\mathbb{Z}}\|_{\ell^2_1}$. In particular, $g^+(p)$ is a Hilbert-Schmidt operator, and so are the other operators $g^-(p)$, $g^+(q)$ and $g^-(q)$, as well as the compositions. For the coefficient series $(c^{(1)}_{\ell m})_{\ell,m\in\mathbb{N}}$ and $(c^{(2)}_{\ell m})_{\ell,m\in\mathbb{N}}$ of g_1 and g_2 we then find (3.1.10) by the Cauchy-Schwarz inequality.

In particular, $L(p,q)$ and $L(q,p)$ are compact operators on $L^2(\mathbb{R}_+)$ so that

$$p_\Omega q_\Omega = I - L(p,q) \quad \text{and} \quad q_\Omega p_\Omega = I - L(q,p)$$

are Fredholm operators (i.e. have finite dimensional nullspace and closed range with finite dimensional complement). Standard arguments in Fredholm theory then show that p_Ω and q_Ω are likewise Fredholm operators in $L^2(\mathbb{R}_+)$, and that the index of p_Ω (the dimension of the nullspace minus the dimension of the range complement) is the opposite of the index of q_Ω . □

When $p \in S^{0,\nu}_{tr}(\Xi\times\mathbb{R}, \overline{\mathbb{R}}^{n+1}_+)$ with $\nu \geq 0$, and is elliptic with a parametrix symbol q according to Theorem 2.1.16, $q \in S^{0,\nu}_{tr}(\Xi\times\mathbb{R}, \overline{\mathbb{R}}^{n+1}_+)$, then the lemma applies to $p^0(X,0,\xi,\mu)$ and its inverse $q^0(X,0,\xi,\mu)$ (for each $|\xi',\mu| \geq c > 0$) ; note that here

$$L(p^0,q^0) \quad \text{and} \quad L(q^0,p^0) \in S^{-1,\nu-\frac{1}{2}}(\Xi, \overline{\mathbb{R}}^n_+, H^+ \hat{\otimes} H^-_{-1})$$

by the general results of Section 2.6; and the $\ell^2_{\frac{1}{2},\frac{1}{2}}$ norm of their coefficient series will be uniformly bounded if $\nu \geq 1$. At each fixed (X,ξ',μ) , the nullspace $Z(I-L(p^0,q^0))$ and range complement $Z(I-L(p^0,q^0)^*)$ in $L^2(\mathbb{R}_+)$ are finite dimensional subspaces of $\mathcal{S}(\overline{\mathbb{R}}_+)$ (since s.g.o.s on $\overline{\mathbb{R}}_+$ range in $\mathcal{S}(\overline{\mathbb{R}}_+)$) , so also for p^0_Ω and q^0_Ω the nullspaces and range complements are finite dimensional subspaces of $\mathcal{S}(\overline{\mathbb{R}}_+)$, cf. (3.1.11). Moreover, since p^0_Ω and q^0_Ω are continuous operators in $\mathcal{S}(\overline{\mathbb{R}}_+)$, these subspaces actually coincide with the nullspaces resp. range complements for the operators considered as acting in $\mathcal{S}(\overline{\mathbb{R}}_+)$. (Since $\mathcal{S}(\overline{\mathbb{R}}_+)$ is merely a Fréchet space, there is not a simple Fredholm theory for operators in this space, and the Fredholm properties we speak of will always be inferred from some Sobolev space situation.)

In view of the continuity properties of our operators, they are also Fredholm in $H^s(\overline{\mathbb{R}}_+)$ for any $s \geq 0$, with the same nullspace and range complement as above.

The statements carry over to operators of order $\neq 0$ by use of certain auxiliary ps.d.o.s. Let $\zeta(t)$ be a C^∞ function on $\mathbb{R}$, vanishing on $[-\frac{1}{2},\frac{1}{2}]$ and equal to 1 for $t \geq 1$ and $t \leq -1$. Let $\varepsilon > 0$, and define the symbols, for $m \in \mathbb{Z}$,

$$\lambda_\pm^m(\xi',\xi_n,\mu) = \left[|\xi',\mu|\zeta\left(\frac{|\xi',\mu|}{\varepsilon|\xi,\mu|}\right) - i\xi_n\right]^m \zeta(|\xi,\mu|) \,. \tag{3.1.14}$$

These functions are pseudo-differential symbols on $\mathbb{R}^{n+1}$; in particular they are ps.d.o symbols on $\mathbb{R}^n$ depending on the parameter μ; and they have the transmission property. An exercise in the calculus shows:

3.1.2 Lemma. $\lambda_\pm^m$ *is elliptic of order* m *and regularity* $+\infty$, *with parametrix symbol* $\lambda_\pm^{-m}$. *For any* $m \in \mathbb{Z}$, ε *can be chosen so small that for all* $|\xi',\mu| \geq 1$, $\lambda_-^m(\xi',\mu,D_n)_\Omega$ *is a homeomorphism of* $\mathscr{S}(\overline{\mathbb{R}}_+)$ *onto itself, and from* $H^s(\overline{\mathbb{R}}_+)$ *to* $H^{s-m}(\overline{\mathbb{R}}_+)$ *for all* $s \geq 0$ *so that* $s-m \geq 0$.

Proof: It is seen from the symbol that λ_-^m is elliptic of order m and regularity $+\infty$, with parametrix λ_-^{-m}. Let $|\xi',\mu| \geq 1$. Since $(|\xi',\mu| - i\xi_n)^{m'} \in H^-$ for any m',

$$L((|\xi',\mu| - i\xi_n)^{m'}, (|\xi',\mu| - i\xi_n)^{-m'}) = 0 \quad \text{for all } m' \in \mathbb{Z}\,, \tag{3.1.15}$$

so $OP_n((|\xi',\mu| - i\xi_n)^m)_\Omega$ is a bijection of $\mathscr{S}(\overline{\mathbb{R}}_+)$ into itself, with inverse $OP_n((|\xi',\mu| - i\xi_n)^{-m})_\Omega$. By the continuity of the operators (cf. Section 2.5), $OP_n((|\xi',\mu| - i\xi_n)^m)_\Omega$ defines a homeomorphism of $H^s(\overline{\mathbb{R}}_+)$ onto $H^{s-m}(\overline{\mathbb{R}}_+)$ for those $s \geq 0$ for which $s-m \geq 0$. Moreover,

$$L((|\xi',\mu| - i\xi_n)^{-m}, \lambda_-^m) = 0\,, \tag{3.1.16}$$

since $(|\xi',\mu| - i\xi_n)^{-m} \in H^-$.

Now an elementary calculation shows that

$$(|\xi',\mu| - i\xi_n)^{-m}\, \lambda_-^m(\xi,\mu) = 1 + p_\varepsilon(\xi,\mu)\,,$$

where $p_\varepsilon \to 0$ uniformly in (ξ,μ) for $\varepsilon \to 0$. Then in view of (3.1.16),

$$(3.1.17)\qquad OP_n((|\xi',\mu| - i\xi_n)^{-m})_\Omega\, OP_n(\lambda_-^m)_\Omega = I + OP_n(p_\varepsilon)_\Omega \quad ,$$

where the norm of $OP_n(p_\varepsilon)$ in $L^2(\mathbb{R})$, equal to $\sup_{\xi_n} |p_\varepsilon(\xi,\mu)|$, goes to 0 for $\varepsilon \to 0$, uniformly in (ξ',μ). For sufficiently small ε, $I + OP_n(p_\varepsilon)_\Omega$ is then a bijection in $L^2(\mathbb{R}_+)$ and in $\mathscr{S}(\overline{\mathbb{R}}_+)$, so $OP_n(\lambda_-^m)$ is a bijection in $\mathscr{S}(\overline{\mathbb{R}}_+)$, and from $H^s(\overline{\mathbb{R}}_+)$ to $H^{s-m}(\overline{\mathbb{R}}_+)$, as asserted. □

In the following, λ_-^m is used with ε taken so small that the above conclusions hold.

Since $(|\xi',\mu| - i\xi_n)^m$ is not a ps.d.o. symbol in ξ (or in (ξ,μ)), we have to use λ_-^m in our compositions, to stay within genuine ps.d.o.s. λ_+^m does not define an isomorphism in the same way; on the contrary one can show that for ε sufficiently small, the system

$$(3.1.18)\qquad \begin{pmatrix}\lambda_+^1\\ \gamma_0\end{pmatrix} : \mathscr{S}(\overline{\mathbb{R}}_+) \to \begin{matrix}\mathscr{S}(\overline{\mathbb{R}}_+)\\ \times\\ \mathbb{C}\end{matrix}$$

defines an isomorphism (λ_+^m for larger $m > 0$ then demands more boundary conditions, and λ_+^m for negative m demands supplementary Poisson operators). Since this useful fact will not be needed in the present theory, we leave the details to the reader (an analysis was also given in the prepublication [Grubb 11], and related statements appear in [Boutet 3], [Rempel-Schulze 1]), see also Remark 3.1.8.

The operators λ_-^m can be used to reduce the treatement of ps.d.o.s of arbitrary order to the zero order case. When for instance p is of order $-d$ $(d > 0)$, we have that

$$(3.1.19)\qquad \lambda_-^d(D_n)_\Omega\, p^0(D_n)_\Omega = OP_n(\lambda_-^d p^0)_\Omega \quad .$$

The symbol $\lambda_-^d p^0$ is of order 0 (and the same regularity as p), and the preceding discussion applies to this; then in view of the bijectiveness of $\lambda_{-,\Omega}^d$ it is seen altogether that p_Ω^0 is a Fredholm operator from $L^2(\mathbb{R}_+)$ to $H^d(\overline{\mathbb{R}}_+)$, with nullspace and range complement in $\mathscr{S}(\overline{\mathbb{R}}_+)$, etc. The dependence on (ξ',μ) can also be analyzed.

When p^0 is elliptic of order $d > 0$, the preceding discussion applies to $q^0 = (p^0)^{-1}$. Moreover, we have for each (ξ',μ)

$$(3.1.20)\qquad p_\Omega^0\, q_\Omega^0 = I - g^+(p^0)g^-(q^0) \ ,$$

where $g^+(p^0)g^-(q^0)$ depends only on $h_{-1}p^0$ and q^0, and is a singular Green operator of class 0, hence compact in $L^2(\mathbb{R}_+)$. Thus $p^0_\Omega q^0_\Omega$ is a Fredholm operator in $L^2(\mathbb{R}_+)$, and since q^0_Ω is a Fredholm operator from $L^2(\mathbb{R}_+)$ to $H^d(\overline{\mathbb{R}}_+)$, it follows that p^0_Ω is a Fredholm operator from $H^d(\overline{\mathbb{R}}_+)$ to $L^2(\mathbb{R}_+)$. Again the nullspace and range complement is in $\mathscr{S}(\overline{\mathbb{R}}_+)$ (since this holds for q^0_Ω and for $I - g^+(p^0)g^-(q^0)$). By Lemma 3.1.1, the $\ell^2_{\frac12,\frac12}$ norm of the coefficient matrix for $g^+(p^0)g^-(q^0)$ in the Laguerre system is bounded in (ξ',μ) (locally in X), when p is of regularity $\nu \geq 1$.

Let us now consider a strictly homogeneous symbol p^h associated with a given elliptic symbol $p \in S^{d,\nu}_{tr}$ at $x_n = 0$. For $\xi' \neq 0$, $p^h(X,\xi',\xi_n,\mu)$ equals a usual symbol, so p^h_Ω defines a Fredholm operator as described above. Assume that $\nu \geq 1$, let $\mu > 0$, and consider the limit $p^h(X,0,\xi_n,\mu)$ of $p^h(X,\xi',\xi_n,\mu)$ for $\xi' \to 0$, in the sense of the first symbol norms as described in Proposition 2.8.3; where we write

$$(3.1.21)\qquad p^h(X,0,\xi_n,\mu) = \sum_{0\leq j\leq d} s^h_{d-j}(X,0,\mu)\xi_n^j + \sum_{k\in\mathbb{Z}} b^h_k(X,0,\mu)\hat{\varphi}_k(\xi_n,\mu)$$
$$= \sum_{1\leq j\leq d} s^h_{d-j}(X,0,\mu)\xi_n^j + \sum_{k\in\mathbb{Z}} a^h_k(X,0,\mu)\hat{\psi}_k(\xi_n,\mu)\quad,$$

(cf. (2.2.54)ff.). Since p is of regularity $\nu \geq 1$, the ℓ^2_1-norm of the series a^h_k is bounded (locally in X) for $\xi' \to 0$, by Theorem 2.2.8. Similar statements hold for the strictly homogeneous symbol q^h associated with $q^0 = (p^0)^{-1}$ (defined as such for $|\xi',\mu| \geq 1$).

If $d = 0$, we can now apply Lemma 3.1.1 directly to $p^h(D_n)_\Omega$ and $q^h(D_n)_\Omega$ at $\xi' = 0$, obtaining that they are Fredholm operators in $L^2(\mathbb{R}_+)$ with opposite indices, and $I - p^h_\Omega q^h_\Omega$ and $I - q^h_\Omega p^h_\Omega$ are Hilbert-Schmidt operators with coefficient matrices in $\ell^2_{\frac12,\frac12}$. In fact, the $\ell^2_{\frac12,\frac12}$ norm of the coefficient matrices is bounded for (ξ',μ) in compact subsets of $\overline{\mathbb{R}}^n_+\setminus 0$, so by homogeneity, they are <u>bounded on</u> $\overline{\mathbb{R}}^n_+\setminus 0$ (locally in $X \in \Xi$).

If $d < 0$, we get the Fredholm property of $p^h(X,0,\mu,D_n)_\Omega$ as an operator from $L^2(\mathbb{R}_+)$ to $H^d(\overline{\mathbb{R}}_+)$ by composing it to the left with $\lambda_-^{-d}(0,\mu,D_n)_\Omega$. If $d > 0$, we get the Fredholm property of $p^h(X,0,\mu,D_n)_\Omega$ as an operator from $H^d(\overline{\mathbb{R}}_+)$ to $L^2(\mathbb{R}_+)$ by using an identity of the form (3.1.20). In this latter case, which is the most interesting for our purposes, the argument following (3.1.20) applies to show that, also here, $g^+(p^h)g^-(q^h)$ is a Hilbert-Schmidt operator whose $\ell^2_{\frac12,\frac12}$-norm depends locally boundedly on $(\xi',\mu) \in \overline{\mathbb{R}}^n_+\setminus 0$, and since the homogeneity degree is zero, it is actually <u>bounded on</u> $\overline{\mathbb{R}}^n_+\setminus 0$ (locally in X).

Since p^0 and p^h are continuous in (X,ξ',μ) in the first symbol norms, the index of p^h_Ω, as well as that of p^0_Ω, is constant for (X,ξ',μ) in connected sets. In particular, $\text{index}\, p^h_\Omega = \text{index}\, p^0_\Omega$ for all (X,ξ',μ), since they are equal for $|\xi'| \geq 1$.

The considerations extend of course to N×N-matrixformed ps.d.o. symbols.

We shall now introduce the ellipticity concept for parameter-dependent boundary value problems. Since p^0_Ω is a Fredholm operator in $\mathcal{S}(\overline{\mathbb{R}}_+)^N$, it can be supplied with finite rank operators and spaces (trace and Poisson operators), such that it becomes bijective. Now the full operator $(P_\mu)_\Omega$ in $C^\infty(E)$ is generally not Fredholm and it must be supplied with infinite dimensional conditions and spaces to become bijective (or to have a parametrix). The boundary operators T_μ, K_μ, G_μ and S_μ (as in (2.4.10)) serve this purpose, and the main point of defining an ellipticity condition is to give criteria on the principal symbols assuring that the full operator A_μ has a good parametrix. Here, the criteria should be as "algebraic" as possible: It is easy to require the existence of an inverse of the boundary symbol operator lying in the right symbol classes, but this means to presuppose convenient estimates of the terms in the inverse symbol, whereby one really just leaves the difficulties in the parametrix construction to the applications. It is preferable to formulate the ellipticity definition in terms of simple bijectiveness properties, and to prove on the general level that these properties imply the existence of an inverse symbol belonging to the calculus (satisfying the estimates).

For the interior symbol $p(X,\xi,\mu)$, we saw in Proposition 2.1.11 2° how Definition 2.1.2 (containing a simple estimate) could be made purely algebraic when the regularity number ν was > 0. When boundary conditions are included, the desired estimates, now for operators on $\overline{\mathbb{R}}_+$ rather than matrices, are much less simple. But also here we shall show that when the regularity is positive, it suffices to assume bijectiveness of certain boundary symbol operators, for then it is possible to deduce the estimates needed for the parametrix construction. (This is a non-trivial point already in the standard Boutet de Monvel calculus, but it is more complicated here.)

In the following definition, we restrict the attention to cases with positive regularity, since this allows the simplest formulation, but we also take up the discussion of more general cases along the way. Actually, we take the regularity ν_1 of the interior symbol ≥ 1, since ν_1 is always integer in the applications to resolvent constructions.

3.1.3 Definition. *Let* Ξ *be open* $\subset \mathbb{R}^{n'}$, *let* $d \in \mathbb{Z}$, *let* $r \in \mathbb{N}$ *with* $r \leq d_+$ *and let* $\nu_1 \geq 1$ *and* $\nu_2 > 0$. *Let* $N \in \mathbb{N}_+$ *and* M *and* $M' \in \mathbb{N}$. *Let*

$$\begin{aligned}
&\text{(i)} \quad p(X,x_n,\xi,\mu) \in S_{tr}^{d,\nu_1}(\Xi\times\mathbb{R}, \overline{\mathbb{R}}_+^{n+1}) \otimes L(\mathbb{C}^N,\mathbb{C}^N),\\
&\text{(ii)} \quad g(X,\xi,\eta_n,\mu) \in S^{d-1,\nu_2}(\Xi, \overline{\mathbb{R}}_+^n, H^+ \hat{\otimes} H_{r-1}^-) \otimes L(\mathbb{C}^N,\mathbb{C}^N),\\
(3.1.22)\quad &\text{(iii)} \quad t(X,\xi,\mu) \in S^{d,\nu_2}(\Xi, \overline{\mathbb{R}}_+^n, H_{r-1}^-) \otimes L(\mathbb{C}^N,\mathbb{C}^{M'}),\\
&\text{(iv)} \quad k(X,\xi,\mu) \in S^{d-1,\nu_2}(\Xi, \overline{\mathbb{R}}_+^n, H^+) \otimes L(\mathbb{C}^M,\mathbb{C}^N),\\
&\text{(v)} \quad s(X,\xi',\mu) \in S^{d,\nu_2}(\Xi, \overline{\mathbb{R}}_+^n) \otimes L(\mathbb{C}^M,\mathbb{C}^{M'}) .
\end{aligned}$$

The associated boundary symbol operators (3.1.1)-(3.1.4) *and the associated Green operator*

$$(3.1.23)\qquad A_\mu = \begin{pmatrix} P_{\mu,\Omega} + G_\mu & K_\mu \\ T_\mu & S_\mu \end{pmatrix} : \begin{matrix} C_{(0)}^\infty(\Omega'\times\overline{\mathbb{R}}_+)^N \\ \times \\ C_0^\infty(\Omega')^M \end{matrix} \rightarrow \begin{matrix} C^\infty(\Omega'\times\overline{\mathbb{R}}_+)^N \\ \times \\ C^\infty(\Omega')^{M'} \end{matrix}$$

(in case $\Xi = \Omega'$ *with* Ω' *open* $\subset \mathbb{R}^{n-1}$*), are said to be* <u>*elliptic*</u> *(of order* d, *class* r *and regularities* ν_1, ν_2*), when the following conditions* (I)-(III) *hold:*

(I) *The strictly homogeneous interior symbol* $p^h(X,x_n,\xi,\mu)$ *is invertible for all* $(X,x_n,\xi,\mu) \in \Xi\times\mathbb{R}\times(\overline{\mathbb{R}}_+^{n+1}\setminus 0)$ *(i.e.,* p *is elliptic on* $\Xi\times\mathbb{R}\times\overline{\mathbb{R}}_+^{n+1}$ *according to Definition 2.1.2 and Proposition 2.1.11* 2°*).*

(II) *The (*x_n*-independent) principal boundary symbol operator* $a^0(X,\xi',\mu,D_n)$ *(cf.* (3.1.1)*) is bijective from* $\mathcal{S}(\overline{\mathbb{R}}_+)^N\times\mathbb{C}^M$ *to* $\mathcal{S}(\overline{\mathbb{R}}_+)^N\times\mathbb{C}^{M'}$ *for all* (X,ξ',μ) *with* $|\xi'| = 1$ *and* $\mu \geq 0$.

(III) *The limit for* $\xi' \to 0$ *(in the first symbol norms) of the associated strictly homogeneous boundary symbol operator* $a^h(X,0,\mu,D_n)$ *(cf.* (3.1.2), (A.4')*) is bijective from* $H^{d_+}(\overline{\mathbb{R}}_+)^N\times\mathbb{C}^M$ *to* $H^{d_-}(\overline{\mathbb{R}}_+)^N\times\mathbb{C}^{M'}$ *for all* $\mu > 0$, *all* $X \in \Xi$.

When U *is a subset of* $\Xi\times\overline{\mathbb{R}}_+^n$, *the set of symbols is said to be* <u>*elliptic on*</u> U *when* (I)-(III) *hold for* $(X,\xi',\mu) \in U$ *(with* (x_n,ξ_n) *running in* $\mathbb{R}^2$ *for* (I)*).*

The existence of the limit of a^h for $\xi' \to 0$ is assured for $\nu_2 > 0$, by Proposition 2.8.3. Note that Definition 3.1.3 is consistent with Definition 1.5.5

for systems $\{P_\Omega+G+\omega\mu^d,T\}$, where P, G and T are necessarily of positive regularity, cf. Lemma 1.5.6, Definition 1.5.14 and Proposition 2.3.14.

(II) and (III) can be taken together in the formulation:

(II+III) *The strictly homogeneous boundary symbol operator*

$$a^h(X,\xi',\mu,D_n) : H^{d_+}(\overline{\mathbb{R}}_+)^N\times\mathbb{C}^M \to H^{d_-}(\overline{\mathbb{R}}_+)^N\times\mathbb{C}^{M'} \tag{3.1.24}$$

is bijective for all $(X,\xi',\mu)\in\Xi\times(\overline{\mathbb{R}}^n_+\setminus 0)$.

We here use that $a^0 = a^h$ for $|\xi'|\geq 1$, and that the bijectiveness of a^h extends to all $\xi'\neq 0$ by the homogeneity in (ξ',μ) ; moreover, the bijectiveness of a^0 going from $\mathscr{S}(\overline{\mathbb{R}}_+)^N\times\mathbb{C}^M$ to $\mathscr{S}(\overline{\mathbb{R}}_+)^N\times\mathbb{C}^{M'}$ is equivalent with bijectiveness from $H^{d_+}(\overline{\mathbb{R}}_+)^N\times\mathbb{C}^M$ to $H^{d_-}(\overline{\mathbb{R}}_+)^N\times\mathbb{C}^{M'}$, in view of the observations on the nullspace and cokernel of p^0 and q^0 prior to Definition 3.1.3 (and the mapping properties $t^0: H^{d_+}(\overline{\mathbb{R}}_+)^N\to\mathbb{C}^{M'}$, $g^0: H^{d_+}(\overline{\mathbb{R}}_+)^N\to\mathscr{S}(\overline{\mathbb{R}}_+)^N$, $k^0:\mathbb{C}^M\to\mathscr{S}(\overline{\mathbb{R}}_+)^N$, $s^0:\mathbb{C}^M\to\mathbb{C}^{M'}$) .

Also the considerations from Section 1.5 concerning the stronger hypothesis (II') hold in general:

<u>3.1.4 Proposition</u>. *Let* p, g, t, k *and* s *be as in* (3.1.22), *with* $\nu_1\geq 1$ *and* $\nu_2>0$. *Assume that* (I) *holds, and define condition* (II') *as follows:*

(II') *The* (x_n-*independent) principal boundary symbol operator* $a^0(X,\xi',\mu,D_n)$ *is bijective from* $\mathscr{S}(\overline{\mathbb{R}}_+)^N\times\mathbb{C}^M$ *to* $\mathscr{S}(\overline{\mathbb{R}}_+)^N\times\mathbb{C}^{M'}$ *for all* $|\xi',\mu|\geq c_0(X)$, *where* $c_0(X)$ *is a continuous non-negative function on* Ξ .

Then (II)-(III) *hold if and only if* (II')-(III) *hold.*

The proof is contained in the proof of our main result Theorem 3.2.3 later. (We note that Proposition 3.1.4 holds for <u>any</u> choice of the smooth symbol a^0 , coinciding with a^h for $|\xi'|\geq 1$ and lying in the mentioned symbol spaces.) On the other hand, (II') alone does not imply (III), cf. Example 1.5.13.

Concerning the relation to definitions in works of other authors, see Section 1.5.

In preparation for the parametrix construction, we shall discuss some auxiliary elliptic operators (that can also serve as examples, besides the examples studied in Section 1.5, where the parameter-dependence is rather special).

One kind of auxiliary operator is the operator in Lemma 3.1.2, which is used to reduce to the case where the system is of order and class 0 .

When $d \leq 0$, one can replace $a(X,x_n,\xi',\mu,D_n)$ by

$$(3.1.25) \qquad a'(X,x_n,\xi',\mu,D_n) = \begin{pmatrix} \lambda_-^{-d}(\xi',\mu,D_n)_\Omega & 0 \\ 0 & \kappa(\xi',\mu)^{-d} \end{pmatrix} a(X,x_n,\xi',\mu,D_n)$$

cf. (3.1.14) and (2.1.1); here λ_-^{-d} and κ^{-d} are of order $-d$ and regularity $+\infty$. The various composition rules in Section 2.7 show that the new system

$$a' = \begin{pmatrix} (\lambda_-^{-d}p)_\Omega - L(\lambda_-^{-d},p) + \lambda_{-,\Omega}^{-d}g & \lambda_{-,\Omega}^{-d}k \\ \kappa^{-d}t & \kappa^{-d}s \end{pmatrix}$$

is elliptic of order and class 0 (recall that a is of class 0) and regularities ν_1,ν_2 ; for the only possible loss of regularity would stem from the term $L(\lambda_-^{-d},p)$, where λ_-^{-d} in fact gives a contribution $g^+(\lambda_-^{-d})$ whose symbol is supported in $|\xi',\mu| \leq 1$ (since $\lambda_-^{-d} \in H^-$ for $|\xi',\mu| \geq 1$), so that $g^+(\lambda_-^{-d})g^-(p)$ is negligible of regularity $+\infty$. The analysis can now be carried out for a' , implying the desired results for a by composition to the left with

$$(3.1.26) \qquad \begin{pmatrix} \lambda_{-,\Omega}^{d} & 0 \\ 0 & \kappa^{d} \end{pmatrix} .$$

In the further analysis of a' , one uses another auxiliary operator like the one studied below in Theorem 3.1.5.

Now let $d > 0$ (the case of special interest to us). Here we shall compose a to the right with an auxiliary elliptic operator b having the ps.d.o. part $q = p^{-1}$ (principally). The existence of a convenient operator b (microlocally) is shown in the following theorem.

<u>3.1.5 Theorem</u>. *Let* Σ *be open* $\subset \mathbb{R}^{n'}$, *let* $d \in \mathbb{N}$ *and let* $\nu \geq 1$. *Let* $q(X,\xi,\mu) \in S_{tr}^{-d,\nu}(\Sigma\times\mathbb{R}, \overline{\mathbb{R}}_+^{n+1}) \otimes L(\mathbb{C}^N,\mathbb{C}^N)$, *and assume that it is elliptic , and homogeneous in* (ξ,μ) *of degree* $-d$ *for* $|\xi| \geq 1$. *Let* $\Xi \subset \Sigma$ *be compact and connected. There exists a finite covering of* $\Xi\times\overline{\mathbb{R}}_+^n$ *by* r_1+r_2 *relatively open sets* U_r $(r=1,\dots,r_1+r_2)$, *where* r_1 *sets are conical:*

(3.1.27) $U_r = \{(X,t\xi',t\mu) \mid X \in \Xi_r, (\xi',\mu) \in W_r \cap \overline{\mathbb{R}}^n_+, t > c_r \geq 0\}$ *for* $1 \leq r \leq r_1$

(with Ξ_r *open* $\subset \Sigma$ *and* W_r *open* $\subset S^{n-1}$*), and* r_2 *sets* $U_{r_1+1},\dots,U_{r_1+r_2}$ *are bounded, such that on each* U_r *there is an elliptic boundary symbol Green operator*

$$(3.1.28)\quad b_r(X,\xi',\mu,D_n) = \begin{pmatrix} q_\Omega & k_r \\ t_r & s_r \end{pmatrix} : \begin{matrix} \mathscr{S}(\overline{\mathbb{R}}_+)^N \\ \times \\ \mathbb{C}^{M_1} \end{matrix} \to \begin{matrix} \mathscr{S}(\overline{\mathbb{R}}_+)^N \\ \times \\ \mathbb{C}^{M_2} \end{matrix}$$

(defined for $(X,\xi',\mu) \in U_r$*), with* k_r*,* t_r *and* s_r *of order* $-d$*, class 0 and regularity* $+\infty$*. Here* M_2 *can be any integer larger than a certain* M_0 *depending on* $\overline{\Xi}$*; and* $M_2-M_1 = \text{index}\; q(X,\xi',\mu,D_n)_\Omega$*, which is constant on* $\overline{\Xi}\times\overline{\mathbb{R}}^n_+$*. On the sets* U_r *with* $r \leq r_1$*, also* b^h_r *is bijective. The* k_r *and* t_r *can be taken to be finite linear combinations of Laguerre functions.*

Proof: The proof will be formulated for the case $N = 1$, since matrix-formed q merely give notational complications. We have already accounted for the Fredholm property of q_Ω as well as q^h_Ω, considered as operators from $L^2(\mathbb{R}_+)$ to $H^d(\overline{\mathbb{R}}_+)$. The family of operators $q^h_\Omega : L^2(\mathbb{R}_+) \to H^d(\overline{\mathbb{R}}_+)$ depends continuously on (X,ξ',μ) in the operator norm, by Proposition 2.1.11 and the general estimate, valid for $OP_n(a(\xi_n))$ with $|a(\xi_n)| \leq c\langle\xi_n\rangle^{-d}$,

$$(3.1.29)\quad \|a(D_n)_\Omega u\|_{H^d(\overline{\mathbb{R}}_+)} \leq \|a(D_n)e^+u\|_{H^d(\mathbb{R})} = \|F^{-1}[a(\xi_n)\widehat{e^+u}(\xi_n)]\|_d$$
$$= (2\pi)^{-\frac{1}{2}}\|\langle\xi_n\rangle^d a(\xi_n)\widehat{e^+u}(\xi_n)\|_0 \leq \sup_{\xi_n} |\langle\xi_n\rangle^d a(\xi_n)|\, \|u\|_{L^2(\mathbb{R}_+)} .$$

In particular, index q^h_Ω is constant in $(X,\xi',\mu) \in \overline{\Xi}\times(\overline{\mathbb{R}}^n_+\setminus 0)$. Let $p^h = (q^h)^{-1}$. We now use that

$$(3.1.30)\quad p^h_\Omega q^h_\Omega = I - g(X,\xi',\mu,D_n) ,$$

where g is a Hilbert-Schmidt operator in $L^2(\mathbb{R}_+)$ with a kernel

$$(3.1.31)\quad \tilde{g}(X,x_n,y_n,\xi',\mu) = \sum_{\ell,m} c_{\ell m}(X,\xi',\mu)\varphi_\ell(x_n,\kappa^h)\varphi_m(y_n,\kappa^h)$$

satisfying

$$(3.1.32)\quad \sum_{\ell,m\in\mathbb{N}} (1+\ell)(1+m)|c_{\ell m}(X,\xi',\mu)|^2 \leq c(X) \quad \text{for} \quad (\xi',\mu) \in \overline{\mathbb{R}}^n_+\setminus 0 ,$$

in view of Lemma 3.1.1. (The hypothesis $\nu \geq 1$ is used here.) Since $\overline{\Xi}$ is compact, there exists for each $\sigma < 1$ an M_0 such that for $M \geq M_0$, the decomposition

$$(3.1.33)\qquad \begin{gathered}\tilde g = \tilde g_M + \tilde g_M^\dagger \quad , \quad \text{with}\\ \tilde g_M = \sum_{0\le \ell, m<M} c_{\ell m}\varphi_\ell\varphi_m \quad \text{and} \quad \tilde g_M^\dagger = \sum_{\ell \text{ or } m\ge M} c_{\ell m}\varphi_\ell\varphi_m \quad ,\end{gathered}$$

satisfies

$$(3.1.34)\qquad \Big(\sum_{\ell \text{ or } m\ge M} |c_{\ell m}(X,\xi',\mu)|^2\Big)^{\frac12} \le \sigma \ , \quad \text{for all} \quad (X,\xi',\mu)\in \Xi\times(\overline{\mathbb{R}}^n_+\setminus 0)\ .$$

This is the Hilbert-Schmidt norm of $g_M^\dagger(D_n)$, and then the operator $I - g_M^\dagger(D_n)$ on $L^2(\mathbb{R}_+)$ is invertible, since the Neumann series converges:

$$(3.1.35)\qquad \begin{gathered}(I - g_M^\dagger)^{-1} = I - g' \ , \quad \text{with} \quad g' = \sum_{k=1}^{\infty} (g_M^\dagger)^{\circ k} \quad ,\\ \|(g_M^\dagger)^{\circ k}\|_{\text{Hilbert-Schmidt}} \le \sigma^{-k} \quad \text{for} \quad k\ge 1 \quad ;\end{gathered}$$

here $(g_M^\dagger)^{\circ k}$ is the composition of k factors $g_M^\dagger$, and we use (2.6.53) and the Cauchy-Schwarz inequality. It follows that

$$(3.1.36)\qquad (I - g'(D_n))p_\Omega^h(D_n)q_\Omega^h(D_n) = I - g''(D_n) \quad , \quad g'' = (I-g')g_M\ .$$

Since g'' has a kernel of the form $\tilde g'' = \Sigma_{\ell\in\mathbb{N},m<M}\, c''_{\ell m}\varphi_\ell\varphi_m$, it can be regarded as the composition $k\circ t$ of the (row vector) Poisson operator k with symbol

$$(3.1.37)\qquad k = \{k_0,\dots,k_{M-1}\} \quad , \quad k_m(X,\xi,\mu) = \sum_{0\le\ell<\infty} c''_{\ell m}(X,\xi',\mu)\hat\varphi_\ell(\xi_n,\kappa^h)\ ,$$

and the (column vector) trace operator with symbol

$$(3.1.38)\qquad t = \begin{pmatrix}\bar{\hat\varphi}_0(\xi_n,\kappa^h)\\ \vdots \\ \bar{\hat\varphi}_{M-1}(\xi_n,\kappa^h)\end{pmatrix} \quad ,$$

so (3.1.36) implies the formula

$$(3.1.39)\qquad (p_\Omega^h + g''' \quad k)\begin{pmatrix} q_\Omega^h \\ \\ t\end{pmatrix} = I \qquad \text{in} \quad L^2(\mathbb{R}_+) \quad ,$$

where $g'''(D_n) = -g'(D_n)p_\Omega^h(D_n)$.

The surjective operator $(p_\Omega^h + g''' \quad k)$ from $H^d(\overline{\mathbb{R}}_+)\times\mathbb{C}^M$ to $L^2(\mathbb{R}_+)$ is a Fredholm operator, since p_Ω^h is a Fredholm operator from $H^d(\overline{\mathbb{R}}_+)$ to $L^2(\mathbb{R}_+)$, g' is compact in $L^2(\mathbb{R}_+)$, and k has finite rank. Then the kernel $Z(X,\xi',\mu)$ of $(p_\Omega^h + g''' \quad k)$ has constant dimension M' when (X,ξ',μ) varies, and

$$M' = M + \text{index}\ p_\Omega\ .$$

Now it follows from the identity (3.1.39) that $H^d(\overline{\mathbb{R}}_+)\times\mathbb{C}^M$ is the direct sum of Z and the range of $\begin{pmatrix} q^h_\Omega \\ t \end{pmatrix}$,

$$H^d(\overline{\mathbb{R}}_+)\times\mathbb{C}^M = Z \dotplus R\left(\begin{pmatrix} q^h_\Omega \\ t \end{pmatrix}\right), \quad \text{for each } (X,\xi',\mu) .$$

Then if $\begin{pmatrix} q^h_\Omega \\ t \end{pmatrix}$ is supplemented with the injection i_Z of Z into $H^d(\overline{\mathbb{R}}_+)\times\mathbb{C}^M$, we get a bijective operator

$$(3.1.40) \qquad \left(\begin{pmatrix} q^h_\Omega \\ t \end{pmatrix} \quad i_Z\right) : \begin{matrix} L^2(\mathbb{R}_+) \\ \times \\ Z \end{matrix} \rightarrow \begin{matrix} H^d(\overline{\mathbb{R}}_+) \\ \times \\ \mathbb{C}^M \end{matrix} .$$

We now have to bring this into the realm of Green operators.

Consider a fixed $(X_0,\xi_0',\mu_0)\in\overline{\Xi}\times(\overline{\mathbb{R}}^n_+\cap S^{n-1})$. Let $\{u_1,\dots,u_{M'}\}$ be a basis of $Z(X_0,\xi_0',\mu_0)$. Since the linear hull of the Laguerre functions $\varphi_k(x_n,\kappa(\xi_0',\mu_0))$ is dense in $H^d(\overline{\mathbb{R}}_+)$ (for it is dense in $\mathscr{S}(\overline{\mathbb{R}}_+)$ which is dense in $H^d(\overline{\mathbb{R}}_+)$), there is a linearly independent set of vectors $\{u_1',\dots,u_{M'}'\}$ lying in the linear hull of the Laguerre functions and spanning a space Z', so that (3.1.40) with Z replaced by Z' is likewise a bijection. Here $i_{Z'}$ can be expressed as a column matrix

$$\begin{pmatrix} k' \\ s' \end{pmatrix} : \mathbb{C}^{M'} \rightarrow \begin{matrix} H^d(\overline{\mathbb{R}}_+) \\ \times \\ \mathbb{C}^M \end{matrix}$$

where k' is a Poisson operator (with symbol in the linear hull of Fourier transformed Laguerre functions) and s' is a multiplication operator (a matrix). t is already on a convenient form. So we have altogether a bijection

$$(3.1.41) \qquad \begin{pmatrix} q^h_\Omega & k' \\ t & s' \end{pmatrix} : \begin{matrix} L^2(\mathbb{R}_+) \\ \times \\ \mathbb{C}^{M'} \end{matrix} \rightarrow \begin{matrix} H^d(\overline{\mathbb{R}}_+) \\ \times \\ \mathbb{C}^M \end{matrix}$$

at (X_0,ξ_0',μ_0). Note that

$$(3.1.42) \qquad M - M' = -\operatorname{index} p^h_\Omega = \operatorname{index} q^h_\Omega .$$

Now we <u>fix</u> k', t and s', but allow q^h_Ω to vary with $(X,\xi',\mu)\in\Sigma\times S^{n-1}$; then since q^h_Ω depends continuously on (X,ξ',μ) in the operator norm (as shown above), there is a neighborhood of (X_0,ξ_0',μ_0) in $\Sigma\times S^{n-1}$, where (3.1.41) is still bijective. If we extend the symbols by homogenity and use the compactness of $\overline{\Xi}\times(\overline{\mathbb{R}}^n_+\cap S^{n-1})$, we obtain a finite cover of $\overline{\Xi}\times(\overline{\mathbb{R}}^n_+\setminus 0)$ by conic sets U_r'

where there exist invertible operators b_r^h of the form (3.1.41). Note that the symbols t and k' are simple finite sums of Laguerre functions depending only on $|\xi',\mu|$, for each U_r'. In particular, they are of regularity $+\infty$.

Finally, we shall show how to replace q_Ω^h by q_Ω itself. This simply uses Lemma 2.1.9 2^0, where the estimate (2.1.20) implies (since $\nu > 0$) that the norm of $q_\Omega^h - q_\Omega$ as an operator from $L^2(\mathbb{R}_+)$ to $H^d(\overline{\mathbb{R}}_+)$ goes to zero for $|\xi',\mu| \to \infty$ in view of (3.1.29), even on the set where $|\xi'| \leq 1$. Then also

$$\begin{pmatrix} q_\Omega & k' \\ t & s' \end{pmatrix} : \begin{matrix} L^2(\mathbb{R}_+) \\ \times \\ \mathbb{C}^{M'} \end{matrix} \to \begin{matrix} H^d(\overline{\mathbb{R}}_+) \\ \times \\ \mathbb{C}^{M} \end{matrix}$$

is bijective in each U_r, for $|\xi',\mu|$ sufficiently large. So we have to remove some compact neighborhoods of 0 from the U_r, but these are covered by finitely many sets where the existence of auxiliary operators is assured by a simpler version of the above arguments. This shows the proposition, with $M_1 = M'$ and $M_2 = M$. □

3.1.6 Remark. The proof shows that the family of kernels $Z(X,\xi',\mu)$ of $(p_\Omega^h + g''' \quad k)$ is a C^0 vector bundle of dimension M'; in fact

$$(3.1.43) \qquad I - \begin{pmatrix} q_\Omega^h \\ t \end{pmatrix} (p_\Omega^h + g''' \quad k)(X,\xi',\mu)$$

provides an isomorphism of $Z(X_0,\xi_0',\mu_0)$ onto $Z(X,\xi',\mu)$, when (X_0,ξ_0',μ_0) and (X,ξ',μ) are close together. Moreover, with $\nu' = \nu - \frac{1}{2}$, $Z(X,\xi',\mu)$ is a $C^{\nu'-}$ vector bundle, since the operators depend $C^{\nu'-}$ on (X,ξ',μ) (cf. Proposition 2.8.3). One could here continue the line of thought in [Boutet 3, p. 43 ff.] defining an index bundle (concerned there with a C^∞ vector bundle), which could be used in an index theory for the parameter-dependent operators.

3.1.7 Remark. Σ can in particular be the set of (x',θ) with x' in an open set $\Omega' \in \mathbb{R}^{n-1}$ and θ in an interval $]-a,a[\subset \mathbb{R}$; then for each compact connected subset $\overline{\Xi}' \subset \Omega'$, we can cover $\overline{\Xi}' \times \{0\} \times \overline{\mathbb{R}}_+^n$ in such a way that for some $\delta > 0$, the full interval $[-\delta,\delta]$ is included in each U_r. In the proof of Theorem 3.1.5, the auxiliary operators k_r, t_r and s_r are taken to be constant in θ, but they could also be chosen to depend in some specific way on θ. In particular, we shall later study the case where q depends analytically on $z = e^{i\theta}\mu^d$ for $\theta \in]-a,a[$, $\mu \geq 0$ and $|\xi',\mu| \geq c_0$. Then we use that for $\theta = 0$, k_r, t_r and s_r are algebraic functions of κ^h, where

(3.1.44) $\kappa^h = |\xi',\mu| = (|\xi'|^2 + \mu^2)^{\frac{1}{2}} = (|\xi'|^2 + z^{2/d})^{\frac{1}{2}}$

(since they are finite linear combinations of the Laguerre functions $\hat{\varphi}_k(\xi_n,\kappa^h)$ and $\bar{\hat{\varphi}}_k(\xi_n,\kappa^h)$, that are algebraic in κ^h cf. (2.2.10)). When this shall be extended from $\theta = 0$ to θ in an interval around 0, it is preferable to extend $|\xi',\mu|$ as the analytic extension of $(|\xi'|^2 + z^{2/d})^{\frac{1}{2}}$ for z near $\mathbb{R}_+$, instead of being constant in θ. Then one can obtain auxiliary operators b_r in (3.1.28) that are defined, and analytic in z, on conical regions $|\operatorname{Im} z| \le \varepsilon_r \operatorname{Re} z$.

A still better construction is where we replace $|\xi',\mu|$ by

(3.1.45) $\sigma(\xi',\mu,\theta) = (|\xi'|^{2d} + z^2)^{1/2d}$ with $z = e^{i\theta}\mu^d$,

for $|\xi',\mu| \ge 1$, extended to a nonzero C^∞ function for $|\xi',\mu| \le 1$; here there is no singularity at $z = 0$ when $|\xi'| \ge 1$. Then b_r can be obtained to be <u>analytic in</u> z <u>on a keyhole region around</u> $\overline{\mathbb{R}}_+$

(3.1.46) $V_{\delta,\varepsilon,\xi'} = \{z \in \mathbb{C} \mid |z| \le \delta|\xi'|^d$ <u>or</u> $|\operatorname{Im} z| \le \varepsilon \operatorname{Re} z\}$,

when $|\xi'|$ is suitably large. (Of course we need not make reference to polar coordinates in the neighborhood of zero.) For smaller $|\xi'|$ there is <u>analyticity in a sector truncated at zero</u>

(3.1.47) $W_{r,\varepsilon} = \{z \in \mathbb{C} \mid |z| \ge r$ <u>and</u> $|\operatorname{Im} z| \le \varepsilon \operatorname{Re} z\}$.

<u>3.1.8 Remark</u>. In connection with the introduction of $\lambda_\pm^m$ in (3.1.14) we observe that the corresponding non-parametrized symbols (regarded as in Definition 1.5.3)

(3.1.48) $\underline{\lambda}_\pm^m(\xi',\xi_n) = [\,|\xi'|\,\zeta\!\left(\frac{|\xi'|}{\varepsilon|\xi|}\right) \pm i\xi_n]^m \zeta(|\xi|)$

furnish examples where one has parameter-ellipticity but not parabolicity. For instance, $\underline{\lambda}_\pm^1(\xi',\xi_n)$ take their values in $\{z \in \mathbb{C} \mid \operatorname{Re} z \ge 0\}$, so for the rays $z = re^{i\theta}$ with argument $\theta \in \,]\pi/2, 3\pi/2[$, one has that $\underline{\lambda}_\pm^1(\xi',\xi_n) - z$ is parameter-elliptic. But this does not extend to the arguments $\theta = \pi/2$ and $3\pi/2$. Parabolicity would also be contradicted by the fact that $\underline{\lambda}_-^1(\xi',D_n)_\Omega + 1$ has index 0, as in Lemma 3.1.2, and $\underline{\lambda}_+^1(\xi',D_n)_\Omega + 1$ has index 1, as in (3.1.18); here $\underline{\lambda}_+^1 + 1$ is obtained from $\underline{\lambda}_-^1 + 1$ by sending ξ_n into $-\xi_n$ (compare also Remark 1.5.10).

3.2 The parametrix construction.

The crucial step in the parametrix construction takes place for boundary symbol operators with ps.d.o. symbol independent of x_n (sometimes called the "model" situation, where coefficients are "frozen" at a boundary cotangent point). We start with a particularly simple case; and we use the notation

$$V_J = \operatorname{span}\{\varphi_j \mid j \in J\} \tag{3.2.1}$$

for an interval J of $\mathbb{R}$. The result is concerned with operators of the form $I + g(X,\xi',\mu,D_n)$, where g is a slightly generalized singular Green operator, in the sense that it is a Hilbert-Schmidt operator on $L^2(\mathbb{R})$ instead of $L^2(\mathbb{R}_+)$, with symbol in $H_{-1} \hat{\otimes} H_{-1}$ and range in $e^+\mathscr{S}(\overline{\mathbb{R}}_+) \dotplus e^-\mathscr{S}(\overline{\mathbb{R}}_-)$. The Laguerre series expansions are then over the full system $(\varphi_j)_{j\in\mathbb{Z}}$ instead of $(\varphi_j)_{j\in\mathbb{N}}$. (See also the remarks preceding Definition 2.3.7.) When g has the expansion

$$g(X,\xi,\eta_n,\mu) = \sum_{\ell,m\in\mathbb{Z}} c_{\ell m}(X,\xi',\mu)\hat{\varphi}_\ell(\xi_n,\kappa)\overline{\hat{\varphi}}_m(\eta_n,\kappa) ,$$

we denote

$$\|g\|_{\ell^2_{N,N'}} \equiv \|(c_{\ell m})_{\ell,m\in\mathbb{Z}}\|_{\ell^2_{N,N'}} = \Big(\sum_{\ell,m\in\mathbb{Z}} |(1+|\ell|)^N(1+|m|)^{N'}c_{\ell m}|^2\Big)^{\frac12} ; \tag{3.2.2}$$

observe here that $\|g\|_{\ell^2_{0,0}}$ is simply the Hilbert-Schmidt norm,

$$\|g\|_{\ell^2_{0,0}} = \|\tilde{g}(X,x_n,y_n,\xi',\mu)\|_{L^2_{x_n,y_n}} = \frac{1}{2\pi}\|g(X,\xi',\xi_n,\eta_n,\mu)\|_{L^2_{\xi_n,\eta_n}} . \tag{3.2.3}$$

The operators $g(D_n)$ are called extended singular Green operators.

3.2.1 Proposition. *Let* $\nu \geq 0$, *and let* $g \in S^{-1,\nu}(\Xi, \overline{\mathbb{R}}^n_+, H_{-1} \hat{\otimes} H_{-1})$, *homogeneous of degree* -1 *for* $|\xi'| \geq 1$ *and having the Laguerre series expansion*

$$g(X,\xi',\xi_n,\eta_n,\mu) = \sum_{\ell,m\in\mathbb{Z}} c_{\ell m}(X,\xi',\mu)\hat{\varphi}_\ell(\xi_n,\kappa)\overline{\hat{\varphi}}_m(\eta_n,\kappa) ; \tag{3.2.4}$$

and denote for any $M \in \mathbb{N}$

$$g_M = \sum_{|\ell|,|m|<M} c_{\ell m}\hat{\varphi}_\ell\overline{\hat{\varphi}}_m \quad \textit{and} \quad g_M^\dagger = \sum_{|\ell| \text{ or } |m|\geq M} c_{\ell m}\hat{\varphi}_\ell\overline{\hat{\varphi}}_m \tag{3.2.5}$$

(so $g = g_M + g_M^\dagger$*) . If there is an* M *such that*

$$\|g_M^\dagger\|_{\ell^2_{0,0}} \leq \sigma < 1 \quad \textit{for all} \quad (X,\xi',\mu) , \tag{3.2.6}$$

then

$$(3.2.7)\qquad I - g_M^{\dagger}(X,\xi',\mu,D_n) : L^2(\mathbb{R}) \to L^2(\mathbb{R})$$

is invertible with inverse $(I - g_M^{\dagger})^{-1} = I + g'$ *, where* g' *is again an extended s.g.o. with symbol* g' *lying in* $S^{-1,\nu}(\Xi, \overline{\mathbb{R}}^n_+, H_{-1} \hat{\otimes} H_{-1})$ *; it is homogeneous of degree* -1 *in* (ξ, η_n, μ) *for* $|\xi'| \geq 1$ *and satisfies*

$$(3.2.8)\qquad \|g'\|_{\ell^2_{0,0}} \leq \sigma/(1-\sigma) \ , \ \textit{for all } (X,\xi',\mu) \ .$$

Moreover, for N *and* $N' \in \overline{\mathbb{R}}_+$ *,*

$$(3.2.9)\qquad \|g'\|_{\ell^2_{N,N'}} \leq \|g_M^{\dagger}\|_{\ell^2_{N,N'}} + \|g_M^{\dagger}\|_{\ell^2_{N,0}} \|g_M^{\dagger}\|_{\ell^2_{0,N'}} (1-\sigma)^{-1} \ .$$

Furthermore, when (3.2.6) *holds,*

$$(3.2.10)\qquad \begin{aligned} (I - g)(I - g_M^{\dagger})^{-1} &= I + g'' \ , \\ (I - g_M^{\dagger})^{-1}(I - g) &= I + g''' \ , \end{aligned}$$

where g'' *and* g''' *lie in* $S^{-1,\nu}(\Xi, \overline{\mathbb{R}}^n_+, H_{-1} \hat{\otimes} H_{-1})$ *, and have the properties:*

$$(3.2.11)\qquad \begin{aligned} &(i)\ \ g''(X,\xi',\mu,D_n) \ \textit{ranges in } V_{]-M,M[} \ ; \\ &(ii)\ \ g'''(X,\xi',\mu,D_n) \ \textit{vanishes on } L^2(\mathbb{R}) \ominus V_{]-M,M[} \ . \end{aligned}$$

The results extend to $S_{1,0}$ *symbol spaces, and to square matrix formed operators. Analogous results hold in the framework of strictly homogeneous symbol spaces.*

Proof: g' is obtained by a Neumann series construction

$$(3.2.12)\qquad g' = \sum_{k=1}^{\infty} (g_M^{\dagger})^{\circ k} \ ,$$

just as in (3.1.35); and since the Hilbert-Schmidt norm of $(g_M^{\dagger})^{\circ k}$ is $\leq \sigma^{-k}$, the Hilbert-Schmidt (or $\ell^2_{0,0}$) norm of g' is $\leq \Sigma_{k\geq 1}\sigma^{-k} = \sigma/(1-\sigma)$. The proof of (3.2.9) is just a refinement of this argument: By the Cauchy-Schwarz inequality, one has for $k \geq 2$,

$$\begin{aligned} \|(g_M^{\dagger})^{\circ k}\|_{\ell^2_{N,N'}} &= \|g_M^{\dagger} \circ g_M^{\dagger} \circ \dots \circ g_M^{\dagger} \circ g_M^{\dagger}\|_{\ell^2_{N,N'}} \\ &\leq \|g_M^{\dagger}\|_{\ell^2_{N,0}} \|g_M^{\dagger}\|_{\ell^2_{0,0}} \dots \|g_M^{\dagger}\|_{\ell^2_{0,0}} \|g_M^{\dagger}\|_{\ell^2_{0,N'}} \\ &\leq \sigma^{-k+2} \|g_M^{\dagger}\|_{\ell^2_{N,0}} \|g_M^{\dagger}\|_{\ell^2_{0,N'}} \ , \end{aligned}$$

and hence

$$\|g'\|_{\ell^2_{N,N'}} = \|\sum_{k=1}^{\infty} (g_M^{\dagger})^{\circ k}\|_{\ell^2_{N,N'}}$$

$$\leq \|g_M^{\dagger}\|_{\ell^2_{N,N'}} + \sum_{k=2}^{\infty} \sigma^{2-k} \|g_M^{\dagger}\|_{\ell^2_{N,0}} \|g_M^{\dagger}\|_{\ell^2_{0,N'}}$$

$$= \|g_M^{\dagger}\|_{\ell^2_{N,N'}} + \|g_M^{\dagger}\|_{\ell^2_{N,0}} \|g_M^{\dagger}\|_{\ell^2_{0,N'}} (1-\sigma)^{-1} .$$

To show that g' has a symbol in the claimed symbol space, we consider the integral operator kernels. $f_k = (g_M^{\dagger})^{\circ k}$ has the kernel

$$\tilde{f}_k(X,x_n,y_n,\xi',\mu) = \int_{\mathbb{R}} \cdots \int_{\mathbb{R}} \int_{\mathbb{R}} \tilde{g}_M^{\dagger}(x_n,w_1)\tilde{g}_M^{\dagger}(w_1,w_2)\cdots\tilde{g}_M^{\dagger}(w_{k-1},y_n) \, dw_1 dw_2 \ldots dw_{k-1} \tag{3.2.13}$$

(the $\tilde{g}_M^{\dagger}$ depending furthermore on (X,ξ',μ)), which is estimated by use of the Cauchy-Schwarz inequality. An application of $x_n^m D_{x_n}^{m'}$ to $\tilde{f}_k$ only affects the first factor in the integral, and application of $y_n^m D_{y_n}^{m'}$ affects only the last factor. When k is large, an application of derivatives $D_X^{\beta} D_{\xi'}^{\alpha} D_{\mu}^{j}$ affects, by the Leibniz formula, only $|\alpha|+|\beta|+j$ of the factors in each term. The remaining factors contribute to the estimate with the factor σ. So, for instance, one has for $k \geq 2$

$$\|x_n^m D_{x_n}^{m'} y_n^{\ell} D_{y_n}^{\ell'} \tilde{f}_k\|_{L^2_{x_n,y_n}} \tag{3.2.14}$$

$$\leq \|x_n^m D_{x_n}^{m'} \tilde{g}_M^{\dagger}\| \, \|(\tilde{g}_M^{\dagger})^{\circ(k-2)}\| \, \|y_n^{\ell} D_{y_n}^{\ell'} \tilde{g}_M^{\dagger}\|$$

$$\leq c(X)\sigma^{k-2}(\rho^{\nu-[m-m']_+} + 1)(\rho^{\nu-[\ell-\ell']_+} + 1)\kappa^{-m+m'-\ell+\ell'}$$

$$\leq c'(X)\sigma^{k-2}(\rho^{\nu-[m-m']_+ - [\ell-\ell']_+} + 1)\kappa^{-m+m'-\ell+\ell'} ,$$

since $\nu \geq 0$; here $c'(X)$ is independent of k. By the summation (3.2.12) we get the desired estimates of the symbol-kernel and symbol of g'. g' is clearly homogeneous of degree -1 when g is so.

The last formulas are obvious: Since $g = g_M + g_M^{\dagger}$,

$$(I-g)(I-g_M^{\dagger})^{-1} = I - g_M(I-g_M^{\dagger})^{-1} = I + g'' ,$$

where g'' ranges in $V_{]-M,M[}$ by the definition of g_M. Moreover,

$$(I-g_M^{\dagger})^{-1}(I-g) = I - (I-g_M^{\dagger})^{-1}g_M = I + g''' \quad ,$$

where g''' vanishes on the space spanned by the φ_j with $|j| \geq M$, since g_M does so. The methods generalize to $S_{1,0}$ and S_{hom} symbol spaces. □

If $g(X,\xi,\eta_n,\mu)$ depends analytically on $z = e^{i\theta}\mu^d$ (in a similar way as in Remark 3.1.7), g' will likewise be analytic in z in the same region, since the Leibniz formula gives:

$$\bar{\partial}_z(I-g)^{-1} = -(I-g)^{-1} \circ \bar{\partial}_z(I-g) \circ (I-g)^{-1} = 0 \; ,$$

when $\bar{\partial}_z g = 0$.

For the case where g is polyhomogeneous with g^0 satisfying (3.2.6), the result is valid in a parametrix sense (as shown more generally in Section 3.3).

3.2.2 Remark. The property (3.2.6) can be obtained in various ways.

First, note that if g is strictly homogeneous, and we furthermore have that g depends continuously on $X \in \Xi$ and $(\xi',\mu) \in \bar{\mathbb{R}}^n_+ \setminus 0$ in the Hilbert-Schmidt norm, then since the expansion coefficients $c_{\ell m}(X,\xi',\mu)$ are likewise continuous in (X,ξ',μ), $g_M^{\dagger}$ is continuous in Hilbert-Schmidt norm, for each fixed M. At each point $(X_0,\xi_0',\mu_0) \in \Xi\times(S^{n-1} \cap \bar{\mathbb{R}}^n_+)$, one can choose M so large that $\|g_M^{\dagger}\| < \sigma$ at that point; then by the homogenity and continuity, the inequality extends to a conical neighborhood (as in (3.1.27) with $c_r = 0$). For $\bar{\Xi}_1$ compact $\subset \Xi$, the set $\bar{\Xi}_1 \times (\bar{\mathbb{R}}^n_+ \setminus 0)$ is covered by a finite system of such neighborhoods, so there is an M that works on the whole set.

Next, note that if $\nu > 0$ and g is strictly homogeneous, it does indeed have the above continuity property, by Proposition 2.8.3.

Now, if $\nu > 0$ and we are considering a smooth g (homogeneous for $|\xi'| \geq 1$ and lying in $S^{-1,\nu}(\Xi, \bar{\mathbb{R}}^n_+, H_{-1} \hat{\otimes} H_{r-1})$), then its deviation from the associated strictly homogeneous symbol satisfies, by Proposition 2.8.2,

$$\|g_M^{\dagger} - g_M^{h\dagger}\|_{\ell^2_{0,0}} \leq \|g - g^h\|_{\ell^2_{0,0}} \leq c(X)\langle\xi',\mu\rangle^{-\nu} \text{ for } |\xi',\mu| \geq 1 \; ,$$

for any $M \in \mathbb{N}$. Then, if we take M so large that $\|g_M^{h\dagger}\| < \sigma/2$ on $\bar{\Xi}_1 \times (\bar{\mathbb{R}}^n_+ \setminus 0)$ according to the preceding description, we can afterwards take R so large that (cf. (A.3))

$$\|g_M^{\dagger}\|_{\ell^2_{0,0}} \leq \|g_M^{h\dagger}\|_{\ell^2_{0,0}} + \|g - g^h\|_{\ell^2_{0,0}} \leq \sigma \text{ for } (X,\xi',\mu) \in \bar{\Xi}_1 \times (\bar{\mathbb{R}}^n_+ \setminus B_R(0)) \; .$$

Here the argument of Proposition 3.2.1 works on the set $\Xi_1 \times (\overline{\mathbb{R}}^n_+ \setminus B_R(0))$, or it works on $\Xi_1 \times \overline{\mathbb{R}}^n_+$ for a modification of g obtained by truncation in $B_{R+1}(0)$.

Finally, another method to obtain (3.2.6) for $g \in S^{-1,\nu}(\Xi, \overline{\mathbb{R}}^n_+, H_{-1} \hat{\otimes} H_{r-1})$ can be used when there is a $\delta > 0$ so that the $\ell^2_{\delta,\delta}$ norm of g is bounded, for then

$$(3.2.15) \quad \|g_M^\dagger\|^2_{\ell^2_{0,0}} = \sum_{|\ell| \text{ or } |m| \geq M} |c_{\ell m}|^2 \leq (1+M)^{-2\delta} \cdot \sum_{|\ell| \text{ or } |m| \geq M} (1+|\ell|)^{2\delta}(1+|m|)^{2\delta}|c_{\ell m}|^2$$
$$\leq (1+M)^{-2\delta} \|(c_{\ell m})_{\ell,m\in\mathbb{Z}}\|^2_{\ell^2_{\delta,\delta}} \to 0 \quad \text{for } M \to \infty .$$

on the full set $\Xi_1 \times \overline{\mathbb{R}}^n_+$. The boundedness of the $\ell^2_{\delta,\delta}$ norm holds when $\nu \geq 1$, in view of (2.3.32) (for $(N,N') = (1,0)$ and $(0,1)$) and the inequality

$$(3.2.15') \qquad (1+|\ell|)^{\frac12}(1+|m|)^{\frac12} \leq (1+|\ell|) + (1+|m|) .$$

If $g = g^+(p_1) \circ g^-(p_2)$ for some p_1 and p_2 of regularity 1 and orders $d \geq 0$ resp. $-d$, then we can only be certain that g has regularity $\frac12$; but here one gets a uniform bound on the $\ell^2_{\frac12,\frac12}$ norm anyway, by use of Corollary 2.6.14. (Lemma 3.1.1 treats a special case.)

We shall now prove the fundamental result, that the inverse of an elliptic boundary problem belongs to the calculus (in the boundary symbol framework, to begin with), having specific regularities deduced from the given regularities. In this proof, we shall use systematically a point of view, where Poisson and trace operators (and matrices over the boundary) are imbedded as special cases of singular Green operators, by identification of their finite dimensional domain or range spaces with subspaces of $L^2(\mathbb{R}_-)$. (The s.g.o. part acts originally in $L^2(\mathbb{R}_+)$, so there is ample space to take from $L^2(\mathbb{R}_-)$, to imbed the whole system into one that acts in $L^2(\mathbb{R})$!). The idea stems from a conversation with L. Boutet de Monvel on the non-parametrized calculus.

Consider a system

$$(3.2.16) \qquad h = \begin{pmatrix} g & k \\ t & s \end{pmatrix} : \begin{matrix} \mathcal{S}(\overline{\mathbb{R}}_+)^N \\ \times \\ \mathbb{C}^M \end{matrix} \to \begin{matrix} \mathcal{S}(\overline{\mathbb{R}}_+)^N \\ \times \\ \mathbb{C}^{M'} \end{matrix}$$

where g, t, k and s are as in (3.1.22), of order and class 0 ; it also maps $L^2(\mathbb{R}_+)^N \times \mathbb{C}^M$ into $L^2(\mathbb{R}_+)^N \times \mathbb{C}^{M'}$.

For simplicity, assume first that $N = 1$. Then we identify $\mathbb{C}^M$ with a subspace of $L^2(\mathbb{R}_-)$, using the notation (3.2.1) ,

(3.2.17) $$\mathbb{C}^M \simeq V_{[-M,0[} \subset L^2(\mathbb{R}_-)$$

(i.e., $\mathbb{C}^M$ is identified with the span of the Laguerre functions $(\varphi_j)_{-M\le j<0}$; and we identify similarly

$$\mathbb{C}^{M'} \simeq V_{[-M',0[} \subset L^2(\mathbb{R}_-) \quad ,$$

note that they are also subspaces of $\mathscr{S}(\overline{\mathbb{R}}_-)$. When $L^2(\mathbb{R}_+)$ and $L^2(\mathbb{R}_-)$ are regarded as subspaces of $L^2(\mathbb{R})$ in the natural way, the Poisson operator

$$k(X,\xi',\mu,D_n) = \{k_1,\dots,k_M\} : \mathbb{C}^M \to L^2(\mathbb{R}_+)$$

identifies with the (extended) s.g.o. $g(k)$ in $L^2(\mathbb{R})$ with kernel

(3.2.18) $$\begin{aligned}\tilde{g}(k)(X,x_n,y_n,\xi',\mu) &= \sum_{-M\le m<0} \tilde{k}_{-m}(X,x_n,\xi',\mu)\varphi_m(y_n,\kappa) \quad . \\ &= \sum_{\substack{\ell\in\mathbb{N}\\ -M\le m<0}} c_{\ell m}(X,\xi',\mu)\varphi_\ell(x_n,\kappa)\varphi_m(y_n,\kappa) \quad .\end{aligned}$$

Similarly, the trace operator

$$t(X,\xi',\mu,D_n) = \begin{pmatrix} t_1 \\ \cdot \\ \cdot \\ \cdot \\ t_{M'} \end{pmatrix} : L^2(\mathbb{R}_+) \to \mathbb{C}^{M'}$$

identifies with the (extended) s.g.o. $g(t)$ in $L^2(\mathbb{R})$ with kernel

(3.2.19) $$\begin{aligned}\tilde{g}(t)(X,x_n,y_n,\xi',\mu) &= \sum_{-M'\le\ell<0} \varphi_\ell(x_n,\kappa)\tilde{t}_{-\ell}(X,y_n,\xi',\mu) \quad , \\ &= \sum_{\substack{-M'\le\ell<0\\ m\in\mathbb{N}}} c_{\ell m}\varphi_\ell(x_n,\kappa)\varphi_m(y_n,\kappa) \quad ;\end{aligned}$$

and the matrix

$$s(X,\xi',\mu) = (s_{\ell m})_{\substack{1\le\ell\le M'\\ 1\le m\le M}} : \mathbb{C}^M \to \mathbb{C}^{M'}$$

identifies with the (extended) s.g.o. $g(s)$ in $L^2(\mathbb{R})$ with kernel

(3.2.20) $$\tilde{g}(s)(X,x_n,y_n,\xi',\mu) = \sum_{\substack{-M'\le\ell<0\\ -M\le m<0}} s_{-\ell,-m}(X,\xi',\mu)\varphi_\ell(x_n,\kappa)\varphi_m(y_n,\kappa) \quad .$$

Altogether, we find, taking the Laguerre expansions of the k_m and t_ℓ into account, that $h(X,\xi',\mu,D_n) : L^2(\mathbb{R}_+)\times\mathbb{C}^M \to L^2(\mathbb{R}_+)\times\mathbb{C}^{M'}$ (cf.(3.2.16)), identifies with the (extended) s.g.o. $g(h)(X,\xi',\mu,D_n)$ in $L^2(\mathbb{R})$, whose kernel has a Laguerre series expansion

$$(3.2.21)\qquad \tilde{g}(h)(X,x_n,y_n,\xi',\mu) = \sum_{\substack{-M'<\ell<\infty\\ -M\leq m<\infty}} c_{\ell m}(X,\xi',\mu)\varphi_\ell(x_n,\kappa)\varphi_m(y_n,\kappa)$$

where the $c_{\ell m}$ are the coefficients of g when ℓ and $m \geq 0$, and are determined from (3.2.18)-(3.2.20) when ℓ or $m < 0$.

If the original s.g.o. g is N×N-matrix formed, with $N > 1$, we identify $\mathbb{C}^M$ and $\mathbb{C}^{M'}$ with subspaces of the first factor in $L^2(\mathbb{R}_-)^N$,

$$(3.2.21')\qquad \mathbb{C}^M \simeq V_{[-M,0[} \times\{0\} \times \dots \times \{0\} \subset L^2(\mathbb{R}_-)^N ,$$

and we get an identification of h with an extended singular Green operator (3.2.21) where the $c_{\ell m}$ are N×N-matrixformed. This gives only notational complications, so details can be filled in whenever necessary.

It is clear from the construction, that (cf. (3.1.22))

$$(3.2.22)\qquad g(h) \in S^{-1,\nu_2}(\Xi, \overline{\mathbb{R}}^n_+, H_{-1} \hat{\otimes} H_{-1}) \otimes \mathcal{L}(\mathbb{C}^N,\mathbb{C}^N) .$$

Note that $g(h)(D_n)$ ranges in $e^+\mathscr{S}(\overline{\mathbb{R}}_+)^N \dotplus e^-\mathscr{S}(\overline{\mathbb{R}}_-)^N$.

<u>3.2.3 Theorem.</u> *Let* ν_1 *be integer* ≥ 1, *and let* $\nu_2 \in \mathbb{R}_+$. *Let* $a(X,\xi',\mu,D_n)$ *satisfy the hypotheses of Definition 3.1.3, with* p *independent of* x_n, *and with* a *equal to its own principal part.*

1^0 *The inverse* $(a^h)^{-1}$ *of the strictly homogeneous boundary symbol operator* a^h *(considered as in* (3.1.24)*) is again a strictly homogeneous boundary symbol operator*

$$(3.2.23)\qquad (a^h)^{-1} = \begin{pmatrix} q^h_\Omega + g'^h & k'^h \\ t'^h & s'^h \end{pmatrix} : \begin{matrix} H^{d_-}(\overline{\mathbb{R}}_+)^N \\ \times \\ \mathbb{C}^{M'} \end{matrix} \to \begin{matrix} H^{d_+}(\overline{\mathbb{R}}_+)^N \\ \times \\ \mathbb{C}^{M} \end{matrix}$$

of order $-d$, *class* $r' = d_-$ *and regularities* ν_1,ν_3 *(*q^h *is of regularity* ν_1 *and the other terms are of regularity* ν_3*), where*

$$\nu_3 = \min\{\nu_1 - \tfrac{1}{2}, \nu_2\} \quad ;$$

here $q^h = (p^h)^{-1}$, *and* $(a^h)^{-1}$ *is again elliptic.*

2^0 *There is a continuous function* $c_0(X)$ *so that the boundary symbol operator* $a(X,\xi',\mu,D_n)\colon \mathscr{S}(\overline{\mathbb{R}}_+)^N\times\mathbb{C}^M \to \mathscr{S}(\overline{\mathbb{R}}_+)^N\times\mathbb{C}^{M'}$ *is bijective for* (ξ',μ) *outside the set* $\{(\xi',\mu) \mid |\xi',\mu| \le c_0(X)\ ,\ |\xi'| \le 1\}$; *and the inverse* a^{-1} *(extended smoothly to all* (ξ',μ) , *e.g., by truncation in* $\{|\xi',\mu| \le c_0(X),\ |\xi'| \le 1\}$ *) is a boundary symbol operator*

$$(3.2.24)\qquad a^{-1} = \begin{pmatrix} q_\Omega + g' & k' \\ t' & s' \end{pmatrix} : \begin{matrix}\mathscr{S}(\overline{\mathbb{R}}_+)^N \\ \times \\ \mathbb{C}^{M'}\end{matrix} \to \begin{matrix}\mathscr{S}(\overline{\mathbb{R}}_+)^N \\ \times \\ \mathbb{C}^{M}\end{matrix}$$

of order $-d$, *class* $r' = d_-$ *and regularities* ν_1,ν_3 *where* ν_3 *is defined by* (3.2.25); *here* $a^{-1} = (a^h)^{-1}$ *for* $|\xi'| \ge 1$. *More precisely,*

$$\begin{aligned}
&\text{(i)} && q \in S_{tr}^{-d,\nu_1}(\Xi\times\mathbb{R},\ \overline{\mathbb{R}}^{n+1}_+) \otimes L(\mathbb{C}^N,\mathbb{C}^N)\ ,\\
&\text{(ii)} && g' \in S^{-d-1,\nu_3}(\Xi,\ \overline{\mathbb{R}}^n_+,\ H^+ \hat{\otimes} H^-_{r'-1}) \otimes L(\mathbb{C}^N,\mathbb{C}^N)\ ,\\
(3.2.25)\ &\text{(iii)} && t' \in S^{-d,\nu_3}(\Xi,\ \overline{\mathbb{R}}^n_+,\ H^-_{r'-1}) \otimes L(\mathbb{C}^N,\mathbb{C}^M)\ ,\\
&\text{(iv)} && k' \in S^{-d-1,\nu_3}(\Xi,\ \overline{\mathbb{R}}^n_+,\ H^+) \otimes L(\mathbb{C}^{M'},\mathbb{C}^N)\ ,\\
&\text{(v)} && s' \in S^{-d,\nu_3}(\Xi,\ \overline{\mathbb{R}}^n_+) \otimes L(\mathbb{C}^{M'},\mathbb{C}^M)\ .
\end{aligned}$$

3^0 *If the Laguerre expansions of the singular Green, trace and Poisson operators of class* 0 *occuring in* a *have uniformly estimated* $\ell^2_{\delta,\delta}$*-norms resp.* ℓ^2_δ*-norms for* $\delta \in\ ]0,\tfrac12[$ *(i.e., the norms are* $O(c(X)\langle\xi',\mu\rangle^\alpha)$ *with* $\alpha = d$ *for* g , $\alpha = d+\tfrac12$ *for* t *and* $\alpha = d-\tfrac12$ *for* k*)* , *then the parts of class* 0 *in* a^{-1} *likewise have uniformly estimated* $\ell^2_{\delta,\delta}$*-norms resp.* ℓ^2_δ*-norms.*

Proof: We may assume $N = 1$, since the cases $N > 1$ only deviate from this by more complicated notation . Restrict the attention to a compact connected subset $\overline{\Xi}'$ of Ξ .

When $d \ge 0$, we compose $a(X,\xi',\mu,D_n)$ to the right with the auxiliary boundary symbol operators $b_r(X,\xi',\mu,D_n)$ constructed in Theorem 3.1.5 for $(X,\xi',\mu) \in U_r$ $(r = 1,\dots,r_1+r_2)$. (To assure uniform estimates, we shrink U_r a little, to compact Ξ_r and W_r .) To make the dimensions fit together, we choose M_2 in Theorem 3.1.5 at least as large as the given M in (3.1.22); then also $M_1 \ge M'$ (since M_2-M_1 = index q_Ω = -index p_Ω = M-M'). Now a can be replaced by the possibly larger system

$$(3.2.26)\qquad a' = \begin{pmatrix} p_\Omega + g & k & 0 \\ t & s & 0 \\ 0 & 0 & I \end{pmatrix} : \begin{matrix} \mathscr{S}(\overline{\mathbb{R}}_+) \\ \times \\ \mathbb{C}^M \\ \times \\ \mathbb{C}^{M_1-M'} \end{matrix} \to \begin{matrix} \mathscr{S}(\overline{\mathbb{R}}_+) \\ \times \\ \mathbb{C}^{M'} \\ \times \\ \mathbb{C}^{M_1-M'} \end{matrix}$$

where I is the identity operator in $\mathbb{C}^{M_1-M'}$. By composition with b_r (on each U_r) we arrive at a system of the form

$$(3.2.27)\qquad a'' = a' \circ b_r = \begin{pmatrix} I + g'' & k'' \\ t'' & s'' \end{pmatrix} : \begin{matrix} \mathscr{S}(\overline{\mathbb{R}}_+) \\ \times \\ \mathbb{C}^{M_1} \end{matrix} \to \begin{matrix} \mathscr{S}(\overline{\mathbb{R}}_+) \\ \times \\ \mathbb{C}^{M_1} \end{matrix} ,$$

where g'', t'', k'' and s'' are of order and class zero, and regularity ν_3 on U_r, according to the rules of calculus.

When $d < 0$, we first compose to the left with an order reducing operator as in (3.1.25), arriving at an operator a' of order and class 0, with regularities ν_1, ν_2 again (as shown after (3.1.25)). The ps.d.o. part of a' is $p' = \lambda_-^{-d} p$, which is of order 0; its inverse $q' = q\lambda_-^d$ being likewise of order 0. Since

$$p'_\Omega = (\lambda_-^{-d} p)_\Omega = \lambda_{-,\Omega}^{-d} p_\Omega$$

for $|\xi',\mu| \geq 1$, where $\lambda_{-,\Omega}^{-d}$ is bijective according to Lemma 3.1.2,

$$\operatorname{index} p'_\Omega = \operatorname{index} p_\Omega = -\operatorname{index} q_\Omega = -\operatorname{index} q'_\Omega .$$

We then apply Theorem 3.1.5 to q' and compose the resulting operators b_r to the left with a' (after an augmentation by the identity matrix on $\mathbb{C}^{M_1-M'}$ as above, if necessary), which leads to an operator family a'' (for each U_r) of the form

$$(3.2.28)\qquad a'' = b_r \circ a' = b_r \circ \begin{pmatrix} \lambda_{-,\Omega}^{-d} & 0 \\ 0 & \kappa^{-d} \end{pmatrix} \circ a$$

$$= \begin{pmatrix} I + g'' & k'' \\ t'' & s'' \end{pmatrix} : \begin{matrix} \mathscr{S}(\overline{\mathbb{R}}_+) \\ \times \\ \mathbb{C}^{M_2} \end{matrix} \to \begin{matrix} \mathscr{S}(\overline{\mathbb{R}}_+) \\ \times \\ \mathbb{C}^{M_2} \end{matrix}$$

(including an extension from $\mathbb{C}^M$ to $\mathbb{C}^{M_2}$ if necessary); here g'', t'', k'' and s'' are of order and class 0 and regularity ν_3 according to the rules of calculus.

There is a quite parallel reduction of the associated strictly homogeneous symbols.

Let us now study the invertibility of a'', for both cases (3.2.27) and (3.2.28) at the same time, denoting M_1 in (3.2.27) resp. M_2 in (3.2.28) by M in the following. We consider the sets U_r with $r \leq r_1$ in the following. The construction we describe now, will be carried out simultaneously for a'' and for the associated strictly homogeneous system a''^h; we write the details for a''. Let us denote

$$(3.2.29) \qquad a'' = \begin{pmatrix} I+g'' & k'' \\ t'' & s'' \end{pmatrix} = I+h \;, \quad \text{where} \quad h = \begin{pmatrix} g'' & k'' \\ t'' & s''-I \end{pmatrix} ,$$

and use the point of view explained before the theorem, identifying h with an extended s.g.o. $g(h)$ acting in $L^2(\mathbb{R})$, or rather, in $V_{[-M,\infty[}$ (cf. (3.2.1)). For brevity, we call $g(h) = f$, it has a symbol-kernel

$$(3.2.30) \qquad \tilde{f}(X,x_n,y_n,\xi',\mu) = \sum_{\substack{-M<\ell<\infty \\ -M\leq m<\infty}} b_{\ell m}(X,\xi',\mu)\varphi_\ell(x_n,\kappa)\varphi_m(y_n,\kappa) ,$$

where the $b_{\ell m}$ for ℓ and $m \geq 0$ are the coeffients in g'', and the other coefficents are determined from k'', t'' and $s''-I$ as in (3.2.18-20).

The strictly homogeneous symbol a''^h is similarly carried over to the form $I+f^h$, mapping $V_{[-M,\infty[}$ into itself, and with f^h being the strictly homogeneous symbol associated with f. (Here we exclude the set $|\xi',\mu| \leq 1$ where κ should be replaced by $\kappa^h = |\xi',\mu|$ in the definition of the Laguerre functions; for this purpose it suffices to assume that $c_r \geq 1$ in the definition of U_r, cf. (3.1.27
By the hypothesis of ellipticity, $I+f^h$ is bijective on $V_{[-M,\infty[}$, and on $L^2(\mathbb{R})$ ($I+f^h$ is the identity on $V_{]-\infty,-M[}$), for $(X,\xi',\mu) \in U_r$.

We shall now use the decomposition

$$(3.2.31) \qquad \begin{aligned} &\tilde{f} = \tilde{f}_{M_1} + \tilde{f}^{\dagger}_{M_1} \;, \quad \text{with} \\ &\tilde{f}_{M_1} = \sum_{\substack{-M<\ell<M_1 \\ -M\leq m<M_1}} b_{\ell m}\varphi_\ell\varphi_m \quad \text{and} \quad \tilde{f}^{\dagger}_{M_1} = \sum_{\ell \text{ or } m\geq M_1} b_{\ell m}\varphi_\ell\varphi_m \;, \end{aligned}$$

and the analogous decomposition for $\tilde{f}^h$. The observations in Remark 3.2.2 show that since $\nu_3 > 0$, one can choose M_1 so large that (after replacing c_r in the definition of U_r by a larger number if necessary)

$$\|\tilde{f}^{\dagger}_{M_1}\|_{\ell^2_{0,0}} \leq \tfrac{1}{2} \quad \text{and} \quad \|\tilde{f}^{h\dagger}_{M_1}\|_{\ell^2_{0,0}} \leq \tfrac{1}{2} \quad \text{for all} \quad (X,\xi',\mu) \in U_r .$$

Then we apply Proposition 3.2.1 to the operator $I+f_{M_1}^{\dagger}$, which shows that it is invertible on $L^2(\mathbb{R})$, with an inverse of the same type. Moreover,

$$(I+f)(I+f_{M_1}^{\dagger})^{-1} = (I+f_{M_1}+f_{M_1}^{\dagger})(I+f_{M_1}^{\dagger})^{-1} = I + f' ,$$

where f' is again a singular Green operator (in the extended sense) of order and class 0 and regularity ν_3 ; and f' ranges in $V_{[-M,M_1[}$, since f_{M_1} maps into this space; it also vanishes on $V_{]-\infty,-M[}$. The analogous statements hold for the associated strictly homogeneous symbols.

Applying a decomposition like (3.2.31) to f' , we can now again find $M_2 \geq M_1$ such that (with a modified U_r)

$$\|\tilde{f}'^{\dagger}_{M_2}\|_{\ell^2_{0,0}} \leq \tfrac{1}{2} \quad \text{for all} \quad (X,\xi',\mu) \in U_r ,$$

and there is a similar estimate for the strictly homogeneous symbol. Again the operator $I+f'^{\dagger}_{M_2}$ is invertible, by Proposition 3.2.1, and we have

$$(I+f'^{\dagger}_{M_2})^{-1}(I+f') = (I+f'^{\dagger}_{M_2})^{-1}(I+f'_{M_2} + f'^{\dagger}_{M_2}) = I + f'' ,$$

where f'' is zero on $V_{[M_2,\infty[}$ and $V_{]-\infty,-M[}$. Recall here that

$$(I+f'^{\dagger}_{M_2}) = I + \sum_{k\geq 1} (-f'^{\dagger}_{M_2})^{\circ k}$$

where $\Sigma(-f'^{\dagger}_{M_2})^{\circ k}$ ranges in $V_{[-M,M_1[}$, since f' and $f'^{\dagger}_{M_2}$ do so. It follows that

$$f'' = (I+f'^{\dagger}_{M_2})^{-1} f'_{M_2}$$

ranges in $V_{[-M,M_1[}$. Since $M_1 \leq M_2$, we have altogether:

$$(I+f'^{\dagger}_{M_2})^{-1}(I+f)(I+f_{M_1}^{\dagger})^{-1} = I + f'' , \tag{3.2.32}$$

where f'' is zero on $V_{[M_2,\infty[}$ and $V_{]-\infty,-M[}$ and ranges in $V_{[-M,M_2[}$.

The parallel construction works for the strictly homogeneous symbols, so that we get

$$(I + f'^{h\dagger}_{M_2})^{-1}(I + f^h)(I + f^{h\dagger}_{M_1})^{-1} = I + f''^h \quad \text{on} \quad U_r ,$$

where f''^h is zero on $V_{[M_2,\infty[}$ and $V_{]-\infty,M[}$ and ranges in $V_{[-M,M_2[}$, for $(X,\xi',\mu) \in U_r$.

On $V_{[-M,M_2[}$, $I+f''$ can be expressed relative to the basis $\{\varphi_m(x_n,\kappa)\}_{m\in[-M,M_2[}$, by a matrix

$$r(X,\xi',\mu) = (r_{\ell m}(X,\xi',\mu))_{\ell,m\in[-M,M_2[} ,$$

whose entries are ps.d.o. symbols in $S^{0,\nu_3}(U_r)$, since f'' is of order 0 and regularity ν_3 . (Recall (2.3.30); the $r_{\ell m}$ are obtained by taking the scalar product of $I+\tilde{f}''$ with $\varphi_\ell(x_n,\kappa)\varphi_m(y_n,\kappa)$.) Similarly, $I+f''^h$ is expressed by the associated strictly homogeneous matrix

$$r^h(X,\xi',\mu) = (r^h_{\ell m}(X,\xi',\mu))_{\ell,m\in[-M,M_2[} .$$

The crucial observation to be made now is, that the ellipticity assumption on a^h implies that r^h is elliptic, as an element of $S^{0,\nu_3}(U_r) \otimes L(\mathbb{C}^{M+M_2},\mathbb{C}^{M+M_2})$. For, we have all the way reduced a^h by compositions with invertible, homogeneous symbols of positive regularity, arriving at $I+f''^h$ which is invertible on $L^2(\mathbb{R})$ for $(X,\xi',\mu) \in U_r$; and then since f''^h vanishes on $V_{]-\infty,-M[} \oplus V_{[M_2,\infty[}$ and ranges in $V_{[-M,M_2[}$, $I+f''^h$ is bijective in $V_{[-M,M_2[}$ (and equals I in $V_{]-\infty,-M[} \oplus V_{[M_2,\infty[}$) . Here it identifies with r^h .

Since $\nu_3 > 0$, it then follows, by Proposition 2.1.11 2°, that $r(X,\xi',\mu)$ is elliptic; more precisely, there is a constant c_0 so that $r(X,\xi',\mu)$ is invertible for $|\xi',\mu| \geq c_0$, $(X,\xi',\mu) \in U_r$, with r^{-1} satisfying the estimate (2.1.5). (Recall that r is its own principal symbol.)

It is at this point that condition (III) in Definition 3.1.3 is used; even if (II') holds, we can only get as far as showing invertibility of r (for $(X,\xi',\mu) \in U_r$, $|\xi',\mu| \geq c_0$) , but that does not assure the estimate of r^{-1} required in Definition 2.1.2 , in the intersection of U_r with $\{|\xi'| \leq 1\}$.

Denote the inverse of r (defined for large enough $|\xi',\mu|$, and extended by truncation to all of U_r) by r' ,

$$r'(X,\xi',\mu) = (r'_{\ell m}(X,\xi',\mu))_{\ell,m\in[-M,M_2[} ;$$

with associated strictly homogeneous symbol r'^h . By Theorem 2.1.16 and Lemma 2.8, $r' \in S^{0,\nu_3}(U_r)$ and $r'^h \in S^{0,\nu_3}_{hom}(U_r)$.

$(a^h)^{-1}$ is now found by retracing all the steps, and we likewise find an inverse a^{-1} (for large $|\xi',\mu|$) , lying in the correct symbol space. Let us explain the procedure for a .

First note that the above considerations imply

$$(I + f'')^{-1} = I + f''' ,$$

where f''' is the finite rank s.g.o. with symbol-kernel defined from the elements in r' :

$$(3.2.34)\quad \tilde{f}'''(X,x_n,y_n,\xi',\mu) = \sum_{\ell,m\in[-M,M_2[} (r'_{\ell m}(X,\xi',\mu)-\delta_{\ell m})\varphi_\ell(x_n,\kappa)\varphi_m(y_n,\kappa)$$

($\delta_{\ell m}$ denotes the Kronecker delta). Here $\tilde{f}'''$ is a s.g.o. symbol-kernel of order 0 and regularity ν_3 . Next

$$(I+f)^{-1} = (I+f^{\dagger}_{M_1})^{-1}(I+f''')(I+f'^{\dagger}_{M_2})^{-1}$$
$$= I + f^{(4)} ,$$

where $f^{(4)}$ is again a s.g.o. (in the extended sense), of order 0 and regularity ν_3 , with Laguerre coefficients that are zero for ℓ or $m < -M$. The interpretation of a'' as $I+f$ corresponds to an interpretation of $(a'')^{-1}$ as $I+f^{(4)}$,

$$(3.2.35)\qquad (a'')^{-1} = \begin{pmatrix} I+g^{(4)} & k^{(4)} \\ t^{(4)} & I+s^{(4)} \end{pmatrix}$$

where $g^{(4)}$, $t^{(4)}$, $k^{(4)}$ and $s^{(4)}$ are determined from the Laguerre series coefficients of $f^{(4)}$ as described before this theorem. We finally get a^{-1} (modulo the trivial augmentation in (3.2.26))

$$(3.2.36)\qquad \begin{aligned} a^{-1} &= b_r \circ (a'')^{-1} && \text{if } d \geq 0 , \\ a^{-1} &= (a'')^{-1} \circ b_r \circ \begin{pmatrix} \lambda^{-d}_{-,\Omega} & \\ 0 & \kappa^{-d} \end{pmatrix} && \text{if } d < 0 , \end{aligned}$$

for $(X,\xi',\mu) \in U_r$ with $|\xi',\mu| \geq c_0$. Similarly, we get $(a^h)^{-1}$ for all $(X,\xi',\mu) \in U_r$, and it extends by homogeneity to the full cone. By the rules of calculus, the resulting systems have q (resp. q^h) as ps.d.o. part; it is of order $-d$ and regularity ν_1 ; and the other symbols in the matrix are of order $-d$ and regularity ν_3 , and class d_- . This shows 1^0 and 2^0 of the theorem. Note also that we have shown that conditions (II) and (III) in Definition 3.1.3 together imply (II'), so that Proposition 3.1.4 is proved. The ellipticity of a^{-1} and $(a^h)^{-1}$ is obvious.

Finally consider the statement on Laguerre summability. Here the contribution $g^+(p)g^-(p)$ (resp. $g^+(q)g^-(q)$) to a'' is in $\ell^2_{\frac12,\frac12}$, by Corollary 2.6.14, since p and q are of regularity ≥ 1 . For the other parts of a'' one uses the composition rules in Proposition 2.6.15 ff., implying altogether that the terms in a'' are in $\ell^2_{\delta,\delta}$ resp. ℓ^2_δ . Then the extended s.g.o. f (cf. (3.2.30)) is in $\ell^2_{\delta,\delta}$. By use of (3.2.9) (and rules like (2.6.53, 55)), one finds that this property is preserved throughout, in the transformations of $I+f$ in the above proof. ▯

Observe that the essence of the proof lies in a reduction to the study of an elliptic pseudo-differential matrix depending on the boundary cotangent variables. This idea of a "reduction to the boundary" has also been a central theme in earlier studies of elliptic boundary value problems. For differential boundary value problems, it is well-known how one can reduce to an ellipticity discussion for matrices of a fixed rank related to the order of the operator (see e.g. the discussion in terms of Calderón projectors, as in [Seeley 2], [Hörmander 2, 8], [Grubb 19]). For pseudo-differential boundary problems (where the above proof of course works, with less efforts, in the non-parametrized case) we have reduced to a matrix with finite rank in a more intricate way; here the rank depends on the "size" of the symbols. The proofs for the non-parametrized case in [Boutet 3, (1.16)] and [Rempel-Schulze 1] exhibit a similar feature. See also the discussion of generalized Calderón projectors in these works.

<u>3.2.4 Remark</u>. The regularity set $\{\nu_1,\nu_2\}$ is preserved under passage to the inverse when $\nu_2 \leq \nu_1 - \frac{1}{2}$, for then $\nu_3 = \nu_2$; note that the Green systems with such regularity sets form an "algebra".

<u>3.2.5. Remark</u>. In some cases, ν_3 can be replaced by $\min\{\nu_1,\nu_2\}$, namely when $L(p,q)$ (in case $d \geq 0$), resp. $L(q',p')$ (in case $d < 0$), has regularity ν_1 instead of $\nu_1 - \frac{1}{2}$. This holds in many instances. For one thing, it holds if p resp. q' has symbol in H^- with respect to ξ_n (so that $L(p,q)$ resp. $L(q',p')$ essentially vanishes), in particular if p resp. q' is (or can be taken to be) a differential operator. It will also hold in the case accounted for in Proposition 2.6.11 where one factor is μ-independent, and hence in the case where $p(X,\xi,\mu) = p_c(X,\xi) + e^{i\theta}\mu^d$, where $L(p,q) = L(p_c,q)$ (this is highly important in the resolvent construction). For such cases, the regularity set is preserved when $\nu_2 \leq \nu_1$, which it is in our main applications.

<u>3.2.6 Remark</u>. By the representation of a^{-1} on the form (3.2.36) , it is seen that if the given symbol a depends analytically on $z = e^{i\theta}\mu^d$ (with respect to the symbol seminorms) in a keyhole region or sector, then so does a^{-1} (with respect to the symbol seminorms of the stated regularity); this is verified by inspection of each composition in the proof (cf. Remark 3.1.7).

Having completed the basic step, we shall now allow x_n-dependent ps.d.o. symbols and lower order terms in each place, applying the complete rules of calculus (taking also the x'-variable into account). This simply boils down to an application of our composition rules, and gives no new difficulties.

To distinguish in a simple way between x_n-dependent and x_n-independent symbols, we temporarily use a lower index 0 to indicate that $p(X,x_n,\xi,\mu)$ is replaced by $p(X,0,\xi,\mu)$, so for instance

$$(3.2.37)\quad \begin{aligned} p_0(X,\xi',\mu,D_n) &= OP_n(p(X,0,\xi,\mu)) ,\\ a_0(X,\xi',\mu,D_n) &= \begin{pmatrix} p_{0,\Omega}+g & k \\ t & s \end{pmatrix}(X,\xi',\mu,D_n) . \end{aligned}$$

3.2.7 Theorem. *Let* Ω' *be open* $\subset \mathbb{R}^{n-1}$, *let* $d \in \mathbb{Z}$, *let* $r \in \mathbb{N}$ *with* $r \leq d_+$, *and let* ν_1 *be integer* ≥ 1 *and* $\nu_2 \in \mathbb{R}_+$. *Let*

$$(3.2.38)\quad A_\mu = \begin{pmatrix} P_{\mu,\Omega}+G_\mu & K_\mu \\ T_\mu & S_\mu \end{pmatrix} : \begin{matrix} C^\infty_{(0)}(\Omega'\times\overline{\mathbb{R}}_+)^N \\ \times \\ C^\infty_0(\Omega')^M \end{matrix} \to \begin{matrix} C^\infty(\Omega'\times\overline{\mathbb{R}}_+)^N \\ \times \\ C^\infty(\Omega')^{M'} \end{matrix}$$

be an elliptic system of order d, *class* r *and regularities* ν_1,ν_2, *defined from symbols* p,g,t,k,s *satisfying the conditions in Definition 3.1.3 with* $\Xi = \Omega'$. *Let*

$$\nu_3 = \min\{\nu_1 - \tfrac{1}{2}, \nu_2\} .$$

There exists a parametrix

$$(3.2.39)\quad B_\mu = \begin{pmatrix} Q_{\mu,\Omega}+G'_\mu & K'_\mu \\ T'_\mu & S'_\mu \end{pmatrix} : \begin{matrix} C^\infty_{(0)}(\Omega'\times\overline{\mathbb{R}}_+)^N \\ \times \\ C^\infty_0(\Omega')^{M'} \end{matrix} \to \begin{matrix} C^\infty(\Omega'\times\overline{\mathbb{R}}_+)^N \\ \times \\ C^\infty(\Omega')^{M} \end{matrix} ,$$

elliptic of order $-d$, *class* $r' = d_-$ *and regularities* ν_1,ν_3, *such that when* φ' *and* $\psi' \in C^\infty_0(\Omega')$, φ_n *and* $\psi_n \in C^\infty_0(\mathbb{R})$ *equal to* 1 *near* $x_n = 0$, *and we set* $\varphi'\varphi_n = \varphi$ *and* $\psi'\psi_n = \psi$, *then*

$$(3.2.40)\quad \begin{pmatrix} \varphi & 0 \\ 0 & \varphi' \end{pmatrix} A_\mu \begin{pmatrix} \psi & 0 \\ 0 & \psi' \end{pmatrix} B_\mu - \begin{pmatrix} \varphi\psi & 0 \\ 0 & \varphi'\psi' \end{pmatrix} \sim 0 ,$$

$$(3.2.41)\quad \begin{pmatrix} \varphi & 0 \\ 0 & \varphi' \end{pmatrix} B_\mu \begin{pmatrix} \psi & 0 \\ 0 & \psi' \end{pmatrix} A_\mu - \begin{pmatrix} \varphi\psi & 0 \\ 0 & \varphi'\psi' \end{pmatrix} \sim 0 ,$$

when considered as Green operators of order 0 *and regularities* ν_1, ν_3 *(of class* r' *in* (3.2.40) *and class* r *in* (3.2.41)*).*

The principal boundary symbol operator for B_μ *is of the form (for* $|\xi',\mu| \geq c_0 > 0$)

$$(3.2.42)\quad b^0(x',x_n,\xi',\mu,D_n) = \begin{pmatrix} q^0(x',x_n,\xi',\mu,D_n)_\Omega + g'^0(x',\xi',\mu,D_n) & k^0(x',\xi',\mu,D_n) \\ t'^0(x',\xi',\mu,D_n) & s^0(x',\xi',\mu) \end{pmatrix}$$

where $q^0(x,\xi,\mu) = p^0(x,\xi,\mu)^{-1}$, *and* g'^0, t'^0, k'^0 *and* s'^0 *equal the corresponding terms in the inverse* $(a_0^0)^{-1}$ *of the principal* x_n*-independent boundary symbol operator* a_0^0 . *The complete boundary symbol operator for* B_μ *is determined, up to negligible modifications, by*

$$(3.2.43)\quad \begin{gathered} b(x,\xi',\mu,D_n) \sim b^0(x,\xi',\mu,D_n) \circ (I + \sum_{k\geq 1} r^{\circ k}) \\ \textit{where}\quad r = I - a \circ b^0 \ , \end{gathered}$$

(rearranged according to degrees of homogeneity), with the compositions described in Theorem 2.7.9.

Proof: By Theorem 3.2.3, the x_n-independent boundary symbol operator $a_0^0(x',\xi',\mu,D_n)$ has (for $|\xi',\mu| \geq c_0 > 0$) an inverse $(a_0^0)^{-1}$ in the stated symbol classes. Its ps.d.o. part is $q_0^0 = (p_0^0)^{-1}$. Let q be a full parametrix symbol for p , according to Theorem 2.1.16; then $q_0^0 = q^0|_{x_n=0}$. Observe now that when we write $\varphi_n p^0 = p_0^0 + x_n p'$, $\psi_n q^0 = q_0^0 + x_n q'$ (where p' and q' are again x_n-dependent ps.d.o. symbols having the transmission property) , then

$$\begin{aligned} (3.2.44)\quad (\varphi_n p^0)_\Omega \circ_n (\psi_n q^0)_\Omega &= (\varphi_n p^0 \psi_n q^0)_\Omega - L(\varphi_n p^0, \psi_n q^0) \\ &= \varphi_n \psi_n - L(p_0^0 + x_n p', q_0^0 + x_n q') \\ &= \varphi_n \psi_n - L(p_0^0, q_0^0) + g_1 \\ &= (p_0^0)_\Omega \circ_n (q_0^0)_\Omega + g_1 + \varphi_n \psi_n - I \end{aligned}$$

where g_1 is a s.g.o. of order -1 and regularity $\nu_1 - 1 - \frac{1}{2}$, in view of Theorem 2.7.7 and Remark 2.7.8 (each factor x_n decreases the order and regularity by 1). Similar results are found for the compositions of ps.d.o. symbols with singular symbols, e.g.

$$g \circ_n (\psi_n q^0)_\Omega = g \circ_n (q^0_{0,\Omega} + x_n q'_\Omega) = g \circ_n q^0_{0,\Omega} + g_2 \;,$$

where g_2 is of order -1 and regularity $m(\nu_1,\nu_2) - 1$. Altogether, it is seen that (for $|\xi',\mu| \geq c_0$)

$$(3.2.45) \quad \begin{pmatrix} \varphi_n & 0 \\ 0 & I \end{pmatrix} a^0 \circ_n \begin{pmatrix} \psi_n & 0 \\ 0 & I \end{pmatrix} b^0 = a^0_0 \circ_n b^0_0 + \begin{pmatrix} \varphi_n\psi_n - I & 0 \\ 0 & 0 \end{pmatrix} + r_1 = \begin{pmatrix} \varphi_n\psi_n & 0 \\ 0 & I \end{pmatrix} + r_1 \;,$$

where r_1 is of order -1 and regularity $\nu_3 - 1$, and contains only singular terms. In view of the composition rules with respect to the full x-variable, this implies that

$$\begin{pmatrix} \varphi & 0 \\ 0 & \varphi' \end{pmatrix} OP'(a^0(x,\xi',\mu,D_n)) \begin{pmatrix} \psi & 0 \\ 0 & \psi' \end{pmatrix} OP'(b^0(x,\xi',\mu,D_n)) = \begin{pmatrix} \varphi\psi & 0 \\ 0 & \varphi'\psi' \end{pmatrix} - R_{1,\mu} \;,$$

where $R_{1,\mu}$ is of order -1 and regularities ν_1-1, ν_3-1 . Now set

$$(3.2.46) \quad B^0_\mu = OP' \begin{pmatrix} q(x,\xi',\mu,D_n)_\Omega + g'^0(x',\xi',\mu,D_n) & k'^0(x',\xi',\mu,D_n) \\ t'^0(x',\xi',\mu,D_n) & s'^0(x',\xi',\mu) \end{pmatrix}$$

then the complete composition rules (Theorem 2.7.9) imply that

$$\begin{pmatrix} \varphi & 0 \\ 0 & \varphi' \end{pmatrix} A_\mu \begin{pmatrix} \psi & 0 \\ 0 & \psi' \end{pmatrix} B^0_\mu = \begin{pmatrix} \varphi\psi & 0 \\ 0 & \varphi'\psi' \end{pmatrix} - R_{2,\mu} \;,$$

where $R_{2,\mu}$ is of order -1 and regularities ν_1-1, ν_3-1 . Here

$$R_{2,\mu} \sim OP'(r_2(x,\xi',\mu,D_n)) \;,$$

for a certain boundary symbol operator r_2 determined from the symbols of A_μ and B^0_μ , and φ', ψ', φ_n and ψ_n , by the composition rules. For any $x \in \Omega' \times \overline{\mathbb{R}}_+$, we can choose φ', ψ', φ_n and ψ_n so that they equal 1 on a neighborhood of x , and here we set

$$(3.2.47) \quad b(x,\xi',\mu,D_n) \sim \begin{pmatrix} q_\Omega + g'^0 & k'^0 \\ t'^0 & s'^0 \end{pmatrix} \circ \sum_{k=0}^{\infty} r_2^{\circ k}$$

(with respect to the complete composition rules); this can be done consistently for all choices of cut-off functions. Then the operator

$$B_\mu = OP'(b(x,\xi',\mu,D_n))$$

satisfies (3.2.40) for any given set $\varphi', \psi', \varphi_n, \psi_n$. So B_μ is a right parametrix. A left parametrix B'_μ (satisfying (3.2.41)) can be constructed by the analogous (transposed) arguments. Finally, a standard argument shows that $B_\mu \sim B'_\mu$: Let φ and φ' be given, and choose ψ, ψ', ψ_1 and ψ'_1 so that ψ and ψ_1 are 1 on supp φ , and ψ' and ψ'_1 are 1 on supp φ' . Then by the already shown equivalences,

$$\begin{pmatrix}\varphi & 0\\ 0 & \varphi'\end{pmatrix}(B_\mu - B'_\mu) \sim \begin{pmatrix}\psi & 0\\ 0 & \psi'\end{pmatrix}B'_\mu\begin{pmatrix}\psi_1 & 0\\ 0 & \psi'_1\end{pmatrix}A_\mu\begin{pmatrix}\varphi & 0\\ 0 & \varphi'\end{pmatrix}B_\mu - \begin{pmatrix}\varphi & 0\\ 0 & \varphi'\end{pmatrix}B'_\mu\begin{pmatrix}\psi_1 & 0\\ 0 & \psi'_1\end{pmatrix}A_\mu\begin{pmatrix}\psi & 0\\ 0 & \psi'\end{pmatrix}B_\mu \sim 0$$

in view of the composition rules. Since φ and φ' can be adapted to be 1 on given subsets of $\Omega' \times \overline{\mathbb{R}}_+$, it follows that $B_\mu - B'_\mu \sim 0$. The formula (3.2.43), that is slightly simpler than (3.2.47), follows from the composition rules for symbols. (3.2.47) is used in Remark 3.2.8 below. □

<u>3.2.8 Remark</u>. As in Remark 3.2.5, the regularity number ν_3 can be improved to be $\min\{\nu_1,\nu_2\}$ when $L(P_\mu,Q_\mu)$ is of regularity ν_1 ; here the improved regularity extends to the full parametrix symbol because there is no further loss of regularity in the parametrix construction for $I - r_2$ above, where the ps.d.o. part has symbol I , cf. (3.2.47).

<u>3.2.9 Remark</u>. The operators in the theorem can of course be allowed to depend on more parameters. Of particular interest for us is the situation described in Remarks 3.1.7 and 3.2.6, and we note that the construction shows that analytic dependence on $z = e^{i\theta}\mu^d$ in a keyhole region or sector carries over to all terms in the complete symbol. When there is an exact inverse, analyticity also carries over to that, e.g. as in the treatment below.

Finally, the arguments extend to the manifold situation. We give the full description for compact manifolds. Here, a Green operator

$$A_\mu = \begin{pmatrix} P_{\mu,\Omega} + G_\mu & K_\mu \\ T_\mu & S_\mu \end{pmatrix} : \begin{matrix} C^\infty(E) \\ \times \\ C^\infty(F) \end{matrix} \to \begin{matrix} C^\infty(E) \\ \times \\ C^\infty(F') \end{matrix}$$

is said to be elliptic, when Definition 3.1.3 holds in local coordinates. Since p^0, a_0^0 and a_0^h are actually functions on $(T^*(\overline{\Omega})\times\overline{\mathbb{R}}_+)\setminus 0$ and $(T^*(\Gamma)\times\overline{\mathbb{R}}_+)\setminus 0$ (matrix resp. operator valued), the inverses $q^0 = (p^0)^{-1}$, $(a_0^0)^{-1}$ and $(a_0^h)^{-1}$

are likewise globally defined. As a first approximation of the parametrix of A_μ, one can construct a Green operator

$$(3.2.48)\qquad \mathcal{B}_\mu^0 = \begin{pmatrix} Q_{\mu,\Omega}^0 + G_\mu'^0 & K_\mu'^0 \\ T_\mu'^0 & S_\mu'^0 \end{pmatrix} : \begin{matrix} C^\infty(E) \\ \times \\ C^\infty(F') \end{matrix} \to \begin{matrix} C^\infty(E) \\ \times \\ C^\infty(F) \end{matrix},$$

whose ps.d.o. part Q_μ^0 has principal symbol q^0, and whose singular terms have the principal symbols g'^0, t'^0, k'^0, s'^0 appearing in $(a_0^0)^{-1}$. For example, $G_\mu'^0$ can be constructed as explained in (2.4.77) ff., when we as operators $G_\mu^{(i)}$ $(i=1,\dots,i_1)$ take operators on $\overline{\mathbb{R}}_+^n$ with symbols equal to the symbols obtained from g'^0 by the trivializations $\zeta_i : F|_{\Gamma_i} \to \Xi_i' \times \mathbb{C}^M$ and $\zeta_i' : F'|_{\Gamma_i} \to \Xi_i' \times \mathbb{C}^{M'}$ (one can take the sets Γ_i so small that F and F' are simultaneously trivialized here). For $i=i_1+1,\dots,i_2$ we set $G_\mu^{(i)} = 0$. Then the local principal symbols add up to g'^0 when (2.4.79) is applied, since $\Sigma\varphi_i = 1$.

The composition rules in Section 2.7 and the various considerations on negligible operators in Section 2.4 imply altogether that

$$(3.2.49)\qquad A_\mu \mathcal{B}_\mu^0 = I - \mathcal{R}_\mu$$

where $\mathcal{R}_\mu$ is of order -1 and regularities ν_1-1, ν_3-1 (and class d_-). A true parametrix is then found by reiteration:

$$(3.2.50)\qquad \mathcal{B}_\mu \sim \mathcal{B}_\mu^0 \sum_{k=0}^{\infty} \mathcal{R}_\mu^{\circ k},$$

where a concrete version of $\mathcal{B}_\mu$ can be constructed by use of a fixed system of local coordinates. It is worth noting here that the series (3.2.50) can be shown to converge in operator norm, for sufficiently large μ. For, the estimates in Section 2.5 show that $\mathcal{B}_\mu$ is continuous

$$(3.2.50')\qquad \mathcal{B}_\mu : \begin{matrix} H^{s,\mu}(E) \\ \times \\ H^{s-\frac12,\mu}(F') \end{matrix} \to \begin{matrix} H^{s+d,\mu}(E) \\ \times \\ H^{s+d-\frac12,\mu}(F) \end{matrix}$$

for $s \geq d_-$, uniformly in μ, and the "error term" $\mathcal{R}_\mu$ even satisfies

$$\|\mathcal{R}_\mu f\|_{H^{s+1,\mu}(E)\times H^{s+\frac12,\mu}(F')} \leq c(\langle\mu\rangle^{1-\nu_3}+1)\|f\|_{H^{s,\mu}(E)\times H^{s-\frac12,\mu}(F')}$$

for $s \geq d_-$, so that, since

$$\mu\|v\|_s + \|v\|_{s+1} \leq c\|v\|_{s+1,\mu}$$

for $s \geq 0$, one has

(3.2.51) $\|R_\mu f\|_{H^{s,\mu}(E)\times H^{s-\frac{1}{2},\mu}(F')} \leq c\mu^{-\nu_3'}\|f\|_{H^{s,\mu}(E)\times H^{s-\frac{1}{2},\mu}(F')}$ for $s \geq d_-$

with $\nu_3' = \min\{\nu_3,1\}$. Since $\nu_3 > 0$, the norm of R_μ in $H^{s,\mu}(E)\times H^{s-\frac{1}{2},\mu}(F')$ goes to zero for $\mu \to \infty$, each s , so for sufficiently large μ , one has a true inverse

(3.2.52) $$A_\mu^{-1} = B_\mu = B_\mu^0 \sum_{k=0}^{\infty} R_\mu^{\circ k} \quad .$$

To see that the true inverse is a Green operator belonging to the calculus, we set it in relation to a parametrix B_μ' constructed from the symbols. Consider the case $d \geq 0$, where B_μ' is of class 0 (if $d \leq 0$, A_μ itself is of class 0 and one uses a left composition instead of the right composition, we now describe). Then

$$A_\mu B_\mu' = I - R_\mu'$$

where R_μ' is a negligible element of the space of Green operators of order 0, class 0 and regularity ν_3 . Just as above, R_μ' has a norm satisfying (3.2.51), so for sufficiently large μ ,

(3.2.53) $$B_\mu = B_\mu' \sum_{k=0}^{\infty} (R_\mu')^{\circ k} = B_\mu' + B_\mu' \sum_{k=1}^{\infty} (R_\mu')^{\circ k} \quad .$$

To show that B_μ belongs to our operator class, we shall show that the second term defines a negligible operator of the right kind. This is achieved, if we show that

(3.2.54) $$S_\mu = \sum_{k=1}^{\infty} (R_\mu')^{\circ k}$$

is negligible as a Green operator of order and class 0 and regularity ν_3 , for then the composition $B_\mu' S_\mu$ is negligible as an operator of order $-d$, class 0 and regularity ν_3 , by the composition rules.

In view of Definition 2.4.4, the negligibility of R_μ' can be expressed by the statement that it is an integral operator in $L^2(E)\times L^2(F')$ with kernel

$$r(x,y,\mu) = \begin{pmatrix} r_1(x,y,\mu) & r_2(x,y',\mu) \\ r_3(x',y,\mu) & r_4(x',y',\mu) \end{pmatrix}$$

satisfying (in local coordinates near the boundary)

$$\| x_n^k D_{x_n}^{k'} y_n^m D_{y_n}^{m'} D^\beta_{x',y'} D^j_\mu r_1(x,y,\mu) \|_{L^2(I_a \times I_a)} \le c(x',y') \langle\mu\rangle^{-\nu_3 + [k-k']_- + [m-m']_- - j}$$

(3.2.55)

$$\| x_n^k D_{x_n}^{k'} D^\beta_{x',y'} D^j_\mu r_2(x,y',\mu) \|_{L^2(I_a)} \le c(x',y') \langle\mu\rangle^{-\nu_3 - \frac{1}{2} + [k-k']_- - j} ,$$

$$\| y_n^k D_{y_n}^{k'} D^\beta_{x',y'} D^j_\mu r_3(x',y,\mu) \|_{L^2(I_a)} \le c(x',y') \langle\mu\rangle^{-\nu_3 + \frac{1}{2} + [k-k']_- - j} ,$$

$$| D^\beta_{x',y'} D^j_\mu r_4(x',y',\mu) | \le c(x',y') \langle\mu\rangle^{-\nu_3 - j} ,$$

for all indices; and uniform estimates in subsets of the interior:

$$| D^\beta_{x,y} D^j_\mu \, r_1(x,y,\mu) | \le c(x,y) \langle\mu\rangle^{-\nu_3 - j} ,$$

(3.2.56) $$| D^\beta_{x,y'} D^j_\mu \, r_2(x,y',\mu) | \le c(x,y') \langle\mu\rangle^{-\nu_3 - \frac{1}{2} - j} ,$$

$$| D^\beta_{x',y} D^j_\mu \, r_3(x',y,\mu) | \le c(x',y) \langle\mu\rangle^{-\nu_3 + \frac{1}{2} - j} ,$$

when $\operatorname{dist}(x,\Gamma)$ and $\operatorname{dist}(y,\Gamma)$ are $\ge b > 0$.

Our task is to show that S_μ satisfies a similar system of estimates. To begin with, we replace R'_μ by

(3.2.57) $$R''_\mu = \begin{pmatrix} I & 0 \\ 0 & \langle\mu\rangle^{-\frac{1}{2}} \end{pmatrix} R'_\mu \begin{pmatrix} I & 0 \\ 0 & \langle\mu\rangle^{\frac{1}{2}} \end{pmatrix} ,$$

observing that

(3.2.58) $$S_\mu = \begin{pmatrix} I & 0 \\ 0 & \langle\mu\rangle^{\frac{1}{2}} \end{pmatrix} \sum_{k=1}^{\infty} (R''_\mu)^{\circ k} \begin{pmatrix} I & 0 \\ 0 & \langle\mu\rangle^{-\frac{1}{2}} \end{pmatrix} ;$$

and R''_μ satisfies the estimates (3.2.55)-(3.2.56) with $\frac{1}{2}$ removed. The Hilbert-Schmidt norm of R''_μ then satisfies (we leave out the bundle notation)

(3.2.59) $$\| R''_\mu \|_{\text{Hilbert-Schmidt}} = \left(\| r_1 \|^2_{L^2(\Omega\times\Omega)} + \| r_2 \|^2_{L^2(\Omega\times\Gamma)} + \| r_3 \|^2_{L^2(\Gamma\times\Omega)} + \| r_4 \|^2_{L^2(\Gamma\times\Gamma)} \right)^{\frac{1}{2}}$$

$$\le c \langle\mu\rangle^{-\nu_3} ,$$

so that for μ so large that $c\langle\mu\rangle^{-\nu_3} \leq \frac{1}{2}$, (3.2.58) converges in the Hilbert-Schmidt norm. The next thing is to apply the various derivatives and multiplications to the kernel of S_μ and observe, just as in the proof of Proposition 3.2.1, that each derivative or multiplication touches only a fixed finite number of the factors in $(R''_\mu)^{\circ k}$, the remaining factors being again estimated by their Hilbert-Schmidt norm which is $\leq \frac{1}{2}$. (The sup norm estimates in x',y' that result from (3.2.55)-(3.2.56) are interchangeable with $L^2_{x',y'}$ estimates, since all derivatives in x',y' respect the same type of bounds.) Summation for $k \to \infty$ then implies, as in the proof of Proposition 3.2.1, that the kernel of the sum S_μ satisfies the desired estimates. Thus S_μ is negligible, as an operator of order and class 0 and regularity ν_3, and B_μ belongs to our calculus.

B'_μ and B_μ can of course be patched together by taking

$$B''_\mu = \chi(\mu)B'_\mu + (1-\chi(\mu))B_\mu$$

where $\chi = 1$ for the small μ where B_μ is not defined, and $\chi = 0$ for large μ

Altogether, we have obtained the following result.

<u>3.2.10 Theorem</u>. *Let* $d \in \mathbb{Z}$, *let* $r \in \mathbb{N}$ *with* $r \leq d_+$, *and let* ν_1 *be integer* ≥ 1 *and* $\nu_2 \in \mathbb{R}_+$. *Let* E, F *and* F' *be* C^∞ *vector bundles over a smooth compact manifold* $\overline{\Omega}$ *and its boundary* Γ, *as described in the Appendix, and let*

$$\text{(3.2.60)} \qquad A_\mu = \begin{pmatrix} P_{\mu,\Omega} + G_\mu & K_\mu \\ T_\mu & S_\mu \end{pmatrix} : \begin{matrix} C^\infty(E) \\ \times \\ C^\infty(F) \end{matrix} \to \begin{matrix} C^\infty(E) \\ \times \\ C^\infty(F') \end{matrix}$$

be an elliptic Green operator of order d, *class* r *and regularities* ν_1,ν_2 *(satisfying Definition 3.1.3 in local coordinates). Then* A_μ *has a* <u>*parametrix*</u> *(for* $\mu \in \overline{\mathbb{R}}_+$*)*

$$\text{(3.2.61)} \qquad B_\mu = \begin{pmatrix} Q_{\mu,\Omega} + G'_\mu & K'_\mu \\ T'_\mu & S'_\mu \end{pmatrix} : \begin{matrix} C^\infty(E) \\ \times \\ C^\infty(F') \end{matrix} \to \begin{matrix} C^\infty(E) \\ \times \\ C^\infty(F) \end{matrix},$$

which is an elliptic Green operator of order $-d$, *class* $r' = d_-$ *and regularities* ν_1,ν_3, *where*

(3.2.62) $$\nu_3 = \min\{\nu_1 - \tfrac{1}{2}, \nu_2\} \quad ,$$

There exists $\mu_0 \geq 0$ *so that for* $\mu \geq \mu_0$, A_μ *has an inverse* B_μ , *which is of the kind* (3.2.61) *ff*. *(the parametrix can here be taken equal to the inverse). Moreover, one has the estimates* (3.2.50') *for* $s \geq d_-$.

3.2.11 Remark. As in the local formulations it is seen that the regularity ν_3 can be improved to

(3.2.63) $$\nu_3 = \min\{\nu_1, \nu_2\}$$

when $L(P_\mu, Q_\mu)$ has regularity ν_1 . Also the remarks on analyticity extend: If A_μ furthermore depends on a parameter θ , such that $A_{\mu,\theta}$ is analytic in $z = e^{i\theta}\mu^d$ in a certain region, then the inverse $B_{\mu,\theta} = (A_{\mu,\theta})^{-1}$ is likewise analytic in z where it exists, with respect to the operator norms and symbol norms that were taken into account in the proof.

3.2.12 Remark. The above arguments apply also to the ps.d.o. P_μ alone. If the manifold Σ , in which $\bar{\Omega}$ is imbedded, is compact, P_μ has a true inverse Q_μ for μ sufficiently large, and it is a ps.d.o. belonging to our calculus. The inverse is found e.g. by taking a parametrix Q'_μ , observing that the norm of

$$I - P_\mu Q'_\mu = \mathcal{P}_\mu$$

in $H^{s,\mu}(\tilde{E})$ goes to zero for $\mu \to \infty$, and setting

(3.2.64) $$Q_\mu = P_\mu^{-1} = Q'_\mu \sum_{k=0}^{\infty} (\mathcal{P}_\mu)^{\circ k}$$

where the sum converges in norm. The convergence also holds in Hilbert-Schmidt norm for μ sufficiently large, and arguments as above show that $Q_\mu - Q'_\mu$ is negligible of regularity $\nu_1 + d$. (See also Remark 3.3.6.)

3.2.13 Remark. The above results of course also have extensions to non-compact manifolds, at least in the parametrix sense.

3.2.14 Example. Let us give a very simple example of an application of Theorem 3.2.10. Let P_μ be the second order ps.d.o.

(3.2.65) $$P_\mu = (P' + \mu^4)(P'' + \mu^2)^{-1} \quad ,$$

where P' and P'' are positive selfadjoint elliptic ps.d.o.s on Σ of order 4 resp. 2, having the transmission property. Since P' and P'' are of regularity 4 resp. 2, when considered as μ-dependent, it follows that P_μ is of regularity 2; and it is clearly parameter-elliptic. On the boundary symbol level,

$$(p^0(x',0,\xi',\mu,D_n)u,u) \geq c\|u\|^2_{1,\mu} \quad \text{for } u \in C_0^\infty(\mathbb{R}) ,$$

and an analysis as in Theorem 1.7.2 shows that the Dirichlet problem

$$(3.2.66) \qquad \begin{aligned} P_{\mu,\Omega}u &= f \quad \text{in } \Omega , \\ \gamma_0 u &= \varphi \quad \text{at } \Gamma , \end{aligned}$$

is parameter-elliptic. Here the trace operator is of regularity ∞ . Theorem 3.2.10 now implies that

$$A_\mu = \begin{pmatrix} P_{\mu,\Omega} \\ \gamma_0 \end{pmatrix}$$

has a parametrix (inverse, for large μ)

$$\mathcal{B}_\mu = (Q_{\mu,\Omega}+G_\mu \quad K_\mu) ,$$

where $Q_\mu = (P''+\mu^2)(P'+\mu^4)^{-1}$ (it has regularity 2), and G_μ and K_μ have regularity 3/2 . In particular,

$$(3.2.67) \qquad \mathcal{B}_\mu : \begin{matrix} H^{s,\mu}(\Omega) \\ \times \\ H^{s+3/2,\mu}(\partial\Omega) \end{matrix} \rightarrow H^{s+2,\mu}(\Omega)$$

is continuous, uniformly in $\mu \in \overline{\mathbb{R}}_+$, for all $s \geq 0$.

In this example, P' and P'' can in particular be taken with rational symbols, in which case the problem can be regarded from the point of view of [Frank-Wendt 1-3], [Wendt 1], that builds on a delicate analysis of roots and poles and factorizations, for boundary value problems depending on a small parameter ε ; here ε^{-1} has the same rôle as our parameter μ . (See also Section 4.7.)

Our ellipticity requirement is more restrictive than that of Frank and Wendt, who allow symbols with a more general behavior at $\xi' = 0$. Frank and Wendt discuss in particular some interesting examples, whereas their general theory is sketchily presented.(For example, it is not clear whether they define s.g.o.s as finite or infinite sums of products of Poisson operators and trace operators, and only some of the composition rules are treated. (Cf. e.g. [Frank-Wendt 2, 2.4.5-7].) There is limited information on how they intend to handle the effects of negative regularity in variable coefficient cases. The difficulties here are illustrated in Example 3.2.16 below.)

3.2.15 Remark. As a special consequence of Theorem 3.2.10, we can obtain operator families $(\Lambda^m_{\pm,\mu})_\Omega$ on E with symbols $\lambda^m_\pm$ as in (3.1.14) and with very convenient properties. We explain this in detail for λ^m_-. Let $\tilde{E}$ be an extension of E over a compact extension Σ of $\overline{\Omega}$. Let $m \in \mathbb{Z}$. By use of a fixed system of coordinate patches, we can define ps.d.o.s $\Lambda^m_{-,\mu}$ in $\tilde{E}$ with symbols $\lambda^m_- I$ near Γ, extended to elliptic symbols on all of Σ (one can use the symbol $\langle\xi,\mu\rangle^m$ at a distance from Γ, as indicated in a related construction in [Boutet 3]). Since λ^m_- is in H^- as a function of ξ_n, we can choose $\Lambda^m_{-,\mu}$ such that

(3.2.68) $$r^+\Lambda^m_{-,\mu}(I - e^+r^+)u = 0 \quad \text{for any} \quad u \in \mathcal{D}'(E) ,$$

(not just ~ 0) ; then

(3.2.69) $$L(\Lambda^m_{-,\mu},P) = 0$$

for any ps.d.o. P. As shown in Lemma 3.1.2, $(\Lambda^m_{-,\mu})_\Omega$ is a parameter-elliptic operator in E of order m and of regularity $+\infty$. Then Theorem 3.2.10 implies the existence of a $\mu_0 \geq 0$ so that for $\mu \geq \mu_0$, the operator defines a homeomorphism

(3.2.70) $$(\Lambda^m_{-,\mu})_\Omega : C^\infty(E) \xrightarrow{\sim} C^\infty(E) ,$$

with an inverse of order $-m$ and class $[-m]_+$. By the continuity properties of these operators (cf. Section 2.5), we have in particular homeomorphisms

(3.2.71) $$(\Lambda^m_{-,\mu})_\Omega : H^s(E) \xrightarrow{\sim} H^{s-m}(E) \quad \text{when} \quad s \geq 0 \ \text{with} \ s-m \geq 0 ,$$

uniformly in $\mu \geq \mu_0$.

These are of interest also in parameter-independent situations, where they can be used for a fixed μ ; one can check (by using Taylor expansions of ζ) that the symbol λ^m_- is polyhomogeneous in ξ for each fixed μ. Parameter-independent operators of a similar kind, with symbol (3.1.48) near Γ, were introduced in [Boutet 3], and the mapping properties were analyzed in detail in [Rempel-Schulze 1, Section 3.1.2]. Parameter-dependent operators related to the $(\Lambda^m_{-,\mu})_\Omega$ above, but not of the classical ps.d.o. type, are used in [Rempel-Schulze 4], where (3.2.71) is extended to all $s \in \mathbb{R}$. - The operators $(\Lambda^m_{-,\mu})_\Omega$ are convenient for certain arguments, but because of the loss of regularity in general parameter-dependent compositions $L(P_\mu,P'_\mu)$, we generally avoid using them with variable μ.

The following additional construction will be useful later on: Let $m \in \mathbb{N}_+$, and let Λ_μ^{2m} be a strongly elliptic, positive selfadjoint polyhomogeneous ps.d.o. of regularity $+\infty$ on $\bar{\Sigma}$ (constructed e.g. with symbol $\langle\xi,\mu\rangle^{2m}$), so that the Dirichlet problem for Λ_μ^{2m} is elliptic and uniquely solvable (cf. Section 1.7). Since $L(\Lambda_{-,\mu}^{-m}, \Lambda_\mu^{2m}) = 0$, so that $(\Lambda_{-,\mu}^{-m})_\Omega \Lambda_{\mu,\Omega}^{2m} = (\Lambda_{-,\mu}^{-m} \Lambda_\mu^{2m})_\Omega$, the reduced Dirichlet problem

$$(3.2.72)\qquad \begin{aligned} (\Lambda_{-,\mu}^{-m} \Lambda_\mu^{2m})_\Omega\, u &= f \\ \rho_m u &= 0 \end{aligned}$$

(cf. the notation (A.55)ff.) is likewise uniquely solvable for $\mu \geq \mu_0$, in view of (3.2.71). The solution operator $R_{-m,\mu}$ is of order $-m$ and class 0; it is of the form

$$(3.2.73)\qquad R_{-m,\mu} = (Q_{-m,\mu})_\Omega + G_{-m,\mu}$$

with a ps.d.o. $Q_{-m,\mu}$ and a s.g.o. $G_{-m,\mu}$, defining homeomorphisms

$$(3.2.74)\qquad R_{-m,\mu} : H^{s,\mu}(E) \xrightarrow{\sim} \mathring{H}^{m,\mu}(E) \cap H^{s+m,\mu}(E) \quad \text{for } s \geq 0 ,$$

uniformly in $\mu \geq \mu_0$. These operators can also be used in the parameter-dependent theory, for fixed μ; and their symbols are polyhomogeneous in ξ for each μ.

<u>3.2.16 Remark</u>. Concerning extensions of the parametrix construction to cases where the hypotheses stated in the above theorems are violated, let us mention that one can make observations quite parallel to those of Remark 2.1.19. Assume that A_μ is a system of order d and regularities ν_1,ν_2, and <u>assume that</u> there exists a principal inverse symbol $b^0(x',\xi',D_n)$ of order $-d$ and regularities ν_1',ν_2'. (Recall that a great deal of the efforts in proving the preceding theorems were devoted to proving the existence of a good principal inverse symbol; when a principal inverse symbol is given on beforehand, one needs less of the other hypotheses.) By Theorem 2.7.9 (note that $D_{\xi'}^\alpha$ falls on the first factor only)

$$(3.2.75)\qquad \begin{aligned} c_r &= I - a^0 \circ b^0 \quad \text{is of order } -1 \text{ and regularities } \nu_{1,r},\nu_{2,r} , \\ c_\ell &= I - b^0 \circ a^0 \quad \text{is of order } -1 \text{ and regularities } \nu_{1,\ell},\nu_{2,\ell} , \end{aligned}$$

where $\nu_{1,r} = m(\nu_1 - 1, \nu_1')$ and $\nu_{1,\ell} = m(\nu_1' - 1, \nu_1)$, and

$$\nu_{2,r} = \min\{\nu_1-3/2,\nu_2-1,\nu_1'-\tfrac{1}{2},\nu_2',\nu_1+\nu_1'-2,\nu_1+\nu_2'-1,\nu_2+\nu_1'-1,\nu_2+\nu_2'-1\}\ ,$$

(3.2.76)

$$\nu_{2,\ell} = \min\{\nu_1-\tfrac{1}{2},\nu_2,\nu_1'-3/2,\nu_2'-1,\nu_1+\nu_2'-2,\nu_1+\nu_2'-1,\nu_2+\nu_1'-1,\nu_2+\nu_2'-1\}\ .$$

In particular

$$\nu_{2,r} = m(\nu_2-1,\nu_2') \quad \text{and} \quad \nu_{2,\ell} = m(\nu_2'-1,\nu_2)\ ,$$

(3.2.77)

$$\text{when} \quad \nu_1 \geq \nu_2 + \tfrac{1}{2} \quad \text{and} \quad \nu_1' \geq \nu_2' + \tfrac{1}{2}\ .$$

The question is now whether c_r or c_ℓ retain enough regularity, and in particular, whether reiteration as in Theorem 3.2.7 gives a reasonable symbol class; here again the composition rules of Section 2.7 are applied. In the case (3.2.77), we have that if $\nu_{2,r} \geq -1$, then $\Sigma_{k=0}^{\infty} c_r^{\circ k}$ defines a symbol of order 0 and regularity $\nu_{2,r} + 1 \geq 0$, so there is a reasonable right parametrix; whereas if $\nu_{2,r} < -1$, $b^0 c_r$ has a less controlled μ-behavior than b^0 (the order is 1 lower, but the regularity is at least $|\nu_{2,r}|$ lower, so the symbol can be a power $\mu^{|\nu_{2,r}|-1}$ higher); and the $c_r^{\circ k}$ have worse estimates in μ, the larger k is. Similar statements hold for c_ℓ.

As an illustration, consider the following modified version of (1.7.104):

$$-\Delta u + \mu^2 u = f \quad \text{in } \Omega$$

(3.2.78)

$$(1-\Delta_{x'})^{-\frac{1}{2}} S_1\gamma_1 u + (1-\Delta_{x'})^{\frac{1}{2}}\gamma_0 u = \varphi \quad \text{at } \Gamma\ ,$$

where S_1 is a nonzero first order ps.d.o. in Γ with real principal symbol $s_1(x',\xi')$. Here the trace operator is of order 1, and generally of regularity 0 and not normal. We multiply the boundary condition with $\Lambda_\mu' = (\mu^2-\Delta_{x'})^{\frac{1}{2}}$ to get a second order trace operator (without changing the regularity), then we have a system with principal homogeneous boundary symbol operator

(3.2.79)
$$a^h = \begin{pmatrix} \kappa^2 + D_n^2 \\ \kappa\sigma^{-1} s_1\gamma_1 + \kappa\sigma\gamma_0 \end{pmatrix}\ ;$$

here we use the notations $|\xi'| = \sigma$ and $|\xi',\mu| = \kappa$. An explicit calculation shows that a^h has the inverse

(3.2.80)
$$b^h = (r_D^h + g^h \quad k^h)\ ,$$

where r_D^h is the solution operator $r_D^h : f \rightsquigarrow v$ for the Dirichlet problem

$$(\kappa^2 + D_n^2)v = f \qquad \text{for} \quad x_n > 0 \ ,$$
$$\gamma_0 v = 0 \qquad \text{at} \quad x_n = 0 \ ;$$

and g^h resp. k^h are singular Green resp. Poisson operators with symbol-kernels

(3.2.81)
$$\tilde{g}^h = \frac{-i\sigma^{-1}s_1}{i\sigma^{-1}s_1\kappa + \sigma} \exp(-\kappa(x_n+y_n)) \ ,$$
$$\tilde{k}^h = \frac{1}{\kappa(i\sigma^{-1}s_1\kappa + \sigma)} \exp(-\kappa x_n) \ , \qquad \text{for} \quad \sigma \neq 0 \ .$$

Since s_1 is real and is $O(\sigma)$, we see that $\|\tilde{g}^h\|$ is $O(\kappa^{-2})$; and calculations of the derivatives show that in fact $\tilde{g}^h$ is of order -2 and regularity 0. However, $\tilde{k}^h$ (which is likewise of order -2) is in general only of regularity -1. (The exception is when $s_1(x',\xi') \neq 0$ for all $\xi' \neq 0$, in which case one gets regularity 0. But then S_1 can essentially be divided out, so that the problem becomes normal.)

The principal resolvent symbol $r^h = r_D^h + g^h$ defines locally an operator R_μ^0 that is continuous from $H^{s,\mu}$ to $H^{s+2,\mu}$ for $s \geq 0$, uniformly in μ. In particular, the norm of R_μ^0 in L^2 is $O(\langle\mu\rangle^{-2})$, consistently with (1.7.109). However, the Poisson operator K_μ^0 defined from k^h will generally not be uniformly continuous from $H^{s-\frac{1}{2},\mu}$ to $H^{s+2,\mu}$, because of the negative regularity, so one does not get the optimal result as in (3.2.50').

With respect to the full composition rules, we have in general that $c_r = I - a^h \circ b^h$ is of order -1 and regularities $+\infty,-1$, so that $\Sigma_{k=0}^{\infty} c_r^{\circ k}$ defines a symbol of order 0 and regularities $+\infty,0$; this gives us a right parametrix symbol b for a^h of order -2 and regularities $+\infty,0$. On the other hand, $c_\ell = I - b^h \circ a^h$ will be of order -1 and regularities $+\infty,-2$, so that $c_r^{\circ k}$ is of order $-k$ and regularities $+\infty,-2k$, and the terms cannot be summed up in our symbol classes.

But even for the right parametrix, we observe that the arguments before Theorem 3.2.10, leading to the exact inverse construction for large μ, cannot be used, since the lower order terms do not have a better μ-behavior than the principal term.

Examples of this kind were studied before in [Rempel-Schulze 3,4], see our comments in Remark 1.7.17.

Another interesting case is the system in (1.5.48), where the trace operator $\gamma_1 R$ is of order -1 and regularity $-\frac{1}{2}$, and certainly not normal. Its symbol is essentially $(|\xi'|-i\xi_n)^{-1}$, so let us now consider the related symbol of order 2

(3.2.82)
$$a^h = \begin{pmatrix} p^h \\ t^h \end{pmatrix} = \begin{pmatrix} |\xi|^2 + \mu^2 \\ \kappa^3(\sigma - i\xi_n)^{-1} \end{pmatrix} \ .$$

An explicit calculation gives that there is an inverse $(q_\Omega^h + g^h \quad k^h)$ with $q^h = (|\xi|^2 + \mu^2)^{-1} = (\kappa+i\xi_n)^{-1}(\kappa-i\xi_n)^{-1}$,

(3.2.83) $$k^h = c\kappa^{-3}\frac{\kappa+\sigma}{\kappa+i\xi_n},$$

and

(3.2.84) $$\begin{aligned} g^h &= -k^h \circ t^h \circ q_\Omega^h \\ &= -c\,\frac{\kappa+\sigma}{\kappa+i\xi_n}\, h^-_{\eta_n}\left[\frac{1}{(\sigma-i\eta_n)(\kappa+i\eta_n)(\kappa-i\eta_n)}\right] \\ &= c\,\frac{\kappa+\sigma}{\kappa+i\xi_n}\left[\frac{1}{\mu^2(\sigma-i\eta_n)} + \frac{\kappa+\sigma}{2\mu^2\kappa(\kappa-i\eta_n)}\right]. \end{aligned}$$

Here k^h is of order -2 and regularity 1, but g^h is μ^{-2} times a symbol of order 0 and regularity $-\frac{1}{2}$ (the negative regularity comes from the factor $(\sigma-i\eta_n)^{-1}$). The fact that g^h is only of order 0, not -2, is compensated for by the factor μ^{-2} (as far as μ-estimates are concerned). But the fact that the regularity of g^h, as well as of t^h, is $-\frac{1}{2}$, is more grave, for it implies that when we calculate $c_r = I - a^h \circ b^h$ and $c_\ell = I - b^h \circ a^h$, we get terms with regularity $-3/2$. For fixed ξ', the symbols $b^h \circ c_r$ and $c_\ell \circ b^h$ are a power $\mu^{\frac{1}{2}}$ higher than b^h itself, and iterated terms will blow up in μ for $k \to \infty$.

Note also that the norm of $OP_n(g^h)$ as an operator in $L^2(\mathbb{R}_+)$ is only $O(\langle\mu\rangle^{-3/2})$ for $\mu \to \infty$ (fixed ξ'), where it should be $O(\langle\mu\rangle^{-2})$ if Theorem 3.2.3 were applicable. This carries over to the full resolvent associated with (1.5.48), in the constant coefficient case with $\Omega = \mathbb{R}^n_+$. (Recall that μ^2 corresponds to $-\lambda$ in (1.5.48), so the L^2 norm of the resolvent is $O(\langle\lambda\rangle^{-3/4})$.)

The example enters in a prominent way in [Frank-Wendt 1, Section 3.2] and in earlier works of these authors. See also [Grubb 15] and [Rempel-Schulze 3, 4], and our observations in Remark 1.5.13 and Section 4.7.

3.3 The resolvent of a realization.

The general parametrix construction will now be applied to the study of the resolvent of (non-parametrized) pseudo-differential elliptic boundary problems, where we shall in particular elaborate the very interesting case of realizations described in Chapter 1.

Let d be an integer > 0 and consider a system of the kind studied in Chapter 1 (polyhomogeneous)

$$(3.3.1)\qquad A = \begin{pmatrix} P_\Omega + G \\ \\ T \end{pmatrix} \; : \; C^\infty(E) \to \begin{matrix} C^\infty(E) \\ \times \\ C^\infty(F) \end{matrix}$$

for given vector bundles E over $\overline{\Omega}$ and F over Γ ; here P is elliptic of order d in a neighboring bundle $\tilde{E}$, G is of order d and class $r \leq d$

$$(3.3.2)\qquad G = \sum_{0 \leq j < r} K_j \gamma_j + G'$$

and T is a normal trace operator:

$$(3.3.3)\qquad T = \begin{pmatrix} T_0 \\ T_1 \\ \cdot \\ \cdot \\ \cdot \\ T_{d-1} \end{pmatrix} \; : \; C^\infty(E) \to C^\infty(F) = C^\infty(\oplus_{0 \leq k < d} F_k) \; ,$$

where $T_k : C^\infty(E) \to C^\infty(F_k)$ is of order k and has the form

$$(3.3.4)\qquad T_k = s_{kk}(x')\gamma_k + \sum_{0 \leq j < k} S_{kj}\gamma_j + T_k' \; ,$$

with surjective morphisms $s_{kk}(x') : E_\Gamma \to F_k$. (T_k is void if $\dim F_k = 0$.)

Let us assume that

$$(3.3.5)\qquad A_\lambda = \begin{pmatrix} P_\Omega + G - \lambda I \\ \\ T \end{pmatrix}$$

is parameter-elliptic on the ray $\lambda = e^{i\theta} r$, as defined in Definition 1.5.5. We write

$$(3.3.6)\qquad -\lambda = \omega\mu^d \; , \qquad \mu \in \overline{\mathbb{R}}_+ \; ,$$

where $\omega = e^{i(\theta-\pi)}$. For the application of the results in Section 3.2, we have to adapt A_λ slightly, to obtain a system where all terms are of order d . This is done by composition of T to the left with the invertible elliptic ps.d.o. in F

$$(3.3.7) \qquad \Theta_\mu = \begin{pmatrix} \Lambda_{\mu,d} I_{F_0} & & \cdots & 0 \\ 0 & \Lambda'_{\mu,d-1} I_{F_1} & \cdots & 0 \\ \vdots & \vdots & \ddots & \vdots \\ 0 & 0 & \cdots & \Lambda'_{\mu,1} I_{F_{d-1}} \end{pmatrix}$$

where

$$(3.3.8) \qquad \Lambda'_{\mu,k} = ((-\Delta_\Gamma)^d + \mu^{2d})^{k/2d} \quad , \qquad I_{F_k} = \text{the identity on } F_k ,$$

$-\Delta_\Gamma$ being a positive Laplacean on Γ . (Note that the symbol of $\Lambda'_{\mu,1}$ is $(|\xi'|^{2d} + \mu^{2d})^{1/2d}$ as in (3.1.45).) Since $\Lambda'_{\mu,k}$ and its inverse are of regularity $+\infty$, $\Theta_\mu T$ and T have the same regularity. We then set

$$(3.3.9) \qquad A'_{\theta,\mu} = \begin{pmatrix} I & 0 \\ 0 & \Theta_\mu \end{pmatrix} A_\lambda = \begin{pmatrix} P_\Omega + G + \omega\mu^d I \\ T'_\mu \end{pmatrix} \quad \text{with} \quad T'_\mu = \Theta_\mu T \ ;$$

it is of order d . The regularity ν_1 of the ps.d.o. part $P_\Omega + \omega\mu^d I$ is d ,

$$(3.3.10) \qquad \nu_1 = d \ ,$$

unless P is differential operator, so that the regularity is $+\infty$. The class r of $A'_{\theta,\mu}$ is the largest of the numbers r_1 = class of G and $r_2 = k_{max} + 1$, where k_{max} is the largest k for which $\dim F_k > 0$ (recall that we allow $\dim F_k = 0$ for convenience of notations). The regularity ν_2 of the "singular part" of $A'_{\theta,\mu}$ is determined as in Definition 1.5.14 (consistently with Proposition 2.3.14); when G = 0 and T is a differential trace operator, it equals $+\infty$, and otherwise it can be any integer or half-integer in the interval between $\frac{1}{2}$ and d ;

$$(3.3.11) \qquad \nu_2 \in [\tfrac{1}{2}, d] \ .$$

Recall that $\nu_2 = \frac{1}{2}$ takes place precisely when the trace operator includes a term

$$(3.3.12) \qquad T_0 = \gamma_0 + T'_0$$

with a nontrivial T_0' , or G includes a nontrivial term $K_{d-1}\gamma_{d-1}$; otherwise ν_2 will be ≥ 1 .

Now parameter-ellipticity of A_λ on the ray $\lambda = e^{i\theta}r$ (Definition 1.5.5) implies ellipticity of $A'_{\theta,\mu}$ in the sense of Definition 3.1.3. Moreover, even in the cases where $\nu_2 = \frac{1}{2}$, the Laguerre series hypothesis in Theorem 3.2.3 3^0 is satisfied (actually with $\delta = \frac{1}{2}$), as is seen by application of the considerations in Remark 2.6.16 to the symbols of T_0' and K_{d-1} (all other terms have regularity ≥ 1 so their Laguerre series have the required estimates with $\delta = 1$).

We observe furthermore, that for $P + \omega\mu^d I$ and its parametrix $Q_{\theta,\mu}$ (which is of order $-d$ and regularity d , by Theorem 2.1.16) ,

$$L(P + \omega\mu^d I,\, Q_{\theta,\mu}) = L(P, Q_{\theta,\mu}) \text{ has regularity } d , \tag{3.3.13}$$

in view of Theorem 2.7.10 (note that $L(I, Q_{\theta,\mu}) = 0$).

Finally, it is worth observing that when A_λ is parameter-elliptic on the ray $\lambda = e^{i\theta}r$, it is likewise parameter-elliptic on the rays $\lambda = e^{i\theta'}r$ for θ' in a neighborhood of θ , since the conditions (I)-(III), in view of the homogeneity, reduce to invertibility of a continuous family of operators for a parameter running in a compact set; such an invertibility extends to a neighborhood. Obviously, A_λ and its symbol terms are analytic in $\lambda \in \mathbb{C}$, and also $A'_{\theta,\mu}$ extends to depend analytically on $\lambda = e^{i\theta}\mu^d$, in a keyhole region around the ray $\{\lambda = e^{i\theta}r \mid r \geq 0\}$. (Concerning the symbol, cf. Remark 3.1.7.)

Then we can conclude from the theorems and remarks in Section 3.2:

<u>3.3.1 Theorem.</u> *Let* $\theta_0 \in [0,2\pi]$, *let* d *be a positive integer, and let*

$$A = \begin{pmatrix} P_\Omega + G \\ \\ T \end{pmatrix} : C^\infty(E) \to \begin{matrix} C^\infty(E) \\ \times \\ C^\infty(F) \end{matrix} \tag{3.3.14}$$

be a Green operator with P *of order* d , G *of order* d *and class* $r \leq d$, *and* T *normal, associated with the order* d . *Let* $\nu \in [\frac{1}{2}, d]$ *stand for the regularity of* G *and* T , *when considered as parameter-dependent operators. Assume that*

$$A_\lambda = \begin{pmatrix} P_\Omega + G - \lambda I \\ \\ T \end{pmatrix} : C^\infty(E) \to \begin{matrix} C^\infty(E) \\ \times \\ C^\infty(F) \end{matrix} \tag{3.3.15}$$

is parameter-elliptic on the ray $\lambda = e^{i\theta_0} r$ *(cf. Definition* 1.5.5*).*

There is a truncated sector

$$(3.3.16) \qquad W_{r_0,\varepsilon} = \{\lambda \in \mathbb{C} \mid |\lambda| \geq r_0 \ \text{ and } \ |\theta_0 - \arg\lambda| \leq \varepsilon\}$$

with $r_0 \geq 0$ *and* $\varepsilon > 0$ *, such that* A_λ *is invertible for* $\lambda \in W_{r_0,\varepsilon}$ *. Writing*

$$\lambda = \exp[i(\theta-\pi)]\mu^d \quad , \quad \mu \geq 0 \quad ,$$

and setting

$$A'_{\theta,\mu} = \begin{pmatrix} P_\Omega + G - \lambda I \\ \Theta_\mu T \end{pmatrix}$$

with Θ_μ *defined by* (3.3.7)*, we have that* A_λ^{-1} *coincides with a parametrix* B_λ *of* A_λ *defined for* λ *in a keyhole region*

$$(3.3.17) \qquad V_{\delta,\varepsilon} = \{\lambda \in \mathbb{C} \mid |\lambda| \leq \delta \ \text{ or } \ |\theta_0 - \arg\lambda| \leq \varepsilon\} \quad ,$$

such that

$$(3.3.18) \quad B_\lambda = B'_{\theta,\mu}\begin{pmatrix} I & 0 \\ 0 & \Theta_\mu \end{pmatrix} = (Q_{\theta,\mu,\Omega} + G_{\theta,\mu} \quad K_{\theta,\mu}) = (Q_{\lambda,\Omega} + G_\lambda \quad K_\lambda) \ ,$$

where $B'_{\theta,\mu}$ *is a parametrix of* $A'_{\theta,\mu}$ *for each* $\theta \in [-\varepsilon,\varepsilon]$ *, parametrized by* $\mu \in \overline{\mathbb{R}}_+$*. More precisely,* $Q_{\theta,\mu}$ *is a (parameter-dependent) ps.d.o. of order* $-d$ *and regularity* d *for each* θ *;* $G_{\theta,\mu}$ *is a (parameter-dependent) s.g.o. of order* $-d$ *, class* 0 *and regularity* ν *for each* θ *, and*

$$(3.3.19) \qquad K_{\theta,\mu} = \{K_{\theta,\mu,0} , \dots , K_{\theta,\mu,d-1}\}$$

is a row vector of (parameter-dependent) Poisson operators $K_{\theta,\mu,k}$ *from* $C^\infty(F_k)$ *to* $C^\infty(E)$ *of orders* $-k$ *and regularity* ν *.*

A_λ^{-1} *depends analytically on* λ *in the truncated sector* $W_{r_0,\varepsilon}$ *, and the terms in the symbols depend analytically on* λ *in keyhole regions*

$$(3.3.20) \qquad V_{\delta,\varepsilon,\xi'} = \{\lambda \in \mathbb{C} \mid |\lambda| \leq \delta|\xi'|^d \ \text{ or } \ |\theta_0 - \arg\lambda| \leq \varepsilon\} \ ,$$

for $|\xi'| \geq c_0 > 0$ *(for small* $|\xi'|$ *there is analyticity in truncated sectors). The principal symbols of* $G_{\theta,\mu}$ *and* $K_{\theta,\mu}$ *have uniformly estimated* $\ell^2_{\delta,\delta}$ *resp.* ℓ^2_δ*-norms, for* $\delta \in \,]0,\tfrac{1}{2}[$ *.*

As explained in (1.5.18)-(1.5.20), the resolvent $R_\lambda = (B - \lambda I)^{-1}$

of the realization $B = (P+G)_T$ is simply the first block in $\mathcal{B}_\lambda$,

(3.3.21) $R_\lambda = Q_{\theta,\mu,\Omega} + G_{\theta,\mu}$, also written $Q_{\lambda,\Omega} + G_\lambda$.

In view of Theorems 2.5.4 and 2.5.5, we have as an immediate consequence:

<u>3.3.2 Corollary.</u> *When* A *satisfies the hypotheses of Theorem 3.3.1, the resolvent* R_λ *of the realization* $(P+G)_T$ *exists for* λ *in a truncated sector* $W_{r_0,\varepsilon}$ (3.3.16), *and it is of the form* (3.3.21), *where* $Q_{\theta,\mu}$ *(also called* Q_λ*) is a parametrix of* $P-\lambda I$, *and* $G_{\theta,\mu}$ *(also called* G_λ*) is of order* $-d$, *class* 0 *and regularity* ν , *as a s.g.o. depending on the parameter* μ *(for each* θ*). Here* R_λ *satisfies the estimates (cf.* (A.59')*)*

(3.3.22)
$$(\langle\mu\rangle^{s+d}\|R_\lambda f\|_0 + \|R_\lambda f\|_{s+d}) \le c_s(\langle\mu\rangle^s\|f\|_0 + |f|_s)$$
for each $s \ge 0$, *uniformly in* $W_{r_0,\varepsilon}$.

Similar results holds for Q_λ *on* $\widetilde{E}$, *when* Σ *is compact.*

<u>3.3.3 Remark.</u> The above formulations do not quite take the regularities $+\infty$ for differential operators into account, so let us supply the information here: If both P and T are differential operators and G is 0 , the regularity of $\mathcal{B}_{\theta,\mu}$ is $+\infty$ in all terms. If only P is differential, the regularity will be ν as stated above; and if P is pseudo-differential, $G = 0$ and T is differential, the regularity will be d (as stated above, assigning the regularity d to G and T). So only the pure differential operator case needed extra mentioning. Anyway, this is the case covered already in [Seeley 3] by polynomial methods, so it is not here that the novelty lies. Let us mention that the symbol-kernel estimates, that result from our treatment, give back the estimates of [Seeley 3], except for a factor $\exp(-c\langle\xi',\mu\rangle(x_n+y_n))$, which is directly linked with the behavior of the roots in the symbol polynomials.

In the following, the presentation is generally simplified by leaving out the special features obtainable in the pure differential operator case.

Note that the resolvent R_λ is a bijection

(3.3.23) $R_\lambda : L^2(E) \xrightarrow{\sim} D(B) = \{u \in H^d(E) \mid Tu = 0\}$,

and satisfies

$$R_\lambda(P_\Omega + G - \lambda I)u = u \qquad \text{for} \quad u \in D(B)\ ,$$
$$(P_\Omega + G - \lambda I)R_\lambda f = f \qquad \text{for} \quad f \in L^2(E)\ ,$$

quite exactly (not just in a parametrix sense; recall the problems around that in Proposition 1.4.2). We also recall that when A_λ^{-1} is written

$$A_\lambda^{-1} = (R_\lambda \quad K_\lambda)\ ,$$

we find in particular the identities

$$\begin{aligned} R_\lambda(P_\Omega + G - \lambda I)u + K_\lambda Tu &= u\ , \quad \text{for} \quad u \in H^d(E)\ , \\ TR_\lambda f &= 0, \ \text{for} \quad f \in L^2(E)\ ; \\ (P_\Omega + G - \lambda I)K_\lambda\varphi &= 0, \quad \text{and} \\ TK_\lambda\varphi &= \varphi, \ \text{for } \varphi \in \prod_{0\le j<d} H^{d-j-\frac12}(F_j)\ . \end{aligned} \tag{3.3.24}$$

It will be of interest to list some conclusions of the general calculus for the special operator R_λ .

The term Q_λ is rather well known (from [Seeley 1] and subsequent other studies on resolvents of ps.d.o.s without boundary conditions), and our results for it are primarily listed for completeness' sake. The novelty lies in the study of the s.g.o. term G_λ .

In the following we assume that the ray of parameter-ellipticity is $\overline{\mathbb{R}}_-$; this can always be obtained by a rotation.

One has for $Q = Q_{\theta,\mu}$:

3.3.4 Theorem. *Let* A_λ *be as in Theorem 3.3.1, parameter-elliptic on the ray* $\overline{\mathbb{R}}_-$. *In each local trivialization* $\Xi\times\mathbb{C}^N$ *for* $\widetilde{E}$, *the symbol* $q(x,\theta,\xi,\mu)$ *of* $Q_{\theta,\mu} = Q_\lambda$ *(with* $\lambda = -e^{i\theta}\mu^d$, $\theta \in [-\varepsilon,\varepsilon]$ *and* $\mu \in \overline{\mathbb{R}}_+$) *has the form*

$$q(x,\theta,\xi,\mu) \sim \sum_{\ell\in\mathbb{N}} q_{-d-\ell}(x,\theta,\xi,\mu) \quad (\textit{also denoted } q_{-d-\ell}(x,\xi,\lambda)\ \textit{etc.})\ ,$$

where q *and the* $q_{-d-\ell}$ *lie in* $C^\infty(\Xi\times[-\varepsilon,\varepsilon]\times\overline{\mathbb{R}}^{n+1}_+) \otimes L(\mathbb{C}^N,\mathbb{C}^N)$, *the latter are homogeneous in* (ξ,μ) *of degree* $-d-\ell$ *for* $|\xi| \geq 1$; *and*

$$(3.3.25)\qquad |D_x^\beta D_\xi^\alpha D_\mu^j[q(x,\theta,\xi,\mu) - \sum_{\ell<M} q_{-d-\ell}(x,\theta,\xi,\mu)]|$$

$$\leq c(x)(\rho(\xi,\mu)^{d-|\alpha|-M} + 1)\langle\xi,\mu\rangle^{-d-|\alpha|-M-j}$$

$$\leq \begin{cases} c'(x)\langle\xi,\mu\rangle^{-d-|\alpha|-M-j} & \textit{when}\quad |\alpha|+M \leq d\ , \\ c'(x)\langle\xi\rangle^{d-|\alpha|-M}\langle\xi,\mu\rangle^{-2d-j} & \textit{when}\quad |\alpha|+M > d\ ; \end{cases}$$

in particular, these estimates are satisfied by q_{-d-M} . *The homogeneous terms are analytic in* λ *in a keyhole region*

$$(3.3.26)\qquad V_{\delta,\varepsilon,\xi'} = \{\lambda \in \mathbb{C} \mid |\lambda| \leq \delta|\xi'|^d \quad \textit{or} \quad |\mathrm{Im}\lambda| \leq -\varepsilon\,\mathrm{Re}\lambda\}$$

around the ray $\overline{\mathbb{R}}_-$ *for* $|\xi'| \geq c_0 > 0$, *and in truncated sectors around* $\mathbb{R}_-$ *for smaller* $|\xi'|$.

The distribution kernel $K_Q(x,y,\theta,\mu)$ *of* $Q_{\theta,\mu}$ *has an asymptotic expansion*

$$(3.3.27)\qquad K_Q(x,y,\theta,\mu) \sim \sum_{\ell\in\mathbb{N}} K_{Q,-d-\ell}(x,y,\theta,\mu)\ \ (\textit{also denoted } K_{Q,-d-\ell}(x,y,\lambda)\ \textit{etc.})\ ;$$

here, when

$$(3.3.28)\qquad d' \equiv -d-|\alpha|+|\beta|+|\gamma|-M-j < -n\ ,$$

the kernel satisfies the estimates

$$(3.3.29)\qquad |(x-y)^\alpha D_x^\beta D_y^\gamma D_\mu^j[K_Q - \sum_{\ell<M} K_{Q,-d-\ell}]|$$

$$\leq \begin{cases} c(x)\langle\mu\rangle^{n+d'} & \textit{if}\quad d-|\alpha|+|\gamma|-M > -n\ , \\ c(x)(\langle\mu\rangle^{n+d'}\log(\mu+2)+\langle\mu\rangle^{-2d-j}) & \textit{if}\quad -n \geq d-|\alpha|+|\gamma|-M \geq -n-|\beta|\ , \\ c(x)\langle\mu\rangle^{-2d-j} & \textit{if}\quad d-|\alpha|+|\gamma|+|\beta|-M < -n\ . \end{cases}$$

When $d > n$, *one has in particular on the diagonal, setting*

$$(3.3.30)\qquad z = (-\lambda)^{1/d}\ , \quad \textit{by analytic continuation from } -\lambda \in \mathbb{R}_+\ ,$$

that

$$(3.3.31)\qquad K_Q(x,x,\theta,\mu) = r_{-d}(x)z^{n-d} + r_{-d-1}(x)z^{n-d-1} + \dots$$

$$+ r_{-2d-n+1}(x)z^{-2d+1} + r'_{-2d-n}(x,\theta,\mu)$$

where the $r_k(x)$ *are* C^∞ *functions and* r'_{-2d-n} *is a* C^∞ *function for* $\mu \neq 0$, *satisfying*

$$(3.3.32)\qquad |D_x^\beta D_\mu^j\, r'_{-2d-n}(x,\theta,\mu)| \le c(x)|\mu|^{-2d-j}\log\langle\mu\rangle \quad \textit{for} \quad \mu \ge c > 0 .$$

When Q_λ *is a precise inverse of* $P-\lambda I$ *(for large* $|\lambda|$*),* $K_Q(x,y,\theta,\mu)$ *and* $r'_{-2d-n}(x,\theta,\mu)$ *are analytic in* λ *in a truncated sector around* $\mathbb{R}_-$. *It follows that the trace of* Q_λ *satisfies*

$$(3.3.33)\qquad \operatorname{tr} Q_\lambda = \int_\Omega \operatorname{tr} K_Q(x,x,\lambda)dx = \underline{r}_{-d} z^{n-d} + \ldots + \underline{r}_{-2d-n+1} z^{-2d+1} + \underline{r}'_{-2d-n}(\lambda) ,$$

with constants $\underline{r}_{-d-\ell} = \int_\Omega \operatorname{tr} r_{-d-\ell}(x)dx$, *and* $|\underline{r}'_{-2d-n}(\lambda)| \le c\langle\lambda\rangle^{-2}\log\langle\lambda\rangle$ *for* $|\lambda| \ge c_0 > 0$, *in a truncated sector around* $\mathbb{R}_-$.

Proof: The estimates (3.3.25) are the symbol estimates for q , valid since q is of order $-d$ and regularity d , cf. Definition 2.1.1. The estimates (3.3.29) are seen by use of the formula (1.2.4) (where the condition (3.3.28) assures that the kernel is a continuous function). More specifically, one has (when $j = M = 0$)

$$\begin{aligned}(3.3.34)\qquad (x-y)^\alpha D_x^\beta D_y^\gamma K_Q(x,y,\theta,\mu) &= (x-y)^\alpha D_x^\beta D_y^\gamma F^{-1}_{\xi\to x-y} q \\ &= (x-y)^\alpha D_x^\beta D_y^\gamma (2\pi)^{-n} \int e^{i(x-y)\cdot\xi} q(x,\theta,\xi,\mu)d\xi \\ &= (2\pi)^{-n} \int e^{i(x-y)\cdot\xi} \overline{D}_\xi^\alpha [\sum_{\theta\le\beta} c_\theta (i\xi)^\theta(-i\xi)^\gamma D_x^{\beta-\theta} q]d\xi \quad ,\end{aligned}$$

where we use the Leibniz formula in the application of D_x^β . Since $D_\xi^\alpha(\xi^\rho q) \in S^{-d+|\rho|-|\alpha|,d+|\rho|-|\alpha|}$ (cf. Lemma 2.1.6 and (2.1.2)), we find by use of the elementary Lemma 3.3.5 below:

$$\begin{aligned} &\left|\int e^{i(x-y)\cdot\xi} \overline{D}_\xi^\alpha [\xi^{\gamma+\theta} D_x^{\beta-\theta} q]d\xi\right| \\ &\le c\int (\langle\xi\rangle^{d+|\gamma+\theta|-|\alpha|} + \langle\xi,\mu\rangle^{d+|\gamma+\theta|-|\alpha|})\langle\xi,\mu\rangle^{-2d}d\xi \\ &= c_1\langle\mu\rangle^{n-d+|\gamma+\theta|-|\alpha|} + c\int\langle\xi\rangle^{d+|\gamma+\theta|-|\alpha|}\langle\xi,\mu\rangle^{-2d}d\xi \\ &\le \begin{cases} c_2\langle\mu\rangle^{n-d+|\gamma+\theta|-|\alpha|} & \text{if } d+|\gamma+\theta|-|\alpha| > -n \\ c_2\langle\mu\rangle^{n-d+|\gamma+\theta|-|\alpha|}\log(\mu+2) & \text{if } d+|\gamma+\theta|-|\alpha| = -n , \\ c_2\langle\mu\rangle^{-2d} & \text{if } d+|\gamma+\theta|-|\alpha| < -n . \end{cases}\end{aligned}$$

Now θ runs thorugh the multiindices between 0 and β , so the estimate for the full expression (3.3.34) is, with $d' = -d-|\alpha|+|\beta|+|\gamma|$,

$$|(x-y)^\alpha D_x^\beta D_y^\gamma K_Q| \le \begin{cases} c\langle\mu\rangle^{n+d'} & \text{if } d+|\gamma|-|\alpha| > -n , \\ c(\langle\mu\rangle^{n+d'}\log(\mu+2)+\langle\mu\rangle^{-2d}) & \text{if } -n \ge d+|\gamma|-|\alpha| \ge -n-|\beta| , \\ c\langle\mu\rangle^{-2d} & \text{if } d+|\gamma|+|\beta|-|\alpha| < -n . \end{cases}$$

Taking M or j different from zero gives a shift in the orders and regularities, where similar calculations can be applied, and we altogether find (3.3.29).

On the diagonal, K_Q is simply determined by

$$K_Q(x,x,\theta,\mu) = (2\pi)^{-n}\int_{\mathbb{R}^n} q(x,\xi,\theta,\mu)d\xi \quad ,$$

when $d > n$. Here we use that q is polyhomogeneous of regularity $d > 0$. This implies, by use of Lemma 2.1.9, that the first d terms are estimated by

$$\begin{aligned}
&|q_{-d-\ell}(x,\theta,\xi,\mu)| \le c(x)\langle\xi,\mu\rangle^{-d-\ell} \quad \text{for} \quad \ell \le d \ , \quad \text{with}\\
(3.3.35)\quad &|q^h_{-d-\ell}(x,\theta,\xi,\mu)| \le c(x)|\xi,\mu|^{-d-\ell} \quad ,\\
&|q_{-d-\ell} - q^h_{-d-\ell}| \le c(x)\langle\xi,\mu\rangle^{-2d} \quad \text{for} \quad |\xi,\mu| \ge c > 0 \quad ;
\end{aligned}$$

and the next $n-1$ terms are estimated by

$$\begin{aligned}
&|q_{-d-\ell}(x,\theta,\xi,\mu)| \le c(x)\langle\xi\rangle^{d-\ell}\langle\xi,\mu\rangle^{-2d} \ , \quad \text{for} \quad d < \ell < d+n \ , \quad \text{with}\\
(3.3.36)\quad &\\
&|q^h_{-d-\ell}(x,\theta,\xi,\mu)| \le c(x)|\xi|^{d-\ell}|\xi,\mu|^{-2d} \quad ;
\end{aligned}$$

so that altogether

$$(3.3.37)\quad q(x,\theta,\xi,\mu) = q^h_{-d}(x,\theta,\xi,\mu) + \ldots + q^h_{-2d-n+1}(x,\theta,\xi,\mu) + q'(x,\theta,\xi,\mu)$$

where $q' = q'' + q'''$, satisfying for $\mu \ge c_0 > 0$,

$$\begin{aligned}
&|q''| \le c(x)|\xi|^{1-n}|\mu|^{-2d}\chi(\xi) \ , \quad \text{with} \quad \chi \in C_0^\infty(\mathbb{R}^n) \ , \quad \chi(\xi) = 1 \quad \text{for} \quad |\xi| \le 1\\
(3.3.38)\quad &\\
&|q'''| \le c(x)\langle\xi\rangle^{-n}\langle\xi,\mu\rangle^{-2d} \ .
\end{aligned}$$

By the homogeneity, we have for $\ell < d+n$ and $\theta = 0$:

$$\begin{aligned}
&(2\pi)^{-n}\int_{\mathbb{R}^n} q^h_{-d-\ell}(x,0,\xi,\mu)d\xi = r_{-d-\ell}(x)\mu^{n-d-\ell} \quad ,\\
(3.3.39)\quad &\\
&\text{with} \quad r_{-d-\ell}(x) = (2\pi)^{-n}\int_{\mathbb{R}^n} q^h_{-d-\ell}(x,0,\xi,1)d\xi
\end{aligned}$$

(which is well-defined since $d-\ell > -n$ and $-d-\ell < -n$) , and the remainder gives, by Lemma 3.3.5,

$$(3.3.40)\quad \left|(2\pi)^{-n}\int_{\mathbb{R}^n}(q'' + q''')d\xi\right| \le c(x)|\mu|^{-2d}\log\langle\mu\rangle \quad \text{for} \quad \mu \ge 1 \ ,$$

with related estimates of derivatives. This shows (3.3.31) for $\theta = 0$, and the statement extends to small θ by continuity.

When Q_λ is a precise inverse of $P-\lambda I$, it depends analytically on λ in the norm of operators from $L^2(E)$ to $H^d(E)$. Since $d > n$, the kernel is continuous in x,y and now furthermore analytic in λ, so r'_{-2d-n} is analytic. Integration in x gives the trace of Q_λ (cf. e.g. [Agmon 3, Theorems 12.21 and 13.5]), showing (3.3.33). □

3.3.5 Lemma. 1° *Let* $a > n$. *Then we have for* $\mu \geq 0$:

(3.3.41) $$\int_{\mathbb{R}^n} \langle\xi,\mu\rangle^{-a} d\xi = \text{const.}\langle\mu\rangle^{n-a} .$$

2° *Let* a *and* $b \in \mathbb{R}$ *with* $a+b > n$. *Then for* $\mu \geq 0$:

(3.3.42) $$\int_{\mathbb{R}^n} \langle\xi\rangle^{-a}\langle\xi,\mu\rangle^{-b} d\xi \leq \begin{cases} c\langle\mu\rangle^{n-a-b} & \text{if } a < n , \\ c\langle\mu\rangle^{-b}\log(\mu+2) & \text{if } a = n , \\ c\langle\mu\rangle^{-b} & \text{if } a > n \text{ and } b \geq 0 . \end{cases}$$

Proof: (3.3.41) is seen by replacing ξ by $\eta = \xi/\langle\mu\rangle$:

$$\int_{\mathbb{R}^n} \langle\xi,\mu\rangle^{-a} d\xi = \langle\mu\rangle^{n-a} \int_{\mathbb{R}^n} \langle\eta,1\rangle d\eta = \text{const.}\langle\mu\rangle^{n-a} .$$

For (3.3.42) we have in the cases $a \leq n$:

$$\begin{aligned} \int_{\mathbb{R}^n} \langle\xi\rangle^{-a}\langle\xi,\mu\rangle^{-b} d\xi &= \int_{|\xi|\leq 1} \langle\xi\rangle^{-a}\langle\xi,\mu\rangle^{-b} d\xi + \int_{|\xi|\geq 1} \langle\xi\rangle^{-a}\langle\xi,\mu\rangle^{-b} d\xi \\ &\leq c_1\langle\mu\rangle^{-b} + c_2\int_1^\infty r^{-a}(r+\mu+2)^{-b} r^{n-1} dr \\ &= c_1\langle\mu\rangle^{-b} + c_2(\mu+2)^{n-a-b}\int_{(\mu+2)^{-1}}^\infty s^{-a}(s+1)^{-b} s^{n-1} ds \\ &\leq \begin{cases} c_3\langle\mu\rangle^{n-a-b} & \text{if } a < n , \\ c_3\langle\mu\rangle^{-b}\log(\mu+2) & \text{if } a = n . \end{cases} \end{aligned}$$

If $a > n$ and $b \geq 0$, then simply

$$\int_{\mathbb{R}^n} \langle\xi\rangle^{-a}\langle\xi,\mu\rangle^{-b} d\xi \leq \langle\mu\rangle^{-b}\int\langle\xi\rangle^{-a} d\xi .$$ □

The proof of Theorem 3.3.4 is fairly standard; it is done in so much detail because we shall use the same arguments in other cases later on. Results of the above kind follow from [Seeley 1] (with the corrections in [Seeley 3]). More refined estimates (for λ close to the spectrum of P) were given in [Grubb 6], with applications to spectral theory. Resolvent kernels have also been studied for more general classes of operators (e.g. in [Robert 1], [Mohamed 1,2]...), going in directions outside the scope of this book. The next remarks give some precisions, and further information on kernels and traces are given at the end of this Section.

3.3.6 Remark. As observed in Remark 3.2.12, Q_λ is an exact inverse of $P-\lambda I$ for large $|\lambda|$, when Σ is compact. Now in the applications, it is usually so that $\overline{\Omega}$ and the bundles E and F are given, and one is free to choose the neighboring manifold Σ and bundle $\widetilde{E}$ over Σ. It is then advantageous to take Σ compact. In fact, we shall now show that when P is given in E and is here parameter-elliptic on the ray $\overline{\mathbb{R}_-}$, say, and we extend E to a bundle $\widetilde{E}$ over a compact boundaryless manifold Σ in which $\overline{\Omega}$ is imbedded (one can take the "double" of E along Γ), then P can always be extended to $\widetilde{E}$ in such a way that the parameter-ellipticity on $\overline{\mathbb{R}_-}$ is preserved. This is achieved by means of a homotopy carrying $p^0(x,\xi)$ into $|\xi|^d I$ (for $(x,\xi)\in T^*(\overline{\Omega})\setminus 0$) while preserving the parameter-ellipticity; for example:

$$(3.3.43)\qquad \begin{aligned} &(1-t)p^0(x,\xi) + t\alpha|\xi|^d I \quad \text{for } t\in[0,\tfrac{1}{2}]\ ,\\ &(1-t)(p^0(x,\xi)+\alpha|\xi|^d I) + t|\xi|^d I \quad \text{for } t\in[\tfrac{1}{2},1]\ ,\end{aligned}$$

where we take $\alpha \geq 0$ so large that $\operatorname{Re} p^0(x,\xi)+\alpha|\xi|^d I$ is positive definite. Now, when $p^0(x,\xi)$ is given as a section in E (considered as a bundle over $T^*(\overline{\Omega})$), we use the homotopy to extend it in a continuous way for $(x',x_n)\in\Gamma\times[-1,0]$ such that $p^0(x,\xi)$ is $|\xi|^d I$ for $x_n=-1$; then we can set $p^0(x,\xi)=|\xi|^d I$ on the rest of Σ. (Actually, p^0 should be C^∞ in x and not just C^0; this is obtained by further smoothing.)

The property of parameter-ellipticity on all rays with arguments θ in an interval $[\pi-\theta_0,\ \pi+\theta_0]$ say, can likewise be preserved under a homotopy to $|\xi|^d I$; so also here (and in particular in the parabolic case) we can take Σ to be compac The cases where $p^0(x,\xi)$ is symmetric but not semibounded can likewise be handled, since p^0 is homotopic to a diagonal matrix with a fixed number of positive resp. negative elements in the diagonal (note that here p^0 is parameter-elliptic in two disjoint sectors $0 < \theta < \pi$ and $\pi < \theta < 2\pi$). Finally, in more pathological cases where $p^0(x,\xi)$ is parameter-elliptic in several disjoint sectors, one can at least take Σ compact for the consideration of each individual sector.

The abovementioned homotopy argument can also be used to modify a given $p_d(x,\xi)$ in the region $|\xi| \leq 1$ such that $p_d(x,\xi) - \lambda I$ is invertible for all $\xi \in \mathbb{R}^n$, all $\lambda \in \overline{\mathbb{R}}_-$.

<u>3.3.7 Remark</u>. The terms $q_{-d-\ell}$ can of course be described explicitly by the construction in Theorem 2.1.16 (cf. [Seeley 1], [Hörmander 4], [Nagase 1]). One has that $q_{-d-\ell}$ for $|\xi| \geq 1$ is a sum of products, whose factors are (λ-independent) derivatives of the homogeneous terms in $p(x,\xi)$, together with $(p_d(x,\xi)-\lambda)^{-1}$; the terms in $q_{-d-\ell}$ contain from 2 to $2\ell+1$ of the latter factors when $\ell \geq 1$. In the scalar case these factors can be grouped together in formulas

$$
(3.3.44) \qquad \begin{aligned} q_{-d} &= (p_d(x,\xi) - \lambda)^{-1}, \\ q_{-d-\ell} &= \sum_{k=1}^{2\ell} p_{\ell,k}(x,\xi)(p_d(x,\xi) - \lambda)^{-k-1} \qquad \text{for } \ell \geq 1, \end{aligned}
$$

where the $p_{\ell,k}$ are λ-independent symbols of order $dk-\ell$ (derived from $p_d,\dots,p_{d-\ell}$).

In the pseudo-differential case, there can be nonzero terms with $k = 1$ no matter how large ℓ is, whereas the sum begins with $k \geq \ell/d$ in the differential operator case (the $p_{\ell,k}$ are here polynomial in ξ). This demonstrates again the finite versus infinite regularity ν in the ps.d.o. versus d.o. case.

When p_d is chosen for $|\xi| \leq 1$ such that $p_d(x,\xi) - \lambda$ is bijective for all $\xi \in \mathbb{R}^n$ and all λ on the ray (see the preceding remark), the formulas in (3.3.44) extend to all $\xi \in \mathbb{R}^n$. The terms are obviously analytic in λ in a keyhole region around the ray.

<u>3.3.8 Remark</u>. The kernel and trace estimates in (3.3.31-33) required $d > n$. In the cases where $d \leq n$, one can still estimate the kernel and trace of the operator

$$
(3.3.45) \qquad Q'_{\lambda,M} = Q_\lambda - \sum_{\ell<M} Q_{\lambda,-d-\ell}
$$

for $d+M > n$; here the $Q_{\lambda,-d-\ell}$ are operators defined from the symbols $q_{-d-\ell}$ by use of local coordinate systems and partitions of unity; we express this briefly by writing

$$
Q_{\lambda,-d-\ell} = OP(q_{-d-\ell}) .
$$

One can here take $Q_{\lambda,-d-\ell}$ to depend analytically on λ in a keyhole region (as an operator from $L^2(\tilde{E})$ to $H^{d+\ell}(\tilde{E})$) ; then $Q'_{\lambda,M}$ is analytic in λ when Q_λ and the $Q_{\lambda,-d-\ell}$ are so. When $M > n-d$, so that $Q'_{\lambda,M}$ is of order $< -n$, it is of trace class and has a continuous kernel, and the analyticity in operator topology implies analyticity of the kernel in sup-norm. We note that (3.3.29) in particular implies the estimates of the kernel $K_{Q'_M}(x,y,\lambda)$ of $Q'_{\lambda,M}$:

$$\text{(3.3.46)}\qquad \sup_{x,y} |D^\beta_{x,y} K_{Q'_M}(x,y,\lambda)| \leq c(\langle\lambda\rangle^{(n-d-M+|\beta|)/d}\log(2+\langle\lambda\rangle) + \langle\lambda\rangle^{-2})$$

$$\text{when } M-|\beta|+d > n \ ;$$

the $\langle\lambda\rangle^{-2}$ estimate wins for large M .

For $n-d < M < n+d$ it is furthermore seen (by the same arguments as in Theorem 3.3.4) that the kernel and trace have asymptotic expansions

$$\text{(3.3.47)}\qquad \begin{aligned} K_{Q'_M}(x,x,\lambda) &= r_{-d-M}(x)z^{n-d-M} + \ldots + r_{-2d-n+1}(x)z^{-2d+1} + r'_{-2d-n}(x,\lambda) \\ \operatorname{tr} Q'_{\lambda,M} &= \underline{r}_{-d-M}z^{n-d-M} + \ldots + \underline{r}_{-2d-n+1}z^{-2d+1} + \underline{r}'_{-2d-n}(\lambda) \ , \end{aligned}$$

the terms being defined as in the theorem. For $M \geq d+n$, the kernel and trace of $Q'_{\lambda,M}$ are estimated like r' and $\underline{r}'$ above.

3.3.9 Remark. One can ask whether Q_λ has a complete symbol that is analytic in λ, but this is a somewhat formal question, since such a complete symbol is only linked with Q_λ in an asymptotic sense. For most purposes, the above remainders $Q'_{\lambda,M}$ with analytic kernels suffice. One could associate with the kernel of $Q'_{\lambda,M}$ a "symbol" q'_M by use of (2.1.41) and a finite version of (2.1.46 ii), such that $q' = \Sigma_{\ell<M} q_{-d-\ell} + q'_M$ is a good analytic approximation to a standard symbol q of Q_λ (the difference operator $Q_\lambda - OP(q')$ being analytic in λ and highly smoothin[g]

We have here worked in local coordinates; for a global treatment see [Widom 2].

We now turn to the singular Green operator part of R_λ ; the term G_λ . Here it is of interest to consider (in local coordinates near Γ) the symbol $g(x',\theta,\xi,\eta_n,\mu)$ of G_λ ; the symbol-kernel $\tilde{g}(x',x_n,y_n,\theta,\xi',\mu)$, and also the integral of the symbol-kernel on the x_n-diagonal

$$\text{(3.3.48)}\qquad \tilde{\tilde{g}}(x',\theta,\xi',\mu) = \int_{\mathbb{R}_+} \tilde{g}(x',x_n,x_n,\theta,\xi',\mu)dx_n$$

(which gives the trace of the boundary symbol operator $g(x',\theta,\xi',\mu,D_n))$; and besides these the kernel of G_λ , and its trace. The most important contribution to G_λ comes from a neighborhood of the boundary, for when φ and ψ are 1 on a neighborhood of Γ and vanish outside a larger neighborhood, then

$$(3.3.49)\qquad G_\lambda = \varphi G_\lambda \psi + (1-\varphi)G_\lambda + \varphi G_\lambda(1-\psi) ,$$

where the two last terms are negligible, as elements in the class of s.g.o.s of order $-d$ and regularity ν . In other words, they are negligible of regularity $\nu+d+1$, which means that their kernels satisfy estimates as in (2.4.33) with ν replaced by $\nu+d$. In particular, the bounds on derivatives of these kernels, together with the uniform bounds on compact subsets of Ω , show that

$$(3.3.50)\qquad \sup_{x,y\in\Omega} |K_{[G_\lambda - \varphi G_\lambda \psi]}(x,y,\theta,\mu)| \le c\langle\mu\rangle^{-d-\nu+1} ,$$

and further estimates can be inferred from Definition 2.4.4 3^0 .

In the following, we list the results for $G_\lambda' = \varphi G_\lambda \psi$ (φ and ψ taken with support in $\Sigma_+' = \Gamma\times[0,1[$) ; we can view the latter as an operator on $\Gamma\times\overline{\mathbb{R}}_+$, and we denote it for simplicity G_λ again.

The next theorem sums up the analysis achieved by means of the thorough study of our parametrized calculus in Chapter 2 and the preceding sections.

3.3.10 Theorem. *Let A_λ be as in Theorem 3.3.1, parameter-elliptic on the ray $\overline{\mathbb{R}}_-$, and consider the s.g.o. term G_λ in A_λ^{-1} ; it is precisely defined and analytic for $|\lambda|$ large, λ in a sector $|\mathrm{Im}\,\lambda| \le -\varepsilon\,\mathrm{Re}\,\lambda$ around $\mathbb{R}_-$; and it is defined in a parametrix sense in a keyhole region around $\overline{\mathbb{R}}_-$, the terms in the symbol being analytic there (in $V_{\delta,\varepsilon,\xi'}$ defined by (3.3.26), for $|\xi'| \ge c_0 > 0$).*

In each local trivialization $(\Xi'\times\overline{\mathbb{R}}_+)\times\mathbb{C}^N$ for $E|_{\Sigma_+'}$ (cf. Appendix and Section 2.4) the symbol $g(x',\theta,\xi,\eta_n,\mu)$ has the form

$$g(x',\theta,\xi,\eta_n,\mu) \sim \sum_{\ell\in\mathbb{N}} g_{-d-1-\ell}(x',\theta,\xi,\eta_n,\mu) ,$$

lying in $S^{-d-1,\nu}(\Xi'\times[-\varepsilon,\varepsilon], \overline{\mathbb{R}}_+^n, H^+ \hat{\otimes} H_{-1}^-) \otimes L(\mathbb{C}^N,\mathbb{C}^N)$ (Definition 2.3.7). More specifically, the associated symbol-kernel $\tilde{g}(x',x_n,y_n,\theta,\xi',\mu)$ (the sesqui-inverse Fourier transform of g , cf. (2.3.25)) has the form

$$\tilde{g}(x',x_n,y_n,\theta,\xi',\mu) \sim \sum_{\ell\in\mathbb{N}} \tilde{g}_{-d-1-\ell}(x',x_n,y_n,\theta,\xi',\mu) ,$$

where $\tilde{g}$ and the $\tilde{g}_{-d-1-\ell}$ lie in $C^\infty(\Xi'\times\overline{\mathbb{R}}_{++}^2\times[-\varepsilon,\varepsilon]\times\overline{\mathbb{R}}_+^n)\otimes L(\mathbb{C}^N,\mathbb{C}^N)$, the latter being quasihomogeneous in (x_n,y_n,ξ',μ)

$$(3.3.51)\quad \tilde{g}_{-d-1-\ell}(x',\frac{x_n}{t},\frac{y_n}{t},\theta,t\xi',t\mu) = t^{-d+1-\ell}\tilde{g}_{-d-1-\ell}(x',x_n,y_n,\theta,\xi',\mu)$$

for $t \geq 1$ *and* $|\xi'| \geq 1$,

and one has for all integer indices

$$(3.3.52)\quad \| D^{\beta}_{x',\theta} D^{\alpha}_{\xi'} D^{j}_{\mu} x_n^k D^{k'}_{x_n} y_n^m D^{m'}_{y_n} [\tilde{g} - \sum_{\ell<M} \tilde{g}_{-d-1-\ell}] \|_{L^2_{x_n,y_n}} \leq$$

$$\leq c(x)(\langle\xi'\rangle^{\nu-M'} + \langle\xi',\mu\rangle^{\nu-M'})\langle\xi',\mu\rangle^{-d-\nu+M''-j}$$

$$\leq \begin{cases} c'(x')\langle\xi',\mu\rangle^{-d-M'+M''-j} , & \textit{when } M' \leq \nu , \\ c'(x')\langle\xi'\rangle^{\nu-M'}\langle\xi',\mu\rangle^{-d-\nu+M''-j} & \textit{when } M' \geq \nu , \end{cases}$$

here

$$(3.3.53)\quad \begin{aligned} M' &= [k-k']_+ + [m-m']_+ + |\alpha| + M , \\ M'' &= [k-k']_- + [m-m']_- \quad ; \quad \textit{so} \quad -M'+M'' = -k+k'-m+m'-|\alpha|-M . \end{aligned}$$

Moreover, the integral (3.3.48) *of the symbol-kernel is a pseudo-differential symbol on the boundary, of order* $-d$ *and regularity* $\nu - 1/4$,

$$(3.3.54)\quad \tilde{\tilde{g}}(x',\theta,\xi',\mu) \in S^{-d,\nu-\frac{1}{4}}(\Xi'\times[-\varepsilon,\varepsilon],\bar{\mathbb{R}}^n_+) \otimes L(\mathbb{C}^N,\mathbb{C}^N) \quad .$$

Proof: The statements on $g(x',\theta,\xi,\eta_n,\mu)$ and $\tilde{g}(x',x_n,y_n,\theta',\xi',\mu)$ are simply restatements of the fact that g is a s.g.o. symbol of order $-d$, class 0 and regularity ν , as proved in Theorem 3.3.1 and Corollary 3.3.2. We shall now prove the statements on $\tilde{\tilde{g}}$. The parameters (x',θ,ξ',μ) will often be omitted in the calculations. For $\varepsilon > 0$ (to be chosen later), we find

$$(3.3.55)\quad |\tilde{\tilde{g}}|^2 \leq \Big(\int_0^\infty \frac{\varepsilon+x_n}{\varepsilon+x_n} |\tilde{g}(x_n,x_n)|dx_n\Big)^2$$

$$\leq c\,\varepsilon^{-1}\int_0^\infty (\varepsilon^2+x_n^2)\,|\tilde{g}(x_n,x_n)|^2 dx_n$$

$$\leq c\int_0^\infty\int_0^\infty \Big[\varepsilon\partial_{y_n}|\tilde{g}(x_n,y_n)|^2 + \varepsilon^{-1}\partial_{y_n}|y_n\tilde{g}(x_n,y_n)|^2\Big]dx_n dy_n$$

$$\leq 2c(\varepsilon\|\tilde{g}\|\,\|\partial_{y_n}\tilde{g}\| + \varepsilon^{-1}\|y_n\tilde{g}\|\,\|\partial_{y_n}y_n\tilde{g}\|) \quad ,$$

with norms over $L^2_{x_n,y_n}(\mathbb{R}^2_{++})$. (A trace estimate as in (A.54) was used to extend the integration to y_n .) Then by (3.3.52)

$$(3.3.56)\quad |\tilde{\tilde{g}}(x',\theta',\xi',\mu)| \leq c(x')(\varepsilon(\rho^\nu+1)^2\kappa^{-2d+1} + \varepsilon^{-1}(\rho^{\nu-1}+1)(\rho^\nu+1)\kappa^{-2d-1})^{\frac{1}{2}} \quad .$$

If $\nu \geq 1$, this gives, by taking $\varepsilon = \kappa^{-1}$

(3.3.57) $$|\tilde{\tilde{g}}| \leq c(x')\kappa^{-d} \quad .$$

If $\nu \leq 0$ (this occurs in lower order parts and derivatives), $\rho^{\nu}+1 \simeq \langle\xi'\rangle^{\nu}\kappa^{-\nu}$, and we find, taking $\varepsilon = \langle\xi'\rangle^{-\frac{1}{2}}\kappa^{-\frac{1}{2}}$,

(3.3.58) $$\begin{aligned} |\tilde{\tilde{g}}| &\leq c(x')(\varepsilon\langle\xi'\rangle^{2\nu}\kappa^{-2\nu-2d+1} + \varepsilon^{-1}\langle\xi'\rangle^{2\nu-1}\kappa^{-2\nu-2d})^{\frac{1}{2}} \\ &\leq c(x')\langle\xi'\rangle^{\nu-\frac{1}{4}}\kappa^{-d-\nu+\frac{1}{4}} \quad . \end{aligned}$$

If $\nu = \frac{1}{2}$ (the value between 0 and 1 that is of interest in the study of realizations), we must appeal to other considerations, for here the use of (3.3.56) only leads to regularity $-\frac{1}{8}$. We take instead the Laguerre series estimates into account.

Note first that the terms in $\tilde{\tilde{g}}$

(3.3.59) $$\tilde{\tilde{g}}_{-d-1-\ell}(x',\theta,\xi',\mu) = \int_0^\infty \tilde{g}_{-d-1-\ell}(x',x_n,x_n,\theta,\xi',\mu)dx_n$$

are homogeneous in (ξ',μ) of degree $-d-\ell$ for $|\xi'| \geq 1$, in view of the quasi-homogeneity of $\tilde{g}$ (3.3.51). Moreover, when ν is integer, the above arguments apply just as well to the derived expressions

$$D^{\beta}_{x,\theta}D^{\alpha}_{\xi'}D^{j}_{\mu}\,[\tilde{\tilde{g}} - \sum_{\ell<M}\tilde{\tilde{g}}_{-d-1-\ell}]$$

(of regularity $\nu-|\alpha|-M$), which completes the proof of (3.3.54) in the case $\nu \in \mathbb{N}$. (There is a slight inconsistency of notations here in the indexation of the terms in $\tilde{\tilde{g}}$, in that the index equals the degree of homogeneity minus 1.)

For the case $\nu = \frac{1}{2}$, the above proof works for $\tilde{\tilde{g}} - \tilde{\tilde{g}}^0$, showing that it is in $S^{-d-1,\nu-\frac{1}{4}-1}$, since $\tilde{g} - \tilde{g}^0$ is of regularity $-\frac{1}{2}$ (here (3.3.58) applies). It also shows that $D^{\alpha}_{\xi'}\tilde{\tilde{g}}^0 \in S^{-d-|\alpha|,\nu-\frac{1}{4}-|\alpha|}$ for $|\alpha| \geq 1$. So it only remains to estimate the term $\tilde{\tilde{g}}^0$ and its derivatives in x' and μ. For $\tilde{\tilde{g}}^0$ itself, we use the last remark in Theorem 3.2.3, that estimates of the $\ell^2_{\delta,\delta}$ resp. ℓ^2_{δ} norms of the Laguerre coefficients in the principal symbol of A_λ by $O(c(x')\kappa^r)$ (for the appropriate r associated with each order and for some $\delta \in]0,\frac{1}{2}[$), imply analogous estimates of $\ell^2_{\delta,\delta}$ resp. ℓ^2_{δ} norms in the principal inverse symbol. Here we recall that in the present case, the given operator has such uniformly bounded $\ell^2_{\frac{1}{2},\frac{1}{2}}$ and $\ell^2_{\frac{1}{2}}$ estimates, so the resulting operator has uniformly bounded $\ell^2_{\delta,\delta}$ and ℓ^2_{δ} estimates for any $\delta < \frac{1}{2}$, cf. (3.3.12) ff. and (3.2.15') (some improvements to $\delta = \frac{1}{2}$ are possible).

The following is written for the case $N = 1$ (otherwise the argument is applied to each element in the N×N-matrix $\tilde{\tilde{g}}^0$).

In view of the identities (Lidskii's theorem)

$$\text{(3.3.60)} \qquad \operatorname{tr} g(D_n) = \int_0^\infty \tilde{g}(x_n,x_n)dx_n = \sum_{\ell\in\mathbb{N}} c_{\ell\ell}$$

where $(c_{\ell m})_{\ell,m\in\mathbb{N}}$ is the system of Laguerre coefficients of g (cf. (2.3.19)), we have

$$\text{(3.3.61)} \qquad |\tilde{\tilde{g}}^0| = |\sum_{\ell\in\mathbb{N}} c^0_{\ell\ell}| \le \sum_{\ell\in\mathbb{N}} (1+\ell)^{-a}(1+\ell)^a |c^0_{\ell\ell}|$$

$$\le c_a(\sum_{\ell\in\mathbb{N}} (1+\ell)^{2a}|c^0_{\ell\ell}|^2)^{\frac{1}{2}} \qquad \text{for } a > \tfrac{1}{2}$$

$$\le c_a(\sum_{\ell,m\in\mathbb{N}} (1+\ell)^a(1+m)^a|c^0_{\ell m}|^2)^{\frac{1}{2}}$$

$$= c_a \, \|(c^0_{\ell m})_{\ell,m\in\mathbb{N}}\|_{\ell^2_{a/2,a/2}}$$

where $a/2$ must be greater than $\frac{1}{4}$ (since $a > \frac{1}{2}$). By the information on $\tilde{\tilde{g}}^0$, the latter expression is $O(c(x')_\kappa{}^{-d})$ when $a/2 \le \delta$, where δ can be arbitrarily close to $\frac{1}{2}$. There is just enough leeway (between $\frac{1}{4}$ and $\frac{1}{2}$!) to conclude that

$$|\tilde{\tilde{g}}^0(x',\theta,\xi',\mu)| \le c(x')_\kappa{}^{-d} \quad ,$$

which is what it should satisfy in the case of regularity ≥ 0 . For the x' and μ-derivatives, the corresponding estimates are shown by investigation of the compositions involved in formula (3.2.50), again giving regularity ≥ 0 (consistently with regularity $\nu - \frac{1}{4} = \frac{1}{4}$). So (3.3.54) can be concluded in this case also.

Finally, the other half-integer cases ($\nu = k+\frac{1}{2}$ with k integer between 1 and d-1) build on the above arguments for the first few terms and derivatives in $\tilde{\tilde{g}}$, in a slightly more elaborated fashion. □

For the various symbols and symbol-kernels and their terms we also use the notation $g_j(x',\xi,\eta_n,\lambda)$, $\tilde{g}_j(x',x_n,y_n,\xi',\lambda)$, $\tilde{\tilde{g}}_j(x',\xi',\lambda)$, etc. (with $\lambda = -e^{i\theta}\mu^d$) , when there is no ambiguity; this can be practical in formulas where the analyticity plays a rôle.

On the basis of Theorem 3.3.10, the behavior of the kernel of G_λ can easily be studied.

3.3.11 Theorem. *Hypotheses as in Theorem 3.3.10. The distribution kernel* $K_G(x,y,\theta,\mu)$ *(also denoted* $K_G(x,y,\lambda)$*) of* $G_{\theta,\mu} = G_\lambda$ *(reduced to an operator on* $\Gamma\times\overline{\mathbb{R}}_+$ *as explained before Theorem 3.3.10) has an asymptotic expansion*

$$K_G(x,y,\theta,\mu) \sim \sum_{\ell\in\mathbb{N}} K_{G,-d-1-\ell}(x,y,\theta,\mu) \tag{3.3.62}$$

satisfying for multiindices with

$$d' \equiv -d-|\alpha|+|\beta|+|\gamma|-M-j-k+k'-m+m' < -n+1 \ , \tag{3.3.63}$$

the following estimates in local coordinates:

$$\|(x'-y')^\alpha D^\beta_{x'} D^\gamma_{y'} D^j_\mu x_n^k D^{k'}_{x_n} y_n^m D^{m'}_{y_n}[K_G - \sum_{\ell<M} K_{G,-d-1-\ell}]\|_{L^2_{x_n,y_n}} \tag{3.3.64}$$

$$\leq \begin{cases} c(x')\langle\mu\rangle^{n-1+d'} & \text{when } \nu+|\gamma|-M' > 1-n \ , \\ c(x')[\langle\mu\rangle^{n-1+d'}\log(2+\mu) + \langle\mu\rangle^{-d-\nu+M''-j}] & \text{when } 1-n \geq \nu+|\gamma|-M' \geq 1-n-|\beta| \ , \\ c(x')\langle\mu\rangle^{-d-\nu+M''-j} & \text{when } \nu+|\gamma|+|\beta|-M' < 1-n \ ; \end{cases}$$

here M' *and* M'' *are as defined in* (3.3.53).

On the diagonal one has, when $d \geq n$, *setting* $z = (-\lambda)^{1/d}$,

$$\int_0^\infty K_G(x',x_n,x',x_n,\theta,\mu)dx_n = s_{-d-1}(x')z^{n-d-1}+\dots+s_{-d-\nu'-n}(x')z^{-d-\nu'}+s'(x',\theta,\mu) , \tag{3.3.65}$$

where the $s_{-d-\ell}(x')$ *are* C^∞ *functions,* ν' *is the largest integer* $< \nu$, *and* s' *is a* C^∞ *function for* $\mu \neq 0$, *satisfying*

$$|D^\beta_{x'} D^j_\mu s'(x',\theta,\mu)| \leq c(x')|\mu|^{-d-\nu+1/4-j} , \quad \textit{for} \ \mu \geq c > 0 \ ; \tag{3.3.66}$$

the estimates are valid in a truncated sector around $\mathbb{R}_-$, *with analyticity in the same way as in Theorem 3.3.4. In particular, when* $d > n$, *the trace of* G_λ *satisfies*

$$\operatorname{tr} G_\lambda = \int_{\Gamma\times\mathbb{R}_+} \operatorname{tr} K_G(x,x,\theta,\mu)dx = \underline{s}_{-d-1}z^{n-d-1}+\dots+\underline{s}_{-d-\nu'-n}z^{-d-\nu'} + s'(\lambda) \ , \tag{3.3.67}$$

with constants $\underline{s}_{-d-\ell} = \int_\Gamma s_{-d-\ell}(x')dx'$, *and*

$$|s'(\lambda)| \leq c|\lambda|^{-1-(\nu-1/4)/d} \quad \textit{for} \quad |\lambda| \geq c_0 > 0 \ ,$$

λ *in a truncated sector around* $\mathbb{R}_-$.

Proof: The kernel estimates are derived from the symbol-kernel estimates (3.3.52) by use of the Fourier inversion formula (in local coordinates)

$$K_G(x,y,\theta,\mu) = F^{-1}_{\xi'\to x'-y'}\,\tilde{g}(x',x_n,y_n,\theta,\xi',\mu) \quad ,$$

that gives a function K_G continuous in (x',y') with values in $L^2_{x_n,y_n}$ when the order is $< -n+1$. One proceeds as in the proof of Theorem 3.3.4; only the estimates are somewhat more complicated because of the special effect of operations with respect to x_n and y_n. One has the typical estimate (by use of Lemma 2.6.3 3^0 and Lemma 3.3.5) for an expression of order $d' = -d-|\alpha|+|\gamma|+|\theta|-k+k'$:

$$\Big\|\int_{\mathbb{R}^{n-1}} e^{i(x'-y')\cdot\xi'}\bar{D}^{\alpha}_{\xi'}[(\xi')^{\gamma+\theta}D^{\beta-\theta}_{x'}x_n^kD^{k'}_{x_n}\tilde{g}]\,d\xi'\Big\|_{L^2_{x_n,y_n}}$$
$$\leq c\int_{\mathbb{R}^{n-1}}(\rho(\xi',\mu)^{\nu-|\alpha|+|\gamma+\theta|-[k-k']_+} + 1)\langle\xi',\mu\rangle^{d'}\,d\xi'$$

$$\leq \begin{cases} c_1\langle\mu\rangle^{n-1+d'} & \text{if } \nu-|\alpha|+|\gamma+\theta|-[k-k']_+ > 1-n\ , \\ c_1\langle\mu\rangle^{n-1+d'}\log(2+\mu) & \text{if } \nu-|\alpha|+|\gamma+\theta|-[k-k']_+ = 1-n\ , \\ c_1\langle\mu\rangle^{-d-\nu+[k-k']_-} & \text{if } \nu-|\alpha|+|\gamma+\theta|-[k-k']_+ < 1-n\ . \end{cases}$$

Estimating x'-derivatives by use of the Leibniz formula, and taking the effect of M and j into account, one finds (3.3.64). $K_G(x,y,\mu)$ is continuous if $d>n$.

For $d\geq n$, the x_n-integral is continuous in x',y' (cf. (3.3.55) ff.), and

$$\text{(3.3.68)}\qquad \int_0^\infty K_G(x',x_n,x',x_n,\theta,\mu)\,dx_n = (2\pi)^{1-n}\int_{\mathbb{R}^{n-1}}\tilde{\tilde{g}}(x',\theta,\xi',\mu)\,d\xi' \quad ,$$

which is the diagonal value of the kernel of the ps.d.o. with symbol $\tilde{\tilde{g}}$, described in the preceding theorem. The properties of the symbol space $S^{-d,\nu-\frac14}(\Omega',\bar{\mathbb{R}}^n_+)$ imply here, similarly to (3.3.35-38), that

$$\text{(3.3.69)}\qquad \tilde{\tilde{g}}(x',\theta,\xi',\mu) = \tilde{\tilde{g}}^h_{-d-1} + \ldots + \tilde{\tilde{g}}^h_{-d-\nu'-n} + \tilde{\tilde{g}}' \ ,$$

where the terms for $\ell < \nu-\frac14$ (i.e. $\ell \leq \nu'$) satisfy

$$\begin{aligned} &|\tilde{\tilde{g}}_{-d-1-\ell}(x',\theta,\xi',\mu)| \leq c(x')\langle\xi',\mu\rangle^{-d-\ell} \ , \\ \text{(3.3.70)}\quad &|\tilde{\tilde{g}}^h_{-d-1-\ell}(x',\theta,\xi',\mu)| \leq c(x')|\xi',\mu|^{-d-\ell} \quad \text{and} \\ &|\tilde{\tilde{g}}_{-d-1-\ell} - \tilde{\tilde{g}}^h_{-d-1-\ell}| \leq c(x')\langle\xi',\mu\rangle^{-d-\nu+\frac14} \quad \text{for } |\xi',\mu| \geq c > 0\ , \end{aligned}$$

and next, for $\nu-\frac14 < \ell < \nu-\frac14+n-1$ (i.e. $\nu' < \ell \leq \nu'+n-1$) satisfy

$$(3.3.71)\quad \begin{aligned} |\tilde{\tilde{g}}_{-d-1-\ell}| &\le c(x')\langle\xi'\rangle^{\nu-1/4-\ell}\langle\xi',\mu\rangle^{-d-\nu+1/4} \\ |\tilde{\tilde{g}}^h_{-d-1-\ell}| &\le c(x')|\xi'|^{\nu-1/4-\ell}|\xi',\mu|^{-d-\nu+1/4} \quad , \end{aligned}$$

with a remainder $\tilde{\tilde{g}} - \sum_{\ell\le\nu'+n-1} \tilde{\tilde{g}}_{-d-1-\ell} = \tilde{\tilde{g}}' = f'' + f'''$, where

$$(3.3.72)\quad \begin{aligned} |f''| &\le c(x')|\xi'|^{\nu-1/4-(\nu'+n-1)}|\mu|^{-d-\nu+1/4}\chi(\xi') \quad , \\ |f'''| &\le c(x')\langle\xi'\rangle^{\nu-1/4-\nu'-n}\langle\xi',\mu\rangle^{-d-\nu+1/4} \quad . \end{aligned}$$

Then (3.3.65-67) follow by integration and analytic extension, as in Theorem 3.3.4. (We assume $d>n$ for (3.3.67) in order to use [Agmon 3], but since G_λ is of trace class for $d=n$ also ([Grubb 17]), $d=n$ can probably be included.) □

3.3.12 Remark. Just as in Remark 3.3.8 one can observe that there are kernel estimates also when $d \le n-1$. Here one can consider

$$(3.3.73)\qquad G'_{\lambda,M} = G_\lambda - \sum_{\ell<M} G_{\lambda,-d-1-\ell} \quad ,$$

where the $G_{\lambda,-d-1-\ell}$ are operators defined from the symbols $g_{-d-1-\ell}$ by use of local coordinate systems and partitions of unity, briefly expressed:

$$G_{\lambda,-d-1-\ell} = \mathrm{OPG}\,(g_{-d-1-\ell}) \quad ,$$

and $G'_{\lambda,M}$ as well as the terms $G_{\lambda,-d-1-\ell}$ are of order $< -n$ when $d+M$ resp. $d+\ell$ is $> n$. Then the kernels are continuous in (x,y) and depend analytically on λ . We have in particular the estimates of the kernel $K_{G'_M}(x,y,\lambda)$ of $G'_{\lambda,M}$

$$(3.3.74)\qquad \sup_{x',y'} \| D^\beta_{x',y'} K_{G'_M}(x,y,\lambda)\|_{L^2_{x_n,y_n}} \le c\langle\lambda\rangle^{-1-\nu/d} \quad ,$$

when $M > n-1+\nu+|\beta|$; this is a special case of (3.3.64), which also lists the estimates for other derived functions. One can show asymptotic expansions of $K_{G'_M}$ when $n-d < M < n+\nu$, containing part of the expressions in (3.3.65) and (3.3.67) (somewhat like in (3.3.47)).

Collecting the results for Q_λ and G_λ one finds, setting $R_{\lambda,-d-\ell} = (Q_{\lambda,-d-\ell})_\Omega + G_{\lambda,-d-\ell}$, the expansion

$$(3.3.74')\quad \mathrm{tr}[R_\lambda - \sum_{\ell<M} R_{\lambda,-d-\ell}] = \sum_{M\le j\le n+\nu'} a_{-d-j}(B)(-\lambda)^{(n-d-j)/d} + O(|\lambda|^{-1-(\nu-1/4)/d})$$

for $M+d > n$; where each $a_{-d-j}(B)$ is a constant determined from the k'th symbols in P and their derivatives up to order $j-k$, for $0 \le k \le j$, and the k'th symbols in G and T and their derivatives up to order $j-k$, for $0 \le k \le j-1$.

Let us end this section with some remarks on possible improvements of the kernel estimates given above. The main outcome will be, that for the case without boundary, one can obtain full asymptotic expansions in a certain sense, by use of recomposition of Q_λ with powers of P ; but there is no analogue for our pseudo-differential boundary problems, since recomposition with powers of B does not increase the regularity here.

Consider first the boundaryless case. Let $Q_\lambda = (P-\lambda)^{-1}$ in a truncated sector around $\mathbb{R}_-$ (permissible in view of Remark 3.3.6). By iteration of the identity

$$Q_\lambda = \lambda^{-1}(\lambda-P+P)Q_\lambda = -\lambda^{-1} + \lambda^{-1}PQ_\lambda$$

one has for $m \in \mathbb{N}_+$,

$$(3.3.75) \qquad Q_\lambda = -\lambda^{-1} - \lambda^{-2}P - \ldots - \lambda^{-m}P^{m-1} + \lambda^{-m}P^mQ_\lambda \; .$$

Similarly, one has for $R_\lambda = (B-\lambda)^{-1}$, any $m \in \mathbb{N}_+$,

$$(3.3.76) \qquad R_\lambda f = (-\lambda^{-1}-\lambda^{-2}B - \ldots - \lambda^{-m}B^{m-1})f + \lambda^{-m}B^mR_\lambda f \quad \text{for} \quad f \in D(B^{m-1}) \; .$$

The identity (3.3.75) is convenient because Q_λ is here written as a sum of terms that are polynomial in λ^{-1} and P , plus a term whose regularity is generally <u>better</u> than that of Q_λ , namely: P^mQ_λ has regularity md , by Lemma 2.1.6 (for P^m has regularity md and is μ-independent). This leads to new representations of the terms in the symbol of q beyond the first $d+n-1$ terms occurring in Theorem 3.3.4. Consider the case where $d > n$ (otherwise one subtracts a few terms from Q_λ) .

Let us denote

$$(3.3.77) \qquad P^mQ_\lambda = Q_\lambda^{(m)}$$

with symbol

$$(3.3.78) \qquad q^{(m)}(x,\theta,\xi,\mu) \sim \sum_{\ell\in\mathbb{N}} q^{(m)}_{(m-1)d-\ell} \quad \text{in } S^{(m-1)d,md}(\Sigma\times[-\varepsilon,\varepsilon], \overline{\mathbb{R}}^{n+1}_+) \; .$$

We also write

$$(3.3.79) \qquad \begin{aligned} Q_\lambda^{(m)} &= \sum_{\ell<M} Q^{(m)}_{\lambda,(m-1)d-\ell} + Q_M^{(m)\prime} \; , \quad \text{where} \\ Q^{(m)}_{\lambda,(m-1)d-\ell} &= OP(q^{(m)}_{(m-1)d-\ell}(x,\xi,\lambda)) \; . \end{aligned}$$

Now for example, taking $m = 2$ in (3.3.75), we can write

$$q = -\lambda^{-1} - \lambda^{-2}p + \lambda^{-2}q^{(2)}$$
$$= -\lambda^{-1} - \lambda^{-2}(p_d + \ldots + p_{-d} + O(\langle\xi\rangle^{-d-1})) + \lambda^{-2}(q_d^{(2)} + \ldots + q_{-d}^{(2)} + O(\langle\xi\rangle^{-1}\langle\xi,\mu\rangle^{-d})),$$

which gives a new representation of q_{-3d} (equal to $q_{-d-\ell}$ with $\ell = 2d$)

$$(3.3.80) \qquad q_{-3d} = -\lambda^{-2}p_{-d} + \lambda^{-2}q_{-d}^{(2)} ,$$

where the strictly homogeneous version of $q_{-d}^{(2)}$ is $O(|\xi,\mu|^{-d})$, hence is locally bounded at $\xi = 0$, in contrast with q_{-3d}^h. Of course, p_{-d}^h carries a big singularity at $\xi = 0$, but the interesting thing about this is that it is λ-independent.

More generally, the ℓ-th term in q satisfies (for $\ell > 0$)

$$(3.3.81) \quad q_{-d-\ell} = -\lambda^{-2}p_{d-\ell} - \ldots - \lambda^{-m}p_{(m-1)d-\ell}^{(m-1)} + \lambda^{-m}q_{(m-1)d-\ell}^{(m)}$$

where $p^{(j)}$ denotes the symbol of P^j, and $q_{(m-1)d-N}^{(m)}$ is of order $(m-1)d-N$ and regularity $md-N$; and the remainders satisfy:

$$(3.3.82) \quad q - \sum_{\ell<N} q_{-d-\ell} = \sum_{j=1}^{m} \lambda^{-j}\cdot(\lambda\text{-independent terms}) + \lambda^{-m}r_{m,N}(x,\theta,\xi,\mu) ,$$

where $r_{m,N}$ is of order $(m-1)d-N$ and regularity $md-N$.

When one tries to imitate the calculations in the proof of Theorem 3.3.4, one finds e.g. that $\lambda^{-2}q_{-d}^{(2)}$ gives a nice contribution to the kernel and its trace (the latter contribution being of the form $c(x)(-\lambda)^{-3+n/d}$ plus a remainder of lower order), whereas $-\lambda^{-2}p_{-d}$ gives a singular contribution to the distribution kernel; however, this has the form of a λ-independent coefficient times λ^{-2}, which is of <u>integer degree</u>.

An elaboration of this point of view leads to some kind of full asymptotic expansion of the kernel, as a series of nice terms plus a singular contribution where λ merely enters as negative integer powers.

For the heat kernel expansion that we shall consider in Section 4.2 (as well as for the fractional powers considered in Section 4.4), <u>the contributions from $-\lambda^{-1} - \lambda^{-2}P - \ldots - \lambda^{-m}P^{m-1}$ vanish</u>, and one can show a satisfactory complete expansion of the trace.

Now it would be nice if one could obtain similar information from (3.3.76) for boundary value problems. Unfortunately, the regularity of B^mR_λ does not in general improve with increasing m, when B is truly pseudo-differential. For here,

(3.3.83) $$B^m R_\lambda = (P_\Omega + G)^m (Q_{\lambda,\Omega} + G_\lambda) \;,$$

where the regularity of G_λ is decisive, in the following sense:

Let G and T have regularity $\nu \in [\frac{1}{2}, d]$, then G_λ has regularity ν by Theorem 3.3.1. The regularity of G itself may possibly be better, equal to some $\nu_1 \in [\nu, d]$, and then the regularity of $(P_\Omega + G)^m Q_{\lambda,\Omega}$ is at least ν_1. In the consideration of $(P_\Omega + G) G_\lambda$, a regularity improvement can occur in the terms resulting from compositions $G' G_\lambda$ where G' is a s.g.o. of class 0 and high order (like in Lemma 2.1.6 and Proposition 2.6.11). But there is a typical contribution where the regularity does not increase, namely

(3.3.84) $$s_d^m(x') D_{x_n}^{dm} G_\lambda \;,$$

stemming from the term of highest order with respect to D_{x_n} in P^m (here $s_d(x')$ is nonzero in view of the ellipticity). The regularity of (3.3.84) is ν (the same as G_λ) , for the symbol-kernel simply equals $s_d^m(x') D_{x_n}^{dm} \tilde{g}(x_n, y_n, \dots)$, which lies in $S^{(m-1)d,\nu}(\mathscr{S}(\overline{\mathbb{R}}_{++}^2))$ (cf. (2.3.28)). Other "bad" contributions will come from s.g.o. terms of the type $K_j \gamma_j$ in $(P_\Omega + G)^m$. (See also Ex.2.2.6.)

For this reason, we do not develop the resolvent kernel in further terms, and the heat trace expansion we obtain in Section 4.2 stops after a finite number of terms, depending directly on ν . (Luckily, there will be enough terms in the expansion to get interesting consequences, e.g. for index theory.)

The kernel of the Poisson operator K_λ can be analyzed in a similar way as in Theorems 3.3.10-11 and exhibits many of the same features.

In fact, the above analysis is a special case of a general asymptotic analysis with a similar flavor, that can be carried out for each of the terms in $\mathcal{B}_\mu$, when $\mathcal{B}_\mu$ is a parametrix (or inverse) of A_μ , as in Theorem 3.2.10.

3.4 Other special cases.

The following kind of parameter-dependent ps.d.o. is a little more general than the case $P + \omega\mu^d$ considered in Section 3.3:

$$(3.4.1) \qquad P_\mu = P^{(m)} + \mu^d P^{(m-1)} + \ldots + \mu^{(m-1)d} P^{(1)} + \mu^{md} I ,$$

with $P^{(j)}$ denoting a (parameter-independent) ps.d.o. in $\widetilde{E}$ of order jd, having, as always, the transmission property at Γ. Here P_μ is of order md, but the regularity is generally only d, since the $P^{(j)}$ are of regularity jd, with $j \geq 1$, cf. (2.1.8). (If some terms are differential operators, the regularity can be higher, but to fix the ideas in the following, we assume that $P^{(1)}$ is a ps.d.o.) (3.4.1) can be said to be polynomially parameter-dependent.

For P_μ we shall consider boundary problems

$$(3.4.2) \qquad \begin{aligned} P_{\mu,\Omega} u + G_\mu u &= f \quad \text{on } \Omega , \\ T_\mu u &= \varphi \quad \text{at } \partial\Omega , \end{aligned}$$

where G_μ and T_μ are of the form

$$(3.4.3) \qquad \begin{aligned} G_\mu &= G^{(m)} + \mu^d G^{(m-1)} + \ldots + \mu^{(m-1)d} G^{(1)} , \\ T_\mu &= T^{(m)} + \mu^d T^{(m-1)} + \ldots + \mu^{(m-1)d} T^{(1)} ; \end{aligned}$$

here the $G^{(j)}$ are s.g.o.s of order jd and class $\leq jd$, and the $T^{(j)}$ are column vectors

$$(3.4.4) \qquad T^{(j)} = \left\{ T_k^{(j)} \right\}_{0 \leq k < md} ,$$

with $T_k^{(j)}$ of order $k-(m-j)d$ and of the form

$$(3.4.5) \qquad \begin{aligned} T_k^{(j)} &= 0 \quad \text{if} \quad k-(m-j)d < 0 , \\ T_k^{(j)} &= S_{r,r}^{(j)}(x')\gamma_r + \sum_{0 \leq \ell < r} S_{r,\ell}^{(j)} \gamma_\ell + T_k'^{(j)} \quad \text{if} \quad r \equiv k-(m-j)d \geq 0 , \end{aligned}$$

the $S_{r,\ell}^{(j)}$ being ps.d.o.s over Γ of order $r-\ell$ and $T_k'^{(j)}$ being a trace operator of order r and class 0.

The hypotheses are general enough to include the case

$$(3.4.6)\qquad \begin{aligned} (P_\Omega+G+\mu^d)^m v &= f \\ Tv &= \varphi_0 \\ T(P_\Omega+G+\mu^d)v &= \varphi_1 \\ &\vdots \\ T(P_\Omega+G+\mu^d)^{m-1}v &= \varphi_{m-1} \end{aligned}$$

obtained by iteration of the problem

$$(3.4.7)\qquad \begin{aligned} (P_\Omega+G+\mu^d)u &= f\ , \\ Tu &= \varphi\ . \end{aligned}$$

But of course, also more general choices of G_μ and T_μ could be considered.

The singular Green operator and the trace operator in (3.4.3) have a certain regularity $\nu \geq \frac{1}{2}$. Observe that (3.4.6) will not in general have a better regularity than (3.4.7) (similarly to the observations at the end of Section 3.3).

Let us assume that the system

$$(3.4.8)\qquad A_\mu = \begin{pmatrix} P_{\mu,\Omega} + G_\mu \\ T_\mu \end{pmatrix}$$

is parameter-elliptic in the sense of Definition 3.1.3; this will in particular require normality of $T^{(m)}$ (which in the case (3.4.6) amounts to normality of T). (Examples where the parameter-ellipticity holds, can be constructed e.g. from strongly elliptic cases.) The ps.d.o. part P_μ then has a parametrix Q_μ of order $-md$ and regularity d, by Theorem 2.1.16; and if Σ is compact, Q_μ is the true inverse, for μ sufficiently large.

The "leftover" singular Green operator $L(P_\mu,Q_\mu)$ is seen to be of regularity d and not just $d-\frac{1}{2}$, by a variant of Proposition 2.6.11: One has that

$$(3.4.9)\qquad L(P^{(j)},Q_\mu) = G^+(P^{(j)})G^-(Q_\mu)\ ,$$

where $G^+(P^{(j)})$ is μ-independent and is of order $jd \geq d$ and class 0, and $G^-(Q_\mu)$ is of order $-md$, regularity $d-\frac{1}{2}$ and class 0; then applications of the Leibniz formula and the Cauchy-Schwarz inequality show that $L(P^{(j)},Q_\mu)$ is of order $(j-m)d$ and regularity

$$\min\{jd+d-\tfrac{1}{2}\ ,\ jd\} = jd\ ,$$

as in Lemma 2.1.6.

The construction in Theorem 3.2.3 therefore gives an inverse symbol with regularity $\nu' = \min\{\nu,d\}$, and A_μ has a parametrix

(3.4.10) $$\mathcal{B}_\mu = (Q_{\mu,\Omega} + G'_\mu \quad K_\mu)$$

with regularity d in the ps.d.o. part and ν' in the other terms.

It is found just as in Section 3.2, that the parametrix can be taken to be an inverse of A_μ for μ sufficiently large; and the construction extends by analyticity to μ in a neighborhood of $\overline{\mathbb{R}}_+$.

In particular, the semi-homogeneous problem

(3.4.11) $$\begin{aligned}(P_{\mu,\Omega} + G_\mu)u &= f\\ T_\mu u &= 0\end{aligned}$$

has the solution operator

(3.4.12) $$R_\mu = Q_{\mu,\Omega} + G'_\mu \ ,$$

where the two terms have kernels as described in Theorems 3.3.4 and 3.3.11 (with modifications due to the fact that the order is md instead of d). Summing up, we have obtained the result for polynomially parameter-dependent systems:

3.4.1 Theorem. *Let* m *and* $d \in \mathbb{N}_+$, *let* P_μ *be of the form* (3.4.1), *of order* md *and regularity* d , *and consider the system* A_μ (3.4.8), *as defined in* (3.4.3-5), *with singular terms of regularity* ν ($\geq \frac{1}{2}$). *When* A_μ *is parameter-elliptic (Definition 3.1.3), it has a parametrix* B_μ *(an inverse for sufficiently large* μ*) of the form* (3.4.10), *where the ps.d.o. part* Q_μ *is of regularity* d *and the singular terms are of regularity* $\nu' = \min\{\nu,d\}$. B_μ *extends analytically to* μ *in a neighborhood of* $\overline{\mathbb{R}}_+$, *and kernel estimates as in Section 3.3 are valid.*

3.4.2 Remark. The theorem applies in particular to the powers $(B+\mu^d)^{-m}$ of the resolvent of a realization B defined as in Section 3.3. Here a direct composition $(B+\mu^d)^{-1} \circ (B+\mu^d)^{-1}$ according to the rules of calculus in Section 2.6-2.7 could lead to a loss of regularity by $\frac{1}{2}$, but when $(B+\mu^d)^{-m}$ is considered as the inverse of the operator

$$(3.4.13)\qquad (B+\mu^d)^m = \sum_{0\le j\le m} \binom{m}{j}\mu^{(m-j)d}B^j$$

(the realization associated with (3.4.6)), then we get from Theorem 3.4.1 that the regularity of $(B+\mu^d)^{-m}$ is no worse than the regularity of $(B+\mu^d)^{-1}$.

Another case that can be studied by our methods is the resolvent of a square matrix-formed Green operator of order $d \in \mathbb{N}$ and class $\le d$

$$(3.4.14)\qquad A = \begin{pmatrix} P_\Omega+G & K \\ T & S \end{pmatrix} : \begin{matrix} H^d(E) \\ \times \\ H^{d-\frac12}(F) \end{matrix} \rightarrow \begin{matrix} H^0(E) \\ \times \\ H^{-\frac12}(F) \end{matrix} .$$

When $d = 0$, the operator $A-\lambda I$ is invertible for all $\lambda \in \mathbb{C}$ with $|\lambda| \ge C$, for some large constant C , since A is bounded in $H^0(E)\times H^{-\frac12}(F)$. For such operators, there is an easy functional calculus allowing quite general functions of A (as in the last section of [Seeley 1]). In this case, the integration parameter λ in the Cauchy integral formulas runs on a compact curve, so the resulting operators have whatever properties are preserved under such integrations. For example, the complex powers A^z are again Green operators of order and class 0 , belonging to the Boutet de Monvel calculus (polyhomogeneous resp. of $S_{1,0}$ type, when A is so, with principal homogeneous symbol resp. symbol having no eigenvalues on $\overline{\mathbb{R}}_-$) . More details are given at the end of this section.

When $d \ge 1$, we need the full parameter-dependent calculus. P and S are of regularity d , G has a certain regularity in $[\frac12,d]$ as explained in Section 3.3, and T is by assumption of the form (and of order d)

$$(3.4.15)\qquad T = \sum_{0\le \ell\le d-1} S_\ell\gamma_\ell + T' ,$$

with T' of class 0 and S_ℓ of order $d-\ell$, so that it has a regularity in $[1,d]$. K has regularity $d-\frac12$, cf. (2.3.51). Altogether, the singular terms have a regularity $\nu \in [\frac12,d]$.

Now consider

$$(3.4.16)\qquad A_\lambda = A_{\theta,\mu} = \begin{pmatrix} P_\Omega+G-\lambda & K \\ T & S-\lambda \end{pmatrix} , \text{ with } \lambda = -e^{i\theta}\mu^d \ (\mu \ge 0) .$$

Definition 3.1.3 can be applied without difficulty, since the principal x_n-independent homogeneous boundary symbol operator does have a limit for $\xi' \to 0$,

$$(3.4.17)\quad a^h(x',0,\mu,D_n) = \begin{pmatrix} s_d(x',0)D_n^d + e^{i\theta}\mu^d & 0 \\ 0 & e^{i\theta}\mu^d \end{pmatrix} : \begin{matrix} H^d(\overline{\mathbb{R}}_+)^N \\ \times \\ \mathbb{C}^M \end{matrix} \to \begin{matrix} H^0(\overline{\mathbb{R}}_+)^N \\ \times \\ \mathbb{C}^M \end{matrix}$$

where $s_d(x',0)$ is the coefficient of D_n^d in P . An example where this boundary symbol operator is invertible is $P = OP(\lambda_-^d)$, cf. (3.1.14) and Lemma 3.1.2.

Note that T is not normal here. One could also take for T an operator of order $d-j$ and for K an operator of order $d+j$, for some integer $j \in [1,d]$, so that one has to do with a system

$$(3.4.18)\quad A = \begin{pmatrix} P_\Omega + G & K \\ T & S \end{pmatrix} : \begin{matrix} H^d(E) \\ \times \\ H^{d+j-\frac12}(E) \end{matrix} \to \begin{matrix} H^0(E) \\ \times \\ H^{j-\frac12}(E) \end{matrix} \quad ;$$

here we assume that T is of the form

$$(3.4.19)\quad T = s'_{d-j}(x')\gamma_{d-j} + \sum_{0\le \ell < d-j} S_\ell \gamma_\ell + T'$$

with S_ℓ of order $d-j-\ell$ and T' of class 0 , to get regularity $\ge \frac12$. (More generally, one could take for T a column of such operators with different orders $d-j$; and for K a corresponding row.) The limit boundary symbol operator for $A + e^{i\theta}\mu^d I$ is here

$$(3.4.20)\quad a^h(x',0,\mu,D_n) = \begin{pmatrix} s_d(x',0)D_n^d + e^{i\theta}\mu^d & 0 \\ s'_{d-j}(x')\gamma_{d-j} & e^{i\theta}\mu^d \end{pmatrix} ,$$

which is invertible e.g. if $p = \lambda_-^d$, as above. Note that normality is not required.

Non-integer orders for T and K could also be used, as long as the operators have positive regularity when considered as μ-dependent.

In each of the abovementioned cases, we can apply Theorem 3.2.3. And also here, the resulting regularity can be improved, by use of the information that by Proposition 2.6.11,

$$L(P,Q_{\theta,\mu}) \quad \text{is of regularity} \quad d \;,$$

where $Q_{\theta,\mu}$ denotes the parametrix of $P + e^{i\theta}\mu^d$. Then there is no extra loss of regularity in the parametrix $B_{\theta,\mu}$ $(= B_\lambda)$ of $A_{\theta,\mu}$. Again one gets analyticity in λ by similar considerations as in Section 3.3. Note that B_λ is of the form

$$B_\lambda = \begin{pmatrix} Q_{\lambda,\Omega}+G'_\lambda & K'_\lambda \\ T'_\lambda & S'_\lambda \end{pmatrix} , \tag{3.4.21}$$

where $Q_{\lambda,\Omega}$ and G'_λ, and their kernels, can be analyzed just as in Section 3.3. A slight generalization of the methods permit similar analyses of the other terms K'_λ, T'_λ and S'_λ (here T'_λ is of class 0). Altogether, we find:

3.4.3 Theorem. *Let* A *be a quadratic system as in* (3.4.14) *or* (3.4.18) *ff. (with the conventions explained there), and assume that* $A-\lambda$ *is parameter-elliptic on the ray* $\lambda = re^{i\theta_0}$. *Let* $\nu > 0$ *denote the regularity of* G,T *and* K. *There is a truncated sector* $W_{r_0,\varepsilon}$ (3.3.16) *such that* $A-\lambda$ *is invertible for* $\lambda \in W_{r_0,\varepsilon}$; *here* $(A-\lambda)^{-1}$ *coincides with a parametrix* B_λ *of* $A-\lambda$, *defined and analytic in a keyhole region* $V_{\delta,\varepsilon}$ (3.3.17). *The ps.d.o. part of* B_λ *has regularity* d, *and the singular part has regularity* $\nu' = \min\{\nu,d\}$. *There are symbol estimates and kernel estimates as in Section 3.3, with generalizations to* K'_λ *and* T'_λ.

It is worth pointing out that the special case where $\dim F = 0$, i.e. where A is of the form

$$A = P_\Omega + G : H^d(E) \to H^0(E) \;, \tag{3.4.22}$$

with $A-\lambda$ parameter-elliptic, is covered by this analysis. Such operators can arize in the study of differential operator systems, see e.g. [Grubb-Geymonat 1].

For future reference, let us here also elaborate the statements on the easy functional calculus for operators of order and class 0. Consider a system A of the form (3.4.14) or (3.4.22) (the case $\dim F = 0$), with all operators of order 0 and class 0. Then A defines a bounded operator (also called A) in $H^0(E)\times H^{-\frac{1}{2}}(F)$, and the spectrum $\mathrm{sp}(A)$ is a compact set in $\mathbb{C}$.

Consider first the polyhomogeneous case. The spectra of the principal interior and boundary symbol operators

$$(3.4.23)\qquad \begin{aligned} sp_{\Omega}(a^0) &\equiv \cup\{sp(p^h(x,\xi)) \mid x \in \overline{\Omega},\ \xi \neq 0\} \\ sp_{\Gamma}(a^0) &\equiv \cup\{sp(a^h(x',\xi',D_n)) \mid x' \in \Gamma,\ \xi' \neq 0\} \end{aligned}$$

are contained in $sp(A)$; in fact they constitute the essential spectrum of A , ess sp(A) . (For, it is shown in [Grubb-Geymonat 1] how the essential spectrum for operators as in (3.4.22) is seen to be equal to $sp_{\Omega}(a^0) \cup sp_{\Gamma}(a^0)$, by use of singular sequences associated with ps.d.o. symbols and s.g.o. symbols; and the treatment extends to (3.4.14) by an inclusion of similar singular sequences for Poisson and trace operators, furnished by W. Höppner in [Rempel-Schulze 1, Section 2.3.4].)

Now let $f(\lambda)$ be a complex function that is holomorphic on a neighborhood ω of $sp(A)$, and let C be a closed smooth curve in ω , going around $sp(A)$ in the positive direction (once). Then we set

$$(3.4.24)\qquad f(A) = \frac{i}{2\pi}\int_C f(\lambda)\ (A-\lambda I)^{-1} d\lambda \quad .$$

Since the integration curve is compact, there are no convergence problems, and the integration can be interchanged with the integrations in the Green operator definitions, showing that $f(A)$ in local coordinates near Γ has the form

$$(3.4.25)\qquad f(A) \sim OP'(f(a(x',\xi',D_n))) \quad ,$$

modulo operators of order $-\infty$ and class 0 , where

$$(3.4.26)\qquad f(a) = \frac{i}{2\pi}\int_C f(\lambda)(a(x',\xi',D_n) - \lambda I)^{(-1)} d\lambda \quad ,$$

$(a(x'\xi',D_n) - \lambda I)^{(-1)}$ being a parametrix symbol for $a(x',\xi',D_n) - \lambda I$. Similarly, $f(A) \sim OP(f(p(x,\xi)))$ in interior coordinate patches. We here use the standard Boutet de Monvel calculus at each λ , where λI is considered as the constant λ times the elliptic zero-order operator I (the identity). The symbol constructions are performed such that the symbol terms depend analytically on λ , and there is a remainder kernel of any high degree of smoothness in (x,y) , depending analytically on λ , by Agmon's kernel theorem [Agmon 4]. For each fixed λ , the symbols in $(a - \lambda I)^{(-1)}$ are Green operator symbols of order and class 0 , and the various symbol estimates are preserved after the integration in λ . Moreover, the transmission property (1.2.12) with $d = 0$ is preserved, and so are homogeneity properties in case f is homogeneous. The ps.d.o. part of $f(A)$ will be equal to $f(P)$, where $f(P)$ is the operator defined by [Seeley 1, Theorem 5].

For example, if C can be chosen such that 0 lies in its exterior set and can be connected to ∞ there (e.g. when $sp(A)$ is disjoint from $\overline{\mathbb{R}}_-$ or another ray), then the complex powers A^z can be defined for all $z \in \mathbb{C}$, by the formula

$$A^z = \frac{i}{2\pi} \int_C \lambda^z \, (A - \lambda I)^{-1} d\lambda \quad ; \tag{3.4.27}$$

and here the principal boundary symbol operator $a^{(z)0}$ is determined by

$$\begin{aligned} a^{(z)0}(x',\xi',D_n) &= \frac{i}{2\pi} \int_C \lambda^z (a^0(x',\xi',D_n) - \lambda I)^{-1} d\lambda \quad , \\ &= (a^0(x',\xi',D_n))^z \;, \quad \text{for } |\xi'| \geq 1 \;. \end{aligned} \tag{3.4.28}$$

The function λ^z is of course taken to be holomorphic on a neighborhood of C and the interior set for C . E.g. if $\overline{\mathbb{R}}_-$ is exterior to C , λ^z is cut along $\overline{\mathbb{R}}_-$ and coincides with positive real powers for λ and $z \in \mathbb{R}_+$.

When A is selfadjoint in $H^0(E)\times H^{-\frac{1}{2}}(F)$, the function $f(A)$ above can be defined consistently with the operator function defined by spectral theory, cf. [Dunford-Schwartz 1].

When A is merely of $S_{1,0}$-type, one can obtain nearly the same results. Here the union of the spectra of $p(x,\xi)$ and $a(x',\xi',D_n)$ (with respect to a fixed choice of local coordinate patches) will be contained in a compact set $K \subset \mathbb{C}$, and for $\lambda \in \mathbb{C}\setminus K$ one can construct parametrix symbols $(p - \lambda I)^{(-1)}$ and $(a - \lambda I)^{(-1)}$ in the following way: Since $(p(x,\xi) - \lambda I)^{-1}$ is bounded, it follows from the Leibniz formula as in (2.1.54) that $q_0 = (p(x,\xi) - \lambda I)^{-1}$ is in $S^0_{1,0}$; a full parametrix q of $p - \lambda I$ is then obtained by imitating the rest of the proof of Theorem 2.1.16. For $a(x',\xi',D_n) - \lambda I$ one first constructs an auxiliary elliptic system (in each coordinate patch)

$$b = \begin{pmatrix} q_\Omega & k \\ t & s \end{pmatrix}$$

by an easy variant of Theorem 3.1.5 (there are uniform Laguerre norm estimates in ξ' , and no unbounded parameter). Then this is used to reduce to the case where the ps.d.o. part is I , and the inverse of the resulting system $I + h(x',\xi',D_n)$ (with a singular Green-like operator h) is seen to be of the form $I + h'$ with a singular Green-like operator h' , by an easy variant of Theorem 3.2.3. We must here take K so large that $s_0(x')-\lambda$ is bijective for $\lambda \notin K$ and all x' , where $s_0(x')$ is the first coefficient in the expansion $p(x',0,\xi) \sim \Sigma_{j\leq 0} s_j \xi_n^j$; this assures that q_0 is in $\mathcal{H}_0$ as a function of ξ_n , and the transmission property carries over to q_0 and q .

Now if C can be taken so large that it contains K in its interior set, the definition (3.4.24) again gives an operator satisfying (3.4.25-26); so f(A) is a Green operator of $S_{1,0}$-type again, and of order and class 0 . (The analyticity of the symbols on C is obtained as in the polyhomogeneous case.) In particular, complex powers A^z are well-defined if there is a curve around K having 0 connected with ∞ in its exterior set.

Altogether, we have found

3.4.4 Theorem. *Let A be a polyhomogeneous square matrix-formed Green operator* (3.4.14) *(or in particular* (3.4.22)*) of order and class* 0 . *Then the spectrum of* A *as an operator in* $H^0(E)\times H^{-\frac{1}{2}}(F)$ *is a compact subset of* $\mathbb{C}$, *and*

$$\text{sp}(A) \supset \text{sp}_\Omega(a^0) \cup \text{sp}_\Gamma(a^0) = \text{ess sp}(A) \ . \tag{3.4.29}$$

When f *is a complex function on* $\mathbb{C}$ *that is holomorphic on a neighborhood* ω *of* sp(A) , *we define the operator* f(A) *by the Cauchy formula*

$$f(A) = \frac{i}{2\pi}\int_C f(\lambda)(A-\lambda I)^{-1}d\lambda \ , \tag{3.4.30}$$

where C *is a smooth closed curve in* ω *going around* sp(A) *in the positive direction. Then* f(A) *is a Green operator of order and class* 0 , *of* $S_{1,0}$*-type, again with the structure* (3.4.14)*(resp.* (3.4.22)*), and its symbols are determined by similar Cauchy integrals applied to the symbols of* A . *In particular, the complex powers* A^z *are defined (and polyhomogeneous), when the curve* C *can be chosen so that* 0 *is in its exterior set and can be connected to* ∞ *there.*

The definition of f(A) *and the description of its symbols extend to the case where* A *is of* $S_{1,0}$*-type, of order and class* 0 , *provided that* f *is holomorphic on a neighborhood of a compact set containing the spectra of* $p(x,\xi)$, $s_0(x')$ *and* $a(x',\xi',D_n)$ *(defined via some choice of local coordinates); then* f(A) *is again of* $S_{1,0}$*-type and of order and class* 0 .

3.4.5 Remark. As we shall see later in Section 4.4, the complex powers B^z of a realization usually fall outside the Boutet de Monvel calculus; the ps.d.o. part need not have the transmission property, and the rest resembles a s.g.o. but lacks the full symbol estimates. It can sometimes be advantageous to appeal to the above theorem instead. For example, if B is the Dirichlet realization of a selfadjoint positive ps.d.o. of order $2m$ as in Section 1.7, then $B^{-\frac{1}{2}}$ is usually not a standard Green operator. However, we can in some questions replace $B^{-\frac{1}{2}}$ by

$$R' = R_{-m}(B')^{-\frac{1}{2}} , \tag{3.4.31}$$

where R_{-m} is an isomorphism of $L^2(E)$ onto $\overset{\circ}{H}{}^m(E)$ and B' is the zero-order operator

$$B' = R_{-m}^* B R_{-m} , \tag{3.4.32}$$

that is a selfadjoint positive bijection in $L^2(E)$ (since B maps $\overset{\circ}{H}{}^m(E)$ isomorphically onto its dual space $H^{-m}(E)$). Here R_{-m} can be chosen as a Green operator of order $-m$ and class 0, cf. Remark 3.2.15, so that Theorem 3.4.4 applies to B', and $R_{-m}(B')^{-\frac{1}{2}}$ is likewise a Green operator of order $-m$ and class 0 (mapping $H^s(E)$ isomorphically onto $\overset{\circ}{H}{}^m(E) \cap H^{s+m}(E)$ for $s \geq 0$). Then

$$\begin{aligned} B^{-1} &= (B^{-\frac{1}{2}})^2 = R'(R')^* , \quad \text{and} \\ (R')^* B R' &= I \quad \text{on } L^2(E) . \end{aligned} \tag{3.4.33}$$

The idea is applied to spectral questions in Section 4.6.

For operators or realizations of positive order, the spectrum is unbounded, so one has to be much more cautious in the definition of operator functions by the Cauchy integral formula (3.4.24). When B is a realization defined from a system satisfying Definition 1.5.5, then $(B-\lambda)^{-1}$ is $O(\langle\lambda\rangle^{-1})$ in $L^2(E)$ in a certain sector V, cf. (3.3.22). Then we have for example:

(i) $f(B)$ is defined on $L^2(E)$ by (3.4.24) if C runs in V and $f(\lambda)$ is $O(\langle\lambda\rangle^{-\varepsilon})$ there;

(ii) $f_1(B)$ is defined as $f(B)B^k$ on $D(B^k)$ if $f_1(\lambda) = \lambda^k f(\lambda)$ with f as in (i), $k \in \mathbb{N}$.

The symbols generally cannot be expected to be as nice as in [Seeley 1,4], cf. the remarks at the end of Section 3.3.

In the next chapter, we concentrate the efforts on two very important cases, the exponential function and the power function, where the study is justified by interesting applications; this may also serve as a model for investigations of other functions.

CHAPTER 4

SOME APPLICATIONS

4.1 Evolution problems.

Chapter 4 presents a variety of applications of the resolvent calculus established in Chapters 2 and 3.

In the present Section 4.1 we give a description of the most immediate consequences of the resolvent estimates for non-local evolution problems, parallel to the results for parabolic differential operator problems.

The central ingredient here, the "heat operator" $\exp(-tB)$ defined from a realization B, is analyzed in more detail in Section 4.2. In particular, the asymptotic properties of the kernel and the trace are studied, and the limitations due to the regularity number ν are discussed.

In Section 4.3, we use the trace formulas of Section 4.2 to derive a new index formula for arbitrary normal elliptic realizations.

Section 4.4 treats another operator function, namely B^z, where we describe the detailed structure for $\operatorname{Re} z < 0$, and show how the trace extends as a meromorphic function into the region $\operatorname{Re} z \geq 0$, limited however by the finite regularity ν.

In Section 4.5, we take up the classical problem of describing asymptotically the spectrum of realizations; we here obtain some new results for the pseudo-differential case, including estimates for B^z.

This is used in Section 4.6, where we extend the spectral results to implicit eigenvalue problems of the type $A_0 u = \lambda A_1 u$.

Finally, we discuss in Section 4.7 a somewhat different application, namely a class of singular perturbation problems, where the analysis of the solution operators for $\varepsilon \to 0$ is shown to be derivable from the analysis of operators and symbols for $\mu \to \infty$, that is established in the rest of the book.

Let us first consider the time-dependent problems. We shall show that the results on the resolvent R_λ of the boundary value problem

$$\begin{aligned} P_\Omega u(x) + Gu(x) - \lambda u(x) &= f(x) \quad \text{for} \quad x \in \Omega \,, \\ Tu(x) &= 0 \quad \text{for} \quad x \in \Gamma \,, \end{aligned} \tag{4.1.1}$$

have immediate consequences for the associated evolution problem

$$\text{(4.1.2)}\quad \begin{aligned} &\text{(i)} && \partial_t u(x,t) + P_\Omega u(x,t) + Gu(x,t) = f(x,t) && \text{for } x \in \Omega,\ t > 0 ,\\ &\text{(ii)} && Tu(x,t) = \varphi(x,t) && \text{for } x \in \Gamma,\ t > 0 ,\\ &\text{(iii)} && u(x,0) = u_0(x) && \text{for } x \in \Omega . \end{aligned}$$

Assume that the system satisfies the definition of parabolicity (see Definition 1.5.5 3^0), i.e., parameter-ellipticity on the rays $\lambda = re^{i\theta}$ for all $\theta \in [\pi/2, 3\pi/2]$. (Recall in particular (1.5.44).) Then by Corollary 3.3.2, the resolvent R_λ exists for all λ in a truncated obtuse sector (with $\delta \geq 0$ and $\varepsilon > 0$)

$$\text{(4.1.3)}\qquad W_{\delta,\pi/2+\varepsilon} = \{\lambda \in \mathbb{C} \mid |\lambda| \geq \delta,\ |\pi\text{-}\arg\lambda| \leq \pi/2 + \varepsilon\} ,$$

satisfying the estimates there:

$$\text{(4.1.4)}\qquad \begin{aligned} &\left(\langle\lambda\rangle^{1+s/d}\|R_\lambda f\|_0 + \|R_\lambda f\|_{s+d}\right) \leq c_s\left(\langle\lambda\rangle^{s/d}\|f\|_0 + \|f\|_s\right) \\ &\qquad \text{for each } s \geq 0 ,\ \text{uniformly for } \lambda \text{ in } W_{\delta,\pi/2+\varepsilon} . \end{aligned}$$

In particular, the norm of R_λ in $L^2(E)$ is $O(|\lambda|^{-1})$ for $|\lambda| \to \infty$, $\lambda \in W_{\delta,\pi/2+\varepsilon}$. (Inside the set $\{|\lambda| \leq \delta\}$, the spectrum consists of a finite number of eigenvalues.)

It follows that the exponential function of the realization $B = (P+G)_T$

$$\text{(4.1.5)}\qquad U(t) \equiv \exp(-tB) = \frac{i}{2\pi}\int_C e^{-\lambda t} R_\lambda \, d\lambda$$

is well defined for $t > 0$, when we take for C a contour in $W_{\delta,\pi/2+\varepsilon}$ beginning and ending with rays with argument $\pi/2-\varepsilon'$ resp. $3\pi/2+\varepsilon'$ for $\varepsilon' \in \,]0,\varepsilon[$; and $\exp(-tB)$ defined in this way is continuous from $L^2(E)$ into $H^s(E)$ for any $s \geq 0$, $t > 0$.

The definition of $\exp(-tB)$ is consistent with the holomorphic semigroup definition (as in [Hille-Phillips 1], [Kato 1], [Friedman 1], [Tanabe 1], ...) and the Laplace transform definitions (as in [Agranovich-Vishik 1] and [Lions-Magenes 2]) and results from these works, based on "operator-theoretic" methods, generalize to the present case.

For example, the holomorphic semigroup theory shows that when $u_0 \in L^2(E)$, and we set $u(t) = U(t)u_0$, then $u(t) \in D(B)$ for $t > 0$ and

$$\text{(4.1.6)}\qquad \begin{aligned} &\partial_t u(t) + Bu(t) = 0 \qquad \text{for } t > 0 ,\\ &u(0) = u_0 ; \end{aligned}$$

so $u(t)$ solves the problem (4.1.2) with f and φ equal to zero. A solution of the problem (4.1.2) with only (ii) homogeneous, i.e. the problem

$$\begin{aligned} \partial_t u(t) + Bu(t) &= f(t) \quad \text{for} \quad t > 0 , \\ u(0) &= u_0 ; \end{aligned} \tag{4.1.7}$$

is now obtained on the form

$$u(t) = U(t)u_0 + \int_0^t U(t-s)f(s)ds ; \tag{4.1.8}$$

and solutions of the completely inhomogeneous problem (4.1.2) can be obtained, when the equation can be reduced to a case with (ii) homogeneous, where the preceding results apply. For precision, we formulate a theorem (using [Kato 1, IX.1.7]):

4.1.1 Theorem. *Let* $\{P_\Omega+G,T\}$ *be as described in Sections 1.5 and 3.3. When the system* $\{\partial_t+P_\Omega+G,T\}$ *is parabolic (Definition 1.5.5 3°), the realization* $B = (P+G)_T$ *generates a holomorphis semigroup* $U(t) = \exp(-tB)$.

One then has, for example, that if $f \in C^\sigma([0,a];L^2(E))$ *(for some* $\sigma \in]0,1]$ *and* $a \in \mathbb{R}_+$*),* $\varphi = 0$, *and* $u_0 \in L^2(E)$, *then the problem (4.1.2) has a unique solution defined by (4.1.8); here* $u \in C^0([0,a];L^2(E)) \cap C^{1,\sigma}(]0,a];L^2(E))$, $u(t) \in D(B)$ *for* $t > 0$ *and* $Bu \in C^0([0,a];L^2(E)) \cap C^\sigma(]0,a];L^2(E))$.

One can also formulate results in spaces defined in terms of fractional powers of B , cf. [Friedman 1], leading to results for nonlinear problems. (Fractional powers of realizations are discussed to some extent in Section 4.4 below.)

In the framework of [Lions-Magenes 2] one finds, by a straightforward adaptation of the proof and notation of Theorem 4.5.3 there:

4.1.2 Theorem. *Let* $\{P_\Omega+G,T\}$ *be as described in Sections 1.5 and 3.3, with* $\{\partial_t+P_\Omega+G,T\}$ *parabolic. Let* r *be an integer* ≥ 0 , *and let* $I = [0,a]$ *for some* $a > 0$. *Let* $f,\varphi = \{\varphi_0,\dots,\varphi_{d-1}\}$ *and* u_0 *be given in the respecitve spaces*

$$\begin{aligned} f &\in \underline{H}^{r,r/d}(\bar\Omega\times I,\underline{E}) , \\ \varphi &\in \prod_{0\leq k<d} H^{r+d-k-\frac12,(r+d-k-\frac12)/d}(\Gamma\times I,\underline{F}_k) , \\ u_0 &\in H^{r+d/2}(\bar\Omega,E) , \end{aligned} \tag{4.19}$$

and satisfying a compatibility condition: There exists $w \in \underline{H}^{r+d,r/d+1}(\bar{\Omega}\times I,\underline{E})$ *such that*

$$(4.1.10)\qquad \begin{aligned} &Tw = \varphi\ , \quad w(x,0) = u_0(x)\ , \quad \textit{and} \\ &\partial_t^j(\partial_t + P_\Omega + G)w(x,0) = \partial_t^j f(x,0) \qquad \textit{for} \quad 0 \le j < r/d - \tfrac{1}{2}\ . \end{aligned}$$

Then the problem (4.1.2) *with these data has a unique solution* $u \in \underline{H}^{r+d,r/d+1}(\bar{\Omega}\times I,\underline{E})$.

Here $\underline{E}$ and $\underline{F}_k$ denote the trivial extensions of E resp. F_k to vector bundles over $\bar{\Omega}\times I$ resp. $\Gamma\times I$, and $\underline{H}^{r,s}(\bar{\Omega}\times I,\underline{E})$ is the space

$$\underline{H}^{r,s}(\bar{\Omega}\times I,\underline{E}) = H^0(I;H^r(\bar{\Omega},E)) \cap H^s(I,H^0(\bar{\Omega};E))\ ,$$

see [Lions-Magenes 2] for further details. The uniqueness statement could also be formulated as an estimate of the norm of the solution by the norms of the data, in the mentioned spaces.

The compatibility condition in the theorem of course deserves a deeper analysis, but in some cases (e.g. when $r/d \le \frac{1}{2}$), the discussion in [Lions-Magenes 2, Section 4.3] applies directly. Extensions of the result can be worked out by the methods of [Lions-Magenes 2].

The considerations in Section 3.4 will likewise lead to solvability theorems for problems with ∂_t in higher powers,

$$(4.1.11)\qquad \begin{aligned} &[\partial_t^m + \partial_t^{m-1}(P_\Omega^{(1)} + G^{(1)}) + \dots + P_\Omega^{(m)} + G^{(m)}]u(x,t) = f(x,t) \quad \text{for} \quad x > 0\ , \\ &\sum_{0\le j\le m-1} \partial_t^j T^{(m-j)} u(x,t) = \varphi(x,t) \quad \text{for} \quad t > 0\ , \\ &\partial_t^j u(x,0) = u_j(x) \quad \text{for} \quad 0 \le j \le m-1 \end{aligned}$$

by Laplace transform methods as in [Agranovich-Vishik 1]; and one can also treat evolution problems for Green systems,

$$(4.1.12)\qquad \begin{aligned} &\partial_t\begin{pmatrix} u \\ v \end{pmatrix} + \begin{pmatrix} P_\Omega + G & K \\ T & S \end{pmatrix}\begin{pmatrix} u \\ v \end{pmatrix} = \begin{pmatrix} f \\ g \end{pmatrix} \quad \text{for} \quad t > 0\ , \\ &\begin{pmatrix} u \\ v \end{pmatrix}\Bigg|_{t=0} = \begin{pmatrix} u_0 \\ v_0 \end{pmatrix}\ , \end{aligned}$$

by the mentioned theories. Also other problems with ∂_t entering in various ways can be treated.

Extensions to t-dependent symbols in all these cases seems quite accessible by the known methods; here one can use our symbol classes with t as an extra space variable, linked with the "dual" variable μ .

The operator $U(t)$ in (4.1.5) will be analyzed more closely in Section 4.2. Before we go on to that, let us make some further general comments:

One can investigate the possible extension to the present framework of many more of the classical L^2 results for differential operators, as presented in the works quoted above and their references and further developments; see also [Solonnikov 1], [Eidelman 1], [Ladyzenskaja-Solonnikov-Uralceva 1], and numerous other works; in particular, the results based on (4.1.4) generalize readily. Recall however, that the present thoery is quite demanding in its requirements on smoothness of coefficients and manifolds. It is possible, but requires a heavy notational apparatus, to keep an account of how much smoothness suffices for how large a number of terms in the approximation of a solution. At any rate, the present work permits the analysis of all derivatives.

The study of parabolic problems beyond L^2 theory has been aimed towards general L^p estimates, estimates in Hölder spaces C^σ, and investigations of the operators in more general spaces (for example weighted Sobolev spaces with weight going to zero for $t \to 0$), leading to sharper results and interesting extensions to nonlinear problems. A development of the present theory in these directions would be of great interest, and we hope to take it up elsewhere. Some informations may be derived from the study of symbols and kernels given below.

Pseudo-differential parabolic boundary problems have been studied earlier in [Vishik-Eskin 2] and in [Eskin-Čan Zui Ho 1]. The paper of Vishik and Eskin considers general scalar operators of a pseudo-differential kind in the place of our ∂_t+P (without the G), and trace operators of the form $\gamma_0 B_j$ where B_j is likewise of a pseudo-differential type. Here the factorization method and Wiener-Hopf calculus of [Vishik-Eskin 1], with nonsmooth symbols, is suitably genralized. The announcement [Eskin-Čan Zui Ho 1] treats matrix formed operators, including potential operators (Poisson operators) and boundary pseudo-differential terms; the details have not been available to us.

For the boundaryless case, [Drin 1,2,3] and [Eidelman-Drin 1,2] have treated initial problems for ∂_t+P in various cases with homogeneous symbols, obtaining fine estimates of the kernels. Degenerate parabolic cases are treated in [Iwasaki 1].

Other works on parabolic pseudo-differential problems, such as the works around [Fabes-Jodeit 1], [Fabes-Rivière 1] and [Jodeit 1], the articles [Lascar 1], [Bove-Franchi-Obrecht 1,2], the announcement [de Gosson 1], and the discussion in [Rempel-Schulze 1, Section 4.3.6], are concerned with anisotropic pseudo-differential problems. But, as mentioned in Section 1.5, the study of ∂_t+P (and its boundary problems) falls <u>outside</u> the scope of these works, when P is not a differential operator, since ∂_t+P is <u>not</u> a genuine ps.d.o. (not even an anisotropic one) in the (x,t)-variable, when p is not polynomial; this is precisely the reason for all our trouble with the regularity numbers etc.

4.2 The heat operator.

The fundamental object in the study of evolution equations (4.1.2) is the semigroup $U(t) = \exp(-tB)$, also called the "heat operator" because of the resemblance with the heat equation (the special case where $P = -\Delta$ and $G = 0$). We devote this section to a study of the detailed structure of $U(t)$, based on the analysis of the resolvent in Chapter 3.

$U(t)$ consists of two terms, $V(t)_\Omega$ (stemming from the pseudo-differential operator) and $W(t)$ (stemming from the boundary condition). Both are negligible operators (have C^∞ kernels) for each fixed $t > 0$, and the interesting thing is to analyze their behavior for $t \to 0$. (An analysis for $t \to \infty$ is certainly also possible, but we do not go into that here.) $V(t)$ has been studied in many other works; we include an analysis here for completeness' sake and because some of the methods for $V(t)$ are used again for $W(t)$.

The main results are: a description of the operator family $V(t)$, its symbol $v(x,\xi,t)$ (in a pseudo-differential sense) and its kernel $K_V(x,y,t)$. A complete asymptotic expansion of the trace $\operatorname{tr} V(t)_\Omega$ for $t \to 0+$. A description of the operator family $W(t)$, its symbol $w(x',t,\xi',\xi_n,\eta_n)$ (in a singular Green operator sense) and its kernel $K_W(x,y,t)$. An asymptotic expansion of the trace $\operatorname{tr} W(t)$ for $t \to 0+$, with a finite number of exact terms, depending on the regularity ν of the given boundary value problem. As a corollary an expansion of the trace of the semigroup $U(t)$.

As in Section 4.1, we assume parabolicity of the polyhomogeneous system $\{\partial_t + P_\Omega + G, T\}$. By addition of a suitable constant a to P (this amounts to replacing $U(t)$ by $e^{-at}U(t)$) , we can obtain that R_λ exists for λ in an obtuse "keyhole region" $V_{\delta',\pi/2+\varepsilon'}$ (for some $\delta' > 0$, $\varepsilon' > 0$), <u>containing</u> $\{\lambda \in \mathbb{C} \mid \pi/2-\varepsilon \leq \arg\lambda \leq 3\pi/2 + \varepsilon\}$:

$$(4.2.1)\qquad V_{\delta',\pi/2+\varepsilon'} = \{\lambda \in \mathbb{C} \mid |\lambda| \leq \delta' \ \underline{\text{or}}\ |\pi-\arg\lambda| \leq \pi/2 + \varepsilon'\} ,$$

so that the contour C in (4.1.5) can be chosen to lie in $\{\lambda \mid \operatorname{Re}\lambda > 0\}$, e.g. as the boundary of a similar region

$$(4.2.2)\qquad C_\delta = \partial V_{\delta,\pi/2+\varepsilon} , \text{ with } \delta \in\,]0,\delta'[\text{ and } \varepsilon \in\,]0,\varepsilon'[,$$

or as the boundary of a sector

$$(4.2.3)\qquad C_0 = \partial V_{0,\pi/2+\varepsilon} = \{\lambda \in \mathbb{C} \mid \arg\lambda = \pi \pm (\pi/2+\varepsilon)\} .$$

4.2.1 Remark. Many of the calculations in the following could also be justified without a multiplication of $U(t)$ by e^{-at}, by using instead that R_λ exists in $V_{\delta,\pi/2+\varepsilon}$ except on a finite set of eigenvalues of B.

As accounted for in Remark 3.3.6, we can assume that the neighboring manifold Σ, in which $\overline{\Omega}$ is imbedded, is compact. Then also P has an exact resolvent $Q_\lambda = (P-\lambda I)^{-1}$ for large λ, and we can modify P if necessary, so that the resolvent Q_λ exists for $\lambda \in V_{\delta,\pi/2+\varepsilon}$. It can also be assumed for the principal symbol $p_d(x,\xi)$, that $(p_d(x,\xi)-\lambda I)^{-1}$ is defined for all $\lambda \in V_{\delta,\pi/2+\varepsilon}$, all $(x,\xi) \in T^*(\Sigma)$. Similar considerations hold for b^0.

Taking C as one of the mentioned contours, we have

$$
\begin{aligned}
\text{(4.2.4)} \qquad U(t) &= \frac{i}{2\pi}\int_C e^{-\lambda t} R_\lambda \, d\lambda \\
&= \frac{i}{2\pi}\int_C e^{-\lambda t} Q_{\lambda,\Omega} \, d\lambda + \frac{i}{2\pi}\int_C e^{-\lambda t} G_\lambda \, d\lambda \\
&= V(t)_\Omega + W(t) ,
\end{aligned}
$$

where

$$
\text{(4.2.5)} \qquad
\begin{aligned}
V(t) &= \frac{i}{2\pi}\int_C e^{-\lambda t} Q_\lambda \, d\lambda , \\
W(t) &= \frac{i}{2\pi}\int_C e^{-\lambda t} G_\lambda \, d\lambda .
\end{aligned}
$$

The semigroup $V(t)$ has been studied before (e.g. in [Duistermaat-Guillemin 1], [Widom 2], [Taylor 1], [Treves 2], under various hypotheses), so the main new interest lies in the analysis of $W(t)$. Let us however analyze $V(t)$ first, completing the known results.

4.2.2 Theorem. *Let* $\widetilde{E}$ *be a vector bundle over a compact manifold* Σ *without boundary, and let* P *be a pseudo-differential operator of order* $d > 0$ *in* $\widetilde{E}$ *such that* $\partial_t + P$ *is parabolic (Definition 1.5.3). Assume (as accounted for above) that the resolvent* $Q_\lambda = (P-\lambda I)^{-1}$ *exists in a region* $V_{\delta',\pi/2+\varepsilon'}$ *with* $p_d(x,\xi)-\lambda I$ *invertible there, for all* (x,ξ). *Let* $V(t) = \exp(-tP)$. *The family of operators* $V(t)$ *is strongly continuous in* $L^2(\widetilde{E})$ *for* $t \geq 0$. *For each* $t > 0$, *the* $V(t)$ *are integral operators with* C^∞ *kernel (i.e. ps.d.o.s with symbol in* $S^{-\infty}$*), and* $V(t)$ *is continuous on* $t \geq 0$ *as a family of ps.d.o.s of order* 0 *(of* $S_{1,0}$ *type), and on* $t > 0$ *as a family of negligible ps.d.o.s.* V *has a* *symbol* $v(x,t,\xi)$ *in local coordinates, satisfying*

$$v(x,t,\xi) \sim \sum_{\ell \in \mathbb{N}} v_{-\ell}(x,t,\xi) \ , \tag{4.2.6}$$

where the terms $v_{-\ell}(x,t,\xi)$ *are defined by*

$$\begin{aligned} v_0(x,t,\xi) &= \exp(-tp_d(x,\xi)) \ , \\ v_{-\ell}(x,t,\xi) &= \frac{i}{2\pi}\int_C e^{-t\lambda} q_{-d-\ell}(x,\xi,\lambda)d\lambda \ . \end{aligned} \tag{4.2.7}$$

They are quasi-homogeneous in (t,ξ) *of degree* $-\ell$

$$v_{-\ell}(x,s^{-1/d}t,s\xi) = s^{-\ell}v_{-\ell}(x,t,\xi) \quad \textit{for} \ |\xi| \geq 1 \ , \ s \geq 1 \ , \tag{4.2.8}$$

and satisfy estimates, with a fixed $c > 0$, *and* $c(x)$ *depending on the indices (cf. also* (4.2.45)ff.*),*

$$|D_x^\beta D_\xi^\alpha v_{-\ell}(x,t,\xi)| \leq c(x)\langle\xi\rangle^{-\ell-|\alpha|}\exp(-ct\langle\xi\rangle^d)[t\langle\xi\rangle^d]^a \ , \ \textit{for any} \ a \leq \min\{1,\ell+|\alpha|\} \ ; \tag{4.2.9}$$

and v *can be chosen so that*

$$|D_x^\beta D_\xi^\alpha[v - \sum_{\ell<M} v_{-\ell}]| \leq c(x)\langle\xi\rangle^{-M-|\alpha|}\exp(-ct\langle\xi\rangle^d)[t\langle\xi\rangle^d]^a \ , \ \textit{for any} \ a \leq \min\{1,M+|\alpha|\} \ . \tag{4.2.10}$$

The operators $V_{-\ell}(t)$ *with symbols* $v_{-\ell}(x,t,\xi)$ *have kernels* $K_{V_{-\ell}}(x,y,t)$, *satisfying (with* $c'' > 0$*)*

$$\begin{aligned} \sup_{x,y}|K_{V_0}(x,y,t)| &\leq c't^{-n/d}\exp(-c''t) \quad \textit{for} \ t > 0 \ , \\ K_{V_0}(x,x,t) &= c_0(V,x)t^{-n/d} + O(t) \ \textit{for} \ t \to 0 \ ; \end{aligned} \tag{4.2.11}$$

and for $0 < \ell < d+n$

$$\begin{aligned} \sup_{x,y}|K_{V_{-\ell}}(x,y,t)| &\leq c'(t^{(\ell-n)/d} + t)\exp(-c''t) \quad \textit{for} \quad t > 0 \ , \\ K_{V_{-\ell}}(x,x,t) &= c_\ell(V,x)t^{(\ell-n)/d} + O(t) \qquad \textit{for} \quad t \to 0 \ ; \end{aligned} \tag{4.2.12}$$

here

$$c_\ell(V,x) = (2\pi)^{-n}\int_{\mathbb{R}^n} v^h_{-\ell}(x,1,\xi)d\xi \ , \quad \textit{when} \quad 0 \leq \ell < d+n \ . \tag{4.2.13}$$

Moreover one has (cf. also (4.2.47)*)*

$$\sup_{x,y} |K_{V_{-d-n}}(x,y,t)| \le c't(1+|\log t|)\exp(-c''t) \text{ when } \ell = d+n ,$$
(4.2.14)
$$\sup_{x,y} |K_{V_{-\ell}}(x,y,t)| \le c't \exp(-c''t), \text{ when } \ell > d+n .$$

The kernels are continuous in x,y *and in* $t \ge 0$ *when* $\ell \ge n$, *and are zero at* $t = 0$ *when* $\ell > n$.

The remainders

$$V'_M(t) = V(t) - \sum_{\ell<M} V_{-\ell}(t)$$

have kernels $K_{V'_M}(x,y,t)$ *satisfying*

(4.2.15) $$\sup_{x,y} |K_{V'_M}(x,y,t)| \le c't(1+|\log t|)\exp(-c''t), \text{ when } M > d+n .$$

One also has that $K_{V'_M}$ *is continuous in* $x,y \in \Sigma$ *and* $t \ge 0$ *for* $M \ge n$, *with*

(4.2.16) $$K_{V'_M}(x,y,0) = 0 \quad \text{for } M > n .$$

<u>Proof</u>: Since p_d is chosen such that q_{-d} can be taken equal to $(p_d-\lambda I)^{-1}$ for all ξ , we have immediately that

(4.2.17) $$v_0(x,t,\xi) = \frac{i}{2\pi}\int_{C_\delta} e^{-\lambda t}(p_d(x,\xi)-\lambda)^{-1}d\lambda = \exp(-t\, p_d(x,\xi)) ,$$

and it is easily checked that the inequalities hold for $\ell = 0$. Also for the other terms, it is advantageous to make explicit calculations. In the scalar case it is obvious what one gets, for here we find from (3.3.44)

(4.2.18) $$v_{-\ell}(x,t,\xi) = \sum_{k=1}^{2\ell} p_{\ell,k}(x,\xi)\,\frac{i}{2\pi}\int_{C_\delta} e^{-\lambda t}(p_d-\lambda)^{-k-1}d\lambda$$
$$= \sum_{k=1}^{2\ell} p_{\ell,k}\, c_k\, t^k \exp(-tp_d) , \quad \text{for } \ell \ge 1 .$$

Here, since $k \ge 1$, we have

$$|p_{\ell,k}t^k| \le c(x)\langle\xi\rangle^{dk-\ell}t^k = c(x)\langle\xi\rangle^{d-\ell}t[\langle\xi\rangle^d t]^{k-1}$$

with $[\langle\xi\rangle^d t]^{k-1} \le 1$ for $\langle\xi\rangle^d t \le 1$, and

$$|[\langle\xi\rangle^d t]^{k-1} \exp(-\tfrac{1}{2}tp_d)| \le c(x) \quad \text{for } \langle\xi\rangle^d t \ge 1 ,$$

where the other factor $\exp(-\frac{1}{2}tp_d)$ provides the exponential estimate;

this shows (4.2.9) with $a = 1$ and $\alpha,\beta = 0$. The estimates with $a < 1$ are simple consequences, and the derivatives are determined and estimated by differentiating under the integral sign in (4.2.7).

(4.2.10) can be obtained by constructing $v(x,t,\xi)$ in a suitable way from the terms $v_{-\ell}$ (by a variant of the usual "reconstruction procedure"), but it is far more interesting that we can estimate the exact kernels $K_{V'_M}$. The $V_{-\ell}(t)$ are constructed from the locally defined symbols $v_{-\ell}(x,t,\xi)$ by coordinate transformations and partitions of unity, as in the construction of operators from their symbols in Section 2.4. For simplicity of the presentation, we leave out the precise formulation, and write as if the operators act in $\mathbb{R}^n$. We first find for $V_0(t)$

$$(4.2.19)\qquad K_{V_0}(x,y,t) = (2\pi)^{-n}\int_{\mathbb{R}^n} e^{i(x-y)\cdot\xi}\exp(-tp_d(x,\xi))d\xi\ ,$$

so that, by comparison with the strictly homogeneous symbol,

$$|K_{V_0}(x,y,t)| \le c_1\int_{\mathbb{R}^n}|\exp(-tp_d^h(x,\xi))|d\xi + c_1\int_{|\xi|\le 1}|\exp(-tp_d)-\exp(-tp_d^h)|d\xi$$
$$\le c_1t^{-n/d}\int_{\mathbb{R}^n}\exp(-c|\eta|^d)d\eta + c_2t \le c_3t^{-n/d} + c_2t\ ,$$

and for the value on the diagonal,

$$K_{V_0}(x,x,t) = (2\pi)^{-n}t^{-n/d}\int_{\mathbb{R}^n}\exp(-p_d^h(x,\eta))d\eta$$
$$+ (2\pi)^{-n}\int_{|\xi|\le 1}[\exp(-tp_d)-\exp(-tp_d^h)]d\xi = c_0(V,x)t^{-n/d} + O(t)\ .$$

It is used here that $|\exp(-tp_d)-1| \le 2|tp_d|$ since $\operatorname{Re} p_d \ge 0$, so that $\exp(-tpd)-\exp(tp_d^h)$ is $O(t)$ for $|\xi| \le 1$. K_{V_0} is $O(\exp(-\frac{1}{2}ct))$, for $t \ge 1$.

To treat $K_{V_{-\ell}}$ for $\ell > 0$, we note that the strictly homogeneous symbol $v_{-\ell}^h$ associated with $v_{-\ell}$ satisfies

$$|v_{-\ell}^h(x,t,\xi)| \le c(x)|\xi|^{d-\ell}t\,\exp(-ct|\xi|^d)\ ,$$

when $\ell > 0$. This is integrable when $d-\ell > -n$, so we find for the kernel of $V_{-\ell}(t)$:

$$(4.2.20)\qquad K_{V_{-\ell}}(x,y,t) = (2\pi)^{-n}\int_{\mathbb{R}^n} e^{i(x-y)\cdot\xi}v_{-\ell}(x,t,\xi)d\xi$$

is $O(\exp(-\frac{1}{2}ct))$ for $t \ge 1$ and satisfies for all t

$$|K_{V_{-\ell}}(x,y,t)| \le c_1\int_{\mathbb{R}^n} |v^h_{-\ell}|\,d\xi + c_1\int_{|\xi|\le 1} (|v_{-\ell}|+|v^h_{-\ell}|)d\xi$$

$$\le c_2 t \int_{\mathbb{R}^n} |\xi|^{d-\ell}\exp(-ct|\xi|^d)d\xi + c_3 t$$

$$= c_2 t^{(\ell-n)/d}\int_{\mathbb{R}^n} |\eta|^{d-\ell}\exp(c|\eta|^d)d\eta + c_3 t = c_4 t^{(\ell-n)/d} + c_3 t \ .$$

A similar argument shows

$$K_{V_{-\ell}}(x,x,t) = (2\pi)^{-n}\int_{\mathbb{R}^n} v^h_{-\ell}(x,t,\xi)d\xi + c_1\int_{|\xi|\le 1}(v_{-\ell}-v^h_{-\ell})d\xi$$

$$= c_\ell(V,x)t^{(\ell-n)/d} + O(t) \ .$$

The estimates in (4.2.14) are seen directly by inserting the estimates (4.2.9) in the formula (4.2.20):

$$|K_{V_{-\ell}}(x,y,t)| \le c'(x)t\int_{\mathbb{R}^n} \langle\xi\rangle^{d-\ell}\exp(-ct\langle\xi\rangle^d)d\xi$$

$$\le c'(x)t\Big(\int_{|\xi|\le 1} + \int_{|\xi|\ge 1}\Big)$$

$$\le c''(x)t\Big(1 + \int_1^\infty r^{d-\ell+n-1}\exp(-ctr^d)dr\Big)$$

$$\le \begin{cases} c''(x)t(c_1+c_2|\log t|) & \text{when } \ell = d+n \ , \\ c''(x)t(c_1+c_2t^{-1+(\ell-n)/d}) & \text{when } \ell > d+n \ . \end{cases}$$

For matrix formed operators one obtains these results by a slightly larger effort, using the results of Lemma 4.2.3 below.

Finally, we estimate the remainder kernel as follows. When $M > d+n$, the operator $Q'_{\lambda,M} = Q_\lambda - \sum_{\ell<M} OP(q_{-d-\ell})$ is of order $-d-M < -2d-n$, and has a continuous kernel $K_{Q'_M}(x,y,\lambda)$ that is analytic in λ , with

$$\text{(4.2.21)} \qquad \sup_{x,y} |K_{Q'_M}(x,y,\lambda)| \le c\langle\lambda\rangle^{-2}$$

cf. Remark 3.3.8. Here

$$V'_M(t) = \frac{i}{2\pi}\int_{C_\delta} e^{-\lambda t}\, Q'_{\lambda,M}\, d\lambda \ ,$$

and the integration carries over to the kernel, giving

$$K_{V'_M}(x,y,t) = \frac{i}{2\pi}\int_{C_\delta} e^{-\lambda t} K_{Q'_M}(x,y,\lambda)d\lambda \quad . \tag{4.2.22}$$

Since $|e^{-\lambda t}| \leq e^{-ct}$ on C_δ for some $c > 0$, the integral converges uniformly for all $t \geq 0$, so $K_{V'_M}(x,y,t)$ is continuous in $t \geq 0$ and is $\mathcal{O}(\exp(-ct))$. Now observe furthermore that

$$K_{V'_M}(x,y,0) = \frac{i}{2\pi}\int_{C_\delta} K_{Q'_M}(x,y,\lambda)d\lambda = 0 \ , \tag{4.2.23}$$

for, in view of (4.2.21), the integral over C_δ can be reduced to an integral over the boundary of $V_{\delta,\pi/2+\varepsilon} \cap \{\lambda \mid |\lambda| \leq R\}$ for a large R, and this integral is zero since the integrand is analytic to the left of C_δ. Moreover, $V'_M(t)$ is C^∞ in $t > 0$ and the derivative of the kernel satisfies, for $t \leq 1$,

$$\begin{aligned} |\partial_t K_{V'_M}(x,y,t)| &= \frac{i}{2\pi}\int_{C_\delta} e^{-\lambda t}(-\lambda)K_{Q'_M}(x,y,\lambda)d\lambda| \\ &\leq c_1 \exp(-c_2 t) + c_3\int_\delta^\infty \exp(-c_2|\lambda|t)\langle\lambda\rangle^{-1}\, d|\lambda| \\ &\leq c_4 + c_5\int_{\delta t}^\infty \exp(-c_2\rho)\rho^{-1}\, d\rho \\ &\leq c_6 + c_7|\log t| \quad . \end{aligned} \tag{4.2.24}$$

Then by Taylor's formula

$$|K_{V'_M}(x,y,t)| \leq ct(1+|\log t|), \quad \text{for} \quad t \in [0,1] \quad .$$

This shows (4.2.15), and (4.2.16) follows by a combination of the results. □

It is also possible to estimate some (x,y)-derivatives, depending on how large ℓ or M is. This just requires keeping check on how the derivatives affect the symbols; it is all based on (3.3.29). The general outcome is that the larger ℓ or M is, the more derivatives have estimates like those above, for example:

$$\sup_{x,y} |D^\beta_{x,y} K_{V'_M}(x,y,t)| \leq c't(1+|\log t|)\exp(-ct) \quad \text{when} \quad M > d+n+|\beta| \ . \tag{4.2.25}$$

The structure of the $V_{-\ell}$ is analyzed in more detail in Theorem 4.2.5 below, which also gives additional results on the V'_M.

4.2.3 Lemma. *Let* $M \in \mathbb{N}$, *let* $\sigma_1,\dots,\sigma_{M+1}$ *be nonnegative integers with*

$$\sigma = \sigma_1 + \dots + \sigma_{M+1} \geq 1 , \tag{4.2.26}$$

and let $f(x,\xi,\lambda)$ *be a (matrix formed) symbol of order* $k \in \mathbb{R}$ *of the form*

$$f(x,\xi,\lambda) = f_1(p_d-\lambda)^{-\sigma_1} f_2(p_d-\lambda)^{-\sigma_2} \dots (p_d-\lambda)^{-\sigma_M} f_{M+1} , \tag{4.2.27}$$

where the $f_j(x,\xi)$ *are ps.d.o. symbols of order* $s_j \in \mathbb{R}$, *homogeneous for* $|\xi| \geq 1$ *and vanishing for* x *outside a ball* B *in* $\mathbb{R}^n$. *Let* $F_\lambda = OP(f(x,\xi,\lambda))$ *on* $\mathbb{R}^n$, *and let* $E(t)$ *be the operator family defined from* F_λ *for* $t > 0$ *by*

$$E(t) = \frac{i}{2\pi} \int_{C_\delta} e^{-t\lambda} F_\lambda \, d\lambda . \tag{4.2.28}$$

Then $E(t) = OP(e(x,t,\xi))$, *where*

$$e(x,t,\xi) = \frac{i}{2\pi} \int_{C_\delta} e^{-t\lambda} f(x,\xi,\lambda)\, d\lambda ; \tag{4.2.29}$$

it satisfies the homogeneity condition and estimates:

$$\begin{aligned} &\text{(i)} \quad e(x,s^{-1/d}t,s\xi) = s^{d+k} e(x,t,\xi) \quad \text{for} \quad |\xi| \geq 1,\ s \geq 1 , \\ &\text{(ii)} \quad |D_x^\beta D_\xi^\alpha e(x,t,\xi)| \leq c'(x) \langle\xi\rangle^{d+k-|\alpha|} \exp(-ct\langle\xi\rangle^d) \end{aligned} \tag{4.2.30}$$

with a fixed $c > 0$, *and* $c'(x)$ *depending on the indices.*

For $t > 0$, $E(t)$ *is a smoothing operator, and its kernel is defined by*

$$K_E(x,y,t) = (2\pi)^{-n} \int_{\mathbb{R}^n} e^{i(x-y)\cdot\xi} e(x,t,\xi)\, d\xi . \tag{4.2.31}$$

When

$$k \equiv s_1 + \dots + s_{M+1} - \sigma d > -n \tag{4.2.32}$$

the kernel satisfies (with $c' > 0$*)*

$$\begin{aligned} &\text{(i)} \quad \sup_{x,y} |K_E(x,y,t)| \leq c_1 t^{-(d+k+n)/d} \exp(-c't) \\ &\text{(ii)} \quad K_E(x,x,t) = e_0(x) t^{-(d+k+n)/d} + O(1) \quad \text{for} \quad t \to 0 , \end{aligned} \tag{4.2.33}$$

where e_0 *is defined from the associated strictly homogeneous symbol*

(4.2.34) $$e_0(x) = (2\pi)^{-n} \int_{\mathbb{R}^n} e^h(x,1,\xi)d\xi \quad .$$

Moreover, if $\sigma \geq 2$ *(cf.* (4.2.26)*), then* $e(x,t,\xi) \to 0$ *for* $t \to 0$, *and one has*

(4.2.35)
$$\text{(i)} \quad |D_x^\beta D_\xi^\alpha e(x,t,\xi)| \leq c'(x)\langle\xi\rangle^{d+k-|\alpha|}(\langle\xi\rangle^d t)\exp(-ct\langle\xi\rangle^d)$$
$$\text{(ii)} \quad \sup_{x,y} |K_E(x,y,t)| \leq c_1(t^{-(k+n)/d} + 1)\exp(-c't)$$
$$\text{(iii)} \quad K_E(x,x,t) = e_1(x)t^{-(k+n)/d} + O(1) \quad \textit{for} \quad t \to 0 \ ,$$

where

(4.2.36) $$e_1(x) = (2\pi)^{-n} \int_{\mathbb{R}^n} \partial_t e^h(x,1,\xi)d\xi \quad .$$

<u>Proof</u>: The operator F_λ is defined as an operator from $\mathscr{S}(\mathbb{R}^n)$ to $C^0(B)$ by

$$F_\lambda u = (2\pi)^{-n} \int_{\mathbb{R}^n} e^{ix\cdot\xi} f(x,\xi,\lambda)\hat{u}(\xi)d\xi \quad ,$$

where the integrand is in $L^1(\mathbb{R}^n)$. Since $f(x,\xi,\lambda)$ is $O(\langle\xi\rangle^s(\langle\xi\rangle^d+|\lambda|)^{-\sigma})$, where $s = s_1+\ldots+s_{M+1}$, F_λ extends to a continuous operator from $H^s(\mathbb{R}^n)$ to $C^0(B)$ (uniformly), and the integration in (4.2.28) can be performed in this operator topology. Using the analyticity of F_λ and its symbol, we can write

$$\begin{aligned} E(t)u &= i(2\pi)^{-1-n} \int_{C_\delta} e^{-\lambda t} \int_{\mathbb{R}^n} e^{ix\cdot\xi} f(x,\xi,\lambda)\hat{u}(\xi)d\xi d\lambda \\ &= (2\pi)^{-n} \int_{\mathbb{R}^n} e^{ix\cdot\xi} e(x,t,\xi)\hat{u}(\xi)d\xi \end{aligned}$$

with e defined by (4.2.29). In the latter integral we observe that since $p_d(x,\xi)$ is homogenous of degree d in ξ for $|\xi| \geq 1$, and has eigenvalues in the complement of $V_{\delta,\pi/2+\varepsilon}$ for all ξ , the integration curve C_δ may for each ξ be replaced by a closed curve $C_{\delta,R}$ (with $R \geq \text{const.}\langle\xi\rangle^d$) consisting of the part of C_δ lying in the circle $\{|\lambda| \leq R\}$ together with the part of the circumference lying to the right of C_δ . Then for $|\xi| \leq 1$, we have with a fixed R

(4.2.37) $$\left|\frac{i}{2\pi}\int_{C_\delta} e^{-\lambda t} f(x,\xi,\lambda)d\lambda\right| = \left|\frac{i}{2\pi}\int_{C_{\delta,R}} e^{-\lambda t} f(x,\xi,\lambda)d\lambda\right| \leq c_1 \exp(-c_2 t) \ ,$$

whereas for $|\xi| \geq 1$ we can make the boundary curve homogeneous in ξ and compare f with the associated strictly homogeneous symbol. Here, with fixed r and r' ,

$$(4.2.38)\quad \frac{i}{2\pi}\int_{C_\delta} e^{-\lambda t} f^h(x,\xi,\lambda)d\lambda = \frac{i}{2\pi}\int_{C_{r|\xi|^d, r'|\xi|^d}} e^{-\lambda t} f^h(x,\xi,\lambda)d\lambda \;,$$

where the contribution from each part of the curve is $\mathcal{O}(\exp(-c|\xi|^d t)|\xi|^{k+d})$; and on the other hand

$$(4.2.39)\quad \left|\frac{i}{2\pi}\int_{C_\delta} e^{-\lambda t}(f(x,\xi,\lambda)-f^h(x,\xi,\lambda))d\lambda\right| = \left|\frac{i}{2\pi}\int_{C_{\delta,R}} e^{-\lambda t}(f-f^h)d\lambda\right| \le c_1\exp(-c_2 t) \;,$$

vanishing for $|\xi| \geq 1$. This shows the estimate in (4.2.30) for α and $\beta = 0$, and the estimates of the derivatives are obtained by similar calculations on the derivatives of f , that are of the same form.

In view of the exponential factor in the estimates (4.2.30), $E(t)$ is smoothing for $t > 0$. Now consider the kernel, defined by (4.2.31), and let $k > -n$. Then we can estimate K_E by comparing e with e^h :

$$K_E(x,y,t) = (2\pi)^{-n}\int_{\mathbb{R}^n} e^{i(x-y)\cdot\xi} e^h(x,t,\xi)d\xi + (2\pi)^{-n}\int_{|\xi|\le 1} e^{i(x-y)\cdot\xi}(e-e^h)d\xi \;,$$

here

$$\left|\int_{|\xi|\le 1} e^{i(x-y)\cdot\xi}(e-e^h)d\xi\right| \le c\int_{|\xi|\le 1}(\langle\xi\rangle^k + |\xi|^k)d\xi = c_1 \;,$$

and

$$\left|(2\pi)^{-n}\int_{\mathbb{R}^n} e^{i(x-y)\cdot\xi} e^h(x,t,\xi)d\xi\right| \le c_2\int_{\mathbb{R}^n} |\xi|^{d+k}\exp(-ct|\xi|^d)d\xi$$

$$= c_2 t^{-1-k/d-n/d}\int_{\mathbb{R}^n} |\eta|^{d+k}\exp(-c|\eta|^d)d\eta = c_3 t^{-(d+k+n)/d} \;.$$

The exponential estimate for $t \to \infty$ is obtained by replacing $\exp(-ct\langle\xi\rangle^d)$ by $[\exp(-\tfrac{1}{2}c\langle\xi\rangle^d t)]^2$, so we get (4.2.33i). The proof of (4.2.33ii) is similar:

$$K_E(x,x,t) = \mathcal{O}(1) + (2\pi)^{-n}\int_{\mathbb{R}^n} e^h(x,t,\xi)d\xi = e_0(x)t^{-(d+k+n)/d} + \mathcal{O}(1) \;,\ \text{for } t \to 0 \;,$$

where $e_0(x)$ satisfies (4.2.34).

Now consider the case where $\sigma \geq 2$. Here we find that the integral defining $e(x,t,\xi)$ converges uniformly in $t \geq 0$, and in fact

$$e(x,0,\xi) = \frac{i}{2\pi}\int_{C_\delta} f(x,\xi,\lambda)d\lambda = 0 \;,$$

since f is $\mathcal{O}(\langle\lambda\rangle^{-2})$ so that the integration can be transferred to a curve in the left half plane where f is analytic. Now we can also use that for $t > 0$,

$$\partial_t e(x,t,\xi) = \frac{i}{2\pi}\int_{C_\delta} e^{-t\lambda}(-\lambda)f(x,\xi,\lambda)d\lambda \quad ;$$

here an insertion of

$$(4.2.40) \qquad \lambda(p_d-\lambda)^{-1} = -1 + p_d(p_d-\lambda)^{-1}$$

in (4.2.27) gives $\lambda f(x,\xi,\lambda)$ as a sum of two terms of the preceding kind (with k augmented by d). Then the preceding arguments show that

$$|\partial_t e(x,t,\xi)| \le c(x)\langle\xi\rangle^{2d+k}\exp(-ct\langle\xi\rangle^d)$$

and Taylor's formula gives for $t \le 1$

$$|e(x,t,\xi)| \le c(x)\langle\xi\rangle^{d+k}(\langle\xi\rangle^d t)\exp(-ct\langle\xi\rangle^d) \ ;$$

for larger t this follows already from (4.2.30). This shows the estimate (4.2.35i) for α and β equal to zero, and the rest of the proof is completed essentially as above. □

4.2.4 Remark. When $\sigma \ge m$ for larger m in the lemma, one can show that $\partial_t^j e(x,0,\xi) = 0$ for $j \le m-1$, and obtain an estimate of $e(x,t,\xi)$ with t in (4.2.30i) replaced by t^{m-1}. Since this is only useful for differential operators (cf. Remark 3.3.7), we do not bother to give the details (which lead to well known kernel and trace expansions, see e.g. [Greiner 1]).

Estimates of the $v_{-\ell}(x,t,\xi)$ can also be found in [Treves 2, Chapter III] for systems with diagonal principal part; the remainders are not analyzed there. [Widom 2] treats $\exp(-tP)$ in a framework of general functional calculus.

Improvements of the results in Theorem 4.2.2 can be obtained by use of the representation (3.3.75) of Q_λ.

4.2.5 Theorem. *Hypotheses and notation as in Theorem 4.2.2. For any* $m \in \mathbb{N}$, *the operators* $V(t)$ *may be represented on the form*

$$(4.2.41) \qquad V(t) = \frac{i}{2\pi}\int_{C_\delta} e^{-\lambda t}\lambda^{-m}P^m Q_\lambda \, d\lambda$$

(integrated with respect to the operator norm from $H^{(m-1)d}(\tilde{E})$ *to* $L^2(\tilde{E})$*),* *and here*

$$(4.2.42)\quad V_{-\ell}(t) = \frac{i}{2\pi}\int_{C_\delta} e^{-\lambda t}\lambda^{-m}OP(q^{(m)}_{(m-1)d-\ell}(x,\xi,\lambda))d\lambda \quad \text{for} \quad \ell \in \mathbb{N}\ ,$$

$$(4.2.43)\quad V'_M(t) = \frac{i}{2\pi}\int_{C_\delta} e^{-\lambda t}\lambda^{-m}Q^{(m)\,'}_{\lambda,M}\,d\lambda \quad \text{for} \quad M \in \mathbb{N}\ ,$$

with the terminology (3.3.77-79). *For any* $m \in \mathbb{N}$, *one has with* $c > 0$,

$$(4.2.44)\quad |D^\beta_x D^\alpha_\xi D^m_t v_{-\ell}(x,t,\xi)| \le c'(x)\langle\xi\rangle^{md-\ell-|\alpha|}\exp(-c\langle\xi\rangle^d t)\ ,$$

and there are expansions, for $\ell > 0$,

$$(4.2.45)\quad v_{-\ell}(x,t,\xi) = tv_{-\ell,1}(x,\xi)+\ldots+t^{m-1}v_{-\ell,m-1}(x,\xi) + t^m v'_{-\ell,m}(x,t,\xi)\ ,$$

where $v_{-\ell,j} = \frac{1}{j!}\,\partial^j_t v_{-\ell}(x,0,\xi)$ *is a ps.d.o. symbol of order* $jd-\ell$, *and*

$$(4.2.46)\quad |D^\beta_x D^\alpha_\xi v'_{-\ell,m}(x,t,\xi)| \le c(x)\langle\xi\rangle^{md-\ell-|\alpha|} \quad \text{for} \quad t \in [0,1]\ .$$

$v_0(x,t,\xi)$ *has a similar expansion with a nonvanishing zero'th term* $v_0(x,0,\xi) = 1$.

The kernels of the corresponding operators $V_{-\ell}(t)$ *for* $t > 0$ *satisfy, for any* $m \in \mathbb{N}$,

$$(4.2.47)\quad \sup_{x,y}|\partial^m_t K_{V_{-\ell}}(x,y,t)| \le \begin{cases} c'(t^{(\ell-md-n)/d}+1)\exp(-ct), & \text{when } \ell-md \ne n\ , \\ c'(1+|\log t|)\exp(-ct), & \text{when } \ell-md = n\ . \end{cases}$$

In particular, if $\ell < (m-1)d+n$, *there is a formula*

$$(4.2.48)\quad \begin{aligned} &\partial^m_t K_{V_{-\ell}}(x,x,t) = c_{\ell,m}(x)t^{-(md-\ell+n)/d} + O(1) \quad \text{for} \quad t \to 0\ , \\ &c_{\ell,m}(x) = (2\pi)^{-n}\int_{\mathbb{R}^n} \partial^m_t v^h_{-\ell}(x,1,\xi)d\xi\ . \end{aligned}$$

Moreover, when $\ell > md+n$ *for some* $m \in \mathbb{N}$, *the* $K_{V_{-\ell}}$ *are* C^m *in* $t \ge 0$ *and satisfy*

$$(4.2.49)\quad K_{V_{-\ell}}(x,y,t) = t\partial_t K_{V_{-\ell}}(x,y,0)+\ldots+\frac{t^{m-1}}{(m-1)!}\,\partial^{m-1}_t K_{V_{-\ell}}(x,y,0) + t^m R(x,y,t)\ ,$$

where the $\partial^j_t K_{V_{-\ell}}(x,y,0)$ *and* $R(x,y,t)$ *are continuous and locally bounded in* x,y *and in* $t \ge 0$.

The remainder operators $V'_M(t)$ *have kernels* $K_{V'_M}(x,y,t)$ *that are* C^m *in* $t \ge 0$ *for* $M > (m+1)d+n$, *satisfying*

(4.2.50) $\sup_{x,y} |\partial_t^m K_{V'_M}(x,y,t)| \le c' \exp(-ct)$ *when* $M > (m+1)d+n$,

and having expansions

(4.2.51) $K_{V'_M}(x,y,t) = t\partial_t K_{V'_M}(x,y,0)+\ldots+\frac{t^{m-1}}{(m-1)!}\partial_t^{m-1}K_{V'_M}(x,y,0)+t^m R(x,y,t)$

when $M > (m+1)d+n$,

with $\partial_t^j K_{V'_M}(x,y,0)$ *and* $R(x,y,t)$ *continuous in* x,y *and in* $t \ge 0$.

Proof: The representation (4.2.41) for $t > 0$ follows immediately by insertion of (3.3.75) in (4.2.5), and the formulas (4.2.42) follow likewise from the definitions (3.3.77-79), when we observe the identity (3.3.81):

(4.2.52)
$$v_{-\ell}(x,t,\xi) = \frac{i}{2\pi}\int_{C_\delta} e^{-\lambda t} q_{-d-\ell}(x,\xi,\lambda)d\lambda$$
$$= \frac{i}{2\pi}\int_{C_\delta} e^{-\lambda t}(\lambda^{-1}\delta_{\ell,0}p_d - \lambda^{-2}p_{d-\ell} - \ldots - \lambda^{-m}p^{(m-1)}_{(m-1)d-\ell} + \lambda^{-m}q^{(m)}_{(m-1)d-\ell})d\lambda$$
$$= \frac{i}{2\pi}\int_{C_\delta} e^{-\lambda t}\lambda^{-m}q^{(m)}_{(m-1)d-\ell}d\lambda \quad ,$$

since the $p_k^{(j)}$ are independent of λ , and

(4.2.53)
$$\int_{C_\delta} e^{-\lambda t}\lambda^{-j}d\lambda = 0 \quad \text{for} \quad j \in \mathbb{N} .$$

Note that when $m \ge 1$, (4.2.41) makes sense for $t \ge 0$ since Q_λ is $\mathcal{O}(\langle\lambda\rangle^{-1})$; similar remarks can be made for (4.2.42) and (4.2.43). Differentiation of (4.2.52) gives for $m > 0$

$$\partial_t^m v_{-\ell}(x,t,\xi) = \frac{i(-1)^m}{2\pi}\int_{C_\delta} e^{-\lambda t}q^{(m)}_{(m-1)d-\ell}d\lambda \qquad \text{for} \quad t > 0 .$$

Since $q^{(m)}_{(m-1)d-1}$ is of the form (4.2.27) with $\sigma \ge 1$ and $k = (m-1)d-\ell$, it follows from Lemma 4.2.3 that the estimates (4.2.44) are valid. Then (4.2.45) is obtained by Taylor expansions (the continuity of $\partial_t^j v_{-\ell}$ at $t = 0$ follows from the boundedness of $\partial_t^{j+1} v_{-\ell}$ for $t \to 0$) . The statement on $v_0(x,t,\xi)$ is obvious from its explicit form (4.2.7).

The symbols $\partial_t^m v_{-\ell}$ define the operators $\partial_t^m V_{-\ell}(t)$ defined in suitable operator norms.

Now consider the kernels. There is first the case where $(m-1)d-\ell > -n$; here it is seen from Lemma 4.2.3 that the kernel of $V_{-\ell}(t)$ satisfies an estimate

$$\sup_{x,y} |\partial_t^m K_{V_{-\ell}}(x,y,t)| \le c' t^{-(md-\ell+n)/d} \exp(-ct) ,$$

which gives (4.2.47) for $\ell < (m-1)d+n$. Moreover, one here has as in (4.2.33ii), that

$$\partial_t^m K_{V_{-\ell}}(x,x,t) = c_{\ell,m}(x) t^{-(md-\ell+n)/d} + O(1) \quad \text{for} \quad t \to 0 ,$$

with the constant determined as in (4.2.34), this shows (4.2.48). On the other hand, if $md-\ell < -n$, we have the general estimate

$$(4.2.54) \quad |\partial_t^m K_{V_{-\ell}}(x,y,t)| = (2\pi)^{-n} \left|\int_{\mathbb{R}^n} e^{i(x-y)\cdot\xi} v_{-\ell}(x,t,\xi)d\xi\right|$$

$$\le c' \int_{\mathbb{R}^n} \langle\xi\rangle^{md-\ell} \exp(-ct\langle\xi\rangle^d) d\xi \le c_1 \exp(-\tfrac{1}{2}ct) ,$$

showing (4.2.47) for $\ell > md+n$. There remain the cases $(m-1)d+n \le \ell \le md+n$, but these can be treated by integration of $\partial_t^{m+2} K_{V_{-\ell}}$. Since $\ell < (m+1)d+n$, the first case of our estimate applies to $\partial_t^{m+2} K_{V_{-\ell}}$, giving

$$|\partial_t^{m+2} K_{V_{-\ell}}(x,y,t)| \le c t^{(\ell-(m+2)d-n)/d} \quad \text{for} \quad t \le 1$$

and hence by integration

$$|\partial_t^{m+1} K_{V_{-\ell}}(x,y,t)| = \left|c_1 + \int_1^t \partial_{t'}^{m+2} K_{V_{-\ell}} dt'\right| \le c_2 + c_3 t^{(\ell-(m+1)d-n)/d} \quad \text{for} \quad t \le 1 .$$

Another integration gives, for $t \le 1$,

$$|\partial_t^m K_{V_{-\ell}}(x,y,t)| = \left|c_4 + \int_1^t \partial_{t'}^{m+1} K_{V_{-\ell}} dt'\right|$$

$$\le \begin{cases} c_5 + c_6 t^{(\ell-md-n)/d} & \text{if} \quad \ell-md \neq n , \\ c_5 + c_6 |\log t| & \text{if} \quad \ell-md = n . \end{cases}$$

The exponential estimates are obtained for $t \ge 1$ as in (4.2.54), and we altogether find (4.2.47).

When $\ell > n$, we know from Theorem 4.2.2 that $K_{V_{-\ell}}(x,y,0) = 0$. Then (4.2.49) follows from the above by use of Taylor's formula, for $\ell > md+n$.

Consider finally the remainders. Let $m \ge 1$, and take $M > md+n$; we shall use the representation (4.2.43)

$$V_M'(t) = \frac{i}{2\pi}\int_{C_\delta} e^{-\lambda t}\lambda^{-m} Q_{\lambda,M}^{(m)'}\,d\lambda \quad .$$

Here $Q_{\lambda,M}^{(m)'}$ is a parameter-dependent ps.d.o. of order $(m-1)d-M$ and regularity $md-M$, so it has a symbol $q_M^{(m)'}$ satisfying

$$(4.2.55) \qquad |q_M^{(m)'}(x,\xi,\lambda)| \le c(x)\langle\xi\rangle^{md-M}\langle\xi,\mu\rangle^{-d} \quad ;$$

and its kernel (which is exactly determined from $Q_\lambda^{(m)} = P^m Q_\lambda$ and its first M terms $Q_{\lambda,(m-1)d-\ell}^{(m)}$ for $\ell = 0,1,\dots,M-1$) is a continuous function satisfying

$$(4.2.56) \qquad \sup_{x,y} |K_{Q_M^{(m)'}}(x,y,\lambda)| \le c\langle\lambda\rangle^{-1}$$

(since it is the sum of a negligible kernel of regularity d and a kernel derived from (4.2.55) by integration). The kernel is analytic in λ where $Q_\lambda^{(m)}$ and the $Q_{\lambda,(m-1)d-\ell}^{(m)}$ are analytic. These considerations allow us to write

$$K_{V_M'}(x,y,t) = \frac{i}{2\pi}\int_{C_\delta} e^{-\lambda t}\lambda^{-m} K_{Q_M^{(m)'}}(x,y,\lambda)\,d\lambda \quad .$$

Because of (4.2.56) the integral remains convergent for $t \ge 0$ if we differentiate in t up to order $m-1$

$$\partial_t^j K_{V_M'}(x,y,t) = \frac{i(-1)^j}{2\pi}\int_{C_\delta} e^{-\lambda t}\lambda^{j-m} K_{Q_M^{(m)'}}(x,y,\lambda)\,d\lambda \quad ,$$

and we have

$$\sup_{x,y}|\partial_t^j K_{V_M'}(x,y,t)| \le c\left|\int_{C_\delta} e^{-\lambda t}\langle\lambda\rangle^{j-m-1}\,d\lambda\right| \le c_1 \exp(-\delta t), \quad \text{for } t \ge 0, \quad \text{when } j \le m-1 \; .$$

Since it is known from Theorem 4.2.2 that $K_{V_M'}(x,y,0) = 0$, one gets the Taylor expansion

$$K_{V_M'}(x,y,t) = t\partial_t K_{V_M'}(x,y,0) + \dots + \frac{t^{m-2}}{(m-2)!}\partial_t^{m-2} K_{V_M'}(x,y,0) + t^{m-1} R(x,y,t)$$

with continuous functions $\partial_t^j K_{V_M'}(x,y,0)$ and $R(x,y,t)$ for $t \ge 0$. Replacing $m-1$ by m, we have shown the remaining part of the theorem. $\square$

Note that by a slight variant of the argument showing (4.2.49), we also have

$$(4.2.57) \quad K_{V_{-md-n}}(x,y,t) = \sum_{j=1}^{m-1} \frac{t^j}{j!}\partial_t^j K_{V_{-md-n}}(x,y,0) + O(t^m \log t)$$

for $t \leq 1$. Similarly,

$$(4.2.58)\quad K_{V_{-md-n+k}}(x,y,t) = \sum_{j=1}^{m-1} \frac{t^j}{j!}\,\partial_t^j K_{V_{-md-n+k}}(x,y,0) + O(t^{m-k/d}) \quad \text{for } 0<k<d .$$

Note also that the estimates (4.2.51) of $K_{V'_M}$ for very large M can be combined with the estimates of the $K_{V_{-\ell}}$, to give an analysis of general $K_{V'_M}$, in view of the formula

$$(4.2.59)\qquad V'_M(t) = \sum_{M \leq \ell < N} V_{-\ell}(t) + V'_N(t) .$$

4.2.6 Remark. The smoothness properties of the kernels $K_{V_{-\ell}}(x,y,t)$ in the variables x,y can be analyzed on the basis of the estimates (4.2.44). As for the kernels $K_{V'_M}(x,y,t)$, we observe that when M is taken even larger than in (4.2.50), there is room for some estimates of (x,y)-derivatives; and also the off-diagonal behavior can be controlled:

$$(4.2.60)\quad \sup_{x,y} |(x-y)^\alpha D^\beta_{x,y} \partial_t^m K_{V'_M}(x,y,t)| \leq c' \exp(-ct) \quad \text{when } M>(m+1)d-|\alpha|+|\beta|+n .$$

To see this, it suffices to observe that the corresponding derivative of $K_{Q_M^{(m+1)'}}(x,y,\lambda)$ satisfies

$$(4.2.61)\qquad \sup_{x,y} |(x-y)^\alpha D^\beta_{x,y} K_{Q_M^{(m+1)'}}(x,y,\lambda)| \leq c\langle\lambda\rangle^{-1}$$

since $Q_M^{(m+1)'}$ is of order $md-M$ and regularity $(m+1)d-M$, so (4.2.60) can be shown as in the end of the proof of Theorem 4.2.5.

(4.2.60) shows on one hand that $\partial_t^m K_{V'_M}$ has more well-behaved x,y-derivatives, the larger M is (take $\alpha = 0$ and β large). On the other hand it shows that $K_{V'_M}$ is very smooth for $x \neq y$, also for small M and large m .

Fine kernel estimates are also given in [Eidelman-Drin 1,2], [Drin 1,2,3].

As an easy consequence of the above results we obtain the famous trace estimate, on a precise form:

4.2.7 Corollary. *Hypotheses and notation as in Theorem 4.2.2. The diagonal value of the kernel of* $V(t)$ *has an asymptotic expansion for* $t \to 0+$

$$(4.2.62)\quad K_V(x,x,t) \sim \sum_{\substack{j\in\mathbb{N}\\ j-n\notin d\mathbb{N}_+}} c_{j-n}(x,V)t^{(j-n)/d} + \sum_{j-n\in d\mathbb{N}_+} c_{j-n}(x,V)t^{(j-n)/d}\log t + \sum_{\ell\in\mathbb{N}_+} r_\ell(x,V)t^\ell \;,$$

in the usual sense: the difference between $K_V(x,x,t)$ *and the sum of terms up to* $(j-n)/d = N \in \mathbb{N}_+$, $\ell = N$, *is* $O(t^{N+1/d})$. *Each coefficient* $c_{j-n}(x,V)$ *is determined from the first* $j+1$ *symbols* $p_d,\dots,p_{d-j}$, *whereas the coefficients* $r_\ell(x,V)$ *depend on the full operator* P . *The coefficients are* C^∞ *in* x .

As a consequence, if we set

$$(4.2.63)\quad \underline{c}_k = \underline{c}_k(V) = \int_\Sigma \operatorname{tr} c_k(x,V)dx \;, \quad \underline{r}_\ell = \underline{r}_\ell(V) = \int_\Sigma \operatorname{tr} r_\ell(x,V)dx \;,$$

the trace of $V(t)$ *has the asymptotic expansion for* $t \to 0+$

$$(4.2.64)\quad \operatorname{tr} V(t) \sim \sum_{\substack{j\in\mathbb{N}\\ j-n\notin d\mathbb{N}_+}} \underline{c}_{j-n} t^{(j-n)/d} + \sum_{j-n\in d\mathbb{N}_+} \underline{c}_{j-n} t^{(j-n)/d}\log t + \sum_{\ell\in\mathbb{N}_+} \underline{r}_\ell t^\ell \;.$$

Proof: Take a large $m \in \mathbb{N}$, and consider $\partial_t^m K_V(x,x,t)$. We have by (4.2.48)

$$(4.2.65)\quad \partial_t^m K_{V_{-\ell}}(x,x,t) = c_{\ell,m}(x)t^{(\ell-md-n)/d} + O(1) \text{ for } t \to 0, \text{ when } \ell<(m-1)d+n \;;$$

and we have for $M = (m-1)d+n$ $(> ((m-3)+1)d+n)$

$$(4.2.66)\qquad K_{V'_M}(x,x,t) = \sum_{j=1}^{m-4} \frac{t^j}{j!}\, \partial_t^j K_{V'_M}(x,x,0) + O(t^{m-3}) \;,$$

by (4.2.51). Repeated integration of (4.2.65) shows (in view of (4.2.12) for $\ell=n$)

$$(4.2.67)\quad K_{V_{-\ell}}(x,x,t) = \begin{cases} c'_{\ell,m}(x)t^{(\ell-n)/d} + p_{\ell,m}(x,t) + O(t^m) & \text{if } \ell-n \notin d \\ c'_{\ell,m}(x)t^{(\ell-n)/d}\log t + p'_{\ell,m}(x,t) + O(t^m) & \text{if } \ell-n \in d \end{cases}$$

with $c'_{\ell,m}$ determined from $c_{\ell,m}$ and with $p_{\ell,m}(x,t)$ and $p'_{\ell,m}(x,t)$ denoting polynomials of degree m in t with coefficients continuous in x . For all ℓ , we find from Theorem 4.2.2 that the zero-th term in the polynomials must vanish. Summation of (4.2.66) and (4.2.67) gives

$$K_V(x,x,t) = \sum_{\substack{0\le\ell<M\\ \ell-n\notin d\mathbb{N}_+}} c'_{\ell,m}(x)t^{(\ell-n)/d} + \sum_{\substack{\ell=dj+n\\ 1\le j<m-1}} c'_{\ell,m}(x)t^j\log t + p_m(x,t) + O(t^{m-3}) \;,$$

where $p_m(x,t)$ is a polynomial in t (for each x) with $p_m(x,0) = 0$. Since m can be taken arbitrarily large, this shows the existence of the asymptotic expansion (4.2.62). It is seen that the coefficients $c'_{\ell,m}$ depend on $V_{-\ell}$, which in turn depends on the symbols in P of orders from d down to $d-\ell$; on the other hand, the coefficients in $p_m(x,t)$ have no such simple description. That the coefficients are C^∞ in x is seen as follows: For the $c_{\ell,m}(x)$ there are explicit formulas (4.2.48) defining the functions in terms of (exponentially) convergent integrals of smooth functions, the $c'_{\ell,m}(x)$ just get some constant factors, and for $p_{\ell,m}$ and $p'_{\ell,m}$ in (4.2.67), the smoothness is seen at each step in the integration by setting $t = 1$. For the coefficients in (4.2.66), we use that the j-th coefficient is in fact in C^{m-3-j} in view of Remark 4.2.6; this regularity goes to ∞ when $m \to \infty$.

The trace estimates (4.2.63) follow by integration in x , since

$$\operatorname{tr} V(t) = \int_\Sigma \operatorname{tr} K_V(x,x,t)dx . \qquad \square$$

The trace estimate (4.2.64) was first shown in [Duistermaat-Guillemin 1] in the case of scalar positive ps.d.o.s, and M. Taylor also shows the formula for such operators of order 1 his book [Taylor 1]. A proof for the general case is given in [Widom 2].

As indicated in Remark 4.2.4, one can apply versions of Lemma 4.2.3 with $\sigma \geq m$, $m \in \mathbb{N}$, to the lower order terms in the differential operator case; this leads to an expansion like (4.2.62) where all logarithmic terms vanish and all the coefficients r_ℓ are determined from the symbol; a complete deduction is given in [Greiner 1] (which also accounts for earlier work on the problem).

The semigroup $V(t)$ enters in boundary value problems as the term $V(t)_\Omega = r_\Omega V(t)e_\Omega$ in

$$U(t) = V(t)_\Omega + W(t) ,$$

and the kernel of $V(t)_\Omega$ is of course here obtained by restriction of the full kernel $K_V(x,y,t)$ on $\Sigma\times\Sigma\times\mathbb{R}_+$ to the "quadrant" $\bar\Omega\times\bar\Omega\times\mathbb{R}_+$. Consequently, the trace of $V(t)_\Omega$ has an asymptotic expansion

$$(4.2.68) \qquad \operatorname{tr} V(t)_\Omega \sim \sum_{\substack{j\in\mathbb{N} \\ j-n\notin d\mathbb{N}_+}} \underline{c}_{j-n,\Omega} t^{(j-n)/d} + \sum_{j-n\in d\mathbb{N}_+} \underline{c}'_{j-n,\Omega} t^{(j-n)/d}\log t + \sum_{\ell\in\mathbb{N}_+} \underline{r}_{\ell,\Omega} t^\ell \quad \text{for} \quad t \to 0+ ,$$

where

$$(4.2.69) \qquad \underline{c}_{k,\Omega} = \int_\Omega \operatorname{tr} c_k(x,V)dx \quad \text{and} \quad \underline{r}_{\ell,\Omega} = \int_\Omega \operatorname{tr} r_\ell(x,V)dx .$$

We now turn to the analysis of the term $W(t)$ stemming from the boundary condition. Recall that

(4.2.70) $$W(t) = \frac{i}{2\pi}\int_{C_\delta} e^{-\lambda t} G_\lambda d\lambda \ ,$$

where G_λ is the singular Green operator part of R_λ . The analysis is formulated in local coordinates near Γ , or rather, with $\bar{\Omega}$ replaced by $\Gamma\times\mathbb{R}_+$, as explained before Theorem 3.3.10. In the formulation, we often disregard the inaccuracies stemming from coordinate changes and cut-off functions, observing once and for all that they give rise to negligible terms whose $L^2_{x_n,y_n}$ norms are $O(t^{\nu/d} + t|\log t|)$ for $t \to 0$; more details are given in the proof of Theorem 4.2.9.

The following argument will be used in many places.

4.2.8 Lemma. *Let* $f(x',x_n,y_n,\xi',\lambda)$ *be analytic in* λ *in a region* $V_{\delta'\langle\xi'\rangle^d,\pi/2+\varepsilon'}$ (4.2.1) *satisfying the estimate (with* $\mu = |\lambda|^{1/d}$*), for some* k *and* $m \in \mathbb{R}$,

(4.2.71) $$\|f(x',x_n,y_n,\xi',\lambda)\|_{L^2_{x_n,y_n}} \leq c(x')\langle\xi'\rangle^k\langle\xi',\mu\rangle^{-d-m} \ .$$

Let $C(\xi') = \partial V_{\delta\langle\xi'\rangle^d,\pi/2+\varepsilon}$ *for some* $\delta \in]0,\delta'[$ *and* $\varepsilon \in]0,\varepsilon'[$. *Then the function*

(4.2.72) $$e(x',x_n,y_n,t,\xi') = \frac{i}{2\pi}\int_{C(\xi')} e^{-\lambda t} f(x',x_n,y_n,\xi',\lambda)d\lambda \ , \quad t > 0 \ ,$$

satisfies

$$\|e(x',x_n,y_n,t,\xi')\|_{L^2_{x_n,y_n}} \leq c'(x')\langle\xi'\rangle^{k-m}\exp(-ct\langle\xi'\rangle^d) \ .$$

(4.2.73) $$\cdot\begin{cases} (t\langle\xi'\rangle^d)^{m/d}(1+|\log(t\langle\xi'\rangle^d)|) & \text{if } m/d \in \mathbb{N} = \{0,1,2,\dots\} \ , \\ (t\langle\xi'\rangle^d)^{m/d} & \text{if } m/d \in \mathbb{R}\setminus\mathbb{N} \ . \end{cases}$$

Proof: Because of the exponential decrease of $e^{-\lambda t}$ on $C(\xi')$, the integral is well-defined, also when considered for continuous functions valued in the Banach space $L^2_{x_n,y_n}(\mathbb{R}^2_{++})$. The norm of e in that space is estimated by

$$\|e\| \leq c_1 \left| \int_{C(\xi')} \exp(-c_1|\lambda|t)c(x')\langle\xi'\rangle^k \langle\xi',\mu\rangle^{-d-m} d\lambda \right| ,$$

where the contribution from the part where $|\lambda|$ equals $\delta\langle\xi'\rangle^d$ is $O(\exp(-c_2\langle\xi'\rangle^d t)\langle\xi'\rangle^{k-m})$, and the contribution from the rays are of the type

$$\begin{aligned}(4.2.74) \qquad I &= \int_{\delta\langle\xi'\rangle^d}^{\infty} \exp(-c|\lambda|t)\langle\xi'\rangle^k \langle\xi',\mu\rangle^{-d-m} d|\lambda| \\ &\leq t^{m/d}\langle\xi'\rangle^k \int_{\delta\langle\xi'\rangle^d t}^{\infty} \exp(-c\rho)(\langle\xi'\rangle^d t+\rho)^{-1-m/d} d\rho \\ &\leq \langle\xi'\rangle^{k-m}(\langle\xi'\rangle^d t)^{m/d} \exp(-\tfrac{c}{2}\delta\langle\xi'\rangle^d t)\int_{\delta\langle\xi'\rangle^d t}^{\infty} \exp(-\tfrac{c}{2}\rho)(\langle\xi'\rangle^d t+\rho)^{-1-m/d} d\rho\end{aligned}$$

When $m/d < 0$, the integral is uniformly bounded in ξ' and t , which shows the lemma in this case. When $m/d = 0$, we get its validity from the estimate

$$\begin{aligned}(4.2.75) \qquad \int_{\delta s}^{\infty} \exp(-c_1\rho)(s+\rho)^{-1} d\rho &= \int_{\delta s}^{1} + \int_{1}^{\infty} \\ &\leq \int_{\delta s}^{1} (s+\rho)^{-1} d\rho + \int_{1}^{\infty} \exp(-c_1\rho)\rho^{-1} d\rho \leq c_1 + c_3|\log s| .\end{aligned}$$

When $m/d \in]0,1]$, we have from (4.2.74) the validity of the estimates (4.2.73) for $\langle\xi'\rangle^d t \geq 1$. To handle the values $\langle\xi'\rangle^d t \leq 1$, we observe furthermore that

$$\frac{i}{2\pi}\int_{C(\xi')} f(x',x_n,y_n,\xi',\lambda)d\lambda = 0 ,$$

as an integral with values in $L^2_{x_n,y_n}$, since $\|f\|$ here is $O(\langle\lambda\rangle^{-1-m/d})$ so that the integration contour can be carried over to a compact contour to the left of $C(\xi')$, where f is analytic. Then

$$\|e(x',x_n,y_n,t,\xi')\| = \left\| \frac{i}{2\pi}\int_{C(\xi')} (e^{-\lambda t}-1)f(x',x_n,y_n,\xi',\lambda)d\lambda \right\| ,$$

where the contribution from the circular part of $C(\xi')$ is $O(\langle\xi'\rangle^{k-m} t\langle\xi'\rangle^d)$, since $|e^{-\lambda t}-1| \leq 2|\lambda|t$; and the other contributions are of the type (where we denote $t\langle\xi'\rangle^d = s$)

$$\begin{aligned}(4.2.76) \qquad &\langle\xi'\rangle^k \int_{\delta\langle\xi'\rangle^d}^{\infty} |e^{-\lambda t}-1| \langle\xi',\mu\rangle^{-d-m} d|\lambda| \leq c_1\langle\xi'\rangle^k t^{m/d}\Big(\int_{\delta s}^{1} \rho(s+\rho)^{-1-m/d} d\rho + \\ &+ \int_1^{\infty} [\exp(-c_2\rho)+1](s+\rho)^{-1-m/d} d\rho\Big) \leq \begin{cases} c_3\langle\xi'\rangle^{k-m}(t\langle\xi'\rangle^d)^{m/d} & \text{if } m/d < 1 \\ c_3\langle\xi'\rangle^{k-m}(t\langle\xi'\rangle^d)(1+|\log(t\langle\xi'\rangle^d)|) & \text{if } m/d = 1 .\end{cases}\end{aligned}$$

Taking this together with the exponential estimate valid for $t\langle\xi'\rangle^d \to \infty$, we find (4.2.73). For the cases $m/d \in \,]N-1,N]$ we use a generalization of the preceding argument, where we subtract from $\exp(-\lambda t)$ the terms of order $< N$ in its Taylor series, to get the needed estimate by $O((t\langle\xi'\rangle^d)^{m/d})$ resp. $O((t\langle\xi'\rangle^d)^{m/d}(1+|\log t\langle\xi'\rangle^d|))$ for $t\langle\xi'\rangle^d$ close to 0. □

4.2.9 Theorem. *Let* E *be a vector bundle over a compact manifold* $\bar\Omega$ *with boundary* Γ, *and let* $A = \{P_\Omega + G, T\}$ *be a Green operator of order* $d \in \mathbb{N}_+$ *and class* $r \le d$, *with* T *normal. Let* $\nu \in [\frac{1}{2}, d]$ *stand for the regularity of* G *and* T, *when considered as parameter-dependent operators. Assume that the system* $\{\partial_t + P_\Omega + G, T\}$ *is parabolic (Definition 1.5.5). Let* B *denote the realization* $(P+G)_T$, *and let* $U(t) = \exp(-tB)$, *written as in* (4.2.4)

$$U(t) = V(t)_\Omega + W(t) ,$$

where we assume that the set-up has been adapted (as described before Theorem 4.2.2) so that $P-\lambda I$ *and* $p_d(x,\xi)-\lambda I$, *as well as* $\{P_\Omega + G - \lambda I, T\}$ *and its principal boundary symbol operator, are invertible for all* λ *in an obtuse keyhole region* $V_{\delta',\pi/2+\varepsilon'}$ *(resp.* $V_{\delta'\langle\xi'\rangle^d,\pi/2+\varepsilon'}$ *for* a^0*), cf.* (4.2.1).

The family of operators $W(t)$ *is strongly continuous in* $L^2(E)$ *for* $t \ge 0$. *For* $t > 0$, $W(t)$ *is a continuous family of singular Green operators of order* $-\infty$ *and class* 0; *and some (but generally not all) of the symbol seminorms are bounded for* $t \to 0$. *More precisely,* $W(t)$ *has a symbol* $w(x',t,\xi',\xi_n,\eta_n)$ *in local coordinates near* Γ, *satisfying*

$$\text{(4.2.77)} \qquad w(x',t,\xi',\xi_n,\eta_n) \sim \sum_{\ell\in\mathbb{N}} w_{-1-\ell}(x',t,\xi',\xi_n,\eta_n) ,$$

where the terms $w_{-1-\ell}$ *are defined by (cf.* (4.2.2)*)*

$$\text{(4.2.78)} \qquad w_{-1-\ell}(x',t,\xi,\eta_n) = \frac{i}{2\pi}\int_{C_{\delta\langle\xi'\rangle^d}} e^{-t\lambda} g_{-d-1-\ell}(x',\xi,\eta_n,\lambda)\,d\lambda .$$

They are quasi-homogeneous in (t,ξ,η_n) *of degree* $-1-\ell$

$$\text{(4.2.79)} \qquad w_{-1-\ell}(x',s^{-1/d}t,s\xi,s\eta_n) = s^{-1-\ell} w_{-1-\ell}(x',t,\xi,\eta_n) \quad \textit{for } |\xi'| \ge 1,\ s \ge 1 ,$$

and the corresponding symbol-kernels $\tilde w_{-1-\ell}(x',x_n,y_n,t,\xi') = F^{-1}_{\xi_n\to x_n} \bar F^{-1}_{\eta_n\to y_n} w_{-1-\ell}$ *satisfy estimates (cf. also* (4.2.90) *below)*

$$\| D^{\beta}_{x'} D^{\alpha}_{\xi'} x_n^k D_{x_n}^{k'} y_n^m D_{y_n}^{m'} \tilde{w}_{-1-\ell}(x',x_n,y_n,t,\xi')\|_{L^2_{x_n,y_n}}$$

(4.2.80)
$$\leq c(x')\langle\xi'\rangle^{-\ell-|\alpha|-k+k'-m+m'}\exp(-ct\langle\xi'\rangle^d)\sigma_a(t^{1/d}\langle\xi'\rangle) ,$$

for $a \leq (M'-M")/d$ *when* $M' \leq \nu$, $M' \neq M"$, *or* $M' < \nu$, $M' = M"$;

and for $a \leq (\nu-M")/d$ *when* $M' \geq \nu$, $M" \neq \nu$;

here M' *and* $M"$ *are defined (consistently with* (3.3.53)*) by*

$$(4.2.81)\qquad \begin{aligned} M' &= [k-k']_+ + [m-m']_+ + |\alpha| + \ell ,\\ M" &= [k-k']_- + [m-m']_- ; \quad \textit{so} \quad M'-M" = k-k' + m-m' + |\alpha| + \ell , \end{aligned}$$

and σ_a *is the function*

$$(4.2.82)\qquad \sigma_a(s) = \begin{cases} s^a(1+|\log s|) & \textit{for} \quad a \in \mathbb{N}_+ = \{1,2,\dots\} ,\\ s^a & \textit{for} \quad a \in \mathbb{R}\setminus\mathbb{N}_+ . \end{cases}$$

When $M' \geq \nu$, $M" = \nu$, *the left side of* (4.2.80) *is estimated by*

$$(4.2.83)\qquad \text{l.s.} \leq c(x')\langle\xi'\rangle^{-\ell-|\alpha|-k+k'-m+m'}\exp(-ct\langle\xi'\rangle^d)(1+|\log(\langle\xi'\rangle^d t)|) .$$

Note that $\nu \leq d$ *implies* $a \leq 1$ *in all cases. The asymptotic expansion* (4.2.77) *holds in the sense that* $\tilde{w} - \Sigma_{\ell<M}\tilde{w}_{-1-\ell}$ *satisfies estimates like those satisfied by* $\tilde{w}_{-1-M}$ *in* (4.2.80-82).

The kernels $K_{W_{-1-\ell}}(x,y,t)$ *of the corresponding operators (defined via local coordinates)*

$$(4.2.84)\qquad W_{-1-\ell}(t) = \mathrm{OPG}(w_{-1-\ell}(x',t,\xi,\eta_n))$$

satisfy the following estimates in a neighborhood of Γ :

For $\nu < d$,

$$(4.2.85)\qquad \sup_{x',y'}\|K_{W_{-1-\ell}}(x,y,t)\|_{L^2_{x_n,y_n}} \leq c'\exp(-ct)\begin{cases} t^{(\ell+1-n)/d} & \textit{if} \quad \ell < \nu+n-1\\ t^{\ell/d}(1+|\log t|) & \textit{if} \quad \ell = \nu+n-1\\ t^{\nu/d} & \textit{if} \quad \ell > \nu+n-1 . \end{cases}$$

For $\nu = d$,

$$(4.2.86)\qquad \sup_{x',y'}\|K_{W_{-1-\ell}}(x,y,t)\|_{L^2_{x_n,y_n}} \leq c'\exp(-ct)\begin{cases} t^{(\ell+1-n)/d} & \textit{if} \quad \ell < d+n-1\\ t(1+|\log t|) & \textit{if} \quad \ell \geq d+n-1 . \end{cases}$$

On compact sets in the interior of Ω , *the kernels and all their* (x,y)-*derivatives are* $O(t^{\nu/d})$ *if* $\nu < d$, *resp.* $O(t(1+|\log t|))$ *if* $\nu = d$. *The error terms stemming from various choices of cut-off functions in the definition of* (4.2.84) *satisfy all these estimates with large* ℓ $(\ell \geq n+d)$.

The remainder kernels $K_{W'_M}$, *namely the kernels of the operators*

$$(4.2.87) \qquad W'_M(t) = W(t) - \sum_{\ell<M} W_{-1-\ell}(t) ,$$

satisfy estimates like those satisfied by W_{-1-M} , *for* $M \in \mathbb{N}$.

Near the boundary, the kernels $K_{W_{-1-\ell}}$ *and* $K_{W'_M}$ *are continuous as functions of* $(x',y',t) \in \Gamma\times\Gamma\times\overline{\mathbb{R}}_+$ *with values in* $L^2_{x_n,y_n}$ *when* ℓ *and* $M \geq n-1$, *and they are zero at* $t = 0$ *when* ℓ *and* $M > n-1$.

Proof: $V(t)_\Omega$ is the restriction to $L^2(\Omega,E)$ of a strongly continuous semigroup in $L^2(\Sigma,\tilde{E})$, and $U(t)$ is a strongly continuous semigroup in $L^2(\Omega,E)$ (in fact they are holomorphic semigroups, cf. e.g. [Friedman 1]), so $W(t)$ is likewise strongly continuous in $L^2(\Omega,E)$. We use for $t>0$ that the Cauchy integral (4.2.70) converges in the norm of bounded operators in L^2, since Q_λ and R_λ are $O(\langle\lambda\rangle^{-1})$, cf. Corollary 3.3.2.

In coordinate patches near the boundary we define the symbols $w_{-1-\ell}$ by (4.2.78), and note that the associated symbol-kernels are defined by

$$(4.2.88) \qquad \tilde{w}_{-1-\ell}(x',x_n,y_n,t,\xi') = \frac{i}{2\pi}\int_{C_{\delta\langle\xi'\rangle^d}} e^{-t\lambda}\tilde{g}_{-d-1-\ell}(x',x_n,y_n,\xi',\lambda)d\lambda$$

(using the notation indicated after Theorem 3.3.10). The homogeneity property (4.2.79) follows easily from the homogeneity of $g_{-d-1-\ell}$, and since

$$(4.2.89) \qquad \tilde{w}_{-1-\ell}(x',x_n,y_n,t,\xi') = \overline{F}^{-1}_{\xi_n\to x_n}\overline{F}^{-1}_{\eta_n\to y_n} w_{-1-\ell}(x',t,\xi,\eta_n) ,$$

$\tilde{w}_{-1-\ell}$ has the quasi-homogeneity property

$$(4.2.90) \qquad \tilde{w}_{-1-\ell}(x',s^{-1}x_n,s^{-1}y_n,s^{-1/d}t,s\xi') = s^{1-\ell}\tilde{w}_{-1-\ell}(x',x_n,y_n,t,\xi') \quad \text{for } |\xi'| \geq 1 ,\ s \geq 1 .$$

To show the estimates (4.2.80-83), we apply Lemma 4.2.8 using (3.3.52-53). Note that (4.2.73) contains a logarithmic term when $m/d = 0$, whereas (4.2.82) only contains logarithms for a integer > 0 . At any rate, we have the following straightforward applications where $a \neq 0$ (it suffices to show the estimates (4.2.80) with the largest indicated value of a):

$$\| D_{x'}^{\beta} D_{\xi'}^{\alpha} x_n^k D_{x_n}^{k'} y_n^m D_{y_n}^{m'} \tilde{w}_{-1-\ell} \| \le c'(x') \langle\xi'\rangle^{-M'+M''} \exp(-ct\langle\xi'\rangle^d) \sigma_a(t\langle\xi'\rangle^d)$$

(4.2.91)

with $a = M'-M''$ if $M' \le \nu$ and $M'-M'' \neq 0$,

and $a = \nu - M''$ if $M' \ge \nu$ and $\nu - M'' \neq 0$.

In the remaining cases, the lemma gives us this estimate with $\sigma_a(t\langle\xi'\rangle^d)$ replaced by $1+|\log(t\langle\xi'\rangle^d)|$. This accounts for (4.2.83), where the estimated symbol is of regularity ≤ 0 ; but when there is positive regularity, the logarithm can in fact be removed. The most important case where this is done is the case of the undifferentiated principal part $\tilde{w}_{-1}$, so let us consider that in detail.

Recall that we have the estimates of the associated strictly homogeneous symbol:

$$\| \tilde{g}^h_{-d-1}(x',x_n,y_n,\xi',\lambda) \| \le c(x') |\xi',\mu|^{-d} \quad \text{for } (\xi',\mu) \in \overline{\mathbb{R}}^n_+ \setminus 0 ,$$

(4.2.92)

$$\| \partial_{\xi_j} \tilde{g}^h_{-d-1} \| \le c(x') |\xi'|^{-\frac12} |\xi',\mu|^{-d-\frac12} \quad \text{for } (\xi',\mu) \in \overline{\mathbb{R}}^n_+ \setminus 0 , \; j=1,\dots,n-1 ,$$

since g_{-d-1} is of regularity $\ge \frac12$ and order $-d$ (degree $-d-1$), cf. Proposition 2.3.17 and (2.1.21). Here $\tilde{g}^h_{-d-1}$ is defined and analytic in λ for $|\lambda| \ge \delta' |\xi'|^d$, $\arg\lambda \in [\pi/2 - \varepsilon', 3\pi/2 + \varepsilon']$, for some $\delta' > 0$, $\varepsilon' > 0$. Lemma 4.2.8 adapts easily to the homogeneous case, and gives

(4.2.93) $$\| \tilde{w}^h_{-1}(x',x_n,y_n,t,\xi') \| \le c'(x') \exp(-ct|\xi'|^d)(1+|\log(t|\xi'|^d)|) ,$$

(4.2.94) $$\| \partial_{\xi_j} \tilde{w}^h_{-1} \| \le c'(x') |\xi'|^{-1} \exp(-ct|\xi'|^d)(t|\xi'|^d)^{1/2d} = c'(x') |\xi'|^{-\frac12} t^{1/2d} \exp(-ct|\xi'|^d) .$$

(4.2.93) shows that $\tilde{w}^h_{-1}$ is uniformly bounded for $t|\xi'|^d \ge 1$, x' in a compact set. (4.2.94) shows that for each <u>fixed</u> $t_0 > 0$ and fixed x'_0 , $\tilde{w}^h_{-1}$ satisfies the hypotheses of Lemma 2.1.10, as a function of ξ' valued in the Banach space $L^2_{x_n,y_n}$ which implies that $\tilde{w}^h_{-1}$ is continuous in ξ' at $\xi' = 0$ and satisfies an estimate

(4.2.95) $$\| \tilde{w}^h_{-1}(x,y_n,t,\xi') - \tilde{w}^h_{-1}(x,y_n,t,0) \| \le c''(x') |\xi'|^{\frac12} \exp(-ct|\xi'|^d) t^{1/2d} \text{ for } |\xi'| \le$$

(We can also apply the argument to $\tilde{w}^h_{-1}$, considered as a function of ξ' valued in the Banach space $C^0(\overline{B}(x'_0,t_0), L^2_{x_n,y_n})$ where $\overline{B}(x'_0,t_0)$ is a small ball with center (x_0',t_0) ; this shows the continuous dependence on (x',t) .). Here the quasi-homogeneity (4.2.90) of $\tilde{w}^h_{-1}$ implies that $\|\tilde{w}^h_{-1}(x,y_n,t,0)\|$ is homogeneous of degree zero in t , i.e. is constant in t ,

$$(4.2.96)\qquad \|\tilde{w}^h_{-1}(x,y_n,t,0)\|_{L^2_{x_n,y_n}} = a(x') \quad \text{for all } t > 0 .$$

We shall also use an adaption of the proof of Lemma 2.1.9 2^0 to the present case, showing that $\tilde{w}' \equiv \tilde{w}_{-1} - \tilde{w}^h_{-1}$ satisfies, on the set $|\xi'| \leq 1$ where it does not vanish, the estimate for $\xi_1 > 0$ (with similar estimates for other halfspaces):

$$(4.2.97)\quad \|\tilde{w}'(x,y_n,t,\xi')\| = \|-\int_{\xi_1}^{1} \partial_{\xi_1}\tilde{w}'(x,y_n,t,s,\xi_2,\ldots,\xi_{n-1})ds\|$$
$$\leq c(x')t^{1/2d}\int_{\xi_1}^{1} (\langle(s,\xi_2,\ldots,\xi_{n-1})\rangle^{-\frac12} + |(s,\xi_2,\ldots,\xi_{n-1})|^{-\frac12})ds \leq c'(x')t^{1/2d},$$

in view of the estimates of $\partial_{\xi_1}\tilde{w}_{-1}$ and $\partial_{\xi_1}\tilde{w}^h_{-1}$, cf. (4.2.94). Then we find altogether:

$$\|\tilde{w}_{-1}(x,y_n,t,\xi')\| \leq \|\tilde{w}'(x,y_n,t,\xi')\| + \|\tilde{w}^h_{-1}(x,y_n,t,\xi')-\tilde{w}^h_{-1}(x,y_n,t,0)\| + \|\tilde{w}^h_{-1}(x,y_n,t,0)\|$$
$$\leq c_1(x')(t^{1/2d}+|\xi'|^{\frac12}t^{1/2d}+1) \leq c_3(x'), \quad \text{for } \langle\xi'\rangle^d t \leq 1 .$$

It follows that (4.2.80) is valid for $\tilde{w}_{-1}$ itself.

Other expressions $D\tilde{w}_{-1-\ell}$, where the estimate of the corresponding expression $D\tilde{g}_{-d-1-\ell}$ is $O(c(x',\xi')\langle\xi',\mu\rangle^{-d})$ and the next derivatives $\partial_{\xi_j}D\tilde{g}_{-d-1-\ell}$ are $O(c(x',\xi')\langle\xi',\mu\rangle^{-d-\sigma})$ for some $\sigma > 0$, are treated in the same way. This gives the remaining estimates in (4.2.80).

Clearly, we have s.g.o.s of order $-\infty$ for each fixed $t > 0$.

The full symbol w can be chosen so that the asymptotic estimates are satisfied; here we can in fact take

$$(4.2.98)\qquad w = \frac{i}{2\pi}\int_{C_{\delta\langle\xi'\rangle^d}} e^{-\lambda t} g \, d\lambda$$

where g is a full symbol of G_λ .

From the symbols $w_{-1-\ell}$ in local coordinates one defines operators $W_{-1-\ell}$ on the manifold by use of coordinate transfomrations and partitions of unity, cf. Section 2.4 (on coordinate patches in the interior of Ω , zero symbols can be used). This is briefly expressed in formula (4.2.84). The inaccuracies contained in this formulation are of the type

$$W'(t) = \frac{i}{2\pi}\int_{C_\delta} e^{-\lambda t} G'_\lambda \, d\lambda$$

where G_λ' is negligible in the s.g.o. class and analytic in λ on the usual set. Since the kernel of G_λ' is $O(\langle\lambda\rangle^{-1-\nu/d})$ in $C^0(\Gamma\times\Gamma, L^2_{x_n,y_n})$-norm near the boundary and in C^0 norm on interior compact sets (cf. Lemmas 2.4.3 and 2.4.8 ff.), the kernel of W' is estimated, by the help of Lemma 4.2.8, by

$$(4.2.99)\qquad \sup_{x',y'\in\Gamma} \|K_{W'}(x,y,t)\|_{L^2_{x_n,y_n}} \leq c'\exp(-ct)\sigma_a(t)\ , \quad \text{for } a\leq \nu/d\ ,$$

near the boundary, and

$$(4.2.100)\qquad \sup_{x,y\in\Xi} |K_{W'}(x,y,t)| \leq c'\exp(-ct)\sigma_a(t)\ , \quad \text{for } a\leq \nu/d\ ,$$

on compact subsets Ξ of Ω.

Consider now the kernels $K_{W_{-1-\ell}}$ defined from the symbol-kernels $\tilde{w}_{-1-\ell}$ in a coordinate patch by

$$(4.2.101)\qquad K_{W_{-1-\ell}}(x,y,t) = F^{-1}_{\xi'\to x'-y'}\,\tilde{w}_{-1-\ell}(x',x_n,y_n,t,\xi')\ .$$

For $t\geq 1$, the estimates (4.2.80) imply easily for the norms in $L^2_{x_n,y_n}(\mathbb{R}^2_{++})$:

$$\begin{aligned}\|K_{W_{-1-\ell}}(x,y,t)\| &\leq c_1(x')\int_{\mathbb{R}^{n-1}} \langle\xi'\rangle^{-\ell}\exp(-ct\langle\xi'\rangle^d)\sigma_a(t\langle\xi'\rangle^d)d\xi' \\ &\leq c_2(x')\exp(-\tfrac12 ct)\int_{\mathbb{R}^{n-1}} \langle\xi'\rangle^{-\ell'}\exp(-\tfrac12 c\langle\xi'\rangle^d)d\xi' \\ &\leq c_3(x')\exp(-\tfrac12 ct)\ .\end{aligned}$$

In the region $t\leq 1$, there is a simple estimate when $\ell > \nu+n-1$. Here one can take $a=\nu/d$ in (4.2.80), so if $\nu < d$,

$$\|K_{W_{-1-\ell}}(x,y,t)\| \leq c_1(x')\int_{\mathbb{R}^{n-1}} \langle\xi'\rangle^{-\ell}\langle\xi'\rangle^{\nu}t^{\nu/d}d\xi \leq c_2(x')t^{\nu/d}\ ,$$

and if $\nu = d$,

$$\begin{aligned}\|K_{W_{-1-\ell}}(x,y,t)\| &\leq c_1(x')\int_{\mathbb{R}^{n-1}} \langle\xi'\rangle^{-\ell}\langle\xi'\rangle^{d}t(1+\log\langle\xi'\rangle+|\log t|)d\xi' \\ &\leq c_2(x')t(1+|\log t|)\ ,\end{aligned}$$

which shows (4.2.85-86) in case $\ell > \nu+n-1$.

When $\ell < \nu+n-1$, we use the estimates (4.2.80) as follows. If $\nu < d$, we have with $a=\nu/d$ if $\ell\geq\nu$, $a=\ell/d$ if $\ell<\nu$,

$$(4.2.102)\quad \|K_{W_{-1-\ell}}(x,y,t)\| \le c_1(x')\int_{\mathbb{R}^{n-1}} \langle\xi'\rangle^{-\ell}\exp(-ct\langle\xi'\rangle^d)t^a\langle\xi'\rangle^{ad}d\xi$$

$$\le c_2(x')t^a\int_{\mathbb{R}^{n-1}}(1+|\xi'|)^{ad-\ell}\exp(-ct|\xi'|^d)d\xi'$$

$$= c_2(x')t^{a-a+\ell/d-(n-1)/d}\int_{\mathbb{R}^{n-1}}(t^{-1/d}+|\eta|)^{ad-\ell}\exp(-c|\eta|)d\eta \quad (\text{for } \eta = \xi' t^{1/d})$$

$$\le c_3(x')t^{(\ell-n+1)/d}, \quad \text{since } ad-\ell > -n+1 .$$

If $\nu = d$, we can use the preceding calculation when $\ell < d$, and we have when $d \le \ell < d+n-1$,

$$(4.2.103)\quad \|K_{W_{-1-\ell}}(x,y,t)\| \le c_1(x')t\int_{\mathbb{R}^{n-1}} \langle\xi'\rangle^{-\ell+d}\exp(-ct\langle\xi'\rangle^d(1+|\log t\langle\xi'\rangle^d|)d\xi'$$

$$\le c_2(x')\Big(t(1+|\log t|) + t\int_{|\xi'|\ge 1}|\xi'|^{-\ell+d}\exp(-ct|\xi'|^d)(1+|\log t|\xi'|^d|)d\xi'\Big)$$

$$= c_2(x')\Big(t(1+|\log t|) + t^{(\ell-n+1)/d}\int_{|\eta|\ge t^{1/d}}|\eta|^{-\ell+d}\exp(-c|\eta|)(1+|\log|\eta||)d\eta\Big)$$

$$\le c_3(x')\Big(t(1+|\log t|) + t^{(\ell-n+1)/d}\Big)$$

$$\le c_4(x')t^{(\ell-n+1)/d}, \quad \text{since } \ell < d+n-1 .$$

This completes the proof of (4.2.85-86) when $\ell < \nu+n-1$.

It remains to consider $\ell = \nu+n-1$. Here the final integrals in (4.2.102) and (4.2.103) are $O(|\log t|)$ for $t \to 0$, so the considerations there lead to estimates with an extra factor $1+|\log t|$. This gives (4.2.85-86) in the remaining cases.

On interior compact sets, we use that the symbols $g_{-d-1-\ell}$ define ps.d.o.s that are here uniformly negligible of regularity $\nu+d$ (cf. Lemma 2.4.8 5°), so that the contributions to the $W_{-1-\ell}$ are as described in (4.2.100).

Consider finally the remainders (4.2.87). Here it suffices to consider the case $M > d+n-1$, where we get estimates like (4.2.99-100) by use of the fact that the corresponding remainder kernels for G_λ have norms that are $O(\langle\lambda\rangle^{-1-\nu/d})$, cf. Remark 3.3.12. For lower values of M, we get the remainders by adding the appropriate terms $W_{-1-\ell}$, using that

$$(4.2.104)\qquad W'_M = \sum_{M\le\ell<N} W_{-1-\ell} + W'_N .$$

This gives estimates as in (4.2.85-86). The continuity follows from the continuity properties of the G_λ remainder kernels when there are uniform estimates. ▯

The techniques in the above proof can of course also be used to show estimates of the derivatives of the above kernels. For instance, one finds

4.2.10 Theorem. *Hypotheses as in Theorem 4.2.9. Let* α, β *and* $\gamma \in \mathbb{N}^{n-1}$, *and let* k, k', m, m' *and* $\ell \in \mathbb{N}$. *Then* $(x'-y')^\alpha D^\beta_{x'} D^\gamma_{y'} x_n^k D_{x_n}^{k'} y_n^m D_{y_n}^{m'} K_{W_{-1-\ell}}(x,y,t)$ *satisfies the estimates in* (4.2.84-85) *with* ℓ *and* ν *replaced by* ℓ_1 *resp.* ν_1 , *where*

$$(4.2.105)\qquad \begin{aligned} \ell_1 &= \ell + k-k' + m-m' + |\alpha| - |\gamma| , \\ \nu_1 &= \nu - [k-k']_+ - [m-m']_+ - |\alpha| + |\gamma| . \end{aligned}$$

Similar results hold for the remainder kernels $K_{W'_\ell}(x,y,t)$.

Proof: In the formula (4.2.101), $(x'-y')^\alpha$ gives rise to a derivation $\overline{D}^\alpha_{\xi'}$ on $\tilde{w}_{-1-\ell}$, $D^\beta_{x'}$ gives a sum of x'-derivatives multiplied by powers of ξ' , and $D^\gamma_{y'}$ gives rise to a multiplication by $(\xi')^\gamma$. Altogether, the derived symbol-kernel is of order and regularity corresponding to the values ℓ_1 and ν_1 , and we get the estimate by use of the techniques in Theorem 4.2.9. ▯

Observe in particular, that applications of $(x_n D_{x_n})^k$ or $(y_n D_{y_n})^m$ for any k and m , do not change the order of magnitude in the estimates. It can also be remarked that the estimates imply estimates in sup-norm in (x,y) , for large ℓ_1 .

Finally, we consider the trace estimates. Here one observes, as pointed out at the end of Section 3.3, that recomposition with high powers of B cannot be expected to improve the regularity, so the tricks involved in Theorem 4.2.5 do not carry over to the case with boundary, and our asymptotic estimate of the trace contains only a finite number of terms. We obtain it from the symbol trace estimates in Theorem 3.3.10.

We note here that since $W(t)$ is an integral operator with C^∞ kernel for each $t > 0$ (as is $V(t)_\Omega$) , it is a compact operator with rapidly decreasing characteristic values (it lies in $\bigcap_{p>0} \mathfrak{S}_{(p)}$, cf. (A.78) ff.), so the trace is well-defined.

4.2.11 Theorem. *Hypotheses as in Theorem 4.2.9. Let* ν' *denote the largest integer* $<\nu$. *The diagonal value of the* x_n*-integral of the kernel of* $W(t)$

(4.2.106) $$k_W(x',x',t) = \int_0^\infty K_W(x',x_n,x',x_n,t)dx_n$$

has an asymptotic expansion for $t \to 0+$ *(uniformly in* x'*)* ,

(4.2.107) $$k_W(x',x',t) = c_{1-n}(x',W)t^{(1-n)/d} + \ldots + c_{\nu'}(x',W)t^{\nu'/d} + O(t^{(\nu-1/4)/d}) ,$$

and the trace of $W(t)$ *has the asymptotic expansion for* $t \to 0+$

(4.2.108) $$\operatorname{tr} W(t) = \underline{c}_{1-n}(W)t^{(1-n)/d} + \ldots + \underline{c}_{\nu'}(W)t^{\nu'/d} + O(t^{(\nu-1/4)/d})$$

where the coefficients satisfy

(4.2.109) $$\underline{c}_k(W) = \int_\Gamma c_k(x',W)dx' \quad ;$$

here the coefficient $\underline{c}_{j-n}(x',W)$ *is determined from the first* j *symbols of* $\{P_\Omega+G,T\}$: $p_d,\ldots,p_{d-j+1}$; $g_{d-1},\ldots,g_{d-j}$; $t_k,\ldots,t_{k-j+1}$ $(k = 0,\ldots,d-1)$.

Proof: In view of the definition of $\tilde{\tilde{g}}$ in Theorem 3.3.10,

$$\tilde{\tilde{g}}(x',\xi',\lambda) = \int_0^\infty \tilde{g}(x',x_n,x_n,\xi',\lambda)dx_n$$

we have that $k_W(x',y',t)$ is the kernel of the Cauchy integral (4.2.70) of the associated ps.d.o. $\tilde{\tilde{G}}_\lambda = OP'(\tilde{\tilde{g}})$ on Γ (modulo the usual techniques to express the operators in local coordinates). The fact that $\tilde{\tilde{g}}$ is of order $-d$ and regularity $\nu-1/4$ (which is noninteger, since ν is either integer or half-integer), implies the estimates (3.3.70-72) for the expansion (3.3.69). By insertion in the Cauchy integral formula (4.2.70) and integration in ξ' , we find that the homogeneous terms in $\tilde{\tilde{g}}$ for $\ell \leq \nu'+n-1$ give the homogeneous terms in (4.2.107). For the remainder $f''+f'''$ which has an L^1-norm in ξ' that is $O(\langle\lambda\rangle^{-1-(\nu-1/4)/d})$ with $\nu-1/4 > 0$, the resulting kernel is estimated as in the proof of Theorem 4.2.9 (with use of Lemma 4.2.8), which shows that it gives a contribution that is $O(t^{(\nu-1/4)/d})$. In this way, we obtain (4.2.107), and (4.2.108) follows by integration in x' . ▯

In the differential operator case there is an asymptotic expansion of $\operatorname{tr} W(t)$ in terms of all the powers $t^{(j-n)/d}$, $j \geq 1$, cf. [Greiner 1].

Corollary 4.2.7 and Theorem 4.2.11 together imply (cf. (4.2.68)):

4.2.12 Corollary. *Assume that the hypotheses of Theorem 4.2.9 are satisfied, and let* ν' *be the largest integer less than* ν . *The trace of* $U(t) = \exp(-tB)$ *has an asymptotic expansion*

$$\text{tr } U(t) = c_{-n}(B)t^{-n/d} + c_{1-n}(B)t^{(1-n)/d} + \dots + c_0(B) + \dots + c_{\nu'}(B)t^{\nu'/d} + O(t^{(\nu-1/4)/d}) \quad \text{for} \quad t \to 0+ , \tag{4.2.110}$$

where c_{-n} *is a constant determined from* p_d ,

$$c_{-n} = (2\pi)^{-n} \int_{T^*(\Omega)} \exp(-p_d^h(x,\xi))d\xi dx , \tag{4.2.111}$$

and the c_{j-n} *are constants determined from the first* $j+1$ *symbols in* P *and the first* j *symbols in* G *and* T .

4.2.13 Remark. Formulas for the coefficients c_{j-n} , $j > 0$, can be derived from the above analysis. It is of interest to note here that the derivatives of symbols involved, are those up to order $j-k$ resp. $j-k-1$ for the k'th term in P resp. G and T. We also observe that in the pseudo-differential case one need not have the symmetries (e.g. $p_{d-\ell}(x,-\xi) = (-1)^{d-\ell} p_{d-\ell}(x,\xi)$) that lead to cancellations of certain coefficients in the differential operator case.

4.3 An index formula.

The trace formula for the heat operator developed in the preceding section can be used to give an explicit formula for the index of normal elliptic boundary problems, as explained in [Atiyah-Bott-Patodi 1] in connection with differential operators. We recall the elementary background (cf. also (A.93-94)):

Let B be a closed, densely defined operator in a Hilbert space H with a compact parametrix C, such that $BC-I$ and $CB-I$ have finite rank. Then B and B^* are Fredholm operators. Moreover, B^*B is a selfadjoint non-negative operator with the compact parametrix CC^*, so $(B^*B+1)^{-1}$ is compact, and the spectrum $sp(B^*B)$ consists of a sequence of eigenvalues $\in \overline{\mathbb{R}}_+$ going to $+\infty$, with finite multiplicities. Similar observations hold for BB^*. Now one has the identities for the nullspaces

$$(4.3.1) \qquad Z(B) = Z(B^*B)\ , \quad \text{and} \quad Z(B^*) = Z(BB^*)\ ;$$

here the inclusion $Z(B) \subset Z(B^*B)$ is obvious, and on the other hand, $B^*B\varphi = 0$ implies $\|B\varphi\|^2 = (B^*B\varphi,\varphi) = 0$. Moreover, for $\lambda > 0$,

$$(4.3.2) \qquad \begin{aligned} &B : Z(B^*B-\lambda I) \to Z(BB^*-\lambda I) \quad \text{and} \\ &B^*: Z(BB^*-\lambda I) \to Z(B^*B-\lambda I) \quad \text{are bijections,} \end{aligned}$$

for they are easily seen to be injective mappings, and the spaces are finite dimensional. In particular, B^*B and BB^* have the same eigenvalue sequence with the same multiplicities;

$$(4.3.3) \qquad \dim Z(B^*B-\lambda) = \dim Z(BB^*-\lambda) \quad \text{for} \quad \lambda > 0\ .$$

The index of B can then be expressed by several formulas

$$(4.3.4) \qquad \begin{aligned} \text{index } B &= \dim Z(B) - \dim R(B)^{\perp} \\ &= \dim Z(B) - \dim Z(B^*) \\ &= \dim Z(B^*B) - \dim Z(BB^*) \\ &= \sum_{\lambda\in sp(B^*B)} [\dim Z(B^*B-\lambda) - \dim Z(BB^*-\lambda)] \\ &= \sum_{\lambda\in sp(B^*B)} \varphi(\lambda)[\dim Z(B^*B-\lambda) - \dim Z(BB^*-\lambda)] \end{aligned}$$

for any function φ on $\overline{R}_+$ with $\varphi(0) = 1$. Taking in particular $\varphi(\lambda) = e^{-\lambda t}$, we have

$$\text{(4.3.5)}\qquad \operatorname{index} B = \sum_{sp(B^*B)} e^{-\lambda t}\dim Z(B^*B-\lambda) - \sum_{sp(BB^*)} e^{-\lambda t}\dim Z(BB^*-\lambda)$$

$$= \operatorname{tr}\exp(-tB^*B) - \operatorname{tr}\exp(-tBB^*) ,$$

provided that the traces exist.

We shall apply this to the realization $B = (P+G)_T$ defined from a normal elliptic system A associated with a ps.d.o. P of order $d > 0$,

$$\text{(4.3.6)}\qquad A = \begin{pmatrix} P_\Omega+G \\ \\ T \end{pmatrix} : \; H^d(E) \to \begin{matrix} L^2(E) \\ \times \\ H_F^d \end{matrix} ,$$

as defined in Section 1.4 (we use the notation (1.4.5)). Recall that B is an unbounded, closed, densely defined operator in $L^2(E)$, acting like $P_\Omega+G$ and with

$$\text{(4.3.7)}\qquad D(B) = \{u \in H^d(E) \mid Tu = 0\} .$$

The Fredholm properties of A and B were explained in Section 1.4, see in particular Proposition 1.4.2. Since T is surjective, they have the same index:

<u>4.3.1 Lemma.</u> *For normal elliptic boundary problems* (4.3.6), *one has that*

$$\text{(4.3.8)}\qquad \operatorname{index} A = \operatorname{index} B .$$

In fact, the nullspaces are the same,

$$\text{(4.3.9)}\qquad Z(A) = Z(B) \;;$$

and when $K : H_F^d \to H^d(E)$ *is a right inverse of* T *(which exists by Proposition 1.6.5), then the row matrix*

$$\text{(4.3.10)}\qquad \Phi = (I \quad -(P_\Omega+G)K)$$

defines a bijection of any complement of $R(A)$ *in* $L^2(E) \times H_F^d$ *onto a complement of* $R(B)$ *in* $L^2(E)$.

<u>Proof:</u> The formula (4.3.9) is obvious, since $Z(A) = Z(P_\Omega+G) \cap Z(T)$. As for the ranges, observe on one hand that if the vectorspace V is linearly independent from $R(B)$ in $L^2(E)$, then $V \times \{0\}$ is linearly independent from $R(A)$ in $L^2(E) \times H_F^d$, so

$$\operatorname{codim} R(A) \geq \operatorname{codim} R(B) .$$

On the other hand, let K be a right inverse of T. If $\{f,\varphi\} \notin R(A)$ then $f - (P_\Omega+G)K\varphi \notin R(B)$, since otherwise a solution u of

$$\begin{aligned}(P_\Omega+G)u &= f - (P_\Omega+G)K\varphi \\ Tu &= 0\end{aligned}$$

would furnish the solution $v = u+K\varphi$ of

$$\begin{aligned}(P_\Omega+G)v &= f \\ Tv &= \varphi .\end{aligned}$$

In particular, $\{f,\varphi\}$ and $f - (P_\Omega+G)K\varphi$ are both nonzero. Thus if W is a vectorspace in $L^2(E) \times H_F^d$ that is linearly independent from $R(A)$, then Φ maps W onto a space V in $L^2(E)$ such that nonzero elements go into nonzero elements (the mapping is injective), and the nonzero elements in V do not lie in $R(B)$ (V is linearly independent from $R(B)$). Hence

$$\operatorname{codim} R(B) \geq \operatorname{codim} R(A) .$$

Altogether, the two dimensions are the same, and Φ defines a bijection as stated in the lemma. □

According to Definition 1.5.14 and Proposition 2.3.14ff., $P_\Omega+G$ and B are of regularity $\nu \geq \frac{1}{2}$. By Theorem 1.6.9, the adjoint B^* in $L^2(E)$ is the realization of a normal elliptic boundary problem associated with P^*, whose regularity ν_1 can be found on the basis of the formulas (1.6.25-26); again $\nu_1 \geq \frac{1}{2}$. By Theorem 1.4.6, the compositions B^*B and BB^* are realizations of normal elliptic boundary problems associated with P^*P resp. PP^* of order $2d$; the regularities ν_2 and ν_3 of those problems can be determined from the formulas for the s.g.o. terms and trace operator, and they are at least $\frac{1}{2}$. In view of the ellipticity of B and B^*, B^*B and BB^* are d-coercive (and selfadjoint ≥ 0), so in particular they satisfy the parabolicity condition in Definition 1.5.5.

The results on the heat operator then apply to B^*B and BB^*, giving trace formulas

$$\begin{aligned}&\operatorname{tr} \exp(-tB^*B) = c_{-n}(B^*B)t^{-n/2d} + \ldots + c_0(B^*B) + O(t^{1/8d}) , \\ (4.3.11)\quad &\\ &\operatorname{tr} \exp(-tBB^*) = c_{-n}(BB^*)t^{-n/2d} + \ldots + c_0(BB^*) + O(t^{1/8d}) ,\end{aligned}$$

cf. Corollary 4.2.12. (Actually, the formulas are found on the form of traces of $e^{ct}\exp(-t(B^*B+c))$ etc., for a suitable choice of $c \geq 0$.) In view of (4.3.5) we then have

$$\text{index } B = f(t) + g(t) + c_0(B^*B) - c_0(BB^*) ,$$

where $f(t)$ is a polynomial in $t^{-1/2d}$ without constant term, and $g(t) \to 0$ for $t \to 0$. Since the formula is valid for all $t > 0$ with the left side independent of t , $f(t)$ must converge to a constant for $t \to 0$; and that can only hold if $f \equiv 0$. Thus altogether,

$$\text{index } B = c_0(B^*B) - c_0(BB^*) . \tag{4.3.12}$$

To sum up, we have obtained

4.3.2 Theorem. *Let* $d > 0$. *Let* $A = \{P_\Omega+G,T\}$ *be a normal elliptic system, as defined in Section 1.4, and let* $B = (P+G)_T$ *be the associated realization in* $L^2(E)$. *Then* A *and* B *have the same index, and it is determined by the formula*

$$\text{index } A = \text{index } B = c_0(B^*B) - c_0(BB^*) , \tag{4.3.13}$$

where $c_0(B^*B)$ *and* $c_0(BB^*)$ *are the coefficients of* t^0 *in the expansions in* t *of the traces of the heat operators* $\exp(-tB^*B)$ *resp.* $\exp(-tBB^*)$. *Here* $c_0(B^*B)$ *and* $c_0(BB^*)$ *are determined from the* k'th *homogeneous terms in* P *and their derivatives up to order* $n-k$, *for* $0 \leq k \leq n$, *together with the* k'th *homogeneous terms in* G *and* T *and their derivatives up to order* $n-1-k$, *for* $0 \leq k \leq n-1$.

Index formulas for pseudo-differential elliptic boundary problems have been established earlier in [Rempel 1] and [Rempel-Schulze 1], for general elliptic systems A as in (1.2.14). In [Rempel 1] there is proved a formula involving the k'th homogeneous terms and their derivatives up to order $n-k$, for $0 \leq k \leq n$, from each entry in A . In comparison with this, (4.3.13) is interesting because it uses one less term and one less derivative from G and T .

Improvements of the index formula are given in [Rempel-Schulze 1, Section 4.1]. For one thing, the index depends only on the principal symbols. This follows readily from the fact that the addition of a lower order operator to A represents a compact perturbation in the Sobolev space setting, and the index of a bounded Fredholm operator from one Hilbert space to another is unchanged by a compact perturbation. Moreover, the index is invariant under homotopies. Rempel and Schulze now show, by a generalization of the method of [Fedosov 1] (see also [Hörmander 7]) for the boundaryless case, how their index formula can be reduced to an explicit differential-geometric expression, under certain further hypotheses. It would be interesting to see which expressions can be obtained on the basis of (4.3.13). However, we have not at present made any efforts in this direction.

4.4 Complex powers

By the standard composition rules and invertibility theory, it is easy to define integral powers of our operators in the following cases:

- the case of a quadratic elliptic system (cf. Section 3.4)

$$(4.4.1) \qquad A_1 = \begin{pmatrix} P_\Omega + G & K \\ T & S \end{pmatrix} : \begin{matrix} H^d(E) \\ \times \\ H^{d-\frac{1}{2}}(F) \end{matrix} \to \begin{matrix} H^0(E) \\ \times \\ H^{-\frac{1}{2}}(F) \end{matrix} ;$$

- the case of a realization $B = (P+G)_T$ defined from an elliptic "column" system (cf. Section 3.3)

$$(4.4.2) \qquad A = \begin{pmatrix} P_\Omega + G \\ T \end{pmatrix} : H^d(E) \to \begin{matrix} H^0(E) \\ \times \\ H_F^d \end{matrix} ,$$

of the kind we have treated with particular interest in the present book.

The resolvent calculus will permit us to treat complex powers (and especially fractional powers) A_1^z resp. B^z for certain values $z \in \mathbb{C}$, as we shall now show. Here as usual, the main efforts will be devoted to the realizations B , since the study of these is closest to the differential operator problems. In either case, the analysis is based on a Cauchy integral formula, as in Seeley's treatments of the differential operator case [Seeley 4] and the boundaryless ps.d.o. case [Seeley 1].

Consider a normal polyhomogeneous elliptic system (4.4.2) and the corresponding realization $B = (P+G)_T$. The powers B^z will be defined by a Cauchy integral, when possible:

$$(4.4.3) \qquad \begin{aligned} B^z &= \frac{i}{2\pi} \int_C \lambda^z R_\lambda \, d\lambda = \frac{i}{2\pi} \int_C \lambda^z (Q_{\lambda,\Omega} + G_\lambda) d\lambda \\ &= \left[\frac{i}{2\pi} \int_C \lambda^z Q_\lambda \, d\lambda \right]_\Omega + \frac{i}{2\pi} \int_C \lambda^z G_\lambda \, d\lambda \\ &= (P^z)_\Omega + G^{(z)} . \end{aligned}$$

We here assume that $\{P_\Omega + G - \lambda I , T\}$ is parameter-elliptic on the ray $\overline{\mathbb{R}}_-$, and defines an invertible realization B . The analytic function λ^z is determined to be positive for $\lambda \in \mathbb{R}_+$ and z real, extended to be analytic in $\lambda \in \mathbb{C} \setminus \overline{\mathbb{R}}_-$; and for C we usually take the "Laurent loop"

(4.4.4) $\{\lambda = re^{+i\pi} \mid \infty > r \geq \delta\} \cup \{\lambda = \delta e^{i\theta} \mid \pi > \theta > -\pi\} \cup \{\lambda = re^{-i\pi} \mid \delta \leq r < \infty\}$,

bounding the set

$$W'_\delta = \{\lambda \in \mathbb{C} \mid |\lambda| > \delta \quad \text{and} \quad |\arg\lambda| < \pi\} .$$

If B happens to have some eigenvalues on $\mathbb{R}_-$ (since $B-\lambda I$ is invertible for large $|\lambda|$ there are at most finitely many), we can use that the system will in fact be parameter-elliptic on rays close to $\mathbb{R}_-$, so λ^z can be cut along such a ray avoiding the eigenvalues, and C be moved to this. C can also be taken as the boundary of a keyhole region around $\mathbb{R}_-$ (or nearby)

$$V_{\delta,\pi_1,\pi_2} = \{\lambda \in \mathbb{C} \mid |\lambda| \leq \delta \quad \text{or} \quad \pi_1 \leq \arg\lambda \leq \pi_2\}$$

with π_1 and π_2 close to π . - One can also define B^z modulo a nullspace of B , as in [Seeley 4]. Similar precautions and conventions hold for the powers of P .

The convergence of the integrals in (4.4.3) is assured by Corollary 3.3.2, for $\operatorname{Re} z < 0$.

Seeley showed (cf. [Seeley 1]) how the operator

(4.4.6) $$P^z = \frac{i}{2\pi} \int_C \lambda^z Q_\lambda \, d\lambda$$

for $\operatorname{Re} z < 0$ is a polyhomogeneous elliptic ps.d.o. of order $d \operatorname{Re} z$, with symbol defined from the symbol of P by the same Cauchy integral formula applied to $q(x,\xi,\lambda)$; and he showed how the definition extends to all $z \in \mathbb{C}$ by recursion formulas. In the present context we note that P^z rarely has the transmission property, when P has it, for it means (cf. (2.1.12)) that when $\xi' = 0$ and $\xi_n = \pm 1$ are inserted in $q(x,\xi,\lambda) \sim (p(x,\xi)-\lambda)^{\circ(-1)}$, then

(4.4.7) $$D_x^\beta D_\xi^\alpha q_{-d-\ell}(x,0,1,\lambda) = (-1)^{-d-\ell-|\alpha|} D_x^\beta D_\xi^\alpha q_{-d-\ell}(x,0,-1,\lambda) ,$$

which by insertion in the Cauchy integral formula (4.4.6) gives

(4.4.8) $$D_x^\beta D_\xi^\alpha \sigma_{d \operatorname{Re} z-\ell}(P^z)(x,0,1) = (-1)^{-d-\ell-|\alpha|} D_x^\beta D_\xi^\alpha \sigma_{d \operatorname{Re} z-\ell}(P^z)(x,0,-1) \quad ;$$

this fits with the transmission property for operators of order $d \operatorname{Re} z$ <u>if and only if</u> $d \operatorname{Re} z + d$ <u>is even</u>;

(4.4.9) $$d \operatorname{Re} z + d \in 2\mathbb{Z} .$$

In the other cases, one gets a more general class (encountered briefly in [Boutet 1,2], and studied in more detail in [Hörmander 8, 18.2] and in the works of Rempel and Schulze on ps.d.o.s without the transmission property). Complex powers have been analyzed for more general ps.d.o. classes by [Nagase-Shinkai 1], [Hayakawa-Kumano-go 1], [Dunau 1], [Robert 1] and others.

Since P^z is so well studied, we shall concentrate the efforts on $G^{(z)}$. For differential boundary value problems, $G^{(z)}$ was studied in [Seeley 4] and [Laptev 1,2].

In the analysis of $G^{(z)}$, one may of course strive to show that it has as many of the nice properties of s.g.o.s as possible. We show (for $\operatorname{Re} z < 0$) that it is defined by a symbol with good properties in (x',ξ'), but with certain limitations on the estimates in the x_n-direction. That this is not just a defect of our methods, is demonstrated by the following example.

4.4.1 Example. Consider the "biharmonic" operator $(1-\Delta)^2$ on a bounded smooth open set $\Omega \subset \mathbb{R}^n$. The Dirichlet realization B of $(1-\Delta)^2$ is defined by the system of boundary conditions

$$\gamma_0 u = 0 , \quad \gamma_1 u = 0 ; \tag{4.4.10}$$

it is selfadjoint positive and 2-coercive. Then $B^{-\frac{1}{2}}$ is well-defined and is of the form

$$\begin{aligned} B^{-\frac{1}{2}} &= [((1-\Delta)^2)^{-\frac{1}{2}}]_\Omega + G^{(-\frac{1}{2})} \\ &= (1-\Delta)_\Omega^{-1} + G^{(-\frac{1}{2})} . \end{aligned}$$

It is known from the investigations of [Grisvard 1] and [Seeley 8] that the range of $B^{-\frac{1}{2}}$, i.e. the domain of $B^{\frac{1}{2}}$, consists of the functions in $H^2(\bar\Omega)$ satisfying those of the boundary conditions for B that are well-defined on $H^2(\bar\Omega)$; here one gets (4.4.10) entirely (this also follows from the variational definition), so

$$R(B^{-\frac{1}{2}}) = D(B^{\frac{1}{2}}) = \{u \in H^2(\bar\Omega) \mid \gamma_0 u = \gamma_1 u = 0\} = \mathring{H}^2(\bar\Omega) .$$

The same considerations on the boundary symbol level give that

$$(b^0)^{-\frac{1}{2}} : L^2(\mathbb{R}_+) \to \mathring{H}^2(\overline{\mathbb{R}}_+) \text{ is a bijection, for each } \xi' \neq 0 .$$

Now if $G^{(-\frac{1}{2})}$ (which is continuous from $L^2(\Omega)$ to $H^2(\bar\Omega)$, since the other operators are so) were a standard singular Green operator (necessarily of order -2 and class 0), then one would have

$$(1-\Delta)_\Omega B^{-\frac{1}{2}} = I + G'$$

with a standard s.g.o. G' of order and class 0 . Similarly, on the boundary symbol level one would have

$$(|\xi'|^2 + D_{x_n}^2)(b^0)^{-\frac{1}{2}} = I + g' \quad \text{on} \quad L^2(\mathbb{R}_+) \ , \quad \text{for each} \quad \xi' \neq 0 \ ,$$

with a Hilbert-Schmidt operator g' . In particular, $I+g'$ would have index 0, but this contradicts the fact that $|\xi'|^2 + D_{x_n}^2$ is <u>not</u> surjective from $\overset{\circ}{H}{}^2(\overline{\mathbb{R}}_+)$ to $L^2(\mathbb{R}_+)$; the range has codimension 1. On the operator level, the corresponding observation is that $B^{-\frac{1}{2}}$ maps $L^2(\Omega)$ into a space with two boundary conditions, which is one too many in order for $G^{(-\frac{1}{2})}$ to be a s.g.o.

So the operators $G^{(z)}$ resulting from the calculus cannot in general be expected to be s.g.o.s; but since they can be defined similarly from symbols, satisfying part of the usual estimates, they will be regarded as <u>generalized singular Green operators</u>. (Of course, $G^{(z)}$ <u>can</u> be a standard s.g.o. in exceptional cases, like when we take $(B^2)^{-\frac{1}{2}}$ for a selfadjoint positive realization B.)

Note that the phenomena observed in the example occur already for differential operators. The structure of these are used to advantage in [Seeley 4,6] and [Laptev 1,2] ; the class of generalized s.g.o.s we get in the general ps.d.o. case is larger than theirs.

Before we study the detailed symbol structure of $G^{(z)}$, let us note that the inequality

$$\langle\lambda\rangle^{1-\theta}\|R_\lambda u\|_{\theta d} \leq c\|u\|_0 \quad , \qquad \text{for} \quad \theta \in [0,1] \ ,$$

which holds since R_λ is of order $-d$ and regularity ≥ 0 (cf. Corollary 3.3.2), implies for the complex powers, by insertion in the Cauchy formula and integration in norm:

$$\text{(4.4.10')} \qquad \|B^z u\|_{\theta d} \leq c'\|u\|_0 \qquad \text{when} \quad \operatorname{Re} z < -\theta \quad \text{and} \quad \theta \in [0,1] \ .$$

The same estimates hold for P_Ω^z and $G^{(z)}$ separately. In some cases where B is positive and z is real, we can include the case $z = -\theta$ in the estimate; in fact we can then characterize the range of $B^{-\theta}$ (or the domain of B^θ) precisely, like in [Grisvard 1]. This hinges on a result on "real" interpolation (as in (A.47)) of spaces defined by certain normal boundary conditions, that we now show.

(Extensions of the characterization of $R(B^z)$ to more general cases might be based on complex interpolation as in [Seeley 8], but the present theory has not been carried far enough to furnish the necessary estimates.)

4.4.2 Theorem. *Let* $T = S\rho + T'$ *be a normal boundary operator associated with the order* d, *and assume that* S *is a differential operator. Here* $T = \{T_0, \dots, T_{d-1}\}$ *with* T_k *of order* k *going from* E *to* F_k. *Denote*

$$H^s(E\,;T) = \{u \in H^s(E) \mid T_k u = 0 \quad \textit{for} \quad k < s-\tfrac{1}{2}\}\ ,$$

considered as a closed subspace of $H^s(E)$.

1^0 *For any* $\theta \in\,]0,1[$ *with* $\theta d-[\theta d] \neq \frac{1}{2}$, *the* $1-\theta$ *interpolation space between* $H^d(E;T)$ *and* $L^2(E)$ *is characterized by*

$$[H^d(E\,;T)\ ,\ L^2(E)]_{1-\theta} = H^{\theta d}(E\,;T)\ . \tag{4.4.11}$$

2^0 *Let* $B = (P+G)_T$ *be a selfadjoint positive realization of order* d, *with* T *as above. For any* $\theta \in\,]0,1[$ *with* $\theta d-[\theta d] \neq \frac{1}{2}$, *the domain of* B^θ *satisfies*

$$D(B^\theta) = H^{\theta d}(E\,;T)\ , \tag{4.4.11'}$$

with equivalent norms.

Proof: Part 1^0 of the theorem was shown for differential normal boundary (without the term T') in the scalar case in [Grisvard 1], and extended to the vector bundle situation in [Seeley 8]. Our proof simply consists of a reduction to their situation by means of Lemma 1.6.8. According to that lemma, there exists a Green operator Λ, which is a bijection in $H^s(E)$ for any $s \geq 0$ and is such that $S\rho + T' = S\rho\Lambda$; in particular, Λ defines a homeomorphism of $H^d(E;T)$ onto $H^d(E;S\rho)$. By the results of Grisvard and Seeley

$$[H^d(E;S\rho),\ L^2(E)]_{1-\theta} = H^{\theta d}(E;S\rho) = \{u \in H^{\theta d}(E) \mid (S\rho)_k u = 0 \quad \text{for} \quad k < \theta d-\tfrac{1}{2}\}\ .$$

Since $\Lambda\colon H^d(E;T) \to H^d(E;S\rho)$ and $\Lambda\colon L^2(E) \to L^2(E)$ are homeomorphisms, Λ likewise defines a homeomorphism between the interpolated spaces, so

$$\begin{aligned}[H^d(E;T),\ L^2(E)]_{1-\theta} &= \Lambda^{-1}[H^d(E;S\rho),\ L^2(E)]_{1-\theta} \\ &= \Lambda^{-1}H^{\theta d}(E;S\rho) = \{u \in H^{\theta d}(E) \mid (S\rho)_k \Lambda u = 0 \quad \text{for} \quad k < \theta d-\tfrac{1}{2}\} \\ &= \{u \in H^{\theta d}(E) \mid T_k u = 0 \quad \text{for} \quad k < \theta d-\tfrac{1}{2}\} = H^{\theta d}(E;T)\ .\end{aligned}$$

The second part of the theorem is now an immediate consequence of the definition of interpolation spaces as domains of powers of positive selfadjoint operators, cf. (A.47). □

4.4.3 Remark. Also in the exceptional cases, where $\theta d = j+\frac{1}{2}$ for an integer j, the interpolation space can be characterized, but here it is not a closed subspace of $H^{\theta d}(E)$ (according to [Seeley 8, §4]). Extend the bundles F_k to bundles $\underline{F}_k$ over $\Sigma_+' = \Gamma\times[0,1[$ in the obvious way, and write the boundary value $(S\rho)_k u$ as $\gamma_0 R_k u$, where R_k is a differential operator from E to $\underline{F}_k$ of order k. Then, according to Grisvard and Seeley,

$$[H^d(E;S\rho),\, L^2(E)]_{1-\theta} = \{u \in H^{j+\frac{1}{2}}(E) \mid (S\rho)_k u = 0 \quad \text{for} \quad k < j,\ R_j u \in L^2_{-\frac{1}{2}}(\underline{F}_j)\}\ ,$$

where $L^2_{-\frac{1}{2}}(\underline{F}_j)$ is the space of measurable v with $x_n^{-\frac{1}{2}} v \in L^2(\underline{F}_j)$. Hence, with Λ defined as in the above proof,

$$(4.4.12)\quad [H^d(E;T),\, L^2(E)]_{1-\theta} = \Lambda^{-1}[H^d(E;S\rho),\, L^2(E)]_{1-\theta}$$
$$= \{u \in H^{j+\frac{1}{2}}(E) \mid T_k u = 0 \quad \text{for} \quad k < j\ ,\quad R_j\Lambda u \in L^2_{-\frac{1}{2}}(\underline{F}_j)\}\ .$$

It follows in particular that

$$(4.4.12')\quad D(B^\theta) = [H^d(E;T),\, L^2(E)]_{1-\theta} \subset H^{j+\frac{1}{2}}(E) \quad \text{when} \quad \theta d = j+\tfrac{1}{2}\ ,$$

when B satisfies the hypotheses of the theorem.

Note that (4.4.10') and (4.4.12') imply that $G^{(-s)}$ maps $L^2(E)$ into $H^{sd}(E)$ when $s \in \mathbb{R}_+$, since P^{-s} is of order $-ds$ [Seeley 1].

Let us now analyze the symbolic structure of the operator $G^{(z)}$ for $\operatorname{Re} z < 0$. As usual, we pass to local coordinates (in particular replacing Ω by $\Gamma\times\mathbb{R}_+$ when considering the behavior of the operators near the boundary). Here we observe once and for all that the "error terms" in the resolvent construction, that are negligible in the class of operators of order $-d$ and regularity ν, give rise to generalized s.g.o. error terms G' here, satisfying estimates of the type (cf. Lemma 2.3.11)

$$(4.4.13)\quad \| D^\beta_{x'} D^\alpha_{\xi'} x_n^k D^{k'}_{x_n} y_n^m D^{m'}_{y_n} \tilde{g}' \|_{L^2_{x_n,y_n}}$$
$$\le c(x')\langle\xi'\rangle^{-M} \Big| \int_C \lambda^z \langle\mu\rangle^{-d-\nu+[k-k']_+ + [m-m']_+}\, d\lambda \Big|$$
$$\le c'(x')\langle\xi'\rangle^{-M}\ , \quad \text{for any } M\ , \quad \text{when}$$

(4.4.13') $\quad d\,\mathrm{Re}\,z + [k-k']_- + [m-m']_- < \nu$.

It follows that the corresponding kernels $K_{G'}(x,y,z)$ satisfy:

(4.4.14) $\quad \sup_{x',y'} \| D^{\theta}_{x',y'} x_n^k D_{x_n}^{k'} y_n^m D_{y_n}^{m'} K_{G'} \|_{L^2_{x_n,y_n}} < \infty$, when (4.4.13') holds.

<u>4.4.4 Theorem.</u> *Let* $B = (P+G)_T$ *be a normal, invertible elliptic realization of regularity* $\nu \in [\frac{1}{2}, d]$, *with* $\{P_\Omega + G - \lambda\,,\ T\}$ *parameter-elliptic on* $\overline{\mathbb{R}_-}$. *Let* $z \in \mathbb{C}$ *with* $\mathrm{Re}\,z < 0$, *and define* $G^{(z)}$ *by*

(4.4.14) $$G^{(z)} = \frac{i}{2\pi} \int_C \lambda^z G_\lambda \, d\lambda \ , \qquad \mathrm{Re}\,z < 0 \ ,$$

as in (4.4.3) *ff. Denote*

(4.4.15) $$\mathrm{Re}\,z = s \ .$$

Then $G^{(z)}$ *is, in local coordinates near* Γ , *a generalized singular Green operator*

(4.4.16) $$G^{(z)}u(x) = (2\pi)^{1-n} \int_{\mathbb{R}^{n-1}} \int_0^\infty e^{ix'\cdot\xi'} \tilde{g}^{(z)}(x',x_n,y_n,\xi') \acute{u}(\xi',y_n)\,dy_n\,d\xi'$$

with $\tilde{g}^{(z)} \sim \sum_{\ell \in \mathbb{N}} \tilde{g}^{(z)}_{ds-1-\ell}$; *here the* ℓ*'th term is quasihomogeneous*

(4.4.16) $$\tilde{g}^{(z)}_{ds-1-\ell}\Big(x', \frac{x_n}{t}, \frac{y_n}{t}, t\xi'\Big) = t^{ds+1-\ell} \tilde{g}^{(z)}_{ds-1-\ell}(x',x_n,y_n,\xi') \quad \textit{for } t \geq 1 \textit{ and } |\xi'| \geq 1 \ ,$$

and the series approximates $\tilde{g}^{(z)}$ *in the sense that*

(4.4.17) $$\| D^\beta_{x'} D^\alpha_{\xi'} x_n^k D_{x_n}^{k'} y_n^m D_{y_n}^{m'} [\tilde{g}^{(z)} - \sum_{\ell < M} \tilde{g}^{(z)}_{ds-1-\ell}] \|_{L^2_{x_n,y_n}} \leq c(x') \langle \xi' \rangle^{ds-|\alpha|-k+k'-m+m'-M}$$

holds for the indices satisfying

(4.4.18) $$\begin{aligned} &-k+k'-m+m'-|\alpha|-M < |ds| \ , \\ &[k-k']_- + [m-m']_- < |ds| + \nu \ . \end{aligned}$$

Proof: We define the symbol-kernels by Cauchy integrals

$$\tilde{g}^{(z)}(x',x_n,y_n,\xi') = \frac{i}{2\pi}\int_{C(\xi')}\lambda^z\tilde{g}(x',x_n,y_n,\xi',\lambda)d\lambda \tag{4.4.19}$$

with similar definitions of the homogenous parts, then the corresponding operators match when the integrals are convergent in symbol norm. The homogeneities are easily verified. In view of the estimates (3.3.52), we have that the integrand in the corresponding Cauchy integral is $O(\langle\lambda\rangle^{-1-\varepsilon})$, when

$$-k+k'-m+m'-|\alpha|-M < |ds| \ , \quad \text{if} \quad [k-k']_+ + [m-m']_+ + |\alpha|+M \le \nu \ , \tag{4.4.20}$$

and when

$$[k-k']_- + [m-m']_- < |ds|+\nu \ , \quad \text{if} \quad [k-k']_+ + [m-m']_+ + |\alpha|+M \ge \nu \ . \tag{4.4.21}$$

Then the integral converges and defines a symbol-kernel satisfying the required estimate. Since

$$-k+k'-m+m'-|\alpha|-M = [k-k']_- + [m-m']_- - ([k-k']_+ + [m-m']_+ + |\alpha|+M) \ ,$$

we see that the conditions "if ..." can be left out in (4.4.20-21), so we get (4.4.18). □

Note that in the above formulas, the indices for which the integrals converge are not only limited by a condition involving ν , but also by a ν-independent condition, that enters even if the regularity is $+\infty$; this was to be expected in view of Example 4.4.1.

Observe on the other hand, that when a (derived) symbol-kernel satisfies the estimates, it also does so after application of $(x_nD_{x_n})^k$ and $(y_nD_{y_n})^m$ to any powers k and m. Further estimates are shown in Proposition 4.5.9.

For the kernels $K_{G^{(z)}}(x,y)$ of the associated operators, one can build up an analysis from the symbol-kernel estimates, like in our analysis of the resolvent and the heat operator, using that (in local coordinates)

$$K_{G^{(z)}}(x,y) = F^{-1}_{\xi'\to x'-y'}\,\tilde{g}^{(z)}(x',x_n,y_n,\xi') \ . \tag{4.4.22}$$

One finds for example readily from (4.4.17-18):

4.4.5 Corollary. *Under the hypotheses of Theorem 4.4.4, the kernel* $K_{G^{(z)}}$ *of* $G^{(z)}$ *satisfies*

$$(4.4.23) \quad \sup_{x',y'} \|(x'-y')^\alpha x_n^k D_{x_n}^{k'} y_n^m D_{y_n}^{m'} K_{G^{(z)}}(x,y)\|_{L^2_{x_n,y_n}} < \infty ,$$

when the following inequalities are satisfied:

$$(4.4.24) \quad \begin{aligned} &-k+k'-m+m'-|\alpha| < d|\mathrm{Re}\, z| + 1-n , \\ &[k-k']_- + [m-m']_- < d|\mathrm{Re}\, z| + \nu . \end{aligned}$$

Estimates of other derivatives, and further details, can be explicited whenever necessary. These types of estimates are not treated in [Seeley 4] and [Laptev 1,2].

4.4.6 Remark. It follows in particular that the kernel is continuous in (x,y), when $\mathrm{Re}\, z < -(n+1)/d$, since the integral converges in sup-norm, with

$$(4.4.25) \quad \sup_{x,y} |K_{G^{(z)}}(x,y)| \leq c \sup_{x',y'} \|K_{G^{(z)}}(x,y)\|_{H^2(\overline{\mathbb{R}}^2_{++})} < \infty$$

then. Moreover, the estimates (4.4.10') ff. for B^z, P^z_Ω and $G^{(z)}$ show that B^z and $G^{(z)}$ (and also their adjoints) map $L^2(E)$ into $H^{|ds|-\varepsilon}(E)$, any $\varepsilon > 0$, so in fact the kernel is continuous for $\mathrm{Re}\, z < -n/d$. For the differential operator case, fine pointwise estimates are obtained in [Laptev 2].

Let us now consider the function

$$(4.4.26) \quad k^{(z)}(x') = \int_0^\infty K_{G^{(z)}}(x,x)dx_n$$

and its integral in x', which is closely related to the trace of $G^{(z)}$; these functions were considered for the differential operator case in [Seeley 4].

In the study of the trace and other spectral features we use the notions and results recalled in the Appendix. Since P^z is a ps.d.o. of order $ds = d\,\mathrm{Re}\, z$, it is continuous from $L^2(\widetilde{E})$ to $H^{|ds|}(\widetilde{E})$, and hence for $\mathrm{Re}\, z \leq 0$, $(P^z)_\Omega$ is continuous from $L^2(E)$ to $H^{|ds|}(E)$. One therefore has, as noted in Lemma A.4, that P^z_Ω is compact for $\mathrm{Re}\, z < 0$, with

$$(4.4.27) \quad s_k(P^z_\Omega) \leq c(P^z_\Omega)k^{-|ds|/n} \text{ for } k=1,2,\dots , \quad \text{when} \quad \mathrm{Re}\, z < 0 .$$

As for B^z, we can use the abstract formula (cf. [Seeley 4])

$$B^z = B^{-j}B^{z+j} \text{ , where } j \in \mathbb{N} \text{ and } \operatorname{Re} z+j \in [-1,0[\text{ ,}$$

to write B^z as the composition of B^{z+j}, that maps $L^2(E)$ into $H^s(E)$ for $s < |\operatorname{Re} z+j|$, cf. (4.4.10'), and B^{-j}, that is continuous from $H^s(E)$ to $H^{s+jd}(E)$; then by Lemma A.4 (cf. also (A.75)),

(4.4.28) $\quad s_k(B^z) \le c_\varepsilon(B^z)k^{-d|\operatorname{Re} z|/n+\varepsilon}$ for $k=1,2,\dots$; for any $\varepsilon > 0$.

In particular, P_Ω^z and B^z, and hence also $G^{(z)}$, are of trace class when

(4.4.29) $$\operatorname{Re} z < -n/d \text{ .}$$

Thus the trace of $G^{(z)}$ is well-defined for sufficiently large negative $\operatorname{Re} z$. (The informations on spectral properties of $G^{(z)}$ and B^z will be substantially improved in Section 4.5 later on.) It is of interest to study analytic extensions of the trace, and also of the "local trace" $k^{(z)}(x')$ defined in (4.4.26).

The expression (4.4.26) can be directly derived from $\tilde{\tilde{g}}(x',\xi',\lambda)$, cf. (3.3.48). First define

(4.4.30) $$\tilde{\tilde{g}}^{(z)}(x',\xi') = \frac{i}{2\pi}\int_{C(\xi')} \lambda^z \tilde{\tilde{g}}(x',\xi',\lambda)d\lambda \text{ ,}$$

and observe that since $\tilde{\tilde{g}}$ is a polyhomogeneous ps.d.o. symbol of order $-d$ and regularity $\nu-\frac{1}{4} > 0$ (Theorem 3.3.10), $\tilde{\tilde{g}}^{(z)}$ is for any z with $\operatorname{Re} z < 0$ a polyhomogeneous ps.d.o. of order ds on $\mathbb{R}^{n-1}$ (and in this localized situation, $\tilde{\tilde{g}}$ and $\tilde{\tilde{g}}^{(z)}$ are compactly supported in x') . If $ds < 1-n$, the corresponding ps.d.o. on $\mathbb{R}^{n-1}$ has a continuous kernel $k^{(z)}(x',y')$, and in particular the diagonal value $k^{(z)}(x')$ is well-defined for each x'. It depends analytically on z in

(4.4.31) $$\{z \in \mathbb{C} \mid \operatorname{Re} z < (1-n)/d\} \text{ ,}$$

since the analyticity of λ^z carries over to the value of the Cauchy integral (one can exchange $\bar{\partial}_z$ with the integration).

Observe here that low order terms in $\tilde{\tilde{g}}(x',\xi',\lambda)$ are $O(\langle\lambda\rangle^{-1-(\nu-\frac{1}{4})/d})$ and therefore contribute to the Cauchy integral (4.4.30) with functions that are analytic in z in the larger set

(4.4.32) $$\{z \in \mathbb{C} \mid \operatorname{Re} z < (\nu-\tfrac{1}{4})d\} \text{ .}$$

The same holds for the error terms stemming from localization; and since these are of arbitrarily low order, the resulting kernels will be C^∞ in x'. In particular, we can replace $k^{(z)}(x')$ by

$$(4.4.33) \qquad k_1^{(z)}(x') = \int_0^1 K_{G^{(z)}}(x,x)\varphi(x_n)dx_n$$

for some function $\varphi \in C_0^\infty(]-1,1[)$ equal to 1 near 0 ; then $k_1^{(z)}(x')$ has a sense on the original manifold $\bar{\Omega}$, the error $k^{(z)}(x') - k_1^{(z)}(x')$ being analytic on (4.4.32) and C^∞ in x'.

We shall now study analytic extensions of $k^{(z)}(x')$ (or $k_1^{(z)}(x')$) beyond (4.4.31).

Up to $\operatorname{Re} z < 0$, one can proceed essentially as in [Seeley 1] (applying the arguments to the ps.d.o. symbol $\tilde{\tilde{g}}(x',\xi',\lambda))$; but for $\operatorname{Re} z \geq 0$ that method would require extensions of the definition of $\tilde{\tilde{g}}^{(z)}$ by recursion formulas, related to recomposition of B^z with positive powers B^k, which can be expected to be problematic, as we have seen at the end of Section 3.3. Actually, we can get above $\operatorname{Re} z = 0$ by a direct argument based on Theorem 3.3.11, so let us do that instead.

The most straightforward case is where $d > n$, which we first assume; then one has the asymptotic formulas (3.3.65) and (3.3.66), from which the Cauchy integral can be directly calculated. Observe that by assumption, the resolvent $(B-\lambda)^{-1}$ exists for λ in a neighborhood of $\overline{\mathbb{R}_-}$, so that we may apply formula (3.3.65) to $[B-(\lambda-\delta)]^{-1}$ for a small positive δ, replacing the complex number called z in that formula by $(\delta-\lambda)^{1/d}$. Then we have for $\operatorname{Re} z < (1-n)/d$,

$$(4.4.34) \quad k^{(z)}(x') = \frac{i}{2\pi}\int_C \lambda^z\Big[\sum_{1\leq j\leq n+\nu'} s_{-d-j}(x')(\delta-\lambda)^{(n-j-d)/d} + s''(x',\lambda)\Big]d\lambda \quad ,$$

where, as we recall, ν' is the largest integer $< \nu - \frac14$ (also equal to the largest integer less than ν). Here $s''(x',\lambda)$ is $O(\langle\lambda\rangle^{-1-(\nu-\frac14)/d})$, so the contribution from that term is analytic in (4.4.32), like the error terms mentioned further above. Note in particular that the Cauchy integrals of this term and the error terms give zero for $z = 0$, since the integrand is then analytic and $O(\langle\lambda\rangle^{-1-(\nu-\frac14)/d})$ in a neighborhood of $\overline{\mathbb{R}_-}$.

The remaining part of $k^{(z)}(x')$, namely the contribution from the sum over j in (4.4.34), is extended into the region $\operatorname{Re} z < (\nu-\frac14)/d$ by analytic continuation. Consider the typical term. When $\operatorname{Re} z > -1$, the Laurent loop C can be replaced by the union of two rays, $C' = \{\lambda = re^{+i\pi}\} \cup \{\lambda = re^{-i\pi}\}$ (taking the respective determinations of λ^z on each ray). Here when z is real and $z+w < -1$,

$$\text{(4.4.35)} \qquad \frac{i}{2\pi}\int_{C'} \lambda^z(\delta-\lambda)^w d\lambda = c_\delta(e^{i\pi z} - e^{-i\pi z})\int_0^\infty x^z(1+x)^w dx$$

$$= c_\delta(e^{i\pi z} - e^{-i\pi z})\,\frac{\Gamma(z+1)\Gamma(-z-w-1)}{\Gamma(-w)} ,$$

by a well-known formula. Note that when z is integer, the expression is zero because of the factor $e^{i\pi z} - e^{-i\pi z}$. Now the last entry in (4.4.35) can be continued analytically to all those complex values of z and w for which the poles of the entering gamma functions are avoided. (Recall that $\Gamma(z)$ is meromorphic on $\mathbb{C}$ with simple poles at $z = 0,-1,-2,\ldots$ and no zeroes.) We need to insert the values

$$w = (n-j)/d - 1 \quad \text{for} \quad j=1,2,\ldots,n+\nu' ,$$

(note that $w < 0$), and as for z , we already know $k^{(z)}(x')$ for $\operatorname{Re} z < (1-n)/d$, where $(1-n)/d > -1$, so it suffices to consider

$$-1 < \operatorname{Re} z < (\nu - 1/4)/d .$$

Here the only poles can occur when

$$z + w + 1 \in \mathbb{N} ,$$

i.e. for

$$z \in \mathbb{N} + (j-n)/d , \qquad j=1,2,\ldots,n+\nu' ,$$

which gives the possible set of poles in the strip $\operatorname{Re} z \in \,]-1,(\nu - 1/4)/d[$,

$$z \in \{(1-n)/d , (2-n)/d ,\ldots, 0, 1/d ,\ldots, \nu'/d\} .$$

The case $z = 0$ deserves special attention. Here the terms in (4.4.34) with $j \neq n$ contribute with zero, in view of the remarks on (4.4.35) (the expression is analytic in the neighborhood of $(z,w) = (0,(n-j-d)/d)$ then). For $j = n$ (i.e. $w = -1$) , a calculation of the integral (4.4.35) for $z < 0$ gives

$$\frac{i}{2\pi}\int_C \lambda^z(\delta-\lambda)^{-1}d\lambda = \delta^z \to 1 \quad \text{for} \quad z \to 0 .$$

It follows that $k^{(z)}(x')$ is analytic at $z = 0$ and, since all other contributions vanish,

(4.4.36) $$k^{(z)}(x') = s_{-d-n}(x') \quad .$$

The above covers the case $d > n$. If $d \leq n$, one can carry out a similar analysis on the basis of the formula (valid for low values of z)

(4.4.37) $$B^z = \frac{i}{2\pi} c_k \int_C \lambda^{z+k-1}(B-\lambda)^{-k} d\lambda \quad ,$$

where k has been taken so large that $(B-\lambda)^{-k}$ is of order $< -n$, so that the kernel and trace have expansions like (3.3.65-66) (again with $n+\nu'$ exact terms, since $(B-\lambda)^{-k} = [(B-\lambda)^k]^{-1}$ is of regularity ν again, cf. Remark 3.4.3).

Note that the same results are valid for $k_1^{(z)}(x')$, since its deviation from $k^{(z)}(x')$ stems from negligible terms.

As in the resolvent analysis, the integral of $k^{(z)}(x')$ or $k_1^{(z)}(x')$ over the boundary gives the trace of $G^{(z)}$ when $\operatorname{Re} z < -n/d$, modulo an error stemming from negligible terms; the error is analytic in z for $\operatorname{Re} z < (\nu-\frac{1}{4})/d$ and vanishes for $z = 0$. The analytic extension of $k^{(z)}(x')$ or $k_1^{(z)}(x')$ then leads to an analytic extension of the trace of $G^{(z)}$. Altogether, we have the following result.

4.4.7 Theorem. *Hypotheses as in Theorem 4.4.4. For sufficiently low values of* $\operatorname{Re} z$, *the generalized s.g.o.* $G^{(z)}$ *is of trace class (cf.* (4.4.29) *and Theorem 4.5.10), and has a continuous kernel* $K_{G^{(z)}}(x,y)$ *(cf. Remark 4.4.6). Here the* x_n*-integral* $k_1^{(z)}(x')$ *of the diagonal value (cf.* (4.4.33)*) extends to a meromorphic function of* z *in the region*

(4.4.38) $$\{z \in \mathbb{C} \mid \operatorname{Re} z < (\nu - \tfrac{1}{4})/d\}$$

with simple poles, contained in the set

(4.4.39) $$\{z = (j-n)/d \mid j=1,2,\dots,n+\nu' \; ; \; j \neq n\} \quad .$$

In particular, the function is analytic at $z = 0$ *and has the value*

(4.4.40) $$k_1^{(0)}(x') = s_{-d-n}(x') \quad ,$$

cf. (3.3.65) *and Remark 3.3.12.*

The trace of $G^{(z)}$ *extends to a meromorphic function of* z, *also denoted*

$\mathrm{tr}\, G^{(z)}$, *on the set* (4.4.38), *with poles contained in* (4.4.39). *In particular, the function is analytic at* $z = 0$ *and has the value*

(4.4.41) $$\mathrm{tr}\, G^{(z)} \Big|_{z=0} = \underline{s}_{-d-n} \, ,$$

cf. (3.3.67) *and Remark 3.3.12.*

Combining the above result with the results of [Seeley 1] for P^z , one has

4.4.8 Corollary. *Hypotheses as in Theorem 4.4.4. The function* $\mathrm{tr}\, B^z$, *which is traditionally defined for* $\mathrm{Re}\, z < -d/n$, *cf. Corollary 4.5.11 below, extends to a meromorphic function of* z *on* (4.4.38), *with simple poles in the set*

(4.4.42) $$\{z = (j-n)/d \mid j=0,1,\dots,n+\nu' \, , \, j \neq n\} \, .$$

In particular, the function is analytic at $z = 0$, *and takes the value*

(4.4.43) $$\mathrm{tr}\, B^z \Big|_{z=0} = a_{-d-n}(B) \, ,$$

cf. (3.3.74').

It can be seen from the analysis of (4.4.34) and the corresponding analysis for P^z , how the coefficients in the resolvent trace expansion (3.3.74') determine the residues at the poles (cf. also [Seeley 1, 4]). Note in particular that the pole at $z = -n/d$ stems from P^z alone; the residue is a nonzero constant determined from p^0 .

Concerning the value at $z = 0$, we furthermore observe that when $\{\partial_t + P_\Omega + G, T\}$ is parabolic (i.e. the parameter-ellipticity holds on all rays $\lambda = re^{i\theta}$ with $\pi/2 \leq \theta \leq 3\pi/2$) , then the coefficient of t^0 in $\mathrm{tr}\exp(-tB)$ (cf. Corollary 4.2.12) coincides with $a_{-d-n}(B)$;

(4.4.44) $$\mathrm{tr}\, B^z \Big|_{z=0} = c_0(B) = a_{-d-n}(B) \, ,$$

just as in the boundaryless case. This can be seen by insertion of (3.3.74') in the Cauchy integral defining $\exp(-tB)$ (with a contour passing to the left of the origin). Also the residues at the poles fit together with the coefficients in (4.2.110), as described in [Duistermaat-Guillemin 1].

In [Rempel-Schulze 4], the resolvent of realizations B is analyzed for a class of pseudo-differential boundary problems permitting ps.d.o.s P not having the transmission property. (See Remark 1.5.16 above concerning the ellipticity hypotheses.) They show for operators of order $d > 1$, that the trace of B^z extends to a meromorphic function of z in

$$\{z \in \mathbb{C} \mid \operatorname{Re} z < -(n-\tfrac{1}{2})/d\} , \tag{4.4.45}$$

with a simple pole at $z = -n/d$, whose residue is determined from p^0 by the usual formula.

Concerning the system (4.4.1), the whole analysis can be carried through in a very similar way. We can for $\operatorname{Re} z < 0$ write

$$A_1^z = \begin{pmatrix} (P^z)_\Omega & 0 \\ 0 & 0 \end{pmatrix} + H^{(z)} , \quad \text{where} \quad H^{(z)} = \begin{pmatrix} G^{(z)} & K^{(z)} \\ T^{(z)} & S^{(z)} \end{pmatrix} \tag{4.4.46}$$

consists of operators generalized from the usual classes (satisfying only a finite number of the usual symbol estimates). In the study, one arrives at conclusions quite parallel to the above, so we shall not go thorugh the details here.

Besides the two operator functions we have consided in this and the preceding section, there are of course many other that could be of interest. For example, one can study $\exp(-tB^{1/d})$, that is used to solve

$$\begin{aligned} \partial_t^d u - Bu &= f \quad \text{for} \quad t > 0 , \\ u|_{t=0} &= u_0 , \end{aligned} \tag{4.4.47}$$

or more general functions. Also the function $\exp(itB^{1/d})$ has an interest, but this is not so simple to define by Cauchy integrals, and requires more delicate considerations.

4.5 Spectral asymptotics.

The spectral terminology and techniques, we use in this section, are explained in Section A.6 of the Appendix.

Let P be a polyhomogeneous ps.d.o. in E of order a, with principal symbol p^0. If $a \geq 0$, we define the spectral coefficient (cf. (A.81)):

$$C(p^0,\Omega) = (2\pi)^{-n} \int_{T^*(\Omega)} N(1\,;p^h(x,\xi)^* p^h(x,\xi))d\xi dx\ , \tag{4.5.1}$$

noting that when P is elliptic, the expression can also be written

$$C(p^0,\Omega) = \frac{1}{n(2\pi)^n} \int_\Omega \int_{|\xi|=1} \operatorname{tr}(p^h(x,\xi)^* p^h(x,\xi))^{-n/2a} d\omega dx \tag{4.5.2}$$

for reasons of homogeneity (both expressions are used in the literature, but (4.5.1) shows best the invariant meaning). Moreover, if p^h is selfadjoint at each (x,ξ), we define

$$C^\pm(p^0,\Omega) = (2\pi)^{-n} \int_{T^*(\Omega)} N^\pm(1\,;p^h(x,\xi))d\xi dx\ ; \tag{4.5.3}$$

note that $C(p^0,\Omega) = C^+(p^0,\Omega)$ when p^h is positive selfadjoint. If $a < 0$, we are instead interested in (cf. (A.80))

$$C'(p^0,\Omega) = (2\pi)^{-n} \int_{T^*(\Omega)} N'(1\,;p^h(x,\xi)^* p^h(x,\xi))d\xi dx \tag{4.5.4}$$

(equal to $C(1/p^0,\Omega)$ when p is elliptic) and, in the selfadjoint case,

$$C'^\pm(p^0,\Omega) = (2\pi)^{-n} \int_{T^*(\Omega)} N'^\pm(1\,;p^h(x,\xi))d\xi dx\ . \tag{4.5.5}$$

One of the interesting questions to study for realizations B is the behavior of their eigenvalues. For <u>differential operator realizations</u> $B = P_T$ of order d, and for ps.d.o.s on compact manifolds without boundary, the question has been answered with very high accuracy, in the form of estimates (under various hypotheses)

$$N(t;B) = C(p^0,\Omega)t^{n/d} + O(t^{(n-1)/d}) \quad \text{for} \quad t \to \infty\ ; \tag{4.5.6}$$

in some cases with further sharpenings, where $O(t^{(n-1)/d})$ is replaced by $c_1 t^{(n-1)/d}$ + remainder; see the works of [Hörmander 5], [Demay 1], [Ivrii 1,2,3], [Melrose 2], [Seeley 7], [Pham The Lai 1], [Métivier 1], [Vasiliev 1,2] and others. (We are not here concerned with ps.d.o.s in $\mathbb{R}^n$, for which we refer to e.g. [Shubin 1], [Helffer 1], [Mohamed 3] and their references.)

In boundary problems for pseudo-differential operators, the second term has not (yet) been estimated with such a high precision. The best results for general realizations defined in the Boutet de Monvel calculus still seem to be those of [Grubb 8-10], where the remainder is $\mathcal{O}(t^{(n-\frac{1}{2}+\varepsilon)/d})$ (or $\mathcal{O}(t^{(n-1+\varepsilon)/d})$ in special cases, see also Remark 4.5.7). We now give an account of these and other estimates for pseudo-differential boundary problems, supplied with some new results obtainable by use of the functional calculus. (There is an overlap with [Levendorskii 2], who considers more abstractly defined operators; here he assumes that the domain contains $C_0^\infty(\Omega)$, which restricts the generality, e.g. the s.g.o. can only be of class 0, cf. Example 1.6.9'.)

First we consider principal asymptotic estimates (with remainder $o(t^{n/d})$). It was shown in [Grubb 5], by use of a resolvent construction (of a more coarse kind than the one in the present book) and a standard Tauberian argument, that the Dirichlet realization P_γ of a strongly elliptic selfadjoint ps.d.o. system P (of a type arizing in connection with Douglis-Nirenberg elliptic systems) satisfies

$$N(t;P_\gamma) = C(p^0,\Omega)t^{n/d} + o(t^{n/d}) \quad \text{for} \quad t \to \infty . \tag{4.5.7}$$

Recall that (4.5.7) can equivalently be written (cf. Lemma A.5)

$$\lambda_k(P_\gamma)k^{-d/n} \to C(p^0,\Omega)^{-d/n} \quad \text{for} \quad k \to \infty . \tag{4.5.8}$$

As noted in [Grubb 10], the resolvent construction works in particular for ps.d.o.s P that act like differential operators near the boundary (modulo a negligible term), and are positive selfadjoint, so (4.5.7-8) likewise holds for these.

Moreover, it is easy to extend the result to Dirichlet realizations of strongly elliptic even-order ps.d.o.s regardless of the transmission property, as indicated in [Grubb 7]. For completeness' sake, we include the simple proof.

4.5.1 Proposition. *Let* $m \in \mathbb{N}_+$, *and let* P *be a strongly elliptic, selfadjoint ps.d.o. in* $\tilde{E}$ *of order* $d = 2m$, *not necessarily having the transmission property at* Γ . *Let* P_F *be the Friedrichs extension of* r^+Pe^+ *defined on* $C_0^\infty(\Omega,E)$; *it has domain in* $\mathring{H}^m(E)$ *and represents the Dirichlet problem in some sense (in a precise sense when* P *has the transmission property, cf. Theorem 1.7.2). Then* P_F *has at most finitely many eigenvalues* ≤ 0 , *and the positive eigenvalues satisfy*

$$N^+(t\,;P_F) = C(p^0,\Omega)t^{n/d} + o(t^{n/d}) \quad \textit{for} \quad t \to \infty , \tag{4.5.9}$$

where $C(p^0,\Omega)$ *is defined by* (4.5.1).

Proof: We use that the result is known in the cases where P is a differential operator near Γ. Since P is strongly elliptic, it satisfies (1.7.2), and we can modify it by addition of a constant such that we get an operator (also called P) satisfying

$$(4.5.10) \qquad C\|u\|_m^2 \geq (Pu,u) \geq c\|u\|_m^2 \qquad \text{for} \quad u \in H^d(\tilde{E}) \quad ,$$

with constants $C \geq c > 0$. Since (4.5.10) holds in particular for $u \in C_0^\infty(\Omega,E)$, the positive symmetric sesquilinear form

$$s_0(u,v) = (Pu,v) \qquad \text{on} \quad C_0^\infty(\Omega,E)$$

extends by continuity to a positive symmetric sesquilinear form $s(u,v)$ on $\mathring{H}^m(E)$; the associated variational operator in $L^2(E)$ (cf. Lemma 1.7.1) is by definition P_F. Now let Λ be the operator in $L^2(\tilde{E})$ associated with the $H^m(\tilde{E})$ scalar product on $H^m(\tilde{E})$, in particular

$$(\Lambda u,u) = \|u\|_m^2 \qquad \text{for} \quad u \in D(\Lambda) \ ;$$

here Λ is an elliptic positive selfadjoint differential operator of order d, and $D(\Lambda) = H^d(\tilde{E})$. Let $\varphi(x_n) \in C_0^\infty(]-\varepsilon,\varepsilon[)$ with values in $[0,1]$ and $\varphi(x_n) = 1$ on $]-\varepsilon/2, \varepsilon/2[$, and let

$$P' = P + \varphi(x_n)(C\Lambda - P)\varphi(x_n) \ ,$$

$$P'' = P - \varphi(x_n)(P - c\Lambda)\varphi(x_n) \ .$$

Then in view of (4.5.10)

$$(P''u,u) \leq (Pu,u) \leq (P'u,u) \qquad \text{for} \quad u \in C^\infty(\tilde{E}) \ ,$$

and in particular

$$(P_F''u,u) \leq (P_Fu,u) \leq (P_F'u,u) \qquad \text{for} \quad u \in C_0^\infty(\Omega,E) \ ;$$

with a similar inequality for the associated sesquilinear forms. Here P_F'' and P_F' are the Dirichlet realizations of P'' resp. P' in $L^2(E)$, so they have the spectral behavior (4.5.1)-(4.5.3). By the maximum-minimum principle, it follows that P_F has a discrete spectrum with

$$\lambda_k(P_F'') \leq \lambda_k(P_F) \leq \lambda_k(P_F') \qquad \text{for all} \quad k \ .$$

Since $P - \varphi P \varphi$ is negligible near Γ, and P' and P'' are strongly elliptic, (4.5.8) applies to P' and P'', which gives

$$\limsup_{k\to\infty} \lambda_k(P_F)k^{-d/n} \le C(p'^0,\Omega)^{-d/n}$$
$$\liminf_{k\to\infty} \lambda_k(P_F)k^{-d/n} \ge C(p''^0,\Omega)^{-d/n}$$

Taking ε small, we can make $C(p'^0,\Omega)$ and $C(p''^0,\Omega)$ arbitrarily close to $C(p^0,\Omega)$, and it follows that in fact

$$\lambda_k(P_F)k^{-d/n} \to C(p^0,\Omega)^{-d/n} \quad \text{for} \quad k\to\infty\,, \tag{4.5.11}$$

which shows the proposition, when P satisfies (4.5.10). More generally, (4.5.10) was obtained by addition of a positive constant, so there are at most finitely many negative eigenvalues, and the positive eigenvalues are estimated like (4.5.11).

Note that the domain of P_F is not assumed to lie in $H^{2m}(E)$.

For certain other types of elliptic boundary problems, where P need not have the transmission property, and is of order > 1, the principal estimate (4.5.7) has been shown in [Rempel-Schulze 4] by a resolvent analysis.

For operators of negative order, one can often obtain asymptotic eigenvalue estimates by perturbation from special cases, even for operators that are not elliptic. Before demonstrating this, we recall a very useful result on operators of singular Green type, showing how the boundary dimension enters instead of the interior dimension. (As usual, the operators are assumed polyhomogeneous.)

4.5.2 Proposition. *Let* $a \in \mathbb{R}_+$, *and let* G *be one of the following operators:*

1^0 $G = B^{z/d} - (P^{z/d})_\Omega$, *where* d *is even* > 0, $\operatorname{Re} z = -a$, *and* $B = P_T$ *is an elliptic realization in* E *of an elliptic differential operator* P *of order* d *together with a differential trace operator* T, *such that* $\{P-\lambda, T\}$ *is parameter-elliptic for* $\lambda \in \overline{\mathbb{R}}_-$.

2^0 $G = G_1^+(Q)$ *or* $G_1^-(Q)$, *where*

$$G_1^+(Q) = r^+Qe^-\,, \qquad G_1^-(Q) = r^-Qe^+\,, \tag{4.5.12}$$

Q *being a ps.d.o. in* $\tilde{E}$ *of order* $-a$, *not necessarily with transmission property.*

3^0 G *is a singular Green operator in* E *of order* $-a$ *and class* 0.

Then (cf. (A.78) *ff.)*

$$G \in \mathfrak{S}_{((n-1)/a)}\,, \tag{4.5.13}$$

and the characteristic values $s_k(G)$ *satisfy an asymptotic estimate*

$$s_k(G)k^{a/(n-1)} \to C(g^0) , \tag{4.5.14}$$

where $C(g^0)$ *is a constant determined from the principal symbols involved. There are similar estimates of* $\lambda_k^+(G)$ *and* $\lambda_k^-(G)$ *in the selfadjoint case.*

(4.5.14) was proved in case 1^0 and 2^0 by [Laptev 1,2] and in case 3^0 by [Grubb 17]. Moreover, [Grubb 17] showed that when Q has the transmission property (in particular $a \in \mathbb{N}_+$), then $G_1^+(Q)$ and $G_1^-(Q)$ are very much like singular Green operators (they become s.g.o.s by composition with a reflection in Γ, locally), and $G_1^+(Q)^*G_1^+(Q)$ and $G_1^-(Q)^*G_1^-(Q)$ are true s.g.o.s in $L^2(E)$ resp. $L^2(\tilde{E}\setminus E)$, so 2^0 is then covered by 3^0. (1^0 is covered by 3^0 when z/d is integer.) See also Theorem 2.6.6.

(4.5.13) alone is simpler to show than (4.5.14); for s.g.o.s it appears in [Grubb 5, 7-10, 16, 17]. We showed in [Grubb 17] how (4.5.13) extends to s.g.o.s of $S_{1,0}$ type, with a detailed estimate of the quasi-norm

$$N_{(n-1)/a}(G) \equiv \sup_k\{s_k(G)k^{a/(n-1)}\} ; \tag{4.5.15}$$

this is recalled further below (Proposition 4.5.8). In applications, it often suffices to have (4.5.13), or estimates of (4.5.15). (A more detailed account of the historical development of (4.5.13-15) for s.g.o.s is given in [Grubb 17].)

The next proposition shows an argument using (4.5.13) together with simple perturbation tricks; and we later take up these estimates for the generalized s.g.o.s $G^{(z)}$ defined from pseudo-differential boundary problems as in Section 4.4.

<u>4.5.3 Proposition</u>. *Let* $a \in \mathbb{R}_+$, *and let* Q *be a polyhomogeneous ps.d.o. of order* $-a$ *in* $\tilde{E}$ *(neither ellipticity nor the transmission property at* Γ *is assumed). Then* Q_Ω *has the spectral behavior*

$$s_k(Q_\Omega)k^{a/n} \to C'(q^0,\Omega)^{a/n} , \quad \text{for} \quad k \to \infty , \tag{4.5.16}$$

where $C'(q^0,\Omega)$ *is defined by* (4.5.4).

Proof: The result is known, when Ω is replaced by the full manifold, cf. [Birman-Solomiak 2] (or [Grubb 17, Lemma 4.5 ff.], where approximation from elliptic cases is used).

Consider first the case where Q is strongly elliptic and selfadjoint, satisfying

$$C\|u\|^2_{-a/2} \geq (Qu,u) \geq c\|u\|^2_{-a/2} \qquad \text{for} \quad u \in C^\infty(\tilde{E}) ,$$

with $C \geq c > 0$. Let χ_1 and $\chi_2 \in C^\infty(\Sigma)$ with values in $[0,1]$, such that χ_1 equals 1 on a large subset of Ω and vanishes on a neighborhood of $\Sigma\setminus\Omega$, whereas χ_2 equals 1 on a neighborhood of $\bar{\Omega}$ and vanishes outside a slightly larger neighborhood. Then

$$Q_\Omega = (\chi_2 Q \chi_2)_\Omega = r^+\chi_2 Q \chi_2 e^+ ,$$

so that, by (A.75),

$$\begin{aligned} \lambda_k(Q_\Omega) &\leq \|r^+\|\lambda_k(\chi_2 Q\chi_2)\|e^+\| \\ &= \lambda_k(\chi_2 Q \chi_2) , \end{aligned}$$

since the operators r^+ and $e^+ = (r^+)^*$ have norm 1. On the other hand,

$$\lambda_k((\chi_1 Q \chi_1)_\Omega) \leq \lambda_k(Q_\Omega)$$

since χ_1 has norm ≤ 1 as an operator in $L^2(E)$. Since (4.5.16) holds in the boundaryless case, and χ_1 and χ_2 can be chosen such that $C'(\chi_1 q^0 \chi_1,\Sigma)$ and $C'(\chi_2 q^0 \chi_2,\Sigma)$ are arbitrarily close to $C'(q^0,\Omega)$, it follows as in Proposition 4.5.1 that (4.5.16) holds for Q_Ω.

When Q is merely selfadjoint ≥ 0, we apply the preceding result to $Q_\Omega + \varepsilon\Lambda_\Omega^{-a}$ for $\varepsilon \to 0$, where Λ^{-a} is a selfadjoint invertible nonnegative ps.d.o. in $\tilde{E}$ with principal symbol $|\xi|^{-a}$; then since $s_k(\varepsilon\Lambda_\Omega^{-a}) = \varepsilon s_k(\Lambda_\Omega^{-a})$, the perturbation argument in Lemma A.7 gives the result for Q_Ω in the limit.

Finally, when Q is arbitrary, we note that with the notation (4.5.12),

$$\begin{aligned} (4.5.17) \qquad Q_\Omega^* Q_\Omega &= r^+Q^*e^+r^+Qe^+ \\ &= r^+Q^*(I - e^-r^-)Qe^+ \\ &= (Q^*Q)_\Omega - G_1^+(Q^*)G_1^-(Q) , \end{aligned}$$

since the orders are ≤ 0. Note also that $(Q^*)_\Omega = (Q_\Omega)^*$. Here $(Q^*Q)_\Omega$ is covered by the preceding case,

$$\lambda_k((Q^*Q)_\Omega)k^{2a/n} \to C'(q^{0*}q^0,\Omega)^{2a/n} . \tag{4.5.18}$$

For $G_1^+(Q^*)$ and $G_1^-(Q)$, we use the result of Laptev quoted in Proposition 4.5.2, which gives, in view of (A.75 ii), that

$$s_k(G_1^+(Q^*)G_1^-(Q)) \leq \text{const. } k^{-2a/(n-1)} .$$

An application of the perturbation argument Lemma A.6 1^0 then gives that

$$s_k(Q_\Omega)^2 = \lambda_k((Q_\Omega)^*Q_\Omega) = \lambda_k((Q^*)_\Omega Q_\Omega)$$

behaves like (4.5.18), and this implies the proposition.

The result may well be known (since it can be deduced as simply as above), but we have not seen a proof for the general case before. ([Widom 3] obtains it for operators of order $\in\ ]-1,0[$, see also [Widom 4].)

We now turn to the more delicate remainder estimates (improvements of $o(t^{n/d})$) , that require more information on the structure of the operator. Here, the result of [Grubb 8, 9, 10] is first recalled; note that it is <u>not</u> a priori restricted to <u>normal</u> boundary conditions (but they are convenient in the definition of examples to which the result applies).

<u>4.5.4 Theorem.</u> *Let* P *be selfadjoint, strongly elliptic of order* $d > 0$ *and having the transmission property (so* d *is even), and let* $B = (P+G)_T$ *be a selfadjoint elliptic, not necessarily lower bounded realization, defined as in Definition 1.4.1 (cf. also Theorem 1.6.11). Then*

$$N^+(t\,;B) = C(p^0,\Omega)t^{n/d} + O(t^{(n-\theta)/d}) \quad \textit{for} \quad t\to\infty , \tag{4.5.19}$$

$$N^-(t\,;B) = O(t^{(n-1)/d}) \quad \textit{for} \quad t\to\infty , \tag{4.5.20}$$

where θ *can be any real number* $< \frac{1}{2}$ *in general, and* θ *can be any real number* < 1 *when* P *is a scalar differential operator (cf. also Remark 4.5.7). When* B *is invertible, the estimates can equivalently be written*

$$\lambda_k^+(B^{-1}) = C(p^0,\Omega)^{d/n}k^{-d/n} + O(k^{-(d+\theta)/n}) \quad \textit{for} \quad k\to\infty , \tag{4.5.21}$$

$$\lambda_k^-(B^{-1}) = O(k^{-d/(n-1)}) \quad \textit{for} \quad k\to\infty . \tag{4.5.22}$$

As shown in [Grubb 8, 9, 10], the proof consists of two ingredients: One is a fine analysis of the resolvent kernel of the simplest Dirichlet realization P_γ in case $d > n$, by a generalization of the techniques of [Agmon 5] and [Beals 1] (commutation with nested cut-off functions), showing the results in this case. The other ingredient is the observation that (assuming invertibility, which can be obtained by a small translation)

$$(4.5.23) \qquad (P+G)_T^{-N} = [(P^N)_\gamma]^{-1} + G_{(N)} \ ,$$

where $G_{(N)}$ is a s.g.o. of order $-dN$ and class 0, and N can be taken as large as we like (in particular larger than n/d). By a coarse version of Proposition 4.5.2 3^0, $G_{(N)}$ is in $\mathfrak{S}_{((n-1)/Nd)}$, and then the perturbation argument in Lemma A.6 2^0 can be applied to (4.5.23). In this application, we can take N so large that $q = q'$, which gives (4.5.19-20) for B^N (cf. Lemma A.5); and we can take N odd so that the positive resp. negative eigenvalues for B^N are the N'th powers of the positive resp. negative eigenvalues for B; this gives the result for B. (Similar techniques are used below, e.g. in (4.5.25)ff.)

The theorem has the following interesting consequence, that sharpens Proposition 4.5.3 in the case of elliptic ps.d.o.s with the transmission property.

<u>4.5.5 Corollary</u>. *Let* $d \in \mathbb{N}_+$, *let* Q *be an elliptic ps.d.o. in* $\tilde{E}$ *of order* $-d$, *having the transmission property, and let* G' *be a s.g.o. in* E *of order* $-d$ *and class* 0. *Then the counting function for the characteristic values of* $Q_\Omega + G'$ *satisfies*

$$(4.5.24) \quad N'(t\,; Q_\Omega+G') = C'(q^0,\Omega)t^{n/d} + O(t^{(n-\theta)/d}) \quad \textit{for} \quad t \to \infty \ ,$$

for any $\theta < \frac{1}{2}$ *in general, and any* $\theta < 1$ *if* Q *is a parametrix of a scalar differential operator (cf. also Remark 4.5.7).*

<u>Proof</u>: Let P be a parametrix of Q on $\tilde{E}$, then P is elliptic of order d. Let $(P^*P)_\gamma$ be the Dirichlet realization of the strongly elliptic operator P^*P, and let $\tilde{B}$ be the operator with domain

$$D(\tilde{B}) = D((P^*P)_\gamma) = H^{2d}(E) \cap \mathring{H}^d(E) \ ,$$

acting like $(P^*P)_\gamma$ on the orthogonal complement of the nullspace $Z((P^*P)_\gamma)$ and like I on $Z((P^*P)_\gamma)$. Then $\tilde{B}$ is a normal, selfadjoint, positive, elliptic realization of P^*P (of the form $(P^*P+\tilde{G})_\gamma$), and $C(p^{0*}p^0,\Omega) = C'(q^0,\Omega)$. To this operator we can apply Theorem 4.5.4, which gives for its N-th power (since $N(t\,;\tilde{B}) = N(t^N\,;\tilde{B}^N))$, $N \in \mathbb{N}_+$:

$$N(t\,;\tilde{B}^N) \equiv N'(t\,;\tilde{B}^{-N}) = C'(q^0,\Omega)t^{n/2dN} + O(t^{(n-\theta)/2dN}) ,$$

for $t \to \infty$. Since $Q_\Omega + G'$ is of negative order and class 0 , it has an adjoint within the calculus, and the rules of calculus imply

$$(4.5.25) \quad [(Q_\Omega+G')^*(Q_\Omega+G')]^N = [(Q^*Q)^N]_\Omega + G_{(N)} = \tilde{B}^{-N} + G'_{(N)} ,$$

where $G_{(N)}$ and $G'_{(N)}$ are s.g.o.s of order $-2dN$ and class 0 . By Proposition 4.5.2 3^0 ,

$$s_k(G'_{(N)}) = O(k^{-2dN/(n-1)}) .$$

We can then apply Lemma A.6 2^0 (cf. also Lemma A.5) to the last sum in (4.5.25), with

$$p = n/2dN \quad , \quad q = n/(2dN+\theta)$$
$$r = (n-1)/2dN \quad \text{and} \quad s = (n-\theta)/2dN .$$

This gives that

$$N'(t\,;[(Q_\Omega+G')^*(Q_\Omega+G')]^N) = C'(q^0,\Omega)t^{n/2dN} + O(t^{s'}) ,$$

where

$$s' = \max\left\{\frac{n-\theta}{2dN} , \frac{n-1}{2dN}\,\frac{n+2dN}{n-1+2dN}\right\} .$$

For any $\theta < 1$, one can here take N so large that the first term in $\{\ \}$ is largest, so we can obtain $s' = (n-\theta)/2dN$. Then (4.5.24) follows, since

$$N'(t\,;Q_\Omega+G') = N'(t^{2N}\,;[(Q_\Omega+G')^*(Q_\Omega+G')]^N) . \qquad \square$$

For realizations, one obtains in particular:

<u>4.5.6 Corollary</u>. *Let* $B = (P+G)_T$ *be an elliptic realization of order* $d > 0$ *, as in Definition 1.4.1. Any parametrix* R *satisfies* (4.5.24). *If* B *is normal, or densely defined and invertible, one has moreover*

$$(4.5.26) \quad N(t\,;B) = C(p^0,\Omega)t^{n/d} + O(t^{(n-\theta)/d}) \quad \textit{for} \quad t \to \infty ,$$

with θ *as in Theorem 4.5.4.*

Proof: Since R is of the form $Q_\Omega + G'$, where Q is a parametrix of P and G' is a s.g.o. or order $-d$ and class 0 , the preceding corollary applies to R ; note here that $C(p^0,\Omega) = C(p^{0*}p^0,\Omega)$. Then when B is invertible, and densely defined so that B^* exists, (4.5.26) follows by application of the preceding result to B^{-1} .

When B is normal (hence densely defined by Lemma 1.6.8), we apply instead Theorem 4.5.4 directly to B^*B , which is a normal, selfadjoint elliptic realization of P^*P , by Theorems 1.4.6 and 1.6.9. □

The last result also gives an estimate of the number of eigenvalues of B in $[-t,t]$, when B is selfadjoint, P not necessarily strongly elliptic. Individual estimates of $N^+(t\,;B)$ and $N^-(t\,;B)$ in that case will be obtained later on, for normal boundary conditions.

4.5.7 Remark. Let us observe that in all these proofs, the restriction to $\theta < \frac{1}{2}$ for general P stems from a similar restriction for its Dirichlet realization, or the Dirichlet realization of P^*P . If the estimate of the Dirichlet realization can be extended to larger values of θ in $[\frac{1}{2},1[$, the same extension will be valid for the other realizations. As observed already in [Grubb 8], the estimate can probably be extended to all $\theta < 1$ whenever P is a scalar ps.d.o., by use of methods like in [Robert 2]; but we have not pursued this idea. Moreover, in matrix cases where P is diagonal or suitably diagonalizable, the improved estimates can be extended from the scalar case.

Let us now turn to the new material obtainable in the framework of operational calculus. The main point will be an analysis of the spectral properties of $G^{(z)}$ for $\operatorname{Re} z < 0$, where $G^{(z)}$ is the generalized s.g.o. part of B^z defined in Section 4.4 from a pseudo-differential realization. We shall show how Laptev's result (Proposition 4.5.2 1°) is generalized, and give some applications.

The proof is a combination of the symbol estimates in Section 4.4 with the detailed analysis in [Grubb 17], that gave a careful account of which symbol estimates are required for each spectral estimate. Rather than repeating all the details on how the analysis is set up by the help of Laguerre expansions, we shall simply quote the results that we need here, and explain their application. We shall use [Grubb 17, Theorem 4.8] and its proof, and we shall also need the observations before Theorem 4.8 on how one avoids to have to deal with the term R in Proposition 4.7. (We allow ourselves to omit the details, for once.)

To quote the results, we first recall the norm notation for a singular Green symbol-kernel of order $-a$ and class 0 (parameter-independent)

$$(4.5.27)\quad |||\tilde{g}|||_{\alpha,\beta,k,k',m,m'} \equiv \sup \langle\xi'\rangle^{a+|\alpha|+k-k'+m-m'} \| D^{\beta}_{x'} D^{\alpha}_{\xi'} x_n^k D^{k'}_{x_n} y_n^m D^{m'}_{y_n} \tilde{g} \|_{L^2_{x_n,y_n}}$$

Recall also the definition of the Laguerre operator (2.2.12), that we use in the following with parameter $\sigma \sim \langle\xi'\rangle$ (dropping the plus),

$$(4.5.28)\quad L_{\sigma}u(x_n) = -\sigma^{-1}\partial_{x_n} x_n \partial_{x_n} u(x_n) + \sigma x_n u(x_n) + u(x_n)\ , \quad \text{for} \quad u \in \mathscr{S}(\overline{\mathbb{R}}_+)\ .$$

(Here we may write L_{σ,x_n} or L_{σ,y_n}, when it is applied to a function of x_n resp. y_n.) The result from [Grubb 17] will be used in the following formulation.

<u>4.5.8 Proposition</u>. *Let* G *be a generalized s.g.o. of order* $-a<0$ *and class* 0 *on* $\mathbb{R}^n_+$, *with a symbol-kernel* $\tilde{g}(x',x_n,y_n,\xi')$ *(of* $S_{1,0}$ *type) that is compactly supported in* x' *and for which the symbol norms in* (4.5.30) *are finite. Then* $G \in \mathfrak{S}_{((n-1)/a)}$, *and the characteristic values satisfy*

$$(4.5.29)\qquad s_k(G) \le c(G)k^{-a/(n-1)}\ ,$$

where $c(G)$ *is estimated by*

$$(4.5.30)\quad c(G) \le C \sup\{|||L^{N_2}_{\sigma,y_n}\tilde{g}|||_{0,\beta,0,0,0,0} \mid |\beta| \le N_1\}$$

for some constant C *depending on* a, n, *the symbol support and* N_1, N_2 ; *here* $\sigma=\sigma(\xi')\sim\langle\xi'\rangle$, *and* N_1 *is even and* N_2 *is real, chosen such that*

$$(4.5.31)\qquad N_1 > a+n-1\ , \qquad N_2 > \max\left\{\frac{1}{2}\ , \frac{a}{n-1}\right\}\ .$$

In the formulation of [Grubb 17], Theorem 4.8, N_2 is taken to be integer, and the norms in (4.5.30) are replaced by a collection of norms $|||\tilde{g}|||_{0,\beta,0,0,\ell,\ell'}$ where $\ell \le N_2$ and $\ell' \le 2N_2$; so the condition there follows from (4.5.30) by insertion of L_{σ,y_n} defined by (4.5.28). However, for minimal hypotheses it is advantageous to keep L_{σ,y_n} itself, and to allow non-integer powers, which we can do in view of the proof of Theorem 4.8. It is seen there (in the last 8 lines) that L_{σ,y_n} enters only because of its rôle in the formula for Laguerre expansions

$$(4.5.32) \qquad \sum_{\ell,m} (1+m)^{2N} |s_{\ell m}(x',\xi')|^2 = 2^{-2N} \| L^N_{\sigma,y_n} \tilde{g} \|^2_{L^2_{x_n,y_n}} ,$$

when $\tilde{g}(x',x_n,y_n,\xi') = \Sigma_{\ell,m\in\mathbb{N}}\, s_{\ell m}(x',\xi')\varphi_\ell(x_n,\sigma)\varphi_m(y_n,\sigma)$ (as in (2.6.41) for $\kappa=\sigma$). In (4.5.32), one can let N be non-integer, using fractional powers of the positive selfadjoint operator L_σ in $L^2(\mathbb{R}_+)$. In the proof of Theorem 4.8, one simply needs the boundedness of the left hand side of (4.5.32) for some $N = N_2 > \max\{1/2, a/(n-1)\}$, along with similar estimates of x'-derivatives up to order N_1.

Now consider $G^{(z)}$, as described in Theorem 4.4.4. Here it is seen that N_1 causes no problems, since differentiation in x' does not affect the boundedness of the symbol norms. On the other hand, the condition on N_2 does lead to some restrictions, since the application of L_{σ,y_n} in general lowers the regularity. In the following proposition, we investigate the effect of applications of powers of the Laguerre operator to $\tilde{g}^{(z)}$ and its derivatives, preparing also for more general applications later on.

<u>4.5.9 Proposition</u>. *Let* $z\in\mathbb{C}$ *with* $s = \operatorname{Re} z < 0$, *and let* $\tilde{g}^{(z)}(x',x_n,y_n,\xi')$ *be the generalized s.g.o. symbol-kernel defined in Theorem 4.4.4, cf.* (4.4.19). *Let* N *and* $N' \in \overline{\mathbb{R}}_+$, *and let* $k,k',m,m',M\in\mathbb{N}$ *and* $\alpha,\beta\in\mathbb{N}^{n-1}$. *Let* $\sigma(\xi')$ *be a positive* C^∞ *function, equal to* $|\xi'|$ *for* $|\xi'|\geq 1$. *Then the estimates*

$$(4.5.33) \qquad \| L^N_{\sigma,x_n} L^{N'}_{\sigma,y_n} x_n^k D_{x_n}^{k'} y_n^m D_{y_n}^{m'} D^\beta_{x'} D^\alpha_{\xi'} (\tilde{g}^{(z)} - \sum_{\ell<M} \tilde{g}^{(z)}_{ds-1-\ell}) \|_{L^2_{x_n,y_n}} \leq c(x')\langle\xi'\rangle^{ds-k+k'-m+m'-|\alpha|-M}$$

hold for the indices satisfying

$$(4.5.34) \qquad \begin{aligned} &-k+k'-m+m'-|\alpha|-M+N+N' < |ds| + \nu'', \quad \textit{where} \\ &\nu'' = \min\{\nu-[k-k']_+ - [m-m']_+ - |\alpha|-M, 0\} \qquad (\leq 0) . \end{aligned}$$

In particular, $\tilde{g}^{(z)}$ *itself satisfies*

$$\| L^{N}_{\sigma,x_n} L^{N'}_{\sigma,y_n} \tilde{g}^{(z)} \|_{L^2_{x_n,y_n}} \le c(x')\langle\xi'\rangle^{-|ds|}$$

(4.5.35)

$$\textit{for any } N \textit{ and } N' \in \overline{\mathbb{R}}_+ \textit{ with } N+N' < |ds| .$$

Proof: For a symbol-kernel $\tilde{f}(x',x_n,y_n,\xi',\mu) \in S^{r,\tau}(\Xi, \overline{\mathbb{R}}^n_+, \mathcal{S}(\overline{\mathbb{R}}^2_{++}))$, the two terms in L_{σ,x_n} in (4.5.28) act as follows (norms in $L^2_{x_n,y_n}(\mathbb{R}^2_{++})$):

$$\|\sigma^{-1}D_{x_n}x_nD_{x_n}\tilde{f}\| \le c(x')\langle\xi'\rangle^{-1}(\langle\xi'\rangle^{\tau} + \langle\xi',\mu\rangle^{\tau})\langle\xi',\mu\rangle^{r-\tau+1}$$

$$= c(x')(\langle\xi'\rangle^{\tau-1}\langle\xi',\mu\rangle^{r-\tau+1} + \langle\xi'\rangle^{-1}\langle\xi',\mu\rangle^{r+1}) ,$$

(4.5.36)

$$\|\sigma x_n\tilde{f}\| \le c(x')\langle\xi'\rangle(\langle\xi'\rangle^{\tau-1} + \langle\xi',\mu\rangle^{\tau-1})\langle\xi',\mu\rangle^{r-\tau}$$

$$= c(x')(\langle\xi'\rangle^{\tau}\langle\xi',\mu\rangle^{r-\tau} + \langle\xi'\rangle\langle\xi',\mu\rangle^{r-1}) ,$$

which indicates that $\sigma^{-1}D_{x_n}x_nD_{x_n}$ lowers the regularity to $\min\{\tau-1,-1\}$ and σx_n lowers the regularity to $\min\{\tau,1\}$. The weakest regularity is here $\min\{\tau-1,-1\} = -\tau_- - 1$. By application of the analogous estimate to derivatives and lower order parts of $\tilde{f}$, it is seen that

$$\tilde{f} \in S^{r,\tau} \quad \text{implies} \quad L_{\sigma,x_n}\tilde{f} \in S^{r,-\tau_- -1} .$$

L_{σ,y_n} acts similarly. Thus we find by iteration

$$L^{N}_{\sigma,x_n}L^{N'}_{\sigma,y_n}\tilde{f} \in S^{r,-\tau_- - N-N'} , \tag{4.5.37}$$

for any nonnegative integers N and N' . Since the regularity is ≤ 0 here, the estimates extend readily by interpolation to noninteger N and N' . For example, the typical estimates (cf. (2.1.17))

$$\|L^{N}_{\sigma,x_n}\tilde{f}\| \le c(x')\langle\xi'\rangle^{-\tau_- - N}\langle\xi',\mu\rangle^{r+\tau_- + N}$$

$$\|L^{N+1}_{\sigma,x_n}\tilde{f}\| \le c(x')\langle\xi'\rangle^{-\tau_- - N-1}\langle\xi',\mu\rangle^{r+\tau_- + N+1}$$

imply, by the interpolation inequality (A.48),

$$\|L^{N+\theta}_{\sigma,x_n}\tilde{f}\| \le \|L^{N}_{\sigma,x_n}\tilde{f}\|^{1-\theta}\|L^{N+1}_{\sigma,x_n}\tilde{f}\|^{\theta}$$

$$\le c(x')\langle\xi'\rangle^{-\tau_- - N-\theta}\langle\xi',\mu\rangle^{r+\tau_- + N+\theta} ,$$

for all $\theta \in]0,1[$. Thus (4.5.37) holds for all N and $N' \in \overline{\mathbb{R}}_+$. (Note that the parameter σ in L_{σ,x_n} is independent of μ ; this is because we shall need to carry the operator outside the Cauchy integral (4.4.19). The estimates (4.5.36) would be quite different if σ were replaced by $\kappa \sim \langle\xi',\mu\rangle$.)

Now let $\tilde{g}$ be the s.g.o. part of the resolvent, and let

$$\tilde{f}(x',x_n,y_n,\xi',\lambda) = x_n^k D_{x_n}^{k'} y_n^m D_{y_n}^{m'} D_{x'}^{\beta} D_{\xi'}^{\alpha}(\tilde{g} - \sum_{\ell<M} \tilde{g}_{-d-1-\ell}) \quad ; \tag{4.5.38}$$

it is of order

$$d' = -d - k+k' - m+m' - |\alpha| - M$$

and regularity

$$\nu' = \nu - [k-k']_+ - [m-m']_+ - |\alpha| - M ,$$

and $L_{\sigma,x_n}^N L_{\sigma,y_n}^{N'} \tilde{f}$ is likewise of order d' , but of regularity

$$\nu'' = \min\{\nu',0\} - N - N' ,$$

for any N and $N' \geq 0$. In particular, for fixed ξ' ,

$$L_{\sigma,x_n}^N L_{\sigma,y_n}^{N'} \tilde{f} \quad \text{is} \quad O(\langle\mu\rangle^{d'-\nu''}) \quad .$$

Thus the Cauchy integral formula

$$L_{\sigma,x_n}^N L_{\sigma,y_n}^{N'} \tilde{f}^{(z)} = \frac{i}{2\pi} \int_{C(\xi')} \lambda^z L_{\sigma,x_n}^N L_{\sigma,y_n}^{N'} \tilde{f} \, d\lambda \tag{4.5.39}$$

is justified (in the $L^2_{x_n,y_n}$-sense), when

$$d \operatorname{Re} z + d' - \nu'' < -d$$

(recall that $\mu = |\lambda|^{1/d}$), and the estimates (4.5.33) are valid then. This shows the proposition.

Besides (4.5.35), we note the special case $D_{x_n}^{k'} \tilde{g}^{(z)}$, which satisfies:

$$\begin{aligned} &\| L_{\sigma,x_n}^N L_{\sigma,y_n}^{N'} D_{x_n}^{k'} \tilde{g}^{(z)} \| \leq c(x') \langle\xi'\rangle^{ds+k'} \quad , \\ &\text{when} \quad N + N' + k < |ds| \quad . \end{aligned} \tag{4.5.40}$$

Let us now see how Proposition 4.5.8 applies to the generalized s.g.o.s defined from these symbols. (The spectral bounds discussed here are preserved under the coordinate transformations used in the definition of the operators on manifolds. Actually, one can define the localization so that also finer asymptotic estimates carry over, as described in [Grubb 17].)

4.5.10 Theorem. *Let* $z \in \mathbb{C}$ *with* $s = \operatorname{Re} z < 0$. *Let* $G^{(z)} = OPG(\tilde{g}^{(z)})$ *and, more generally, let*

$$F^{(z)} = OPG(b(x',\xi')x_n^k D_{x_n}^{k'} y_n^m D_{y_n}^{m'} D_{x'}^{\beta} D_{\xi'}^{\alpha} (\tilde{g}^{(z)} - \sum_{\ell<M} \tilde{g}_{ds-1-\ell}^{(z)})) \tag{4.5.41}$$

for some $b \in S^r(\mathbb{R}^{n-1})$.

1°. *Let* $\operatorname{Re} z < -1/2d$. *Then* $G^{(z)} \in \mathfrak{S}_{((n-1)(|ds|-\delta))}$, *where* $\delta = 0$ *if* $n > 2$, *and* δ *is any positive number if* $n = 2$. *In other words, the characteristic values satisfy (for* $k \in \mathbb{N}_+$*)*

$$\begin{aligned} s_k(G^{(z)}) &\le c(G)k^{-|ds|/(n-1)} && \text{if } n > 2\,; \\ s_k(G^{(z)}) &\le c(\delta,G)k^{-|ds|+\delta} && \text{if } n = 2\,, \quad \delta > 0\,; \end{aligned} \tag{4.5.42}$$

and the constants $c(G)$ *and* $c(\delta,G)$ *are estimated as in Proposition 4.5.8, with* $a = |ds|$ *resp.* $|ds|-\delta$. *In particular,* $G^{(z)}$ *is of trace class, when*

$$\operatorname{Re} z < -(n-1)/d\,. \tag{4.5.43}$$

2°. *With* $-a_0$ *denoting the "normal order" of* $F^{(z)}$

$$-a_0 \equiv ds - k+k' - m+m' - |\alpha| - M\,,$$

assume that

$$a_0 + \nu'' > \tfrac{1}{2}\,, \tag{4.5.44}$$

where $\nu'' \le 0$ *is defined in* (4.5.34), *and that the order* $-a = -a_0+r$ *of* $F^{(z)}$ *is negative,*

$$a = a_0 - r > 0\,. \tag{4.5.45}$$

Then $F^{(z)} \in \mathfrak{S}_{(p)}$ *with* $p = \max\{(n-1)/a, 1/(a_0+\nu''-\delta)\}$, *i.e., the characteristic values satisfy*

$$\begin{aligned} s_k(F^{(z)}) &\le c(F^{(z)})k^{-a/(n-1)} && \textit{if} \quad a < (a_0+\nu'')(n-1) \\ s_k(F^{(z)}) &\le c(F^{(z)},\delta)k^{-a_0-\nu''+\delta} && \textit{if} \quad a \ge (a_0+\nu'')(n-1)\,, \end{aligned} \tag{4.5.46}$$

for any $\delta > 0$.

Proof: We have that $F^{(z)} = OPG(\tilde{f}^{(z)})$, where

$$\tilde{f}^{(z)}(x',x_n,y_n,\xi') = b(x',\xi')\frac{i}{2\pi}\int_{C(\xi')} \lambda^z x_n^k D_{x_n}^{k'} y_n^m D_{y_n}^{m'} D_{x'}^{\beta} D_{\xi'}^{\alpha}(\tilde{g} - \sum_{\ell<M} \tilde{g}_{-d-1-\ell})d\lambda \quad ;$$

so by Proposition 4.5.9,

$$(4.5.47) \qquad \| D_{x'}^{\gamma} L_{\sigma,y_n}^{N} \tilde{f}^{(z)} \| \leq c\langle\xi'\rangle^{-a} \quad ,$$

for $N < a_0+\nu''$ and all γ .

Consider first $F^{(z)}$ as an operator of order $-a$. In the special case $G^{(z)}$, where $a_0 = a = |ds|$ and $\nu'' = 0$, the application of Proposition 4.5.8 requires the existence of $N \in \mathbb{R}_+$ so that

$$(4.5.48) \qquad \max\{\tfrac{1}{2} , |ds|/(n-1)\} < N < |ds| \ .$$

For $n > 2$ this is satisfied when $|ds| > \frac{1}{2}$, i.e.,

$$\operatorname{Re} z < -1/2d \ ,$$

showing 1° in this case. When $n = 2$, (4.5.48) cannot be obtained! We then regard $G^{(z)}$ as an operator of higher order, see further below. For the general operator $F^{(z)}$, Proposition 4.5.8 can be applied when there exists N satisfying

$$\max\{\tfrac{1}{2}, (a_0-r)/(n-1)\} < N < a_0+\nu'' \ ,$$

which can be obtained if $a_0+\nu'' > \frac{1}{2}$ and $a_0+\nu'' > (a_0-r)/(n-1)$; in that case one gets the optimal estimate in the first line of (4.5.46), corresponding to the true order $-a$ of $F^{(z)}$.

Now consider $F^{(z)}$ as an operator (of $S_{1,0}$-type) of order $-a'$ for some $a' \in]0,a[$ where a' can be adapted to our purposes. In the consideration of $G^{(z)}$ for $n = 2$ we need merely take $a' = |ds|-\delta$ for some $\delta > 0$, then there exists N satisfying

$$\max\{\tfrac{1}{2}, |ds|-\delta\} < N < |ds| \ ,$$

so Proposition 4.5.8 applies (always provided $|ds| > \frac{1}{2}$); this proves the rest of 1°. For more general $F^{(z)}$ we must now find N so that

$$\max\{\tfrac{1}{2}, a'/(n-1)\} < N < a_0+\nu'' \ .$$

Again $a_0+\nu'' > \frac{1}{2}$ is necessary; and we see that when $a = a_0-r \geq (a_0+\nu'')(n-1)$, we can now take

$$a' = (a_0 + \nu'' - \delta)/(n-1)$$

for any $\delta > 0$. Proposition 4.5.8 then implies the second line in (4.5.46). ▯

As an example, let

(4.5.49) $$F^{(z)} = D_{x'}^{\alpha'} D_{x_n}^{\alpha_n} G^{(z)} \quad ;$$

here $D_{x'}^{\alpha'}$ carries over to the symbol as a sum of multiplications with $(\xi')^{\beta}$ $(\beta \leq \alpha')$ combined with differential operators in x' . We can then apply the theorem with $r = |\alpha'|$ and $k' = \alpha_n$. The normal order is

$$-a_0 = ds + \alpha_n$$

and the full order is

$$-a = -a_0 + r = ds + |\alpha| .$$

As long as $ds + |\alpha| < 0$, the theorem applies, provided

$$\alpha_n < |ds| - \tfrac{1}{2} ,$$

giving the first estimate in (4.5.46) if $a < a_0\ (n-1)$ and the second estimate if $a \geq a_0\ (n-1)$ (in particular if $n = 2$) .

The consideration of such derived operators gives a step towards the treatment of composition rules involving the generalized s.g.o.s $G^{(z)}$; but these rules will in general be much less satisfactory than the rules for s.g.o.s in Boutet de Monvel's calculus.

The restriction $\operatorname{Re} z < -1/2d$ was overlooked in the first formulation of results in [Grubb 12]. It seems plausible that the estimate (4.5.42) should be extendible to all $\operatorname{Re} z < 0$ (as in Proposition 4.5.2 1° for differential operators). Here are some weaker results:

Note that since $(P^z)_\Omega$ for $\operatorname{Re} z < 0$ maps into $H^{|\operatorname{Re} z| d}(E)$ by [Seeley 1]), one has in view of Lemma A.4,

(4.5.50) $$s_k(P_\Omega^z) \leq c_z\, k^{-d|\operatorname{Re} z|/n} , \qquad \text{for} \quad k \in \mathbb{N}_+ .$$

The fact that B^z maps into $H^{\theta d}(E)$ for $\theta < |\operatorname{Re} z|$, cf. (4.4.10'), implies such an estimate for B^z with $|\operatorname{Re} z|$ replaced by θ , so we get by use of (A.75 i)

(4.5.51) $$s_k(G^{(z)}) \leq c_z\, k^{-d\theta/n} \quad \text{for} \quad k \in \mathbb{N}_+ , \quad \text{whenever} \quad \operatorname{Re} z < -\theta < 0 .$$

Moreover, if B is as in Theorem 4.4.2 and $-1/2d \leq \operatorname{Re} z < 0$,

(4.5.52) $\quad s_k(G^{(z)}) \leq c_z\ k^{-d|\mathrm{Re}z|/n}$, for $k \in \mathbb{N}_+$,

since $B^z = B^{\mathrm{Re}z}B^{z-\mathrm{Re}z}$ maps into $H^{|\mathrm{Re}z|d}$ then.

(4.5.50) and Theorem 4.5.10 give information on B^z in general, when $\mathrm{Re}z < -1/2d$:

4.5.11 Corollary. *Hypotheses as in Theorem 4.4.4. When* $z \in \mathbb{C}$ *with* $\mathrm{Re}\, z < -1/2d$, *then* $B^z \in \mathfrak{S}_{(n/d|\mathrm{Re}\, z|)}$, *and*

(4.5.53) $$s_k(B^z)k^{d|\mathrm{Re}\, z|/n} \to C'((p^0)^z,\Omega)^{d|\mathrm{Re}\, z|/n} \quad \textit{for} \quad k \to \infty .$$

In particular, B^z *is of trace class, when* $\mathrm{Re}\, z < -n/d$.

Proof: One applies Lemma A.6 1^0 with $T = (P^z)_\Omega$ and $T' = G^{(z)}$.

4.5.11' Remark. With some more efforts, one can moreover apply the method of [Grubb 17, Theorem 4.10] to find $\lim_{k\to\infty} s_k(G^{(z)})k^{d|\mathrm{Re}\, z|/(n-1)}$, under slightly more restrictive hypotheses.

The above theorem opens up for more results obtained by manipulations with operational calculus and perturbation techniques. We give two such applications here, further applications are given in Section 4.6.

The first one sharpens Proposition 4.5.3 in some particular cases not covered by Corollary 4.5.5.

4.5.12 Corollary. *Let* P *be a strongly elliptic, positive selfadjoint ps.d.o. of order* d *in* $\tilde{E}$ *having the transmission property. For* $a > 1/2d$, *the truncated ps.d.o.* $(P^{-a})_\Omega$ *(which is selfadjoint* ≥ 0*) satisfies*

(4.5.54) $$N'(t\,;(P^{-a})_\Omega) = C(p^0,\Omega)t^{n/da} + O(t^{(n-\theta')/da}) \quad \textit{for} \quad t \to \infty ,$$

where $\theta' = \min\{\theta, da/(n-1+da) - \delta\}$; *here* θ *is as in Theorem 4.5.4, and* $\delta = 0$ *if* $n > 2$, δ *arbitrary* > 0 *if* $n = 2$.

Proof: Consider $B = P_\gamma$, the Dirichlet realization of P. By Theorem 4.5.4,

$$N'(t;B^{-1}) = N(t;B) = C(p^0,\Omega)t^{n/d} + O(t^{(n-\theta)/d}) \quad \text{for} \quad t \to \infty ,$$

and hence, since $\lambda_k(B^{-a}) = \lambda_k(B^{-1})^a$,

$$N'(t;B^{-a}) = N'(t^{1/a};B^{-1}) = C(p^0,\Omega)t^{n/da} + O(t^{(n-\theta)/da}) \quad \text{for} \quad t \to \infty ,$$

which is equivalent with the statement on eigenvalues (cf. Lemma A.5)

$$(4.5.55) \quad \lambda_k(B^{-a}) = C(p^0,\Omega)^{da/n}\, k^{-da/n} + O(k^{-(da+\theta)/n}) \quad \text{for} \quad k \to \infty .$$

For $G^{(-a)}$ we have by Theorem 4.5.10 (with $\delta' = 0$ if $n > 2$, $\delta' > 0$ if $n = 2$),

$$(4.5.56) \quad s_k(G^{(-a)}) = O(k^{-(da-\delta')/(n-1)}) \quad \text{for} \quad k \to \infty .$$

Since $(P^{-a})_\Omega = B^{-a} - G^{(-a)}$, we get from (4.5.55) and (4.5.56), by Lemma A.6 2^0,

$$(4.5.57) \quad \lambda_k((P^{-a})_\Omega) = C(p^0,\Omega)^{da/n}\, k^{-da/n} + O(k^{-(da+\theta')/n}) ,$$

where we have taken (in case $n > 2$; for $n = 2$ there is a small loss)

$$p = n/da , \qquad q = n/(da+\theta) , \qquad r = (n-1)/da ,$$

and θ' is determined by

$$\frac{n}{da+\theta'} = q' = \max\{q, p\,\frac{r+1}{p+1}\} = \max\{\frac{n}{da+\theta}, \frac{n}{da}\,\frac{da+n-1}{da+n}\} ,$$

i.e.

$$\theta' = \min\{\theta, da\,\frac{da+n}{da+n-1} - da\} = \min\{\theta, da/(da+n-1)\} .$$

Here (4.5.57) is equivalent with (4.5.54), by Lemma A.5.

The corollary gives for example for $P = -\Delta$ on a bounded domain in $\mathbb{R}^3$, $a > 1/4$,

$$(4.5.58) \quad N'(t;(-\Delta)_\Omega^{-a}) = C(|\xi|^2,\Omega)t^{3/2a} + O(t^{(3-a/(1+a))/2a}) \quad \text{for} \quad t \to \infty ,$$

we think this precison is new. ($(-\Delta)^{\frac{1}{2}}$ is a prominent example of a ps.d.o. without the transmission property, cf. (4.4.9), and it is not covered by the spectral analysis in [Rempel-Schulze 4], since it is only of order 1. Here $(-\Delta)^{-\frac{1}{2}}$ is of course the ps.d.o. part of any parametrix R of a ps.d.o. boundary problem for $(-\Delta)^{\frac{1}{2}}$. As far as we know, the spectral properties of the Green and Mellin terms in $R - (-\Delta)_\Omega^{-\frac{1}{2}}$ have not been analyzed.)

The second application is an estimate of $N^+(t;B)$ and $N^-(t;B)$ individually, when B is selfadjoint, without P being strongly elliptic. In that case there is a principal term of order $t^{n/d}$ both for the positive and for the negative eigenvalues. (All odd order selfadjoint cases with the transmission property will be of this kind, and an even order example is

(4.5.59) $$P = \begin{pmatrix} -\Delta & P_{12} \\ P_{12}^* & \Delta \end{pmatrix}$$

with P_{12} of second order, chosen so that P is elliptic.)

Let P be selfadjoint, elliptic but not strongly elliptic, and invertible in $L^2(E)$, and let $B = (P+G)_T$ be a normal elliptic realization; selfadjoint and not lower bounded. In the study of the counting functions $N^{\pm}(t\,;B)$ it suffices to determine the asymptotic properties of an odd power B^{2N+1}, so we can assume that the order d is as large as we please. The positive and negative parts $\pm R_{\pm}$ of B^{-1} are defined by operational calculus:

(4.5.60) $$\begin{aligned} R_{\pm} &= \tfrac{1}{2}\,(|B|^{-1} \pm B^{-1}) \\ &= \tfrac{1}{2}((B^2)^{-\frac{1}{2}} \pm B^{-1})\ , \end{aligned}$$

here $\pm R_{\pm}$ act like B^{-1} on the positive resp. negative eigenspace, being zero on its orthogonal complement. (One can assume that B is invertible.)

Now R_+ and R_- have the ps.d.o. parts

(4.5.61) $$Q_{\pm} = \tfrac{1}{2}(|P|^{-1} \pm P^{-1})\ ,$$

that are not elliptic, so they are not directly covered by the analysis of parametrices. However, we can reconstruct $R_{\pm}$ from the operators

(4.5.62) $$R'_{\pm} = |B|^{-1} \pm \tfrac{1}{2}\,B^{-1}\ ,$$

which have elliptic ps.d.o. parts

(4.5.63) $$Q'_{\pm} = |P|^{-1} \pm \tfrac{1}{2}\,P^{-1}\ ;$$

here

(4.5.64) $$\begin{aligned} R'_+ &= \tfrac{3}{2}\,R_+ + \tfrac{1}{2}\,R_-\ , \\ R'_- &= \tfrac{1}{2}\,R_+ + \tfrac{3}{2}\,R_-\ , \end{aligned}$$

which can be solved for R_+ and R_-.

In case d is <u>even</u>, $|P|$ has the transmission property, for if $d = 2m$, the order of P^2 is $4m$, so its square root satisfies (4.4.9):

$$4m \cdot \tfrac{1}{2} + 4m = 6m \in 2\mathbb{Z}\ .$$

Then Q'_+ also has the transmission property, and since it is positive and strongly elliptic, the Dirichlet realization B_1 of $(Q'_+)^{-1}$ is elliptic and invertible, so we can compare R'_+ with that. Here we find

$$\begin{aligned} R'_+ &= |P|_\Omega^{-1} + G^{(-\frac12)} + \tfrac12 P_\Omega^{-1} + \tfrac12 G_B \\ &= Q'_{+,\Omega} + G^{(-\frac12)} + \tfrac12 G_B \\ &= B_1^{-1} + G^{(-\frac12)} + \tfrac12 G_B - G_{B_1} \quad , \end{aligned}$$

where $G^{(-\frac12)}$ is the generalized s.g.o. term in $(B^2)^{-\frac12}$, satisfying by Theorem 4.5.10,

$$s_k(G^{(-\frac12)}) \le ck^{-d/(n-1)+\delta} \text{ for } k \in \mathbb{N}_+ \quad ,$$

where $\delta = 0$ if $n > 2$ and δ is a small positive number if $n = 2$; and G_B and G_{B_1} are ordinary s.g.o.s of order $-d$ and class 0 , satisfying

$$s_k(G_B) \text{ and } s_k(G_{B_1}) \text{ are } \mathcal{O}(k^{-d/(n-1)}) \quad \text{for } k \in \mathbb{N}_+ \ .$$

Since we also have, by Theorem 4.5.4,

$$N'(t;B_1^{-1}) = C'(q_+'^0,\Omega)t^{n/d} + \mathcal{O}(t^{(n-\theta)/d}) \quad ,$$

we can apply the perturbation argument Lemma A.6 2° in the same way as in Corollary 4.5.12, obtaining

$$N'(t\,;R'_+) = C(q_+'^0,\Omega)t^{n/d} + \mathcal{O}(t^{(n-\theta')/d}) \quad , \tag{4.5.65}$$

where

$$\theta' = \min\{\theta,\ d/(d+n-1) - \delta'\} \quad ;$$

here $\delta' = 0$ if $n > 2$, and δ' is an arbitrary small positive number if $n = 2$. Since $\theta < 1$, and d can be assumed to be as large as we need, we can in fact take

$$\theta' = \theta \ .$$

R'_- satisfies similar estimates.

Now (4.5.64) implies

$$\begin{aligned} N'(t;R'_+) &= \tfrac23 N'(t;R_+) + 2N'(t;R_-) \\ N'(t;R'_-) &= 2N'(t;R_+) + \tfrac23 N'(t;R_-) \end{aligned}$$

that can be solved for $N'(t;R_+)$ and $N'(t;R_-)$, showing that they likewise satisfy estimates

$$N'(t;R_\pm) = C(p_\pm^0,\Omega)t^{n/d} + \mathcal{O}(t^{(n-\theta)/d}) \ .$$

Here $N'(t;R_\pm) = N'^{\pm}(t;B^{-1})$, and $C(p_\pm^0,\Omega) = C^\pm(p^0,\Omega)$ (cf. (4.5.3)), so we have obtained the following theorem.

4.5.13 Theorem. *Let* P *be selfadjoint, elliptic and invertible on* $\tilde{E}$, *of even order* d *and having the transmission property at* Γ. *Let* $B = (P+G)_T$ *be a normal realization of* P *that is likewise selfadjoint. The positive resp. negative eigenvalues satisfy*

$$N^{\pm}(t;B) = C^{\pm}(P^0,\Omega)t^{n/d} + O(t^{(n-\theta)/d}) \quad \text{for } t \to \infty , \tag{4.5.66}$$

with θ *defined as in Theorem 4.5.4, and* $C^{\pm}(p^0,\Omega)$ *defined by* (4.5.3).

For the odd order case, we do not have as fine an estimate of $N'(t;R'_{\pm})$, since $Q'_{\pm}$ do not have the transmission property (and are not fractional powers of operators having it). However, we do have the principal estimates, by Proposition 4.5.3

$$N'(t;Q_{\pm})t^{-n/d} \to C^{\pm}(p^0,\Omega) ,$$

which, in view of the lower order estimates of the (generalized) ps.d.o. parts of $|B|^{-1}$ and B^{-1} lead to similar principal estimates of $N'(t;R_{\pm})$, by Lemma A.6 1^0. Thus we have

4.5.14 Corollary. *When* $B = (P+G)_T$ *is a normal selfadjoint realization of a ps.d.o.* P *(of odd order) within the Boutet de Monvel calculus, then*

$$N^{\pm}(t;B)t^{-n/d} \to C^{\pm}(p^0,\Omega) \quad \text{for } t \to \infty . \tag{4.5.67}$$

The results of Theorem 4.5.13 (which was formulated in [Grubb 12]) and Corollary 4.5.14 overlap with [Kozlov 1,2] in the differential operator case, see also Section 4.6, and [Levendorskii 2].

The theorems have consequences for Douglis-Nirenberg elliptic systems (differential as well as pseudo-differential) in a similar way as in [Grubb 5, 10]. It would of course be interesting to improve the remainder estimates. For one thing, the perturbation methods used above may not be optimal (improvements may possibly be inferred from some indications at the end of [Métivier 1]). Moreover, the study of propagation of singularities for hyperbolic boundary problems, that has been a very succesful tool in differential operator cases, has not (as far as we know) been aborded yet in the case of boundary problems for pseudo-differential operators.

4.6 Implicit eigenvalue problems.

In this and the following section we present two more applications of our theory, that on one hand shed new light on some classical problems for differential operators, and on the other hand give results on generalizations of those problems to pseudo-differential settings. The first application is the investigation of eigenvalue problems of Pleijel-type. One considers problems of the form

$$(4.6.1)\qquad \begin{aligned} \lambda A_1 u &= A_0 u \quad &&\text{in } E\,,\\ Tu &= 0 &&\text{at } \Gamma\,, \end{aligned}$$

where the eigenvalues are the values λ for which there exist nonzero solutions u (eigenfunctions). Here A_0 and A_1 are Green operators in E of order r resp. $r+d$ ($r \geq 0$ and $d > 0$) and T is a trace operator, such that $(A_0)_T$ is symmetric and $(A_1)_T$ is selfadjoint.

Pioneering work on this problem was done by Å. Pleijel who treated the case $A_1 = \Delta^2$, $A_0 = D_x^{\,2} - D_y^{\,2}$ in [Pleijel 1]; and important contributions giving the principal asymptotic behavior of λ in general cases where A_1 is strongly elliptic and the domain and coefficients are allowed to be increasingly nonsmooth, were given by M.S. Birman and M.Z. Solomiak, see their survey [Birman-Solomiak 3] where the history of the problem is also explained. Remainder estimates have been obtained in [Kozlov 1,2]; and [Levendorskii 1 ,2] announces extensions to problems for Douglis-Nirenberg elliptic systems, allowing pseudo-differential operators A_1 and A_0 . (The definition of explicit realizations determined by boundary conditions is not discussed there.) [Ivrii 3] has sharper estimates in very special cases.

We begin by observing that it is very easy to discuss "characteristic values" associated with (4.6.1) in the present framework, in the sense that (4.6.1) reduces to the Green operator problem

$$(4.6.2)\qquad Rv = \lambda v \quad , \qquad R = A_0 (A_1)_T^{-1} \quad ,$$

by composition with $(A_1)_T^{-1}$ in the invertible case; and here the characteristic values $s_k(R)$ can easily be estimated. One uses Corollary 4.5.5 if A_0 is elliptic, or Proposition 4.5.3 combined with Proposition 4.5.2 3^0 if A_0 is more general, which gives the asymptotic behavior

$$(4.6.3)\qquad N'(t;R) = C'(q^0,\Omega)t^{n/d} + o(t^{(n-\theta)/d})$$

with $\theta < \frac{1}{2}$ when A_0 is elliptic and $\theta = 0$ if A_0 is not elliptic; here q^0 is the principal symbol of the ps.d.o. part of R . However, such an estimate of

$s_k(R) = \lambda_k(R^*R)^{\frac{1}{2}}$ only gives a rough information; it is quite far from an estimate of the true eigenvalues of the nonselfadjoint operator R .

To get hold of the eigenvalues themselves, we need a more symmetric reduction of (4.6.1). Here, when $(A_1)_T$ is selfajdoint positive, the obvious thing to do could be to set $u = (A_1)_T^{-\frac{1}{2}}v$, and compose the first line with $(A_1)_T^{-\frac{1}{2}}$; this gives the eigenvalue problem

(4.6.4) $$(A_1)_T^{-\frac{1}{2}}A_0(A_1)_T^{-\frac{1}{2}}v = \lambda v \quad ,$$

where the operator on the left hand side is of negative order, with a structure determined by our operational calculus. However, the composition properties of the complex powers are not sufficiently well developed to give a very precise result here; recall that $(A_1)_T^{-\frac{1}{2}}$ is usually not a standard Green operator.

On the other hand, there is for the interesting case of the Dirichlet boundary condition another method, where $(A_1)_T^{-\frac{1}{2}}$ is replaced by a Green operator, using the zero order operational calculus, as described in Remark 3.4.5; leading to a straightforward application of our spectral results for Green operators in Section 4.5.

We here use an auxiliary operator R_{-m} as defined in Remark 3.2.15; in fact we take a fixed $\mu \geq \mu_0$ and set $R_{-m} = R_{-m,\mu}$ in (3.2.74). Then R_{-m} defines a homeomorphism

(4.6.5) $$R_{-m} : H^s(E) \xrightarrow{\sim} \mathring{H}^m(E) \cap H^{s+m}(E)$$

for all $s \geq 0$, and

(4.6.6) $$R_{-m} = Q_{-m,\Omega} + G_{-m}$$

where Q_{-m} is a ps.d.o. of order $-m$ and G_{-m} is a s.g.o. of order $-m$ and class 0 ; both are polyhomogeneous elements of the standard Boutet de Monvel calculus.

Consider now the problem (4.6.1) with the Dirichlet condition (cf. the notation (A.55) ff.)

(4.6.7) $$\begin{aligned} \lambda A_1 u &= A_0 u \quad , \\ \rho_m u &= 0 \quad , \end{aligned}$$

where we assume that $A_1 = P_{1,\Omega} + G_1$ is of even order $2m$ and class $\leq m$, positive and formally selfadjoint; such that the Dirichlet realization $(A_1)_\gamma$ is positive selfadjoint in $L^2(E)$, with domain $D((A_1)_\gamma) = \mathring{H}^m(E) \cap H^{2m}(E)$. As for $A_0 = P_{0,\Omega} + G_0$, we assume that it is of order $r < 2m$ and class $\leq m$, and is such that its restriction to $\overset{\circ}{H}^m(E) \cap H^{2m}(E)$ is a symmetric operator in $L^2(E)$.

The considerations in Remark 3.4.5 can now be applied to $B = (A_1)_\gamma$: We set

$$B' = R^*_{-m}(A_1)_\gamma R_{-m} \quad , \tag{4.6.8}$$

it is positive selfadjoint and of order and class 0 , so that we can define

$$R' = R_{-m}(B')^{-\frac{1}{2}} \quad ; \tag{4.6.9}$$

then an insertion of $u = R'v$ in (4.6.7) and composition with $(R')^*$, leads to the eigenvalue problem

$$\begin{aligned} \tilde{R}v &= \lambda v \ , \\ \text{with} \quad \tilde{R} &= R'^*A_0R' \quad . \end{aligned} \tag{4.6.10}$$

Here $\tilde{R}$ is a Green operator of order $r-2m = -d$, and of class 0 since it is composed of continuous mappings $R' = L^2(E) \to \overset{\circ}{H}^m(E)$, $A_0 : \overset{\circ}{H}^m(E) \to H^{m-r}(E) \subset$ $\subset H^{-m}(E)$ and $R'^* : H^{-m}(E) \to L^2(E)$. Then the results of Section 4.5 can be applied to $\tilde{R}$. It is of the form

$$\begin{aligned} \tilde{R} &= \tilde{Q}_\Omega + \tilde{G} \ , \quad \text{with} \\ \tilde{Q} &= (Q^*_{-m}P_1Q_{-m})^{-\frac{1}{2}}\ Q^*_{-m}P_0Q_{-m}(Q^*_{-m}P_1Q_{-m})^{-\frac{1}{2}} \end{aligned} \tag{4.6.11}$$

(one can assume that P_1 and Q_{-m} are bijective in $\tilde{E}$). For the description of the constants $C'^{\pm}(\tilde{q}^0,\Omega)$ we note that the eigenvalues of $\tilde{q}^0(x,\xi)$ can be identified with the eigenvalues of the equation

$$\lambda p_1^0(x,\xi) = p_0^0(x,\xi) \tag{4.6.12}$$

by retracing the steps in the above construction; so the j-th positive resp. negative eigenvalue of $\tilde{q}^0(x,\xi)$ satisfies

$$\lambda_j^{\pm}(\tilde{q}^0(x,\xi)) = \lambda_j^{\pm}(p_1^0(x,\xi)^{-\frac{1}{2}}p_0^0(x,\xi)p_1^0(x,\xi)^{-\frac{1}{2}}) \ . \tag{4.6.13}$$

Then we also have

$$\begin{aligned} C'^{\pm}(\tilde{q}^0,\Omega) &= (2\pi)^{-n}\int_{T^*(\Omega)} N'^{\pm}(1;\tilde{q}^0(x,\xi))\,d\xi dx \\ &= (2\pi)^{-n}\int_{T^*(\Omega)} N'^{\pm}(1;p_1^0(x,\xi)^{-\frac{1}{2}}p_0^0(x,\xi)p_1^0(x,\xi)^{-\frac{1}{2}})\,d\xi dx \\ &= C'^{\pm}((p_1^0)^{-\frac{1}{2}}p_0^0(p_1^0)^{-\frac{1}{2}},\ \Omega) \quad , \end{aligned} \tag{4.6.14}$$

and similarly

$$C'(\tilde{q}^0,\Omega) = C'((p_1^0)^{-\frac{1}{2}}p_0^0(p_1^0)^{-\frac{1}{2}}, \Omega) \ . \tag{4.6.15}$$

When there are no ellipticity conditions on P_0, we now simply get, by application of Proposition 4.5.3 to $\tilde{Q}_\Omega$ and Proposition 4.5.2 3^0 to $\tilde{G}$, combined by use of Lemma A.6,

$$N'(t;\tilde{R}) = C'(\tilde{q}^0,\Omega)t^{n/d} + o(t^{n/d}) \qquad \text{for} \quad t \to \infty \ , \tag{4.6.16}$$

which gives a principal asymptotic estimate of the number of eigenvalues with absolute value $\geq 1/t$.

Now let P_0 be elliptic. Then the estimate (4.6.16) can be drastically improved by use of Corollary 4.5.5, which shows:

$$N'(t;\tilde{R}) = C'(\tilde{q}^0)t^{n/d} + O(t^{(n-\theta)/d}) \qquad \text{for} \quad t \to \infty \ , \tag{4.6.17}$$

for any $\theta < \frac{1}{2}$. The estimate extends to $\theta < 1$ if the parametrix $\tilde{P}$ of $\tilde{Q}$,

$$\tilde{P} = \tilde{Q}^{(-1)} = (Q_{-m}^* P_1 Q_{-m})^{\frac{1}{2}} \ (Q_{-m})^{-1} P_0^{\ (-1)} (Q_{-m}^*)^{-1} (Q_{-m}^* P_1 Q_{-m})^{\frac{1}{2}} \tag{4.7.18}$$

is equivalent with a scalar differential operator, which will of course rarely be the case; see however also Remark 4.5.7. Recall that the study of the problem (4.6.1) with $A_0 = I$ is covered already by Section 4.5.

If $\tilde{R}$ is ≥ 0 (or ≤ 0), (4.6.17) counts the positive (negative) eigenvalues. However, if $\tilde{R}$ has both positive and negative eigenvalues, the function $N'(t;\tilde{R})$ counts the positive and negative eigenvalues with absolute value $> 1/t$ together, and one can ask for a separate estimate. Here, when P_0 is strongly elliptic (in particular r is even), the ps.d.o. $\tilde{P}$ in (4.6.18) is strongly elliptic selfadjoint of order d (when we for $P_0^{(-1)}$ take a selfadjoint parametrix), and a comparison of $\tilde{R}$ with the solution operator for the Dirichlet realization $\tilde{P}_\gamma$ will show that

$$\begin{aligned} N'^{+}(t;\tilde{R}) &= C'(\tilde{q}^0,\Omega)t^{n/d} + O(t^{(n-\theta)/d}) \ , \\ N'^{-}(t;\tilde{R}) &= O(t^{(n-1)/d}) \ , \end{aligned} \tag{4.6.19}$$

just as in the proof of Theorem 4.5.4.

When P_0 is elliptic of even order but not strongly elliptic, we can get fine estimates by appealing to Theorem 4.5.13. Here we observe that if $\tilde{P}$ has a selfadjoint realization $\tilde{A}_{T_1}$, say, then $\tilde{A}_{T_1}^{-1} - \tilde{R}$ is a s.g.o. of order $-d$ and class 0 , so the estimates (4.5.66) for $N^{\pm}(t;\tilde{A}_{T_1})$ imply similar estimates for $N'^{\pm}(t;\tilde{R})$ by the usual perturbation argument:

$$(4.6.20)\qquad \begin{aligned} N'^{+}(t;\tilde{R}) &= C'^{+}(\tilde{q}^0,\Omega)t^{n/d} + O(t^{(n-\theta)/d}) , \\ N'^{-}(t;\tilde{R}) &= C'^{-}(\tilde{q}^0,\Omega)t^{n/d} + O(t^{(n-\theta)/d}) , \qquad \text{for } t \to \infty . \end{aligned}$$

The conditions for this is somewhat implicit, but we leave it at that, to stay within the most immediate applications of the results of Section 4.5. We have shown:

<u>4.6.1 Theorem</u>. *Let* $r \in \mathbb{N}$ *and* $d \in \mathbb{N}_+$, *with* $r+d$ *even, equal to* $2m$. *Let* $A_1 = P_{1,\Omega} + G_1$ *be elliptic of order* $2m$ *and class* m, *and assume that the Dirichlet realization* $(A_1)_\gamma$ *in* E *is a positive selfadjoint operator in* $L^2(E)$. *Let* $A_0 = P_{0,\Omega} + G_0$ *be of order* r *and class* m , *and assume that the restriction of* A_0 *to* $H^{2m}(E) \cap \mathring{H}^m(E)$ *is symmetric in* $L^2(E)$. *Then the problem*

$$(4.6.21)\qquad \begin{aligned} \lambda A_1 u &= A_0 u \\ \rho_m u &= 0 \end{aligned}$$

has the same set of eigenvalues as the selfadjoint operator $\tilde{R}$ *of order* $-d$ *and class* 0 *defined by* (4.6.8-10), *such that the eigenfunctions* u *of* (4.6.21) *correspond to the eigenfunctions* v *of* $\tilde{R}$ *by* $u = R'v$. *These eigenvalues have the asymptotic behavior* (4.6.16), *where the constant satisfies* (4.6.15).

When P_0 *is elliptic, the estimate can be sharpened to* (4.6.17). *Here, if* P_0 *is strongly elliptic, the positive and negative eigenvalues are estimated separately by* (4.6.19). *If* P_0 *is elliptic of even order but not strongly elliptic, and the selfadjoint parametrix* $\tilde{P}$ (4.6.18) *of the ps.d.o. part of* $\tilde{R}$ *has a selfadjoint realization, then the positive and negative eigenvalues of* $\tilde{R}$ *are estimated separately by* (4.6.20), *where the constants are defined by* (4.6.14).

The result is contained in the result of [Kozlov 2] when A_1 and A_0 are even-order <u>differential</u> operators (in fact Kozlov can then get $\theta = \frac{1}{2}$, and he obtains some remainder estimates also when A_0 is non-elliptic); but we think the result is new for odd order A_0 . [Ivrii 3]has obtained sharper remainder estimates (with $\theta=1$) in cases where A_1 resp. A_0 are principally equal to $(-\Delta)^p$ resp. $(-\Delta)^q$, $p > q \geq 0$. [Levendorskii 1,2] gives some even-order results where A_1 and A_0 are allowed to be nonlocal. The transmission property is not required there, but the existence of the realization is abstractly assumed, without further analysis.

For general operators in the Boutet de Monvel calculus, our result is completely new. We also trust that the proof is of interest, because of the simple way in which it is built up from results that are known in the case $A_0 = I$.

Very similar considerations can be applied to the problem (4.6.7), when A_0 is strongly elliptic instead of A_1 . More precisely, we assume that $(A_0)_\gamma$ is selfadjoint positive, and that $(A_1)_\gamma$ is selfadjoint, elliptic and bijective, but not necessarily positive. A_0 is of even order $r = 2r'$, and A_1 is of order $2m = 2(r'+d')$, where $2d' = d$. Since the functions u in the Dirichlet domain for A_1 also lie in the Dirichlet domain for A_0 , we can define

$$B'' = R^*_{-r'}(A_0)_\gamma R_{-r'} \tag{4.6.22}$$

as a positive selfadjoint Green operator of order and class 0 , and we set

$$R'' = R_{-r'}(B'')^{-\frac{1}{2}} \quad , \tag{4.6.23}$$

so that (4.6.7) by insertion of $u = R''w$ and application of R''^* reduces to the problem

$$\begin{aligned} &\lambda\tilde{A}w = w \ , \\ &\tilde{T}w = 0 \ , \\ &\text{with } \tilde{A} = R''^*A_1R'' \ , \text{ and} \\ &\tilde{T} = \{\tilde{T}_0,\dots,\tilde{T}_{d'-1}\} \ , \qquad \tilde{T}_j = \gamma_{r'+j}R''v \ . \end{aligned} \tag{4.6.24}$$

The conditions $\gamma_j R''w = 0$ for $j < r'$ are automatically satisfied since R'' maps $L^2(E)$ into $\overset{\circ}{H}{}^{r'}(E)$, so they can be omitted. The remaining conditions are normal boundary conditions of order ≥ 0 (since the ps.d.o. part of R'' is elliptic), hence of regularity $\geq \frac{1}{2}$. The realization $\tilde{B} = \tilde{A}_{\tilde{T}}$ is bijective because of the way it is constructed from bijective operators, and it is elliptic, since the same bijectiveness holds on the boundary symbol level. It is also selfadjoint, since we have for $w \in D(\tilde{B})$,

$$(\tilde{B}w,w) = (R''^*A_1R''w,w) = \langle A_1R''w,\overline{R''w}\rangle \in \mathbb{R} \ , \tag{4.6.25}$$

(duality between $H^{-m}(E)$ and $\overset{\circ}{H}{}^m(E)$) , using that $\rho_m R''w = 0$ then (note that $D(\tilde{B}) \subset H^d(E)$) . We can then apply Theorem 4.5.13 to $\tilde{B}$, obtaining

4.6.2 Theorem. *Let* $r' \in \mathbb{N}$ *and* $d' \in \mathbb{N}_+$, *let* $m = r'+d'$, *and let* $r = 2r'$, $d = 2d'$. *Let* $A_1 = P_{1,\Omega} + G_1$ *be of order* $2m$ *and class* m, *with* P_1 *elliptic but not necessarily strongly elliptic, and assume that the Dirichlet realization* $(A_1)_\gamma$ *is invertible, and selfadjoint as an operator in* $L^2(E)$. *Let* $A_0 = P_{0,\Omega} + G_0$ *be of order* r *and class* r', *with* P_0 *strongly elliptic selfadjoint, and assume that the Dirichlet realization* $(A_0)_\gamma$ *is selfadjoint positive. Then the problem* (4.6.21) *has the same set of eigenvalues as the inverse of* $\widetilde{B} = \widetilde{A}_{\widetilde{T}}$, *defined by* (4.6.22-24), *such that the eigenfunctions* u *of* (4.6.21) *correspond to the eigenfunctions* w *of* (4.6.24) *by* $u = R''w$. *These eigenvalues have the asymptotic behavior for* $t \to \infty$:

$$(4.6.26) \qquad N^{\pm}(t;\widetilde{B}) = C^{\pm}((p_0^0)^{-\frac{1}{2}}p_1^0(p_0^0)^{-\frac{1}{2}}, \Omega)t^{n/d} + O(t^{(n-\theta)/d}) \quad ;$$

for $t \to \infty$, *where* θ *is any number* $< \frac{1}{2}$ *(and* θ *can be taken* < 1 *if the ps.d.o. part* $\widetilde{P}$ *of* $\widetilde{A}$ *is a scalar differential operator, see also Remark 4.5.7). When* P_1 *is strongly elliptic,* $N^-(t;\widetilde{B})$ *is* $O(t^{(n-1)/d})$ *or finite.*

V.A. Kozlov has obtained such results for <u>differential</u> operators A_1 and A_0, cf. [Kozlov 1], where A_1 is also allowed to be of odd order, and more general boundary conditions are included (without an exact analysis of these, only some hypotheses on selfadjointness and domain regularity); he can get $\theta = \frac{1}{2}$. The result also has an overlap with [Levendorskii 1,2]. [Ivrii 3] has $\theta = 1$ in special cases.

Finally, the methods extend to Douglis-Nirenberg elliptic systems in a very easy way. Let $m_1 \le m_2 \le \dots \le m_N$ and $r_1' \le r_2' \le \dots \le r_N'$ be nonnegative integers, and let

$$(4.6.27) \qquad \begin{aligned} A_0 &= (A_{0,st})_{s,t=1,\dots,N} = P_{0,\Omega} + G_0 \,, \\ A_1 &= (A_{1,st})_{s,t=1,\dots,N} = P_{1,\Omega} + G_1 \,, \end{aligned}$$

where the $A_{0,st}$ are Green operators from bundles E_t to bundles E_s over $\bar{\Omega}$ of order $r_s' + r_t'$ and the $A_{1,st}$ are Green operators from E_t to E_s of order $m_s + m_t$, respectively, all singular Green terms being of class $\le m_1$. We assume that $m_s - r_s' = d_s' \ge d'$ for all s, for some $d' > 0$, and we set $2d' = d$. Consider the eigenvalue problem

$$(4.6.28) \qquad \begin{aligned} \lambda A_1 u &= A_0 u \qquad && (u = (u_1,\dots,u_N)) \,, \\ \rho_{m_t} u_t &= 0 \qquad \text{for} && t = 1,\dots, N \,. \end{aligned}$$

The concepts of ellipticity and strong ellipticity are here defined in the Douglis-Nirenberg sense, where e.g. the principal symbol of A_1 is defined as the matrix of principal symbols $\sigma_{m_s+m_t}(A_{1,st})$ corresponding to the indicated orders of the elements in A_1. (Cf. e.g. [Grubb 5].) The boundary condition in (4.6.28) is the Dirichlet condition for A_1; the corresponding realization will be denoted $(A_1)_\gamma$.

When $(A_1)_\gamma$ is positive selfadjoint, we proceed as in the proof of Theorem 4.6.1, with some modifications: Instead of R_{-m} we take the diagonal matrix

$$(4.6.29)\qquad R = \begin{pmatrix} R_{-m_1} & 0 & \dots & 0 \\ 0 & R_{-m_2} & & 0 \\ \vdots & \vdots & \ddots & \vdots \\ 0 & 0 & \dots & R_{-m_N} \end{pmatrix},$$

with R_{-m_t} acting like (4.6.5) in E_t. Then

$$(4.6.30)\qquad B' = R^*(A_1)_\gamma R$$

is positive selfadjoint and of order and class 0, so that we can define

$$(4.6.31)\qquad R' = R(B')^{-\frac{1}{2}},$$

satisfying

$$R'^*(A_1)_\gamma R' = (B')^{-\frac{1}{2}}R^*(A_1)_\gamma R(B')^{-\frac{1}{2}} = I .$$

Insertion of $u = R'v$ in (4.6.28) and composition with R'^* gives the problem

$$(4.6.32)\qquad \begin{aligned} &\tilde{R}v = \lambda v , \\ \text{with}\quad &\tilde{R} = R'^*A_0R' , \end{aligned}$$

here $\tilde{R} = \tilde{Q}_\Omega + \tilde{G} = (\tilde{R}_{st})_{s,t=1,\dots,N}$, with each $\tilde{R}_{st} = \tilde{Q}_{st,\Omega} + \tilde{G}_{st}$ being a Green operator of order $-d_s' - d_t'$ and class 0.

Now let J be the subset of the indices $\{1,\dots,N\}$ for which $d_s' = d'$, and let J' be the complement $\{1,\dots,N\}\setminus J$ (where $d_s' \geq d'+1$). Then we can write $\tilde{R}$ in blocks

$$(4.6.33)\qquad \tilde{R} = \begin{pmatrix} \tilde{R}_{JJ} & 0 \\ 0 & 0 \end{pmatrix} + \begin{pmatrix} 0 & \tilde{R}_{JJ'} \\ \tilde{R}_{J'J} & \tilde{R}_{J'J'} \end{pmatrix},$$

after a rearrangement of rows and columns preserving the selfadjointness.

If $\tilde{R}_{JJ}$ has elliptic ps.d.o. part $\tilde{Q}_{JJ}$, and if the parametrix $\tilde{P}_{JJ}$ of $\tilde{Q}_{JJ}$ has a selfadjoint elliptic realization, then we find by the techniques preceding Theorem 4.6.1 the estimates

$$N'^{\pm}(t;\tilde{R}_{JJ}) = C'^{\pm}(\tilde{q}^0_{JJ},\Omega)t^{n/d} + O(t^{(n-\theta)/d}) \quad \text{for} \quad t \to \infty , \tag{4.6.34}$$

with $\theta < \frac{1}{2}$ in general ($\theta < 1$ if $\tilde{P}_{JJ}$ is a scalar differential operator, etc.). This is in particular obtained if P_0 is Douglis-Nirenberg strongly elliptic, for then $\tilde{Q}_{JJ}$ is likewise strongly elliptic (of order -d), and the Dirichlet realization of $\tilde{P}_{JJ}$ is selfadjoint; in this case the negative eigenvalues of $\tilde{R}_{JJ}$ even satisfy the sharper estimate

$$N'^{-}(t;\tilde{R}_{JJ}) = O(t^{(n-1)/d}) \quad \text{for} \quad t \to \infty . \tag{4.6.35}$$

When merely $\tilde{Q}_{JJ}$ is elliptic, we have at least by Corollary 4.5.5

$$N'(t;\tilde{R}_{JJ}) = C'(\tilde{q}^0_{JJ},\Omega)t^{n/d} + O(t^{(n-\theta)/d}) \quad \text{for} \quad t \to \infty ; \tag{4.6.36}$$

and if no ellipticity is assumed, we have the principal estimate, by Propositions 4.5.2-3,

$$N'(t;\tilde{R}_{JJ}) = C'(\tilde{q}^0_{JJ},\Omega)t^{n/d} + o(t^{n/d}) \quad \text{for} \quad t \to \infty . \tag{4.6.37}$$

Propositions 4.5.2-3 also give that the remaining part of $\tilde{R}$, which is of order $\leq$ -d-1, satisfies

$$N'\left(t; \begin{pmatrix} 0 & \tilde{R}_{JJ'} \\ \tilde{R}_{J'J} & \tilde{R}_{J'J'} \end{pmatrix}\right) = O(t^{n/(d+1)}) \quad \text{for} \quad t \to \infty . \tag{4.6.38}$$

We can now apply Lemmas A.5-6 to the sum (4.6.33). When $\tilde{Q}_{JJ}$ is not elliptic, one gets an estimate like (4.6.37) for $\tilde{R}$. When $\tilde{Q}_{JJ}$ is elliptic, the combination of (4.6.36) with (4.6.38) gives, with

$$p = \frac{n}{d} , \qquad s = \frac{n-\theta}{d} , \qquad r = \frac{n}{d+1} ,$$

that s' in Lemma A.6 is determined by

$$s' = \max\left\{\frac{n-\theta}{d} , \frac{n}{d+1}\,\frac{n/d+1}{n/(d+1)+1}\right\} = \frac{n-\theta'}{d} ,$$

where

$$\theta' = \min\left\{\theta , \frac{n}{n+d+1}\right\} . \tag{4.6.39}$$

The same exponents are used when (4.6.34) is taken together with (4.6.38). Finally, (4.6.35) and (4.6.38) together give the exponent $(n-\theta'')/d$, cf. (A.75 i),

$$\theta'' = \min\left\{1\ ,\ \frac{n}{d+1}\right\} . \tag{4.6.40}$$

The estimates can of course be improved if the last matrix in (4.6.33) is of lower order $-\ell \leq -d-1$; then

$$\theta' = \min\left\{\theta,\ n\,\frac{\ell-d}{n+\ell}\right\} \quad \text{and} \quad \theta'' = \min\left\{1,\ n\,\frac{\ell-d}{\ell}\right\} . \tag{4.6.41}$$

In all cases, $\theta'' \geq \theta'$, and θ is any value $< \frac{1}{2}$ in general and any value < 1 if $\tilde{P}_{JJ}$ happens to be a scalar differential operator (cf. also Remark 4.5.7).

The final result is

4.6.3 Theorem. *Consider the eigenvalue problem* (4.6.28) *for multi-order systems* (4.6.27) *with entries* $A_{0,st}$ *of order* $r'_s + r'_t$ *and* $A_{1,st}$ *of order* $m_s + m_t$, *for nonnegative integers* $m_1 \leq m_2 \leq \dots \leq m_N$ *and* $r'_1 \leq r'_2 \leq \dots \leq r'_N$ *with*

$$m_s = r'_s + d'_s \quad \textit{for all} \quad s\ ; \tag{4.6.42}$$

here it is assumed that all s.g.o. terms are of class $\leq m_1$ *and*

$$d' = \min\{d'_s \mid s = 1,\dots,N\} > 0 . \tag{4.6.43}$$

Let $(A_1)_\gamma$ *be positive selfadjoint. Then the eigenvalues (and eigenfunctions) of* (4.6.28) *can be identified with the eigenvalues (and images of eigenfunctions) of* $\tilde{R}$ *constructed in* (4.6.29-32). *With* $\tilde{Q}_{JJ}$ *denoting the ps.d.o. part of the block* $\tilde{R}_{JJ}$ *in* $\tilde{R}$ *of maximal order* $-d = -2d'$, *one has the following asymptotic estimates for* $t \to \infty$: *In general*

$$N'(t;\tilde{R}) = C'(\tilde{q}^0_{JJ},\Omega)t^{n/d} + o(t^{n/d}) . \tag{4.6.44}$$

When $\tilde{Q}_{JJ}$ *is elliptic, then furthermore*

$$N'(t;\tilde{R}) = C'(\tilde{q}^0_{JJ},\Omega)t^{n/d} + O(t^{(n-\theta')/d}) , \tag{4.6.45}$$

and when $\tilde{Q}_{JJ}$ *is strongly elliptic, then*

$$\begin{aligned} N'^{+}(t;\tilde{R}) &= C'(\tilde{q}^0_{JJ},\Omega)t^{n/d} + O(t^{(n-\theta')/d}) \\ N'^{-}(t;\tilde{R}) &= O(t^{(n-\theta'')/d}) . \end{aligned} \tag{4.6.46}$$

When $\widetilde{Q}_{JJ}$ *is elliptic, not strongly elliptic, and its parametrix* $\widetilde{P}_{JJ}$ *has a selfadjoint realization, then*

$$(4.6.47)\qquad N'^{\pm}(t;\widetilde{R}) = C'^{\pm}(\tilde{q}^0_{JJ},\Omega)t^{n/d} + O(t^{(n-\theta')/d}) \ .$$

Here θ' *and* θ'' *are defined by* (4.6.41) *if the blocks in* $\widetilde{R}$ *besides* $\widetilde{R}_{JJ}$ *are of order* $\leq -\ell$; *in particular if* $\ell = d+1$ *they are defined by* (4.6.39) *and* (4.6.40).

A comparable result is announced in [Levendorskii 1], however, he assumes $\widetilde{Q}_{JJ}$ elliptic, he does not allow A_0 to be a multi-order system, and his exponents on the remainder terms are somewhat weaker than ours, valid for ps.d.o.s of a general kind but without singular Green terms.

Also the proof of Theorem 4.6.2 extends to the multi-order case: If A_0 is strongly elliptic instead of A_1 , one can reduce A_0 to order and class 0 by use of

$$(4.6.48)\qquad S = \begin{pmatrix} R_{-r_1'} & 0 & \cdots & 0 \\ 0 & R_{-r_2'} & \cdots & 0 \\ \vdots & \vdots & \ddots & \vdots \\ 0 & 0 & \cdots & R_{-r_N'} \end{pmatrix} ;$$

here we have that

$$(4.6.49)\qquad B'' = S^*(A_0)_\gamma S$$

is positive selfadjoint and of order and class 0, so that we can define

$$(4.6.50)\qquad R'' = S(B'')^{-\frac{1}{2}} \ ,$$

satisfying

$$R''^*(A_0)_\gamma R'' = I \ .$$

Insertion of $u = R''w$ in (4.6.28) now gives the problem

$$(4.6.51)\qquad \begin{aligned} &\lambda\widetilde{A}w = w \ , \\ &\widetilde{T}w = 0 \ , \\ &\text{with } \widetilde{A} = R''^*A_1R'' \ , \quad \text{and} \\ &(\widetilde{T}w)_s = \{\gamma_{r_s'}, \ldots, \gamma_{m_s-1}\}\,(R''w)_s \qquad \text{for} \quad s = 1,\ldots,N \end{aligned}$$

(column vector); here we note that $\gamma_0(R''w)_s, \ldots, \gamma_{r'_s-1}(R''w)_s$ are zero because of the properties of R''. The realization $\tilde{B} = \tilde{A}_{\tilde{T}}$ is seen to be elliptic selfadjoint and bijective, when $(A_1)_\gamma$ is so, just as in the proof of Theorem 4.6.2. Here $\tilde{A}$ is a multi-order system $\tilde{A} = (\tilde{A}_{st})_{s,t=1,\ldots,N}$ with $\tilde{A}_{st}$ of order $d'_s + d'_t$. The solution operator $\tilde{R}' = \tilde{B}^{-1}$ has a structure very similar to that of $\tilde{R}$ in (4.6.33), with a block $\tilde{R}'_{JJ}$ of highest order $-d$ and the other blocks of order $-\ell \leq -d-1$. The ps.d.o. part $\tilde{Q}_{JJ}$ of $\tilde{R}_{JJ}$ need not be elliptic, but it certainly is so, if A_1 is strongly elliptic. An investigation similar to that preceding Theorem 4.6.3 gives

4.6.4 Theorem. *Consider the eigenvalue problem* (4.6.28) *for multi-order systems, where the orders are as in Theorem 4.6.3. Assume that* A_0 *is strongly elliptic, with a positive selfadjoint Dirichlet realization, and assume that* $(A_1)_\gamma$ *is elliptic, selfadjoint and bijective, but not necessarily strongly elliptic. Then the eigenvalues (and eigenfunctions) of* (4.6.28) *can be identified with the inverse eigenvalues (and the images of eigenfunctions) of the realization* $\tilde{B} = \tilde{A}_{\tilde{T}}$ *constructed in* (4.6.48-51). *With* $\tilde{Q}'_{JJ}$ *denoting the ps.d.o. part of the block* $\tilde{R}'_{JJ}$ *in the solution operator* $\tilde{R}' = \tilde{B}^{-1}$, *one has similar asymptotic estimates for* $t \to \infty$ *as in Theorem 4.6.3, with* $\tilde{R}_{JJ}$ *replaced by* $\tilde{R}'_{JJ}$ *and* $\tilde{Q}_{JJ}$ *replaced by* $\tilde{Q}'_{JJ}$ *(also in the hypotheses).*

The result has a certain overlap with [Levendorskii 1].

4.6.5 Remark. The estimates of θ' and θ'' in Theorem 4.6.3 and 4.6.4, resulting from perturbations by lower order terms, can possibly be improved by means of the perturbation theorems of [Markus and Matsaev 1] (instead of Lemmas A.5-6), as used by [Kozevnikov 1] for Douglis-Nirenberg elliptic systems on boundaryless manifolds.

4.7 Singular perturbations.

The parametrix construction in this work depends very clearly on the property of positive regularity ν of the entering systems (see e.g. Theorem 3.2.3, Remarks 2.1.19, 3.2.16 etc.). But there also exist problems with negative regularity that are of interest, Example 1.5.13 describes one of them. Now, although the problem apparently falls outside the present theory, we can in fact find some reductions that transform it to a problem with positive regularity, so that it can be treated as an application after all. In this final section we describe in a systematic way a class of problems containing Example 1.5.13; one could say that these problems are regularizable. Another case of regularizable problems will be treated in a joint work with V. Solonnikov [Grubb-Solonnikov 1].

The application of our theory that we shall describe now is in a way related to the preceding application, because it has a parameter in the same place as the eigenvalue parameter there. It is the study of the solution u_ε of a "perturbed" boundary problem

$$(4.7.1) \qquad \begin{aligned} \varepsilon^d A_1 u_\varepsilon + A_0 u_\varepsilon &= f \quad && \text{in } E\,, \\ T_0 u_\varepsilon &= \varphi_0 \quad && \text{at } \Gamma\,, \\ T_1 u_\varepsilon &= \varphi_1 \quad && \text{at } \Gamma\,; \end{aligned}$$

and especially its behavior for $\varepsilon \to 0$, in relation to the solution of the "unperturbed" problem

$$(4.7.2) \qquad \begin{aligned} A_0 u &= f \quad && \text{in } E\,, \\ T_0 u &= \varphi_0 \quad && \text{at } \Gamma\,. \end{aligned}$$

Here A_0 is of order $r \geq 0$ (the case $r = 0$ is quite simple, so the main efforts are directed towards the case $r > 0$), and A_1 is of order $r+d$, $d > 0$.

It is assumed that the system $\{A_0, T_0\}$ is elliptic, that $\{A_1, T_0, T_1\}$ is elliptic, and that $\{\varepsilon^d A_1 + A_0, T_0, T_1\}$ has an ellipticity property for $\varepsilon \in]0,1]$, all the operators belonging to the Boutet de Monvel calculus. (More assumptions will be made below.) Note that when the first line in (4.7.1) is multiplied by μ^d, where $\mu = \varepsilon^{-1}$, we arrive at a problem where the ps.d.o. part is like the non-parameter-elliptic cases discussed in Remark 2.1.19.

Problems like (4.7.1-2) were treated in the fundamental paper [Vishik-Lyusternik 1] and have been further studied by many other people; let us mention the works [Huet 1-6] and [Greenlee 1] that take advantage of abstract principles of functional analysis, and the works of [Demidov 1], [Eskin 1], [Frank 1], [Frank-Wendt 1-3] and [Wendt 1], that use symbolic calculus of pseudo-differential operators; see also the references in the mentioned works.

We shall now show how the solution operator can be represented in a very simple way by use of a combination of the Boutet de Monvel calculus and the parameter-dependent calculus of the present work, allowing in particular a thorough discussion of the behavior for $\varepsilon \to 0$. Here the problems (4.7.1-2) can be rather general pseudo-differential boundary problems, but the interest of the results lies not only in this extension but also in the application to more classical problems. (Besides the discussion in [Frank 1] of coerciveness and limiting properties of symbols for differential boundary problems, Frank and Wendt have discussed the invertibility properties more generally by use of a factorization method for ps.d.o.s with rational symbols, obtaining interesting results particularly on the principal symbol level ([Frank-Wendt 1-3], [Wendt 1]). However, their presentation of the needed parameter-dependent version of the Boutet de Monvel calculus is far from complete - lacking e.g. a precise definition and analysis of singular Green operators - and their description of the variable coefficient case gives only sparse information on how the effects of negative regularity are handled.)

The present method consists of a reduction of (4.7.1), by use of parameter-independent operators, to a parameter-dependent problem of positive regularity, where the results of Chapter 3 can be applied, somewhat like in the quoted works of D. Huet. We neither need smooth factorizations of symbols nor rationality.

The method can be generalized to situations with a more complicated ε-dependence than (4.7.1); this will be taken up elsewhere, since we here just want to give a brief account of the main ideas. We also restrict the attention here to the most important parameter-independent Sobolev norm estimates (with powers of ε in front), that can be straightforwardly deduced. (The method will also allow estimates in the mixed parameter-dependent Sobolev norms

(4.7.3) $$\|(1-\Delta)^{s/2}(1-\varepsilon^2\Delta)^{s'/2}u\|_{L^2},$$

that are used in the works of A. Demidov, G.I. Eskin, L.S. Frank and W.D. Wendt, but it is not necessary to go via these spaces to get control of remainders.)

Our basic hypotheses are the following.

4.7.1 Assumption. *All the mentioned operators belong to the Boutet de Monvel calculus and are polyhomogeneous, as presented in Chapter 1.*

1° $A_0 = P_{0,\Omega} + G_0$ *is of order* $r \in \mathbb{N}$, *with the s.g.o.* G_0 *of class* 0. T_0 *is a trace operator associated with the order* r, *the largest order in* T_0

being denoted k_0' $(< r)$; *more precisely*

(4.7.4) $\quad T_0 = \{T_{0,j}\}_{0\le j\le k_0'}$ *with* $T_{0,j}\colon H^s(E) \to H^{s-j-\frac12}(F_j^0)$, $\dim F_j^0 = M_j$,

for $s > j+\frac12$; *and* T_0 *maps sections in* E *into sections in* $F^0 = \oplus_{0\le j\le k_0'} F_j^0$ *(of dimension* $M^0 = \Sigma_{0\le j\le k_0'} M_j$*). The system* $\{A_0, T_0\}$ *is elliptic and invertible, with the inverse denoted*

$$(4.7.5) \qquad \begin{pmatrix} A_0 \\ T_0 \end{pmatrix}^{-1} = \begin{pmatrix} R_0 & K_0 \end{pmatrix} ;$$

here $K_0 = \{K_{0,j}\}_{0\le j\le k_0'}$ *with* $K_{0,j}$ *of order* $-j$.

2^0 $A_1 = P_{1,\Omega} + G_1$ *is of order* $r+d$ $(d \in \mathbb{N}_+)$, *with the s.g.o.* G_1 *of class* 0 . T_1 *is a* <u>*normal*</u> *trace operator associated with the order* $r+d$; *the smallest resp. largest order of nontrivial terms in* T_1 *are denoted* k_1 *resp.* k_1' $(< r+d$, *cf. also* (4.7.7)*)*, *so*

(4.7.6) $\quad T_1 = \{T_{1,j}\}_{k_1\le j\le k_1'}$ *with* $T_{1,j}\colon H^s(E) \to H^{s-j-\frac12}(F_j^1)$, $\dim F_j^1 = M_j$,

for $s > j+\frac12$; *and* T_1 *maps section* E *into sections in* $F^1 = \oplus_{k_1\le j\le k_1'} F_j^1$ *(of dimension* $M^1 = \Sigma_{k_1\le j\le k_1'} M_j$*). The system* $\{A_1, T_0, T_1\}$ *is elliptic.*

3^0 *The orders of the trace operators satisfy*

$$(4.7.7) \qquad 0 \le k_0' < k_1 \le r \quad \textit{and} \quad 0 \le k_1' - k_1 < d ,$$

and the system $\{\varepsilon^d A_1 + A_0, T_0, T_1\}$ *has the following ellipticity properties for* $\varepsilon > 0$ *: The principal interior symbol*

(4.7.8) $\quad \varepsilon^d p_1^0(x,\xi) + p_0^0(x,\xi)$ *is bijective for* $|\xi| = 1$, $\varepsilon > 0$, *all* x .

The principal boundary symbol operator

$$(4.7.9) \qquad \begin{pmatrix} \varepsilon^d a_1^0(x',\xi',D_n) + a_0^0(x',\xi',D_n) \\ t_0^0(x',\xi',D_n) \\ t_1^0(x',\xi',D_n) \end{pmatrix} : \mathscr{S}(\overline{\mathbb{R}}_+)^N \to \begin{matrix} \mathscr{S}(\overline{\mathbb{R}}_+)^N \\ \times \\ \mathbb{C}^{M^0} \\ \times \\ \mathbb{C}^{M^1} \end{matrix}$$

is bijective for $|\xi'| = 1$, $\varepsilon > 0$, *all* x' .

The assumptions are obviously satisfied in the case where P_0 and P_1 are strongly elliptic ps.d.o.s in Σ (of orders $r = 2r'$ resp. $r+d = 2m$) , with

$P_0 + P_0^*$ positive, G_0 and G_1 are zero (or have a suitably small principal symbol), and

$$T_0 = \{\gamma_0, \dots, \gamma_{r'-1}\}$$
(4.7.10)
$$T_1 = \{\gamma_{r'}, \dots, \gamma_{m-1}\},$$

defining Dirichlet conditions (or T_0 and T_1 are close to this), cf. Sections 1.5 and 1.7.

Observe that when $r = 0$, T_0 is void, so that A_0 itself must be a bijective elliptic operator in $L^2(E)$. The procedure described further below amounts in this case to the replacement of u_ε by $w = A_0^{-1} u_\varepsilon$, giving a simple reduction to a resolvent construction.

4.7.2 Example. Let $A_0 = -\Delta$ and $A_1 = \Delta^2$, and consider the two problems

(4.7.11)
$$\begin{aligned} \varepsilon^2\Delta^2 u_\varepsilon - \Delta u_\varepsilon &= f && \text{in } \Omega, \\ \gamma_0 u_\varepsilon &= 0 && \text{at } \Gamma, \\ \gamma_1 u_\varepsilon &= 0 && \text{at } \Gamma; \end{aligned}$$

and

(4.7.12)
$$\begin{aligned} \varepsilon^2\Delta^2 u'_\varepsilon - \Delta u'_\varepsilon &= f && \text{in } \Omega, \\ \gamma_0 u'_\varepsilon &= 0 && \text{at } \Gamma, \\ \gamma_2 u'_\varepsilon &= 0 && \text{at } \Gamma; \end{aligned}$$

that are perturbations of the Dirichlet problem for $-\Delta$

(4.7.13)
$$\begin{aligned} -\Delta u &= f \\ \gamma_0 u &= 0. \end{aligned}$$

By use of the inverse

(4.7.14)
$$\begin{pmatrix} -\Delta \\ \gamma_0 \end{pmatrix}^{-1} = (R_0 \quad K_0),$$

we can easily reduce (4.7.12) to a system where our resolvent theory applies: One can write u'_ε as $R_0 w$ for some w because of the condition $\gamma_0 u'_\varepsilon = 0$; by insertion of this and multiplication by $\mu^2 = \varepsilon^{-2}$, (4.7.12) reduces to

(4.7.15)
$$\begin{aligned} -\Delta w + \mu^2 w &= g \\ \gamma_2 R_0 w &= 0, \end{aligned}$$

where $g = \mu^2 f$. Here $\gamma_2 R_0 = \gamma_0 + T_0'$ for some trace operator T_0' of order and class 0 . It is seen e.g. by consideration of sesquilinear forms, that

$$\begin{pmatrix} \varepsilon^2(D_n^2 + |\xi'|^2)^2 + D_n^2 + |\xi'|^2 \\ \gamma_0 \\ \gamma_2 \end{pmatrix} : H^4(\overline{\mathbb{R}}_+) \to \begin{matrix} L^2(\mathbb{R}_+) \\ \times \\ \mathbb{C} \\ \times \\ \mathbb{C} \end{matrix}$$

is bijective for $\xi' \neq 0$ and $\varepsilon > 0$; then it follows by composition with $r_0(x',\xi',D_n)$ that the boundary symbol operator for (4.7.15) satisfies (I) and (II) in Definition 1.5.5. Also (III) holds, since the limit for $\xi' \to 0$ of the strictly homogeneous boundary symbol operator for (4.7.15) is

$$\begin{pmatrix} D_n^2 + \mu^2 \\ \gamma_0 \end{pmatrix} , \tag{4.7.16}$$

which is invertible for $\mu > 0$. Then the theorems of Section 3.3 can be applied to (4.7.15); we get unique solvability for $\mu \geq \mu_0$ (for some $\mu_0 \geq 0$) and μ-dependent estimates of the solution w in terms of g , that can be used to give estimates of u_ε' and $u_\varepsilon' - u$ in terms of f , with precise informations on the ε-dependence. This will be done in a general setting further below.

When the same procedure is applied to (4.7.11), one arrives at the problem

$$\begin{aligned} -\Delta w + \mu^2 w &= g , \\ \gamma_1 R_0 w &= 0 , \end{aligned} \tag{4.7.17}$$

where $\gamma_1 R_0$ is of order -1 , hence <u>not normal</u>; in fact it is of <u>negative regularity</u> $-\frac{1}{2}$. To this, our resolvent construction cannot be directly applied, as noted also in Example 1.5.13 and Remark 3.2.16. (Problems of the kind (4.7.17) are discussed in [Frank-Wendt 1,2] by use of operators acting in the mixed parameter-dependent Sobolev spaces with norms (4.7.3).) However, a little more use of Boutet de Monvel's calculus in the case (4.7.11) will lead to a new reduction, where we do obtain a problem with positive regularity, solvable with the methods of the present book. We now turn to the general explanation of the method; the particular example will be taken up again in Examples 4.7.10 and 4.7.16 below.

The main idea is to compose A_0 with an auxiliary operator, a "regularizing factor", before the elimination of A_0 in the first line of (4.7.1), such that the trace operator that results from T_1 in this procedure has <u>nonnegative</u> order and remains <u>normal</u>. The regularizing factor is of the following kind:

Recall from Remark 3.2.15 (taking a large fixed μ there) that there exists, for each $\ell \in \mathbb{N}$, an elliptic ps.d.o. $\Lambda_-^{-\ell}$ of order $-\ell$ in $\tilde{E}$ which defines a homeomorphism (by restriction to Ω)

$$(4.7.18) \qquad \Lambda_{-,\Omega}^{-\ell} : H^s(E) \xrightarrow{\sim} H^{s+\ell}(E) \qquad \text{for} \quad s \geq 0$$

and satisfies

$$(4.7.19) \qquad L(\Lambda_-^{-\ell}, P) = 0 \quad \text{for any ps.d.o. } P .$$

Let us for brevity denote

$$(4.7.20) \qquad \Lambda_{-,\Omega}^{-\ell} = \Lambda_{-\ell} \quad .$$

Now consider (4.7.1-2), satisfying Assumption 4.7.1. Take an integer $\ell \in [0,r]$ such that

$$(4.7.21) \qquad k_0' < r-\ell \leq k_1 \quad \text{and} \quad k_1' - r + \ell < d$$

(one can in particular take $\ell = r - k_1$, since (4.7.7) holds). An application of $\Lambda_{-\ell}$ to the first line in (4.7.1) gives the equivalent problem

$$(4.7.22) \qquad \begin{aligned} \varepsilon^d \Lambda_{-\ell} A_1 u_\varepsilon + \Lambda_{-\ell} A_0 u_\varepsilon &= \Lambda_{-\ell} f \;, \\ T_0 u_\varepsilon &= \varphi_0 \;, \\ T_1 u_\varepsilon &= \varphi_1 \;. \end{aligned}$$

The associated unperturbed problem is

$$(4.7.23) \qquad \begin{aligned} \Lambda_{-\ell} A_0 u &= \Lambda_{-\ell} f \;, \\ T_0 u &= \varphi_0 \;, \end{aligned}$$

whose solution operator will be denoted

$$(4.7.24) \qquad \begin{pmatrix} \Lambda_{-\ell} A_0 \\ T_0 \end{pmatrix}^{-1} = (R_0' \quad K_0') \;.$$

Since $k_0' < r-\ell$, and $\Lambda_{-\ell} A_0$ is of class 0 (by (4.7.19)), the system $\{\Lambda_{-\ell} A_0, T_0\}$ is well-defined on $H^{r-\ell}(E)$, and R_0' is of class 0 (and order $-r+\ell$). Note moreover, that

$$(4.7.25) \qquad R_0 = R_0' \Lambda_{-\ell} \quad \text{and} \quad K_0 = K_0' \;,$$

in view of (4.7.5).

Setting $\Lambda_{-\ell}A_0u_\varepsilon = w$, we can write u_ε on the form

$$u_\varepsilon = R_0'w + K_0\varphi_0 \; . \tag{4.7.26}$$

We insert this in (4.7.22) and multiply the first line with μ^d , where $\mu = \varepsilon^{-1}$; then we get the equations for w

$$\begin{aligned} \Lambda_{-\ell}A_1R_0'w + \mu^d w &= \mu^d\Lambda_{-\ell}f - \Lambda_{-\ell}A_1K_0\varphi_0 \; , \\ T_1R_0'w &= \varphi_1 - T_1K_0\varphi_0 \; . \end{aligned} \tag{4.7.27}$$

Let us denote

$$\Lambda_{-\ell}A_1R_0' = A = P_\Omega + G \tag{4.7.28}$$

and observe that it is of order d , with ps.d.o. part

$$P = \Lambda_{-\ell}P_1\,(\Lambda_{-\ell}P_0)^{-1} \tag{4.7.29}$$

(in the parametrix sense) and s.g.o. part G <u>of class</u> 0 in view of (4.7.19), so that A is well-defined on $L^2(E)$. We furthermore denote

$$T_1R_0' = T \; , \tag{4.7.30}$$

which is <u>normal</u>, since T_1 is normal of order $\geq k_1$, and R_0' is of order $-r+\ell \geq -k_1$ (cf. (4.7.21)) with elliptic ps.d.o. part $(\Lambda_{-\ell}P_0)^{-1}$,cf. (1.2.13) ; here

$$\begin{aligned} &T = \{T_{j-r+\ell}\}_{k_1\leq j\leq k_1'} \; , \text{ with} \\ &T_{j-r+\ell} \text{ of order } j-r+\ell \, , \text{ going from } E \text{ to } F_j^1 \; , \end{aligned} \tag{4.7.31}$$

and the largest order $k_1' - r+\ell$ is less than d by (4.7.21). Altogether, we have arrived at a problem

$$\begin{aligned} (A + \mu^d)w &= g \; , \\ Tw &= \psi \; , \end{aligned} \tag{4.7.32}$$

with A of order d and class 0 , and T normal, associated with the order d . (A larger order of T could be allowed, at the cost of augmenting the order of the first line; we leave that out for simplicity.) In particular, $\{A,T\}$ <u>is of regularity</u> $\geq \frac{1}{2}$.

We now want to apply the resolvent construction of Chapter 3 to the reduced problem (4.7.32), so let us check how many of the assumptions in Definition 3.1.3 (or Definition 1.5.5) are verified.

4.7.3 Lemma. *Consider the statements in Definition 3.1.3, applied to* $p^h(x,\xi) + \mu^d$ *and*

$$a^h(x',\xi',\mu,D_n) = \begin{pmatrix} p^h(x',0,\xi',D_n)_\Omega + g^h(x',\xi',D_n) + \mu^d \\ t^h(x',\xi',D_n) \end{pmatrix} , \tag{4.7.33}$$

the strictly homogeneous interior symbol, resp. boundary symbol operator associated with (4.7.32).

Property (I) *follows for* $\mu > 0$ *from* (4.7.8) *and the ellipticity of* P_0 , *and it follows for* $\mu = 0$ *from the ellipticity of* P_1 *and* P_0 .

Property (II) *follows for* $\mu > 0$ *from* (4.7.9) *and the ellipticity of* $\{A_0,T_0\}$, *and it follows for* $\mu = 0$ *from the ellipticity of* $\{A_1,T_0,T_1\}$ *and* $\{A_0,T_0\}$.

Normality of T *follows from the normality of* T_1, *the ellipticity of* $\{A_0,T_0\}$ *and the fact that* $r-\ell \leq k_1$; *a formula for the coefficients of the top normal order terms is given in* (4.7.34) *below.*

Proof: One simply goes through the reductions mentioned above, on the interior symbol level for (I), and on the boundary symbol level for (II). The normality was already observed further above; for precision we note that the coefficient $s_{j-r+\ell,j-r+\ell}$ of $\gamma_{j-r+\ell}$ in $T_{j-r+\ell}$ is

$$\begin{aligned} s_{j-r+\ell,j-r+\ell}(x') &= s_{1,jj}(x')(\lambda_-^{-\ell}(x',0,0,1)p_0^0(x',0,0,1))^{-1} , \\ &= c\, s_{1,jj}(x')p_0^0(x',0,0,1)^{-1} , \end{aligned} \tag{4.7.34}$$

where $s_{1,jj}(x')$ is the coefficient of γ_j in $T_{1,j}$, and c is a nonzero constant (equal to $(\mu_1-i)^{\ell}$, when $\lambda_-^{-\ell}$ is chosen as in (3.1.14) with μ fixed $= \mu_1 \geq 1$ and $\varepsilon \leq 1$, cf. Remark 3.2.15); one may here also consult (1.2.10-11). □

As observed in Proposition 1.5.9, the properties (I) and (II) together with normality imply (III) in the scalar case and in certain vectorial cases, so in these cases $\{A+\mu^d,T\}$ is parameter-elliptic without further hypotheses. For the remaining cases, we add the following assumption.

4.7.4 Assumption. *The following differential boundary symbol operator is bijective for all* x' , *all* $\mu > 0$:

$$(4.7.35)\quad a^h(x',0,\mu,D_n) = \begin{pmatrix} p^h(x',0,0,1)D_n^d + \mu^d I \\ \\ \{s_{j-r+\ell,j-r+\ell}(x')\gamma_{j-r+\ell}\}_{k_1\le j\le k_1'} \end{pmatrix} : \mathcal{S}(\overline{\mathbb{R}}_+)^N \to \begin{matrix}\mathcal{S}(\overline{\mathbb{R}}_+)^N \\ \times \\ \mathbb{C}^{M^1}\end{matrix} ,$$

where $p^h(x',0,0,1) = p_1^h(x',0,0,1)p_0^h(x',0,0,1)^{-1}$, *and* $s_{j-r+\ell,j-r+\ell}(x')$ *is described in* (4.7.34).

That $a^h(x',0,\mu,D_n)$ has this form, was shown in Section 1.5 (around (1.5.37)). We use that

$$(4.7.36)\qquad p^h(x,\xi) = p_1^h(x,\xi)p_0^h(x,\xi)^{-1} ,$$

which holds since $\lambda_-^{-\ell}$ can be regarded as a scalar factor.

Since (4.7.35) is a differential operator, the condition for invertibility can of course be formulated as the invertibility of a certain matrix function (of x' and μ) , as in the classical theory.

Altogether we have, by Lemma 4.7.3 and Proposition 1.5.9:

4.7.5 Proposition. *Let the systems* $\{A_0,T_0\}$ *and* $\{A_1,T_0,T_1\}$ *satisfy Assumption 4.7.1, choose* ℓ *so that* (4.7.21) *holds, and define* $A = P_\Omega+G$ *and* T *by* (4.7.28-31). *When* $N = 1$, *and when* $N > 1$ *and* $p_1^0(x',0,0,1)p_0^0(x',0,0,1)^{-1}$ *is diagonalizable and each fiber dimension* M_j *is either* N *or* 0 *for* $k_1 \le j \le k_1'$, *one can then conclude that* $\{A+\mu^d,T\}$ *is parameter-elliptic (for* $\mu \ge 0$) ; *in these cases Assumption 4.7.4 automatically holds. In general* $\{A+\mu^d,T\}$ *is parameter-elliptic when Assumption 4.7.4 is included in the hypotheses.*

4.7.6 Remark. In the treatment [Frank 1] of differential operator problems (4.7.1-2) (and other ε-dependent problems, with details for constant coefficient cases), a Coerciveness Condition was defined, consisting of certain hypotheses (i)-(iv). These appear to be related to, respectively: (i) the ellipticity of $\{A_0,T_0\}$, (ii) the normality and invertibility of (4.7.35), (iii) the ellipticity of $\{A_1,T_0,T_1\}$, and (iv) the invertibility property (4.7.9) for $\{\varepsilon^d A_1+A_0,T_0,T_1\}$.

4.7.7 Remark. A slight relaxation of the class hypotheses in Assumption 4.7.1 can be allowed, namely, we could take:

(4.7.37) $\quad$ class of $G_0 \leq r-\ell$; $\quad$ class of $G_1 \leq d$.

Then we arrive at a problem (4.7.32) where G is of class $\leq d$ (instead of class 0), so that A is still of regularity $\geq \frac{1}{2}$; and the considerations from the resolvent construction can be applied to this problem too.

When all the assumptions hold, the system defining (4.7.32) is invertible for $\mu \geq \mu_0$ when μ_0 is sufficiently large, by Theorem 3.3.1, with inverse

$$\begin{pmatrix} A+\mu^d \\ T \end{pmatrix}^{-1} = (R_\mu \quad K_\mu) \qquad \text{for } \mu \geq \mu_0 \tag{4.7.38}$$

of regularity $\frac{1}{2}$ (at least). Here

$$\begin{aligned} &K_\mu = \{K_{\mu,j-r+\ell}\}_{k_1 \leq j \leq k_1'} \quad \text{, with} \\ &K_{\mu,j-r+\ell} \text{ of order } -j+r-\ell \text{ , going from } F_j^1 \text{ to } E \text{ .} \end{aligned} \tag{4.7.39}$$

Note that R_μ is the inverse of $B+\mu^d$ (is the resolvent of B), where B is the <u>realization</u> $B = A_T$.

We can then find the solution of (4.7.1) by retracing the steps in the above construction: (4.7.32) must be solved with g and ψ equal to the right hand terms in (4.7.27), which gives

$$w = R_\mu(\mu^d \Lambda_{-\ell} f - \Lambda_{-\ell} A_1 K_0 \varphi_0) + K_\mu(\varphi_1 - T_1 K_0 \varphi_0) \quad ,$$

and then u_ε is determined from (4.7.26) by

$$\begin{aligned} u_\varepsilon = &\; R_0' R_\mu \mu^d \Lambda_{-\ell} f \\ &+ [K_0 - R_0'(R_\mu \Lambda_{-\ell} A_1 + K_\mu T_1) K_0] \varphi_0 \\ &+ R_0' K_\mu \varphi_1 \;, \quad \text{for } \varepsilon \in \,]0,\varepsilon_0] \;, \end{aligned} \tag{4.7.40}$$

where $\varepsilon_0 = \mu_0^{-1}$ (understood as $+\infty$ when $\mu_0 = 0$).

So it is seen that (4.7.1) is uniquely solvable for $\varepsilon \in \,]0,\varepsilon_0]$, and the solution operator is

$$\begin{aligned} &\begin{pmatrix} \varepsilon^d A_1 + A_0 \\ T_0 \\ T_1 \end{pmatrix}^{-1} = (M_\varepsilon \quad N_{\varepsilon,0} \quad N_{\varepsilon,1}) \text{ , with} \\ &M_\varepsilon = R_0' \mu^d R_\mu \Lambda_{-\ell} \;, \\ &N_{\varepsilon,0} = K_0 - R_0'(R_\mu \Lambda_{-\ell} A_1 + K_\mu T_1) K_0 \;, \\ &N_{\varepsilon,1} = R_0' K_\mu \;, \qquad \varepsilon = \mu^{-1} \;. \end{aligned} \tag{4.7.41}$$

This solution operator will now be compared with the solution operator for the unperturbed problem (4.7.2), cf. (4.7.5). Let us write

$$M_\varepsilon = R_0 + M'_\varepsilon \ ,\qquad N_{\varepsilon,0} = K_0 + N'_{\varepsilon,0} \ , \tag{4.7.42}$$

and observe the following formulas, where $\varepsilon = \mu^{-1} > 0$ throughout:

$$N'_{\varepsilon,0} = -R'_0(R_\mu \Lambda_{-\ell} A_1 + K_\mu T_1)K_0 \tag{4.7.43}$$

and (cf. (4.7.25))

$$M'_\varepsilon = R'_0 \mu^d R_\mu \Lambda_{-\ell} - R_0 = R'_0(\mu^d R_\mu - I)\Lambda_{-\ell} \ . \tag{4.7.44}$$

Here (4.7.38) implies on one hand

$$\mu^d R_\mu - I = \mu^d R_\mu - (A+\mu^d)R_\mu = -AR_\mu \ , \tag{4.7.45}$$

and on the other hand

$$\mu^d R_\mu - I = \mu^d R_\mu - R_\mu(A+\mu^d) - K_\mu T = -R_\mu A - K_\mu T \ , \tag{4.7.46}$$

applied to functions, for which the operators have a sense (note that (4.7.46) requires more smoothness than (4.7.45)). This gives the two representations of M'_ε :

$$M'_\varepsilon = -R'_0 A R_\mu \Lambda_{-\ell} \ , \tag{4.7.47}$$

$$M'_\varepsilon = -R'_0 R_\mu A \Lambda_{-\ell} - R'_0 K_\mu T \Lambda_{-\ell} \ . \tag{4.7.48}$$

In view of the definitions of A and T , we have

$$A\Lambda_{-\ell} = \Lambda_{-\ell} A_1 R_0 \quad \text{and} \quad T\Lambda_{-\ell} = T_1 R_0 \ , \tag{4.7.49}$$

so the formula (4.7.48) can also be written

$$M'_\varepsilon = -R'_0 R_\mu \Lambda_{-\ell} A_1 R_0 - R'_0 K_\mu T_1 R_0 \ . \tag{4.7.50}$$

In particular,

$$M_\varepsilon = R_0 - R'_0(R_\mu \Lambda_{-\ell} A_1 + K_\mu T_1)R_0 \ . \tag{4.7.51}$$

Jointly with (4.7.43), this gives the following formula linking the "perturbed solution" (4.7.40) directly with the "unperturbed solution" $u = R_0 f + K_0 \varphi$:

$$(4.7.52) \qquad u_\varepsilon = u - R_0'(R_\mu\Lambda_{-\ell}A_1 + K_\mu T_1)u + R_0'K_\mu\varphi_1 \ ,$$

showing how u_ε deviates from u by one term depending on u and another term depending on the extra boundary condition.

We call M_ε', $N_{\varepsilon,0}'$ and $N_{\varepsilon,1}'$ the deviation operators. They can be represented by still other formulas in special cases.

Altogether, we have shown:

4.7.8 Theorem. *Let* A_0, A_1, T_0 *and* T_1 *be given, satisfying Assumption 4.7.1, and add Assumption 4.7.4 in the relevant cases accounted for in Proposition 4.7.5, such that* $\{A+\mu^d,T\}$ *is parameter-elliptic. Then there exists an* $\varepsilon_0 > 0$ *such that the problem* (4.7.1) *is uniquely solvable for all* $\varepsilon \in \,]0,\varepsilon_0]$, *with the solution operator* $(M_\varepsilon \quad N_{\varepsilon,0} \quad N_{\varepsilon,1})$ *described explicitly by* (4.7.41-52).

4.7.9 Example. A case that has often been studied is where A_0 and A_1 are of even order $r = 2r'$ resp. $r+d = 2(r'+d') = 2m$. (In fact, for scalar differential operators, r and d must be even when $n > 2$, in order to have ellipticity.) In particular, the Dirichlet problem, where (cf. (A.55) ff.)

$$(4.7.53) \qquad \begin{aligned} T_0 &= \rho_{r'} = \{\gamma_0,\gamma_1,\dots,\gamma_{r'-1}\} \\ T_1 &= \{\gamma_{r'},\dots,\gamma_{r'+d'-1}\} \ , \end{aligned}$$

has been studied. Here our Assumption 4.7.1 is satisfied when P_0 and P_1 are strongly elliptic and G_0 and G_1 are zero, by Theorem 1.7.2; in particular, (4.7.8) and (4.7.9) hold since $\mathrm{Re}(\varepsilon^d p_1^0 + p_0^0) > 0$ for $\varepsilon > 0$, $|\xi| \geq 1$. Note that $k_0' = r'-1$, $k_1 = r'$ and $k_1' = r'+d'-1$, so the only possible choice of ℓ is $\ell = r'$. With this choice, A defined in (4.7.28) is of order $d = 2d'$, and T defined in (4.7.30) is of the form

$$(4.7.54) \qquad T = \{T_{j-r'}\}_{r' \leq j < r'+d'} = S\rho_{d'} + T' \ ,$$

with T' of class 0 and S a bijective triangular ps.d.o. in $(E_\Gamma)^{d'}$, (with bijective morphisms in the diagonal), i.e. T is of Dirichlet-type associated with A .

Now Assumption 4.7.4 is automatically satisfied if $N = 1$, and for $N > 1$ it likewise holds, thanks to the strong ellipticity of the limit operator $a^h(x',0,\mu,D_n)$. So Theorem 4.7.8 applies.

In this example, one can also include nonzero singular Green operators G_0 and G_1, and one can include lower normal order terms and nonlocal terms in T_0 and T_1 ; here the parameter-ellipticity is satisfied if the resulting systems are strongly elliptic (as discussed in Section 1.7) or are suitable perturbations from strongly elliptic cases to more general systems.

Other boundary problems for strongly elliptic ps.d.o.s can likewise be treated in this way, when they define strongly elliptic (m-coercive) realizations. This includes operators in N-dimensional bundles, for which the trace operators T_0 and T_1 may have entries mapping into lower dimensional bundles (a specific example is given below). The m-coerciveness assures the validity of (III) which is then not covered by Proposition 1.5.9.

One can of course also give non-strongly elliptic and odd order examples (e.g. with ps.d.o.s constructed from $\lambda_\pm^m$, cf. Remarks 3.1.8 and 3.2.15).

<u>4.7.10 Example</u>. We here continue Example 4.7.2. The singular perturbation problem (4.7.11), that was problematic there, is now covered by the above considerations on Dirichlet problems, here $\ell = 1$. Note that in (4.7.12), one has the choice between taking $\ell = 0$ or 1 , giving two different representations of the solution. - We can just as well let u be vector-valued in these examples.

Let us include a case where u is an N-vector ($N > 1$) and the trace operators have rank $< N$: Let $\bar\Omega \subset \mathbb{R}^n$ and let $Q(x')$ be a morphism in the trivial bundle $\Gamma\times\mathbb{C}^N$, of the form of an orthogonal projection with rank $N-1$ for each $x' \in \Gamma$. (For example, if $N = n$ and u is <u>real</u>, so that $u|_\Gamma$ is a section in $\Gamma\times\mathbb{R}^n$, there is an interesting case where one takes for Q the projection onto the tangential vectors; this range is usually a nontrivial bundle.) In view of the formula

$$(4.7.55)\quad (\Delta^2 u,v)_{L^2(\Omega)^N} = (\Delta u,\Delta v)_{L^2(\Omega)^N} - i(\gamma_1\Delta u,\gamma_0 v)_{L^2(\Gamma)^N} - i(\gamma_0\Delta u,\gamma_1 v)_{L^2(\Gamma)^N} ,$$

the system of trace operators

$$(4.7.56)\quad T_0 u = \begin{pmatrix} Q\gamma_0 u \\ (I-Q)\gamma_1 u \end{pmatrix} , \qquad T_1 u = \begin{pmatrix} Q\gamma_0 \Delta u \\ (I-Q)\gamma_1 \Delta u \end{pmatrix} ,$$

defines a 2-coercive realization of Δ^2 as well as of $\varepsilon^2\Delta^2-\Delta$ for any $\varepsilon > 0$; and Theorem 4.7.8 applies to the singular perturbation problem for $\varepsilon^2\Delta^2-\Delta$ with this choice of trace operators. See also Example 1.6.16.

The solution operators are in each case given by the appropriate version of formulas (4.7.41-52).

4.7.11 Remark. It is possible to give a series development of the solution operator, based on the exact formula (3.2.52) and the fact that the μ-dependent operators $\{A+\mu^d, T\}$ and $(R_\mu \quad K_\mu)$ are of regularity $\geq \frac{1}{2}$:

Let $(R_\mu^0 \quad K_\mu^0)$ be an operator constructed from the principal parametrix symbols for $\{A+\mu^d, T\}$, and let

$$(4.7.57) \qquad S_\mu = I - \begin{pmatrix} A+\mu^d \\ T \end{pmatrix} (R_\mu^0 \quad K_\mu^0) \ ;$$

then S_μ is of order -1 and regularity $d-1$ in the ps.d.o. part, regularity $\geq -\frac{1}{2}$ in the other part, and class 0 . It is seen, like in the consideration after (3.2.49), that the norm of S_μ as an operator in $H^{s,\mu}(E)\times H^{s-\frac{1}{2},\mu}(F^1)$, considered for a fixed $s \geq 0$, is $O(\mu^{-\frac{1}{2}})$ (or satisfies a better estimate, if T is of strictly positive order). Now one has as in (3.2.52)

$$(4.7.58) \qquad (R_\mu \quad K_\mu) = (R_\mu^0 \quad K_\mu^0) \sum_{k=0}^{\infty} S_\mu^{\circ k} \qquad \text{for} \quad \mu \geq \mu_0$$

(μ_0 sufficiently large), where the norm of $S_\mu^{\circ k}$ in $H^{s,\mu}(E)\times H^{s-\frac{1}{2},\mu}(E)$ is $\leq (c\mu^{-\frac{1}{2}})^k$, for each k , and $S_\mu^{\circ k}$ is of order $-k$. This gives an expansion of $(R_\mu \quad K_\mu)$ in a series with terms of decreasing size compared to μ and decreasing order. Expressed in ε , the terms are $\leq (c\varepsilon^{\frac{1}{2}})^k$. For M_ε, M'_ε, $N'_{\varepsilon,0}$ and $N_{\varepsilon,1}$, the expansions of R_μ and K_μ can be inserted in the formulas, giving expansions in terms of decreasing order, and decreasing size $\leq c_1(c_2\varepsilon^{\frac{1}{2}})^k$, when considered as mappings between suitable ε-dependent Sobolev space families. The expansions are truly convergent, not just in an asymptotic sense. (This is an exact, global construction. We recall on the other hand, that the negligible terms in the symbolic calculus are at best $O(\mu^{-d-\nu})$, where ν is the regularity, cf. Lemma 2.4.3. Note however, that the full symbolic calculus gives the operators as sums of operators with homogeneous symbols, whereas S_μ is more complicated.)

We use the above representations to show some estimates of the operators. For brevity, we concentrate on estimates in non-parametrized Sobolev spaces, that are fundamental for applications. The finest estimates are obtained for the deviation operators M'_ε , $N'_{\varepsilon,0}$ and $N_{\varepsilon,1}$; then these imply estimates for M_ε and $N_{\varepsilon,0}$ when the constant terms R_0 and K_0 are added.

Since R_μ is of order $-d$, class 0 and regularity $\geq \frac{1}{2}$, it satisfies the following estimates for any $s \geq 0$ and $t \in [0,d+s]$ (by Theorems 2.5.4-5 or by Corollary 3.3.2):

$$\langle\mu\rangle^{s+d-t}\|R_\mu f\|_t \leq c_1\|R_\mu f\|_{s+d,\mu} \leq c_2\|f\|_{s,\mu} \leq c_3(\langle\mu\rangle^s\|f\|_0 + \|f\|_s) ,$$

and hence in particular:

(4.7.59) $\|R_\mu f\|_t \leq c(\langle\mu\rangle^{t-d}\|f\|_0 + \langle\mu\rangle^{t-d-s}\|f\|_s)$ for $s \geq 0$, $0 \leq t \leq d+s$.

As for K_μ , we have that it is of the form (4.7.39) and of regularity $\geq \frac{1}{2}$, so in view of Theorem 2.5.1,

$$\langle\mu\rangle^{j-r+\ell+\frac{1}{2}+s-t}\|K_{\mu,j-r+\ell}\psi\|_t \leq c_1\|K_{\mu,j-r+\ell}\psi\|_{j-r+\ell+\frac{1}{2}+s,\mu}$$

$$\leq c_2\|\psi\|_{s,\mu} \leq c_3(\langle\mu\rangle^s\|\psi\|_0 + \|\psi\|_s) \quad \text{for } s \geq 0,\ 0 \leq t \leq j-r+\ell+\tfrac{1}{2}+s ,$$

and hence in particular,

(4.7.60) $\|K_{\mu,j-r+\ell}\psi\|_t \leq c(\langle\mu\rangle^{t-\frac{1}{2}-j+r-\ell}\|\psi\|_0 + \langle\mu\rangle^{t-\frac{1}{2}-j+r-\ell-s}\|\psi\|_s)$

for $s \geq 0$, $0 \leq t \leq j-r+\ell+\frac{1}{2}+s$.

The first representation (4.7.47) of M'_ε gives

$$\|M'_\varepsilon f\|_t = \|R'_0 A R_\mu \Lambda_{-\ell} f\|_t \leq c_1\|R_\mu\Lambda_{-\ell}f\|_{t-r+\ell+d} , \quad \text{for } t-r+\ell > -\tfrac{1}{2}$$

$$\leq c_2(\langle\mu\rangle^{t-r+\ell}\|\Lambda_{-\ell}f\|_0 + \langle\mu\rangle^{t-r+\ell-s}\|\Lambda_{-\ell}f\|_s)$$

$$\text{when } 0 \leq t-r+\ell+d \leq d+s ,\quad s \geq 0 ,$$

where the first requirement on t stems from the fact that the s.g.o. part of R'_0A can be of class d (cf. (1.2.46)), and the last requirement stems from (4.7.59). With the notation

(4.7.61)
$$\|f\|^+_s = \|f\|_{H^s(E)} \quad \text{when } s \geq 0 ,$$
$$\|f\|^+_s = \|e^+f\|_{H^s(\tilde{E})} \quad \text{when } s < 0 ,$$

we have in particular, since $\varepsilon = \mu^{-1}$,

(4.7.62) $\|M'_\varepsilon f\|_t \leq c(\varepsilon^{r-\ell-t}\|f\|^+_{-\ell} + \varepsilon^{r-\ell-t+s}\|f\|^+_{s-\ell})$ for $r-\ell-\frac{1}{2} < t \leq r-\ell+s$, $s \geq 0$.

Observe that the exponent on ε can be positive here only when $t < r-\ell$. A better result is obtained from the other representations (4.7.48) and (4.7.50), under stricter smoothness assumptions on f .

The first term in (4.7.50) gives

$$(4.7.63)\quad \|R_0'R_\mu\Lambda_{-\ell}A_1R_0f\|_t \le c_1\|R_\mu\Lambda_{-\ell}A_1R_0f\|_{t-r+\ell}$$
$$\le c_2(\langle\mu\rangle^{t-r+\ell-d}\|\Lambda_{-\ell}A_1R_0f\|_0 + \langle\mu\rangle^{t-r+\ell-d-s}\|\Lambda_{-\ell}A_1R_0f\|_s)$$
$$\le c_3(\varepsilon^{d+r-\ell-t}\|R_0f\|_{d+r-\ell} + \varepsilon^{d+r-\ell-t+s}\|R_0f\|_{d+r-\ell+s})$$
$$\text{when}\quad 0 \le t-r+\ell \le d+s\ ,\quad s \ge 0\ ,$$

and this gives furthermore, since $R_0 = R_0'\Lambda_{-\ell}$,

$$(4.7.64)\quad \|R_0'R_\mu\Lambda_{-\ell}A_1R_0f\|_t \le c_4(\varepsilon^{d+r-\ell-t}\|f\|^+_{d-\ell} + \varepsilon^{d+r-\ell-t+s}\|f\|^+_{d-\ell+s})\ .$$

The second term in (4.7.50) consists of contributions from each $K_{\mu,j-r+\ell}$, where we find

$$(4.7.65)\quad \|R_0'K_{\mu,j-r+\ell}T_{1,j}R_0f\|_t \le c_1\|K_{\mu,j-r+\ell}T_{1,j}R_0f\|_{t-r+\ell}$$
$$\le c_2(\langle\mu\rangle^{t-\frac12-j}\|T_{1,j}R_0f\|_0 + \langle\mu\rangle^{t-\frac12-j-s}\|T_{1,j}R_0f\|_s)$$
$$\le c_{3,\delta}(\varepsilon^{j+\frac12-t}\|R_0f\|_{j+\frac12+\delta} + \varepsilon^{j+\frac12-t+s}\|R_0f\|_{j+\frac12+s})$$
$$\le c_{4,\delta}(\varepsilon^{j+\frac12-t}\|f\|^+_{j-r+\frac12+\delta} + \varepsilon^{j+\frac12-t+s}\|f\|^+_{j-r+\frac12+s})$$
$$\text{when}\quad 0 \le t-r+\ell \le j+\tfrac12-r+\ell+s\ ,\quad s \ge 0\ ,\quad \delta > 0\ ;$$

here we have again used that $R_0 = R_0'\Lambda_{-\ell}$, and moreover that $j \ge k_1 \ge r-\ell$ so that the continuity of R_0' from $H^\sigma(E)$ to $H^{\sigma+r-\ell}(E)$ for $\sigma \ge 0$ can be used. Summing over $k_1 \le j \le k_1'$, and estimating intermediate terms by interpolation, we obtain

$$(4.7.66)\quad \|R_0'K_\mu T_1R_0f\|_t \le c \sum_{k_1\le j\le k_1'} (\varepsilon^{j+\frac12-t}\|T_{1,j}R_0f\|_0 + \varepsilon^{j+\frac12-t+s}\|T_{1,j}R_0f\|_s)$$
$$\le c_\delta(\varepsilon^{k_1+\frac12-t}\|R_0f\|_{k_1+\frac12+\delta} + \varepsilon^{k_1'+\frac12-t+s}\|R_0f\|_{k_1'+\frac12+\delta'})\ ,$$
$$\le c_\delta'(\varepsilon^{k_1+\frac12-t}\|f\|^+_{k_1-r+\frac12+\delta} + \varepsilon^{k_1'+\frac12-t+s}\|f\|^+_{k_1'-r+\frac12+\delta'})\ ,$$

when the indices satisfy

$$(4.7.67)\quad 0 \le t-r+\ell \le \min\{k_1+\tfrac12-r+\ell, k_1'+\tfrac12-r+\ell+s\}\ ,\quad s \ge 0,\quad \delta > 0,\ \delta' = \max\{s,\delta\}\ .$$

We can similarly estimate the contributions from u in (4.7.52):

$$\|R_0'R_\mu\Lambda_{-\ell}A_1u\|_t \le c(\varepsilon^{d+r-\ell-t}\|u\|_{d+r-\ell} + \varepsilon^{d+r-\ell-t+s}\|u\|_{d+r-\ell+s}) \tag{4.7.68}$$
$$\text{when } 0 \le t-r+\ell \le d+s\ , \quad s \ge 0$$

follows as in (4.7.64), and the proof of (4.7.66) also shows

$$\|R_0'K_\mu T_1u\|_t \le c_\delta(\varepsilon^{k_1+\frac{1}{2}-t}\|u\|_{k_1+\frac{1}{2}+\delta} + \varepsilon^{k_1'+\frac{1}{2}-t+s}\|u\|_{k_1'+\frac{1}{2}+\delta'}) \tag{4.7.69}$$
$$\text{when the indices satisfy (4.7.67).}$$

The estimates obtained in (4.7.64) and (4.7.66) give together an estimate of $M_\varepsilon'f$ (by the formula (4.7.50)), that will be summed up in the theorem below; let us here just note that when f is sufficiently smooth, $\|M_\varepsilon'f\|_t$ is $O(\varepsilon^{k_1+\frac{1}{2}-t})$:

$$\|M_\varepsilon'f\|_t \le c(f)\varepsilon^{k_1+\frac{1}{2}-t} \quad \text{for } t \ge r-\ell, \text{ when } f \in H^a(E)\ , \tag{4.7.70}$$
$$a \ge \max\{t-\ell, t-r+\delta, t-r\} \qquad (\text{for some } \delta > 0)\ .$$

This can be used to derive convergence statements for nonsmooth f, from the uniform estimates in (4.7.62). On one hand, (4.7.62) implies, for $t = r-\ell+s$, $s \ge 0$,

$$\begin{aligned} \varepsilon^{t-r+\ell}\|M_\varepsilon'f\|_t &\le c(\|f\|_{-\ell}^+ + \varepsilon^{t-r+\ell}\|f\|_{t-r}^+)\ , \\ &\le c_1\|f\|_{t-r}^+\ , \end{aligned} \tag{4.7.71}$$

and on the other hand, (4.7.70) shows that

$$\varepsilon^{t-r+\ell}\|M_\varepsilon'f\|_t \le c(f)\varepsilon^{\frac{1}{2}} \quad \text{for } f \in H^a(E)\ ,\ a \ge \max\{t-\ell, t-r+1\}\ , \tag{4.7.72}$$

since $r-\ell \le k_1$. Then one has, with $f_k \in H^a(E)$

$$\varepsilon^{t-r+\ell}\|M_\varepsilon'f\|_t \le c_1\|f-f_k\|_{t-r}^+ + c(f_k)\varepsilon^{\frac{1}{2}}\ . \tag{4.7.73}$$

Since $H^a(E)$ is dense in $\mathring{H}^{t-r}(E)$ for $t-r \le 0$, resp. dense in $H^{t-r}(E)$ for $t-r \ge 0$, one can for each $\varepsilon' > 0$ and each $f \in \mathring{H}^{t-r}(E)$, resp. $H^{t-r}(E)$, choose f_k in $H^a(E)$ so that the first term in the right hand side of (4.7.73) is $\le \varepsilon'/2$. Then one can find ε_0 so that the second term in the right hand side is $\le \varepsilon'/2$ for $\varepsilon \le \varepsilon_0$, so altogether

$$\varepsilon^{t-r+\ell}\|M_\varepsilon'f\|_t \le \varepsilon' \quad \text{for} \quad \varepsilon \le \varepsilon_0 \quad (\text{depending on } f)\ . \tag{4.7.74}$$

This shows that $\varepsilon^{t-r+\ell}\|M_\varepsilon'f\|_t \to 0$ for $\varepsilon \to 0$, when $t \ge r-\ell$ (strong convergence, not necessarily uniform convergence).

For the full analysis, it remains to consider $N'_{\varepsilon,0}$ and $N_{\varepsilon,1}$. Here we use the notation

$$(4.7.75)\qquad (\sum_j \|\varphi_j\|^2_{s_j})^{\frac{1}{2}} = \|\varphi\|_{\{s_j\}} .$$

For $N'_{\varepsilon,0}$ described by (4.7.43), we find by use of (4.7.59-60) and the known continuity properties of the other operators:

$$(4.7.76)\qquad \begin{aligned} \|R_0'R_\mu\Lambda_{-\ell}A_1K_0\varphi_0\|_t &\le c_1\|R_\mu\Lambda_{-\ell}A_1K_0\varphi_0\|_{t-r+\ell} \\ &\le c_2(\langle\mu\rangle^{t-r+\ell-d}\|\Lambda_{-\ell}A_1K_0\varphi_0\|_0 + \langle\mu\rangle^{t-r+\ell+d-s}\|\Lambda_{-\ell}A_1K_0\varphi_0\|_s) \\ &\le c_3(\varepsilon^{d+r-\ell-t}\|\varphi_0\|_{\{d+r-\ell-j-\frac{1}{2}\}} + \varepsilon^{d+r-\ell-t+s}\|\varphi_0\|_{\{d+r-\ell-j-\frac{1}{2}+s\}}) \\ &\qquad \text{when } 0 \le t-r+\ell \le d+s ,\quad s \ge 0 , \end{aligned}$$

since $\Lambda_{-\ell}A_1K_{0,j}$ is a Poisson operator of order $-\ell+d+r-j$ for each j, cf. (1.2.33). Moreover, since K_μ is of the form (4.7.39), and T_1K_0 is a ps.d.o. matrix

$$(4.7.77)\qquad \begin{aligned} &S = T_1K_0 = (T_{1,i}K_{0,j})_{k_1\le i\le k_1',0\le j\le k_0'} , \text{ with} \\ &S_{ij} = T_{1,i}K_{0,j} \text{ going from } F_j^0 \text{ to } F_i^1 , \text{ of order } i-j , \end{aligned}$$

we have that

$$(4.7.78)\qquad \begin{aligned} \|R_0'K_\mu T_1K_0\varphi_0\|_t &\le c_1\|K_\mu S\varphi_0\|_{t-r+\ell} \le c_2 \sum_{\substack{k_1\le i\le k_1' \\ 0\le j\le k_0'}} \|K_{\mu,i-r+\ell}S_{ij}\varphi_{0,j}\|_{t-r+\ell} \\ &\le c_3 \sum_{i,j} (\langle\mu\rangle^{t-\frac{1}{2}-i}\|S_{ij}\varphi_{0,j}\|_0 + \langle\mu\rangle^{t-\frac{1}{2}-i+s}\|S_{ij}\varphi_{0,j}\|_s) \\ &\le c_4 \sum_{i,j} (\varepsilon^{i+\frac{1}{2}-t}\|\varphi_{0,j}\|_{i-j} + \varepsilon^{i+\frac{1}{2}-t+s}\|\varphi_{0,j}\|_{i-j+s}) \\ &\qquad \text{for } r-\ell \le t \le k_1+\tfrac{1}{2}+s ,\quad s \ge 0 . \end{aligned}$$

Finally, we have for $N_{\varepsilon,1}$,

$$(4.7.79)\quad \|R_0'K_\mu\varphi_1\|_t \le c_1\|K_\mu\varphi_1\|_{t-r+\ell} \le c_2 \sum_{k_1\le i\le k_1'} (\langle\mu\rangle^{t-\frac12-i}\|\varphi_{1,i}\|_0 + \langle\mu\rangle^{t-\frac12-i-s}\|\varphi_{1,i}\|_s)$$

$$\le c_3 \sum_{k_1\le i\le k_1'} (\varepsilon^{i+\frac12-t}\|\varphi_{1,i}\|_0 + \varepsilon^{i+\frac12-t+s}\|\varphi_{1,i}\|_s)$$

$$\text{for}\quad r-\ell \le t \le k_1+\tfrac12+s\ ,\quad s\ge 0\ .$$

The various estimates are collected in the following theorem, where we have also used that $d+r-\ell \ge k_1'+1$ by (4.7.21), so that some intermediate norms can be omitted from the expressions.

<u>4.7.12 Theorem</u>. *Assumptions as in Theorem 4.7.8.*

1^0 *The deviation operator* M_ε' *satisfies*

$$(4.7.80)\quad \|M_\varepsilon' f\|_t \le c(\varepsilon^{r-\ell-t}\|f\|^+_{-\ell} + \varepsilon^{r-\ell-t+s}\|f\|^+_{s-\ell})$$

$$\textit{when}\quad r-\ell-\tfrac12 < t \le r-\ell+s\ ,\quad s\ge 0\ .$$

Moreover, it satisfies

$$(4.7.81)\quad \begin{aligned}\|M_\varepsilon' f\|_t &\le \|R_0'R_\mu\Lambda_{-\ell}A_1R_0f\|_t + \|R_0'K_\mu T_1R_0f\|_t\\ &\le c\varepsilon^{d+r-\ell-t+s}\|R_0f\|_{d+r-\ell+s} + c_\delta\varepsilon^{k_1+\frac12-t}\|R_0f\|_{k_1+\frac12+\delta}\\ &\le c'\varepsilon^{d+r-\ell-t+s}\|f\|^+_{d-\ell+s} + c_\delta'\varepsilon^{k_1+\frac12-t}\|f\|^+_{k_1-r+\frac12+\delta}\ ,\end{aligned}$$

$$\textit{when}\quad r-\ell\le t\le d+r-\ell+s\ ,\quad s\ge 0\ ,\quad \delta>0\ .$$

2^0 *When* $t\ge r-\ell$ *and* $\|f\|^+_{t-r}<\infty$, *one has that*

$$(4.7.82)\quad \varepsilon^{t-r+\ell}\|M_\varepsilon' f\|_t \to 0 \quad \textit{for}\quad \varepsilon\to 0\ ;$$

in particular,

$$(4.7.83)\quad \|M_\varepsilon' f\|_{r-\ell}\to 0 \quad \textit{for}\quad \varepsilon\to 0\ ,\quad \textit{when}\quad f\in \mathring{H}^{-\ell}(E)\ .$$

3^0 *The deviation operator* $N_{\varepsilon,0}'$ *satisfies the estimates*

$$(4.7.84)\quad \|N_{\varepsilon,0}'\varphi_0\|_t \le c(\varepsilon^{d+r-\ell-t+s}\|\varphi_0\|_{\{d+r-\ell-j-\frac12+s\}} + \varepsilon^{k_1+\frac12-t}\|\varphi_0\|_{\{k_1-j\}})$$

$$\textit{when}\quad r-\ell\le t\le d+r-\ell+s\ ,\quad s\ge 0\ .$$

4^0 *The deviation operator* $N_{\varepsilon,1}$ *satisfies*

$$\|N_{\varepsilon,1}\varphi_1\|_t \le c \sum_{k_1\le i\le k_1'} (\varepsilon^{i+\frac12-t}\|\varphi_{1,i}\|_0 + \varepsilon^{i+\frac12-t+s}\|\varphi_{1,i}\|_s) \tag{4.7.85}$$

when $r-\ell \le t \le k_1+\frac12+s$, $s \ge 0$.

5^0 *The deviation operator* $-R_0'(R_\mu\Lambda_{-\ell}A_1+K_\mu T_1)$ *in formula* (4.7.52) *is estimated by*

$$\|R_0'(R_\mu\Lambda_{-\ell}A_1+K_\mu T_1)u\|_t \le c\varepsilon^{d+r-\ell-t+s}\|u\|_{d+r-\ell+s} + c_\delta\varepsilon^{k_1+\frac12-t}\|u\|_{k_1+\frac12+\delta} \tag{4.7.86}$$

when $r-\ell \le t \le d+s$, $s \ge 0$ *and* $\delta > 0$.

By combination of the preceding estimates with standard estimates of R_0 and K_0 we find for the full solution operator $(M_\varepsilon \quad N_{\varepsilon,0} \quad N_{\varepsilon,1})$, using again that $R_0 = R_0'\Lambda_{-\ell}$:

4.7.13 Corollary.

1^0 *The entry* $M_\varepsilon = R_0 + M_\varepsilon'$ *in the solution operator satisfies*

$$\|M_\varepsilon f\|_t \le \|R_0 f\|_t + \|M_\varepsilon' f\|_t \le c(\|f\|^+_{\max\{t-r,-\ell\}} + \varepsilon^{r-\ell-t+s}\|f\|^+_{s-\ell}) \tag{4.7.87}$$

when $r-\ell-\frac12 < t \le r-\ell+s$, $s \ge 0$.

One also has

$$\begin{aligned}\|M_\varepsilon f\|_t &\le \|R_0 f\|_t + c\varepsilon^{d+r-\ell-t+s}\|R_0 f\|_{d+r-\ell+s} + c_\delta\varepsilon^{k_1+\frac12-t}\|R_0 f\|_{k_1+\frac12+\delta}\\ &\le c'\|f\|^+_{\max\{t-r,-\ell\}} + c_\delta'\varepsilon^{d+r-\ell-t+s}\|f\|^+_{d-\ell+s} + c_\delta'\varepsilon^{k_1+\frac12-t}\|f\|^+_{k_1-r+\frac12+\delta}\end{aligned} \tag{4.7.88}$$

when $r-\ell \le t \le d+r-\ell+s$, $s \ge 0$, $\delta > 0$.

2^0 *The entry* $N_{\varepsilon,0} = K_0 + N_{\varepsilon,0}'$ *in the solution operator satisfies*

$$\|N_{\varepsilon,0}\varphi_0\|_t \le c(\|\varphi_0\|_{\{t-j-\frac12\}} + \varepsilon^{d+r-\ell-t+s}\|\varphi_0\|_{\{d+r-\ell-j-\frac12+s\}} + \varepsilon^{k_1+\frac12-t}\|\varphi_0\|_{\{k_1-j\}}) \tag{4.7.89}$$

when $r-\ell \le t \le d+r-\ell+s$, $s \ge 0$.

3^0 *For* u_ε *expressed in terms of* u *and* φ_1 *as in* (4.7.52), *one has*

$$\|u_\varepsilon\|_t \le \|u\|_t + c\varepsilon^{d+r-\ell-t+s}\|u\|_{d+r-\ell+s} + c_\delta\varepsilon^{k_1+\frac12-t}\|u\|_{k_1+\frac12+\delta}$$

(4.7.90)
$$+ c \sum_{k_1\le i\le k_1'} (\varepsilon^{i+\frac12-t}\|\varphi_{1,i}\|_0 + \varepsilon^{i+\frac12-t+s'}\|\varphi_{1,i}\|_{s'}) ,$$

when $r-\ell \le t \le \min\{d+s,\ k_1+\frac12+s'\}$, $s \ge 0$, $s' \ge 0$, $\delta > 0$.

4.7.14 Remark. In special cases one can use other estimates of R_0 than the above, where we have used the representation $R_0 = R_0'\Lambda_{-\ell}$. For example, when A_0 is of order $r = 2r'$ and T_0 is the Dirichlet trace operator $\{\gamma_0,\gamma_1,\ldots,\gamma_{r'-1}\}$, then R_0 is continuous

(4.7.91) $$R_0 : H^s(E) \to H^{s+2r'}(E) \cap \mathring{H}^{r'}(E) \quad \text{for} \quad s \ge -r' ,$$

which can be used in the development of estimates going via R_0 .

4.7.15 Example. Consider a Dirichlet problem or Dirichlet-type problem, as described in Example 4.7.9. Since $\ell = r' = k_1$ here, we get for instance (going back to (4.7.62-79) in some cases):

(4.7.92)

(i) $\|M_\varepsilon' f\|_t \le c\varepsilon^{r'-t}\|f\|^+_{-r'}$ for $r' - \frac12 < t \le r'$;

(ii) $\varepsilon^{t-r'}\|M_\varepsilon' f\|_t \to 0$ for $\varepsilon \to 0$, when $\|f\|^+_{t-2r'} < \infty$ and $t \ge r'$,

(iii) $\|M_\varepsilon' f\|_t \le c \sum_{r'\le j<r'+d'} \varepsilon^{j+\frac12-t}\|T_{1,j}R_0 f\|_0 + c_1\varepsilon^{d+r'-t}\|R_0 f\|_{d+r'}$

$\le c_\delta\varepsilon^{r'+\frac12-t}\|f\|^+_{-r'+\frac12+\delta} + c_2\varepsilon^{d+r'-t}\|f\|^+_{d-r'}$ for $r' \le t \le r'+\frac12$, $\delta>0$,

(iv) $\|M_\varepsilon' f\|_t \le c(f)\varepsilon^{r'+\frac12-t}$ for $f \in C^\infty(E)$, $t \ge r'$.

In these formulas, the best exponent on ε is $\frac12$; but better exponents can be obtained when f is subject to boundary conditions. For example, if $T_{1,j}R_0 f = 0$ for $r' \le j < j_0$, some $j_0 \le r'+d'$ then by (iii), (and (4.7.63, 65)),

(4.7.93) $$\|M_\varepsilon' f\|_t \le c \sum_{j_0\le j<r'+d'} \varepsilon^{j+\frac12-t}\|T_{1,j}R_0 f\|_0 + c_1\varepsilon^{d+r'-t}\|R_0 f\|_{d+r'} ,$$
$$\le c(f)\varepsilon^{j_0+\frac12-t} , \quad \text{for} \quad r' \le t \le j_0 + \tfrac12 ,$$

and when $j_0 = r'+d'$ (by (4.7.63) alone)

(4.7.94) $$\|M_\varepsilon' f\|_t \le c_1\varepsilon^{d+r'-t}\|R_0 f\|_{d+r'} \quad \text{for} \quad r' \le t \le d+r' .$$

In the case of a pure Dirichlet condition (4.7.53), the condition $T_{1,j}R_0f = 0$ for $r' \leq j < r'+d$ means precisely that

(4.7.95) $$u = R_0 f \in \overset{\circ}{H}{}^{r'+d'}(E) .$$

One can possibly obtain intermediate estimates by interpolation, improving (iv).

The results are well in accordance with those of [Huet 4,5], [Greenlee 1], [Frank 1] and others, for differential operators (where in some special cases the requirements on f are milder, and lower values of t are included); and it seems that we avoid a factor $|\log \varepsilon|$ in comparison with some consequences of [Frank-Wendt 1]. The strong estimate (4.7.94) has some relations to an early result [Huet 1] for $u \in C_0^\infty(\Omega)$, pertaining to the case $A_0 = I$. See also [Huet 6].

4.7.16 Example. As a "test" case, we consider the boundary problems for $\varepsilon^2\Delta^2 - \Delta$, described in Examples 4.7.2 and 4.7.10. Here $r' = d' = 1$.

For the Dirichlet problem (4.7.11), where $\ell = 1$, we find e.g.:

$$\|M'_\varepsilon f\|_t \leq c\varepsilon^{3-t+s}\|f\|_{1+s} + c_\delta \varepsilon^{3/2-t}\|f\|^+_{-\frac{1}{2}+\delta} \quad \text{for } 1 \leq t \leq 3+s,\ s \geq 0,\ \delta > 0 .$$

(4.7.96) $$\|N'_{\varepsilon,0}\varphi_0\|_t \leq c(\varepsilon^{3-t+s}\|\varphi_0\|_{5/2+s} + \varepsilon^{3/2-t}\|\varphi_0\|_1) \quad \text{for } 1 \leq t \leq 3+s,\ s \geq 0 .$$

$$\|N'_{\varepsilon,1}\varphi_1\|_t \leq c(\varepsilon^{3/2-t}\|\varphi_1\|_0 + \varepsilon^{3/2-t+s}\|\varphi_1\|_s) \quad \text{for } 1 \leq t \leq 3/2+s,\ s \geq 0 .$$

(4.7.94) here gives

(4.7.97) $$\|M'_\varepsilon f\|_t \leq c\varepsilon^{3-t}\|R_0 f\|_3 \quad \text{for } 1 \leq t \leq 3, \text{ when } R_0 f \in \overset{\circ}{H}{}^2(E) \cap H^3(E) .$$

For the problem (4.7.12), we can take either $\ell = 1$ or $\ell = 0$. Here $\ell = 1$ gives the estimates:

$$\|M'_\varepsilon f\|_t \leq c\varepsilon^{3-t+s}\|f\|_{1+s} + c_\delta \varepsilon^{5/2-t}\|f\|_{\frac{1}{2}+\delta} \quad \text{for } 1 \leq t \leq 3+s,\ s \geq 0,\ \delta > 0 .$$

(4.7.98) $$\|N'_{\varepsilon,0}\varphi_0\|_t \leq c(\varepsilon^{3-t+s}\|\varphi_0\|_{3/2+s} + \varepsilon^{5/2-t}\|\varphi_0\|_2) \quad \text{for } 1 \leq t \leq 3+s,\ s \geq 0 .$$

$$\|N'_{\varepsilon,0}\varphi_1\|_t \leq c(\varepsilon^{5/2-t}\|\varphi_1\|_0 + \varepsilon^{5/2-t+s}\|\varphi_1\|_s) \quad \text{for } 1 \leq t \leq 5/2+s,\ s \geq 0 .$$

The choice $\ell = 0$ gives the estimates (which do not add very much to the picture):

$$\|M'_\varepsilon f\|_t \le c\varepsilon^{4-t+s}\|f\|_{2+s} + c_\delta \varepsilon^{5/2-t}\|f\|_{\frac{1}{2}+\delta} \quad \text{for} \quad 2 \le t \le 4+s,\ s \ge 0, \delta > 0\ .$$

(4.7.99) $$\|N'_{\varepsilon,0}\varphi_0\|_t \le c(\varepsilon^{4-t+s}\|\varphi_0\|_{5/2+s} + \varepsilon^{5/2-t}\|\varphi_0\|_2) \quad \text{for} \quad 2 \le t \le 4+s,\quad s \ge 0\ .$$

$$\|N_{\varepsilon,1}\varphi_1\|_t \le c(\varepsilon^{5/2-t}\|\varphi_1\|_0 + \varepsilon^{5/2-t+s}\|\varphi_1\|_s) \quad \text{for} \quad 2 \le t \le 5/2+s,\quad s \ge 0\ .$$

Finally, consider the case with the trace operators (4.7.56). Here $k'_0 = 1$, $k_1 = 2$, $k'_1 = 3$, so the only possible choice of ℓ is $\ell = 0$. We then get

$$\|M'_\varepsilon f\|_t \le c\varepsilon^{4-t+s}\|f\|_{2+s} + c_\delta \varepsilon^{5/2-t}\|f\|_{\frac{1}{2}+\delta} \quad \text{for} \quad 2 \le t \le 4+s,\ s \ge 0,\ \delta > 0\ .$$

$$\|N'_{\varepsilon,0}\varphi_0\|_t \le c\varepsilon^{4-t+s}(\|\varphi_{0,0}\|_{7/2+s} + \|\varphi_{0,1}\|_{5/2+s}) + c'\varepsilon^{5/2-t}(\|\varphi_{0,0}\|_2 + \|\varphi_{0,1}\|_1)$$

(4.7.100) $$\text{for} \quad 2 \le t \le 4+s,\ s \ge 0\ .$$

$$\|N_{\varepsilon,1}\varphi_1\|_t \le c(\varepsilon^{5/2-t}\|\varphi_{1,2}\|_0 + \varepsilon^{5/2-t+s}\|\varphi_{1,2}\|_s) + c'(\varepsilon^{7/2-t}\|\varphi_{2,2}\|_0 +$$
$$+ \varepsilon^{7/2-t+s}\|\varphi_{2,2}\|_s) \text{ for } 2 \le t \le 5/2+s,\quad s \ge 0\ .$$

The treatment of this last case appears to be new.

All the results of course extend to nonlocal perturbations (by suitably small terms).

We end with some observations on improvements of the above results. It was seen in the proof of Theorem 4.7.12 how the two different representations of M_ε, (4.7.47) and (4.7.48), give different estimates (either with very general f or with improved decrease in ε), and it is natural to search for still other representations. One possibility would be to work with versions of $\Lambda^{-\ell}_{-,\Omega}$ with noninteger ℓ; this would allow a greater freedom in the choice of t. Operators $\Lambda^{-s}_{-,\Omega}$ with $s \in \mathbb{R}$ are defined in the calculus of [Eskin 1] (which is somewhat coarser than the calculus used here, in its concept of ps.d.o.s), and this is further developed in [Rempel-Schulze 2]. The use of such operators should be coupled with a generalization of the resolvent construction of [Rempel-Schulze 4] to the needed cases.

Another possibility is to search for other representations of the middle term $\mu^d R_\mu - I$ in (4.7.44) than (4.7.45) and (4.7.46), in order to get better estimates than (4.7.80) without requiring quite as much smoothness as in (4.7.81). For example, it can happen that $B = A_T$ is selfadjoint positive, or can be obtained with these properties by a more clever choice of reduction operator than $\Lambda_{-\ell}$; then if $D(B^\theta) = H^{\theta d}(E)$ for $\theta d < \frac{1}{2}$ (by our Theorem 4.4.2 or an extension), one can obtain improvements of (4.7.80-81) by use of spectral considerations as in [Huet 2,3],[Greenlee 1].

The above study gives the most immediate results that are available from our general calculus; they were announced in [Grubb 18]. It would of course be interesting to make a deeper analysis, possibly combining the above methods with those of the other mentioned authors, and we hope to return to this elsewhere.

APPENDIX: VARIOUS PREREQUISITES

A.1 *General notation.*

We denote by $\mathbb{R}$ the real numbers, $\mathbb{C}$ the complex numbers, $\mathbb{Z}$ the integers and $\mathbb{N}$ the nonnegative integers $0,1,2,\ldots$ (that are used in most indexations). $\mathbb{N}_+$ stands for the positive integers $1,2,3,\ldots$. For $n \in \mathbb{N}_+$, $\mathbb{R}^n$ and $\mathbb{C}^n$ denote the vector spaces of real or complex n-tuples $x=(x_1,\ldots,x_n)$, $\xi=(\xi_1,\ldots,\xi_n)$, etc. Here we set

$$\mathbb{R}^n_\pm = \{x \in \mathbb{R}^n \mid x_n \gtrless 0\} ,$$

and its boundary $\{x \in \mathbb{R}^n \mid x_n=0\}$ is identified with $\mathbb{R}^{n-1}$, whose points are then denoted x',ξ', etc., so that $x = (x',x_n)$. (In particular, $\mathbb{R}^1_\pm = \mathbb{R}_\pm$.). The closure of $\mathbb{R}^n_\pm$ is denoted $\overline{\mathbb{R}}^n_\pm$. We write

$$x \cdot y = x_1 y_1 + \ldots + x_n y_n \quad \text{(sometimes written } xy) ,$$
$$|y| = (y \cdot \bar{y})^{\frac{1}{2}} ,$$
$$\langle y \rangle = (1 + |y|^2)^{\frac{1}{2}} ,$$

where the upper bar stands for complex conjugation. In particular, when $y = (\xi,\mu)$ with $\xi \in \mathbb{R}^n$ and $\mu \in \mathbb{C}$, we write for short

$$\begin{aligned} |(\xi,\mu)| &= |\xi,\mu| = (|\xi|^2 + |\mu|^2)^{\frac{1}{2}} , \\ \langle(\xi,\mu)\rangle &= \langle\xi,\mu\rangle = (1 + |\xi|^2 + |\mu|^2)^{\frac{1}{2}} ; \text{ in particular} \\ \langle\xi\rangle &= \langle\xi',\xi_n\rangle = (1 + |\xi'|^2 + |\xi_n|^2)^{\frac{1}{2}} , \ \langle\xi'\rangle = (1+|\xi'|^2)^{\frac{1}{2}} . \end{aligned} \tag{A.1}$$

This notation will be needed for the parameter-dependent calculus, where we also use

$$\rho(\xi,\mu) = \langle\xi\rangle\langle\xi,\mu\rangle^{-1}, \text{ or } \rho(\xi',\mu) = \langle\xi'\rangle\langle\xi',\mu\rangle^{-1} . \tag{A.2}$$

We denote by $B_r(x)$ the ball with center x and radius r , and by $\overline{B}_r(x)$ its closure

$$B_r(x) = \{y \in \mathbb{R}^n \mid |x-y| < r\} ; \quad \overline{B}_r(x) = \{y \in \mathbb{R}^n \mid |x-y| \le r\} . \tag{A.3}$$

The set $\mathbb{R}_+ \times \mathbb{R}_+$ is denoted $\mathbb{R}^2_{++}$, and $\overline{\mathbb{R}_+} \times \overline{\mathbb{R}_+} = \overline{\mathbb{R}}^2_{++}$. We write

$$\mathbb{C}_\pm = \{z \in \mathbb{C} \mid \operatorname{Im} z \gtrless 0\} ; \tag{A.4}$$

(A.4') $a_\pm$ (or $[a]_\pm$) $= \max\{\pm a,0\}$; $[a]$ = largest integer $\le a$, when $a \in \mathbb{R}$.

The derivative d/dt is also denoted ∂_t, and

$$-i\partial_t = D_t , \qquad +i\partial_t = \overline{D}_t , \qquad \text{with} \quad i = (-1)^{\frac{1}{2}} .$$

For partial derivatives, we write $\partial/\partial x_j = \partial_{x_j}$ or ∂_j, and $D_{x_j} = -i\partial_{x_j}$, $\overline{D}_{x_j} = +i\partial_{x_j}$. We use multiindex notation

$$\text{(A.5)} \qquad D_x^\alpha = D_{x_1}^{\alpha_1} \cdots D_{x_n}^{\alpha_n} , \qquad \overline{D}_x^\alpha = \overline{D}_{x_1}^{\alpha_1} \cdots \overline{D}_{x_n}^{\alpha_n} , \quad \text{for} \quad \alpha \in \mathbb{N}^n ;$$

here the order $\alpha_1 + \dots + \alpha_n$ is denoted $|\alpha|$, and

$$\alpha \pm \beta = (\alpha_1 \pm \beta_1, \dots, \alpha_n \pm \beta_n) ,$$

$\alpha \leq \beta$ means that $\alpha_1 \leq \beta_1, \dots, \alpha_n \leq \beta_n$,

$\alpha!$ stands for $\alpha_1! \dots \alpha_n!$.

A.2 *Functions and distributions.*

When Ω is an open subset of $\mathbb{R}^n$, and $k \in \mathbb{N}$, $C^k(\Omega)$ denotes the space of functions on Ω with continuous derivatives of all orders $\leq k$. When $u \in C^k(\Omega)$ for some $k > 0$, and $x \in \Omega$, one has for y in a ball around x the Taylor formula

$$\text{(A.6)} \qquad u(x+y) = \sum_{|\alpha|<k} \frac{y^\alpha}{\alpha!} \partial^\alpha u(x) + \sum_{|\alpha|=k} \frac{k}{\alpha!} y^\alpha \int_0^1 (1-h)^{k-1} \partial^\alpha u(x+hy)dh .$$

For φ and $u \in C^k(\Omega)$ and $|\alpha| \leq k$ one has the Leibniz formula

$$\text{(A.7)} \qquad D^\alpha(\varphi(x)u(x)) = \sum_{\beta \leq \alpha} \frac{\alpha!}{\beta!(\alpha-\beta)!} D^\beta \varphi(x) D^{\alpha-\beta} u(x) .$$

The support of a function u, $\operatorname{supp} u$, is the complement of the largest open set where u is zero; this notion extends to functions defined "almost everywhere" and distributions. The space $C^\infty(\Omega)$ equals $\cap_{k\in\mathbb{N}} C^k(\Omega)$, and $C_0^\infty(\Omega)$ is the subspace of C^∞ functions with compact support in Ω. More generally one can for any set $M \subset \mathbb{R}^n$ denote by $C_0^\infty(M)$ the C^∞ functions on $\mathbb{R}^n$ with compact support in M. We have more need for the following notation that it should not be confounded with:

$$\text{(A.8)} \qquad C_{(0)}^\infty(M) = \{u|_M \mid u \in C_0^\infty(\mathbb{R}^n)\} .$$

The formulas (A.6) and (A.7) hold also when u takes its values in a Banach space X; $u \in C^k(\Omega,X)$. For $\sigma \in \,]0,1]$, the Hölder space of functions u with $\|u(x)-u(y)\| \leq c(x)|x-y|^\sigma$ for y near x, is denoted $C^\sigma(\Omega,X)$, and $C^{k,\sigma}(\Omega,X) = \{u \mid D^\alpha u \in C^\sigma$ for $|\alpha| \leq k\}$.

We shall need various spaces of testfunctions and distributions, and various Sobolev spaces; here the notations are chosen to be very close to those of [Hörmander 1], in order to avoid lengthy repetitions. So we shall merely list the definitions and refer to [Hörmander 1] and other works ([Schwartz 1], [Lions-Magenes 1],...), for the full explanation. The functions we consider will always be assumed Lebesgue measurable.

$C^\infty(\Omega)$ is provided with the Fréchet topology defined by the sequence of seminorms

$$\sup\{|D^\alpha u(x)| \mid |\alpha| \leq N\ ,\ x \in K_N\}, \quad \text{for} \quad N \in \mathbb{N}\ ,$$

where K_N is a sequence of compact sets in Ω such that

$$\text{(A.9)} \qquad K_0 \subset \overset{\circ}{K}_1 \subset K_1 \subset \dots \subset K_N \subset \overset{\circ}{K}_{N+1} \subset \dots \quad , \quad \text{with} \quad \bigcup_{N\in\mathbb{N}} K_N = \Omega\ .$$

Letting $C_K^\infty(\Omega)$ denote the closed subspace of functions in $C^\infty(\Omega)$ with support in K (compact $\subset \Omega$), one provides $C_0^\infty(\Omega)$ with the inductive limit topology (*LF*-topology)

$$\text{(A.9')} \qquad C_0^\infty(\Omega) = \bigcup_{N\in\mathbb{N}} C_{K_N}^\infty(\Omega)\ .$$

$\mathcal{D}'(\Omega)$ is then defined as the dual space to $C_0^\infty(\Omega)$, i.e. the space of continuous linear functionals on $C_0^\infty(\Omega)$, with the notation $\langle u,\varphi\rangle$ for the value of the functional u on the testfunction $\varphi \in C_0^\infty(\Omega)$. (The notation $\langle u,\varphi\rangle$ will only be used explicitly on very few occasions, where it is easily distinguishable from the functions in (A.1).)

Besides $C_0^\infty(\Omega)$, we need the testfunction space $\mathcal{S}(\mathbb{R}^n)$ (or $\mathcal{S}$) of rapidly decreasing functions on $\mathbb{R}^n$, with the Fréchet topology defined by the seminorms

$$\text{(A.9'')} \qquad \sup\{\langle x\rangle^N |D^\alpha u(x)| \mid x \in \mathbb{R}^n\ ,\ |\alpha| \leq N\}\ , \quad \text{for} \quad N \in \mathbb{N}\ ,$$

and its dual space $\mathcal{S}'(\mathbb{R}^n)$ (or $\mathcal{S}'$), the (Schwartz) space of temperate distributions; here $\mathcal{S}'(\mathbb{R}^n)$ identifies with a subspace of $\mathcal{D}'(\mathbb{R}^n)$. The dual space of $C^\infty(\Omega)$ is $\mathcal{E}'(\Omega)$, the space of distributions with compact support (a subspace of $\mathcal{D}'(\Omega)$). Distributions here are often considered as "extended by zero" to distributions on $\mathbb{R}^n$.

Locally L^p integrable functions ($p \in [1,\infty]$) belong to the distributions. We consider in particular $L^2(M)$ (for measurable sets $M \subset \mathbb{R}^n$) , the space of (equivalence classes of) functions f on M with finite L^2 norm

$$\text{(A.10)} \qquad \|f\|_{L^2(M)} \equiv \left(\int_M |f(x)|^2 dx\right)^{\frac{1}{2}} .$$

$L^2_{loc}(M)$ consists of the functions f on M for which $f|_{M'} \in L^2(M')$ for any compact subset M' of M, and $L^2_{comp}(M)$ consists of the functions in $L^2(M)$ with compact support in M. When u and v are functions, we denote by $(u,v)_M$ the integral

$$\text{(A.11)} \qquad (u,v)_M = \int_M u(x)\bar{v}(x)dx ,$$

whenever it has a sense. It coincides with the sesquilinear distribution duality $\langle u,\bar{v}\rangle$ when $M = \Omega$, $u \in \mathcal{D}'(\Omega)$ is a function, and v is a testfunction in $C_0^\infty(\Omega)$; then

$$\langle u,\bar{v}\rangle = (u,v)_\Omega .$$

Also other generalizations of (A.11) will be denoted $\langle u,\bar{v}\rangle$, with indication (if necessary) of which spaces are being considered. For example, if V is a dense subspace of $L^2(M)$ with a stronger topology, then there is an identification of $L^2(M)$ with a subspace of V^*, so that

$$\text{(A.12)} \qquad \langle u,\bar{v}\rangle_{V^*,V} = (u,v)_{L^2(M)} \quad \text{when} \quad u \in L^2(M) \text{ and } v \in V ;$$

here $\langle u,\bar{v}\rangle_{V^*,V}$ stands for the value of the conjugate linear functional $u \in V^*$ on $v \in V$.

The Fourier transform and conjugate Fourier transform (co-Fourier transform) are defined by

$$\text{(A.13)} \qquad \begin{aligned} (Fu)(\xi) &= \hat{u}(\xi) = \int_{\mathbb{R}^n} e^{-ix\cdot\xi}u(x)dx , \\ (\bar{F}u)(\xi) &= \int_{\mathbb{R}^n} e^{+ix\cdot\xi}u(x)dx , \end{aligned}$$

for functions in $L^1(\mathbb{R}^n)$ or $\mathcal{S}(\mathbb{R}^n)$; and it extends to $\mathcal{S}'(\mathbb{R}^n)$ by use of the formulas (verified first on $\mathcal{S}(\mathbb{R}^n)$)

$$\text{(A.14)} \qquad \begin{aligned} F[x^\alpha D_x^\beta u] &= \overline{D}_\xi^\alpha \xi^\beta Fu , \\ F^{-1} &= (2\pi)^{-n}\bar{F} . \end{aligned}$$

In particular, $(2\pi)^{-n/2}F$ defines an isometry of $L^2(\mathbb{R}^n)$ onto itself (the Parseval-Plancherel theorem):

$$(A.15) \qquad (2\pi)^{-n}\,(\hat{u},\hat{v})_{\mathbb{R}^n} = (u,v)_{\mathbb{R}^n} \qquad \text{for} \quad u,v \in L^2(\mathbb{R}^n)\ .$$

Another important property is

$$(A.16) \qquad F(u * v) = \hat{u}(\xi)\cdot\hat{v}(\xi)\ ,$$

valid for u and $v \in L^2(\mathbb{R}^n)$ and having numerous generalizations. The following inequality is often useful:

$$(A.17) \qquad \langle\xi+\eta\rangle^s \le 2^{|s|}\langle\xi\rangle^s\langle\eta\rangle^{|s|} \quad \text{for any} \quad s \in \mathbb{R}\ .$$

The Fourier transform (A.13) is sometimes for precision denoted $F_{x\to\xi}$, to distinguish it from the "partial Fourier transforms"

$$(A.18) \qquad \begin{aligned} F_{x'\to\xi'} &: u(x) \sim \int_{\mathbb{R}^{n-1}} e^{-ix'\cdot\xi'}\,u(x',x_n)dx' \equiv \acute{u}(\xi',x_n)\ ; \\ F_{x_n\to\xi_n} &: u(x) \sim \int_{\mathbb{R}} e^{-ix_n\xi_n}\,u(x',x_n)dx_n \equiv \grave{u}(x',\xi_n)\ . \end{aligned}$$

There are similar conventions for the inverse Fourier transforms.

A.3 *Sobolev spaces.*

For any $s \in \mathbb{R}$, one defines the Sobolev space $H^s(\mathbb{R}^n)$ by

$$(A.19) \qquad H^s(\mathbb{R}^n) = \{u \in \mathcal{S}'(\mathbb{R}^n) \mid \langle\xi\rangle^s\hat{u}(\xi) \in L^2(\mathbb{R}^n)\}\ ,$$

provided with the norm

$$(A.20) \qquad \|u\|_s = (2\pi)^{-n/2}\,\|\langle\xi\rangle^s\hat{u}(\xi)\|_{L^2(\mathbb{R}^n)}\ .$$

In other words, $H^s(\mathbb{R}^n)$ is the isometric image by the mapping $[(2\pi)^{-n/2}F]^{-1}$ of the weighted L^2 space $L^2_s = L^2(\mathbb{R}^n, \langle\xi\rangle^{2s})$ (here $L^2(\mathbb{R}^n, f(\xi))$ is the space of locally integrable functions $u(\xi)$ with norm $(\int|u(\xi)|^2f(\xi)d\xi)^{\frac{1}{2}} < \infty$);

$$(A.21) \qquad H^s(\mathbb{R}^n) = (2\pi)^{n/2}\,F^{-1}L^2_s\ .$$

Since L^2_s is a Hilbert space, so is $H^s(\mathbb{R}^n)$. One has that $\mathcal{S}$ is dense in $L^2(\mathbb{R}^n)$ and hence in any of the spaces L^2_s and $H^s(\mathbb{R}^n)$.

When $s \geq 0$, L^2_s is a dense subspace of $L^2_0 = L^2(\mathbb{R}^n)$ with a stronger topology, and its dual space identifies with L^2_{-s} in such a way that the duality is consistent with the scalar product in $L^2(\mathbb{R}^n)$ (as in (A.12)), and with the distribution duality,

$$\langle u,\overline{v}\rangle_{L^2_{-s},L^2_s} = (u,v)_{\mathbb{R}^n} \quad \text{if} \quad u \in L^2_0 \text{ and } v \in L^2_s, \quad s \geq 0 .$$

By inverse Fourier transformation one gets a similar identification of the dual space of $H^s(\mathbb{R}^n)$ with $H^{-s}(\mathbb{R}^n)$, such that

(A.22) $$\langle u,\overline{v}\rangle_{H^{-s},H^s} = (u,v)_{\mathbb{R}^n} \quad \text{if} \quad u \subset L^2 \text{ and } v \in H^s, \quad s \geq 0 .$$

$H^s(\mathbb{R}^n)$ likewise identifies with the dual space of $H^{-s}(\mathbb{R}^n)$.

There are some variants of the H^s spaces that we also need. For one thing, we need the <u>anisotropic spaces</u> $H^{(s,t)}(\mathbb{R}^n)$ defined for $s,t \in \mathbb{R}$ by

(A.23) $$H^{(s,t)}(\mathbb{R}^n) = \{u \in \mathscr{S}'(\mathbb{R}^n) \mid \langle\xi\rangle^s\langle\xi'\rangle^t\hat{u}(\xi) \in L^2(\mathbb{R}^n)\} ;$$

they are Hilbert spaces with the norm

(A.24) $$\|u\|_{(s,t)} = (2\pi)^{-n/2} \|\langle\xi\rangle^s\langle\xi'\rangle^t\hat{u}(\xi)\|_{L^2(\mathbb{R}^n)} .$$

$\mathscr{S}$ is dense in this space (since it is dense in $L^2(\mathbb{R}^n, \langle\xi\rangle^{2s}\langle\xi'\rangle^{2t})$), and there is an identification of $H^{(-s,-t)}(\mathbb{R}^n)$ with the dual of $H^{(s,t)}(\mathbb{R}^n)$ for any $s,t \in \mathbb{R}$. Secondly, we sometimes need to study the estimates in their dependence on a parameter; for this we denote, for $\mu \geq 0$ (recall (A.1)),

(A.25) $$\|u\|_{(s,t),\mu} = (2\pi)^{-n/2}\|\langle\xi,\mu\rangle^s\langle\xi',\mu\rangle^t\hat{u}(\xi)\|_{L^2(\mathbb{R}^n)} ,$$

and we denote the space provided with this norm by $H^{(s,t),\mu}(\mathbb{R}^n)$. For $t = 0$ we get the <u>isotropic</u> case, where we simply write

(A.26) $$H^{(s,0),\mu}(\mathbb{R}^n) = H^{s,\mu}(\mathbb{R}^n) ,$$
$$\|u\|_{(s,0),\mu} = \|u\|_{s,\mu} .$$

Observe that $\|u\|_{s,\mu} \simeq (\|u\|^2_s + \langle\mu\rangle^s\|u\|^2_0)^{\frac{1}{2}}$ when $s \geq 0$, and that

(A.27) $\|u\|_{(s,t),\mu} \leq \|u\|_{(s',t'),\mu}$ holds precisely when $s' \geq s$ and $s'+t' \geq s+t$,

so in particular,

$$\|u\|_{s+t,\mu} \le \|u\|_{(s,t),\mu} \quad \text{when} \quad t \le 0 \,, \tag{A.28}$$
$$\|u\|_{s+t,\mu} \ge \|u\|_{(s,t),\mu} \quad \text{when} \quad t \ge 0 \,.$$

There are corresponding inclusions between the various spaces.

A.4 *Spaces over subsets of* $\mathbb{R}^n$.

We now turn to subsets of $\mathbb{R}^n$. For one thing, there are local variants of the preceding spaces. Let Ω be open $\subset \mathbb{R}^n$, and set

$$H^s_{comp}(\Omega) = \{u \in H^s(\mathbb{R}^n) \mid \text{supp } u \quad \text{compact} \subset \Omega\} \,, \tag{A.29}$$
$$H^s_{loc}(\Omega) = \{u \in \mathcal{D}'(\Omega) \mid \varphi u \in H^s(\mathbb{R}^n) \quad \text{for} \quad \varphi \in C_0^\infty(\Omega)\} \quad ;$$

analogous definitions are made for the $H^{(s,t),\mu}$ spaces. $H^s_{comp}(\Omega)$ is topologized as the inductive limit of the Hilbert spaces $H^s_{K_N}(\Omega)$ of distributions in $H^s(\mathbb{R}^n)$ supported in K_N, cf. (A.9). $H^s_{loc}(\Omega)$ is a Fréchet space with respect to the family of seminorms $\|\varphi_N u\|_s$, where φ_N is a sequence of functions in $C_0^\infty(\Omega)$ with $\varphi_N = 1$ on K_N (one can show that multiplication with $\varphi \in C_0^\infty(\mathbb{R}^n)$ is continuous in $H^s(\mathbb{R}^n)$). Similar notions hold for the spaces based on $H^{(s,t),\mu}$.

Secondly, and more central to the present work, there are the spaces over Ω where the behavior at the boundary of Ω is more specified. Here we generally need smoothness of Ω, permitting diffeomorphisms that reduce the questions to the "flat" case where $\Omega = \mathbb{R}^n_+$. We consider this case first. (Much of the analysis could be extended to cases of domains with less smoothness; one can see this from the norms involved. However, we have not made any efforts in this direction.)

Denote by $r^\pm$ the restriction operators (from $\mathcal{D}'(\mathbb{R}^n)$ to $\mathcal{D}'(\mathbb{R}^n_\pm)$)

$$r^\pm v = v|_{\mathbb{R}^n_\pm} \qquad \text{for} \quad v \in \mathcal{D}'(\mathbb{R}^n) \,, \tag{A.30}$$

and by $e^\pm$ the "extension by zero" operators (from $L^1_{loc}(\overline{\mathbb{R}}^n_\pm)$ to $L^1_{loc}(\mathbb{R}^n)$)

$$e^\pm u = \begin{cases} u & \text{on} \quad \mathbb{R}^n_\pm \\ 0 & \text{on} \quad \mathbb{R}^n \setminus \mathbb{R}^n_\pm \end{cases} \,. \tag{A.31}$$

We shall later also need the reflection operator J ,

$$J : u(x',x_n) \sim u(x',-x_n) \tag{A.32}$$

(generalized to distributions if needed), it sends spaces over $\mathbb{R}^n_{\pm}$ into spaces over $\mathbb{R}^n_{\mp}$. The following notation will be used:

$$\begin{aligned} C^k(\overline{\mathbb{R}}^n_{\pm}) &= r^{\pm}C^k(\mathbb{R}^n) \quad , \; k \in \mathbb{N} \;\text{ or }\; k = \infty \, , \\ C^k_{(0)}(\overline{\mathbb{R}}^n_{\pm}) &= r^{\pm}C^k_0(\mathbb{R}^n) \quad , \; k \in \mathbb{N} \;\text{ or }\; k = \infty \, , \\ \mathscr{S}(\overline{\mathbb{R}}^n_{\pm}) &= r^{\pm}\mathscr{S}(\mathbb{R}^n) \quad , \end{aligned} \tag{A.33}$$

where the spaces are topologized by the earlier mentioned seminorms, restricted to $K_N \cap \overline{\mathbb{R}}^n_{\pm}$. One can define continuous extension operators

$$\begin{aligned} &\ell^{\pm}_k : C^k(\overline{\mathbb{R}}^n_{\pm}) \to C^k(\mathbb{R}^n) \, , \quad \ell^{\pm}_k : C^k_{(0)}(\overline{\mathbb{R}}^n_{\pm}) \to C^k_0(\mathbb{R}^n) \, , \quad k \leq \infty \, , \\ &\ell^{\pm} : \mathscr{S}(\overline{\mathbb{R}}^n_{\pm}) \to \mathscr{S}(\mathbb{R}^n) \, , \end{aligned} \tag{A.34}$$

such that $r^{\pm}\ell^{\pm}_k = I$ and $r^{\pm}\ell^{\pm} = I$, as in [Lions-Magenes 1], [Seeley 5] .

Although they will not play a great rôle, we also list the definitions

$$\begin{aligned} \overset{\circ}{\mathscr{D}}'(\overline{\mathbb{R}}^n_{\pm}) &= \{ u \in \mathscr{D}'(\mathbb{R}^n) \mid \operatorname{supp} u \subset \overline{\mathbb{R}}^n_{\pm} \} \, , \\ \overset{\circ}{\mathscr{S}}'(\overline{\mathbb{R}}^n_{\pm}) &= \{ u \in \mathscr{S}'(\mathbb{R}^n) \mid \operatorname{supp} u \subset \overline{\mathbb{R}}^n_{\pm} \} \, , \\ \overset{\circ}{\mathscr{E}}'(\overline{\mathbb{R}}^n_{\pm}) &= \{ u \in \mathscr{E}'(\mathbb{R}^n) \mid \operatorname{supp} u \subset \overline{\mathbb{R}}^n_{\pm} \} \end{aligned} \tag{A.35}$$

(closed subspaces), and

$$\begin{aligned} \mathscr{D}'(\overline{\mathbb{R}}^n_{\pm}) &= r^{\pm}\mathscr{D}'(\mathbb{R}^n) \quad , \\ \mathscr{S}'(\overline{\mathbb{R}}^n_{\pm}) &= r^{\pm}\mathscr{S}'(\mathbb{R}^n) \quad , \\ \mathscr{E}'(\overline{\mathbb{R}}^n_{\pm}) &= r^{\pm}\mathscr{E}'(\mathbb{R}^n) \quad ; \end{aligned} \tag{A.36}$$

the latter can be identified with the quotient spaces $\mathscr{D}'(\mathbb{R}^n)/\overset{\circ}{\mathscr{D}}'(\overline{\mathbb{R}}^n_{\mp})$, $\mathscr{S}'(\mathbb{R}^n)/\overset{\circ}{\mathscr{S}}'(\overline{\mathbb{R}}^n_{\mp})$ resp. $\mathscr{E}'(\mathbb{R}^n)/\overset{\circ}{\mathscr{E}}'(\overline{\mathbb{R}}^n_{\mp})$. (Our notation for $\overset{\circ}{\mathscr{S}}'$ is consistent with [Hörmander 1], but $\overset{\circ}{\mathscr{E}}'(\overline{\mathbb{R}}^n_{\pm})$ is there called $\mathscr{E}'(\overline{\mathbb{R}}^n_{\pm})$. The notation is further modified in [Hörmander 8].) One has that $C^\infty_0(\mathbb{R}^n_{\pm})$ is dense in each of the spaces (A.35), and $C^\infty_{(0)}(\overline{\mathbb{R}}^n_{\pm})$ is dense in each of the spaces (A.36). There are various natural dualities, e.g. $\overset{\circ}{\mathscr{E}}'(\overline{\mathbb{R}}^n_{\pm})$ identifies with the dual space of $C^\infty_{(0)}(\overline{\mathbb{R}}^n_{\pm})$, with respect to an extension of the scalar product (u,v) . The reason for writing $\overline{\mathbb{R}}^n_\pm$ in (A.36), in spite of the restriction to the open set $\mathbb{R}^n_\pm$, is that the elements in these spaces have to be "extendible across the boundary".

For the Sobolev spaces, one similarly defines two groups of spaces over $\overline{\mathbb{R}}^n_\pm$, for s and $t \in \mathbb{R}$,

(A.37)
$$\overset{\circ}{H}{}^s(\overline{\mathbb{R}}^n_\pm) = \{ u \in H^s(\mathbb{R}^n) \mid \operatorname{supp} u \subset \overline{\mathbb{R}}^n_\pm \} ,$$
$$\overset{\circ}{H}{}^{(s,t),\mu}(\overline{\mathbb{R}}^n_\pm) = \{ u \in H^{(s,t),\mu}(\mathbb{R}^n) \mid \operatorname{supp} u \subset \overline{\mathbb{R}}^n_\pm \} ,$$

provided with the norm of H^s resp. $H^{(s,t),\mu}$; and

(A.38)
$$H^s(\overline{\mathbb{R}}^n_\pm) = r^\pm H^s(\mathbb{R}^n) , \quad H^{(s,t),\mu}(\overline{\mathbb{R}}^n_\pm) = r^\pm H^{(s,t),\mu}(\mathbb{R}^n) ;$$
$$\text{with norm } \|u\|_{(s,t),\mu} = \inf\{\|v\|_{(s,t),\mu} \mid u = r^+v,\ v \in H^{(s,t),\mu}(\mathbb{R}^n)\} , \text{ etc.}$$

For $t = 0$ or $\mu = 0$, one gets as a special case the corresponding $H^{s,\mu}$ and $H^{(s,t)}$ spaces. Again it can be seen that $C_0^\infty(\mathbb{R}^n_\pm)$ is dense in the spaces (A.37) and $C^\infty_{(0)}(\overline{\mathbb{R}}^n_\pm)$ is dense in the spaces (A.38) (cf. [Hörmander 1]), so that there are dualities:

(A.39)
$$H^s(\overline{\mathbb{R}}^n_\pm) \text{ and } \overset{\circ}{H}{}^{-s}(\overline{\mathbb{R}}^n_\pm) \text{ are dual spaces ,}$$
$$H^{(s,t),\mu}(\overline{\mathbb{R}}^n_\pm) \text{ and } \overset{\circ}{H}{}^{(-s,-t),\mu}(\overline{\mathbb{R}}^n_\pm) \text{ are dual spaces ,}$$

consistently with the sesquilinear distribution duality.

An inspection of the extension procedure given in [Lions-Magenes 1] shows that there are extension operators $\ell^\pm_s$ from $H^{(s,t),\mu}(\overline{\mathbb{R}}^n_\pm)$ to $H^{(s,t),\mu}(\mathbb{R}^n)$ that are continuous, uniformly in μ , such that $r^\pm \ell^\pm_s = I$.

When m is integer ≥ 0 , the spaces $H^m(\mathbb{R}^n)$ and $H^{(m,t)}(\mathbb{R}^n)$ can also be described in terms of derivatives (cf. (A.18)):

(A.40)
$$H^m(\mathbb{R}^n) = \{u \in L^2(\mathbb{R}^n) \mid D^\alpha u \in L^2(\mathbb{R}^n) \quad \text{for} \quad |\alpha| \leq m\} .$$
$$H^{(m,t)}(\mathbb{R}^n) = \{u \in \mathscr{S}'(\mathbb{R}^n) \mid \langle\xi'\rangle^{t+m-j} D^j_{x_n} \acute{u}(\xi',x_n) \in L^2(\mathbb{R}^n) \quad \text{for} \quad 0 \leq j \leq m\} ;$$

here t is arbitrary in $\mathbb{R}$. Then moreover, for $m \in \mathbb{N}$ and $t \in \mathbb{R}$,

$$H^m(\overline{\mathbb{R}}^n_\pm) = \{u \in L^2(\mathbb{R}^n_\pm) \mid D^\alpha u \in L^2(\mathbb{R}^n_\pm) \quad \text{for} \quad |\alpha| \le m\},$$

(A.41)

$$H^{(m,t)}(\overline{\mathbb{R}}^n_\pm) = \{u \in \mathscr{S}'(\overline{\mathbb{R}}^n_\pm) \mid \langle\xi'\rangle^{t+m-j} D^j_{x_n} \acute{u}(\xi',x_n) \in L^2(\mathbb{R}^n_\pm) \quad \text{for } 0 \le j \le m\} \quad ,$$

with generalizations to $H^{(m,t),\mu}$, the norms satisfying

(A.42) $$\|u\|_{(m,t),\mu} \simeq \Big(\sum_{0\le j\le m} \|\langle\xi',\mu\rangle^{t+m-j} D^j_{x_n} \acute{u}(\xi',x_n)\|_0^2\Big)^{\frac{1}{2}} \; , \quad \text{uniformly in } \mu \; ;$$

in particular

(A.43) $$\|u\|_{m,\mu} \simeq (\|u\|_m^2 + \mu^{2m}\|u\|_0^2)^{\frac{1}{2}} \quad , \qquad \text{uniformly in } \mu \text{ for } \mu \ge 1 \; .$$

Here one has an identification as subspaces

$$\overset{\circ}{H}{}^m(\overline{\mathbb{R}}^n_\pm) \subset H^m(\overline{\mathbb{R}}^n_\pm) \quad ,$$

(A.44)

$$\overset{\circ}{H}{}^{(m,t),\mu}(\overline{\mathbb{R}}^n_\pm) \subset H^{(m,t),\mu}(\overline{\mathbb{R}}^n_\pm) \quad , \qquad \text{for} \quad m \in \mathbb{N} \; .$$

(In contrast, the spaces $\overset{\circ}{H}{}^{-m}(\overline{\mathbb{R}}^n_\pm)$ are not subspaces of $H^{-m}(\overline{\mathbb{R}}^n_\pm)$ for $m > 0$.)

For the values $s \in \,]-\frac{1}{2}, \frac{1}{2}[$, one can show that $H^s(\overline{\mathbb{R}}^n_\pm)$ and $\overset{\circ}{H}{}^s(\overline{\mathbb{R}}^n_\pm)$ coincide (and hence identify with subspaces of $H^s(\mathbb{R}^n)$), see [Lions-Magenes 1]. This can be used to extend some of the estimates we show later on for $s \ge 0$ to values $s > -\frac{1}{2}$, but we shall not make much effort in that direction.

We shall occasionally need the spaces

$$H^{(s,t)}_{\mathrm{loc}}(\overline{\mathbb{R}}^n_\pm) = r^\pm H^{(s,t)}_{\mathrm{loc}}(\mathbb{R}^n), \quad H^{(s,t)}_{\mathrm{comp}}(\overline{\mathbb{R}}^n_\pm) = r^\pm H^{(s,t)}_{\mathrm{comp}}(\mathbb{R}^n) \quad ,$$

(A.45)

$$\overset{\circ}{H}{}^{(s,t)}_{\mathrm{loc}}(\overline{\mathbb{R}}^n_\pm) = \{u \in H^{(s,t)}_{\mathrm{loc}}(\mathbb{R}^n) \mid \operatorname{supp} u \subset \overline{\mathbb{R}}^n_\pm\} \quad ,$$

$$\overset{\circ}{H}{}^{(s,t)}_{\mathrm{comp}}(\overline{\mathbb{R}}^n_\pm) = \{u \in H^{(s,t)}_{\mathrm{comp}}(\mathbb{R}^n) \mid \operatorname{supp} u \subset \overline{\mathbb{R}}^n_\pm\} \quad ,$$

and corresponding $H^{(s,t),\mu}$ versions; here there is duality between $H^{(s,t)}_{\mathrm{loc}}(\overline{\mathbb{R}}^n_\pm)$ and $\overset{\circ}{H}{}^{(-s,-t)}_{\mathrm{comp}}(\overline{\mathbb{R}}^n_\pm)$, etc. Also restrictions to other subsets of $\mathbb{R}^n$ can occur.

A.1 Remark. For $s \ge 0$, the spaces $\overset{\circ}{H}{}^s(\overline{\mathbb{R}}^n_\pm)$ coincide with the spaces $H^s_0(\mathbb{R}^n_\pm)$ of [Lions-Magenes 1], except when $s = \frac{1}{2}$ + integer; in that case $\overset{\circ}{H}{}^s(\overline{\mathbb{R}}^n_\pm)$ here is the same as $H^s_{00}(\mathbb{R}^n_\pm)$ there. The present spaces $H^s(\overline{\mathbb{R}}^n_\pm)$ and $\overset{\circ}{H}{}^s(\overline{\mathbb{R}}^n_\pm)$ have, unrestrictedly, the interpolation properties investigated in

[Lions-Magenes 1] (with exceptional cases there), because the interpolation properties can be derived directly from interpolation properties of the spaces $H^s(\mathbb{R}^n)$. More precisely, we have e.g. for $s > t \geq 0$ and $\theta \in [0,1]$,

$$\text{(A.46)}\qquad \begin{aligned} [H^s(\overline{\mathbb{R}}^n_\pm), H^t(\overline{\mathbb{R}}^n_\pm)]_\theta &= H^{(1-\theta)t+\theta s}(\overline{\mathbb{R}}^n_\pm) , \\ [\mathring{H}^s(\overline{\mathbb{R}}^n_\pm), \mathring{H}^t(\overline{\mathbb{R}}^n_\pm)]_\theta &= \mathring{H}^{(1-\theta)t+\theta s}(\overline{\mathbb{R}}^n_\pm) , \end{aligned}$$

where, as we recall from [Lions-Magenes 1, Ch. 1.2], $[X,Y]_\theta$ denotes the domain of $\Lambda^{1-\theta}$, when Λ is a selfadjoint positive operator in the Hilbert space Y with domain $D(\Lambda) = X \subset Y$; here $[X,Y]_\theta$ is a Hilbert space provided with the norm

$$\text{(A.47)}\qquad \|u\|_{[X,Y]_\theta} = (\|u\|_Y^2 + \|\Lambda^{1-\theta}u\|_Y^2)^{\frac{1}{2}} .$$

Recall also the well-known inequality for a selfadjoint positive operator Λ in a Hilbert space Y

$$\text{(A.48)}\qquad \|\Lambda^{1-\theta}u\|_Y \leq \|\Lambda u\|_Y^{1-\theta}\|u\|_Y^\theta \quad \text{on} \quad D(\Lambda) ,$$

and the resulting interpolation inequality for $u \in X = D(\Lambda)$:

$$\text{(A.49)}\qquad \|u\|_{[X,Y]_\theta} \leq C\|u\|_X^{1-\theta}\|u\|_Y^\theta , \quad \text{for} \quad \theta \in [0,1] ,$$

here $\|u\|_X$ is the graph-norm $(\|u\|_Y^2 + \|\Lambda u\|_Y^2)^{\frac{1}{2}}$.

When $s > \frac{1}{2}$, the distributions in $H^{(s,t)}(\overline{\mathbb{R}}^n_\pm)$ have boundary values (traces) on $\partial\overline{\mathbb{R}}^n_\pm = \mathbb{R}^{n-1}$, and the distributions in $H^{(s,t)}(\mathbb{R}^n)$ have restrictions to the submanifold $\mathbb{R}^{n-1}$. For precision, let us here denote, for $j \in \mathbb{N}$,

$$\text{(A.50)}\qquad \begin{aligned} \gamma_j &: C^\infty(\mathbb{R}^n) \ni u \sim D^j_{x_n} u(x',0) \in C^\infty(\mathbb{R}^{n-1}) , \\ \gamma_j^\pm &: C^\infty(\overline{\mathbb{R}}^n_\pm) \ni u \sim D^j_{x_n} u(x',0) \in C^\infty(\mathbb{R}^{n-1}) ; \end{aligned}$$

observe that γ_j^- is not quite analogous to γ_j^+ since the x_n-axis is directed into $\overline{\mathbb{R}}^n_+$ but out of $\overline{\mathbb{R}}^n_-$. It is well known that the γ_j extend to continuous mappings from $H^{(s,t)}(\mathbb{R}^n)$ to $H^{s+t-j-\frac{1}{2}}(\mathbb{R}^{n-1})$ when $s > j + \frac{1}{2}$, cf. e.g. [Hörmander 1] and [Lions-Magenes 1]. For completeness, we give the simple proof in the parameter-dependent version we need.

A.2 Lemma. *Let* $j \in \mathbb{N}$. *For each* $s > j + \frac{1}{2}$ *and* $t \in \mathbb{R}$, *the mappings*

(A.51)
$$\gamma_j : H^{(s,t),\mu}(\mathbb{R}^n) \to H^{s+t-j-\frac{1}{2},\mu}(\mathbb{R}^{n-1})$$
$$\gamma_j^{\pm} : H^{(s,t),\mu}(\overline{\mathbb{R}}^n_{\pm}) \to H^{s+t-j-\frac{1}{2},\mu}(\mathbb{R}^{n-1})$$

are continuous, uniformly in $\mu \in \overline{\mathbb{R}}_+$.

Proof: Let $u \in \mathscr{S}(\mathbb{R}^n)$. Using that for $v \in \mathscr{S}(\mathbb{R})$,

(A.52)
$$|v(0)| \leq \frac{1}{2\pi} \|\hat{v}\|_{L^1(\mathbb{R})} ,$$

we have, by the elementary properties of the Fourier transform,

(A.53)
$$\|\gamma_j u\|^2_{s+t-j-\frac{1}{2},\mu} = c\int_{\mathbb{R}^{n-1}} |D^j_{x_n} \acute{u}(\xi',0)|^2 \langle\xi',\mu\rangle^{2s+2t-2j-1} d\xi'$$
$$\leq c_1 \int_{\mathbb{R}^{n-1}} \langle\xi',\mu\rangle^{2s+2t-2j-1} \Big(\int_{\mathbb{R}} |\xi_n^j \hat{u}(\xi)|\, d\xi_n\Big)^2 d\xi'$$
$$\leq c_2 \int_{\mathbb{R}^{n-1}} \langle\xi',\mu\rangle^{2s+2t-2j-1} \Big(\int_{\mathbb{R}} |\xi_n^j \hat{u}(\xi)|^2 \langle\xi,\mu\rangle^{2s-2j} d\xi_n\Big)\Big(\int_{\mathbb{R}} \langle\xi,\mu\rangle^{-2s+2j} d\xi_n\Big)$$
$$= c_3 \int_{\mathbb{R}^n} \langle\xi',\mu\rangle^{2t} \langle\xi,\mu\rangle^{2s-2j} |\xi_n^j \hat{u}(\xi)|^2 d\xi \leq c_3 \|u\|^2_{(s,t),\mu} .$$

Since $\mathscr{S}(\mathbb{R}^n)$ is dense in $H^{(s,t),\mu}(\mathbb{R}^n)$, this implies the assertion for γ_j by extension by continuity. For the $\gamma_j^{\pm}$ one can now deduce the result by passing via the earlier mentioned continuous extension operators $\ell_s^{\pm}$. When s is an integer m, an alternative method is to work directly with the characterization (A.41), using the inequality

(A.54)
$$|v(0)|^2 = -\int_0^\infty \partial_t [v(t)\bar{v}(t)]dt \leq 2\|v\|_{L^2(\mathbb{R}_+)} \|\partial_t v\|_{L^2(\mathbb{R}_+)}$$

instead of (A.52). □

The lemma shows in particular that for $m \in \mathbb{N}$,

(A.55)
$$\rho_m^{\pm} = \{\gamma_0^{\pm}, \ldots, \gamma_{m-1}^{\pm}\}$$

is continuous from $H^{(s,t),\mu}(\overline{\mathbb{R}}^n_{\pm})$ to $\Pi_{0 \leq j < m} H^{s+t-j-\frac{1}{2},\mu}(\mathbb{R}^{n-1})$ for $s > m - \frac{1}{2}$, uniformly in μ. One can show that this mapping is surjective, cf. [Lions-Magenes 1], [Hörmander 1]; the details are carried out when needed, namely in Lemma 1.6.4. Furthermore, one has that the kernel of

$$\rho_m^{\pm} : H^{(m,t)}(\overline{\mathbb{R}}^n_{\pm}) \to \Pi_{0\leq j<m} H^{m+t-j-\frac{1}{2}}(\overline{\mathbb{R}}^{n-1}) \tag{A.56}$$

is precisely $\overset{\circ}{H}{}^{(m,t)}(\overline{\mathbb{R}}^n_{\pm})$.

Actually, we drop the + in ρ_m^+ or γ_j^+, when this causes no confusion.

In order to define the various concepts on manifolds, we recall that the spaces have certain invariance properties. For one thing, multiplication by a function $\varphi \in C_0^\infty(\mathbb{R}^n)$, resp. $\varphi \in C_{(0)}^\infty(\overline{\mathbb{R}}^n_{\pm})$, is continuous in all these spaces (uniformly in μ). Secondly, the isotropic "local" Sobolev spaces (A.29), and (A.45) with $t = 0$, are invariant under diffeomorphisms (cf. [Hörmander 1, Section 2.6]). For the anisotropic local spaces (A.45) with $t \neq 0$, there is invariance under coordinate changes in x' alone, or in x_n alone (one can use that for $s \geq 0$, $\langle\xi\rangle^s \simeq \langle\xi'\rangle^s + \langle\xi_n\rangle^s$, and for $s < 0$ there is a characterization by duality).

Now let Ω be an open subset of $\mathbb{R}^n$. Then we set

$$\begin{aligned} H^s(\overline{\Omega}) &= r_\Omega H^s(\mathbb{R}^n) \\ H^{(s,t),\mu}(\overline{\Omega}) &= r_\Omega H^{(s,t),\mu}(\mathbb{R}^n) \end{aligned} \tag{A.57}$$

where r_Ω is the restriction operator

$$r_\Omega = u \rightsquigarrow u|_\Omega \; ; \tag{A.58}$$

the spaces are Hilbert spaces with rorm e.g.

$$\|u\|_{H^{(s,t),\mu}(\overline{\Omega})} = \inf\{\|v\|_{(s,t),\mu} \mid v \in H^{(s,t),\mu}(\mathbb{R}^n) \;, \; u = r_\Omega v\} \; . \tag{A.59}$$

Again we have that

$$\|u\|_{s,\mu} \simeq (\|u\|_s^2 + \langle\mu\rangle^s \|u\|_0^2)^{\frac{1}{2}} \;, \text{ uniformly in } \mu \;, \text{ when } s \geq 0 \; . \tag{A.59'}$$

The reason for writing $\overline{\Omega}$ here is that the distributions in the space are required to have an extension property across $\partial\Omega$, which is not always satisfied by the elements in $H^s(\Omega)$ as defined e.g. in [Lions-Magenes 1], when Ω is not smooth. But in the smooth case, the definitions do coincide.

$\Omega \subset \mathbb{R}^n$ is said to be smooth, when every point $x \in \partial\Omega$ has a neighborhood ω for which there is a diffeomorphism κ of ω onto $B_1(0)$ (cf. (A.3)) such that

$$\begin{aligned}
&\kappa(x) = 0 \quad , \\
\text{(A.60)} \qquad &\kappa(\omega \cap \Omega) = B_1(0) \cap \mathbb{R}^n_+ \quad , \\
&\kappa(\omega \cap \partial\Omega) = B_1(0) \cap \{x_n = 0\}.
\end{aligned}$$

In the smooth case, $H^s(\bar{\Omega})$ is closely related to the $H^s(\bar{\mathbb{R}}^n_+)$ spaces. By an investigation of the invariance properties of the spaces under diffeomorphisms (cf. e.g. [Lions-Magenes 1], [Hörmander 1]), one can show the localization principle:

A.3 Lemma. *Let* Ω *be a smooth, bounded open subset of* $\mathbb{R}^n$, *and let* $\omega_1,\dots,\omega_N$ *be a system of open sets as above, with diffeomorphisms* $\kappa_1,\dots,\kappa_N$, *such that the sets* $\kappa_i^{-1}(B_{\frac{1}{2}}(0))$ *cover* $\partial\Omega$. *Let* $\varphi \in C_0^\infty(B_1(0))$ *with* $\varphi = 1$ *on* $B_{\frac{1}{2}}(0)$. *Then* $u \in H^s(\bar{\Omega})$ *if and only if* $u \in H^s_{loc}(\Omega)$ *and* $\varphi(y)u(\kappa_i^{-1}(y))$ *is in* $H^s(\bar{\mathbb{R}}^n_+)$ *for* $i=1,\dots,N$.

The anisotropic spaces $H^{(s,t),\mu}(\bar{\Omega})$ are not as simple, and here it is in fact more interesting to work with spaces obtained from the anisotropic spaces $H^{(s,t),\mu}(\bar{\mathbb{R}}^n_+)$ by diffeomorphisms as above. This can be done when we have a fixed normal coordinate defined in the neighborhood of $\partial\Omega$. We explain this in more detail below, where we describe the manifold situation.

A.5 *Spaces over manifolds.*

Let $\bar{\Omega}$ denote an n-dimensional C^∞ manifold with interior Ω and boundary Γ ; we assume for simplicity that $\bar{\Omega}$ is compact and connected, although much of the theory holds in more general cases. The structure of $\bar{\Omega}$ is then described by a finite system of coordinate mappings $\kappa_i : \Omega_i \to \Xi_i \subset \mathbb{R}^n$ $(i=1,\dots,i_0)$, where the Ω_i are relatively open in $\bar{\Omega}$, $\cup_{1\le i\le i_0}\Omega_i = \bar{\Omega}$, the Ξ_i are of the form $B_r(0)$ (for $\Omega_i \subset \Omega$) or $B_r(0) \cap \bar{\mathbb{R}}^n_+$ (for $\Omega_i \cap \partial\Omega \neq \emptyset$) , and the κ_i have the property: $\kappa_i \circ \kappa_j^{-1}$ is a diffeomorphism from $\kappa_j(\Omega_i \cap \Omega_j)$ to $\kappa_i(\Omega_i \cap \Omega_j)$ for all $i,j=1,\dots,i_0$. (The κ_i must furthermore be compatible with any other choice of coordinate mappings $\{\kappa_i' : \Omega_i' \to \Xi_i' \mid i=1,\dots,i_0'\}$.) We can assume that $\bar{\Omega}$ is smoothly imbedded in an n-dimensional C^∞ manifold Σ without boundary (the final results do not depend on how Σ is chosen; it should just be considered as a neighborhood of $\bar{\Omega}$).

For points in Σ resp. Γ we use the notation x resp. x' . For convenience, we work under the following hypothesis, although the operator classes are invariant under diffeomorphisms preserving the boundary: We assume that a normal coordinate x_n has been chosen near Γ, in the sense that a fixed neighborhood Σ_2' of Γ in Σ is represented as

$$\Sigma_2' = \Gamma \times]-2,2[$$

with points $x = (x',x_n)$ ($x' \in \Gamma$ and $x_n \in]-2,2[$), and we denote

$$\Sigma' = \Gamma \times]-1,1[\quad ,$$

(A.61) $$\Sigma_+' = \Sigma' \cap \overline{\Omega} = \Gamma \times [0,1[\quad ,$$

$$\Sigma_-' = \Sigma' \cap (\Sigma \setminus \Omega) = \Gamma \times]-1,0] \ .$$

(One can in fact replace Σ by $\Omega \cup \Sigma_-'$, unless a compact Σ is desired.) We assume that Σ and Γ are provided with positive C^∞ densities dx resp. dx' such that $dx = dx'dx_n$ on Σ_2'. We use again the conventions (A.10-11) for functions on Σ or Γ. (If Σ is provided with a Riemannian metric, this gives such a normal coordinate.)

The local coordinate maps describing Σ are chosen to fit with the above. Cover Γ by open sets $\Gamma_1,\dots,\Gamma_{i_1}$, such that Σ_2' is covered by the open sets

$$\Sigma_i = \Gamma_i \times \ ]-2,2[\qquad i=1,\dots,i_1$$

and assume that there are coordinate maps $\kappa_i' : \Gamma_i \to \Xi_i'$ (open $\subset \mathbb{R}^{n-1}$) such that the maps $\kappa_i : (x',x_n) \rightsquigarrow (\kappa_i'(x'),x_n)$ from Σ_i to $\Xi_i = \Xi_i' \times]-2,2[$ are coordinate maps for Σ. Next, cover $\overline{\Omega}\setminus\Sigma_+'$ by open sets Σ_i for $i=i_1+1,\dots,i_2$, with associated coordinate maps $\kappa_i : \Sigma_i \to \Xi_i \subset \mathbb{R}^n_+$, and cover $\Sigma_2' \setminus (\Gamma \times]-1,2[)$ by open sets Σ_i for $i=i_2+1,\dots,i_3$, with associated coordinate maps $\kappa_i : \Sigma_i \to \Xi_i \subset \mathbb{R}^n_-$. If not all of Σ is covered, one can cover the rest by a finite or infinite set $(\Sigma_i)_{i=i_3+1,\dots,}$ with coordinate maps $\kappa_i : \Sigma_i \to \Xi \subset \mathbb{R}^n_-$. It can be assumed that the Ξ_i are bounded, with mutually disjoint closures, and that the κ_i are restrictions of coordinate maps defined on neighborhoods of the $\overline{\Sigma}_i$.

Instead of functions on our manifolds, we may consider sections in vector bundles over them. Here we let $\widetilde{E}$ denote an N-dimensional complex C^∞ vector bundle (a generalization of vector valued functions) over Σ, with restrictions

(A.62) $$E = \widetilde{E}|_{\overline{\Omega}} \quad , \qquad E_\Gamma = \widetilde{E}|_\Gamma \quad ;$$

usually, E is given, and $\widetilde{E}$ is some convenient extension. We identify E over Σ_2' with the lifting of E_Γ, which makes differentiation of sections with respect to x_n well-defined here. (Detailed explanations of vector bundles can be found e.g. in [Atiyah 1] or [Steenrod 1].) The coordinate maps are now assumed to be so conveniently chosen that $\tilde{E}|_{\Sigma_i}$ is trivial for each i, with trivializations $\psi_i : \tilde{E}|_{\Sigma_i} \to \Xi_i \times \mathbb{C}^N$. (The linear map at each $x \in \Sigma_i$ is denoted $\psi_{i,x} : \tilde{E}_x \to \mathbb{C}^N$, and the restriction to Γ_i for $i=1,\dots,i_1$ is denoted $\psi_i' : \tilde{E}|_{\Gamma_i} \to \Xi_i' \times \mathbb{C}^N$.) The sections u in $\tilde{E}$ can over each Σ_i be represented by N-vector valued functions u_i on Ξ_i

$$u_i(y) = \psi_{i,x} u(x) \quad \text{when} \quad y = \kappa_i(x) ,$$

that we express briefly as

$$u_i = \psi_i \circ u .$$

For the symbolic calculus, we need to consider the cotangent bundles $T^*(\Sigma)$, $T^*(E)$ and $T^*(E_\Gamma)$ also. They are n-dimensional real vector bundles, and we can assume that the sets Σ_i are chosen so small that the local coordinate maps $\kappa_i : \Sigma_i \to \Xi_i$ are associated with trivializations $\tilde{\kappa}_i : T^*(\Sigma_i) \to \Xi_i \times \mathbb{R}^n$, such that at each $x \in \Sigma_i \cap \Sigma_j$, the linear map $\tilde{\kappa}_{i,x}\tilde{\kappa}_{j,x}^{-1}$ in $\mathbb{R}^n$ (strictly speaking: from $\{\kappa_j(x)\}\times\mathbb{R}^n$ to $\{\kappa_i(x)\}\times\mathbb{R}^n$) equals precisely the transpose of the Jacobian of $\kappa_j\kappa_i^{-1}$ at $\kappa_i(x)$. This is one way of describing the structure of $T^*(\Sigma)$, that fits directly with the purpose of defining principal symbols invariantly (cf. Lemma 2.1.17 etc.). For Γ, the cotangent bundle $T^*(\Gamma)$ can be described in this way in relation to the coordinate systems $\kappa_i' : \Gamma_i \to \Xi_i'$ for $i=1,\dots,i_1$; and since $\Sigma_2' = \Gamma \times]-2,2[$, one has an identification of $T^*(\Sigma_2')$ with the lifting of $T^*(\Gamma) \oplus \mathbb{R}$ to $\Gamma \times]-2,2[$.

For $s \geq 0$, the Sobolev space $H^s_{loc}(\tilde{E})$ is now defined as the space of sections u in $\tilde{E}$ such that

(A.63) $\quad \psi_i \circ u \in H^s_{loc}(\Xi_i)^N = H^s_{loc}(\Xi_i) \times \dots \times H^s_{loc}(\Xi_i) \quad$ (N components) ;

and the spaces $H^s_{comp}(\tilde{E})$ consists of the elements in $H^s_{loc}(\tilde{E})$ that have compact support in Σ. The topology is defined by the seminorms on each Ξ_i, taken together. For negative s, the corresponding spaces can be defined by sesquilinear duality - either directly (taking $H^{-s}_{loc}(\tilde{E})$ as the dual of $H^s_{comp}(\tilde{E})$, etc.) or via distributions (defining $\mathscr{D}'(\tilde{E})$ first and taking $H^{-s}_{loc}(\tilde{E})$ as the distributions that trivialize to elements in $H^{-s}_{loc}(\Xi)^N$).

They are Hilbert spaces if Σ is compact. (Truly coordinate-free descriptions can be obtained by use of half-densities, as in [Hörmander 6, 8].)

Defining restriction and extension operators by

$$\text{(A.64)}\qquad \begin{aligned} & r_\Omega v = v|_\Omega \qquad\qquad , \text{ also denoted } r^+v \ , \\ & e_\Omega u = \begin{cases} u & \text{on } \Omega \\ 0 & \text{on } \Sigma\setminus\Omega \end{cases} , \text{ also denoted } e^+u \ , \end{aligned}$$

we define the Sobolev spaces over $\overline{\Omega}$ by

$$\text{(A.65)}\qquad \begin{aligned} & H^s(E) = r^+H^s_{loc}(\widetilde{E}) \\ & \mathring{H}^s(E) = \{u \in H^s_{loc}(\widetilde{E}) \mid \operatorname{supp} u \subset \overline{\Omega}\} \ , \end{aligned}$$

they are Hilbert spaces. Lemma A.3 extends to show that the elements in $H^s(E)$ are characterized near Γ by carrying over to elements of $H^s(\overline{\mathbb{R}}^n_+)^N$ in local trivializations. Also the parameter-dependent spaces are needed, here $H^{s,\mu}(E)$ and $\mathring{H}^{s,\mu}(E)$ are provided with norms such as

$$\text{(A.66)}\qquad \|u\|_{s,\mu} = \Big(\sum_{i=1}^{i_3} \|\psi_i \circ (\varphi_i u)\|^2_{s,\mu}\Big)^{\frac12}$$

where the φ_i , $i=1,\dots,i_3$ are in $C_0^\infty(\Sigma_i)$, respectively, with $\Sigma\,\varphi_i = 1$. The standard trace operators are defined near Γ (cf. (A.61)) by

$$\gamma_j u = \Big(\frac{1}{i}\frac{\partial}{\partial x_n}\Big)^j u|_\Gamma$$

for smooth sections, and Lemma A.2 implies the uniform boundedness of e.g.

$$\text{(A.67)}\qquad \gamma_j : H^{s,\mu}(E) \to H^{s-j-\frac12,\mu}(E_\Gamma) \quad \text{when} \quad s > j + \tfrac12 \ ,$$

so that the trace operators can be extended by continuity.

Now the anisotropic spaces $H^{(s,t)}$ do not have a global meaning, but they can be defined over e.g. Σ_2' , Σ' and Σ_+' by use of the local coordinate maps we have described (as in (A.63), (A.66)), leading to spaces of Fréchet or *LF*-type

$$\text{(A.68)}\qquad \begin{aligned} & H^{(s,t),\mu}_{loc}(\widetilde{E}|_{\Sigma_2'}) \quad \text{and} \quad H^{(s,t),\mu}_{loc}(\widetilde{E}|_{\Sigma'}) \ , \\ & H^{(s,t),\mu}_{loc}(E|_{\Sigma_+'}) = r^+H^{(s,t),\mu}_{loc}(\widetilde{E}|_{\Sigma'}) \ , \\ & \mathring{H}^{(s,t),\mu}_{loc}(E|_{\Sigma_+'}) = \{u \in H^{(s,t),\mu}_{loc}(\widetilde{E}|_{\Sigma'}) \mid \operatorname{supp} u \subset \Sigma_+'\} \ , \end{aligned}$$

and Hilbert spaces

$$(A.69)\qquad H^{(s,t),\mu}(\tilde{E}|_{\Sigma'})\ ,\quad H^{(s,t),\mu}(E|_{\Sigma'_+})\ ,\quad \mathring{H}^{(s,t),\mu}(E|_{\Sigma'_+})$$

by restriction to Σ' resp. Σ'_+ . For $s > j+\frac{1}{2}$, the trace operators γ_j have a meaning on these spaces and are continuous e.g. as follows

$$(A.70)\qquad \gamma_j : H^{(s,t),\mu}(E|_{\Sigma'_+}) \to H^{s+t-j-\frac{1}{2}}(E_\Gamma) \quad \text{when} \quad s > j+\tfrac{1}{2}\ .$$

For m integer > 0 , the space $\mathring{H}^{(m,t)}(E|_{\Sigma'_+})$ consists of the sections $u \in H^{(m,t)}(E|_{\Sigma'_+})$ with $\gamma_j u = 0$ for $0 \leq j \leq m-1$. ($E|_{\Sigma'_+}$ is also written $E_{\Sigma'_+}$.)

A.6 *Notions from spectral theory.*

The functional calculus we shall develop, has applications to spectral theory. We here collect some definitions and well-known facts from the spectral theory of compact operators and their inverses.

Let T be a compact operator from a Hilbert space H to another H_1 . When $H = H_1$ and T is selfadjoint ≥ 0 , we denote by $\lambda_k(T)$ the positive eigenvalues, arranged in a nondecreasing sequence and repeated according to multiplicity; they are labelled by $k \in \mathbb{N}_+ = \{1,2,\dots\}$ or by a finite subset $\{1,2,\dots,N\}$. When T is merely selfadjoint, we denote by $\lambda_k^{\pm}$ the positive resp. negative eigenvalues (with $\pm\lambda_k^{\pm}$ arranged monotonically as above). For general compact operators $T\colon H \to H_1$, we denote by $s_k(T)$ the characteristic values, that are defined as the eigenvalues of

$$(A.71)\qquad |T| = (T^*T)^{\frac{1}{2}}$$

so that

$$(A.72)\qquad s_k(T) = \lambda_k((T^*T)^{\frac{1}{2}}) = (\lambda_k(T^*T))^{\frac{1}{2}} \quad \text{for} \quad k=1,2,\dots\ ;$$

in particular, $s_k(T) = \lambda_k(T)$ when T is selfadjoint ≥ 0 .

For a selfadjoint operator T one can define the selfadjoint nonnegative operators

$$(A.73)\qquad T_{\pm} = \frac{1}{2}\,(|T| \pm T)$$

where T_+ and $-T_-$ are the positive resp. negative part of T . (T_+ coincides with T on the space spanned by the eigenvalues ≥ 0 , being zero on its complement, and T_- has the same rôle relative to $-T$.) In particular,

$$(A.74)\qquad \lambda_k^+(T) = \lambda_k(T_+) \quad \text{and} \quad \lambda_k^-(T) = -\lambda_k(T_-) \quad \text{for} \quad k=1,2,\dots\ .$$

The characteristic values have various properties that can be derived on the basis of the minimum-maximum property, cf. [Gohberg-Krein 1]:

$$\text{(A.75)}\qquad \begin{aligned} &\text{(i)}\quad s_{j+k-1}(T+T') \le s_j(T) + s_k(T') \\ &\text{(ii)}\quad s_{j+k-1}(TT') \le s_j(T)s_k(T') \\ &\text{(iii)}\quad s_k(ATB) \le \|A\| s_k(T)\|B\| \end{aligned}$$

for compact operators T and T' (from H to H_1 or H_2 to H) and bounded operators A and B (from H_1 to H_3 resp. from H_2 to H); similar estimates hold for $\lambda_k^{\pm}(T)$ in selfadjoint cases. We also recall that

$$\text{(A.76)}\qquad \lambda_k(T^*T) = \lambda_k(TT^*) \quad \text{for all } k .$$

For $p > 0$, the Schatten class $\mathfrak{C}_p$ is defined as the space of compact operators T in H for which

$$\text{(A.77)}\qquad \|T\|_{\mathfrak{C}_p} \equiv \Big(\sum_k s_k(T)^p\Big)^{1/p} < \infty ;$$

it is a Banach space with this norm. In particular, $\mathfrak{C}_1$ is the space of trace class operators and $\mathfrak{C}_2$ is the space of Hilbert-Schmidt operators. For the analysis of asymptotic properties in spectral theory, it is also interesting to consider the class $\mathfrak{S}_{(p)}$ of compact operators for which

$$\text{(A.78)}\qquad N_p(T) \equiv \sup_k s_k(T)k^{1/p} < \infty$$

here $N_p(T)$ is a quasi-norm (satisfying a non-standard triangle inequality deduced from (A.75 i)); a thorough description of these and other classes is given in [Birman-Solomiak 1]. Note that $\mathfrak{S}_{(p)} \subset \mathfrak{C}_{p+\varepsilon}$ for any $\varepsilon > 0$. The following result is well known (cf. [Agmon 2-4], [Beals 2], [Paraska 1]):

A.4 Lemma. *Let* Ξ *be a smooth compact* n*-dimensional manifold (with or without boundary), and let* T *be an operator in* $L^2(\Xi)$ *that is continuous from* $L^2(\Xi)$ *to* $H^a(\Xi)$ *for some* $a \in \mathbb{R}_+$. *Then* T *is compact, and there is a constant* $c(T)$ *so that*

$$\text{(A.79)}\qquad s_k(T) \le c(T)k^{-a/n} \quad \textit{for all } k ;$$

in other words, $T \in \mathfrak{S}_{(n/a)}$. *The statement also holds for a continuous operator* T *from* $L^2(E)$ *to* $H^a(F)$, *where* E *and* F *are* C^∞ *vector bundles over* Ξ.

Besides the eigenvalues or characteristic values of T, it is often convenient to consider the "counting function", indicating the number of eigenvalues or characteristic values outside an interval around zero. We here use the notation

$$\text{(A.80)}\quad \begin{aligned} N'(t\,;T) &= \text{number of } s_k(T) \geq 1/t \\ N'^{\pm}(t\,;T) &= \text{number of } \lambda_k^+(T) \text{ resp. } \lambda_k^-(T) \text{ outside }]-1/t,\, 1/t[\ , \end{aligned}$$

for any $t > 0$. The definition is chosen so that it matches the usual definition for $A = T^{-1}$ in case T is invertible. Here we use the conventions:

Let $A = T^{-1}$ where T is compact, injective and has dense range. When A is selfadjoint positive, the eigenvalues are arranged nondecreasingly in a sequence

$$\lambda_1(A) \leq \lambda_2(A) \leq \dots \rightarrow \infty$$

with repetitions according to multiplicity. In general, the k'th characteristic value is defined as $s_k(A) = \lambda_k(A^*A)^{\frac{1}{2}}$; and when A is selfadjoint, we also consider the monotone sequences

$$\begin{aligned} 0 &< \lambda_1^+(A) \leq \lambda_2^+(A) \leq \dots \\ 0 &> \lambda_1^-(A) \geq \lambda_2^-(A) \geq \dots \end{aligned}$$

of positive resp. negative eigenvalues, repeated according to multiplicity. (Finite sequences can occur.) So, with a slight abuse of notation,

$$\text{(A.80')}\quad \begin{aligned} \lambda_k(A) &= \lambda_k(T)^{-1} \quad , \\ s_k(A) &= s_k(T)^{-1} \quad , \\ \lambda_k^{\pm}(A) &= \lambda_k^{\pm}(T)^{-1} \quad , \end{aligned}$$

in the respective cases. The counting functions are here defined by

$$\text{(A.81)}\quad \begin{aligned} N(t\,;A) &= \text{number of } s_k(A) \text{ in }]0,t] \\ N^{\pm}(t\,;A) &= \text{number of } \lambda_k^+(A) \text{ resp. } \lambda_k^-(A) \text{ in } [-t,t]\ . \end{aligned}$$

Note that

$$\text{(A.82)}\quad \begin{aligned} N(t\,;A) &= N'(t\,;A^{-1}) \quad , \\ N^{\pm}(t\,;A) &= N'^{\pm}(t\,;A^{-1}) \quad ; \end{aligned}$$

in the respective cases. The definitions extend easily to operators A with a non-trivial nullspace.

The counting functions are more convenient than the characteristic values in certain considerations involving powers of the operators. Note that the functions $t \sim N(t\,;A)$ and $k \sim s_k(A)$ are essentially inverse functions of one another. Moreover, one has:

A.5 Lemma. *Let* T *be a compact operator in* H. *Let* $p > q > 0$, *and let* $C_0 > 0$. *Then* T *has one of the properties, for* $k \to \infty$,

$$\text{(A.84)}\qquad \begin{aligned} &\text{(i)} && s_k(T) = O(k^{-1/p}) \;, \\ &\text{(ii)} && s_k(T) = C_0^{1/p}\, k^{-1/p} + o(k^{-1/p}) \;, \\ &\text{(iii)} && s_k(T) = C_0^{1/p}\, k^{-1/p} + O(k^{-1/q}) \;; \end{aligned}$$

if and only if the counting functions has the corresponding property, for $t \to \infty$,

$$\text{(A.85)}\qquad \begin{aligned} &\text{(i)} && N'(t\,;T) = O(t^p) \;, \\ &\text{(ii)} && N'(t\,;T) = C_0 t^p + o(t^p) \;, \\ &\text{(iii)} && N'(t\,;T) = C_0 t^p + O(t^s) \;; \end{aligned}$$

here s *and* q *are related by*

$$\text{(A.86)}\qquad \begin{aligned} s &= p + 1 - \frac{p}{q} \;, \\ q &= \frac{p}{p - s + 1} \;. \end{aligned}$$

In particular, when $p = n/a$, *then* $q = n/(\theta + a)$ *corresponds to* $s = (n-\theta)/a$.

Similar statements hold for the positive resp. negative eigenvalues, when T *is selfadjoint.*

When T *is injective with dense range, and* A *equals* T^{-1}, *we can replace* $N'(t\,;T)$ *by* $N(t\,;A)$ *in* (A.85) *(and* $N'^{\pm}(t\,;T)$ *by* $N^{\pm}(t\,;A)$ *in the statements for the selfadjoint case), using* (A.80').

In this lemma, the equivalence of the statements (i), and of the statements (ii), is fairly obvious; and since the equivalence of the statements (iii) was shown in [Grubb 6, Lemma 6.2], we shall not go into details with it here.

One has the following perturbation results:

A.6 Lemma. *Let* T *and* T' *be compact operators in* H , *and let* $p > q > 0$, $p > r > 0$. *Let* $C_0 \geq 0$.

1^0 *If* $s_k(T)k^{1/p} \to C_0$ *and* $s_k(T')k^{1/p} \to 0$ *for* $k \to \infty$, *then* $s_k(T+T')k^{1/p} \to C_0$.

2^0 *If* T *and* T' *satisfy, for* $k \to \infty$,

$$\text{(A.87)}\qquad \begin{aligned} &s_k(T) - C_0^{1/p}\, k^{-1/p} \quad \text{is} \quad O(k^{-1/q}) , \\ &s_k(T') \quad \text{is} \quad O(k^{-1/r}) , \end{aligned}$$

then $T+T'$ *satisfies*

$$\text{(A.88)}\qquad s_k(T+T') - C_0^{1/p}\, k^{-1/p} \quad \text{is} \quad O(k^{-1/q'}) ,$$

where

$$\text{(A.89)}\qquad q' = \max\left\{q \;,\; p\,\frac{r+1}{p+1}\right\} .$$

For the counting function $N'(t\,;T+T')$, (A.88) *with* $C_0 > 0$ *means that, with* s *defined by* (A.86),

$$\text{(A.90)}\qquad \begin{aligned} &N'(t\,;T+T') - C_0 t^p \quad \text{is} \quad O(t^{s'}) , \quad \text{where} \\ &s' = \max\left\{s,\; r\,\frac{p+1}{r+1}\right\} . \end{aligned}$$

In the selfadjoint case, one can replace $s_k(T)$ *and* $s_k(T+T')$ *by* $\lambda_k^{\pm}(T)$ *and* $\lambda_k^{\pm}(T+T')$ *in these statements.*

Here 1^0 is the Weyl-Ky Fan theorem (cf. e.g. [Gohberg-Krein 1, Theorem II.2.3]), and 2^0 is a sharpening shown in [Grubb 6, Proposition 6.1]; the statement on the counting function follows immediately from Lemma A.5.

When T and $T_1 = T+T'$ are injective with densely defined inverses $A = T^{-1}$ and $A_1 = (T+T')^{-1}$, the statements translate by (A.82) and Lemma A.5 to statements on $N(t\,;A)$ and $N(t\,;A_1)$ (with $T' = A_1^{-1} - A^{-1}$).

The asymptotic estimates are sometimes found by a passage to the limit; here the following lemma can be useful (for the proof see e.g. [Grubb 17, Lemma 4.2]).

<u>A.7 Lemma</u>. *Let* T_M *and* T'_M $(M \in \mathbb{N})$ *be sequences of compact operators in* H , *and let* $p > 0$.

1° *If* $T_M \to T$ *in the operator norm, and* $N_p(T_M) \leq C$ *for all* M , *then* $N_p(T) \leq C$.

2° *If* $T = T_M + T'_M$ *for each* M , *where*

$$(A.91) \qquad \begin{aligned} s_k(T_M)k^{1/p} &\to C_M \qquad \text{for} \quad M \to \infty , \\ s_k(T'_M)k^{1/p} &\leq \varepsilon_M \qquad \text{for all} \quad k , \end{aligned}$$

with $C_M \to C_0$ *and* $\varepsilon_M \to 0$ *for* $M \to \infty$, *then*

$$(A.92) \qquad s_k(T)k^{1/p} \to C_0 \qquad \text{for} \quad k \to \infty .$$

For operators in general (always linear in the present book), we use the notation

$$(A.93) \qquad \begin{aligned} D(S) &= \text{the domain of } S , \\ R(S) &= \text{the range of } S , \\ Z(S) &= \text{the nullspace (kernel) of } S . \end{aligned}$$

When such precision is not needed, we may use the same formal expression for operators acting similarly but with various domains (e.g. differential or pseudo-differential operators on function or distribution spaces).

An operator $S: H \to H_1$ is called a Fredholm operator, when $\dim Z(S) < \infty$ and $R(S)$ is closed, with a finite dimensional complement V ; then the index is defined by

$$(A.94) \qquad \text{index } S = \dim Z(S) - \dim V .$$

The spectrum of an operator S in H is denoted $sp(S)$; and the essential spectrum is denoted $ess\ sp(S)$, it is the complement of the set of λ for which $S-\lambda$ is Fredholm. We recall that $ess\ sp(S)$ is invariant under compact perturbations.

BIBLIOGRAPHY

M.S. Agranovich and M.I. Vishik:

1. Elliptic problems with a parameter and parabolic problems of a general type, Usp. Mat. Nauk 19 (1963), 53-161 = Russ. Math. Surveys 19, 53-159.

S. Agmon:

1. The coerciveness problem for integro-differential forms, J. Analyse Math. 6 (1958), 183-223.
2. On the eigenfunctions and on the eigenvalues of general elliptic boundary value problems, Comm. Pure Appl. Math. 15 (1962), 119-147.
3. *Lectures on Elliptic Boundary Value Problems*. Van Nostrand Math. Studies, D. Van Nostrand Publ. Co., Princeton 1965.
4. On kernels, eigenvalues, and eigenfunctions of operators related to elliptic problems, Comm. Pure Appl. Math. 18 (1965), 627-663.
5. Asymptotic formulas with remainder estimates for eigenvalues of elliptic operators, Arch. Rat. Mech. An. 28 (1968), 165-183.

M.F. Atiyah:

1. K-*Theory*. Benjamin, Amsterdam 1967.

M.F. Atiyah and R. Bott:

1. The index problem for manifolds with boundary, Coll. Differential Analysis, Tata Institute, Bombay, Oxford University Press 1964, 175-186.

M.F. Atiyah, R. Bott and V.K. Patodi:

1. On the heat equation and the index theorem, Inventiones Math. 19 (1973), 279-330.

M.F. Atiyah and I.M. Singer:

1. The index of elliptic operators, Ann. of Math. 87 (1968), 434-530.

R. Beals:

1. Asymptotic behavior of the Green's function and spectral function of an elliptic operator, J. Functional Analysis 5 (1970), 484-503.
2. Classes of compact operators and eigenvalue distribution for elliptic operators, Am. J. Math. 89 (1967), 1056-1072.

M.S. Birman:

1. On the theory of self-adjoint extensions of positive definite operators, Mat. Sb. 38: 90 (1956), 431-450 (Russian).

M.S. Birman and M.Z. Solomiak:

1. Estimates of singular numbers of integral operators, Russian Math. Surveys 32 (1977), 15-89.
2. Asymptotics of the spectra of pseudodifferential operators with anisotropic-homogeneous symbols, Vestnik Leningrad Univ. 13 (1977), 13-21.
3. Asymptotic behavior of the spectrum of differential equations, Itogi Nauk. Tekh., Mat. An. 14 (1977), 5-58 = J. Soviet Math. 12 (1979), 247-283.

J.M. Bony, P. Courrège and P. Priouret:

1. Semi-groupes de Feller sur une variété à bord compacte et problèmes aux limites intégro-différentiels du second ordre donnant lieu au principe du maximum, Ann. Inst. Fourier (Grenoble) 18 (1968), 369-521.

L. Boutet de Monvel:

1. Comportement d'un opérateur pseudo-différentiel sur une variéte à bord, I-II, J. d'Analyse Fonct. 17 (1966), 241-304.
2. Opérateurs pseudo-differentiels et problèmes aux limites elliptiques, Ann. Inst. Fourier 19 (1969), 169-268.
3. Boundary problems for pseudo-differential operators, Acta Math. 126 (1971), 11-51.
4. *A course on Pseudo-differential Operators and their Applications*. Duke University Mathematics Series II, 1976.

A. Bove, B. Franchi and E. Obrecht:

1. A boundary value problem for elliptic equations with polynomial coefficients in a half space. I Pseudodifferential operators and function spaces, Boll. Un. Mat. Ital. B 18 (1981), 25-45. II The boundary value problem, Boll. Un. Mat. B 18 (1981), 355-380.

Čan Zui Ho and G. Eskin:

1. Boundary value problems for parabolic systems of pseudo-differential equations, Dokl. Akad. Nauk SSSR 198 (1971), 50-53 = Soviet Math. Dokl. 12 (1971), 739-743.

C. Cancelier:

1. Problèmes aux limites pseudodifférentiels donnant lieu au principe de maximum. Thesis, Univ. de Paris-Sud, Orsay 1984.

H.O. Cordes:

1. *Elliptic Pseudo-differential Operators - An Abstract Theory*. Lecture Note 756, Springer Verlag, Berlin 1979.

M. de Gosson:

1. Paramétrix de transmission pour des opérateurs de type parabolique et applications au problème de Cauchy microlocal, C.R. Acad. Sci. Paris (Sér.A) 292 (1981), 51-53.

Y. Demay:

1. Paramétrix pour des systèmes hyperboliques du premiér ordre à multiplicité constante, J. Math. pures et appl. 56 (1977), 393-422.

A. Demidov:

1. Asymptotic behavior of the solution of a boundary value problem for elliptic pseudodifferential equations with a small parameter multiplying the leading operator, Trudy Moskov. Mat. Obsv. 32 (1975), 119-146 = Trans. Moscow Math. Soc. 32 (1975), 115-142.

Ja.M. Drin:

1. Study of one class of parabolic pseudodifferential operators in the spaces of Hölder's functions,Notes of the Ukrainian Academy of Sciences, Series A no. 1 1974, 19-22. (Russian, English summary.)

2. On the structure of pseudodifferential operators with generalized-homogeneous characters. Notes of the Ukrainian Academy of Sciences, Series A (Physical-technical and mathematical sciences) no.7 1974, 592-595. (Russian, English summary.)

3. Fundamental solution of the Cauchy problem for one class of parabolic pseudo-differential equations, Notes of the Ukrainian Academy of Sciences, Series A (physical-mathematical and technical sciences) no.3 1977, 197-203. (Russian.)

J.J. Duistermaat and V. Guillemin:

1. The spectrum of positive elliptic operators and elliptic bicharacteristics, Inventiones Math. 29 (1975), 39-79.

J. Dunau:

1. Fonctions d'un opérateur elliptique sur une variété compacte. J. Math. pures et appl. 56 (1977), 367-391.

N. Dunford and J. Schwartz:

1. *Linear Operators, Part I: General Theory.* Interscience Publishers, New York 1957.

S.D. Eidelman:

1. *Parabolic Systems.* North-Holland Publ. Co. Amsterdam 1969.

S.D. Eidelman and Ja.M. Drin:

1. Necessary and sufficient conditions for the stability of solutions of the Cauchy problem for parabolic pseudodifferential equations. Approximation Methods in Mathematical Analysis, Kiev State Pedagogical Institute, Kiev 1974, 60-71. (Russian.)

2. Construction and investigation of classical fundamental solutions for one-dimensional Cauchy problems for parabolic pseudodifferential equations. Mathematical Investigations, Academy of Science of the Moldavian Republic 1981, 18-33. (Russian.)

G.I. Eskin:

1. *Boundary Value Problems for Elliptic Pseudodifferential Equations.* Transl. Math. Monogr. Vol.52, Amer. Math. Soc. Providence 1981.

2. Asymptotics near the boundary of spectral functions of elliptic self-adjoint boundary problems, Israel J. Math. 22 (1975), 214-246.

E.B. Fabes and M. Jodeit:

1. Boundary value problems for second order parabolic equations, Amer. Math. Soc. Proc. Symp. Pure Math. 10 (1967), 82-105.

E.B. Fabes and N.M. Rivière:

1. Symbolic calculus of kernels with mixed homogeneity, Amer. Math. Soc. Proc. Symp. Pure Math. 10 (1967), 106-137.

B.V. Fedosov:

1. Analytical formulas for the index of elliptic operators, Trudy Mosk. Mat. Obsv. 30 (1974), 159-241 = Trans. Moscow Math. Soc. 30 (1974), 159-240.

L.S. Frank:

1. Coercive singular perturbations I: A priori estimates, Ann. Mat. pura appl. 119 (1979), 41-113.

L.S. Frank and W.D. Wendt:

1. Coercive singular perturbations II: Reduction to regular perturbations and applications, Comm. Part. Diff. Equ. 7 (1982), 469-535.

1'. Coercive singular perturbations II: Reduction and convergence, J. Math. Anal. Appl. 88 (1982), 463-545. [Much of the material here is repeated verbatim in 1.]

2. Coercive singular perturbations III: Wiener-Hopf operators, J. d'Analyse Math. 43 (1984), 88-135.

3. Erratum, Comm. Part. Diff. Equ. 10 (1985), 1227.

A. Friedmann:

1. *Partial Differential Equations.* Holt, Rinehart and Winston, New York 1969.

D. Fujiwara and N. Shimakura:

1. Sur les problèmes aux limites stablement variationnels, J. Math. pures et appl. 49 (1970), 1-28.

Y. Giga:

1. The Stokes operator in L_r spaces, Proc. Japan Acad. 57, Ser.A (1981), 85-90.

I.C. Gohberg and M.G. Krein:

1. *Introduction to the Theory of Linear Nonselfadjoint Operators.* Amer. Math. Soc. Transl. Math. Monogr. 18, Rhode Island, 1969.

W.M. Greenlee:

1. Rate of convergence in singular perturbations, Ann. Inst. Fourier 18 (1969), 135-192.

P. Greiner:

1. An asymptotic extension for the heat equation, Arch. Rat. Mech. Anal. 41 (1971), 163-218.

P. Grisvard:

1. Caractérisation de quelques espaces d'interpolation, Arch. Rat. Mech. Anal. 25 (1967), 40-63.

G. Grubb and G. Geymonat:

1. The essential spectrum of elliptic systems of mixed order, Math. Ann. 227 (1977), 247-276.

2. Eigenvalue asymptotics for selfadjoint elliptic mixed order systems with non-empty essential spectrum, Boll. U.M.I. 16-B (1979), 1032-1048.

G. Grubb:

1. On coerciveness and semiboundedness of general boundary problems, Isr. J. Math. 10 (1971), 32-95.

2. Weakly semibounded boundary problems and sesquilinear forms, Ann. Inst. Fourier Grenoble, 23 (1973), 145-194.

3. Inequalities for boundary value problems for systems of partial differential operators, Astérisque 2-3 (1973), 171-187.

4. Properties of normal boundary problems for elliptic even-order systems, Ann. Scuola Norm. Sup. Pisa 1 (ser.IV) (1974), 1-61.

5. Spectral asymptotics for Douglis-Nirenberg elliptic and pseudo-differential boundary problems, Comm. Part. Diff. Equ. 2 (1977), 1071-1150.

5'. Les problèmes aux limites généraux d'un opérateur elliptique, provenant de la théorie variationnelle, Bull. Sc. Math. 94 (1970), 113-157.

(G. Grubb:)

6. Remainder estimates for eigenvalues and kernels of pseudo-differential elliptic systems, Math. Scand. 43 (1978), 275-307.

7. Sur les valeurs propres des problèmes aux limites pseudo-différentiels, C.R. Acad. Sci. Paris (Sér.A) 286 (1978), 199-201.

8. Estimation du reste dans l'étude des valeurs propres des problèmes aux limites pseudo-différentiels auto-ajdoints, C.R. Acad. Sci. Paris (Sér.A) 287 (1978), 1017-1020.

9. Estimation du reste dans l'étude des valeurs propres des problèmes aux limites pseudo-différentiels elliptiques, Sém. Goulaouic-Schwartz 1978-79, no.14, Palaiseau 1979.

10. On the spectral theory of pseudo-differential elliptic boundary problems, Partial Differential Equations 1978, Banach Center Publications Vol.10, PWN-Polish Scientific Publishers, Warsaw 1983, 147-168.

11. On Pseudo-Differential Boundary Problems. Reports no.2 (1979), 1, 2, 7 and 8 (1980), Copenhagen University Publication Series.

12. A resolvent construction for pseudo-differential boundary value problems, with applications, 18th Scandinavian Congr. Math. Proceedings, 1980, Birkhäuser Boston, 307-320.

13. Problèmes aux limites dépendant d'un paramètre, C.R. Acad. Sci. Paris (Sér.I) 292 (1981), 581-583.

14. La résolvante d'un problème aux limites pseudo-differéntiel elliptique, C.R. Acad. Sci. Paris (Sér.I) 292 (1981), 625-627.

15. The heat equation associated with a pseudo-differential boundary problem, Seminar Analysis 1981-82, Akad. Wiss. Berlin, 27-41. [Also available as Preprint no.2, 1982, Copenhagen Univ. Math. Dept.]

16. Remarks on trace estimates for exterior boundary problems, Comm. Part. Diff. Equ. 9 (1984), 231-270.

17. Singular Green operators and their spectral asymptotics, Duke Math. J. 51 (1984), 477-528.

18. Une méthode pseudo-différentielle pour les perturbations singulières elliptiques, C.R. Acad. Sci. Paris (Sér.I) 301 (1985), 427-430.

19. Boundary problems for systems of partial differential operators of mixed order, J. Functional Analysis 26 (1977), 131-165.

G. Grubb and V.A. Solonnikov:

1. Boundary value problems for the nonstationary Navier-Stokes equation treated by pseudo-differential methods (in preparation).

L. Gårding:

1. Dirichlet's problem for linear elliptic partial differential equations, Math. Scand. 1 (1953), 55-72.

2. On the asymptotic distribution of the eigenvalues and eigenfunctions of elliptic differential operators, Math. Scand. 1 (1953), 237-255.

K. Hayakawa and H. Kumano-go:

1. Complex powers of a system of pseudo-differential operators, Proc. Japan Acad. 47 (1971), 359-364.

B. Helffer:

1. *Théorie spectrale pour des opérateurs globalement elliptiques.* Asterisque 112, Société Mathématique de France, 1984.

E. Hille and R. Phillips:

1. *Functional Analysis and Semi-Groups.* Amer. Math. Soc. Colloq. Publ. 31, Providence 1957.

D. Huet:

1. Remarque sur un theorème d'Agmon et applications à quelques problèmes de perturbation singulière, Boll. Un. Mat. Ital. 21 (1966), 219-227.
2. Perturbations singulières et régularité, C.R. Acad. Sci. Paris (Sér.A) 266 (1968), 1237-1239.
3. *Décomposition spectrale et opérateurs*. Presses universitaires de France, Paris 1977.
4. Approximation d'un espace de Banach et perturbations singulières, C.R. Acad. Sci. Paris 289 (Sér.A) (1979), 595-596.
5. Proper approximation of a normed space and singular perturbations, *Analytical and Numerical Approaches to Asymptotic Problems in Analysis*, North Holland Publ. Co. 1981, 87-98.
6. Convergence propre et perturbations singulières coercives, C.R. Acad. Sci. Paris 301 (Sér.I) (1985), 435-438; Stabilité inverse d'une famille d'opérateurs opérant dans des espaces normés et dépendant d'un paramètre, ibid., 675-677.

L. Hörmander:

1. *Linear Partial Differential operators.* Springer Verlag, New York 1963.
2. Pseudo-differential operators, Comm. Pure Appl. Math. 18 (1965), 501-517.
3. Pseudo-differential operators and non-elliptic boundary problems, Ann. of Math. 83 (1966), 129-209.
4. On the Riesz means of spectral functions and eigenfunction expansions for elliptic differential operators, Belfer Grad. School Sci. (Yeshiva Univ.) Conf. Proceedings, Vol.2 1966, 155-202.
5. The spectral function of an elliptic operator, Acta Math. 121 (1968), 193-218.
6. Fourier integral operators I, Acta Math. 127 (1971), 79-183.
7. The Weyl calculus of pseudo-differential operators, Comm. Pure Appl. Math. 32 (1979), 359-443.
8. *The Analysis of Linear Partial Differential Operators*, Springer Verlag, Berlin 1983 and 1985. [The present work mostly uses Volume III.]
9. On the regularity of solutions of boundary problems, Acta Math. 99 (1958), 225-264.

V.Ja. Ivrii:

1. Second term of the spectral asymptotic expansion of the Laplace-Beltrami operator on manifolds with boundary and for elliptic operators acting in fiberings, Dokl. Akad. Nauk SSSR 250 (1980), 1300-1302 = Soviet Math. Dokl. 21 (1980), 300-302.
2. Precise spectral asymptotics for elliptic operators acting in fiberings, Funkts. Anal. Prilozhen. 16 No.2 (1982), 30-38 = Funct. Analysis Appl. 16 (1983), 101-108.
3. *Precise spectral asymptotics for elliptic operators.* Lecture Notes in Math. 1100, Springer Verlag, Berlin 1984.

C. Iwasaki and N. Iwasaki:

1. Parametrix for a degenerate parabolic equation and its application to the asymptotic behavior of spectral functions for stationary problems, Publ. R.I.M.S. Kyoto 17 (1981), 577-655.

M. Jodeit:

1. Symbols of parabolic singular integrals, Amer. Math. Soc. Proc. Symp. Pure Math. 10 (1967), 184-195.

T. Kato:

1. *Perturbation theory for linear operators.* Springer Verlag, Berlin 1966.

A. Kozevnikov:

1. Remainder estimates for eigenvalues and complex powers of the Douglis-Nirenberg elliptic systems, Comm. Part. Diff. Equ. 6 (1981), 1111-1136.

V.A. Kozlov:

1. Spectral asymptotics for non-semibounded elliptic systems, Vestnik Leningrad, No.1 (1979), 112-113.
2. Estimates of the remainder in formulas for the asymptotic behavior of the spectrum for linear operator bundles, Funkts. Anal. Prilozhen. 17 No.2 (1983), 80-81 = Funct. Analysis Appl. 17 (1983), 147-149.

M.G. Krein:

1. Theory of self-adjoint extensions of symmetric semi-bounded operators and applications I, Mat. Sb. 20: 62 (1947), 431-495 (Russian).

O.A. Ladyzenskaja, V.A. Solonnikov and N.N. Uralceva:

1. *Linear and Quasilinear Equations of Parabolic Type.* "Nauka", Moscow 1967, Amer. Math. Soc. Translation, Providence 1968.

A.A. Laptev:

1. Spectral asymptotics of a composition of pseudo-differential operators and reflections from the boundary, Dokl. Akad. Nauk SSSR 236 (1977), 800-803 = Soviet Math. Dokl. 18 (1977), 1273-1276.
2. Spectral asymptotics of a class of Fourier integral operators, Trudy Mosk. Mat. Obsv. 43 (1981), 92-115 = Trans. Moscow Math. Soc. 1983, 101-127.

R. Lascar:

1. Propagation des singularités des solutions des équations pseudo-différentielles quasi-homogènes, Ann. Inst. Fourier 27 (1977), 79-193.

S.Z. Levendorskii:

1. Asymptotic behavior of the spectrum of problems of the form $Au = tBu$ for operators that are elliptic in the Douglis-Nirenberg sense, Funkts. Anal. Prilozhen. 18 No.3 (1984), 84-85 = Funct. Analysis Appl. 18 (1984), 253-255.
2. Asymptotics of the spectrum of linear operator pencils, Mat.Sbornik 124 (166) (1984), 251-270 = Math. USSR Sbornik 52 (1985), 245-266.

J.L. Lions:

1. *Perturbations Singulières dans les Problèmes aux limites et en Contrôle Optimal.* Lecture Notes in Math. 323, Springer Verlag, Berlin 1973.

J.L. Lions and E. Magenes:

1. *Problèmes aux limites non homogènes et applications*, vol.1. Editions Dunod, Paris 1968.
2. *Problèmes aux limites non homogènes et applications*, vol.2. Editions Dunod, Paris 1968.

A.S. Markus and V.I. Matsaev:

1. Comparison theorems for spectra of linear operators, and spectral asymptotics, Trudy Mosk. Mat. Obsv. 45 (1982), 139-188 = Trans. Moscow Math. Soc. 1984, 139-185.

R. Melrose:

1. Transformation of boundary problems, Acta Math. 147 (1981), 149-236.
2. The trace of the wave group. Amer. Math. Soc. Contemporary Math. vol.27 "Microlocal Analysis" (1984), 127-167.

G. Métivier:

1. Estimation du reste en théorie spectrale, Rend. Sem. Mat. Univ. Politec. Torino 1983, Special Issue, 158-180.

A. Mohamed:

1. Etude spectrale d'opérateurs hypoelliptiques à caractéristiques multiples I, Ann. Inst. Fourier (Grenoble) 32 (1982), 39-90.
2. Etude spectrale d'opérateurs hypoelliptiques à caractéristiques multiples II, Comm. Part. Diff. Equ. 8 (1983), 247-316.
3. Comportement asymptotique, avec estimation du reste, des valeurs propres d'une classe d'opérateurs pseudo-différentiels sur $\mathbb{R}^n$, C.R. Acad. Sci. Paris (Sér.I) 299 (1984), 177-180.

L. Nirenberg:

1. *Lectures on Linear Partial Differential Equations*. Amer. Math. Soc. Regional Conf. Series 17, Providence 1973.

T. Nambu:

1. Feedback stabilization for distributed parameter systems of parabolic type, J. Diff. Equ. 33 (1979), 167-188.

M. Nagase:

1. On the asymptotic behavior of resolvent kernels for elliptic operators, J. Math. Soc. Japan 25 (1973), 464-474.

M. Nagase and K. Shinkai:

1. Complex powers of non-elliptic operators, Proc. Japan Acad. 46 (1970), 779-783.

V.I. Paraska:

1. On asymptotics of eigenvalues and singular numbers of linear operators which increase smoothness, Mat. Sb. 68 (110) (1965), 623-631. (Russian.)

Pham The Lai:

1. Meilleures estimations asymptotiques des restes de la fonction spectrale et des valeurs propres relatifs au laplacien, Math. Scand. 48 (1981), 5-31.

R.S. Phillips:

1. Dissipative operators and hyperbolic systems of partial differential equations, Trans. Amer. Math. Soc. 90 (1959), 193-254.

Å. Pleijel:

1. Certain indefinite differential eigenvalue problems - the asymptotic distribution of their eigenfunctions, Part. Diff. Equ. and Continuum Mech., Wisconsin Press, Madison 1961, 19-37.

S. Rempel:

1. An analytical index formula for elliptic pseudo-differential boundary problems I, Math. Nachr. 94 (1980), 243-275.

S. Rempel and B.-W. Schulze:

1. *Index Theory of Elliptic Boundary Problems.* Akademie-Verlag, Berlin 1982.
2. Parametrices and boundary symbolic calculus for elliptic boundary problems without the transmission property. Math. Nachr. 105 (1982), 45-149.
3. Complex powers for pseudo-differential boundary problems I, Math. Nachr. 111 (1983), 41-109.
4. Complex powers for pseudo-differential boundary problems II, Math. Nachr. 116 (1984), 269-314.

D. Robert:

1. Propriëtés spectrales d'opérateurs pseudo-différentiels, Comm. Part. Diff. Equ. 3 (1978), 755-826.
2. Developpement asymptotique du noyau résolvant d'opérateurs elliptiques, Osaka J. Math. 15 (1978), 233-243.

B.-W. Schulze:

1. Adjungierte elliptische Randwertprobleme, Math. Nachr. 89 (1979), 225-245.

L. Schwartz:

1. *Théorie des Distributions.* Hermann, Paris 1966.

R. Seeley:

1. Complex powers of an elliptic operator, Amer. Math. Soc. Proc. Symp. Pure Math. 10 (1967), 288-307.
2. Topics in pseudo-differential operators, C.I.M.E. *Conf. on Pseudo-Differential Operators*, Edizioni Cremonese, Roma 1969, 169-305.
3. The resolvent of an elliptic boundary problem, Amer. J. Math. 91 (1969), 889-920.
4. Analytic extension of the trace associated with elliptic boundary problems, Amer. J. Math. 91 (1969), 963-983.
5. Extension of C^∞ functions, Proc. Amer. Math. Soc. 15 (1964), 625-626.
6. Norms and domains of the complex powers A_B^z, Amer. J. Math. 93 (1971), 299-309.
7. A sharp asymptotic remainder estimate for the eigenvalues of the Laplacian in a domain of R^3, Adv. in Math. 29 (1978), 144-269.
8. Interpolation in L^p with boundary conditions, Studia Math. 44 (1972), 47-60.

M.A. Shubin:

1. *Pseudodifferential Operators and Spectral Theory*, "Nauka", Moscow 1978 (Russian).

V.A. Solonnikov:

1. *On the Boundary Value Problems for Linear Parabolic Systems of Differential Equations of General Form.* Proc. Stekl. Inst. of Math. 83, Leningrad 1965.

N. Steenrod:

1. *The Topology of Fibre Bundles.* Princeton University Press 1951.

R.S. Strichartz:

1. A functional calculus for elliptic pseudo-differential operators, Amer. Math. 94 (1972), 711-722.

H. Tanabe:

1. *Equations of Evolution*. Pitman Publishing Ltd., London 1979.

M.E. Taylor:

1. *Pseudodifferential Operators*. Princeton University Press, New Jersey 1981.

F. Treves:

1. *Topological Vector Spaces, Distributions and Kernels*. Academic Press, New York 1967.
2. *Introduction to Pseudodifferential and Fourier Integral Operators*, 1-2. Plenum Press, New York 1980.

R. Triggiani:

1. On Nambu's stabilization problem for diffusion processes, J. Diff. Eq. 33 (1979), 189-200.

D.G. Vasiliev:

1. Binomial asymptotics of the spectrum of a boundary value problem, Funkts. Analysis Prilozhen. 17 No.4 (1983), 79-81 = Funct. Anal. Appl. 17 (1983), 309-311.
2. Two-term asymptotics of the spectrum of a boundary-value problem under an interior reflection of a general form, Funkts. Analys. Prilozhen. 18 No.4 (1984), 1-13 = Funct. Analysis Appl. 18 (1984), 267-277.

M.I. Vishik:

1. On general boundary value problems for elliptic differential equations, Trudy Moskov. Mat. Obsv. 1 (1952), 187-246 = Amer. Math. Soc. Transl. 24 (1963), 107-172.

M.I. Vishik and G.I. Eskin:

1. Elliptic equations in convolution in a bounded domain and their applications, Uspehi Mat. Nauk 22 (1967), 15-76 = Russian Math. Surveys 22 (1967), 13-75.
2. Parabolic convolution equations in a bounded domain, Mat. Sb. 71 (1966), 162-190 = Amer. Math. Soc. Transl. 95 (1970), 131-162.

M.I. Vishik and L. Lyusternik:

1. Regular degeneration and boundary layer for linear differential equations with small parameters, Uspekhi Mat. Nauk 12 (1957), 3-122 = Amer. Math. Soc. Transl. (2), 20 (1962), 239-264.

W.D. Wendt:

1. *Coercive Singularly Perturbed Wiener-Hopf Operators and Applications*. Doctoral dissertation, Nijmegen 1983.

H. Widom:

1. A Szegö theorem and complete symbolic calculus for pseudo-differential operators, Ann. Math. Studies Princeton 91 "Singularities" (1978), 261-283.
2. A complete symbolic calculus for pseudo-differential operators, Bull. Sc. Math. 104 (1980), 19-63.
3. Asymptotic expansions for pseudo-differential operators on bounded domains. (Preprint, Univ. of California, Santa Cruz 1984.)
4. *Asymptotic Expansions for Pseudodifferential Operators on Bounded Domains*. Lecture Note 1152, Springer Verlag, Berlin 1985.

INDEX

The numbers x.y.z, or A.x, refer to *statements* (Definition, Theorem, Lemma, Remark etc.), and the numbers (x.y.z), or (A.x), refer to *formulas*, often including the surrounding text. *See the Appendix for explanations of notation.*

Progress in Mathematics

41 STANLEY. Combinatorics and Commutative Algebra
42 JOUANOLOU. Théorèmes de Bertini et Applications
43 MUMFORD. Tata Lectures on Theta II
44 KAC. Infinite Dimensional Lie Algebras
45 BISMUT. Large Deviations and the Malliavin Calculus
46 SATAKE/MORITA. Automorphic Forms of Several Variables Taniguchi Symposium, Katata, 1983
47 TATE. Les Conjectures de Stark sur les Fonctions L d'Artin en s = 0
48 FRÖHLICH. Classgroups and Hermitian Modules
49 SCHLICHTKRULL. Hyperfunctions and Harmonic Analysis on Symmetric Spaces
50 BOREL, ET AL. Intersection Cohomology
51 BERTIN/GOLDSTEIN. Séminaire de Théorie des Nombres, Paris 1982-83
52 GASQUI/GOLDSCHMIDT. Déformations Infinitesimales des Structures Conformes Plates
53 LAURENT. Théorie de la Deuxième Microlocalisation dans le Domaine Complèxe
54 VERDIER/LE POTIER. Module des Fibres Stables sur les Courbes Algébriques Notes de l'Ecole Normale Supérieure, Printemps, 1983
55 EICHLER/ZAGIER. The Theory of Jacobi Forms
56 SHIFFMAN/SOMMESE. Vanishing Theorems on Complex Manifolds
57 RIESEL. Prime Numbers and Computer Methods for Factorization
58 HELFFER/NOURRIGAT. Hypoellipticité Maximale pour des Opérateurs Polynomes de Champs de Vecteurs
59 GOLDSTEIN. Séminaire de Théorie des Nombres, Paris 1983-84
60 PROCESI. Geometry Today: Giornate Di Geometria, Roma, 1984
61 BALLMANN/GROMOV/SCHROEDER. Manifolds of Nonpositive Curvature
62 GUILLOU/MARIN. A la Recherche de la Topologie Perdue
63 GOLDSTEIN. Séminaire de Théorie des Nombres, Paris 1984-85
64 MYUNG. Malcev-Admissible Algebras
65 GRUBB. Functional Calculus of Pseudo-Differential Boundary Problems